# Enamels of Limoges
## 1100–1350

# Enamels of Limoges
## 1100–1350

THE METROPOLITAN MUSEUM OF ART, NEW YORK

Distributed by Harry N. Abrams, Inc., New York

This publication is issued in conjunction with *Enamels of Limoges, 1100–1350*, held at the Musée du Louvre, Paris, October 23, 1995– January 22, 1996, and at The Metropolitan Museum of Art, New York, March 5–June 16, 1996.

The exhibition was organized by The Metropolitan Museum of Art and the Réunion des Musées nationaux / Musée du Louvre.

The exhibition is made possible in part by the National Endowment for the Arts and the William Randolph Hearst Foundation.

We are grateful to the Robert Lehman Foundation for making the Lehman galleries available for the exhibition.

Published by The Metropolitan Museum of Art

John P. O'Neill, Editor in Chief
Teresa Egan, Editor, with the assistance of Margaret Aspinwall, Cynthia Clark, Kathleen Howard, Mary E. D. Laing, Jacolyn A. Mott, and Ellen Shultz
Barbara Cavaliere, Production Editor
Bruce Campbell, Designer, after the French edition by Christophe Ibach
Jay Reingold, Production Manager
Robert Weisberg, Computer Specialist
Alarik Skarstrom, Bibliographer

Photography of Metropolitan Museum objects by Patricia Mazza, The Photograph Studio, The Metropolitan Museum of Art

Color separations for entry illustrations by Prographics, Rockford, Illinois
Printed and bound by Imprimerie Mame, Tours

Translations from the French by Sophie Hawkes, Joachim Neugroschel, and Patricia Stirneman ("Transatlantic Crossings of the Art of Limoges")

Jacket/cover: Details of the reverse of the Chasse of Champagnat (cat. 10). The Metropolitan Museum of Art

Details: Page 64, Chasse of Bellac (cat. 9); page 102, Chasse with the Adoration of the Magi (cat. 22); page 196, Chasse from Ambazac (cat. 55); page 242, Tabernacle (cat. 70b); page 336, Coffret of Saint Louis (cat. 123); page 396, Funerary Plaque of Guy de Meyos (cat. 152)

Library of Congress Cataloging-in-Publication Data is available.

ISBN: 0-87099-758-0; 0-87099-759-9 (pbk); 0-8109-6500-3 (Abrams)

# Contents

# Directors' Foreword

Enamels made by goldsmiths of Limoges constitute the richest surviving corpus of medieval metalwork, both in number and in range of objects, from book covers to reliquaries, from candlesticks to garnitures for horses. Their survival is due, in no small measure, to their sturdiness. But this was not the sole quality that made Limoges enamels so valued in the Middle Ages; even more, they were prized for their rich colors, highly polished surfaces, glimmering gold, inventive designs, and appealing narratives. *Opus lemovicense* adorned the shrine of Saint Peter in Rome and enshrouded the tombs of the children of Saint Louis at the abbey of Royaumont. The universal taste for enamels of Limoges in the Middle Ages was an international phenomenon of unrivaled proportions.

The brilliant color of Limoges enamels has long been a source of fascination. As early as the 1600s their characteristic palette was thought to be the result of the chemical composition of the Vienne River, which flows through the city. Aesthetic and historic interest in Limoges enamels grew in the eighteenth and nineteenth centuries. Antiquarians prized their narrative qualities—the depiction of "druidic" culture and the documentation of ancient costume and manners and of the rich fabric of chivalric culture. Among recusant families in England, Limoges enamels at times represented the last vestiges of a spiritual heritage that had been wrenched from them, while in the wake of the Wars of Religion and of the Revolution in France, Limoges enamels were jealously guarded by the local communities that had owned them since their creation.

Curiously, early auction catalogues often called these works "Byzantine." The active market for them suggests that the enamels' "Byzantine" quality, ancient and thus perhaps exotic, may have been part of their allure. It may have been the exotic decoration and inventive form of enameled crosiers, combined with the mystical appeal of all things medieval, that led Eugène Delacroix to make three drawings of Limousin pastoral staffs. The ensemble of enamelwork brought together for this exhibition makes evident how attractive the rich palette dominated by blues must have been to many nineteenth-century collectors, among them French painters and stained-glass restorers.

Many of the same enamels that were acquired by French antiquarians, collectors, and painters now form the nucleus of the exceptional holdings of the Louvre, which benefited from the acquisition of the Durand and Révoil Collections in 1825 and 1828, respectively. The most important of the later acquisitions, bequests, and gifts was that of Victor Martin Le Roy in 1914. In New York the strength of the Metropolitan Collection is due largely to the gift of J. Pierpont Morgan, the great American collector who seems to have consciously tailored his acquisitions to mirror the taste of an erudite prince. He bought the collection of Georges Hoentschel, *en bloc*, at auction in Paris and purchased equally fine pieces from Paris and London dealers. His legacy to the Metropolitan is a collection of Limoges enamelwork unrivaled outside France.

Yet while avidly collected and widely exhibited in the nineteenth century, enamels created by Limoges artists are today little known outside academic circles. Since the turn of the century the only exhibitions devoted to them have been those at Limoges in 1948 and at the Vatican in 1963. The collections of the Louvre and of the Metropolitan, similar in overall number and serendipitously complementary in strengths, form the basis for our collaboration and provide the core of this exhibition. To present a more complete history of the enameling of Limoges, we turned to sister institutions and private collections in Europe and the United States, and we have been astonished by the generosity of their response. But we owe a special debt to the churches and communities that have safeguarded the heritage of Limoges for the last eight hundred years and that have entrusted their works of art to us for a time. These loans and our own collections form an ensemble of extraordinary works, the taste of Europe for better than two centuries and a precious heritage today.

Philippe de Montebello
*Director,*
*The Metropolitan Museum of Art*

Pierre Rosenberg
*Président-directeur,*
*Musée du Louvre*

# Curators and Contributors to the Catalogue

*Curators of the Exhibition*

Barbara Drake Boehm
Associate Curator, Department of Medieval Art,
The Metropolitan Museum of Art

Elisabeth Taburet-Delahaye
Conservateur en chef au Département
des Objets d'art, Musée du Louvre

*Assisted by*

Catherine Gougeon
Paris

Christina Nielsen
Jennie Raab
New York

*Contributors to the catalogue*

Barbara Drake Boehm (BDB)
*and*
Elisabeth Taburet-Delahaye (ET-D)

*with*

Elisabeth Antoine (EA)
Conservateur au Musée National du Moyen Âge, Thermes
de Cluny

Françoise Baron (FB)
Conservateur général honoraire du Patrimoine

Bernadette Barrière
Professeur à l'Université de Limoges

Dom Jean Becquet, O.S.B.

Isabelle Biron
Ingénieur de Recherche au Laboratoire de Recherche
des Musées de France

Béatrice de Chancel-Bardelot (BC-B)
Conservateur au Département des Sculptures
du Musée du Louvre

Pete Dandridge
Conservator, The Sherman Fairchild Center for Objects
Conservation, The Metropolitan Museum of Art

Geneviève François (GF)
Ingénieur d'étude au Centre National de la Recherche
Scientifique

Jean-René Gaborit
Conservateur général chargé du Département des Sculptures
du Musée du Louvre

Marie-Madeleine Gauthier
Directeur honoraire de Recherche au Centre National de la
Recherche Scientifique, Directeur du Corpus des Émaux
Méridionaux

Michel Pastoureau (MP)
Professeur à l'Ecole Pratique des Hautes Etudes, Paris

William D. Wixom (WDW)
Michel David-Weill Chairman, Department of Medieval Art
and The Cloisters, The Metropolitan Museum of Art

Mark T. Wypyski
Assistant Research Scientist, The Sherman Fairchild Center for
Objects Conservation, The Metropolitan Museum of Art

# Lenders to the Exhibition

AUSTRIA

Vienna: Österreichisches Museum für angewandte
Kunst 56a, 91

BELGIUM

Antwerp: Musée Mayer van den Bergh 2b

CZECH REPUBLIC

Prague: Uméleckoprùmyslové Muzeum 60

FRANCE

Ambazac: Church of Saint-Anthony 55

Amiens: Musée de Picardie 121, 153a

Angers: Musées d'Angers 150b

Apt: Church of Saint-Anne 21

Bellac: Church of Notre-Dame 9

Brive-la-Gaillarde: Collegiate Church of Saint-Martin
142, 153b

Cahors: Treasury of the Cathedral 32

Conques: Treasury of the Abbey of Saint-Foy 1, 7

Fontaine-Chaalis: Musée Jacquemart-André 61

Gimel-les-Cascades: Church of Saint-Pardoux 16

La Sauvetat: Church of Saint-John-the-Baptist 156

Les Billanges: Parish Church of the Nativity-of-Saint-
John-the-Baptist 59

Le Mans: Musée de Tessé 15

Limoges: Musée Municipal de l'Évêché 19

Longpont: Treasury of the Abbey 133

Lyons: Musée des Beaux-Arts 18

Mozac: Church of Saint-Peter 45

Moulins: Bibliothèque Municipale 34

Nexon: Church of the Beheading-of-Saint-John-
the-Baptist 15

Paris: Bibliothèque Nationale de France
Cabinet des Médailles 137

Département des Manuscrits 147, 148

Cabinet des Estampes 149, 150

Direction du Patrimoine

Musée National du Moyen Âge, Thermes de
Cluny 12, 57, 99, 100, 104, 113a,b, 119a,c,
127, 141

Musée du Petit Palais 58d–e

Poitiers: Musées de la Ville de Poitiers et de la Société
des Antiquaires de l'Ouest (Musée Sainte-Croix) 23

Rouen: Musée des Antiquités de la Seine-Maritime 2a

Saint-Denis: Abbey Church 146

Saint-Marcel: Church of Saint-Marcel 75

Saint-Sulpice-les-Feuilles: Church of Saint-Sulpice 54

Saint-Viance: Parish Church of Saint-Viance 71, 118

Toulouse: Treasury of Saint-Sernin 40

ITALY

Florence: Museo Nazionale del Bargello 8a–d, 56b, 58f

POLAND

Warsaw: National Museum 140

RUSSIA

Saint Petersburg: State Hermitage Museum 20, 58c

UNITED STATES OF AMERICA

Baltimore: Walters Art Gallery 119b

Boston: Museum of Fine Arts 122

Bryn Athyn, Pa.: The Glencairn Museum, Academy of
the New Church 38

Minneapolis: Institute of Arts 119d

Oberlin, Ohio: Allen Memorial Art Museum, Oberlin
College 53

Washington, D.C.: National Gallery of Art 22

# Acknowledgments

We would like to express our gratitude to all those who have encouraged and assisted in this exhibition.

The project of bringing together the collections of The Metropolitan Museum of Art and the Musée du Louvre could never have been accomplished without the direction of Michel Laclotte, Philippe de Montebello, and Pierre Rosenberg, who believed in this idea and encouraged us throughout the course of its realization, or without the support of Irène Bizot and Mahrukh Tarapor. We are indebted to the research of Mme Marie-Madeleine Gauthier, who has dedicated herself to the study of the enamels of Limoges for more than forty years. We are especially honored that she agreed to be associated with this project.

We are most grateful to our department heads, Daniel Alcouffe and William D. Wixom, for their abiding commitment to this project. In the Département des Objets d'Art at the Louvre and in the Medieval Department and Lehman Collection of the Metropolitan Museum, Danielle Gaborit-Chopin, Charles T. Little, Jannic Durand, and Laurence B. Kanter shared with us their knowledge, opinions, advice, and constant encouragement.

The exhibition could not have taken place without the help and understanding of the staff at the Direction du Patrimoine, to whom we would like to express our appreciation: Colette di Matteo, Bernard Brochard, Christian Prévost-Marcilhacy, inspecteurs généraux des Monuments Historiques; Caroline Piel, Marie-Anne Sire, Jean-Louis Aurat, Michel Caille, Denis Lavalle, Bruno Mottin, Thierry Zimmer, inspecteurs; Paul-Edouard Robinne, conservateur en chef; and Martine Chavent, conservateur à l'Inventaire général, Direction régionale des Affaires cultureles du Limousin. The generous help of the curators of Antiquities and Art Objects—Claire Delmas, Marie-Claude Léonelli, and Etienne d'Alençon—was also vital.

We would also like to thank the mayors and the priests of the communes who agreed to lend important works to this exhibition: the staffs of the treasuries who gave us access to examine the objects; the curators of the participating public collections in the United States, France, and other European countries; and the private lenders who prefer to remain anonymous.

We are particularly pleased to express our gratitude to the specialists who agreed to write introductions or entries for this catalogue: Elisabeth Antoine, Françoise Baron, Bernadette Barrière, Dom Jean Becquet, Béatrice de Chancel-Bardelot, Geneviève François, Jean-René Gaborit, Marie-Madeleine Gauthier, and Michel Pastoureau. We would also like to thank our colleagues in the laboratories of the Metropolitan Museum and the Musées de France—Isabelle Biron, Pete Dandridge, Michel Vandevyver, and Mark T. Wypyski—for their exemplary spirit of collaboration.

The authors of the catalogue address their thanks to all of their colleagues, museum and library curators, archivists, researchers, and specialists who generously furnished their opinions, information, or photographs: A. Allcorn, J.-M. André, Fr. Arquié-Bruley, Fr. Avril, I. Ballandre, S. Baratte, A. Blumka, M. Bernus-Taylor, M. Borel, Chr. Briend, Th. Calligaro, J.-P. Callu, M. Campbell, S. Castronovo, R. Chanaud, Kitty Chibnik of the Avery Library, Columbia University, G. Costa, K. Czerwoniak, L. Deletre, Cl. Delmas, A. Dion-Tenebaum, A.-Ch. Dionnet, J. Domenge y Mesquida, C. Dumont, P. Ennès, H. Evans, M. Eveno, Father J. Fanning, L. Flavigny, S. Fliegel, D. Foy, J. H. Frantz, D. Germain-Bonne, J. Glattauer, C. Gros, Father P.-M. Gy, J.-P. Hervieu, L. Jackson-Beck, S. Kenfield, R. J. Koestler, L. Komaroff, D. Kötzsche, M. Lasnier, M. H. Lavallée, G. Lebègue, J.-B. Lechat, J.-L. Lemaître, Fr. Lernout, C. Leroy, A. Luchs, K. Lydecker, G. Mabille, F. Martin, Ch. Meslin-Perrier, B. Mille, P. Mironneau, C. de Montalivet, E. de Montesquiou, E. Morales, D. Morel, G. Moyse, A. H. Nielsen, H. Nieuwdorp, V. Notin, N. Petit, V. Pomarède, L. Randall, T. Rapé, S. Rebuttini, H. Richard, M. Rérolle, S. Romanelli, B. Rosasco, A. M. Rosenbaum, A. Rosso, L. Roth, Ch. Rouquet, J. Salomon, L. Saulnier-Pernuit, E. Schmuttermeier, L. Seckelson, M. B. Shepard, G. Socié, J. Soultanian, C. Steiner, P. Stirneman, R. E. Stone, K. Szczepkowska-Naliwajek, N. Stratford, J. Taralon, G. Vikan, H. Westermann-Angerhausen.

The majority of the objects benefited from the patient and meticulous care of specialized restorers, A. Amarger, B. Beillard, P. Branch, F. Dall'Ava, S. Crevat, P. Dandridge, J.-F. DeLaperouse, and A. Oron, to whom we express all our appreciation. We are grateful as well to the photographers D. Arnaudet, D. Bagault, M. Beck-Coppola, and P. Mazza.

In the Medieval Department and the Département des Objets d'art we benefited from the constant aid of M.-C.

Bardoz, C. Brennan, B. Ducrot, D. Dutruc-Rosset, S. Gasnier, A. Gauthier, L. Hebert, M. Huyn, M. Mac Ilwain, R. T. Margelony, F. Néguiral, E. Nesterman, O. Pevny, E. Riand, Ch. Riou, F. Tasso, I. Tordjmann, K. Ulemek-Paunac, J. de Villoutreys, and T. C. Vinton.

We express our thanks also to Linda Sylling, Nina Maruca, and David Harvey, who were responsible for the manifold aspects of the transportation to and presentation of the exhibition in New York.

We are grateful to John P. O'Neill, who oversaw the publication of the catalogue with Anne de Margerie. We are especially thankful for the talents and considerable efforts of Teresa Egan (with the assistance of Margaret Aspinwall, Barbara Cavaliere, Cynthia Clark, Kathleen Howard, Mary E. D. Laing, Jacolyn A. Mott, Tonia Payne, Peter Rooney, Ellen Shultz, Alarik Skarstrom, Mary Smith, and Robert Weisberg), Bruce Campbell, and Jay Reingold.

Catherine Gougeon, Pascale Néraud, Christina Nielsen, and Jennie Raab assisted us with enthusiasm and devotion throughout the preparation of the exhibition and the catalogue. They alone know how much we have depended on them.

Finally, we are thankful and mindful of the support of F., L., T., and P. Delahaye and M., A., and D. Porcelli.

To all those involved we extend our deep appreciation.

E.T.-D. and B.D.B.

# Transatlantic Crossings of the Art of Limoges

Marie-Madeleine Gauthier

Could it be that the transatlantic friendships of yesteryear, which grew from a common love and study of *l'Oeuvre de Limoges*, the remarkable enameling produced in Limoges during the Middle Ages, were sending a message to younger readers of this catalogue? Was it a mere coincidence of publishing schedules that special sets of color photographs from the Biblioteca Apostolica Vaticana reached the Corpus des Émaux Méridionaux on the very day in July 1995 that I was completing my biographical portrait of Frederick Stohlman in these same pages? Is it only by chance that he is our protagonist among the actors playing scholarly roles in the historiographic production of which the organizers have solicited a plot summary in the guise of an introduction?

On October 25, 1938, in Princeton, Frederick Stohlman signed his preface to the second volume of the *Catalogo del Museo Sacro della Biblioteca Apostolica Vaticana* entitled *Gli Smalti del Museo Sacro Vaticano* (Enamels in the Museo Sacro Vaticano), published by order of His Holiness Pope Pius XI. The work appeared in 1939. This elegant quarto volume, whose printing was completed on April 30 of that year by Sansaini & Co., catalogued 144 enameled objects. With sixty-seven pages of text, the book included thirty-seven comparative illustrations on thirty-one plates, and among the 144 enameled works in the catalogue, fifty-three were of Limousin champlevé, including thirteen previously unpublished fragments. The young American scholar first acknowledged his mentor, Professor Charles Rufus Morey, "founder and organizer of the department of medieval studies at Princeton University," and then thanked the eminent French scholar and curator Jean-Joseph Marquet de Vasselot for the "help and advice" he had provided over the years.

Even more striking today is the recognition he gave, on the eve of World War II, to two specialists in the study of early medieval art, André Grabar and Fritz Wolfgang Volbach. The former, originally from Kiev, had studied in Prague and was then a professor at the University of Strasbourg; the latter, having been forced to leave the Berlin Museum, became an assistant at the Museo Sacro Vaticano. Ten years later these two specialists confirmed their European vocation by taking charge of the international cataloguing project which had begun spontaneously in 1948 in the museum and library at Limoges, and which would become the Corpus des Émaux Méridionaux.

It was in New York, in the Print Department at the Metropolitan Museum, that Stohlman found important documentary evidence concerning engraved models for Renaissance Limousin enamels. In London he also obtained documents and advice on Limousin enamels from the collector W. L. Hildburgh, although he thoroughly disagreed with the Spanish attributions made by that author in 1936. His style of citing works in abridged form in the text inaugurated the economical bibliographic practice that was first adopted by the *Art Bulletin*, and later by other reviews publishing academic research in the form of innovative articles.

Half of Stohlman's forty references are to books and articles by scholars from the Limousin: Ernest Rupin, Charles de Linas, and those who, since 1845, had published in the *Bulletin de la Société archéologique et historique du Limousin*, notably the prolific Msgr. Barbier de Montault. An exemplary printed and illustrated book from the first half of this century, *Gli Smalti del Museo Sacro Vaticano*, portrays the tormented journeys of these medieval treasures to their present abodes, a panorama that stretches from the most venerable religious sanctuary of Christian Europe (the Vatican) to the sanctuaries of the fine arts built in the United States toward the end of the second millennium, and, at their forefront, The Metropolitan Museum of Art in New York.

## National Treasures and the Fate of Limousin Enamels of the Middle Ages

The art of enameling on copper, perfected in southwestern Europe from the twelfth to the fourteenth century and practiced with astonishing continuity in Limoges, has been qualified in our time as either "decorative" or "industrial," because of the remarkable volume of production of these southern workshops.

It was only after the French Revolution, after the waves of iconoclastic vandalism, irreverent abandonment, dismantling, salvaging of metal, in short, after so much inexpiable destruction, that, little by little, connoisseurs, scholars, and the inquisitive began to realize just how numerous these works were. These losses notwithstanding, the Corpus numbers nearly eight thousand individual specimens that are localizable today.

The maintenance of written legal records has served as a link, a sort of spiritual genealogy, between the authors of medieval inventories and the scholars of modern times, from the inventories made by the dean of the chapter of Saint Paul's in London in 1298, by the secretary of Msgr. Massillon during the episcopal visit to his diocese of Clermont in 1723, and by the notary of Clement XIII in 1761, right up to Abbé Jacques-Rémi Texier, who in 1835 began scouring the rural churches of the dioceses of Limoges and Tulle, and to Louis Guibert, who chronicled the destruction of the order of Grandmont.

For that precious patrimony, later generations continued to provide faithful trustees who traveled through the country, opening cupboards in rectories and drawers in sacristies, counting liturgical vases, reliquaries, and candlesticks. Antiquarians or connoisseurs, experts or wayfarers, they observed, took notes, and made drawings, as did Auguste Bosvieux, archivist of the Creuse.

They entrusted their accounts and detailed descriptions of archaeological discoveries to the journals of scholarly societies, first in Limoges and then in the Poitou, Charente, Auvergne, Anjou, Normandy, and Morinie. Regardless of the prevailing forms of government, the nineteenth century embraced the revival in Romantic Europe of the taste for enameled objects considered "Gothic" or "Byzantine." These were the words used to designate Limousin enamels sold at auction, in London and especially in Paris. Then, in 1827, the painter Pierre Révoil offered the king of France the enamels that he had collected in Provence; it was his donation to the Palais du Louvre that conferred national recognition on the Alpais ciborium (cat. 69), "made in Limoges" and recovered from the former Benedictine abbey of Montmajour, near Arles.

During the nineteenth century there were nearly four hundred public exhibitions in Europe that included champlevé enamels from Limoges. The earliest, in Lyons in 1827 and 1837, were charitable in intent, while the later exhibitions were directed toward the technical instruction of artists in the leading industrial cities of England.

Beginning in 1863 in Paris, these events became increasingly sumptuous and ambitious in scope and were consequently entitled "universal" exhibitions.

It should be recalled that the Church had always offered pilgrims an opportunity to gaze upon and at times to touch its sacred vases, altars, crosses, and reliquaries—treasures made of gold or silver, precious stones, ivory, and enamels rich with narrative, which concealed even more precious treasures, the relics of saints. From 1180 to 1220 two or three of the principal workshops in Limoges had to respond to the largest demand ever stimulated by the political, royal, and papal cult of a martyr-saint: fifty-one reliquary chasses of Saint Thomas Becket exist and have been fully catalogued. The Corpus of enamels has added twenty examples to the series catalogued by Tancred Borenius in 1931. For the clientele of the enamelers of Limoges, in about the year 1200, the tragically successful, and almost contemporaneous, murder of the archbishop of Canterbury in his cathedral far surpassed in importance the memory of the passion of Saint Stephen, the protomartyr and patron of the cathedral in Limoges (Stephen's story was no less dramatic [see cat. 16], but it had occurred at the time of the evangelists). Becket's history eclipsed even the edifying legend, promoted by Eleanor of Aquitaine and the Plantagenets, of Saint Valerie, herself the daughter of a legendary duke, who devoutly carried her own severed head to her bishop, Martial of Limoges, thus becoming the protomartyr of Aquitaine (see cats. 20, 85).

The liberality required of a prince demanded that the gold and silver plate, candelabra, heraldic achievements, and precious finery of the lord and his lineage be displayed on festive occasions. Similarly, by exhibiting the fine liturgical vessels or secular furnishings used in ancient ceremonies, the democratic twentieth century means to show the workings of art and to inspire a taste for beautiful objects, while making each visitor a citizen of the world.

In Paris, about 1820, the visual impressions of the first connoisseurs to espouse the "troubador" fashion were, alas, as fleeting as those retained from a pilgrimage, as ephemeral as those made on social visits to the *cabinet de curiosités* in the abbeys of Sainte-Geneviève or Saint-Germain-des-Prés, as unstable as the parcels of objects that were exchanged, bid upon, or dispersed through public auction or the division of estates. The erudite antiquarian and inquisitive layman, however, longed for the permanence of what republican and Neoclassical convictions designated as a *museum*.

The hero of the Gothic Revival was Alexandre Du Sommerard, who in his youth volunteered as a soldier in the French Revolution, then left military service for the

Cour des Comptes in 1807, and distinguished himself as a royalist after the fall of Napoleon. Long enamored of the Middle Ages, he acquired the Parisian hôtel (residence) of the abbots of Cluny. This lodging, built over the ruins of the Roman Baths of Lutetia (*Thermes*) between 1485 and 1498, was in 1834 the most appropriate site for the collection of this informed archaeologist and shrewd connoisseur. The most famous spoils of the treasury of Grandmont—the plaques from the main altar, which had been dismantled in 1792—came into his possession in 1841 (cat. 57). Upon Du Sommerard's death in 1842, the city of Paris bought the hôtel and his collection, which, ceded to the state with the ruins of the baths, became the Musée de Cluny. The first curator was the architect Édmond Du Sommerard, son of the founder. For the sovereigns and magistrates in many Western capitals, the Musée de Cluny came to represent the model of a historical and archaeological repository of the arts of the Middle Ages, from Late Antiquity to the Renaissance. The collection of some two hundred Limoges champlevé enamels on copper remains the most archaeologically representative of its kind.

The initial presentation, although modified by later additions and reinstallations, was didactic in inspiration. Technical considerations, stylistic criteria, and chronological classification served alternately as standards of arrangement. But could the visitor without instruction commit this information to memory? In their descriptions of objects, in lithographs, and in printed catalogues accompanied by colorplates, other museums would follow the example set by Alexandre Du Sommerard and his son Édmond, from 1834 onward, in their publications concerning the Hôtel de Cluny and the Palais des Thermes and in the six volumes of the *Arts au Moyen Âge*, which appeared from 1838 to 1846.

### From the Belle Époque to the Great War

The "mobilization" of Limoges enamelwork in the nineteenth century soon brought a call to action from the "bourgeois conquerors," as Fernand Braudel called them. Two of note were the reputed author Ferdinand de Lasteyrie and as of 1842, the future industrialist David Haviland, who established by sailing ship or steamer a propitious coming-and-going for the "arts of the kiln" (enamelwork and porcelain) between Limoges and the East Coast of the United States. As early as 1884 the Wadsworth Atheneum in Hartford, Connecticut, had acquired a Limousin champlevé enamel medallion. About the same time the Haviland factory made plaster casts of the plaques of the chasse of Le Chalard (fig. 118*a*) in the interests of the exhibition of enamels at Limoges in 1886.

Knowledge of the Limousin patrimony would increase under the Third Republic: with the use of early inventories and liturgical and iconographic commentaries as sources, the history of the enamels began to unfold. Well-traveled scholars, such as Charles de Linas and Xavier Barbier de Montault, compiled lists of collections throughout Europe. Their surveys and discoveries provided the foundation for the decisive work carried out by Ernest Rupin. A competent scholar and skilled draftsman, photographer, and engraver, Rupin endowed French archaeology with the first illustrated monograph on the cloister of Moissac. He applied the experimental methods of the naturalist to art objects, classifying works in enamel according to categories determined by their original purpose, decoration, and destination. Along these extremely useful lines he conceived his fundamental *L'Oeuvre de Limoges*, which was issued in Paris in 1890 by Alphonse Picard, publisher for the National Archives and for the Society of the École des Chartes, the prestigious training school for French archivists established by Napoleon. As only two hundred copies of this massive volume were printed in Brive-la-Gaillarde (Corrèze), under the daily supervision of the author, this book is a treasure for both the bibliophile and the connoisseur of enamels.

At the end of the nineteenth century one could admire in Saint Petersburg the collection of enamels given to the czar for the Hermitage Museum by Alexandre Basilewsky, the Russian mining industrialist who had acquired them in Paris in 1861 at the sale of the collection of Prince Soltykoff. Émile Molinier was preparing his monumental *Histoire de l'orfèvrerie* (History of goldsmiths' work), which would appear in 1903. He traveled to Poland to study and describe the enamels that had been brought back after 1871 by the Czartoryski princes from the Hôtel Lambert, their Parisian residence. The works acquired by Isabelle Czartoryski, Countess Dzialinska, especially in 1858, at the sale of the collection of Albert Germeau, the late former prefect of the Limousin, and also in 1865, were displayed at Goluchow Castle near Poznań; the part of the collection acquired by Prince Czartoryski, head of the family in Kraków, benefited from its legal status as an inalienable foundation, which had been conferred on it by the Austro-Hungarian Empire.

The descriptions of enamels written by Émile Molinier, curator at the Louvre, were models of the structural precision taught by Alexandre de Laborde and then Alfred Darcel, his predecessors in the Département des Objets d'Art. They recorded the exact artistic and technical appearance of the objects that were later subject to tragic depredations when Poland was occupied during the two world wars. Beginning in 1939, the difficulties of protecting

Goluchow Castle led to the dispersal—to Boston, Minneapolis, and Paris—of the beautiful and enigmatic appliqué groups by the Master of the Passion (cats. 119d, 122).

The pivotal Exposition Universelle of 1900 in Paris inaugurated a new phase in the political and financial organization of such economic and didactic events. New works of art came to light; the tastes of many collectors were revealed through exhibitions, and the generosity of some patrons was unveiled—not only did they collect choice objects, but they also sponsored the study of these works, fostering the vocations of historians of the art of enameling. At the Villa Madrid, the residence of the Martin Le Roy family in Neuilly, outside Paris, there were many remarkable enamels, including those acquired at the sale of the Frédéric Spitzer Collection in 1893. The master of the house knew the young historian Jean-Joseph Marquet de Vasselot; he admired the soundness of Marquet de Vasselot's ideas, as expressed in the chapter on Oriental influences in André Michel's *Histoire de l'Art*, published in 1905. He noted the intellectual rigor that, in 1904, had led Marquet de Vasselot to choose vermiculé work as a suitable element in the "paleography of ornament" for characterizing and dating a significant group of Limousin reliquaries.

In 1906 Marquet de Vasselot published *Orfèvrerie et émaillerie* (Goldwork and enamelwork), the first volume of the *Catalogue raisonné de la collection Martin Le Roy*. Then, as if in a romance by Chrétien de Troyes, the great collector gave his daughter Jehanne in marriage to the young scholar. Marquet de Vasselot was then honing his expertise at the Bibliothèque Nationale by compiling in his own hand an extensive chronological card index of catalogues of public auctions; until that time, this invaluable resource—an unclassified section of the main reading room, or *Salle des Imprimés* (printed books division)—had been sorely neglected. During these same years he began to inventory the archaeological and historical journals in the Département des Périodiques, systematically classifying his findings in a subject index. At the Musée National du Louvre, to which he was assigned, he published his *Catalogue sommaire de l'orfèvrerie, de l'émaillerie et des gemmes, du Moyen Âge au XVII<sup>e</sup> siècle* (Summary catalogue of goldsmiths' work, enameling, and precious stones from the Middle Ages to the seventeenth century) in 1914, in which he classified works according to schools. Limoges champlevé enamels are included in it and sometimes illustrated (nos. 51–111).

During the Belle Époque the fate of the oldest and rarest enamels on copper which were attributable to southern France and still on the market was sealed between Paris and New York. In 1911 the decorator and architect Georges Hoentschel underwrote the publication of a series of lavish volumes devoted to his holdings, *Collection Georges Hoentschel*. The introduction and commentaries by André Pératé were accompanied by plates with hand-colored illustrations. Was it this refined presentation that convinced the American financier J. Pierpont Morgan to buy the complete set of objects that appeared in volume two, *Émaux du XII<sup>e</sup> au XIV<sup>e</sup> siècle* (Enamels of the twelfth to the fourteenth century)? In any case, in 1917, that fateful year in world history, Morgan's splendid donation of enamels was received at the Metropolitan Museum. This lot of some one hundred cloisonné and champlevé enamels on copper included the oldest and rarest examples of southern European enamelwork: the incunabula attributable to the monastic workshop of Saint-Foy of Conques in the Rouergue, active from the early twelfth century (cat. 8e–i). These plaques and medallions confirmed the hypothesis that there were distinct masters; their works would propagate two different graphic and chromatic modes, one of which found root in Limoges and the other in Silos and Castile.

*Enamels as the Furtive Tokens of Peace*
Place and movement, these are the antithetical yet complementary principles that instilled a fertile tension between the sacred sites where the holy bodies of the saints reposed, attracting pilgrims, and the routes by which these travelers brought back fragments of venerated relics to far-flung destinations, where new cults of worship developed. For the precious treasures of the Old World, the same proverbial couple, *motus / locus*, characterized, during the years 1919–40, the dynamic relationship that existed between the attractions of the marketplace and the stable and secure havens represented by the newly founded museums of the United States. The dealer was called upon to be a connoisseur, and the expert was required to provide documentary and textual references: Stora, Brimo de Laroussilhe, Landau, and Ratton, those modern pilgrims of enamels, carried letters, notes, history books, and catalogues in their steamer trunks.

In 1925 Picard published Marquet de Vasselot's treasure trove of references, the *Bibliographie de l'orfèvrerie et de l'émaillerie française* (Bibliography of French goldsmiths' work and enameling) for the Société Française de Bibliographie. The author donated copies both to Limoges and to Princeton, initiating a protocol for the transatlantic exchange of information, whether in the form of card files or printed books. Marquet de Vasselot established himself in London and Paris as the leading European expert in the ceramic arts and enamelwork. A fellow of the Society of Antiquaries in London and resident member of the Société Nationale des Antiquaires de France, he divided his labors as connoisseur and author between formal classification and stylistic attribution, between cataloguing groups of works and writing monographs on specific objects. Finally, by studying the evidence of funerary archaeology, he attempted to reconstruct the complete provenances of objects, tracing them back to their original owners. The last work to be published during his lifetime was the 1941 *Crosses limousines du XIII⁰ siècle* (Limousin crosiers of the thirteenth century). Here, priority is given to classification by iconography, which not only facilitates identifications but also helps to establish chronologies of related objects by means of comparisons; furthermore, it sheds incidental light on ecclesiastical history and the aspirations of a medieval clientele that was dispersed throughout Europe.

The dire international situation and the tragic circumstances of World War II prompted Marquet de Vasselot to express, as if in a will, his final scholarly opinions in the introductory chapters of *Crosses*. He had established a definitive rearrangement for five arched plaques inscribed with the names of the enthroned saints. These remnants of an apostolic program (cat. 58) were probably adornments of the main altar at Grandmont and would have included among their number a plaque now in Florence bearing the image of Saint Martial, inscribed *Apostolus Christi*, making up the baker's dozen so typical of Limousin works (see Barrière, below p. 20, and cat. 58).

Similar considerations governed the cataloguing of the Limousin gemellions, which Marquet de Vasselot undertook with the heraldry expert Max Prinet; their publication was entrusted to Pierre Verlet in 1950 by Mme Jehanne Marquet de Vasselot. To the Département des Objets d'Art at the Louvre would fall the task of evaluating how Limousin enamelwork was received by lay society in the thirteenth century, as revealed by secular imagery, courtly literature, and heraldry. During the 1930s Limoges inspired a new American scholar, Marvin Chauncey Ross. He was welcomed into the fold by his compatriot William Haviland, head of the industrial dynasty that had founded the famous porcelain factory in 1858. Ross assiduously explored the churches of the Auvergne and the Limousin and the archives of the Haute-Vienne; in the latter he pored through the *carnets* compiled by the sedulous nineteenth-century archivist Bosvieux and discovered, for example, that the dismantled plaques reassembled as a chasse at the Metropolitan Museum (cat. 10) had belonged to the church at Champagnat in the Creuse in the nineteenth century. He gleaned historical data from the perusal of hundreds of pages in the bulletins of the scholarly societies of Limoges, Guéret, Tulle, and Brive. Elected a member of the Société Archéologique et Historique du Limousin, he argued that Limousin art had spread to Lower Saxony as well as to Aragon, Galicia, and Catalonia during the time of Henry the Lion (r. 1142–80), and he discussed the question of Spanish workshops at length with W. L. Hildburgh.

By the 1950s many of the most beautiful enamels had crossed the Atlantic and were in museums on the East Coast of the United States. The Walters Art Gallery in Baltimore appointed Ross curator of its collection of early Christian and medieval antiquities, which included many important Limousin enamels; the collection formed a counterpart to the admirable trove of fine bindings and illuminated manuscripts long under the custodianship of Dorothy Miner. For a while Ross diverted his attention to supervise the collection of nineteenth-century sculpture by Antoine-Louis Barye. He returned to his predilection for Byzantine art with the Council of Europe exhibition held at Athens in 1964 and with his catalogues for Dumbarton Oaks in Washington, D.C. Faithful to the Limousin "Byzantine enamels," he was among the first to visit the memorable exhibition "Limousin Enamels of the Twelfth, Thirteenth, and Fourteenth Centuries," held in the Musée Municipal de l'Évêché at Limoges in 1948.

International economic crises, bloody dictatorships, spoliations, persecutions: peace was precarious in Europe, in contrast to the polite exchanges between the defenders of monumental archaeology. Nationalistic claims regarding the first Romanesque art were made by France and Spain at Yale and Harvard. The guests from Paris, Henri Focillon and Marcel Aubert, entered into competition with Kenneth John Conant and Arthur Kingsley Porter. José Gudiol, who had managed to photograph most of the enamels preserved in Catalonia for the Amatller Institute of Barcelona during the Spanish Civil War, had been condemned to death; in exile he frequented the Hispanic Society of America.

Physicists, archaeologists, historians, and philosophers were expelled from Germanic countries, where they were threatened by Nazism, and immigrated to Anglo-Saxon nations; most were Jews; all were confirmed democrats. The Warburg Institute was established in London. Art historians from Berlin, Frankfurt, Hamburg, Leipzig, Dresden, and Vienna left Germany and Austria forever. To the passengers landing in New York, the Statue of Liberty loomed as a monumental icon of political and intellectual hope.

They transmitted their knowledge to many in America, who soon became their disciples, and they brought to the universities proven scientific methods and to the museums curatorial expertise and experience with collectors. Their books and articles were catalogued and read in model libraries: Columbia, Harvard, Yale, and Princeton provided studious readers free access to their stacks.

The career of Georg Swarzenski, always linked with that of his son Hanns, seems exemplary in this context. Scholars of rare stature, captivated by the beauty of works of art, they loved to share their knowledge and appreciation and did so with great enthusiasm and irony. Georg was born in Dresden in 1876, received a doctorate in art history in Heidelberg, moved to Berlin to become an assistant at the Kaiser-Friedrich Museum, and was made director of the Städelsches Kunstinstitut in Frankfurt in 1906 and of the municipal museums in 1928. The wealth of his experience was reflected in his personal photographic library, which Hanns continued to expand. Forced to resign in 1937, Georg became a research fellow at the Boston Museum of Fine Arts, where he undertook the challenging task of forming a collection, based on European models, that would rival those at the other new American museums.

During the war the invasions of European countries resulted in the plundering of collections and the dispersal of royal treasures, such as the *Welfenschatz* (Guelph Treasure). In Boston, however, Georg Swarzenski succeeded in bringing together works evoking the princely patronage of the Guelphs and the Hohenstaufen and that of their allies or enemies, the Plantagenets and the Capetians. But he did not forget the southern charm of the Limousin Eucharistic dove, formerly in Erfurt (now in a private collection), and in 1949 he wrote to Limoges to inquire about the Baptism of Christ (cat. 122) and later acquired this exceptional relief. I was able to identify the earlier provenance of this Baptism plaque in the Goluchow catalogue, not without a certain pride.

A *trésor imaginaire* of objects from the Romanesque period—enamels, bronzes, ivories, and goldsmiths' work—took shape in Hanns Swarzenski's *Monuments of Romanesque Art: The Art of Church Treasuries in North-Western Europe* (1954). He carried his vision of the treasury everywhere—to exhibitions, colloquia, and friendly gatherings—mischievously questioning his best-informed colleagues about the true antiquity of famous masterpieces.

The Institute for Advanced Study, founded in 1930 in Princeton, was a village of scholars a short distance from the Neo-Gothic buildings of the university, where graduate students might meet Jacques and Raïssa Maritain engaged in sacred dialogue with Thomas Aquinas. The members of this learned community, trained in a broad range of liberal arts, resolved themselves by cooptation into a new order. The major modes of learning distributed the adept in three schools—mathematics, physics, and history—but the statutes allowed friendships to flourish independently—witness the example of Albert Einstein and Erwin Panofsky. Members of the Institute participated in research seminars conducted by other university faculties, such as that of art history, where Professor Morey had laid the foundations for the Index of Christian Art.

At the Index, *realia*, works in diverse arts, were understood as corporeal embodiments that gave images and their conceptual content not only form but also beauty. This inventory of medieval Christian art was intended to document every photograph with relevant bibliographical references. Kurt Weitzmann, who with Adolf Goldschmidt had just completed the multivolume *Elfenbeinskulpturen* (Ivory sculpture), ensured a place for the study of ivories at the Institute for Advanced Study. Frederick Stohlman was entrusted with the international repository of enamels.

*Transatlantic Shuttles Between Paris and New York*
The roughly two hundred Limousin enamels exhibited during the summer of 1948 at the Palais de l'Ancien Évêché in Limoges bore witness to the renewal of an Atlantic alliance in the arts. The works seemed like the tokens of a newfound peace. Pierre Verlet had responded in 1948 to the proposal of Serge Gauthier, curator of the municipal museum in Limoges. Visitors from Great Britain and the United States attended the opening of the exhibition. Over the summer photographers from France, the United States, and the Netherlands took pictures in color and in black and white. The finest were published by Gérard Le Prat in 1950 in *Émaux limousins des XIIᵉ, XIIIᵉ, XIVᵉ siècles*, with a preface by Pierre Verlet; the book introduced the practice of using Ektachromes for color illustrations (the proofs were prepared by Mourlot Frères). All the proofs and the slides were catalogued in the municipal library of Limoges, complemented by indispensable local and international bibliographical references. Charles Oman and William King in London, Erich Meyer in Berlin and Hamburg, and Léon Dewey in Liège graciously provided full sets of photographs of the Limoges works in their museums. In 1952 a new UNESCO regulation allowed the municipality of Limoges to purchase pictures from photographic archives not only in Paris but also in Brussels and in Munich. Lastly, Marvin Chauncey Ross provided a list of the American museums that owned enamels.

Frederick Stohlman traveled every year through a now-peaceful Europe, examining and photographing enamels, which were his special interest at Princeton University. It was from Princeton that his disciples set forth to museums in New England and the Midwest; to France, England, Italy (his adopted country), Spain, Switzerland, and Scandinavia—all were methodically explored. In Sweden the publication in 1951 of *Reliker och relikvarier fran Svenska Kyrkor, tillfällig utstallaing* by M. Rydbeck, M. Thordeman, and Q. Kallström prompted Stohlman to praise the objective of the Swedish Nationalmuseum, which was to reunite and conserve those objects that long ago had strayed north of the Baltic (Norway, although Lutheran as well, had left its Limousin treasures in its *Stavekirke* [stave churches]). In Copenhagen the task was easier, for in 1696 the Danish king had the royal collections consolidated and housed in the Kunstkammeret of the Royal Library, and described in *Museum regium seu, catalogus rerum tam naturalium quam artificialium.*

"Freddy" Stohlman retired in 1961 but continued to contribute new documentary photographs to the Index of Christian Art. He traveled to Switzerland with his family. He had met his wife in Rome, about the time that Cardinal Spellman, archbishop of New York, donated the first file-card copy of the Index to the Vatican Library. Stohlman later went on to Limoges to check the documentation of the photographic collection of enamels. Julien Cain, general administrator of the Bibliothèque Nationale, with grants from the Centre Nationale de Recherche Scientifique (CNRS), had arranged for the endowment of the nascent enterprise that was already being called the Corpus des Émaux (Corpus of enamels), by analogy with the Corpus Vitrearum Medii Aevi (CVMA) and with Jean Porcher's grand project of photographing illuminated manuscripts in France.

Absorbed by classifications, subject indexes, and the analysis of periodicals—in short, fascinated by catalogues—the librarian Julien Cain enthusiastically accepted the cooperative project. In a studious alliance begun in 1957, we would carry forward together the Census developed at Princeton.

In 1948 James J. Rorimer of the Metropolitan Museum, who had played a vital role in the restitution of works of art after World War II, visited Limoges for the first time, spending almost a week at our exhibition. He was eager to discover an unknown masterpiece in our newly liberated provinces but only, he joked, if it were on the scale of the Ambazac chasse (cat. 55). The Morgan Wing at the Metropolitan Museum had been open since 1925, and visitors consulted the handbook by Breck and Rogers to guide them through it. In 1941, the third volume of the *Catalogue of the Collection of George and Florence Blumenthal* was published. In that same year, Rorimer complemented the Morgan Collection of enamels with several rare and early vermiculé pieces received as gifts

from that collection. He had pushed forward the construction of The Cloisters in Fort Tryon Park in order to have a place to display the collections of French monumental sculpture and tapestries, as well as a small treasury of Limoges enamels.

William Forsyth, who had completed his studies at the university in Princeton, where his family lived, marked his arrival at the Metropolitan Museum in 1946 with an article on the new installation of enamels. With the flair of an archaeologist, he raised the long-debated question of a Celtic origin for the technique of enameling in red and sometimes in white or blue on small bronze champlevé vases, at the same time treating an unpublished object that was as rare as it was representative. But his predilection was for late medieval sculpture. Throughout his career he made discoveries and astute attributions. His work often took him to France, where, as a member of the Société Française d'Archéologie and a close friend of its director, Francis Salet, he continually added to our knowledge of sculptural monuments in Champagne and in Lorraine. To this day he regularly contributes commentaries and aesthetic exegeses on these sculptures. In *The Pietà in French Late Gothic Sculpture, Regional Variations* (1995), he devoted a long chapter to Gascony and the Limousin.

On the banks of the Hudson Bill Forsyth directed The Cloisters, that great theater of the arts of the Middle Ages, with serenity during the administrations of James Rorimer and Thomas Hoving. He generously encouraged European visitors to make their own discoveries. Medievalists, racing back and forth from the Morgan Wing to the Pierpont Morgan Library, could meet leisurely in his office. The courteous camaraderie that prevailed under the Metropolitan's auspices brought together the critical tradition of scholarly exchange and the discreet generosity of the connoisseur. The rigorous confidentiality of the art market in New York conformed to the customs of London, Paris, and Munich. Vera Ostoia and Carmen Gómez-Moreno shared responsibility for the Limousin collection: the former took charge of painted enamels and the latter, of champlevé enamels, in what was an amiable sharing of personal experience and technical investigation. Philippe Verdier, working at the Walters Art Gallery, often came from Baltimore to join his compatriots, for, together with André Chastel, Georges Gaillard, and Louis Grodecki, he was among the most loyal disciples of Henri Focillon, the illustrious historian of medieval art and professor at the Collège de France, who took a chair at Yale in 1939. He died in New Haven in 1943, but his doctrine, "The Life of Forms," which continued to be taught at the university, radiated throughout the United States.

Over the last twenty-five fruitful years art historians have coordinated professional committees under the auspices of UNESCO. The CIHA (Congrès International d'Histoire de l'Art) met in Paris in 1958 and in New York in 1961—both propitious occasions. At that time the Comité Français (CFHA), under the aegis of the new Ministry of Culture, recently founded by André Malraux, requested that all archives, libraries, museums, historic monuments, and public institutions that held source material indispensable to art and archaeology coordinate their initiatives. Curators on all levels were asked to work in concert with university professors and researchers, regardless of their national affiliations, be they "foundations," "centers," or "academies." The Zentralinstitut für Kunstgeschichte in Munich endowed the history of Western art with a kind of epistemological balance bar, linked on the one hand to the Institute of Fine Arts at New York University and on the other to Princeton and the Institute for Advanced Study, thanks to the generous ubiquity of Willibald Sauerländer. The Metropolitan Museum invited visiting German scholars to work on cataloguing the collection in their areas of specialization.

Erich Steingräber, former director of the Germanisches Nationalmuseum in Nuremberg, participated in 1946 in the twenty-first CIHA Congress in Bonn, the subject of which was *Westeuropäisches Kunst im 1200* (Western European art in 1200); in 1967 he published his decisively synthetic study of enamels in the article "Émail" (Enamel) in the *Reallexikon zur deutschen Kunstgeschichte* (Lexicon of German art history). For a year he worked in depth on the Metropolitan's collection, after which Dietrich Kötzsche succeeded him in 1967 for a semester. Hermann Fillitz came from Vienna to study the insignia of the Holy Roman Empire and other regalia that had immigrated to the New World. At the time the concepts of "rupture" and "continuity" in the life of forms dominated research, and this approach shaped the unforgettable exhibition "The Year 1200," for which Florens Deuchler was engaged as curator in 1970. Teaching in Geneva, he had just published his thesis on the Ingeborg Psalter, which established important chronological signposts for the years 1180 to 1215. "*Homo helveticus,*" as he jokingly characterized himself, Deuchler was instrumental in forming in Switzerland a melting pot of multilingual scholars from Bern, Zurich, Basel, and Riggisberg, who knew how to coordinate, as Robert Hahnloser did for the CVMA, their intellectual and financial resources.

Semiprecious yet sturdy, Limousin enamels regularly made European exhibitions sparkle with their azure and gold. In 1961, they were present at "Arte Romanico," organized in Barcelona in a heroic wager undertaken by Catalonia and the Council of Europe (the catalogue, however, was not published until 1963). In France enamels were exhibited almost every year, but especially in 1965, when they were included in the exhibition of the royal collections of Denmark at the Louvre and in that of the treasures of the churches of France. In 1963 France displayed enamels, with their brilliant and almost otherworldly polychromy, at the Vatican. Under the aegis of the Council of Europe, the exhibition "L'Europe Gothique," which opened in May 1968 in Paris, dared to juxtapose, albeit on a modest scale, the metal "hardware" of Limousin tomb effigies with dazzling marble sculptures from Italy, a confrontation similar to the subtle counterpoint that Cesare Gnudi orchestrated for Duecento painting. At all these events scholars working with the Corpus of Enamels studied, observed, compiled lists, and drafted or translated entries in exchange for good photographs. The annual Enamel Symposium, which since 1979 has brought together specialists in medieval enamelwork at the British Museum, has afforded the Department of Medieval and Later Antiquities the opportunity—on the initiative of Neil Stratford—to exhibit the Keir Collection, the most important modern collection of champlevé enamels. In 1963, I relinquished my post as librarian in Limoges and became affiliated with CNRS; my mission was to complete the photographic and documentary inventories of the Corpus, country by country, in preparation for a catalogue raisonné in which all the pieces would be described and grouped in Cartesian fashion.

Upon the recommendations of Francis Wormald, André Grabar, and Sirarpie der Nersessian, I was invited to Princeton by the Institute for Advanced Study for a semester in 1964. The day after my arrival on January 6, I was taken by Millard Meiss to the Green House, where Frederick Stohlman then had an office, affiliated with the Index of Christian Art. In 1971 Suzan Stohlman, executor of her father's will, together with Rosalie Green, decided to send to the Corpus des Émaux, then housed in the Manufacture de Sèvres, the chest containing the papers, courses, list, and census of the cofounder of the Corpus.

After observation and documentation, the transition to description was easy in Princeton's intensely studious atmosphere. Frequent sessions at the Metropolitan Museum and trips to Chicago, Detroit, Cleveland, Cincinnati, St. Louis, and Toledo made it possible for me to edit and make drawings on standardized index cards, according to French norms. My visits to Washington and Baltimore often lasted a week.

Neither the agricultural wealth and industrial power of Ohio, nor the pleasant harborside setting on the shore of Lake Erie suffices to explain the Cleveland Museum's enduring mission: the superb balance of this collection of fine art from all over the world is proclaimed by the sure provenance and verifiable origin of each piece—in short, by the excellence of the works of art. The talents of the curators are equal to the rarity of the objects entrusted to their care. The collections come from five continents. The originality of the choices, whatever the medium, is sufficient to transport the visitor on imaginary journeys to China, Central Asia, medieval Europe, and even to the Limousin. In a joint venture inspired by shared friendship and profession, Hubert Landais and William Wixom enlivened the winter of 1966–67 with the exhibition "Treasures from Medieval France." One could admire the appliqué groups by the Master of the Passion (cat. 119), brought together for the first time since their dismantling in the wake of the French Revolution.

As the annual meeting of the College Art Association of America was being held that winter in Cleveland, I thought it would be fitting to present a plausible and definitive reconstruction of the altar frontal, most likely a retable, from which these reliefs, datable to 1240–50, had been detached. My hypothetical drawing relied on the comparative resources provided by the inexhaustible Index of Christian Art, which I examined with Rosalie Green. It was also a perfect opportunity for William Wixom to restore the Spitzer Cross (fig. 63a)—attributed by the Corpus to the Plantagenet royal workshop—to its original, slender proportions. Together we separated it from the thick, garnet-colored velvet support to which it had been so awkwardly attached since the days of Frédéric Spitzer.

Over the course of the semester I spent at Princeton in 1967, following a second invitation from the Institute for Advanced Study, the realization of the Corpus in book form benefited from a number of scholarly discussions with Henri Seyrig, Ernst Kitzinger, and Richard Ettinghausen. A program for future work was sketched out, combining archaeology and the history of enamelwork in Limoges. In fact, as I was preparing a lecture to be given at the Metropolitan Museum, discussions were held regarding different ways of organizing the presentation of the Corpus in book form. The convincing and well-founded suggestion put forth by Florentine Mütherich, who was teaching at Columbia University that winter, prevailed over my "geographic" prejudices. The work would be divided into chronological phases, technical stages, and historical periods—in short, it would be published in a series of definitive volumes incorporating observations and analyses worked out in monographs that had been written while the catalogue was being prepared.

It was from this perspective that the Musée du Louvre and the Metropolitan Museum organized the present exhibition, which was accompanied by a colloquium in Paris. At the same time, *L'École de Limoges, 1190–1216* (Southern Enamels, international catalogue of *L'Oeuvre de Limoges*, vol. 2) is about to go to press in Limoges. It is hoped that Danielle Gaborit-Chopin, zealous champion of Saint Martial of Limoges, apostle of Aquitaine, will sign with the founders of the Corpus des Émaux Méridionaux a new alliance for the Corpus Smaltorum Medii Aevi.

# The Limousin and Limoges in the Twelfth and Thirteenth Centuries

BERNADETTE BARRIÈRE

It would be ideal to begin with a discussion of the place of enamelwork in the economic and social fabric of the city of Limoges as well as in the Limousin region as a whole. But the written sources, entirely lacking in documents of a financial or economic character, do not allow for such an approach—champlevé enamel appears only in the most incidental and fleeting manner. We are left with the paradox that, while we possess a great number of enameled objects that bear witness to the importance of Limousin production, we can learn almost nothing from archival documents. We have real information about the topography of the Limousin and its religious and political world in the twelfth and thirteenth centuries, but we are ignorant of the economic and socioeconomic life of Limoges. And so we invite the reader to join us in exploring the significance and implications of what is known.

## The Limousin and Limoges

Occupying the western part of the Massif Central, the Limousin is a geographically and historically coherent area. It is a region of plateaus (elevation 200–500 meters [650–1650 feet]), deeply furrowed by an extensive river system. Well before the year 1000 there was a relatively dense rural population. During the twelfth and especially the thirteenth century several dozen towns sprang up, stimulated by the presence of a prosperous clerical or monastic community, a powerful lord's château, or a well-traveled road (or a combination of these factors). The high plateaus to the east did not experience this urban activity but maintained a significant rural population.

The Limousin corresponds to the present-day departments of Haute-Vienne, Creuse, and Corrèze, as well as Charente and Nontronnais (Dordogne). In the Middle Ages it was described, in religious and secular terms, as the diocese and the county of Limoges. The boundaries of the diocese remained virtually unchanged until the Revolution, but by the tenth century the administrative division of the county was replaced by a succession of great seigneuries: the viscounties of Limoges, Comborn, Turenne, Aubusson, Rochechouart, Ventadour, and Bridiers and the county of the Marches. These seigneuries, of unequal importance, were part of the duchy of Aquitaine, with the counts of Poitiers/dukes of Aquitaine as suzerains. In the middle of the twelfth century their lineage was linked to the Plantagenets, and their activity sometimes crossed over into Berry, Poitiers, Angoulême, Périgord, and Quercy. The term "Limousin" was applied only to the traditional territory of the diocese, itself heir to the Roman city Augustoritum Lemovincensium (later Lemovices). Medieval Limoges was built at the principal site of this Roman city. In 1000 there were three bases of power in this urban settlement on the gentle northern slopes of the Vienne Valley: the Cité (the episcopal district), the Château of Saint-Martial (originally consisting of just the abbey with its immediate dependencies, and the château of the viscount of Limoges (on the monastic estate, perched on a hill 150 meters [500 feet] above the abbey).

Over the course of the eleventh and twelfth centuries the population around the abbey and the château grew considerably, and the newly settled areas were gradually enclosed by defensive walls. On the eve of the thirteenth century the Château of Saint-Martial (or the Château of Limoges) was well on the way to becoming an important urban complex. The town's burghers, who had already established a magistrate, were determined to make the Château a walled city worthy of the name, able to protect not only its inhabitants but also newly arrived settlers and to demand proper notice from the abbot, the lord and proprietor of the area. In 1212 they began the Château's great wall, which endured for centuries but of which nothing now remains. This wall was 12 meters (40 feet) high and had a circumference of 2,000 meters (6,500 feet); the area enclosed amounted to 32 hectares (80 acres). The wall's eight gates had towers and drawbridges. Outside the wall was a moat 20 meters (65 feet) wide and

Fig. 1. Map of Limoges in the thirteenth century (after B. Barrière)

1. Cathedral of Saint-Étienne
2. Baptistery of Saint-Jean
3. Abbey of Sainte-Marie-de-La-Règle
4. Church of Saint-Maurice
5. Church of Saint-André
6. Saint-Étienne Bridge
7. Abbey of Saint-Martial
8. Church of Saint-Pierre-du-Queyroix
9. Church of Saint-Michel-des-Lions
10. Viscount's Residence
11. Place des Bancs
12. Les Combes Quarter
13. Creux des Arènes
14. Church of Sainte-Marie-des-Arènes
15. Church of Saint-Cessateur
16. Abbey of Saint-Martin

17. Church of Saint-Paul
18. Abbey of Saint-Augustin
19. Church of Saint-Christophe
20. Church of Saint-Julien
21. Franciscan Monastery
22. Butchers' Quarter
23. Manigne Quarter
24. Church of Saint-Gérald
25. Hospital of Saint-Gérald
26. Dominican Monastery
27. Church of Saint-Michel-de-Pistorie
28. Church of Sainte-Valérie
29. Church of Saint-Symphorien/Sainte-Félicité
30. Saint-Martial Bridge Quarter
31. Saint-Martial Bridge
32. Ford of La Rouche-au-Gôt

33. Mireboeuf Gate
34. Montmailler Gate
35. Arènes Gate
36. Lansecot Gate
37. Pissevache Gate
38. Manigne Gate
39. Vieille-Monnaie Gate
40. Butchers' Gate
41. Panet Gate
42. Saint-Maurice Gate
43. Scutari Gate
44. Traboreu Gate
45. Chêne Gate
46. Naveix Gate

7 meters (23 feet) deep. The areas enclosed were: the monastery of Saint-Martial, the parishes of Saint-Pierre-du-Queyroix and of Saint-Michel-des-Lions and their dependencies, and the viscount's residence, as well as older residential sections and newer peripheral neighborhoods. This ambitious project bore ample witness to the hopes the burghers had for their city's future.

Yet this was not the oldest complex: some 500 meters (1,650 feet) east of the Saint-Martial enclosure stood the defensive walls of the Cité, the most ancient part of Limoges. This precinct was dominated by the bell tower of the cathedral of Saint-Étienne and was marked by more traditional, if less dynamic, development. Situated on a steep slope, it had its roots in the ancient Roman settlement. Its bishop, the most eminent man in the Limousin, was its sole lord, and the Cité was dominated by the presence of the Church. The cathedral, whose Gothic reconstruction would not begin until the late 1200s, was located roughly at the center of the Cité. To the south were the episcopal palace and the residences of the canons of the cathedral chapter, and to the north was the baptistery of Saint-Jean. The walls of the Cité were apparently rebuilt in the thirteenth century following a plan similar to that of the Château's. Nothing remains of these walls, but old ground plans and texts indicate a fortification with five gates. At the east this defensive wall was joined to that of the Benedictine abbey of Sainte-Marie de la Règle, protecting the route to the Saint-Étienne bridge, whose completion in the thirteenth century gave the Cité its own way across the river. A large part of the Cité's population was part of the world of the clerics and their households. Burghers, however, also lived there, and by the early thirteenth century a magistrate and a council of elders are known to have existed (unfortunately nothing is known of who they were or what they did).

Any description of Limoges in the Middle Ages must include the life that developed *extra muros* (outside the walls). Easily accessible because of its situation on the Vienne River and long the main town of the Limousin, Limoges was set at the crossroads of major trade and travel routes. Until recent times most southeast-northwest traffic passed over the Roman bridge of Saint-Martial. A lively quarter grew up near the bridge: houses were pressed together along the thoroughfare, and the area had its own church near the ruins of a Roman theater. Many tracks, leading either toward the Cité or toward Saint-Martial, originated at the bridge. The earliest route went directly northwest, avoiding the urban settlement. Half way up the slope it crossed an important southwest-northeast route, which continued on to pass between Saint-Martial and Saint-Pierre-du-Queyroix: it was at this crossroad that

the viscount's château was located. The building of the city wall, however, blocked the crossroad, and these routes had to deviate to use the city's new gates.

To the west of the Château of Saint-Martial lay the Arènes quarter, dominated by the Creux des Arènes, the imposing ruins of the Roman amphitheater, which were often used for fairs or as a site for preaching. Between the Château and the Cité was found the Manigne quarter, to which in about 1240 the Dominicans moved from across the river, and the Boucherie (butchers' quarter), an area of tanners and artisans, where the Franciscans established themselves in 1243, moving from their previous inadequate quarters. Also located outside the walls were the Benedictine abbeys of Saint-Martin, around which a small town grew up, and Saint-Augustin, a traditional burial place of the Limousin bishops.

This overview demonstrates that Limoges underwent profound changes during the period when its enamel workers were most active. In the early twelfth century the physical layout of Limoges had been shaped by the ups and downs of a long and complex history. The Limoges of the late thirteenth century bore little relationship to this earlier town; it was instead marked by the deliberate ordering of urban space so characteristic of rich and powerful cities of the time.

## Saint-Martial

The abbey of Saint-Martial was built on the site of the necropolis where Martial, the fourth-century apostle to the Limousin, was buried. The cult that arose around the tomb of the figure presumed to be the first bishop of the diocese continued to grow, as Gregory of Tours reported in the late sixth century. During these early centuries the bishop of Limoges appointed priests to minister to the cults centered on the tombs of missionary saints; the tomb of Martial was honored by being served by several cathedral canons. In 848, during Charles the Bald's efforts to regain control over Aquitaine, the religious community devoted to Martial adopted monastic rule for both political and religious reasons. Thus the Benedictine abbey of Saint-Martial was formed under the sponsorship of the monks of Saint-Savin-sur-Gartempe; this establishment was intended not only to become an important abbey but also to promote devotion to Saint Martial.

In the twelfth and thirteenth centuries Martial was regarded throughout Christendom as a contemporary of Christ. The story of his life (*Vita prolixior* [Fortunate Life]) seems to have first appeared in the early eleventh century, and copies were to be found in many monastic libraries in the West. This text affirmed that he was a relative of Peter, that as a child he had lived among the apostles, that he had followed Peter to Rome, and that Peter had charged him to spread the Gospel in the Limousin and Aquitaine as a whole. This account, with its complementary tales of the virgin Valerie and the duke Stephen, was in official circulation (although it was locally contested in the eleventh century) and helped confirm the importance of the abbey that held Martial's remains.[1] Like Santiago de Compostela—that distant church whose fame rested on the presumed presence of the body of the apostle James—or the abbey of Saint-Jean-d'Angély—which claimed to possess the skull of John the Baptist—Saint-Martial of Limoges was among the most highly regarded sanctuaries of the West, especially given the remarkable creative, artistic, liturgical, and cultural life it had fostered since the tenth century.

The status of the abbey had changed in 1062–63 when it was annexed by Cluny, the reformed Benedictine monastery whose abbot was then Hugh (1049–1109). The local monks found this transition difficult, and late-twelfth-century witnesses testify that the takeover had not been forgotten. After the death in 1114 of Adémar, the sovereign abbot of Cluny, Saint-Martial constantly sought to loosen its ties to Cluny. Thus, until the mid-twelfth century the abbot of Saint-Martial was chosen from the monks of Cluny, but thereafter the abbey gradually became more autonomous. Even after Saint-Martial began to choose its abbot more freely, it remained part of the great Cluniac network, open to Cluny's influence and to its web of relations throughout the Christian world.

Beginning in the eleventh century, the abbey's buildings were frequently rebuilt, restored, or embellished. One element remained stable: the crypt of Saint Martial and the sanctuaries associated with it (these, now destroyed, were, to the east, the church of Saint-Pierre-du-Sépulcre and the chapel of Saint-Benoît; and, to the south, the pilgrimage church of the Savior, the great Romanesque structure more than 100 meters [325 feet] long). Little is known about the cloister and the other buildings that once stood south of the abbey, except that during the thirteenth century La Clautre (the old cloister) became a secular site housing the grain market and perhaps enamel workshops, which were probably run by lay people. Such a location would have been most advantageous for the enamelers. The abbey had long been a source of commissions for liturgical and devotional objects,[2] wrought in gold as well as enameled. In addition, the illuminated manuscripts in the abbey's library could provide a wealth of subjects and images. The many pilgrims who came to Saint-Martial were potential customers. Furthermore, some of the eighty priories that were associated with the abbey were situated on routes in southern France and Catalonia; commissions and deliveries could be arranged through this network.

## Other Limousin Saints

The Christianization of the Limousin occurred gradually during the Merovingian era. This work was initiated by Martial, his companions Alpinian and Austriclinian, and his immediate successor Aurelian. They were followed by a number of other evangelists, whom their contemporaries perceived as saints and whom the people of the Limousin, especially devoted to the concepts of relics and pilgrimages, long venerated. Among those whose tombs became the sites of cults were: Junien (located in Saint-Junien), Victurnien (Saint-Victurnien), Leonard (Saint-Léonard), Psalmet (Eymoutiers), Yrieix (Saint-Yrieix), Marien (Évaux), Pardoux (Guéret), and Martin l'Espagnol (Brive). The episcopacy entrusted most of these sanctuaries to chapters of canons that encouraged devotion to their respective saints by holding ceremonies and by undertaking the upkeep and decoration of their buildings. During a widespread renovation that appears to have taken place in the second half of the eleventh century, these communities became involved with periodic embellishment or replacement of the coffrets and chasses that housed the saints' relics. As far as was financially possible, they wanted to improve on what was already there: champlevé enamel must have been a possible stage in the movement from wood and lead to gold and silver.

The list of Limousin saints did not end with the Merovingian era; in the twelfth and thirteenth centuries new arrivals were welcomed into the celestial cohort. The Gregorian reforms of the eleventh and twelfth centuries seem to have been well received in the diocese of Limoges, where a number of churchmen became known for their goodness: Israël and Théobald in the chapter of Le Dorat; Gauthier in Lesterps; Geoffrey, founder of the chapter at Le Chalard; Gaucher, founder of the chapter and the small canonical congregation of Aureil; and Marc and Sébastien, founders of the small eremitical order of Lartige. Others followed different paths. Stephen of Muret, choosing a hermit's life, paradoxically attracted many disciples, who, after his death in 1124, organized themselves into an order, whose seat was Grandmont. On the other hand, Stephen of Obazine (d. 1159) went from hermit to monk; the founder of a number of monastic houses, he brought all of them into the Cistercian order in 1147.

The veneration of these figures soon after their deaths was often as ardent as that of saints from centuries past. Their remains were treated with great reverence. The renowned late-twelfth-century enameled and jeweled chasse of Stephen of Muret, today in the church of Ambazac, may have been the first container for the saint's remains. About 1260 a sculpted stone monument—a gisant (recumbent figure) within a pierced chasse—was placed above the burial place of Stephen of Obazine. Interestingly, this was probably made by the royal workshops of the Île-de-France, during a period when the king was commissioning enameled plaques from the Limousin workshops for the sepulchre of two of his children.

Hagiographical documents and monastic chronicles, from Gregory of Tours to Bernard Gui, record the great popularity of relics, both contemporary and ancient, in the Limousin. These were necessarily housed in reliquaries of greater or lesser value. Aside from the relics of the properly Limousin saints, there were other relics: those of Roman martyrs, those brought back from the East, and those housed there at the time of the Norman invasion. In his chronicle of 1183 Geoffrey of Breuil, a monk of Saint-Martial and sometime prior of Vigeois, was the first to draw up a list of the saints whose relics were, as he wrote, the "jewels" of the churches of the Limoges diocese. Some of the relics belonging to Saint-Martial had been transferred to other churches, but the abbey still held the sepulchres and remains of fourteen reputed saints. Six other Limoges churches sheltered the bones of local saints. As for the rest of the diocese, the chronicler does not claim to be exhaustive but, deliberately omitting "relics of the Sepulchre or the Cross of the Lord or those of his mother," lists another forty localities that prided themselves on possessing relics of a martyr or confessor.[3] The long lists of relics from Solignac, Grandmont, Saint-Étienne, and Saint-Augustin, among other Limoges churches, are further evidence of this extraordinary enthusiasm. There were also a number of small parish churches that had their own enameled chasses containing relics of saints whose names were not always preserved. This abundance of relics must have created a demand for all sorts of chasses and coffrets to preserve, transport, and display these treasures.

### Enamelwork in Limoges

It is probably not overly audacious to speculate that the multiplication of enamel workshops in Limoges—with few in other Limousin sites—was in response, initially and mainly, to a regional demand rooted in the piety of the monastic or canonical milieu. The technical and artistic mastery achieved by the Limoges enamelers soon created a broader demand—for a more diversified range of objects from buyers from a wider geographic area. Other factors also influenced the growth of this industry.

Although enamel cost much less than gold, an enameled work could be more beautiful than one in gold. Thanks to the presence of Saint-Martial and of other Limousin religious establishments such as Uzerche, Tulle, and Grandmont, Limoges had an important place in the communication networks of southern Europe.

There is an obscure area in this discussion: copper, an essential element in enameling, was not found in the Limousin. The central question—where did the enamelers get this metal?—remains unanswered. It is evident, however, that the expertise necessary to produce champlevé enamel was drawn from a long tradition of goldsmithery in Limoges. During the twelfth century wealth had increased and was more widely held; both religious and secular objects were commissioned either in gold or in enameled copper, gilded or not. The very sparse thirteenth-century accounting records preserved from Saint-Martial bear eloquent witness to the modest cost of enameled pieces. Bernard Itier, the abbey's treasurer and later its librarian, noted that in 1208 two enamel (*d'esmaus*) candlesticks were bought for four livres and that in 1211 two iron crosses were replaced by two gilded (*dorées*) crosses whose price was twenty-three sous, and that a silver coupe (*ciborium*), gilded inside and out, made in 1212 by the "very famous goldsmith (*aurifex*) Chatard," was valued at sixteen livres.[4] Although workshop production was diversified, the surviving texts mention only the goldsmith and not the enameler, perhaps indicating the more valued skill.

### Power of Princes

In the late eleventh and early twelfth century the people of the Limousin were ruled by their local princes: the viscounts or counts of the Marches. These nobles, who were still under the real but not rigorous control of the last dukes of the Guillaume line in Aquitaine, were absorbed by the refined life at court. In Limoges itself economic issues had long made relations difficult between the Cité and the Château; the bishop of Limoges and the abbot of Saint-Martial were often at loggerheads. Thus the political and cultural horizon of the Limousin was for the most part regional, even local, although pilgrims and crusaders provided a breath of fresh air.

The death of Guillaume X of Aquitaine in 1137 marked the beginning of a new era. He betrothed his daughter Eleanor of Aquitaine to Louis VII of France, and their marriage united France and Aquitaine, although the king took only a slight interest in Aquitaine. After he divorced Eleanor, she married in 1152 Henry Plantagenet, duke of Normandy and count of Anjou (king of England from

1154 to 1189 as Henry II). This marriage made Henry ruler of about half the area of present-day France, and it precipitated recurrent warfare between him and Louis VII. From the beginning Henry exerted his power over the lords of Aquitaine, demanding homage and strict fidelity and taking special interest in prosperous towns, among them Limoges. He made great and much-resented demands of the monks of Saint-Martial and the burghers of the Château; he named two regents from his own entourage for the young viscount of Limoges, Adémar V, whom he later married to Sarah of Cornwall.

About 1158–59 the king-duke undertook to claim the sovereignty of the county of Toulouse and demanded assistance from a number of his vassals in Aquitaine. On his way back from this campaign, which was not fruitful, he stopped again in Limoges.

Shortly thereafter, in 1161, three Premonstratensian canons from England, sent by the bishop of Lincoln, came to Saint-Martial and asked for the saint's relics for their monastery, of which Martial was also patron. Geoffrey of Vigeois relates: "During the All Saints' Day procession, the abbot Peter . . . asked the congregation to acquiesce to their request. . . . After the Gospel reading, the abbot presented the head of the apostle to us all. Then, setting apart Saint Martial's relics as well as those of the Blessed Valerie, he gave [the canons] the joy of a satisfied desire." (The box given to transport these relics was a "magnificent chasse," albeit made of ivory.)[5]

The strongest evidence of English (Plantagenet) presence in the Limousin began in 1169 when Louis VII recognized Henry II's transfer of the title to his French possessions to his three eldest sons: Henry the Younger received Normandy, Maine, and Anjou; Richard (the Lion-Hearted; king of England from 1189 to 1199 as Richard I), Aquitaine; and Geoffrey, Brittany. They were not given real authority, however, and in 1173–74, supported by Eleanor of Aquitaine and Louis VII of France, they rebelled against Henry. This uprising did not succeed, but the king's sons continued to conspire against him. Until the end of the twelfth century the continental fiefs of the Plantagenets were racked by conflict.

Richard had been entrusted with Aquitaine in a solemn investiture in Poitiers. His mother, Eleanor of Aquitaine, then accompanied him to Limoges where he was enthroned by the bishop and where, in a gesture symbolizing his unity with the Limousin soil, he received the "ring of Saint Valerie."[6] Henceforth, the Limousin would be harried by the ambitious and prolific Plantagenet family and its agents. The Limoges viscounts, wary of Richard, took a resolute stand against their overlord in 1174 when, led by Adémar V, they harassed his emplacements.

Richard retaliated by calling up bands of mercenaries. The tense situation continued for a number of years, with frequent hostile encounters. Limoges long remained in conflict with either Richard or with his father, Henry II, whose authority remained preeminent.

Henry II, Eleanor, Henry the Younger, and Richard the Lion-Hearted revered the Limousin's regional sanctuaries, especially Grandmont, and demonstrated their devotion by donations of money and works of art. This concern, however, was offset by harsh family conflicts—father against son, brother against brother, king against queen—which led the Château and the Cité of Limoges to take opposing sides and in consequence to suffer hostile actions, levies, and reprisals.

When Richard razed the Château's defensive walls in 1181, the Limousin viscounts allied themselves with Henry the Younger. The Château's inhabitants were called upon by Adémar V to support Henry the Younger, thus making the Château a major stake in the struggle among the Plantagenets. In 1182–83 the burghers restored and reinforced their fortifications, demolishing nearby churches that might serve as strongholds for the "enemy." At one point, when Henry the Younger was closed up in the Château, his father, Henry II, was fortifying himself in the Cité, and his brother Richard was camped outside the walls.

Needing money to pay his mercenaries, Henry the Younger extorted twenty thousand sous from the burghers of the Château. He then forced the monks of Saint-Martial to "lend" him the abbey's treasury. An inventory was drawn up, and Henry pledged in writing to restore the abbey's goods. Thus disappeared forever "the frontal of the Holy Sepulchre altar on which five figures were represented, as well as the frontal of the main altar on which were depicted Christ in Majesty with the twelve apostles, all in the purest gold; a gold chalice, as well as a silver vase of admirable workmanship . . . , the cross from the altar of Saint Peter, as well as half his reliquary; the chasse of the Blessed Austriclinian and the great cross of Bernard." Geoffrey of Vigeois continues: "All this represented a weight in gold of 50 *marcs* and a weight in silver of 103 *marcs*, but without weighing it or judging it fairly, they estimated the price at twenty-thousand sous. In fact, it was worth considerably more. . . . And neither the cost for the work of the goldsmiths or for the gold used for gilding the silver were taken into account. . . . Certainly, if I had not seen it with my own eyes, I would never have believed that such an unprecedented crime could have taken place."[7] This eyewitness account confirms how highly works in gold and silver were valued; enameled-copper pieces, made of far less costly materials, were of interest only for their intended purpose and their fine appearance.

Henry the Younger soon made off with the treasury of Grandmont, including the Eucharistic dove given by his father. Shortly afterward he fell gravely ill, and from his deathbed he wrote to Henry II, asking him to forgive his faults, as well as those of the inhabitants of Limoges, and to restore the treasures he had stripped from the churches. It seems unlikely that there was any immediate restitution for Saint-Martial, for soon after his son's death Henry II forced the Château into submission and destroyed its fortifications. Grandmont, however, fared differently. This abbey had often served as a meeting ground for the combatants and was used on many occasions by Henry II as a retreat. In the late twelfth century it appears to have benefited from great generosity, allowing it to undertake construction and to commission a magnificent altar frontal combining goldwork and enameling, perhaps to celebrate the canonization of Stephen of Muret, the founder of the Grandmont order, in Rome in 1189 (cats. 54–68).

In the early thirteenth century Philip Augustus recovered much of the English-held lands in France, and the Limousin ceased for a time to be the site of princely battles, although it did begin to feel the authority of the French crown. For the Limousin towns, however, and for Limoges in particular, life continued to be difficult. There was fierce fighting, both open and clandestine, between the powers competing for influence. The burghers of the Château were often in conflict with the abbot of Saint-Martial. And when the burghers of the Cité and the Château found support from the king of England, who confirmed the validity of their municipal institutions, the viscount and the bishop of Limoges were assured of the support of the opposition—the king of France.

The years from 1225 to 1250 were relatively quiet, but the treaty of Paris of 1259 revived hostilities. According to this pact, the Limousin was one of the regions ceded in fief by the king of France to the king of England. However, certain lords and towns, such as the bishop of Limoges and his Cité, were retained by the French crown. The Château of Saint-Martial again became an English fief, an arrangement that the abbey and the burghers willingly accepted but that the viscount resolutely rejected, having been unofficially assured of the French king's support. This complex situation, where there was no right or wrong, soon deteriorated into violent conflict. From 1260 to 1276 the War of the Viscounty was waged between the

militia of the burghers of the Château and the viscount's men. Victory finally went to the viscountess of Limoges, who had been strongly supported by the king of France, and the viscount's authority was increased by the suppression of the Château's municipal institutions. The English crown lost all of its rights in the Château, and the abbey of Saint-Martial, its powers severely curtailed, appears to have been crippled by debt and forced to abandon renovations of the monastery. In the Cité the disagreements between the bishop and the magistrates were continuous; the bishop appealed to the French king, who in 1307 imposed a feudal agreement that abolished the magisterial offices.

Throughout these turbulent times the production of enameled objects continued, demonstrating that Limoges maintained its economic activity and commercial life. Kept in political silence for long periods, burghers of the Château devoted themselves to business. This internal dynamic was also fed by external factors. Indeed, political disturbances in the Limousin caused much misery but did not ravage the enamelwork ateliers or the commercial network that sustained them. As the king and the kingdom of France assumed authority in the Midi, the city of Limoges found itself, as so often in its history, at a critical access point. Geographically well situated for the political interests of the king, the city was also poised to allow Limoges work to be better known and therefore to move into a wide market.

NOTES
1. *Vita prolixior* by Saint Martial; see Landes and Paupert 1991, pp. 45–103.
2. Adémar de Chabannes, "Commemoratio abbatum Lemovicensium sancti Martialis apostoli"; see Duplès-Agier 1874, pp. 1–8.
3. Geoffrey de Vigeois, *Chronique*, book 1, chaps. 14, 15; the publication, edited and translated by P. Botineau and B. Barrière, is in preparation.
4. Bernard Itier, "Chronique," in Duplès-Agier 1874, pp. 73, 81, 83; an edition and translation are in preparation by J.-L. Lemaître.
5. Geoffrey de Vigeois, *Chronique*, chap. 61.
6. Ibid., chap. 67.
7. Geoffrey de Vigeois, *Chronique*, in Labbé 1657, vol. 2, pp. 279–342, book 2, chaps. 13, 14.

BIBLIOGRAPHY: Rupin 1890; C. Lasteyrie 1901; Gauthier 1972b; Gaborit 1976a, pp. 231–46; Becquet 1977–81; Aubrun 1981; Barrière 1984; Gauthier 1987; Lemaître 1989; Barrière 1989; Gauthier 1992, pp. 91–105; Barrière 1994, pp. 42–47.

# Religious Life in the Limousin in the Twelfth and Thirteenth Centuries

Dom Jean Becquet, O.S.B.

In 1189 Stephen of Muret (d. 1124), to whom the abbey of Grandmont owed its origins,[1] was canonized by the pope, having been previously canonized by the bishop of Limoges about 1167. The same liturgical pomp accompanied both ceremonies, with a procession in which prelates and clergy escorted the reliquaries of saints amid a large turnout of the populace. Earlier, in 1182, the monks of Grandmont who had gone to Cologne in search of relics were greeted by the bishop on their return with comparable solemnities.[2] Such events are evidence of the religious culture of the Limousin at the end of the twelfth century at its most spectacular.

The religious structure of the Limousin at the time covered an area of about 2,250 square miles, over which some one thousand churches were scattered, each with the privileges and personnel necessary for its functioning.[3] The bishop of Limoges had the spiritual and administrative responsibility for the whole region, and exercised it with the help of rights analogous to those of the churches within his jurisdiction. In the seventy years between 1106 and 1177, only two bishops—Eustorge de Scorailles and Gérard du Cher, uncle and nephew, respectively, scions of the local gentry—had exercised this responsibility, and the cathedral chapter had been able to elect them in complete freedom with regard to both the monks of Saint-Martial and the duke of Aquitaine.[4]

In the Carolingian period this shrine of the first known local bishop had come under the monastic rule of Saint Benedict, an observance revived by Cluny during the time of Saint Hugh (1062).[5] For two centuries the abbey had distinguished itself in the fields of liturgical music and drama and manuscript illumination. The monks, who came to believe that their patron saint was ranked with the apostles, built a library of over five hundred manuscripts.[6] At the start of the thirteenth century, the chronicler of the community drew up a tally of its working strength: ninety monks in the service of the shrine and twice that number in some forty Limousin dependencies, which provided the

abbey with supplies. Some of these dependencies were themselves sizable: the provostship of Chambon, in particular, which honored the body of the "protomartyr" of Aquitaine, Saint Valerie, numbered a hundred monks, one-third of them present at the imposing consecration in 1212 of the church that is still in use today. Again, in the Limousin, the twenty monks of the ancient abbey of Vigeois were subordinate to Saint-Martial; the priories of Arnac, Saint-Vaulry, and La Souterraine each had more than a dozen monks, as did four other dependencies in the neighboring dioceses.

In Limoges itself, the women's cloister that had always been under the protection of the cathedral had likewise adopted the Benedictine rule, hence its name: Sainte-Marie de la Règle, Our Lady of the Rule. In addition, the bishop had founded Saint-Augustin and Saint-Martin about the year 1000 by calling on monks from Poitiers. Located close to the episcopal precinct, or Cité, each of these two abbeys made do with only a few dependencies. Solignac, which was not far from Limoges, had followed the Benedictine rule from the moment of its foundation by Saint Eligius in the seventh century; indeed, it had even imposed the rule at Beaulieu in the ninth century. Endowed at an earlier date, Solignac and Beaulieu had maintained a respectable heritage.

Farther away in the Limousin, two large Benedictine monasteries had controlled a widening network of dependencies since the tenth century: Tulle, which extended its authority over some sixty establishments all the way to Quercy; and, in Rocamadour, Uzerche, which governed even the small Limousin abbeys of Ahun and Meymac.[7] During the twelfth century the Benedictine rule was so common that an abbey could, if necessary, choose an abbot from a different house; an impressive system of worship in a richly decorated church was the foundation of this observance.

Other large Limousin churches were served by communities of canons,[8] chief among them the cathedral

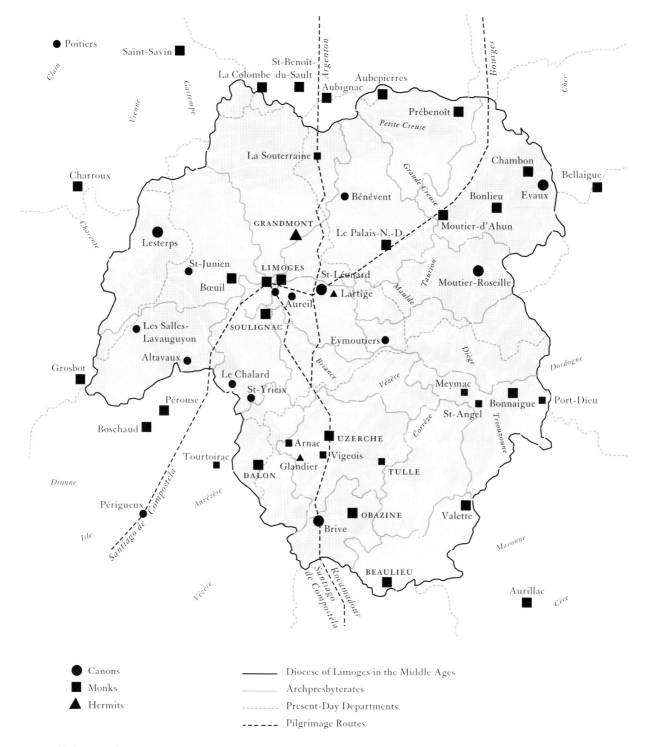

Fig. 2. Establishments of monks, hermits, and canons in the Limousin in the first half of the thirteenth century. (After *Atlas du Limousin*, Limoges, 1994)

dedicated to the protomartyr, Saint Stephen. These canons followed the rather flexible directives laid down in the past by the Carolingian prelates and their emperor. The same directives were adopted by old monastic communities such as the ones at Saint-Yrieix and Saint-Junien, and also by the collegiate churches of Le Dorat, Lesterps, and Eymoutiers (the latter founded about the year 1000) and

by the ancient shrines of Saint-Martin in Brive, Saint-Léonard-de-Noblat, and Évaux. These various chapters of canons numbered some two or three dozen members each and had, like the abbeys, rural dependencies centered on churches of which they were the ecclesiastical patrons with smaller or larger benefices.

During the second half of the eleventh century orders

for a general reform circulated throughout the Latin West; the success of this reform was facilitated by a demographic boom, by the consolidation of feudal structures, and by the direct support of the Holy See in Rome and its legates. Limousin chapters showed their adherence to the reform by adopting the Augustinian rule, which added to the normal requirement of clerical celibacy an individual renunciation of property controlled by the local community. Thus it went at Évaux and Lesterps, Brive and Saint-Léonard, the bishop sometimes intervening in the adoption of the new rule. By the same token, the reform invited laymen with jurisdiction over rural churches and their revenues to relinquish it in favor of the monks and canons by way of the bishop; they complied all the more willingly in that they could appreciate the morality and disinterest of the beneficiaries. As a result, new chapters of canons regular were created almost everywhere, endowed by the bishops with churches handed over by the laity, churches that became both the canons' direct pastoral responsibility and a source of income. This development was so well received that by the beginning of the fourteenth century there were practically no churches under lay authority in the Limousin.

In 1088 a Limousin priest founded Le Chalard. Without adopting, as he did, the new rule, the canons of Saint-Junien and those of the cathedral simultaneously established canons regular at Salles-Lavauguyon and at Bénévent and Aureil, respectively. Aureil was a real success: by about 1200 this priory owned nearly two hundred manuscripts; a dozen canons were in residence, as well as children in its school, and it was in a position to serve some twenty parish churches under its control. Naturally, the holy men who had reformed or founded these chapters had been raised to the altar by the bishop or the pope:[9] Gautier of Lesterps (d. 1070), Geoffrey of Le Chalard (d. 1125), Gaucher of Aureil (d. 1140). The enameled reliquary of Le Chalard (fig. 118a) testifies to a growing treasury of local relics, realized in imitation of the ancient shrines and inspired by devotion to a saint whose presence was still felt.

The monastic order itself, reformed by the Cistercians, was not to be outdone. The tomb of Stephen of Obazine in the great church consecrated in 1177 is evidence to this day of the popular cult that arose following the saint's funeral in 1159.[10] The new monks, however, refrained from assuming parish responsibilities and their profits.

In the course of the twelfth century the Limousin witnessed an astonishing surge in the number of small fraternities of clerics and laymen, inspired by hermit saints and settled in remote places. Each of these rustic sites had its church, which, though plainly built, was spacious enough to welcome friends and neighbors. The monks of Grandmont and of Artige had some forty such houses in the diocese.[11] Even more than the Cistercians, these fraternities rejected the responsibility for and revenues of the country churches; they supported themselves by tilling the soil and with the aid of donations from their benefactors. By the early thirteenth century Grandmont had multiplied itself into some hundred institutions in the domains of the Capetian Louis VII and the Plantagenet Henry II (the successive husbands of Eleanor of Aquitaine), and delegates would convene once a year at the motherhouse.

In 1166 the new, vast church of Grandmont was consecrated by two archbishops and five bishops; the primate of Bourges had donated the relics of eleven martyrs for the new altar, and the monks of Saint-Martial had come with the reliquary of their patron, raised to apostolic status. Possibly this was the occasion for placing the founder's body on the altar, as had been done with Gautier of Lesterps a century earlier. On such an occasion, and under a new prior, the clerics of the motherhouse owed it to themselves to make every effort in the practice of their vocation: the cult of saints. Thus, relics had been brought from Cologne in earthenware jugs—relics clearly destined for a presentation that was worthier of popular veneration. Indeed, in his account of the journey, one of the monks sent to Cologne declared: "It is obvious that in churches where the bodies of saints are entombed in peace, not only the religious communities but also the secular clergy are more attentive to the divine service and willingly come to pray with more devotion. What benefits are assured to Christians by the presence and the merits of one who has shed his blood for Christ or endured a long and bloodless martyrdom."[12]

It was in this spirit that two years earlier the canons regular of La Couronne in the Angoumois had endowed their new establishment in the Limousin with a prestigious assortment of relics. And the chronicler of Saint-Martial wrote about the relics in his abbey: "Their number and their quality brilliantly honor our apostolic and imperial church . . . whose walls, so long as prayers to and praises of God are offered on this site, will benefit from their great blessings. However, after calling attention to the saints of a single holy place, I will now review the names of places that are considered the most illustrious in the entire diocese."[13] Before resuming his chronicle, the author then listed some forty abbeys, chapters, and churches famous for their relics.

As we have said above, the evangelic and apostolic simplicity preached by the new orders could not ignore for long the principal customs of popular devotion, even if their clergy eased the liturgical practices of the older shrines. The new religious could not, any more than their predecessors, disappoint the faithful who attended divine services that were made more splendid because the relics attracted pilgrims. Now these relics needed reliquaries; and just as the old orders had done, the new ones found the wherewithal to acquire them in the late twelfth century. In Grandmont itself, the clerics had established a treasury, which was seized in 1183 by Henry the Younger—in rebellion against his father, Henry II—to pay his mercenaries. In 1185 those same clerics saw their prior, William of Treignac, resign under pressure from the lay brothers and then die in exile. Retaking power, William's followers brought back his remains, which were subsequently preserved in an enameled casket (see cat. 60),[14] while Hugh Lacerta (d. 1157), a model for the lay brothers, remained buried in situ.

The total wherewithal, therefore, that was available to the Limousin churches, large and small, was considerable, for we can count between two and three thousand various canons, monks, and brothers in the diocese of Limoges in about 1200, not to mention a few hundred nuns; half of these were scattered in rural priories, which, despite their modest resources, were always provided with churches. This number clearly fails to take into account later arrivals: the Dominicans and the Franciscans, who were summoned to Limoges in 1221 and 1223,[15] even before the deaths of their respective founders; it was in Limoges that Anthony of Padua gathered crowds before founding a second friary in Brive.

What can be said about the hundreds of rural churches, each with the means to acquire a small treasury—a few liturgical manuscripts, for example—which has now disappeared? Even a modest cantonal seat such as Lapleau (Corrèze) still owns five or six enameled objects that come from neighboring churches. The church in Saint-Hilaire-Foissac, which was large enough to afford a golden "sun," or monstrance, in the eighteenth century, already had the wherewithal to acquire a reliquary and a pyx in the thirteenth. The works of art obtained for larger houses were proportionately more imposing, as the objects gathered here attest. By taking into account the large number of religious establishments that existed, the extent of their membership, and the wealth accrued by some of them, we can begin to comprehend the vast artistic patrimony of the Limousin.

Notes

1. Becquet 1985a, cols. 414–26; *L'Ordre de Grandmont* 1992; *Les Cahiers grandmontains* 1990–95.
2. Becquet 1968, pp. 249–64.
3. Becquet 1979–82: "Les Effectifs des monastères," pp. 105–9; "Trésor et mobilier," pp. 188–89; "Les Reliques," pp. 399–406.
4. Becquet 1987, pp. 85–87.
5. Sohn 1989.
6. Chailley 1960; Gaborit-Chopin 1969; Landes 1994, pp. 23–54, with reference to the work in progress on the apostolicity of Saint Martial, particularly at the universities of Boston and Newark (New Jersey).
7. Aubrun 1981.
8. Becquet 1985b (concerns chiefly the Limousin).
9. Becquet 1991, pp. 26–57; Paris 1993.
10. Barrière 1977.
11. Becquet 1970, pp. 83–142; Artige was devastated by the Wars of Religion in the sixteenth century.
12. Becquet 1968, p. 251.
13. Labbé 1657, vol. 2, pp. 184–87.
14. Souchal 1964, pp. 65–80.
15. By the end of the thirteenth century Limoges had four convents of the mendicant orders already known in the large towns; see Le Goff 1968, pp. 335–52.

# Beginnings and Evolution of the Oeuvre de Limoges

Elisabeth Taburet-Delahaye

## Beginnings of Enamelwork in Aquitaine

The history of the origins of enamelwork in the south of France is, in effect, linked to the treasury of Conques. The binding of enamel to copper may have been attempted in other centers (a few scattered pieces from the tenth and eleventh centuries seem to indicate this), and our knowledge is incomplete, but it is those works in Conques today that allow us to trace the early history of Romanesque enamelwork in southwestern Europe.

In about 1100 enamelwork seems to have been given its first impetus by the patronage of Bégon III, abbot of Conques from 1087 to 1107, who "enshrined numerous relics in reliquaries of gold and silver" and whose name appears on several extant works. No text or inscription attributes the portable altar of Saint Foy (cat. 1) to this abbot, but the object's ornamentation and iconography leave little doubt that it was one of the group of works made for Conques during his abbacy, and that it very probably came from a workshop within the abbey itself. While the style of the enamels reveals ties to eleventh-century art in Aquitaine, their technique is derived, at least in part, from Ottonian or, more likely, Italian examples of the tenth and eleventh centuries. But these have been adapted to create a new form of cloisonné enamel, a major step toward the invention of champlevé enamel on copper. Several pieces similar in style and technique to the altar of Saint Foy (cats. 2–6) attest that this work was not an isolated creation but was made by an active and innovative workshop, whose production continued beyond Bégon's abbacy into that of his successor, Boniface (r. 1107–after 1121). Boniface's name is engraved on a medallion of a coffret found in the abbey church in 1875 (cat. 7), thus suggesting that the oldest surviving champlevé enamels from southern France were made at his instigation. The medallions still in Conques and others in the Metropolitan Museum, the Bargello, and the Louvre (cats. 7, 8) decorated coffrets made for Boniface but can be dated less precisely than is generally believed since the year

in which this abbot died is not known. The medallions' luxuriant decoration of fantastic animals—either isolated, in confrontation, or entwined in ornamental foliage—corresponds to illuminations and sculptures from 1110 to 1140 in Aquitaine, the vast province of southwestern France that was so important in the development of Romanesque art. In their dazzling draftsmanship and color, these Conques medallions prefigure later Aquitaine enamelwork.

Toward the middle of the twelfth century there appear to have been, for however short a time, workshops producing champlevé enamel on copper in a number of places in central and southwestern France, the region that was politically unified in 1152 under the Plantagenet princes. Interrelationships among enamels produced at this time in Conques, Angers, Le Mans (cats. 7, 8, 15), and Limoges (cats. 9, 10) have been observed; there is also clear but complex evidence linking Limousin works to those destined for the monastery at Silos in northern Spain, which was the beneficiary of Plantagenet patronage.

In the mid-1100s Limoges was only one of the important towns in the Plantagenet domain, where workshops engaged in the still-new process of champlevé enamel on copper. The angel of Saint-Sulpice-les-Feuilles (cat. 54), the chasses from Bellac and Champagnat (cats. 9, 10), the fragment of a crosier (?) unearthed at the site of Saint-Augustin-lès-Limoges (fig. 3), attest not only to the similarity between works attributable to Limoges and those produced in Conques under Boniface but also to the technical and artistic mastery attained by Limousin enamel workers by 1150.

## Romanesque Enamels

The rapid spread of workshops in the Limousin during the last thirty or forty years of the twelfth century remains in part without explanation. The geographical setting and the historical circumstances as well as the quality of the

Fig. 3. Crosier fragment, found on the site of Saint-Augustin-lès-Limoges. Formerly Limoges, Musée Municipal de l'Évêché

Limousin enamels may have favored the rapid expansion discussed by Barrière and Boehm. The success of Limousin production derived in part from a quick and appropriate response to the needs of a vast European clientele. The evidence suggests that the years 1150 to 1175 saw the decisive development of Limousin enamelwork and its first international diffusion. And it was at this time that the term *opus lemovicense* first appears (about 1167–69; see Boehm, below). Nevertheless, among the works that have survived, those from this period are puzzlingly rare. Furthermore, there is no strong continuity between the few pieces that date from before 1150 and the important ensemble of works "with vermiculé grounds,"[1] which are regarded as the start of the Limousin workshops' advancement.

The products of these workshops displayed all the creative energy of a still-fledgling art. Narrative scenes often depicted local saints or those especially venerated in the Limousin, such as Stephen, Martial, and Valerie (cats. 16, 17, 20, and fig. 4), with striking animation and color. In contrast, more universal themes—evocations of divine transcendence and majesty (cats. 21, 24, 27)—are presented in a static but still expressively forceful manner. (These tendencies appeared at times in the work of a single artist [cat. 20]). Variations in inventiveness and quality point to activity over several decades by several distinct workshops. Although chasses have a sacred function, some (cats. 21, 22) bear on the reverse a colored decoration of fantastic flora and fauna. This type of decoration, usually associated with scenes of secular life, also appears on candlesticks and coffrets (cats. 33–38), where it may be combined with ornamental motifs and stylistic elements characteristic of works made for altars or for relics.

None of the vermiculé works can be dated exactly. However, a relatively precise chronology can be established for other Limousin enamelwork from 1170 to 1200 on the basis of historical circumstances such as the date when the saint depicted was canonized or when particular relics arrived in the Limousin or when a civic event took place. One group can be assembled around the earliest chasses (cat. 39) for the relics of Thomas Becket (d. 1170), who was canonized in 1173, and the reliquary of the True Cross from Saint-Sernin de Toulouse (cat. 40), made under Abbot Pons (r. 1176–98). The style of these works is still dominated by a taste for animated narration and by a strong Romanesque stylization, but the enameling procedure is the opposite of that used earlier—that is, the figures are in reserve and engraved on an enameled ground. Great use is made of small appliqué heads, standardized to some extent and very likely produced by stamping (see Biron et al. below). Termed "classical heads" by M.-M. Gauthier, these elements were also used extensively by other workshops producing enamels for the abbey of Grandmont (cats. 56, 57).

Unusual in size, iconography, and, to some extent, style, the chasse of Mozac was executed after the death of Abbot Peter in 1181 (figs. 5, 14, and cat. 45). It may be placed at the center of a second group of works in which the use of appliqué figures in half relief reveals an interest in plasticity (cats. 46–48). To this group may be added the important ensemble executed for Orense at the behest of Bishop Alfonso (r. 1174–1213), probably about 1188 (cat. 51), and the Saint Barontus destined for Pistoia (cat. 53).

A homogeneous ensemble of works executed for the order of Grandmont is composed principally of large crosses. They exhibit the same technical and stylistic characteristics as the two plaques from the main altar of the motherhouse of the order very probably executed in 1189–90 (cat. 57). Made by a strongly individualistic workshop active in the last ten or fifteen years of the twelfth century, these works are distinguished not only by the complete enameling of the figures but also by a monumentality, a "classical" equilibrium, and a softening of Romanesque animation and stylization.

Study of the spectacular development of Limousin enamelwork from about 1160–70 to 1200 reveals several fundamental characteristics. The first is its rootedness in the artistic tradition of Aquitaine and more particularly that of the Limousin. Only in rare cases (cat. 16) has a direct tie been established to illuminations from the scriptorium of Saint-Martial of Limoges, whose brilliance during the tenth and eleventh centuries seems dimmed in the early twelfth century. It is, however, certain that Limousin enamelworkers drew from this tradition. Stylistic elements in Limousin illuminations (based for the most part in the art of Aquitaine)[2] as well as particular motifs that appear in enamelwork include monticules (small mounds) covered with spark-shaped imbrications (scalelike decorations) that suggest the ground on which the figures stand (cats. 15, 16, 18), backgrounds strewn with rosette decoration (cats. 29, 55), and orphreys ornamented with geometric motifs or clothing patterned with dots (cats. 19, 20). Friezes or carpets of rosettes inscribed within circles decorate numerous works and entirely cover the reverse sides of some chasses (cats. 24, 27). These were also taken from early medieval art: the reverse of the chasse of Mumma in Saint-Benoît-sur-Loire has this type of decoration.[3]

These borrowings do not, however, underlie all Limousin Romanesque enamelwork. Close ties to artistic centers in central France, notably the Auvergne, were among the factors contributing to the "severe" style of a number of works (cats. 25, 45, 69).

A question remains: Were there ties between Limousin enamels and works, enameled or otherwise, produced in other artistic centers during the second half of the twelfth century? M.-M. Gauthier (1958) showed how the fine ornamental foliage, which gave its name (vermiculé) to the oldest known group of works made in Limoges, had spread through the West—as well as the East—before it was taken up by the goldsmith–enamel-workers of Aquitaine. Stylistically the Limousin enamels seem to demonstrate a familiarity with English art, principally illuminations (cats. 21, 39). This, however, might have been a reaction to Continental works that were important in the formation of the "damp fold" style, which dominated English art in the second half of the twelfth century. It is probable that Limousin artists had some knowledge of proto-Gothic art, with its concern for verisimilitude that developed north of the Loire, notably in the Île-de-France, about 1150. On the other hand contacts with works from the Rhine and Meuse Valleys or with the Byzantine or Byzantinizing creations of Sicily or elsewhere, which are sometimes posited to explain tendencies in Limousin enamel, seem to have been for the most part overestimated (cats. 21, 22, 62).

The location and organization of the first Limousin enamel workshops remain unknown. The consistency of most of the production suggests a concentrated activity—in the same town (Limoges) and perhaps even in the same neighborhood. But were works of exceptional technical virtuosity—such as the Poitiers crosier (cat. 23) or even the appliqués on the Souvigny Bible (cat. 34)—made in

Fig. 4. Chasse of Saint Stephen: detail, Saint Stephen Led from the Town. Gimel, Church of Saint-Pardoux

Fig. 5. Chasse of Mozac: detail, Abbot Peter Celebrating Mass. Mozac, Church of Saint-Peter

Limoges itself or in another center? In some cases a strongly individualized style may raise doubts about the location of what appears to be a Limousin-based workshop. Did the artists who made the Mozac chasse (cat. 45) work in the Aquitanian city of Limoges, or had they moved to the Auvergne from the Limousin? This issue of location is even more relevant to discussions of the authors of the Orense ensemble (cat. 51) or of works destined for Italy (cat. 53). Moreover, to what extent does the concept of workshop aid in the assessment of these artists' activity? ("Workshop" is a convenient term that appears often in these pages.) Similarities may make it possible to recognize with certainty the hand of the same artist (cats. 21, 22). But would this artist have worked in one of a small number of important establishments, rigorously organized

Fig. 6. Eucharistic coffret from the Treasury of Grandmont. Formerly Limoges, Musée Municipal de l'Évêché

half-relief heads of the "classical" type that appeared in Limousin production around 1180, while the figures on the foot and the angel engraved inside the lower coupe show the influence of early Gothic art. The inscription— Magister G. Alpais Lemovicarum—is of great importance since it gives in exceptional fashion not only the artist's name but also where he worked. Attempts have been made to attribute other works to this artist, among them the Eucharistic coffret from Grandmont, an extraordinary piece that must have been made by a very gifted artist (fig. 6).[4]

If the first Gothic infusion seems less evident in other works, it is nonetheless manifest in, for example, the Holy Women at the Tomb and in Saint Peter's drapery on the Louvre and Metropolitan tabernacles (cat. 70).

The quality of work in the early thirteenth century varies more widely than it did in the decades before. A strong and little-diversified demand for certain types of objects, such as book covers (cats. 86, 87) or crosiers (cats. 109, 110), encouraged an almost systematic use of established models, which sometimes led to stereotypical repetition. Innovation, however, was also more common during these years; some artists moved away from Romanesque art by adopting new approaches. A number of works cast old themes in Gothic shapes; among these are a coffer decorated with fantastic creatures and secular scenes, which belonged to Cardinal Bicchieri, bishop of Vercelli, and was mentioned in a 1227 inventory (cat. 88), and medallions from similar coffrets (cat. 92).

The supple and more natural spirit of early Gothic art is evident especially in the enamel reliefs that decorated altars, tabernacles, and chasses. First seen on small but unusually fine works, such as the Annunciation to the Shepherds from the Louvre (cat. 94), this style was used in important ensembles, imposing examples of which are the apostles from the main altar at Grandmont (cat. 58) and the tabernacle from Cherves (cat. 98). A few other very fine works—the Angel of Saint Matthew in the Musée de Cluny (fig. 7), the Virgin and Child in the Cleveland Museum of Art (fig. 8), and the evangelists once in the Spitzer Collection (cited in cat. 95)—confirm the full participation of some Limousin artists in this first Gothic flowering. The broad foliate rinceaux with enameled fleuron supports derive directly from those made at the end of the twelfth century, but the simplification and amplification of the floral motifs give a decidedly Gothic flavor. A rare chronological signpost is provided by the Grandmont apostles, which were very likely completed before 1231.

During these same years other artists did not conform to Gothic conventions. The fine reliquary of Saint Francis of Assisi (cat. 101), for example, is exceptional in Limousin enamelwork, since it can be dated in relation to historical

and perhaps specialized, or would he have worked in one of a number of small shops run by a single master? The complex relationships between the works shown here leads me to prefer the second hypothesis. It does seem certain, however, that for certain works, iconographical themes or decorative motifs were taken from established models and soon became standardized.

## Limoges Between Romanesque and Gothic

The evolution of the *Oeuvre de Limoges* is not marked by disruptions or strongly differentiated stages. There is an exceptional persistence of forms and decorative themes, with some use of earlier treatments. Innovations in nearby centers were accepted only gradually and within limits. The history of Limoges enamel does not appear to be a straight line but a stream in which different currents merge in each period. Thus in the early thirteenth century the abundant production of the Limousin workshops was marked only to a certain degree by the spirit of early Gothic art, or the style of the 1200s, which then dominated the regions north of the Loire. The artists of Limoges had adopted a simplified version of the new manner, incorporating supple draperies with fine, tight folds and full, rounded faces, while preserving the late Romanesque taste for lively narration and figures in motion.

The largest group of works from the early thirteenth century shows an attachment to ideas of the preceding era. However, a certain softening of the forms, a freer treatment of scenes or figures, and the occasional use of a new iconographical theme or decorative motif indicate a slow evolution. The ciborium of Master Alpais (cat. 69) is one of the first and most beautiful creations to exhibit this opening up. The figures decorating the upper and lower coupes are still conceived in Romanesque terms and have events (Saint Francis died in 1226 and was canonized in 1228). Its quadrilobed form and its iconography are new, but its style and decoration seem inspired by themes and practices that had been popular several decades earlier, perhaps at the request of the person who commissioned it. Other related though less imposing works—such as the Metropolitan plaque depicting Saint William of Bourges, canonized in 1218 (cat. 102), and the Bonneval cross in the Musée de Cluny (cat. 104)—present similar enameled figures on a ground of reserved and gilt copper and thus demonstrate a comparable stylistic independence from Gothic impulses.

At this time Limousin workshop production was very diversified, and some objects were untouched by stylistic trends. These include crosses, book covers, and chasses that have a very stylized decoration of half-relief enameled figures (cats. 112–15). There are also a few pieces whose lack of figurative decoration gives them a timeless quality—for example, Eucharistic doves (cats. 105, 106).

Fig. 7. Angel, Symbol of Saint Matthew. Paris, Musée National du Moyen Âge, Thermes de Cluny

### The Last Flowering: From 1250 to 1350

An important new period in Limousin enamelwork began about 1240–50. The impetus, like that earlier in the century, seems to have come from contacts with the art of regions north of the Loire; this time there were also commissions from these same regions. There were changes not only in the style but also in the form and the type of works produced. Some objects, such as book covers, seem to have been less frequently produced; others, such as gemellions (basins used to wash the hands), found new favor. The funerary arts made their appearance in Limoges. The oldest examples of this genre that can be dated with certainty are the tombs of Blanche (d. 1243) and John (d. 1248), the children of Saint Louis—the first and only surviving royal commission to the Limousin goldsmiths (cat. 146). Not unrelated to the taste for strongly repoussé copper gisants (recumbent figures of deceased persons) was that for the half-relief appliqués, very sculptural in style, that decorated large chasses and altars (cats. 61, 116–22). This evolution toward a more monumental treatment, one more solid and somber, reflects stylistic changes in the Île-de-France in the 1240s.

In enameled ornaments, vegetal themes survived but diminished during the second half of the thirteenth century. Heraldic decoration, probably introduced through funerary commissions, became more common, especially on coffrets, gemellions, and candlesticks. These armorial bearings sometimes allow precise dating of a work. For example, the renowned coffret of Saint Louis (cat. 123) bears the escutcheons of the great families during Louis IX's

Fig. 8. Virgin and Child. Cleveland Museum of Art

Fig. 9. The coffret of Richard of Cornwall. Aachen Cathedral Treasury

reign, and its execution can be placed between 1234 and 1237, and similar elements allow the coffret "of Richard of Cornwall" in the cathedral treasury at Aachen to be attributed a very probable date of 1258 (fig. 9).[5] On most objects, however, Limousin artists used heraldic ornament simply as a colorful device without its original personal and familial references (see Pastoureau, below, and cats. 128–35).

In the second half of the thirteenth century, Limousin production displayed diverse tendencies, among which are a strong attachment to the traditions and practices surviving from previous decades. Thus the supple, tight-folded drapery and the rounded faces of early Gothic art are still found on works that appear to have been realized after 1250. One of the most eloquent examples of this phenomenon is the chasse from the butchers' chapel at Saint-Aurelian in Limoges, the only Limousin chasse whose origin is connected to a Limoges building (cat. 141). Despite their sometimes modest stature, the few datable works—such as the dedicatory plaque from the altar of the priory at Artige, which bears the date 1267, and the mandorla reliquary from Saint-Martin of Brive, which held the relics of Saint Clare and thus can be dated after her canonization in 1255—are important evidence in the establishment of a more certain chronology.

Important qualitative differences evident in the first part of the thirteenth century—such as the emergence of stereotypical products—continued and perhaps increased as the century wore on. These conservative and standardizing tendencies seem to have become stronger at the turn of the century. Production appears to have diminished,

and enamelwork seems to owe its survival to a mostly modest and above all local clientele. The only exceptions to this constriction are perhaps funerary pieces: in the gisant of William of Valence (of about 1292; fig. 10) in Westminster Abbey,[6] the face of Blanche of Champagne of about 1306 (cat. 151), and the 1307 effigy of Guy de Meyos (cat. 152), one can see the effects of new trends during the time of Philip the Fair (r. 1285–1314). Responsiveness to innovation is even more evident in the very beautiful enameled plaques in the Wallace Collection,[7] whose original destination is unknown. The attribution of the plaques to Limousin workshops is also uncertain, a problem peculiar to this period, when centers in northern and southern France began to fabricate copper objects with champlevé enamel decoration.

Sculptural works with little or no enamel, such as head reliquaries, seem to have enjoyed an ongoing popularity. In fact, the last known piece of Limousin production is the head reliquary of Saint Ferreolus in Nexon, made in 1346 by Aymeric Chrétien, "goldsmith of the château of Limoges" (cat. 157). Like the much more somber inscription of a century and a half earlier on the ciborium of Master Alpais (cat. 69), the long enameled inscription on this reliquary's reverse states that the goldsmith, a lay person, was established in Limoges, here giving the specific quarter. In the fourteenth century, accounts of the Limoges goldsmith-enamelers became sporadic, indicating the slow decay of the workshops. Traditionally attributed to the town's sacking by the Black Prince in 1370, the demise of enamelwork seems to have taken place gradually over the preceding decades.

Fig. 10. Tomb effigy of William of Valence. London, Westminster Abbey

## NOTES

1. This term, designating the fine ornamental motifs with tight scrolls engraved on grounds of gilt copper, was made current by the fundamental studies on these works published by Linas (1883) and Marquet de Vasselot (1905).
2. In particular, in the *Sacramentaire de Saint-Etienne de Limoges* (Sacramentary of Saint Stephen of Limoges), Paris, BNF, ms. lat. 9438; see Gaborit-Chopin, 1969, figs. 95–98, 169, 171.
3. Paris, 1965, no. 192. Such friezes with rosettes are also found in eleventh century illumination, for example in the *Homélies sur l'Évangile de Jean* by Saint Augustine in the Bibliothèque de Tours (ms. 291, fol. 96), the manuscript in which one also notices other decorative motifs common to Limousin production as well as western Romanesque art, which tends to confirm the recent theory that sees the art of the region around Tours in the tenth and elventh centuries as one of the principal sources of this art.
4. Gauthier 1964a, p. 83.
5. See Vaivre 1974, and cats. 123, 133.
6. See Chancel-Bardelot, Appendix, list no. 16.
7. See Gauthier 1972b, no. 141.

# Opus Lemovicense: *The Taste for and Diffusion of Limousin Enamels*

Barbara Drake Boehm

*De opere lemovicensi*: In an era before advertising, before international telecommunications, this phrase—of Limoges work—was a virtual trademark, recognized throughout Europe, for enamels made in the city of Limoges.[1] Its first recorded use is in a letter of 1167–69, written by John, a religious of the church of Saint-Satyre, to Richard, prior of Saint-Victor in Paris, about a financial transaction. With his superior's permission Jean had left his community to follow Thomas Becket. In his haste he had to borrow money from a friend and promised that the loan would be repaid by the prior of Saint-Victor. To prove that his letter was authentic, Jean reminded Richard that while he had been in the infirmary at Saint-Victor, they had discussed a book cover of Limoges work, which Jean wished to send to the abbot of Witgam (Wingham, Kent, a dependency of Canterbury).[2]

This reference is remarkable not only because of the early date—several decades before the earliest sizable body of surviving Limoges work—but also because of the assumption that seeing a Limoges enamel book cover was

Fig. 11. Enameled vase. Gallo-Roman (3rd century). Discovered at La Guierche in the 19th century. New York, The Metropolitan Museum of Art

memorable enough to be cited as the identifying moment of their visit together. What had so impressed this churchman in the circle of Thomas Becket? What were the distinctive qualities that caused him to value this work of art and to label this particular type of powdered glass fused onto a sheet of copper as *de opere lemovicino*? How did Limoges work become such a sought-after commodity, and how was its use spread throughout most of western Europe before the mid-thirteenth century?

The enrichment of metal with colored decoration had long been considered a remarkable skill. Writing in the mid-third century A.D., the Greek Sophist Philostratus remarked: "It is said that the barbarians of the [Atlantic] ocean pour these colours into bronze moulds, that the colours become as hard as stone, preserving the designs."[3] He was probably describing enameling of the type found on the vase excavated at La Guierche, some thirty-five miles west of Limoges, in the nineteenth century (fig. 1). Brilliant red, blue, and green enamels define the foliate ornament encircling the body of this sturdy but elegant vase. Roman coins found inside the vase provide a date between 260 and 273.

In early Western medieval art, however, it is not this sort of Gallo-Roman enameling, with its overall foliate patterns of color, that was adopted; rather, it was enameling on gold. The esteem in which these enamels on precious metal were held is suggested by their use, like gems, set in combination with, or in place of, colored stones. On the Carolingian altar frontal of San Ambrogio, Milan, enamel strips frame the figurative scenes. On the tenth-century reliquary of the nail of the True Cross at Trier, figurative enamels and gems cluster around the nail's head.[4] In his twelfth-century treatise Theophilus describes a chalice on whose rim enamels alternate with gems and pearls.[5]

The earliest enamels produced in central France belong to this same type. On the portable altar from Conques (cat. 1), figurative enamels alternate with gems in a regular pattern; among the stones are small enameled "gems"

Figs. 12 a and b. Book-cover plaques. Vienna, Museum für angewandte Kunst

bearing a floral design. On the chasse of Bellac (cat. 9) the figurative enamels are again set on disks, which are in turn secured to the chasse by means of bezels, in the same manner as the jewels of the chasse. With the chasse from Champagnat (cat. 10), the enamel decoration begins to break out of the confines imposed by its association with gems: the enameling floats freely on the gilt-copper surface. Still, the placement of the figurative elements in relation to one another seems somewhat unresolved, as on the reverse, where areas of foliate ornament fill the space between the symbols of the evangelists.

An equation of enameling with gems continues even after enameling became the primary medium rather than a jewel-like accent. The funerary image of Geoffrey Plantagenet (cat. 15) was described in a twelfth-century chronicle as "ex auro et lapidibus" (of gold and stones), when in fact the entire monument was fabricated from gilt copper and champlevé enamel.[6]

Though the book cover mentioned in the letter of 1167–69 to Richard of Saint-Victor is not known to have survived, its form probably corresponded to preserved examples like the pair in Lyons (cat. 18) or the pair in the Museum für angewandte Kunst, Vienna (figs. 12a,b). The colored glass is no longer used exclusively for ornamental accents but for the definition of figures, such as those in the Crucifixion and the Christ in Majesty. The enamel has become integral to the work of art itself, and the metal surface has become the support for "painting" with glass.

The expertise of the enamelers of Limoges is as remarkable today as it was in the twelfth century. The channels in the copper are precisely cut. The glass powder is carefully laid in, often with multiple shades in a single field. The palette is rich and the shades are subtly differentiated. The highly polished enamel surfaces are free from bubbles and flaws.

From the end of the twelfth century and into the fourteenth, Limoges enamels were characterized by a predominantly blue palette, an aesthetic shared with Gothic manuscript illumination and stained glass. The interior of a Gothic cathedral seems to be bathed in blue light, and pages of the *Bible moralisée* are dominated by a deep blue field. Sapphire was the color associated with the doors of the Heavenly Jerusalem and with the feet of God himself in Scripture. In his writings Hugh of Saint-Victor (1096–1141) often explores the mystical significance of the color blue; following the tradition of earlier theologians like Bede, he associates sapphire, hyacinth, and lapis with the Old Testament and with heaven. Blue becomes a metaphor for the sacral realm.[7] This kind of association, articulated in the writings of the celebrated scholar from the abbey of Saint-Victor, may well underlie the taste for the blue of Limoges enamels.

A distinguishing feature of Limoges enameling, by comparison with manuscript and stained glass, is its simultaneous use of painting, through the fusion of glass powder and metal, and of sculpture, through its inventive

Fig. 13. Apostle. Present whereabouts unknown

Fig. 14. Chasse of Saint Calminius: detail, Construction of the Monastery of Le Puy. Mozac, Church of Saint-Peter

and highly skilled manipulation of copper. As early as 1160–70, the heads of figures were being worked separately and set against an enameled ground.[8] The overall effect was dazzling: in the sixteenth century F. de Lagarde, a brother of Grandmont, described the abbey's altar frontal as "fort bien ouvré et excellent, aultant ou plus riche que si le tout estoyt d'argent" (very well worked and excellent, as rich or richer than if the whole thing were of silver).[9] Today the visual effect of an altar of Limoges work can only be suggested by surviving elements, like the plaques from Orense (cat. 51); those from Grandmont itself (cat. 58); and those from a third series, of unknown provenance, which are published here together for the first time (cat. 46, figs. 46 *a*, *b*, and fig. 13.)[10]

Enamels of Limoges, whether in the form of book covers (like that mentioned by John of Saint-Satyre), candlesticks, boxes, buckles, or incense burners, were as suited to their function as they were aesthetically appealing. The copper-plate ground is relatively thick and strong. Unlike gold or silver, it is not easily misshapen by handling. Nor does the enameled surface fracture easily. Sturdiness was surely one of the many desirable aspects of Limousin book covers and candlesticks, which were in daily use. An enameled pyx or chrismatory, unlike a silver one, does not tarnish with handling. An enameled reliquary is gilded and colorful but less expensive than one of gold—and therefore less likely to be stolen. Enamel decoration is not only more colorful but also more legible than repoussé.

Limoges was not, however, the only center to produce enamels with these qualities. The Meuse and Rhine Valleys were also renowned for enameling. Abbot Suger of Saint-Denis (r. 1122–51) summoned seven Lorraine goldsmiths to create the great cross for his abbey, and at the end of the twelfth century Nicholas of Verdun was called to work at Klosterneuburg, near Vienna, and at Tournai. Why then was Limoges work more widely diffused than that of the Meuse Valley? A number of historical circumstances favored the development of *opus lemovicense* and allowed it to avoid becoming dependent on a single patron or group of patrons.

The city of Limoges was located at the intersection of major trade routes that dated from Roman times.[11] There was a strong local tradition of goldwork, evidenced in the legend of Saint Eligius (ca. 588–660), patron of goldsmiths, who was born in Chaptelat, near Limoges; by the names of goldsmiths recorded at Limoges; and by the inventories of the abbey of Saint-Martial and other churches.[12] Limoges was situated along ecclesiastical and pilgrimage routes to both northern Spain and Rome. Thus its enamelers were poised to accommodate local ecclesiastical and monastic communities, like Saint-Michel-des-Lions, for which the chasse of Saint Loup was made in 1158,[13] or Bellac and Champagnat (cats. 9, 10), properties of the counts of La Marche, as well as affiliated Benedictine communities like Ruffec (where the chasse of Saint Alpinian was commissioned by Isembert, monk of Saint-Martial who was named prior of its dependency there),[14] Mozac (cat. 45; fig. 14), Saint-Sernin of Toulouse (cat. 40) and dependencies of Cluny, like La Voûlte near Le Puy (fig. 70*b*).

Such monastic associations were not confined to France. The church at Frassinoro in northern Italy preserves important examples of Limoges metalwork. The first is a gilt-copper candlestick without enameling, which bears the name of its maker, Constantine of Limoges (fig. 15). The animals inscribed in roundels are closely related in form to those found on the coffret of Abbot Boniface (cat. 7) of about 1110–30 and constitute an important link between the metalwork executed at Conques and Limoges. The fabrication of the Frassinoro candlestick probably coincides with the church's becoming a dependency in 1107 of the monastery of La Chaise Dieu in the Auvergne, a community that had strong ties to Saint-Martial at Limoges.[15] (Frassinoro was also the monastery in which Guinamundus, the artist who made the late-eleventh-century sepulchre of Saint Front, Périgueux, was resident.)[16] The treasury at Frassinoro also includes a Eucharistic dove of Limoges work, witness to the continuing relationship with central France. The movement of monastic personnel seems to correspond to the diffusion of enamels, as Gauthier has suggested for Roda de Isabena (see cat. 7).

Monasteries typically maintained close links to the seigneurial families of the region, many of whose members entered religious orders and contributed both lands and funding to the communities.[17] It is likely that the creation of works such as the chasse of Saint-Viance (cat. 118) and the equestrian plaque from Gourdon (cat. 93) as well as the late-thirteenth-century coffret of John of Montmirail (cat. 133) involved the direct patronage of the local nobility.

The political relations of France and England in the second half of the twelfth century coincidentally favored the development of enameling at Limoges. Following the marriage in 1152 of Eleanor of Aquitaine to Henry II, the Plantagenet king of England, Limoges became a Plantagenet stronghold, a satellite capital in the heart of medieval Aquitaine. As the stained glass of Poitiers reflects the presence of Eleanor and Henry,[18] so, at Limoges, enamelwork indicates the patronage and favor of the Plantagenets.[19]

The surest manifestation of Plantagenet interest in the Limousin is witnessed at the abbey of Grandmont, which Henry II wanted to make his royal necropolis.[20] His interest may have first been sparked by Eleanor, who had been taught by the Grandmontain Geoffrey de Loroux, former scholar of Angers and archbishop of Bordeaux (r. 1136–58).[21] But Henry's own belief in the value of his association with Grandmont was fervent. During a life-threatening storm at sea, he is recorded as saying: "Let us proceed courageously, for the Grandmontain brothers, in whose prayers we trust, have risen at midnight and are praying for us at Matins. It is impossible for us to perish while the brethren are watching and praying for us."[22] Henry visited Grandmont in the 1170s and in 1182, when he and his son Geoffrey ate in the refectory. In 1177 Henry was at Grandmont for a meeting with the count of La Marche, an encounter that led to the king's purchase of the county (see cat. 10). When his errant son Henry died in 1183, a monk of Grandmont was given the task of informing the king of the tragic news. Henry II was unfailing in his financial support for the community at

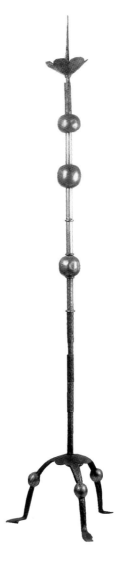

Fig. 15.
Candlestick.
Frassinoro,
Church

43

Grandmont: many payments and shipments of building materials are recorded in the 1170s. When the king died in 1189, he left the abbey of Grandmont two thousand pounds in his will.[23]

The importance of the Grandmontain order in the diffusion of Limoges enameling in France has rightly been emphasized.[24] A number of dependencies of Grandmont are recorded as having crosses whose style indicates that they come from the same workshop as the celebrated Grandmont plaques in the Musée de Cluny (see cat. 57a, b). The treasure excavated at Cherves (cats. 98–100) may represent the holdings of a single nearby Grandmontain foundation. Similarly, the priory of Haie-aux-Bonshommes near Angers, founded in 1178 by Henry II, recorded in its inventory a "very old" gilt-copper and enamel tabernacle, perhaps similar to the one excavated at Cherves, as well as coffrets and reliquaries of Limoges work.[25]

The Grandmont order's influence in the spread of Limoges enamel seems to have been limited to France. (there were but four Grandmontain priories in England and only two in Spain.)[26] The links of Limoges to the Plantagenets themselves, rather than to the Grandmontain order, seem to have encouraged the movement of Limoges enamelers to England and to Spain. The brief but exceptional flowering of enamelwork at Silos was fostered by the royal patronage of Alfonso VIII of Castile and Eleanor of England, daughter of Henry II and Eleanor of Aquitaine, whose gifts to Silos are recorded between 1177 and 1202.[27]

The connections of Limoges with Rome were enhanced via Grandmont at the end of the twelfth century. Gérard Ithier, seventh prior of Grandmont (r. 1188–96), was charged with submitting papers for the canonization of the founder of the Grandmontain order, Stephen of Muret; a priest and lay brother went to Rome with letters from the General Chapter and testimonials from Henry II.[28] In 1189 Pope Clement III (r. 1187–91) sent delegates to the ceremony held at Grandmont on August 30.[29] Pope Innocent III (r. 1198–1216) first delegated William of Bourges (see cat. 102) and then twice visited Grandmont himself to settle disputes concerning authority within the community. This same pope became a patron of Limoges enameling. About 1215 he presented a reliquary of Limoges work in the form of a basilica to the church of Saints Sergius and Bacchus in Rome, his titular church as a cardinal.[30] More important, he undertook to renovate the decoration of the Confessio in Saint Peter's Basilica with a campaign of Limoges work.[31] The Confessio marked the burial place of the Apostle Peter, one of the most sacred spots in Western Christendom, and the importance of this

commission as an indicator of the status of Limoges enameling cannot be overestimated. The Limoges-work decoration followed commissions and embellishments in gold and silver by five popes, from the fifth century through the ninth.[32] The taste for Limoges work on the part of the popes continued, as the inventories of 1295 and of Avignon attest.[33] The importance of Limoges work was validated both through Innocent's commissions and through a proclamation by the Lateran Council, which was convened at Rome in 1215. Attended by more than four hundred archbishops and bishops and more than eight hundred abbots and priors, the council, and Innocent himself, wanted to consolidate and strengthen church authority, partly but not wholly in response to problems of heresy. Liturgical objects of Limoges work were one of the practical means used in establishing universal standards for the practice of the faith. The council established that all churches must have two pyxes for reserving the Sacrament and specified that one of them could be of Limoges work, perhaps because that was already the practice or perhaps because Limoges work was more affordable than precious metal, yet suitable for its precious contents (see cat. 74).

While the enamelers of Limoges produced works of art whose decoration was specific to a particular commission, they were equally adept at fabricating works that could be used in any ecclesiastical context. Chasses rich with vermiculé and representing the Crucifixion and Christ in Majesty were preserved in churches in France, England, and Denmark. Crosiers representing Saint Michael and the Dragon have been found in the tombs of bishops of Germany, France, and Spain (see cats. 109, 110).

The Limoges work in the churches of present-day Denmark and Sweden probably dates from the consolidation of church authority in that region under the archbishops of Lund beginning at the end of the twelfth century. Ecclesiastics such as Absalon of Lund (1128–1201), who established monasticism in the Scandinavian territories, and his successors were trained at the university in Paris. Considerable correspondence between Innocent III and the archbishops of Lund survives,[34] and churchmen from the region were present at the Lateran Council in 1215. It was in the context of this kind of contact with Rome and with France that they acquired objects of Limoges work like the chasse of Uppsala (fig. 16) in the same period.[35]

The coffrets of Cardinal Guala Bicchieri (see cats. 88, 92) were taken to Italy prior to his foundation of a new Augustinian church at Vercelli. With such gifts, he was apparently following an already established example: in his will of 1218 the bishop of Paris gave two Limoges work coffrets to the church of La Chapelle-en-Brie.[36] There are numerous instances of Limoges objects in communities of Augustinian canons, such as Saint-Maurice-d'Agaune, Klosterneuburg, and Roskilde.[37] The use of Limoges objects by Augustinian communities began in the Limousin—the chasse of Saint Geoffrey at Le Chalard was the property of an Augustinian community founded in 1089 (fig. 118a).[38] The spread of Limoges work so far afield seems to have been facilitated by the presence of foreign churchmen in Paris and in particular by their contacts with the abbey of Saint-Victor—where John of Saint-Satyre first saw a Limoges book cover in 1167–69—as the examples of Guala Bicchieri and Absalon of Lund suggest.

The patronage of the Knights Templar and of the Hospitallers seems also to have played a role in the diffusion of Limousin enamels. Their taste for Limoges works is demonstrated by the Virgin of La Sauvetat (cat. 156), a Virgin of the same type preserved at Breuilaufa,[39] the lost altar at Bourganeuf (see cat. 122), and the funerary plaque of Guy de Meyos (cat. 152). The inventory of the commandery of Joigny of 1313 mentions numerous objects "de Limoges."[40]

Much has been made of the dissemination of Limoges work by means of the pilgrimage roads. The appearance of an important altar frontal at Orense (cat. 51), not far from Santiago de Compostela, appears to argue for this hypothesis, at least insofar as individual churches' increased revenues from pilgrims provided the means for the acquisition of enamelwork. The paths followed by the pilgrims were in fact the same routes established between monastic communities and, as the example of Orense and of the eleventh-century monument at Périgueux attest, were well established even among sites that were not primarily associated with pilgrimages.

Church and political connections and individual gifts contributed to the importation of Limoges enamelwork to the Holy Land. An early example of contact between Aquitaine and the Holy Land runs east to west, represented by the relic of the True Cross sent by Amaury, Latin king of Jerusalem, to Grandmont in 1174. But several Limoges work candlesticks excavated at Jerusalem date to about the same period.[41] The single rectangular thirteenth-century Limoges plaque of Christ in Majesty at the monastery of Saint Catherine, Mount Sinai, suggests an offering left by a Crusader. A rough-cut keyhole allows it to serve as a lock plate on the left panel of the sixth-century carved wood door (fig. 17). The decoration of some pyxes with armorial motifs and, in one instance, with the image of knights traveling by boat (fig. 18) suggests that they were made for Crusaders.

Fig. 16. Chasse. Uppsala, Cathedral Treasury

Aristocratic taste for Limoges work seems to have revived in the mid-thirteenth century, perhaps following the example of Saint Louis, two of whose children were interred with brilliantly colored funerary effigies of Limoges work set over them (cat. 146). In Brittany, Champagne, and England, Limoges work tombs enjoyed a vogue not only among nobles and bishops but also among wealthy bourgeois; today these are documented largely through pre-Revolutionary watercolors and surviving contracts (see chapter VI). Princely taste for Limoges enameling from the mid-thirteenth century onward was not confined to tomb sculpture. Coffrets decorated with heraldic motifs, one of which held relics of Saint Louis himself (cat. 124), were also valued. In addition there were objects apparently associated with diplomatic overtures. In 1317 King Philip the Tall sent a *chanfrein*, or a horse's ornamental headdress, to Armenia, during a period of considerable diplomatic effort in the Western Church and kingdoms toward the Armenian kingdoms.[42] A crosier representing Saint Michael belongs to the treasury of the Armenian patriarch of Jerusalem.[43]

In general, we need to admit the possibility of more communication than our sense of the difficulties of road travel and correspondence in the Middle Ages allows us to suppose. It is estimated that between two hundred thousand and five hundred thousand people traveled the pilgrim's route to Santiago de Compostela each year.[44] Nobles and ecclesiastics traveled frequently and communicated by messenger.[45] The peregrinations of the Plantagenets in the Limousin testify to the mobility of political figures in the twelfth century; Louis IX and Blanche of

Fig. 17. Book-cover plaque. Mount Sinai, Monastery of Saint Catherine

Castile visited Limoges in 1244 as they were journeying to the shrine at Rocamadour.[46] The chronicles of Saint-Martial, Limoges, record the visits of other princes to the city: Philip the Fair in 1272 and 1285, Pope Clement V in 1306, James of Majorca in 1307.[47] In 1306–7, when Pope Clement V was in Poitiers, six "Tartares" passed through Limoges, perhaps Mongols whom the pope was hoping to convert to Christianity.[48]

Book covers of Limoges enamel are preserved at Novgorod, possibly sent there along the well-established trade route through Scandinavia.[49] There are even references to what may be Limoges work in China, perhaps among the liturgical objects sent by the Franciscans during their missionary efforts from about 1250 to 1400. This hypothesis is not contradicted by the survival of Limoges work with Franciscan iconography and by the recorded presence of Franciscans at Limoges beginning in the early thirteenth century.[50] In summary, the many aspects of religious and secular life held in common in Europe from the late twelfth through the fourteenth century allowed for the development of a truly international taste and market. Even with the advent of a European community in the late twentieth century, such a phenomenon still surprises us and must be credited finally to the aesthetic appeal of the objects themselves.

Fig. 18. Pyx. Hartford, Wadsworth Atheneum

1. Texts using this designation have been published especially by Texier (1857, cols. 1143–44), Arbellot (1888), Ross (1941), and Lehmann-Brockhaus (1955–60). There are no comparable terms used for enamels from other regions.

2. ("Venerabili domino et amico suo R. priori Sancti Victoris Parisiensis,) suus Joannes eamdem quam sibi desiderat salutem. Quoniam, accepta licentia, exivi de ecclesia Sancti Satyri, ut irem cum domino Cantuariensi archiepiscopo, quidam amicus noster, pro magna necessitate, commodavit mihi decem solidos Andegavenses. Cui promisi quod per manus vestras ei redderem. Ideo precor ut latori praesentium eos consignes, et hoc vobis signum, quod ostendi vobis in infirmario tabulas texti de opere lemovicino, quod mittere volebam abbatiae de Witgam. Et vos mihi respondistis, quod magister Andreas non in tantum respexerat obedientiam, et locum suum de sancto Paulo" (To the venerable lord and friend R, prior of Saint Victor of Paris, his friend John sends him greetings which he himself would like to receive. Having been allowed to leave the church of Saint-Satyre to go with our lord Archbishop of Canterbury, a friend of ours out of great necessity lent to me ten Angevin solidi [coins]. I promised him that I would return them by your hand. This is why I ask you to consign them to the bearer of this. As a guarantee of my word, I remind you that I showed you in the infirmary of your monastery a Gospel cover of Limoges work, which I wished to send to the abbey of Witgam); Du Chesne 1636–49, vol. 4, p. 746, epist. 69. This is a frequently cited text; see, especially, Gauthier 1987, no. 160, pp. 104–5.

3. *Icones*, I, 38, p. 109; see also *Imagines* 1614.

4. Lasko 1972, fig. 91.

5. Chap. 53, Hawthorne and Smith 1976, p. 125.

6. Rupin 1890, p. 7, n. 3.

7. Noted by Texier (1843, pp. 99–106), without specific reference. See, for example, "Allegoriae in Vetus Testamentum," *PL* 175, cols. 661–62; "De bestiis et alies rebus liber secundum," *PL* 177, col. 115–16. For the symbolism of gems and colors before the twelfth century, see Jülich 1986–87.

8. Gauthier 1987, no. 98, pp. 103–4.

9. Texier 1843, p. 73, quoting Abbé Nadaud.

10. The plaque illustrated here as figure 13, which belonged to, or was in the hands of, Arnold Seligmann, Rey and Co. in 1933, was offered to the Busch-Reisinger Museum, Harvard University. An appliqué figure of Christ in Majesty in the Schnütgem Museum, Cologne (Inv. 6560), may likewise come from the same ensemble.

11. See Boehm 1991, p. 115.

12. See Guibert 1885.

13. See Gauthier 1987, no. 75, pp. 83–84.

14. See ibid., p. 103; Texier 1846. Isembert later served as abbot of Saint-Martial from 1174 to 1198.

15. Texier 1843, p. 73, quoting Abbé Nadaud.

16. Gauthier 1987, no. 20, pp. 51–52; Texier 1846.

17. Bull 1993.

18. Grodecki 1977, pp. 72–73, and 286, no. 73.

19. The importance of Plantagenet patronage for Limoges has been especially developed in the publications by Marie-Madeleine Gauthier. See especially Gauthier 1967b.

20. The Grandmontain monks were known for their devotion to the dead. Hutchison 1989, p. 53.

21. Ibid., pp. 63–64.

22. Ibid., p. 57.

23. Ibid., p. 60.

24. Gaborit 1976a; Souchal 1962–1967; Landais 1976.

25. Port 1874–78, vol. 2, p. 342, cited in Landais 1976.

26. For England, see Graham and Clapham 1926, pp. 159–210.

27. Férotin 1897, pp. 103–25.

28. *PL* 204, cols. 1048–49; Hutchison 1989, p. 27.

29. Hutchison 1989, p. 28.

30. "Basilicam de factura lemovica," *PL* 219, col. 207, and Arbellot 1888, p. 240, both cited in Rupin 1890, p. 87, n. 8. On Innocent III's patronage of churches in Rome, see Krautheimer 1980, pp. 203–4.

31. See the publications by Gauthier, notably 1968a, pp. 237–46.

32. See the summary in Gauthier 1968a, pp. 237–38, n. 10.

33. See Molinier 1882–88; Hoberg 1944.

34. See *PL* 215, cols. 200, 223.

35. See B. M. Andersson 1980.

36. Cited in Gay 1887–1928, vol. 1, p. 618.

37. Noted in Castronovo 1992, p. 167.

38. Cottineau 1935–39, vol. 1, col. 671.

39. See Arminjon et al. 1995, p. 73.

40. See Arbellot 1888.

41. Gauthier 1987, nos. 70, p. 81, and 125, p. 120.

42. Oral communication from Helen C. Evans.

43. Marquet de Vasselot 1941, p. 159, nos. 289, 290.

44. See Öhler 1995, p. 187.

45. Ibid., especially p. 101. In the fourteenth century, express messengers could travel up to 200 kilometers a day.

46. Ruben et al. 1872, p. 196.

47. This sampling is taken from Duplès-Agier (1874, pp. 133–34, 136, 142). For additional information on the historical place of the city of Limoges, see Barrière, above, p. 20

48. Duplès-Agier 1874, p. 141. The identification with Mongols was suggested, in an oral communication, by Helen C. Evans.

49. On the Hanseatic routes between East and West, see Schildhauer 1988.

50. Rupin (1890, p. 88) mentions the enameled vessels recorded in China. For Franciscan objects, see below, cats. 101 and 142. For the presence of Franciscans at Limoges, see Duplès-Agier (1874, pp. 130–31). I am grateful to Helen C. Evans for suggesting the possible circumstances of their appearance in the Far East.

# Techniques and Materials in Limoges Enamels

Isabelle Biron, Pete Dandridge, and Mark T. Wypyski,
with the Collaboration of Michel Vandevyver

## I. Introduction

This exhibition has provided a unique opportunity to investigate the techniques and materials used to create the enameled objects attributed to central France—specifically those from Conques and Limoges of the twelfth through the fourteenth century—as well as several key objects that bear directly on the evolution of enameling in this locale. Within Europe, the artists of Limoges were unique in sustaining the practice of enameling throughout the medieval and Renaissance periods. By documenting the origins and development of this tradition, we can establish a framework for the technical evaluation of the artistic output of Limoges relative to its antecedents, its contemporaneous counterparts, and its own evolution.[1] Our study draws on five specific resources: the historical texts that discuss workshop practices; observations drawn from our examination and treatment of the objects in the exhibition; earlier studies of enamels similar to those being analyzed or contemporaneous with them; available analytic capacities; and modern practitioners who have adopted the style and techniques of medieval Limoges.

Three medieval treatises describe different aspects of the arts of glassmaking and enameling. Two of these, the *Mappae Clavicula*[2] from northern Europe and Eraclius's *De Coloribus et Artibus Romanorum*,[3] are twelfth-century compilations that draw on ancient manuscripts and traditions as well as contemporaneous sources. Indeed, the first two books of *De Coloribus* are thought to be of Italian origin, whereas the third book is considered to be French or Norman.[4] Only Theophilus's *De Diversis Artibus*[5] represents the writings of a practitioner, generally thought to be the twelfth-century Benedictine monk Roger of Helmarshausen.[6] His treatise is the sole document that addresses enameling directly,[7] and although he discusses the cloisonné technique on gold, the process is very similar to enameling on copper. The other two texts discuss the related practice of glassmaking,[8] and Eraclius describes the creation of glass "gems," or paste cabochons, from Roman glass.[9] The numerous manuals from the sixteenth, seven-teenth, and eighteenth centuries are of interest as well, in that they describe in much greater detail the materials and the procedures employed by metalworkers and enamelers whose technical origins are in the medieval period.[10] Of these, the most important is René François's *La Façon de l'émaillerie recueillie des anciens émailleurs*, thought to date from the beginning of the sixteenth century, which outlines the techniques and materials used by Limousin enamelers of the time.[11]

In extant technical studies of Limoges enameled objects, the number of pieces examined has been limited, and only isolated aspects of fabrication have been addressed.[12] In the catalogue of northern Romanesque enamels in the British Museum, a wealth of comparative information has been provided about the materials and practices used in the twelfth and thirteenth centuries to produce the champlevé enamels in the museum's collection, predominantly from the Meuse Valley, that are contemporaneous with those in the first half of our study.[13]

The varied analytical capacities of The Metropolitan Museum of Art (MMA) and the Laboratoire de Recherche des Musées de France (LRMF), combined with standard protocols of examination, have allowed us to characterize the working methods of the metalworkers and enamelers of the Limousin and to quantify many of the materials used.[14]

## II. The Techniques of Fabrication

When forming a picture of metalworking practices in Limoges, it is important to realize that they did not develop in a vacuum. Workshops within Limoges and its environs had been producing ecclesiastical and secular objects in silver, gold, and copper well before the period addressed here;[15] and further, Limoges had been a locus for the minting of coins at various times during the preceding five

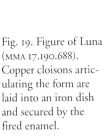

Fig. 19. Figure of Luna (MMA 17.190.688). Copper cloisons articulating the form are laid into an iron dish and secured by the fired enamel.

hundred years.[16] The discussion below adopts a chronology of fabrication suggested by the objects, the historical sources, and discussions with practitioners.

### The Mining and Processing of the Copper

Beginning in the tenth century, the Harz Mountains in Lower Saxony became the major center for mining the copper used in northern Europe throughout the medieval period.[17] Equally productive sites in southern Spain provided copper ore from Roman times until the fifteenth century.[18] Whether either of these was the source for the Limousin, or whether local ore or a combination of both was used, is unclear; however, trace-element analyses of a limited number of objects suggest that there were multiple sources.[19] In what form or at what level of purity the copper was sent from the mines to the ateliers is also difficult to specify.[20] Undoubtedly, refining was an integral part of workshop practice, since only then could the metalworker be assured of attaining an alloy whose physical properties would be compatible with the techniques of fabrication. Indeed, Theophilus describes the need for the further refining of copper that was to be worked in a repoussé style and gilded.[21] By reducing the impurities present in the copper it would become less embrittled and could be more readily hammered, chased, engraved, struck, and scraped;[22] moreover, the potential disruption at the interface of the gold and copper by the migration of lead and zinc to the surface during heating could be avoided.[23] Our analyses seem to confirm this practice, since the composition of the alloys averaged 98.8 percent copper, 0.5 percent lead, with iron/arsenic making up the bulk of the difference (see Appendix, table 1, p. 445).[24]

### The Evolution of Champlevé Enameling

The champlevé method of enameling, which requires the gouging out of the metal substrate to create cells, was the predominant decorative technique associated with the metalwork of medieval Limoges. It is an ancient practice that can be traced to both Celtic and Gaelic traditions; however, the Limousin enameler used incised, hammered sheet rather than the earlier cast forms or cast-in cells. The style that developed in Limoges may be derived more from the cloisonné technique—of equal antiquity and practiced in the West into the tenth century[25]—whereby the cells were created from thin strips of metal, or cloisons, either of gold, silver, or a copper alloy, often soldered to thin supporting sheets of the same metal. The transition from cloisonné to champlevé enameling in the early Romanesque period in the Limousin can be illustrated within the context of the exhibition; however, it should be kept in mind that the progression was not strictly chronological, and that there were some centers, such as Conques, that were working in a champlevé style at a very early date. In the plaque with the figure of Luna (fig. 19),[26] the cloisonné technique has been preserved with copper strips overlaid on an iron dish. Each of the enameled plaques from the portable altar with the symbols of the evangelists (cat. 6) and the figure of Christ (cat. 5) have been fabricated from two copper sheets of identical dimensions:[27] the image was cut out of the top sheet, which was then soldered to the backing sheet that supported the enameled image and the single-width copper cloisons defining it.[28] In the symbol of Saint Luke from the portable altar, the outline of the ox's proper right rear thigh was adjusted with several thicknesses of wire. A variant of this same technique was used for the plaques depicting the Elder of the Apocalypse (cat. 4) and the female saint (cat. 3), where a single copper sheet was employed; the interior of the sheet has been cut away to a uniform depth, creating a large open cell with copper cloisons delineating the design. The final step in the transition seems to be the use of both cloisonné and champlevé techniques within one image, apparent in the Champagnat chasse (cat. 10) and the reliquary of Saint-Sulpice-les-

Fig. 20. Detail of an empty cell from an enameled plaque (Louvre, MRR249). The loss of enamel reveals the tooling associated with a carved cell: a round-nosed scorper around the perimeter and a flat, chisel-like graver in the interior.

Fig. 21. The heterogeneous color and particle size of Limoges enamels are apparent in this dark blue example (Louvre, MRR294). Individual particles are no greater than 0.5 micrometers in diameter. (Photo: D. Bagault)

Feuilles (cat. 54), where cloisons have been selectively: on the Champagnat chasse, in the nimbuses of the angels on either side of the Hand of God; in the horizontal band across the censer swung by the angel on the left; in the nimbuses of the two Marys; in the drapery of Christ; and in the reliquary, for the circular, decorative motifs on the angel's wings. An appealing aspect of the cloisonné technique was the ease with which one could shape the wire to affect the quality of the line. Initially, that same visual effect seems to have been imitated by the Limousin metalworkers, but the refinement achievable in champlevé enameling was restricted somewhat by the inherently subtractive process of carving.

### The Preparation of the Plates for Enameling

With only a few exceptions, all the objects in the exhibition were fabricated from copper sheet, normally between two and five millimeters in thickness, hammered out from cast ingots and periodically annealed, or heated, to relieve the stresses induced by cold-working. The indentations on the backs of the objects, the lack of any casting texture or porosity,[29] the scattered areas of delamination within the metal, and the variation in the thickness of the metal across each plaque are all indications of such working.[30] The sheet would be cut to size with a chisel, and a slight dome, or a curve, would be worked into the reverse. The thickness of the copper precluded the possibility of warpage during subsequent heating, which suggests that the doming might have been an aesthetic choice.

The method (or methods) used to draw or transfer the design onto the metal substrate is unclear; however, care was taken in the layout to assure the successful integration of decorative and functional elements. The systematic placement of reserved spaces for rivets or brads, hinges, and clasps or locks suggests that they were laid out first.[31] Major design elements such as appliqués or enameled figures and symbols would then be positioned.[32] Where enameled elements intersected previously reserved spaces, their cells would be carved to avoid the reserved areas. Lastly, engraved patterns and designs and enameled borders would be sketched out to fill the remaining ground.

The X-rays of the plaques and the losses in the enamels make it apparent that initially the outlines of the cells were cut—typically with a round-nosed graver—[33] and then the bulk of the metal was removed from the interior with either a scorper or a graver, generally to a depth shallower than that of the perimeter (fig. 20).[34] Shaped objects with complex curves and enameled decoration such as the *poupée* figures (cats. 112, 115) and the Eucharistic doves (cats. 105, 106) would have been shaped before the cells were cut. During the working process, the forms or plaques would have been supported, possibly in a bowl of pitch, to maintain stability and provide a resistant surface against which to work.[35]

Prior to the application of the enamel, it would have been necessary to remove from the plaque and the cells any dirt, grease, or corrosion that could inhibit the flow or adhesion of the enamel to the copper. René François mentions the use of salt, urine, vinegar, and wine as potential cleaning agents, which would then be rinsed off with successive washes of water.[36] More likely, an abrasive such as sand would have been sufficient to remove any surface contaminants.[37]

### Enameling

Enamels are made of glass that was either produced

specifically for the purpose or recycled. (See section III, below, for a complete discussion of the possible sources for and the analyses of the enamel.) Theophilus fully elucidates the steps required to create a cloisonné enamel plaque: preparing the enamel powder, filling the cells, firing, and polishing;[38] the techniques are similar to those adopted in the Limousin for champlevé enamel and correspond to François's later description as well.[39] The artist selected the required shards of colored glass, heated them, and dropped each into a dish of cold water, fracturing them. Grinding in water with a pestle reduced the glass to the appropriate size. Too fine a powder would cause air bubbles to be trapped within the fired enamel, thus diminishing the richness of the color.[40] After repeated rinsing to remove potential contaminants, the prepared sample was set aside in a covered dish. Because of the heterogeneous nature of the glass used in the Limousin for enameling, it is possible to gauge the size of the particles with the naked eye (fig. 21).

Using a sharpened quill, the enameler placed the slightly damp powdered glass in the appropriate cells; different colors of varying opacity were often applied within the same recess (fig. 22).[41] Once filled, the plaque and its supporting iron plate were placed near the fire to be warmed slowly and to drive off any moisture retained within the powdered glass. The object was then covered by a domed iron lid pierced with many small holes. By mounding charcoal around the iron muffle and by selective use of the bellows, the temperature within the oven could be raised sufficiently to fuse the enamel.[42] The color of the glowing iron indicated the temperature within the oven and the progress of the firing. When the enamel softened, the charcoal was removed from around the oven and the plaque was allowed to cool slowly, minimizing the risk of the enamel cracking because of the different rates of expansion and contraction for copper and for glass. Since the volume of the enamel was much greater when powdered than when melted, it was necessary to refill the cells and repeat the firing process several times—depending upon the complexity of the design—to achieve an enameled surface that would be without any depressions or voids.

The coloristic effects achieved by the enamelers were limited by their choice of raw materials and by their use of copper as the substrate. The heat required to fuse the enamel caused the formation of a rich layer of red copper oxide—cuprite—over the surface of the exposed copper as well as the surface under the enamel (fig. 22). The use of a predominantly opaque palette precluded much of the visual distortion caused by this underlayer; however, in the case of the translucent blues and greens, the effect of

Fig. 22. Detail from the chasse of Saint Thomas Becket (cat. 39) with a variety of different colored enamels contained within individual cells and encircled by decorative stippling. A thin layer of reddish cuprite formed at the interface of the copper and the enamel. (Photo: D. Bagault)

the cuprite was diminished by increasing the depth of the cells. In the symbol of Saint Luke (cat. 6), an intermediate layer of opaque white was used to isolate the cuprite from the overlaying translucent magenta coloring the ox's body and the book. Other examples suggest that artists made use of the tinting power of the red underlayer to create or enhance a translucent color. For instance, in the suppedaneum of Christ (cat. 43) and the book held by the Lamb of God (cat. 27), a clear enamel has been applied over the cuprite to create a red unusual in Limoges enamels. For a characterization of the enamel colors, see Appendix, Table 7, Munsell notations.

*Polishing*

Theophilus describes a series of steps to reduce the undulating quality of the fired enamel to a finished, polished surface, using a range of different abrasives, beginning with a flat, coarse limestone, with water as a lubricant, and ending with a mixture of saliva and finely ground potsherd rubbed over the surface with a goatskin.[43] Undoubtedly, similar materials were used in the Limousin. While the degree of finishing varies greatly, the quality achievable is apparent in the well-preserved, smooth, reflective surfaces of the Conques medallions (cats. 7, 8).[44] Under magnification one can see that even the most well-preserved surfaces are marred: by linear striations—caused by the coarser abrasives—that span both the enamels and the flat, gilded, reserved areas of their surround; by circular voids where gas trapped within the volume of the softened enamel has partially migrated during fusion;[45] and by specks of metal oxides.[46]

Fig. 23. Detail of a stamped head (cat. 72). The lip around the perimeter of the appliqué indicates the depth of the die used. (Photo: Patricia Mazza)

carry over to the metal and can be seen under the gilding.

## Appliqués

Part of the decorative vocabulary of the Limousin workshops was the use of raised appliqué figures and discrete, solid heads to add a strong three-dimensional aspect. It has been proposed that they were mass-produced, using dies or forms of some sort.[48] Since Limoges had a history of serving as a center for minting coins,[49] it seems likely that the capacity to form dies[50] and the disposition to use them were present. Examination of the heads indicates that in fact matrices were used frequently. On the chrismatory (cat. 71) and the ciborium of Master Alpais (cat. 69), for example, it is possible to find identical heads within each group as well as being able to find a single head—the bearded, balding figure (Saint Paul?)—replicated on each.[51] The depth of the dies is apparent when one looks at the edges of those heads in shallow relief where there is a lip, or shoulder, around the perimeter, just beneath the articulated details (fig. 23), below which excess metal has been cut away. Similarly shallow dies were used for the heads in higher relief, with details and undercuts engraved into the additional, unarticulated metal below the shoulder (cat. 31). Microscopic examination of all the heads indicates that struck details in the hair and beards and around the eyes and mouth were often reinforced or articulated further by engraving and chasing. The extent to which dies were used is unclear. Were all the heads on each object stamped or were some so individualized, requiring so much working after stamping, that it would have been more efficient to carve them entirely? Certainly, the solid, carved heads of the crucified Christ on the cross and on the book cover are unique (cats. 25, 26).

The appliqué figures, like the pierced medallions and crosier knops, were created from copper sheet. Their initial rough forms were raised from the back and then finished from the front, whether in a traditional repoussé technique or with the aid of negative forms. Round-headed peening hammers and tracers, or chasing tools, were used to raise the more massive forms, whereas a tracer sufficed for distinct elements like a slightly projecting arm or a drapery fold. In repoussé work, the metal being raised was normally embedded in a bowl of a yielding medium like pitch. With each strike of the tool the metal would expand; however, if the copper had been hammered into a matrix, it would have compressed and moved laterally when it came into contact with the unforgiving surface of the form. In the appliqués examined, the metal appears to have been stretched during its working.[52] Most of the articulation of the form and the decorative effects were

## Engraving

All the incised details on the plaques, such as the vermiculé grounds and the reserved figures within enamel fields, were defined initially by pushing a graver through softened, annealed copper. Depending upon the depth desired, it may have been necessary to recut the line several times. Whether the engraving was done prior to or after the fusing and polishing of the enamel is an open question. By waiting, the artist could eliminate the necessity of removing either the oxidized layer of cuprite from within the thin engraved lines[47] or the potential overflow, or spillage, of fused enamel in those same details. Of further benefit would be the fact that the engraving of the designs and figures could be initiated in a freshly annealed and polished surface. Alternately, by completing all the engraving prior to enameling, any danger of inadvertent fracturing of the enamel would be avoided.

Certainly, the subsequent scraping of the metal surface was done after the enamel was polished. By pulling either a curved-edge or a straight-edge tool with rounded corners across the surface, the thin lip of metal, or burr, pushed up on either side of an engraved line could be removed or the edges themselves rounded off. The modeling of the broader figural planes was achieved with the same curved tool, the marks of which are apparent on the surface as a series of thin parallel troughs. By cutting back the metal, the scrapers eliminated any trace of the marks left from polishing the enamels. In the flat, reserved areas within enameled figures, where no metal was removed, or in engraved details like the foliate pattern of the cross on the chasse (cat. 24), where the reserved metal was so slight as to make scraping impracticable, the striations from the abrasives achieved from the front. The rounded surfaces of the chasing tools refined the forms and compressed the metal; gravers were used to remove metal, sharpen details, and create decorative patterns. The visual effects attainable with gravers are apparent in the incised lines of drapery patterns where the artist varied the depth and the quality of the V-shaped line by accentuating the rocking motion of his wrist as the point of the graver was pushed through the metal. The undulating, rippled appearance of the broader surfaces of the appliqués derives from the extensive use of the scraper to smooth out rounded forms, reduce high spots, and remove burrs left from engraved lines. Frequently, ring punches were used to simplify the production of *serpenses'* scales or eagles' feathers and to create decorative or textural patterns. The simplicity of the appliqués' shapes, the appearance of the tool marks on the reverse, and the degree to which these objects were finished from the front all suggest a traditional repoussé technique.

If matrices were used, it is more likely to have been with those objects composed of mirror images, such as the doves and the crosiers. In the crosier with the image of Saint Michael and the dragon (cat. 109), half of the volute, its crocketing, and the figures have been cut and shaped from a single sheet joined to its opposite half along a vertical seam; the wings are separate pieces entirely (fig. 24). The low relief of the figural group, the extensive use of files and chisels to articulate the figures and the crocketing along the spine, and the inherent difficulty in transferring surface detail from a negative form onto thick copper make it probable that even these pieces were wrought as unique objects.[53]

The desire to create three-dimensional or other visual effects resulted in several variants to the procedures described above. Cabochons, glass eyes, and the ears and horns of the evangelists' symbols were often added as separate elements. Usually, their settings were pierced through the metal from the front with a pointed tool or, as is the case with some of the cabochons, set into punched sockets. In either instance, the surrounding metal was worked up around the circumference of the insert to secure it in position. The sculptural quality of the heads of the Grandmont apostles (cat. 58) was enhanced by the artist's soldering on an additional sheet of copper to the backs of the heads to allow them to project forward and be fully in the round. Alternately, the heads of the encircling angels on the finial of Master Alpais's ciborium (cat. 69) and the appliqué Virgin on the tabernacle (cat. 70b) have solid heads with bodies modeled in thinner sheet, either carved from a single cast blank or hammered out from thick stock.

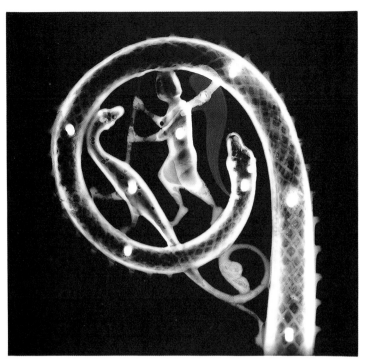

Fig. 24. X-ray of the crosier head (cat. 109) reveals the porous texture of the solder along the single vertical seam that joins the two halves of the crocketing, the volute, and the figures of Saint Michael and the dragon. (Photo: Pete Dandridge)

### Assemblage

The Limousin metalworker usually depended on mechanical joins to assemble separate elements. Most of the appliqué heads were attached with a round-sectioned post that projects from their reverses; it was inserted through a hole in the plaque and hammered over.[54] Copper rivets were used exclusively with the appliqué figures or groups; their heads either stand proud of the surface or have been incorporated into the surrounding design.[55] In some of the chasses, coffrets, and pyxes, mortises and tenons and tabs and slots function as the means of attachment for the sides and/or the bases. Frequently, more complex methods were required, as exemplified by the crosiers and the doves where copper/tin brazes, or hard solders, were used in tandem with mechanical joins (see Appendix, table 2). In the crosier head with the image of Saint Michael, referred to above (cat. 109), the two halves of the head were riveted to each other prior to soldering and enameling; the rivets were then overlaid by enamel and now are visible only in the X-rays (fig. 24). The radiographs also help to identify the location of the solder, which is seen as an opaque, formerly fluid material with a lacunal outline. Potential warpage and movement of the copper were restricted during the subsequent soldering and firing of the enamel by securing the crosier initially with rivets. Similarly, the doves were joined predominantly with rivets or mechanical joins (cats. 105, 106); hard solder was used selectively to attach the bottom half of the tail to the body and, in some instances, to close the seam encircling the body.[56]

## Gilding

The gilding of the metal surfaces was almost the last step in the fabrication of a Limoges enameled object, occurring after the attachment of the appliqué heads and figures. In accordance with our analyses and the historical texts,[57] a paste, or amalgam, of mercury and gold was brushed onto the freshly cleaned surfaces that were to be gilded.[58] The object was then brought to a low heat to volatilize the mercury, leaving a spongy, porous, matte layer of gold overlaying the metal. The gilded surface was burnished with a highly polished, rounded tool, which gave it a reflective, lustrous appearance.[59] However, the matte texture is still preserved in the interiors of the engraved lines, where it was inaccessible to the burnisher. The gold did not adhere to the surface of the enamel, so it could be brushed off and recycled.[60]

## Stippling

Only after gilding were unadorned gilt surfaces enlivened by the linear, repetitive strikes of a small domed punch to encircle the enameled cells or to create delicately stippled patterns across reserved areas (see fig. 22). Under the microscope, it is possible to see that the strikes occasionally trail off into enamels, fracturing their surfaces; that the raised lips around each punch mark have been retained rather than polished or scraped off; and that the gilding within the spherical depressions, unlike that in the engraved lines, has a compact or compressed surface, which would not have been achieved in burnishing.

## Supports

Wood cores serve as supports for many of the chasses; their ends, side elements, and bases are all single planks secured to one another with nails. Generally, the feet are extensions of the gabled ends, whose vertical sides have been cut back to the level of the top of the feet to allow for the long rectangular side elements to be set in flush. Access to the interior could be through a door in one of the gabled ends or in the base. The exposed portions of the core that were not clad with an enameled plaque or embossed sheet were gessoed with either a calcium-carbonate or a calcium-sulfate ground and then painted with minium (red lead) or vermilion (mercury sulfide).[61] A single cross was often painted onto an interior gabled end either in white or in red.

There has been some discussion as to whether the marks that are found on the backs of many Limoges objects are assembly marks.[62] Linear scratches or marks clearly related to joinery have been observed on the interiors of chasses where visual access was possible (cats. 31, 83). Other marks, such as those on the reverses of the series of roundels from Conques (cat. 8), do not relate to assemblage so much as to sequence, and they appear to be either numerals or signs.[63]

## III. Enamel Compositions

### Introduction

Throughout history, many different compositional types of glass have been used, and several different types were theoretically available for use by medieval enamelers. Most of the glass produced in the West up to the end of the Middle Ages is thought to have been made with only two basic raw materials: a source of silica as the matrix former and a fluxing agent to lower the melting temperature. Studies of ancient glassmaking recipes have shown that the siliceous component was usually sand, flint, or ground quartz pebbles, and the fluxing agent was usually an alkali from a plant ash or from a mineral source such as natron, from salt deposits found in some desert regions.[64]

The use of natural, unpurified raw materials by early glassmakers meant that other components were often unintentionally added along with those specifically needed to produce a particular type of glass, resulting in a very complex composition with many elements present. In addition to the silica, sand, for example, may also add some calcium, aluminum, and iron to the glass. Plant ashes—depending on the type of plant used and the method of preparation—may provide the sodium or potassium alkali flux needed, as well as large amounts of calcium and magnesium, and some chlorine, sulfur, and phosphorus.

Colorants and opacifiers were also often added to the basic glass mixture. A commonly used colorant in glasses was copper oxide, which can produce a range of colors, from blue to turquoise to green, depending on the overall composition of the glass and the conditions of manufacture. Copper oxide in its reduced form, cuprite, was used to produce red glass. Opacifiers are dispersed particles or micro-crystals within the glass matrix which diffuse and absorb light, rendering the glass opaque (see figs. 25, 26). Historically, the most commonly used opacifiers were calcium antimonate and tin oxide. Crystalline lead antimonate yellow was often used to produce an opaque

| | Sodium | Magnesium | Aluminum | Silicon | Potassium | Calcium | Tin | Antimony | Lead |
|---|---|---|---|---|---|---|---|---|---|
| | $Na_2O$ | $MgO$ | $Al_2O_3$ | $SiO_2$ | $K_2O$ | $CaO$ | $SnO_2$ | $Sb_2O_3$ | $PbO$ |
| Early Type: | 10–20 | 0.2–2.0 | 1.5–4.5 | 55–70 | 0.5–1.5 | 5–10 | <0.5 | 0.5–10 | 0.5–10 |
| Late Type: | 10–18 | 1.5–5 | 0.5–3.5 | 45–55 | 1.5–2.5 | 2–8 | 5–15 | <0.5 | 5–20 |

Approximate Ranges of Major Element Oxides for Early- and Late-type Enamels (weight percent)

yellow glass; when mixed with copper oxide, it was used to produce an opaque green glass.

Due to the complexity of glasses of this period—which may actually contain twenty or more important components—quantitative chemical analyses of the elements can provide a great deal of information for the classification and history of vitreous materials and document the changes in the materials used over time within a particular industry.[65] However, the complexity of the data may also make interpretation of the results difficult. To better understand the fabrication techniques used by the Limousin enamelers and attempt to define the raw materials used, quantitative elemental analyses were done on enamels from sixty objects from Limoges and its environs, dating from the tenth century to the fourteenth.

## Analytical Techniques

The enamel compositions in this study were analyzed by one of two methods. Those done at The Metropolitan Museum of Art were analyzed with the energy dispersive X-ray spectrometer (EDS) and the scanning electron microscope (SEM).[66] This allowed for analyses of overall compositions of the enamels, with high magnification analyses of individual crystals to identify the opacifier compounds. This type of analysis requires small samples of the enamels, about a cubic millimeter in size. The selection of objects for EDS analysis was therefore limited to those having areas of previous enamel loss, which allowed samples to be taken without compromising undamaged surfaces. Not all the colors on each object could be sampled because of the localized nature of the losses.

The enamels analyzed at the LRMF were done by proton-induced X-ray and gamma-ray emission (PIXE and PIGME), using the AGLAE acclerator (Accélérateur du Grand Louvre par Analyse Élémentaire). The use of this ion beam technique allowed for quantitative elemental chemical analyses directly on the enamels without the need to sample. This method is nondestructive and allows for major, minor, and trace element analyses.[67] It was verified on some samples that the surface analysis of sodium by gamma-ray emission correlated with analysis of the unweathered glass interior.

## Compositions

The majority of the enamel compositions analyzed, with the exception of the reds, fall into two groups that divide chronologically roughly before and after the beginning of the thirteenth century. The differences in the general compositional ranges found for the major elements of the early- and the late-type enamels are listed in the table above. (Complete results of the analyses are listed in the Appendix.) Phosphorus, sulfur, and manganese levels for both types were generally less than one percent, with chlorine less than two percent.

The enamels dated up to about the beginning of the thirteenth century were generally found to be soda-lime-silica glass containing small amounts of magnesium and potassium, generally less than about one and one-half percent each. Many of the early white enamels, however, were found to have higher levels of magnesium, up to more than three percent, while still having low levels of potassium. Aluminum levels in these glasses were found to be generally greater than about one and one-half percent, with calcium levels generally about five to ten percent. Some antimony was noted in almost all the early enamels. Small amounts of lead, up to a maximum of about ten percent, were also seen in some of these enamels, with the highest lead levels found in the yellow and green enamels.

The enamels on objects dating from about the second quarter of the thirteenth century and later were also found to be mainly soda-lime glasses, but unlike the earlier enamels, they were found to have relatively higher levels of magnesium and potassium, with relatively less aluminum and calcium, and they contained large amounts of tin and lead.

## Opacifiers and Colorants

The early-type opaque enamels, the vast majority of those analyzed, appeared to contain calcium antimonate as the main opacifying agent, although trace quantities of tin

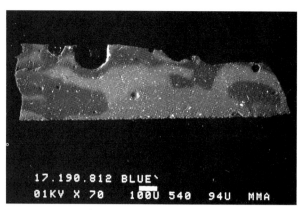

Fig. 25. Scanning electron micrograph of opaque blue enamel from the Saint Peter plaque (cat. 47). The bright areas are rich in lead oxide and contain crystals of tin oxide, whereas the darker areas contain calcium antimonate and very little lead. (Photo: Mark Wypyski)

were also found in several. All the yellow and green enamels examined, with only two exceptions from the later thirteenth century, appeared to contain crystals of the yellow opacifier lead antimonate.[68] Only a single definite use of lead-tin yellow was identified, from a situla (MMA 17.190.509; not exhibited), where it was found to be mixed with lead antimonate yellow.[69] The one flesh-color enamel also appeared to be mainly colored by the white opacifier; it is uncertain what provides the faint pink color—possibly a small amount of cuprite, as is found in red enamel.

The late-type opaque enamels were found to contain tin oxide as the opacifier rather than the calcium antimonate. Substantial amounts of lead, ranging from about ten to more than twenty percent, were also found in the opaque enamels. Lead oxide is nearly always present with tin in vitreous materials, as it was apparently used to facilitate the conversion of metallic tin to tin oxide.[70]

The colorants found in all the enamels are the same as those commonly found in most ancient glasses.[71] Cobalt oxide was used in the blues, generally less than one-half of one percent, although several were found with levels approaching one percent. A strong linear correlation was seen with iron and cobalt. Small amounts of zinc were also found associated with many of the later cobalt blue enamels and may be evidence that the cobalt source used for most of the early blues was different from that used for the late-type enamels. Copper appears to have been rarely added to the cobalt blues intentionally. Copper oxide, in amounts varying from about two to over ten percent, was used for the turquoise. Copper was also used for the greens, where it was found combined with the lead antimonate yellow opacifier, and in two instances, possibly

with lead-tin yellow. Two examples of translucent black—actually dark brown—enamels were seen, apparently colored by iron oxide, and one example was seen of a translucent amethyst colored enamel, which was found to contain a large amount of manganese.

Nine of the objects examined, mainly from the first half of the thirteenth century but including two from the end of the twelfth century, displayed an interesting phenomenon. These objects were found to have both early- and late-type enamels. Three of the enamel colors on the Grandmont Saint James (cat. 58b), for example, the white, blue, and the turquoise, have late-type compositions, whereas the light blue, green, and yellow enamels have early-type compositions. Several of these objects were found to contain individual enamels that appear to be a mixture of early and late compositional types: for example, the white and the blue enamels from the Saint Peter (cat. 47; see also fig. 25). In these mixed enamels, lower levels of potassium were found to correlate with the areas containing antimony, whereas higher levels of potassium were associated with areas containing high lead-tin.

## Red Enamels

All the red enamels analyzed, with one exception, were found to have soda-glass compositions. The red enamels do not really fit into the grouping of early- and late-type compositions, as they generally had somewhat higher levels of potassium and magnesium, similar to those found in the late-type enamels, whereas the levels of aluminum and calcium were generally closer to those of the early-type enamels. Most of the red enamels also contained small amounts of lead, ranging from less than one percent to a maximum of about eleven percent.

The majority of the red enamels were colored and opacified by copper oxide, mainly in the form of red cuprite crystals, dispersed in a clear, uncolored glass matrix. These were also found to have relatively high levels of iron oxide, apparently added as a reducing agent for the cuprite in the glass.[72] Two of the reds were more translucent than those seen on the majority of objects. These were both from the same object, a book cover (cat. 25) from the late twelfth century. Only traces of copper were seen in these two reds, but large amounts of iron were found in both, presumably the source of the red color. Although many of the reds contained traces of tin, possibly from the copper source used for the colorant, only a single red enamel was found to contain a large amount of tin, possibly in the form of crystalline tin oxide (cat. 135). None of the reds were found to contain calcium antimonate as an additional opacifier.

## Potassium Glasses

Potassium glass, like that used for contemporaneous stained-glass windows being produced throughout much of medieval Europe, apparently would have been available for use by the Limousin artisans. However, it appears to have been rarely used, as only four examples of potassium glass were found in this study. One of the red enamels, from the Champagnat chasse (cat. 10), dating from the mid-twelfth century, was unique in being a potash-lime glass of the type found in medieval window glass. This type of glass is characterized by having potassium rather than sodium as the main alkali flux, and it contains a large amount of calcium, usually between ten and twenty percent; magnesium, usually between three and six percent; and phosphorus, usually between about two and five percent. As in most of the other reds, cuprite was the colorant for this glass. One other use was found of a potash-lime glass, in the form of a clear glass cabochon from a mid-thirteenth-century chasse (cat. 115). This glass, however, was actually an applied, inlaid decoration, not an enamel.

Two other enamels analyzed were also found to have very high potassium values, but do not otherwise fit the general characteristics seen in medieval potash-lime glasses. The first one was the white enamel from the Conques medallion no. 21 (cat. 7), dated to the early twelfth century. Unlike most medieval potash glass, this one was found to have very little calcium—less than one percent—and no significant phosphorus. The sodium content was also much higher than what is usually seen in a potash glass. As in the other early-type white enamels, a large amount of antimony was found. A much later enamel, the opaque green from the miter of the reliquary of Nexon (cat. 157), dated to the mid-fourteenth century, was similar to the white in containing a very small amount of calcium, no appreciable phosphorus, and relatively high levels of both sodium and potassium. Like most of the late-type enamels, a large amount of tin rather than antimony was found.

## Discussion

In *De Diversis Artibus*, Theophilus described the production of "white glass"—which is actually clear, uncolored glass—using a mixture of two parts beechwood ash as the source of the alkali flux and one part sand for the silica source.[73] Glass made using wood ash—as well as fern ash—is also described by Eraclius.[74] Compositional analyses of beechwood and fern ashes have found them to contain high levels of potassium and calcium as well as relatively large amounts of magnesium and phosphorus and varying, but generally small, amounts of sodium. During the medieval period, sodium and potassium were both used as alkali fluxes for glass production. A "Mediterranean" composition with a sodium flux, mostly in the form of vessel glass, can be distinguished from a "Continental" composition with a potassium flux, used for most window glass, as described above.[75] In both cases, the alkali is thought to come from plant ash. Analyses of medieval European window glasses have found that the vast majority of these can be characterized as potash-lime glass, with compositions exactly as expected from the use of wood or fern ash, as described by the medieval authors. Studies on the origins and development of the Venetian glassmaking industry have shown that soda glass, made with a flux of marine-plant ash imported from the Levant, was being produced as early as the eighth century.[76] In France itself, particularly southern France, there existed a large industry for the production of soda from marine-plant ash for the making of soap and glass as early as the thirteenth century.[77]

Examination of analytical results on some French window and vessel glasses shows that, although most of the window glass has a potash-lime composition similar to that seen throughout medieval Europe, most of the vessel glass, as well as some of the window glass, appears to have been produced using a soda-based flux made of plant ash. Several examples were seen of red potash-lime window glass, which are very similar in composition to the potash-lime red enamel found in this study. Some of the glasses also appear to have been made by mixing a soda glass with potash-lime glass.[78] In contrast to medieval European window glass, most early glass has been found to be soda glass.[79] Soda glass from the ancient Near East, Egypt, and the Near East in the Islamic period is thought to have been made using a marine or desert plant-ash source of alkali that also contained significant amounts of magnesium and potassium.[80] Roman glass, however, with few exceptions, has been found to contain low levels of magnesium and potassium and is thought to have been produced using a mineral source of alkali, natron, which is nearly pure sodium carbonate, with sand as the source of silica.[81] Roman glass remained relatively constant in composition over a broad geographic area and period of time.

The Limoges enamels of the tenth to the fourteenth century are soda glasses similar to ancient glass and to those of the Mediterranean region during the medieval period, and they are different from the Continental window glass, which is primarily potassium based. The fabrication recipes of the glasses used by the Limoges enamelers during the Middle Ages do not correspond with the texts of Theophilus and Eraclius but appear to continue the ancient tradition of soda-glass production, with a change around the beginning of the thirteenth century in the type of flux and opacifiers used. The early-type enamels appear to be very close to Roman glass in the use of antimony opacifiers and the low levels of magnesium and potassium—probably from a mineral source of sodium—whereas the late-type enamels, with higher levels of magnesium and potassium, are similar to contemporaneous soda glasses made with plant ash and tin oxide as the opacifier.

When specifically discussing the enameling of metal rather than the actual production of glass from raw materials, Theophilus stated that enamelwork was made from "different kinds of glass, namely white, black, green, yellow, blue, red, and purple . . . found in mosaic work in ancient pagan buildings. These are not transparent, but are opaque like marble, like little square stones, and enamelwork is made from them on gold, silver, and copper . . ." —a seemingly clear reference to the use of Roman glass tesserae.[82] Eraclius also recommended the use of Roman glass, although for the making of glass gems, not specifically for enameling.

Recent analytical studies have shown that Roman enamels and glass tesserae were basically similar in composition to that generally seen for Roman translucent glasses.[83] Roman white enamels, however, unlike typical Roman glass, often contain elevated magnesium levels.[84] Antimony compounds have been found in opaque glass from ancient times through the Roman period.[85] The use of these compounds is thought to have persisted until about the middle of the fourth century A.D., when there appears to have been a general changeover to the use of white tin oxide and lead-tin yellow.[86] The Roman red glasses studied, while all being soda glasses, have been divided into four different compositional types, on the basis of differences in the lead content as well as the magnesium, aluminum, and potassium levels.[87] All the red enamels analyzed in this study, with the exception of the one potash red and the two reds from the book cover (cat. 25), were found to be very similar to one or the other of the two low lead types of Roman red glass identified. None of the Limoges red enamels could be described as a high lead "sealing-wax" red type, which is characterized by lead contents of about thirty percent or more. Celtic opaque red glass from Continental Europe, used for enamels and inlays, has also been found to have high lead soda-lime compositions, although with some differences from the Roman high lead red glasses.[88]

Several earlier studies of medieval European glass and enamel centers have found what may be evidence of the reuse of Roman tesserae, as described by Theophilus. Roman glass tesserae have been identified from some eighth-century Scandinavian beadmaking sites.[89] X-ray diffraction and qualitative elemental analyses revealed the use of antimony compounds in some medieval enamels from the Meuse Valley, Cologne, and Limoges.[90] A recent study done at the British Museum Research Laboratory provided quantitative elemental analyses of some Romanesque Mosan, English, and Limoges enamels, showing that these, with the exception of the Mosan reds, appeared to be very similar in composition to Roman glass.[91] Unlike the vast majority of Limoges red enamels, all the Mosan reds were found to have potash-lime or "Continental" glass compositions.[92]

In support of the argument that Roman tesserae were reused, it is important to note that in antiquity and in particular in the Middle Ages, glassmakers in general, and French glassmakers in particular, reworked old glass for other purposes.[93] The ancient recipes for glass production show that colored glass from elsewhere was often used to add color to a "white," or uncolored, base glass. The reuse of colored tesserae would be an effective way to economize on the use of metallic oxide colorants such as cobalt blue, which was very expensive to import, as well as on opacifiers.[94] The mixture of glasses of slightly different color is evident in low-power magnification studies of the Limoges enamels (see fig. 21). The heterogeneity of the grains of powdered glass used in the enamel can also be seen with the electron microscope (see fig. 26).

While the analyses of most of the enamels from Limoges that date to the beginning of the thirteenth century appear to show many similarities to Roman vitreous material and may indicate the reuse of Roman tesserae, it seems hard to believe that there were sufficient quantities of Roman tesserae to supply the enameling industry of Limoges—as well as that of the Meuse Valley—over several centuries of production. Also, it seems hard to understand why locally produced glass was not used, as the fabrication of glasses of different color was apparently well understood, as seen in the texts of the medieval authors. It seems likely that dependence on Roman tesserae for enamel production may even have required the import of tesserae from elsewhere, possibly from Italy, which would probably have been more costly than local production. In addition, the coherence and constancy of the early-type Limoges enamel compositions studied tends to show that very precise recipes were followed for the different colors, which is difficult to reconcile with an uncertain supply of recycled ancient glass over several centuries.

It could be argued that the presence of enamels with Roman-type compositions at this period is actually due to a late survival of Roman glassmaking technology. Roman-type natron glass is thought to have been produced in many areas until the ninth century. Some natron glasses from Italy, however, have been dated from between the ninth and the thirteenth century, although they may have been made by simply reworking older glass.[95] It is possible, therefore, that in some areas, or for specific crafts, the Roman tradition of glassmaking may have continued longer than usually thought. Certainly the stability and quality of Roman-type glasses would have been apparent to later glass workers, and would have been a persuasive argument to continue use of the Roman traditions. There is one problem with this hypothesis: the apparent lack of available natron for the production of this type of soda glass. Glass of the Roman period is generally believed to have been made with natron imported from Egypt.[96] By the late ninth century it appears that Egyptian natron had become unavailable as a raw material for glass production, as even Islamic glass made in Egypt at this time was apparently being produced with a plant-ash source of alkali.[97] The use of Egyptian natron for glassmaking apparently did not begin again until about the eighteenth century.[98] The possibility remains, however, that some other as yet undiscovered source of natron or other natural sodium source available to the Limoges enamel industry of the period did exist, which would produce glass of the same overall composition. A source of antimony for the opacifiers would also be required.

The later-type enamels appear to be closer in composition to the soda-lime plant-ash glasses generally found after about the ninth century. The compositions of the soda glasses from this period in southern France and Venice suggest that the later-type enamels could have been made using these kinds of glass, with tin oxide added as an opacifier, as described in the *Mappae Clavicula*. Analyses of Byzantine glass tesserae found that, at least as early as the tenth century, these had high-magnesium, high-potassium soda-glass compositions.[99] Interestingly, though, analyses of a Byzantine enameled gold enkolpion dated to the mid-tenth century showed that the enamels appear very similar to the early-type enamels in this study and may also represent a reuse of earlier Roman age material or a continuing tradition of Roman glassmaking techniques.[100] However, it is not known how much, if any, of this type of enamel was made for export.

The change in the type of soda glass used in the thirteenth century might be explained by a change in the relative availability of the raw materials—either ancient Roman tesserae or unknown natron and antimony sources—needed for the production of the early-type enamels, thus requiring exclusive use of other contemporary glasses and opacifiers. The change in compositions seen might also be explained by the import around this time of new enameling technology from elsewhere in

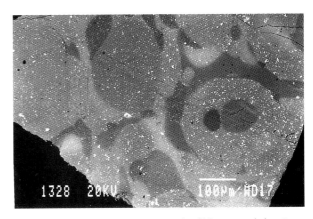

Fig. 26. Scanning electron micrograph of blue enamel showing mixture of translucent and opaque glass containing tin oxide particles (MRR249). (Photo: Isabelle Biron)

Europe. The nine objects that were found to contain both types of enamels or mixtures of both types appear to be from a short transitional period from about the end of the twelfth to the early thirteenth century, when the tin-containing types of enamels were being introduced, but the older antimony-containing types had not yet been discontinued. Several enamels were also found which appear to fit the early-type profile, except for the presence of some tin. These might represent the intentional addition of tin oxide to some early-type enamels, perhaps to alter slightly the final appearance of the enamel.

*Conclusion*

These compositional analyses revealed that, to about the early thirteenth century, enamelers associated with Limoges and those in the Limoges workshops themselves used glasses very similar in composition to ancient Roman glass as the basic material for their enamels, possibly from reworking ancient mosaic glass tesserae as suggested in some medieval texts. If Roman glass tesserae were not actually used, the enamels would represent an extremely late continuation of Roman glassmaking traditions, with as yet unknown sources of raw materials. In the late twelfth century, the Limoges enamelers began to use a different type of glass for the enamels, similar in composition to other soda glasses produced in Europe during this period. The change in the glass composition is also associated with a change in the opacifier compounds used. For a short period from the end of the twelfth century to the beginning of the thirteenth, both types of glass were sometimes used on the same object, sometimes even mixed together in the same enamel color. By about the middle of the thirteenth century it appears that most if not all the enamel decoration produced at Limoges was being made with the late-type enamel. Both the early- and late-type enamel compositions are very different from the glass generally used for medieval stained-glass windows.

Work is continuing on these enamels and on other objects to further characterize the enamels of the period, and to gain a better understanding of the relationships between different types of glass and the development of enamel technology. (See Appendix, tables 5 and 6, for analyses of later objects done in the style of Limoges.) Trace element analyses in particular may help to determine what were the sources of the glass used for the enamels. These kinds of quantitative analyses, with identification of colorants and opacifiers used, as well as analysis of the ratios and correlation of various elements to others, can benefit art historical studies with the ability to apply compositional criteria to help place the material in a historical context.

Notes

1. A preliminary study indicated the efficacy of an extensive survey; see Dandridge and Wypyski 1992, pp. 817–26.
2. Smith and Hawthorne 1974.
3. Merrifield 1967 vol. 1.
4. Merrifield 1967, vol. 1, p. 174.
5. Hawthorne and Smith 1976. Depending upon interpretation, different translations of Theophilus will be cited.
6. Theoplhilus 1961 and 1986, pp. xxxiii–xliv.
7. Hawthorne and Smith 1976, pp. 59, 126–28.
8. Smith and Hawthorne 1974, pp. 50, 62, 67; Merrifield 1967, vol. 1, pp. 208–10, 212–16.
9. Ibid., p. 196.
10. See *Imagines* 1614, and Dominique Mouret (1583) and A. Nouailher, both in Ardant 1855.
11. René François in Ardant 1855.
12. Gaborit-Chopin and Lahanier 1982, pp. 7–27; Oddy, La Niece, and Stratford 1986, pp. 16–17; Distelberger et al. 1993, pp. 19–24, 33–40, 104–18; England 1986, pp. xviii–xxvi, 78–111; Newman 1991, pp. 18–41, 86–87; Otavsky 1973, pp. 37–74; Leone de Castris 1981, pp. 13–30.
13. Stratford 1993.
14. Radiography was employed at the MMA to document joins, reveal tool marks, and provide a relative indication of the lead in the enamels; energy dispersive X-ray spectrometry was used to quantify the compositions of the enamels, solders, and base metals; and X-ray diffraction was used to identify metallic oxides opacifying the enamels. At the LRMF, ion beam analysis was used to evaluate the enamels.
15. Boehm 1991, pp. 127, 134–35.
16. Ibid., pp. 128–29, with earlier citations; Rickard 1932, vol. 2, pp. 510–11.
17. Gregory 1980, pp. 91–93; Craddock 1985, p. 36; Stratford 1993, p. 20, and earlier citations.
18. Gregory 1980, p. 93; Hildburgh 1936, p. 19.
19. According to proton-induced X-ray emission (PIXE) analyses of the surfaces of selected objects undertaken by the LRMF. Possible sources of copper within Limousin monastic dependencies are being explored by B. Barrière. By the sixteenth century many mines existed for extracting copper; see Agricola 1950, note, p. 84.
20. Theophilus discusses the smelting of copper-rich ores, but his information appears to be secondhand; see Hawthorne and Smith 1976, pp. 139–40 and note 1. The clearest and most coincident explanation is the sixteenth-century manuscript of Georgius Agricola, but it is unclear which practices were traditional and which were recently introduced; see Agricola 1950, pp. 388–90, n. 42; pp. 402–5, 407. For a history of the smelting and refining of copper in medieval Europe, see Forbes 1972, vol. 9, pp. 17–24.
21. Hawthorne and Smith 1976, pp. 144–45.
22. Brown and Blin-Stoyle 1959, p. 193.
23. Oddy, La Niece, and Stratford 1986, pp. 6–8.
24. For comparative analyses see Newman 1991, p. 25; Oddy, La Niece, and Stratford 1986, pp. 16–17; Distelberger et al. 1993, pp. 21, 33, 36.
25. See Buckton 1985, pp. 1–6.
26. Dated by Gauthier to Aquitaine in the tenth century; see Gauthier 1987, no. 1, p. 43, ill. 45.
27. The portable altar of Saint Foy (cat. 1) and the medallion of Hosea (cat. 2a) were examined by Elisabeth Taburet-Delahaye, who indicates that they are technically similar to the portable altar and the figure of Christ.
28. Energy dispersive X-ray spectrometry analysis of the hard solder used to join the two sheets on the symbol of Saint Mark (cat. 6) indicated that it is an 85/15 alloy of Cu/Ag (weight percent) with a melting point of about 1000°C. The melting point of the enamel is estimated to be between 800° and 900°C; therefore, it would be possible to join the plaques with solder and then, in a second firing, fuse the enamel without affecting the solder.
29. X-rays taken of the objects revealed none of the porosity associated with casting.
30. On average, the plaques measured 2.4 mm in thickness, with as much as an 0.8 mm fluctuation across the surface of a single piece.
31. Only two plaques in the exhibition do not make use of rivets for attachment—the two images of Christ in Majesty (cats. 12, 43), which would have been secured by separately attached decorative strips.
32. The close correspondence between the shapes of the appliqués and their spaces on the supporting plaques suggests that the artist had them in hand while the design was being laid out. Possible exceptions occur on those chasses where the appliqués extended beyond the surrounding applied borders (cat. 115), necessitating selective filing to accommodate the varying planes.
33. Tools are defined using the terminology of Lowery, Savage, and Wilkins (1971, pp. 167–82).
34. The depth of the cells often varies within the same object depending upon the relative opacity of the enamel. The more translucent enamels required a greater depth to mask the effects of the cuprite. The cells measured ranged from 0.4 to 1.0 mm with the ratio of depth of plaque to depth of cell averaging between three to one and four to one, with the extremes closer to two to one and five to one.
35. Hawthorne and Smith 1976, pp. 129–30.
36. François, in Ardant 1855, p. 24. The distillation of mineral acids does not appear to have been in practice during the period of our study; see Agricola 1950, note, pp. 460–61.
37. Hawthorne and Smith 1976, pp. 138, 152.
38. Ibid., pp. 126–28.
39. François, in Ardant 1855, pp. 30–31.
40. Maryon 1971, pp. 178.
41. To ease the enameling of three-dimensional objects, an adhesive such as plum juice could be added to the powdered enamel; however, firing would destroy any trace of an organic compound, pre-

venting confirmation of its presence; see François, in Ardant 1855, p. 31.

42. The melting point of pure copper is 1083°C and the softening point for the majority of the enamels studied is estimated to be at least 800°C and not greater than 900°C.

43. Hawthorne and Smith 1976, pp. 128.

44. One medallion was not polished, but its enamels are compositionally allied with the originals.

45. The even dispersion of bubbles throughout the depth of the sectioned enamel is apparent under magnification.

46. Possible iron oxides from the interior of the muffle oven (see Maryon 1971, p. 178), or, as the authors found, copper filings from polishing, identified by XRD analyses as cuprite.

47. Stratford 1993, p. 26.

48. Boehm 1991, p. 147; Swarzenski and Netzer 1986, pp. 80, 98; Newman 1991, pp. 26, 86–87; Otavsky 1973, pp. 46–48; Leone de Castris 1981, pp. 22–26.

49. Boehm 1991, pp. 128–29, 131; Spufford 1988, pp. 42, 45, 58, 102; Rickard 1932, vol. 2, pp. 510–11; Boussard 1956, pp. 305–6, 308.

50. Theophilus describes engraving iron rods as stamps for silver or copper sheet; Theophilus 1961 and 1986, chapt. 75, pp. 135–37. The carving of steel dies for striking coins is outlined by Cellini in his sixteenth-century treatise; see Ashbee 1967, pp. 67–78.

51. The centralized production of dies for mints is discussed in Spufford (1988, pp. 43–44, 63). If such a tradition existed in Limoges, a single master could have created the matrices for the various workshops, resulting in a stylistic uniformity in objects from different studios. Whatever the source, the heads were formed by shallow dies struck onto T-shaped blanks. By varying the angle of the strike an upward or downward glance was achieved. There are various options for the fabrication of the blanks: a hammered disk soldered to a drawn or wrought post, an integrally cast disk and post, a drawn or wrought post with one end hammered into a disk similar to a nail head. X-rays indicate the heads are solid, with no visible casting porosity, and with no increased opacity or fluid outlining around the join of disk and post that would indicate the use of solder; however, the central head in one of the appliqués (cat. 79) is clearly raised from a separate sheet and attached to the plaque by a post soldered to the back of the head after its shaping with no indication of solder in the radiograph. Laue back-reflection X-ray diffraction photographs of areas on the heads that had lost their gilding were taken to differentiate a cast or wrought structure; however, subsequent deformation to the surfaces after stamping—chasing, gilding, heating, and burnishing—altered the structure of the surface and rendered the evidence inconclusive. Exposures were five minutes at thirty-seven kilovolts and thirty milliamps at a distance of three centimeters, using the technique described by Wharton (1984, pp. 88–100); and Cullity (1978, pp. 296, 233–47).

52. The authors are grateful to Terry Drayman-Weisser, Director, Department of Conservation and Technical Research, the Walters Art Gallery, Baltimore, for bringing their attention to this qualitative variation in tooling.

53. Texier's examination of the Saint Michael crosiers led him to believe that they were closely related but distinct objects made without the use of molds; see Texier 1843, p. 142.

54. The appliqué heads on the cross (cat. 26) are secured by square posts with the heads set into shallow recesses.

55. An angel on one book cover and an apostle (cats. 48, 96) have exposed rivets that appear original and were cut back from square-sectioned stock, except for their shoulders, which maintain the dimensions of the original and which abut the back of the object to assure a tight seating.

56. Hammered-over tenons attached feet to legs, legs to bodies, and beaks to heads. The wings, the tops of the tails, and the two halves of the bodies were secured with solid transverse rivets; hollow tubes through the heads served as both rivets and as settings for glass eyes. In the MMA dove (cat. 105), copper tabs were soldered to one side of the body, extending across the seam and keying the two sides together before riveting. No solder was used in the central horizontal seam.

57. In the course of our in situ surface analyses of the enamels with X-ray diffractometer (Philips 1710 Open Architecture) and PIXE, traces of mercury were found in association with the gold. No mercury was detected on the Conques medallions fabricated in the nineteenth century and presumed to have been electrolytically gilded (see section on later enamels in Appendix 2). Amalgam gilding is described by Theophilus; see Hawthorne and Smith 1976, pp. 110–14; see also Oddy 1981, pp. 75–79.

58. Theophilus describes using sand to clean metal surfaces prior to gilding; see Hawthorne and Smith 1976, pp. 138.

59. Theophilus describes a steel burnisher; see Hawthorne and Smith 1976, pp. 91, 185. An agate burnisher was excavated from a fourteenth-century goldsmith's atelier in Saint-Denis; see Meyer, Meyer, and Wyss 1990, pp. 90–91.

60. The amalgam was trapped occasionally in dimples in the enamel surface and now appears as opaque dots in the radiographs.

61. 17.190.514 (cat. 24):$CaCO_3$ overlaid by cinnabar. 17.190.523 (cat. 27): $CaCO_3$ overlaid by minium. 41.100.184 (cat. 70b) and 1974.228.1 (cat. 115): $CaSO_4$ overlaid by minium. Samples were analyzed using a Debye-Scherrer camera with a Phillips 1840 X-ray generator and a copper tube at 37Kv and 20Ma for two hours. Patterns were identified with the Fein-Marcourt Search-Match Program using a μPDSM database.

62. Stohlman 1934, pp. 14–18; Stohlman 1935, pp. 390–94.

63. There are also unique marks on discrete objects, such as book covers, that would seem to serve an entirely different function, possibly as a form of identification.

64. Brill 1988, pp. 258–67; Brill, Barag, and Oppenheim 1970, pp. 105–28.

65. Hreglich and Verità 1986, pp. 485–90.

66. Kevex model Delta IV energy dispersive X-ray spectrometer attached to a modified Amray model 1100 (1600 T) scanning electron microscope. (This EDS model does not detect elements below the atomic weight of sodium.) All samples were analyzed at an accelerating voltage of 15 KV to determine the amounts of the lighter elements such as sodium and silicon (the bulk of the glass substance), and also at 30 KV to determine the percentages of heavier elements such as iron and copper.

The enamels were sampled by flaking off very small pieces from areas of previous damage, each on the order of a cubic millimeter in size, with the use of a diamond-edged scribe or a steel scalpel. The samples were prepared for analysis by embedding them in epoxy resin and grinding them with silicon carbide paper to expose the sample interiors. They were then polished with cerium oxide and given a high-vacuum carbon coating to provide conductivity.

Studies of EDS analysis of glass determined that the relative variation in the calculated weight percentages for the major-element oxides is less than two percent for silicon; about five percent for sodium, potassium, and calcium; and about ten percent for magnesium, aluminum, and metals such as copper and iron. The minimum detection limits for the elements with atomic numbers 22 to 30 (titanium to zinc) were found to be under one-tenth of one percent. The minimum detection limits for the oxides of phosphorus, lead, antimony and tin, however, were found to be much higher, about one-half of one percent by weight, mainly due to peak overlap problems. Sulfur is often found in small amounts in many glasses, but cannot be detected when lead is present, also because of peak overlap. For details see Verità, Basso, Wypyski, and Koestler 1994, pp. 241–51.

67. The different color enamels, as well as the metal of the plaques, were surface analyzed using an ion beam of 2.8 MeV provided by the AGLAE particle accelerator. A simultaneous detection system utilized two X-ray detectors, a gamma-ray detector, and charge measurement. Proton-induced X-ray emission spectroscopy (PIXE) was used to measure the major, minor, and trace elements. The sodium values reported here were measured by proton-induced gamma-ray emission (PIGME) spectroscopy, which gives a subsurface analysis representative of the unweathered volume of the glass. The results reported are an average of three points of analysis, which were done to test the homogeneity and the reproducibility of the measurements.

Analyses of trace elements such as cadmium and silver were also done for the enamels but are not reported here because of space restrictions.

68. The crystalline forms of some of the opacifying agents were identified with X-ray diffraction; see Dandridge and Wypyski 1992, p. 822.

Two of the late green enamels in the present study were found to contain large amounts of tin oxide, but it was not possible to definitely identify the presence of either lead-tin yellow or lead antimonate yellow.

69. One other apparent use of lead-tin yellow ($Pb_2SnO_4$) was identified by XRD analysis in the green enamel on the symbol of Saint Matthew (cat. 6). Dandridge and Wypyski 1992, p. 822.

70. Freestone, Bimson, and Buckton 1990, p. 275.

71. Freestone 1991, pp. 41–44.

72. Freestone 1993, p. 41.

73. Hawthorne and Smith 1976, pp. 49–54; Freestone 1992, pp. 739–40.

74. Merrifield 1967, vol. 1, pp. 212–16.

75. See Foy and Sennequier, in Rouen 1989, p. 35.

76. Verità and Toninato 1990, pp. 173–74.

77. Amouric and Foy 1985, pp. 157–71; Foy 1990, pp. 32–38.

78. See Foy 1985, pp. 18–71; Foy 1990, pp. 407–19; Barrera and Velde 1989, pp. 48–54.

79. Sayre and Smith 1961, pp. 1824–1826; Sayre and Smith 1967, pp. 279–311.

80. Brill, Barag, and Oppenheim 1970, pp. 105–28; Lilyquist, Brill, Wypyski, and Koestler 1993, pp. 23–58; Verità 1985, pp. 17–22 .

81. Brill 1988, pp. 258–67. Pliny described the origin of glass as an accidental firing of cakes of natron and beach sand from the river Belus; see Plimy 1962, p. 65.

82. Hawthorne and Smith 1976, p. 59.

83. Henderson 1991, pp. 601–7; Freestone 1990, pp. 103–7.

84. Henderson 1991, p. 606.

85. Sayre and Smith 1961, pp. 1824–1826; Sayre and Smith 1967, pp. 279–311; Turner 1956, pp. 277T–300T.

86. Turner and Rooksby 1961, pp. 1–6. Several instances have, however, been found of the earlier use of tin opacifiers. See Henderson 1985, pp. 267–91; Brill 1969, pp. 47–68.

The use of lead antimonate yellow is not known of again until about the beginning of the sixteenth century, whereas calcium antimonate may not have been used again until the eighteenth century. Wainwright, Taylor, and Harley 1986, pp. 220–25; EDS analyses of *Verre de Nevers* figurines by M. T. Wypyski in Lanmon and Whitehouse 1993, pp. 250–54.

87. Henderson 1991, p. 604.

88. Brun and Pernot 1992, pp. 235–52.

89. Lundstrom 1976, pp. 4–7.

90. Bimson 1981, pp. 161–64; England 1986, pp. xxi–xxii.

91. Freestone 1993, pp. 37–45.

92. This may be the type of red glass known as "Gallien," as discussed by Theophilus; see Buckton 1985, p. 2.

93. See Foy and Sennequier, in Rouen 1989, p.38. Gustavs 1993, pp. 200–201.

94. No cobalt source is known in France; see Foy 1985, p. 38.

95. Verità and Toninato 1990, pp. 169–75.

96. Sayre and Smith 1967, pp. 279–311.

97. Sayre and Smith 1974, pp. 47–70; Gratuze and Barrandon 1990, pp. 155–62.

98. Amouric and Foy 1991, pp. 40, 57.

99. Freestone, Bimson, and Buckton 1990, pp. 271–79.

100. Becker, Schorsch, Williams, and Wypyski 1994, pp. 410–15.

# Catalogue

# I

## The Birth of Enameling in Aquitaine

### 1100–1160

To evoke the birth of Romanesque enamels, this exhibition brings together for the first time two ensembles whose creation is connected to the successive patronage of two abbots of Conques, Bégon III (r. 1087–1107) and Boniface (r. 1107–after 1121).

Grouped with the portable altar of Saint Foy—attributed to an atelier working for Bégon III—are the most beautiful enamels presenting the same technical and stylistic characteristics: the Moses from Antwerp and the Hosea from Rouen (cat. 2), which definitely come from the same work; the female saint from the Louvre (cat. 3); and the Elder of the Apocalypse from the Metropolitan Museum (cat. 4)—which were all part of the collection of the Parisian antiquarian and architect Victor Gay at the end of the nineteenth century—and the stunning symbols of the evangelists, dazzling in their technical virtuosity, which have only recently been linked to this workshop but which are undeniably among its finest works (cat. 6).

Similarly, thanks to the reuniting of the coffret discovered in 1875 in the church of Saint-Foy in Conques, bearing a medallion with the name of Boniface (cat. 7), and the ten medallions formerly in the Carrand Collection, Lyons (cat. 8), the entire collection of extant champlevé enamels on copper from the coffrets created for Abbot Boniface can now be seen together for the first time.

The next stage, which witnessed the appearance of champlevé enamelwork in other centers of southwestern France and northern Spain, and, in particular, the emergence of the Limousin, is illustrated by the earliest works whose execution can be firmly linked to Limoges: the chasses from Bellac (cat. 9) and Champagnat (cat. 10), both created for the counts of La Marche—also exhibited side by side for the first time—as well as the angel of Saint-Sulpice-les-Feuilles, presented in the section devoted to the treasury of Grandmont (cat. 54). The only lacuna is the ornamental fragment of a shaft from a crosier excavated at the site of Saint-Augustin-lès-Limoges, a more modest piece but of great importance nevertheless, due to the nature of its ornamentation and the site of its discovery. It was stolen from the Musée Municipal de l'Évêché, Limoges, in 1981. From these works we may infer a general picture of the first Limousin clientele—both lordly and ecclesiastical, and principally local.

The delicate problem of ties between these first Limousin creations and those executed for the monastery of Santo Domingo de Silos is most powerfully suggested by the *Urna* from the Museo de Burgos, a monumental work whose loan could not be contemplated (detail, fig. 10*a*). Still, the chasse of Champagnat, the book cover from the former Spitzer Collection (cat. 12), and the three medallions recently acquired by the Louvre (cat. 11) bear witness to the close ties between the earliest Limousin works and those destined for Spain. They are key pieces suggesting that their creator could have been trained in Limoges.

Finally, the effigy of Geoffrey Plantagenet (cat. 15), probably executed in Le Mans, seemed crucial to this exhibition because of the many links it presents with Limousin works as well as its historical importance in heralding the role played by Plantagenet patronage in the development of Limousin production.

BDB and ET-D

# 1. Portable Altar of Saint Foy

Conques, ca. 1100

Alabaster, gilt silver, filigree

Six copper medallions, cut and gilt; cloisonné enamel: opaque medium blue, lavender blue, turquoise, almond green, red, white, and pinkish white. Four gold plaques; cloisonné enamel: translucent blue, green, and garnet red; opaque blue and white

28.5 x 19.7 cm (11³⁄₁₆ x 7¾ in.)

INSCRIBED: A and ω (around Christ); S[ANCTA] / MA / RI / A (Holy Mary); S[ANCTA] / FI / D / ES (Saint Foy)

PROVENANCE: First mentioned in Conques in 1846. Classified, Monuments Historiques, 1895

CONDITION: Early repairs on the silver borders around the alabaster plaque; borders with rinceau pattern of trefoil leaves (second half of the fourteenth century) and a fragment decorated with fleurs-de-lis inscribed in lozenges (fourteenth century?).
   (Old?) remounting of bands with decorative filigree; at the bottom of the left vertical border a filigree motif has been roughly cut off. The same is true of the lateral edges of the lower border.

Treasury of the Abbey of Saint-Foy, Conques

The portable altar of Saint Foy is not among the six objects of goldsmiths' work from the treasury at Conques described by Prosper Mérimée (1803–1870) in his *Notes d'un voyage en Auvergne*. Published in 1838, the book prompted the rediscovery of that rich treasury, which then became better known through studies by Abbé Texier (1846) and especially by Darcel (1856). Dated by Darcel "between the ninth and the eleventh centuries," the altar was placed by de Verneilh in the group of works commissioned by Bégon III, the abbot at Conques (1087–95 and 1097–1107). This attribution has since been accepted by all scholars.

Bégon's importance in enriching his abbey's treasury is documented in the Chronicle of Conques, which was written during his tenure as abbot.[1] Furthermore, among the works still preserved in the treasury, three, all of silver, actually bear his name: the portable altar of Bégon (decorated with niello), the lantern of Saint Vincent, and the *A* of Charlemagne.[2] It must, however, be pointed out that, contrary to what is generally stated, there is no clear relationship between these works and the portable altar of Saint Foy, which comprises a rectangular alabaster plaque set in a double silver frame. The first frame, which is narrow, is decorated with repoussé ovals in bas-relief; in the second frame, which is wider, filigree is enhanced by precious stones and by four small plaques of cloisonné enamel on gold and six enameled and gilt-copper medallions depicting Christ and the Paschal Lamb surrounded by the four symbols of the evangelists, the Virgin, Saint Foy, and two other saints.

Thus, this decorative scheme is very different from that of the three objects bearing Bégon's name; not even the fili-gree is identical with the type adorning the *A* of Charlemagne. On the other hand, iconographic comparison with Bégon's portable altar is justified; both pieces include images of Christ Blessing flanked by the Virgin and Saint Foy.[3] Extrapolating from this comparison, we might assume that the two other figures of saints may be the martyrs of the Agen region, Caprais and Vincent.[4]

The four small plaques of cloisonné enamel on gold, similar to the silver plaques on the crown and the throne of the reliquary of Saint Foy and the *A* of Charlemagne, are older elements—quite probably Byzantine or Byzantinizing; they are especially close to those in the icon of Khobi in Georgia.[5]

The importance of the portable altar of Saint Foy in the history of Western art is due chiefly to the presence of the six small medallions of cloisonné enamel on copper: aside from a few isolated pieces (see Gauthier 1987, nos. 1–21), they are the earliest examples of southern enamel work and, at the same time, of the application of enamel to copper in western Europe. They constitute the first ensemble of this type of enameling—organized according to a coherent program and characterized by perfect technical and artistic mastery.

First described by Molinier (1891), the technique of these enamels is extraordinary. Two plaques of the same dimensions are superimposed; the cloisons are applied to the first, and the second is cut in an openwork outline of the person or animal portrayed and then soldered to the first. The superimposition is quite visible on the edges of the medallions depicting the symbols of the evangelists Mark and Luke, where the framing band is broken. The inventive nature of this procedure undoubtedly points to an innovative but isolated workshop that tried to reproduce in copper the effect of cloisonné enamel on gold: since the copper could not be "beaten in" like gold, the base is created by the lower piece and the reserve by the superimposed cut piece.

Certain decorative aspects might have been influenced by works that were executed in Italy during the eleventh century and inspired by Byzantine works such as the enamels on the cover of the book box in the treasury of the cathedral of Vercelli.[6]

However, the human face is treated with a stylization verging on geometric abstraction: the regular oval of the face, the eyes indicated by circles, the chin by a half circle, the eyebrows and the nose by a U-shaped line. This style is characteristic of art in Aquitaine in the late eleventh and early twelfth century. The Apocalypse of Saint-Sever (before 1072) or even the Bibles of Saint-Yrieix (fig. 1*b*) and the

Fig. 1*a*. The letter I. Works of Saint Paterius. Bibliothèque Nationale de France, Paris, ms. lat. 2303, fol. 63v

Saint Foy (detail)

Linked specifically to Conques by the iconographic parallels of Saint Foy and the Virgin to Bégon's portable altar, the portable altar of Saint Foy was most certainly executed for Conques and probably in Conques itself during the abbacy of Bégon III.

As recently pointed out by Delmas,[9] certain details might bear out a hypothesis advanced by Bouillet in 1892 but generally rejected since: the awkward arrangement of the roughly cut filigreed plaques and the restorations done with repoussé silver plaques decorated with small flowers—typical of goldsmiths' work in Rodez during the second half of the fourteenth century—could well mean that this portable altar represents a remounting of the framing border of a Gospel book around an alabaster plaque. Bouillet's theory is further buttressed by the mention of a Gospel book among the works attributed by the chronicler of Conques to the patronage of Bégon III.

ET-D

Fig. 1b. Wisdom. Bible. Town Hall, Saint-Yrieix, fol. 228

NOTES
1. "Bego venerabilis qui claustrum construxit, multas reliquias in auro posuit, textus evangeliorum fieri fecit" (Bégon the Venerable, who built the Cloister, enshrined numerous relics of gold and ordered the making of a Gospel). Cited in Desjardins 1879, p. xliv.
2. See Paris 1965, nos. 540, 541, 543; Gauthier 1963b, pp. 140–43.
3. Hence there is no reason to regard the parallel between the Virgin and Saint Foy as impossible or to deduce that the figure designated by the inscription S / MA / RI / A could be Saint Mary Magdalen (see Gauthier 1987).
4. As suggested by Bouillet (1892) and recently again by Delmas (Toulouse 1989, no. 347).
5. See Amiranashvili 1962, p. 18.
6. See Gauthier 1972b , no. 28, pp. 62–64.
7. Paris, BNF, ms. lat. 8878; Saint-Yrieix (Haute-Vienne), town hall, and Paris, Bibliothèque Mazarine, ms. 1–2. For these manuscripts, see Gaborit-Chopin 1969, pp. 130ff., 176, 212–13; Zaluska 1979; Avril 1983, pp. 182–87.
8. Paris, BNF, ms. lat. 2303, fol. 63v; see Gaborit-Chopin 1969, p. 197.
9. Toulouse 1989, no. 347.

EXHIBITIONS: Paris 1867, no. 1952; Paris 1889, no. 277; Paris 1900, no. 2405; Limoges 1948, no. 1 (not exhibited); Vatican City 1963, no. 1; Paris 1965, no. 544, pl. 42; Toulouse 1989, no. 347.

BIBLIOGRAPHY: Texier 1846 ; Darcel 1856, pp. 84–87; Texier 1857, col. 204; Verneilh 1863, pp. 18–19; Labarte 1865,vol. 3, pp. 433–34; Rupin 1890, pp. 71–73, fig. 137, pl. XI; Molinier 1891, pp. 126–28; Bouillet 1892, pp. 60–62, engraving, p. 61; Bouillet and Servières 1900, pp. 203–5; Molinier 1901, pp. 116–17; Braun 1924, pp. 447–50, pl. 78; Hildburgh 1936, pp. 62–63; Aubert 1954, p. 63, pl. XXI; Gauthier 1963b, pp. 114, 144, pls. 62–64, p. 183; Taralon 1966, p. 297, pl. 218; Lasko 1972, p. 228, pl. 261; Gauthier 1972b, no. 34, pp. 75–76, 326, ill., p. 74; Gaborit-Chopin 1983b, p. 317, pl. 280; Favreau et al. 1984, vol. 9, pp. 39ff.; Gauthier 1987, no. 22, pp. 53–54, ills. 69, 74, pls. IX, X.

Bibliothèque Mazarine, Paris (early twelfth century) include examples of busts inscribed in medallions that are treated in an altogether similar way.[7] The lozenge-shaped halo-crowns of the Virgin and Saint Foy, probably derived from Ottonian or Italian models, also appear in these two Limousin Bibles.

The Paschal Lamb, for its part, shows the same round shapes, the same elongated muzzle extending an inordinately long neck, the same naive expression as that of the Lamb at the center of the letter I in a manuscript, the Works of Saint Paterius, illuminated at Saint-Martial, Limoges, in the late eleventh century (fig. 1a).[8]

# 2. Two Medallions

## a. *Hosea*

Conques, late 11th or early 12th century

Copper: cut and gilt; cloisonné enamel: black, medium and lavender blue, turquoise, white, and pinkish white

Diam.: 6.4 cm (2½ in.)

INSCRIBED: OSE / AE (Hosea)

PROVENANCE: Acquired by the Musée des Antiquités de la Seine-Maritime before 1852

Musée des Antiquités de la Seine-Maritime, Rouen (R 91.21)

a.

## b. *Moses*

Conques, late 11th or early 12th century

Copper: cut and gilt; cloisonné enamel: black, medium blue, turquoise, almond green, white, and pinkish white

Diam.: 6.4 cm (2½ in.)

INSCRIBED: MOY / SES (Moses)

EX COLLS.: Alfred Gérente (sale, Paris, January 22, 1869, no. 100); Carlo Micheli (1809–95, cat. ms., 1898, Musée Mayer van den Bergh, no. 358), sold by Marie Micheli to Fritz Mayer van den Bergh in 1898; Fritz Mayer van den Bergh (d. 1901) Collection

CONDITION: Light wear to the gilding.

Musée Mayer van den Bergh, Antwerp (no. 546)

The Rouen museum's medallion was mentioned for the first time by Laborde in 1852; the medallion in the Antwerp museum appears in the catalogue of the Gérente sale in 1869.

Regarded as a "Byzantine enamel" by Cochet (1868 and 1875), the Rouen medallion was linked to the Louvre plaque (cat. 3) by Darcel (1878a) and de Linas (1885); it was then related to the portable altar of Bégon in Conques and to the medallion in the former Gérente and Micheli collections by Bouillet and Servières (1900) and then Molinier (1901).

There is a very close technical and stylistic kinship with the plaques adorning the portable altar of Saint Foy (cat. 1). These medallions, depicting two Old Testament figures (the patriarch Moses and the prophet Hosea), reveal the same distinctive technique: two plaques are superimposed; the top plaque is cut to form the outline of the figure, and the

cloisons are placed on the bottom plaque. We also find the same palette and a similar placement of the figure in the medallion. Most of all, the heads are drawn in the same way, with olive-shaped faces and perfectly round eyes. All these features justify an attribution to the same workshop. However, certain other details are specific to these two figures: the wavy line defining the hair of the top of the head,

b.

the less rigid draftsmanship, and the coloring of the mouth with a touch of enamel—perhaps allowed by the larger size—are indications of painstaking execution. They confirm that the technique was probably inspired by Italian examples: indeed, such details can be observed on the cross of Velletri.[1] On the other hand, the stylistic sources can be found in Aquitainian art of the second half of the eleventh century: particularly striking here is the kinship with the figures in the Apocalypse of Saint-Sever.[2]

Like the plaques on the portable altar of Conques, these two medallions have no holes for attachment. This indicates that they were to be mounted on an object with—as stressed by Gauthier (1987)—an iconographic program large enough to include both patriarchs and prophets.

ET-D

NOTES
1. See Gauthier 1972b, no. 27.
2. Paris, BNF, ms. lat. 8878; see Avril 1983, pp. 182–87.

EXHIBITIONS: (a): Paris 1878; Paris 1889, no. 69.

BIBLIOGRAPHY: (a): L. Laborde 1852, p. 39, n. 1; Cochet 1868, p. 40, Cabinet 28; Cochet 1875, p. 63, Cabinet 28; Darcel 1878a, p. 566; Linas 1885, p. 471; Linas 1886b, p. 13, fig.; Molinier 1891, p. 128; Vernier 1923, p. 17, fig. 20; Hildburgh 1936, pp. 62–63; Gauthier 1972b, no. 35, p. 76; Gauthier 1987, no. 25, p. 54, ill. 72, pl. X; Flavigny 1992, no. 71; (b): Coo 1965, no. 358, fig. 46, p. 364; Gauthier 1972b, no. 36, p. 526; Gauthier 1987, no. 26, p. 55, ill. 73, pl. X; (a and b): Bouillet and Servières 1900, p. 205 n. 1; Molinier 1901, p. 117; Migeon 1909, p. 420; Gauthier 1972c, p. 630; Niewdorp 1979, p. 20.

# 3. Female Saint

Conques, late 11th or early 12th century

Copper: gilt; champlevé and cloisonné enamel: dark and lavender blue, turquoise, green, white, and pinkish white

9.8 x 5.7 cm (3⅞ x 2¼ in.)

Ex coll.: Victor Gay, Paris (before 1878; inventory, Musée du Louvre Archives, December 1, 1908, vol. 2, no. 65). Gift of friends of Victor Gay, 1909

Condition: Slight wear to the gilding.

Musée du Louvre, Paris; Département des Objets d'art (OA 6273)

Remarked upon at the time of the Exposition Universelle of 1878, when it was already in the possession of Victor Gay, the plaque was linked by Bouillet and Servières (1900) and then by Molinier (1901) to the plaques of the portable altar of Saint Foy (cat. 1).

Yet its technique is slightly different, since only a single copper plaque was used: champlevé for the silhouette of the figure and cloisonné for the other lines.

On the other hand, the faces in both works show the same highly stylized draftsmanship, characteristic of Aquitainian art during the late eleventh and early twelfth century. Certain details in the Louvre figure, such as the double line of the eyebrows or the "trefoil" drawing of the tip of the nose, lend weight to a comparison with the illuminations in the Apocalypse of Saint-Sever (see cat. 1). The luxurious costume is delineated with great care: a gemmed orphrey falls from the waist to the hem of the robe; small, commalike cloisons on the upper part of the body seem to imitate fur; the movement of the leg and the flow of the drapery are suggested by cloisons ending in hooks—recalling certain eleventh-century Italian enamels such as the Pax of Chiavenna.[1] The ample, wide-sleeved robe and the veil covering the head and shoulders recur in western French illumination of about 1100—for example, in the Life of Saint Aubin, folio 1, and the Life of Saint Radegonde, folio 25.[2] The movement of the drapery and the broad gestures of the large hands confirm the kinship with these manuscripts. The crown is studded with jewels and surmounted with fleurons—aspects that distinguish it from the halo-crowns of the Virgin and Saint Foy on the portable altar of Conques (cat. 1). Nevertheless, the crown is comparable to the one worn by Saint Radegonde on certain leaves of the Life of Saint Radegonde at Poitiers (fol. 25v).

The main problem posed by this work is the identity of the crowned but unhaloed figure holding a palm—symbol of eternal life for the elect. The portrayal relates to an Aquitainian tradition: the Lectionary for the Use of Saint-Martial of Limoges, a manuscript dating from the late tenth

or early eleventh century, is illustrated with several figures of male and female saints, each holding a palm in the left hand. The posture of one of the figures, a Saint Agatha (fig. 3a), is very similar to the posture of our saint.[3]

The catalogue of the exhibition mounted in Limoges in 1948 proposed that the Louvre figure was a Saint Valerie—a tempting hypothesis but one that was not supported by any decisive argument. More recently, in 1987, Gauthier suggested that this might be a Saint Foy, which is more likely, given the close ties between the plaque and the Conques enamels. Moreover, the figure of Saint Foy is crowned in both portable altars from the Conques treasury. The absence of a halo in this case is surprising but not exceptional: Saint Radegonde has no halo in the various scenes of her life portrayed in the Poitiers manuscript.

The plaque, which has no attachment holes, was probably mounted, like those of the portable altar of Saint Foy. Its size, its shape, and the frontality of the gesturing figure might point to its being the center of a composition. How then could we not recall the "numerous reliquaries" commissioned by Abbot Bégon?[4] However it may be, this exceptionally well-preserved work reveals great beauty in the coloring of the enamel and great sureness of the draftsmanship—both in the facial features and in the animated drapery of the costume. All these aspects provide dazzling evidence of the accomplishments of the atelier of goldsmiths and enamelers working at Conques in the late eleventh and early twelfth century.

ET-D

Notes

1. See Speyer and Mainz 1992, pp. 378–79.
2. Vie de Saint Aubin, Paris, BNF, ms. nouv. acq. lat. 1390. Probably illuminated under Abbot Gérard (1082–1108); see Porcher 1959a, pp. 179–83, pl. 5. Vie de Sainte Radegonde, Poitiers, Bibliothèque Municipale, ms. 254, last quarter of the eleventh century; see Avril 1983, pp. 165–66.
3. Paris, BNF, ms. lat. 5301, fol. 44; see Gaborit-Chopin 1969, p. 206.
4. See Desjardins 1879, p. xliv, text cited above (cat. 1, n. 1).

Exhibitions: Paris 1878; Paris 1889, no. 442; Limoges 1948, no. 5.

Bibliography: Darcel 1878a, p. 566; Linas 1881, p. 130; Linas 1885b, p. 47, pl. xix, no. 2; Gay 1887–1928, p. 619, fig.; Bouillet and Servières 1900, p. 205, no. 1; Molinier 1901, pp. 117–18; *Bulletin des musées de France* 5 (1909), p. 69; Migeon 1909, pp. 420–21, fig.; Marquet de Vasselot 1914, no. 86; Hildburgh 1936, pp. 62–63; Salet 1946, fig. 179; Gauthier 1950, pp. 28ff., pl. 2 bis; Gauthier 1955, p. 63; Gauthier 1957, p. 159; Gaborit-Chopin 1983b, p. 317, fig. 281; Gauthier 1987, no. 23, p. 54, ill. 70 pl. x.

Fig. 3*a*. Saint Agatha. Lectionary for the use of Saint-Martial of Limoges. Bibliothèque Nationale de France, Paris, ms. lat. 5301, fol. 44

# 4. Elder of the Apocalypse

Conques, early 12th century

Copper: gilt; champlevé and cloisonné enamel: black, medium and lavender blue, turquoise, almond green, white, and pinkish white

Diam.: 7.8 cm (3⅟₁₆ in.)

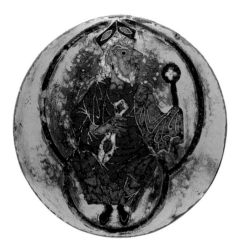

Ex colls.: Victor Gay (inventory, Musée du Louvre Archives, December 1, 1908, vol. 2, no. 66); sale, Paris, March 23–26, 1909, no. 48, pl. 1); Alphonse Kann, Paris; J. Hunt, Dublin (1959); Ronald S. Lauder

Condition: Some losses to the enamel (especially from the mandorla); slight hammer strokes on the surface of the copper are probably modern; gilding perhaps modern.

The Metropolitan Museum of Art, New York; Gift of Mr. and Mrs. Ronald S. Lauder, 1983 (1983.38)

The figure, holding a viol in his right hand, has the white hair of advanced age; he is undoubtedly an Elder of the Apocalypse, as indicated by Brown (1983) and Gauthier (1987). Gauthier identified the object in the left hand as a flagon, linking it to those held by the elders in a manuscript leaf from the end of the eleventh century (originating in Tours and preserved in Auxerre).[1] However, as in most depictions of the tenth and eleventh centuries, those elders hold goblets, some of which have extremely tall stems. The object depicted here looks more like a scepter with a circular knob, the kind held by the figure of the Elder of the Apocalypse in a much later fresco of the same subject in Montmorillon (Vienne).[2]

The medallion is incontestably related to the objects produced at the atelier of enamelers working for the abbot Bégon III at Conques. We find the same range of colors and the combination of champlevé and cloisonné enameling on a single copper plaque as in the Louvre female saint (cat. 3). We also find characteristic details such as the wavy line of

the hair, the circular eye, the double—"forked"—cloisons designating the eyelid and eyebrow, and cloisons with hooked ends defining the drapery. Yet, this piece is distinguished from the earlier ones by its presentation: the figure is frontal, whereas the head is in profile instead of rigorously full face.

This posture ties the Elder to Aquitainian images of the eleventh and twelfth centuries, such as those in the Apocalypse of Saint-Sever (fols. 121v–122) or the tympanum of the abbey church of Moissac,[3] which are very specifically evoked by the dynamic gesture and the sharp profile with the long, thin nose. The pose of the figure, with knees parted, and the shape of the musical instrument lend weight to the parallel drawn with the renowned figures of Moissac. Other ties to Aquitainian art of about 1100 or the first few decades of the twelfth century can be emphasized. For instance, the frequent use of the double mandorla, which, deriving from Carolingian art, occurs particularly in the Bible of Saint-Aubin of Angers,[4] the frescoes of Notre-Dame of Poitiers, and the Christ at the Musée Fénaille in Rodez.[5] This Elder of the Apocalypse, like the female saint at the Louvre (cat. 3), once belonged to Victor Gay, one of the late nineteenth century's most knowledgeable collectors of medieval enameling. More than the earlier works, this Elder shows a certain independence from the enamels of the portable altar of Saint Foy (cat. 1), and it testifies both to the inventiveness of the enamelers working for Bégon III and to the fruitful relations they maintained with the other artistic workshops of western and southwestern Aquitaine.

ET-D

Notes
1. Schapiro 1977, pp. 306–27.
2. Late twelfth century or about 1200. See Deschamps and Thibout 1963, pp. 71–73, pl. xxviii, fig. 3; Demus 1970, pp. 145–46.
3. Barral i Altet 1983, p. 109, fig. 91 (ca. 1120–40).
4. Angers, Bibliothèque Municipale, ms. 4; see Porcher 1959a, pl. p. 193, pp. 182–83.
5. Demus 1970, p. 135; Schapiro 1977, pp. 285ff., fig. 1; see also Cook 1923, pp. 47ff.

Exhibition: Manchester 1959, no. 93.

Bibliography: Brown 1983, p. 17; Gauthier 1987, no. 24, p. 54, ill. 71, pl. x.

# 5. Christ Blessing

Conques(?), beginning of the 12th century

Copper: cut and gilt; cloisonné and champlevé enamel: black, lapis and lavender blue, turquoise, green, white, and pinkish white

8.9 x 5.57 cm (3½ x 2³⁄₁₆ in.)

INSCRIBED: IH [ESU]S / XPC (Christus)

EX COLLS.: Octave Homberg (sale, Paris, May 11–16, 1908, no. 510); Claudius Côte, Lyons (in 1912); Bequest of Mme Claudius Côte, 1960

CONDITION: Gilding deteriorated; losses to enamel; holes pierced later, as indicated by awkward asymmetrical placement.

Musée du Louvre, Paris; Département des Objets d'art (OA 10015)

This standing Christ was linked by Marie-Madeleine Gauthier (1987) to depictions of the Transfiguration and to Byzantine sources. However, since the Carolingian era, such portrayals were frequently found in the West. Sometimes they are associated with the theme of the Ascension—for example, in the frescoes at Saint-Hilaire of Poitiers (about 1100)[1] or on the tympanum of the cathedral of Cahors (about 1140–50)[2]—but often they are isolated evocations of Christ Teaching or Blessing, as in the Carolingian ivories on the covers of the Codex Aureus of Lorsch[3] or the Gospels of Saint-Denis.[4] The gestures of Christ—blessing with his right hand, holding a book in his left—particularly relate the Louvre plaque to these figures of Christ Teaching.

The style and technique of this work connect it to the enamels produced by the atelier working for Bégon III. The almond-shaped plaque is made up of two superimposed copper sheets: the image was cut out of the upper one, and the cloisons were placed on the lower one. The drawing of the nose and the brows and the perfectly round eyes are, moreover, characteristic of those enamels. On the other hand, the shape of the face and the stylized hair and beard are less rigorously geometric, and the lines of the drapery are arranged in a manner less abstract than that of the portable altar of Saint Foy (cat. 1). The draping of the mantle, especially the large V-shaped section under the left arm, specifically recalls western French paintings and sculptures such as the Christ of Saint-Hilaire or the Christ of the Poitiers baptistry.[5]

We do not know whether Bégon III's workshop of enamelers survived his abbacy or whether its particular techniques were adopted by other artists or workshops active during the first half of the twelfth century. Thus, it is difficult to assign a precise date to this plaque, whose deviations from the small earlier group point to a slightly later time.

As in the preceding works, the absence of any original piercings implies that the plaque was set into a mounting, probably as the center of an ensemble.

ET-D

NOTES
1. See Demus 1970, p. 134.
2. See Barral i Altet 1983, p. 115, fig. 97.
3. Vatican Museum; see Goldschmidt 1914, no. 13.
4. Paris, BNF; see Paris 1991, no. 9.
5. See Demus 1970, no. 96, p. 134.

BIBLIOGRAPHY: Bertaux 1912, pl. XXI; Landais 1961a, p. 131, fig. 4; Gauthier 1987, no. 67, pp. 77–78, ill. 166, pl. XXXVI.

# 6. Four Plaques: Symbols of the Evangelists

Conques(?), beginning of the 12th century

Copper: cut and gilt; champlevé and cloisonné enamel: black, lapis and lavender blue, turquoise, green, red, white, pinkish white, and, on the ox, semitranslucent wine red

10–10.4 x 6.1–6.2 cm (3¹⁵/₁₆–4⅛ x 2⅜–2⁷/₁₆ in.)

INSCRIBED: MATHEU[S] (Matthew) / MARCVS (Mark) / LVCAS (Luke) / IOH[ANNE]S (John)

EX COLLS.: Sale, Paris, Hôtel Drouot, November 10, 1853, no. 120(?); Count of Clermont (sale, Paris, April 18, 1864, no. 10); purchased by Frédéric Spitzer but not listed in the catalogue of 1890; Sigismond Bardac, Paris (in 1900); Georges Hoentschel (before 1911); J. Pierpont Morgan, London and New York

CONDITION: Slight wear to gilding; small breaks on the plaque with the eagle of Saint John.

The Metropolitan Museum of Art, New York; Gift of J. Pierpont Morgan, 1917 (17.190.426–29)

One side of each of the four plaques was cut away to form an arc, which enables us to determine their approximate arrangement in the four corners of a composition, perhaps a portable altar or a book cover, that originally must have had a figure of Christ in Majesty or the Paschal Lamb at its center. The representation of the four evangelists by their symbols—a winged man, a lion, an ox, and an eagle—illustrates the vision of the Apocalypse (4:6) in which the Lamb of God is seen surrounded by "four creatures." Since the time of Irenaeus of Lyons Christian tradition has regarded these figures as symbols of the evangelists.

The plaques have been attributed to Trier by Pératé (1911) and to a Spanish-born artist working in Trier by Hildburgh (1936). Souchal (1967) suggested that they might be a "milestone on the road" to the champlevé enamels of the Limousin; and Gauthier (1972b and 1987) attributed them definitively to the "southern area."

Actually, the chief characteristics of the works from Bégon's atelier can be observed in this group—especially the superimposed plaques, one cut out, the other with cloisons. The winged man's face resembles those of the young saints in the portable altar of Saint Foy (cat. 1) and, even more so, those of the Hosea in Rouen and the Moses in Antwerp (cat. 2): the prophet and the patriarch have the same mass of stylized blue-black hair confined within a wavy line. However, the extraordinary technical quality of these plaques has probably not been sufficiently emphasized: the fineness of the cloisons, the skill and imagination of the drawing, the richness of the palette—particularly the semitranslucent wine red on the body of the ox—have no precedent in southern French enameling.

The technical and stylistic sources seen in the works attributed to Conques also appear here—sometimes even more clearly. The decorative treatment of the pelts and the plumage of the three animals, where beaded bands, undulating borders, and "hook" or palmette motifs are juxtaposed, vividly evokes Italian enameling of the eleventh century; the eagle's plumage is comparable to that of the animals on the Vercelli book cover.[1] However, this highly stylized decorative scheme might also have been inspired by textiles from the Orient or Islamic Spain, which employ a similar ornamental repertoire—especially palmettes and alternating beaded and colored bands that define the bodies and wings. The griffins of the Sudarium of Saint Chaffre (Monastier-sur-Gazeilles [Haute-Loire])—executed in the Near East or in Spain during the tenth century—are among the finest earlier examples of such enameling.[2]

Moreover, these decorative tendencies were widespread in the art of southwestern France during the eleventh century, as proved, for instance, by the animals in the initials of the book of tropes from Moissac (Paris, BNF). The entwined birds forming the letter *S* (fol. v verso) are, like these symbols, from the former Hoentschel Collection, treated with great decorative abstraction that is dominated by scalloped borders and beaded bands.[3] Here again, the Aquitainian sources were probably the most influential. Hildburgh (1936) and Gauthier (1987) have shown how the iconographic peculiarities are linked to the tenth- and eleventh-century art of western and southwestern France and ultimately to Carolingian models: this is the case with the eagle clutching a scroll rather than a codex in his talons—a feature that certainly appeared in Spain (for example, on folio 2 of the Beatus of Gerona) as well as in the Gospels illuminated for Fleury about 820 and in the Apocalypse of Saint-Sever (before 1072).[4] The motif of the book carried between the wing tips of the lion and the ox also dates back to Carolingian art. It can be observed on an ivory book cover of the early ninth century (originating in Bourges and preserved in Paris at the Cabinet des Médailles, BNF);[5] then, about 1100, on the canon tables of the Bible of Saint-Aubin of Angers.[6]

Above all, the dynamism of the figures–the dancing posture and dramatic gesture of the winged man of Saint Matthew–echoes the pictorial achievements of western France from about 1100, when the "animated" style flourished, dominating the artistic creations of the area for nearly a century.

So, because of their technical perfection and the refinement of their decorative abstraction, these symbols of the evangelists represent the peak of the enamelers' art attained at the workshop of Bégon III. But they are also, in the expressive power of their style, the direct precursors of the

76

images on the chasse of Champagnat (cat. 10) or the *Urna* at Silos. They stand as the unique milestone between those creations and the first champlevé enamels produced in southern France. Perhaps executed by an exceptional artist trained in Bégon III's enameling workshop, they are, no doubt, of slightly later date than the preceding works. Consequently, they represent both the highest point and the finest extant example of the technique perfected in this workshop.　　　　　　　　　　　　　　　ET-D

NOTES
1. See Gauthier 1972b, no. 28.
2. See Paris 1992, no. 284.
3. Paris, BNF, ms. nouv. acq. lat. 1871; see Porcher 1959c, pl. 20.
4. Beatus of Gerona, 975, Gerona, treasury of the cathedral; see Schapiro 1977, p. 294, fig. 17; Gospels of Fleury, Bern, Bürgerbibliothek, Cod. 348, fol. 8v; see Porcher 1968, pl. 179 (ca. 820); Apocalypse of Saint-Sever, BNF, Paris, ms. lat. 8878, fols. 121–22;. see Avril 1983, pls. 129, 153.
5. See Goldschmidt 1914, no. 19; Little 1985, p. 17, fig. 5, p. 15.
6. Angers, Bibliothèque Municipale, ms. 4, fol. 205v, 206;. see Porcher 1959a, pp. 182–83.

EXHIBITION: Paris 1900, no. 2410.

BIBLIOGRAPHY: Pératé 1911, nos. 1–4, pl. I; Breck and Rogers 1925, p. 58; Hildburgh 1936, pp. 60–62, pl. XI; Souchal 1967, pp. 45–46, fig. 13, p. 44; Gauthier 1972b, no. 33, pp. 73–75, 326, pl. p. 73; Gauthier 1987, nos. 30–33, p. 56, ills. 80–83, pl. XII.

# 7. Coffret of Abbot Boniface

Conques, ca. 1110–30

Thirty-one gilt-copper medallions; champlevé enamel: lapis and medium blue, turquoise, dark green, emerald green, meadow green, yellow, and white; leather-covered wood core with copper studs

31.5 x 57.6 x 31.5 cm (12⅜ x 22⅝ x 12⅜ in.)

Medallions: Diams.: 9.2–9.95 cm (3⅝–3¹⁵⁄₁₆ in.) for those with attachments; 8.2–8.9 cm (3¼–3½ in.) for those without attachments

Almonds: H.: 9.6–10.9 cm (3¾–4¼ in.); W.: 6.2–6.9 cm (2⁷⁄₁₆–2¹¹⁄₁₆ in.)

INSCRIBED:

Medallion no. 7: + SCRINIA : CONCHARUM : MONSTRAN[T] OPVS : VNDIQ[UE] : CLARUM (In all respects, the coffrets of Conques demonstrate brilliant workmanship.)

Fig. 7a. The coffret soon after its discovery in 1875, before the restoration of 1878

Medallion no. 20: + HOC ORNAMENTVM : BONE : SIT : FACII : MONIMENTVM (May this ornament be a memento of Boniface.)

PROVENANCE: The coffret was buried in a wall built between the columns of the choir of the abbey of Saint-Foy in 1590 or shortly thereafter. (The date was inscribed on a double tournois, a coin minted at Tours, found with the coffret; see Bouillet and Servières 1900, pp. 186–88, ill.) It was found on April 21, 1875, when the columns were removed (fig. 7a). The coffret, which had been enclosed in a chestnut box, contained various relics. Twenty-four enamels were still affixed to it: "eight on the lid, six on the front, five on each of the sides" (see Bourret 1880).

Four more medallions were "discovered in Conques, where they had served as bobeches" (Bouillet 1892, p. 73); these four were immediately linked with those on the coffret, since one of them (no. 7) was cited by de Cougny in 1875 (p. 480). They were attached to the coffret during its restoration in 1878 (see below). Classified, Monuments Historiques, 1895.

CONDITION/RESTORATIONS: The coffret was restored in 1878 by Placide Poussielgue-Rusand: the wood core was replaced, the handles were eliminated, the leather repaired in several places, and a large number of studs were removed. During this restoration, the twenty-four enameled medallions were detached, then reattached, but not in their original positions. The four medallions that had served as bobeches were also reattached: the center of each had been pierced and the lugs filed down. At the same time, three newly fabricated medallions were added by Poussielgue.[1]

It was recently proposed that Poussielgue might have gone further than indicated by the 1878 invoice (see note 1). Accordingly, only twelve medallions would be authentic and the remaining nineteen either entirely modern or reenameled.[2]

Examination of the face and the back, confirmed by the analysis of the Laboratoire de Recherche des Musées de France, has allowed an amendment of this judgment: twenty-eight of the thirty-one medallions now in place are original, at least as to their copper supports. They have the same brownish red color and the same type of attachment marks.

The three medallions placed in the center of the back (nos. 8, 10, 11) are Poussielgue's restorations. The modern execution, evident to the naked eye, is revealed by the different composition of the copper—an alloy containing arsenic—the gilding, which contains no mercury, and especially the enameling, which has low levels of calcium and sulfur and a high percentage of aluminum, potassium, and lead. But the invoice of September 2, 1878 (see above and note 1), mentions the engraving and enameling of only two medallions. It is possible that Poussielgue did not charge for the third medallion since it was a reproduction of one of the first two. Indeed, the goldsmith furnished only two drawings—a bird turned toward the right and two confronted birds—which copy two of the four "pierced" medallions that had functioned as bobeches (nos. 7 and 12). That would explain the presence of the gilded circle at the center.[3]

It is more difficult to define the nature and extent of the restoration of the damaged enamels mentioned in the same invoice. Examination of the medallions and study of their laboratory analyses enable us to distinguish two types of intervention in the geometrically decorated medallions on the ends:

1. On two of them (medallion no. 29 and the almond-shaped plaque, no. 31) only the copper is old; the enamels and the gilding, identical in composition with those on the three medallions produced by Poussielgue (nos. 8, 10, 11), are entirely modern.

2. Seven others (medallion no. 24, almond nos. 22, 25–28, 30) have gilding and enamels of the same composition as those of the definitely original medallions. The irregular aspect and the grittiness of their surfaces may be due to an interruption of the work before the polishing—a fairly implausible explanation. More likely than not, Poussielgue refired them but then refrained from polishing so as to preserve the original gilding.

All the other medallions—there are nineteen (eighteen with bird decorations [nos. 1–7, 9, 12–21] and one almond-shaped with lugs [no. 23])—are original and have never been restored. Their condition varies. The four medallions that had served as bobeches (nos. 7, 9, 12, 13) suffered the most damage. Among the others, those that were on the front of the coffret when it was found (nos. 1, 5, 6, 14, 16, 21) are particularly well preserved.

The arrangement of the enamels was altered again before 1924–25, which resulted in the moving of the Poussielgue restorations from the principal face to the roof (see photos, Fonds Marquet de Vasselot, Musée du Louvre, Département des Objets d'art).

The coffret was finally restored by Lucien Toulouse in 1953–54: the medallions, after cleaning, were attached in a fourth arrangement—with the restorations of Poussielgue this time being set on the reverse of the coffret (Paris, Archives des Monuments Historiques, dossier Conques, mémoire du 30 octobre 1952).

N.B. For ease of understanding, the numbering of enamels adopted here is the one proposed by Marie-Madeleine Gauthier and Geneviève François (1987).

Treasury of the Abbey of Saint-Foy, Conques

Principal face, with medallions 1 to 6. On the lid, medallions 14 to 21

The coffret of Abbot Boniface occupies a key position in the history of medieval enameling. The inscriptions on two of the medallions confirm that they were indeed done for the abbey of Saint-Foy at Conques at the behest of Boniface. In order to execute them the enamelers abandoned the technique of cloisonné enameling on copper used by their predecessors; instead, they tried something completely new for them: the champlevé process.

The wood coffret, covered with leather, studded with copper, and enriched with enameled medallions, was found in 1875. It is akin to dated examples from the thirteenth and fourteenth centuries, such as the one in the treasury of Namur,[4] but the support is probably not the original one executed for Abbot Boniface. The twenty-eight original, or partly original, enamels subdivide into two distinct groups. Ten of them—the two medallions and eight almond-shaped plaques on the ends—have geometric decoration; the other eighteen, all circular, are distributed on the face, back, and top and are ornamented with birds. Corresponding to these different themes, there are also distinctions in the palette, the form, and the method of attachment.

In the first group, defined by geometric decoration, one almond-shaped piece (no. 23) can be singled out because of

its state of preservation, the presence of attachment lugs, and the appearance of yellow among the enamels. The nine other enamels are the ones that have undergone the most extensive restorations. Their palette is restrained: one or two blues, one green, and, on two of them (nos. 25 and 29), white; they are without lugs and have only four holes for attachment, which, however, might have been pierced at a later date. The motifs, combining hearts and palmettes in a rigorous, symmetrical arrangement, could be larger-scale adaptations of the design of earlier cloisonné enamels, such as the plaques that frame the cover of the Codex Aureus of Echternach (Germanisches Nationalmuseum, Nuremberg) or the medallion in the center of the Townley Brooch (British Museum, London).[5]

These medallions are thus important witnesses to the evolution of champlevé from cloisonné enameling. The only comparable champlevé enamels are the ones adorning a coffret called the reliquary of Saint Valerius, formerly preserved at Roda de Isábena in Aragon (stolen in 1979), which Gauthier attributed to Abbot Boniface's workshop.[6]

The eighteen medallions decorated with birds are not unprecedented in Western enameling. There is a plaque of cloisonné enamel on iron at the Louvre: decorated with two confronted birds and generally dated to the eleventh century, it is no doubt the best example of all.[7] But here, the conception of the decoration and the technique are typically Romanesque. Furthermore, these medallions have certain characteristics in common that make them a homogeneous ensemble: the same method of attachment,[8] approximately the same dimensions, the same striated rims. Finally, the central motif is always surrounded by an enameled border—an alternation of circles and rectangles on seventeen medallions, a Greek frieze on the last (no. 4). The palette has a total of eight colors: three blues, three greens, a yellow, and a white, with five or six different colors occurring on each medallion. The yellow, absent from the two sets closest in style—the ten medallions from the former Carrand Collection (cat. 8) and the Bellac chasse (cat. 9)—appears on nine medallions.

The main variation from one medallion to another is in the central motif. The most frequent design (eight medallions: nos. 1, 2, 7, 17–21) is a fantastic, rapacious bird. It is seen in either right or left profile, its wings outspread, its plumed head turned either to the front or to the tail. On two medallions (nos. 14 and 16), the bird is seen frontally. On three others (nos. 12, 13, and 15), two birds are confronted, pecking a central stalk. In one instance (no. 9), the bird occupies only half the medallion; the other half is filled with the foliated scroll emerging from its beak. On the last four medallions, the animal is a fantastic creature, a winged quadruped with a bird's head, its tail terminating in a scroll (nos. 5 and 6), or a winged dragon with an ibex's head, also with a tail that terminates in a scroll (nos. 3 and 4).

This decoration is especially captivating because of its variety, the exceptional vitality of the birds, and the quality of the stylization. Its sources have most often been located in oriental art—chiefly textiles or ivories. However, as pointed out by Hildburgh (1936), the kinship is not close enough to suggest a direct influence. A very few specific parallels have been mentioned—for instance, the so-called Saint Thomas Becket chasuble, embroidered in Almería in 1117[9] and now preserved at the cathedral of Fermo—but

1        4

without convincing proof. The motif used most frequently, the bird in profile with outspread wings, is the element that most concretely recalls certain Eastern fabrics, especially the Sassanian textiles decorated with birds inscribed in medallions, such as a fragment preserved in Washington, D.C., the *tiraz* at the Cleveland Museum of Art, and the fabric with haloed pheasants at Jouarre in France.[10]

However, the concept of animal decoration, as seen in works from Conques, is characterized above all by the dynamism and variety of the figures, with their strongly stylized contours and broad colors; this is the very opposite of what is seen in oriental textiles, where identical motifs, decoratively treated, are endlessly repeated. Two other motifs, the two confronted birds and the frontal eagle with outspread wings, are among the fundamental themes of oriental iconography. Yet this same concept at Conques, at once dynamic and stylized, marks a break with the Islamic and Byzantine representations; instead, it is closer to that of works created not so far away.

It must be emphasized that the ornamental use of birds, whether inscribed in circles or not, spread quickly throughout Western art. From the Franco-Saxon illuminations of the seventh and eighth centuries to the Carolingian manuscripts of the School of Tours and those produced by the scriptoria of Angers and Limoges during the eleventh century, the borders and initial letters offer numerous examples of similar illustrations exhibiting animated postures, stylized drawing, broad areas of color, and, in the more recent works, the use of rinceau decoration. All these features allow us to recognize the most immediate sources of the medallions executed for Abbot Boniface. Thus, the dragons with ibex heads (nos. 3 and 4) recall the goats in the Capricorn depictions from the Carolingian zodiacs of the School of Tours—for example, in the Vivian Bible.[11] The two confronted birds (nos. 12, 13, 15) can likewise be linked to the same motif in several manuscripts: Saint Augustine's Commentary on the Psalms, created by the scriptorium of Saint-Aubin of Angers in the eleventh century;[12] and the New Testament attributed to the Agen-Moissac region at

Reverse, with medallions 7 to 13; in the center, the three new medallions added by Poussielgue

6    9    16    20

7    15    18

the beginning of the eleventh century.[13] And finally, the winged quadruped (nos. 5 and 6) has close parallels in Limousin illuminations of the early twelfth century, especially the Bible of Saint-Yrieix (fig. 7*b*).[14]

It is also possible to invoke the few known Romanesque mosaics, particularly the one at the church of Saint-Genès in Thiers (Puy-de-Dôme),[15] which comprises various animals inscribed in medallions, among them a bird in profile with an elongated eye, comparable to the figures on eight of the Conques medallions.

However, the closest ties are with the medallions on the Bellac chasse (cat. 9) and, especially, with the decorative band from the shaft of the crosier that was unearthed in 1940 on the site of the former abbey of Saint-Augustin-lès-

Limoges (fig. 3),[16] where the bird in profile is very close to the birds adorning the Conques medallions (nos. 1, 2, 7, 17–21).

Although the creator (or creators) of Abbot Boniface's coffret probably drew upon various sources, the work is closely bound to the art of western and central France: the rigorous yet relaxed inscribing within circular frames, the extreme but never dry or dull stylization, the sureness and liveliness of the lines, and the color set in broad areas characterize all these medallions—executed in the new technique of champlevé enamel on copper—as exemplary creations of Romanesque art. The creative ingenuity of Abbot Boniface's enamelers is thus borne out by the highly innovative use of enameled decoration: the artists, resisting

Left side, with appliqués 27 to 31

Right side, with appliqués 22 to 26

Fig. 7*b*. The letter B (detail). Bible. Town Hall, Saint-Yrieix, fol. 176v

any temptation to imitate the effects of Byzantine or Western cloisonné enamels, achieved a perfect symbiosis with the new Romanesque interest in color, expressive delineation, and a sense of the vitality of earthly creatures.

It is nevertheless true that we cannot date these enamels precisely. Most scholars have agreed that Boniface's abbacy, which certainly began in 1107 or 1108, ended in 1119[17]—but that is an approximation. Actually, most of the documents concerning Boniface date from the reigns of Henry I of England (1100–1135) and Louis VI of France (1108–37), but no more precisely than that. Boniface's name figures again in a charter issued by Henry I after 1121, and the earliest specific mention of Boniface's successor did not appear until 1154.[18] Nor is it altogether certain that the two quite distinct groups of enamels—those with geometric motifs and those with animal decoration—were exactly contemporaneous.

Furthermore, we should probably take into account the fact that cloisonné enameling on copper blossomed fully at Conques about 1100–1110 (see cats. 1–6). The reliquary of Saint Valerius, formerly preserved at Roda de Isábena—which has champlevé enamels with geometric motifs similar to those on the coffret of Abbot Boniface—may have been executed shortly after 1121. In that year, Ramón II, bishop of Roda (r. 1104–26), exhumed the relics of the bishop-saint Valerius of Saragossa in order to donate an arm to the cathedral there.[19] On that basis, a date of 1110–30 is proposed here.

ET-D

NOTES

1. The invoice of September 2, 1878, reads: "Restoration of wood coffret, repairs of the preserved leather part, special fabrication of leather and studs similar to the old ones; restoration of damaged enamels, engraving, and enameling of two lost medallions, fitting of old fastener to close the chasse." This restoration was implemented "contrary to the advice of the Commission on Historic Monuments." Report by the publisher, Du Sommerard, October 6, 1878. See also, letter from Poussielgue-Rusand to Msgr. Bourret, March 22, 1878, provided by Claire Delmas, to whom I am very grateful: "For the chasse, we are doing what you have written to me, and I already have the specifications to complete the decorated leather. Mr. Gay does not have the enamels, and I will make them anew, which poses no problem (Rodez, Archives diocésaines, série des paroisses, Conques [2], liasse B; for original text, see French edition of this catalogue).

2. See Gauthier 1987, no. 36, p. 59.

3. Incidentally, these same two models—each reproduced three times—can be seen on the six medallions adorning the copy—very free and simplified—of the coffret executed by Poussielgue, preserved at Conques, and corresponding to one of the "coffrets" offered in the catalogue of the house of Poussielgue-Rusand (no. 2233 in the 1893 catalogue).

4. See Namur 1969, no. 3.

5. Westermann-Angerhausen 1973; Speyer and Mainz 1992, pp. 378–89.

6. Gauthier (1987, no. 49) justifies this attribution on the basis of historical relations: a monk from Conques, Pierre d'Andouque, was bishop of Pamplona from 1083 to 1115, and another, Ponce, was bishop of Roda from 1097 to 1104.

7. See Gauthier 1987, no. 67.

8. Except for the four medallions that had served as bobeches, the lugs of which had been filed down. These are now attached by means of two flat strips that cross in an *X* at the back, with a circle at the crossing that fills the hole pierced for their use as bobeches.

9. Gauthier 1972b.

10. See Sheperd 1960; Laporte 1988a.

11. Paris, BNF, ms. lat. 1, fol. 7a (letter D); see Köhler 1930, vol. 1 pl. 79.

12. Angers, Bibliothèque Municipale, ms. 169, fol. 217v; see Angers 1985, p. 58.

13. Paris, BNF, ms. lat. 254, fol. 10; see Porcher 1959c, pl. 2, p. 124; Avril 1983, p. 187.

14. Town hall of Saint-Yrieix, Canons, fol. 303v and letter B, fol. 176v; see Gaborit-Chopin 1969 and Zaluska 1979.

15. See Stern 1954.

16. Gauthier 1987, no. 59. The object was stolen from the Musée Municipal in 1980.

17. See especially, Aubert 1954; Gauthier 1963b, p. 130, and 1987, p. 38; Lasko 1972, p. 154.

18. Desjardins 1879, p. xlv; Lambert 1937.

19. Ross 1933c, pp. 8–9; Gauthier 1987, no. 49.

EXHIBITIONS: Paris 1900, no. 2406; Paris 1965, no. 545.

BIBLIOGRAPHY: Barrau 1875; Gonzague 1875; Cougny 1875; Darcel 1878a, pp. 565–66; Servières 1878, pp. 98–101; Bourret 1880, pp. 12ff.; Molinier 1881, p. 98; Barbier de Montault 1886a, pp. 180–81; Linas 1887, pp. 37–38; Molinier 1889, p. 20; Rupin 1890, pp. 59–62; Bouillet 1892, pp. 73–78; Bouillet and Servières 1900, pp. 186–91; Guibert 1901, p. 12; Ross 1933c; Hildburgh 1936, pp. 64–66; Gauthier 1950, p. 24; Aubert 1954, p. 66, pl. XXII; Gauthier 1963b, pp. 130–31, 145, pls. 66–67; Gauthier 1972b, p. 78; Lasko 1972, pp. 154–55, 228–29, fig. 158; Gaborit-Chopin 1983b, p. 318; Favreau et al. 1984 vol. 9 , no. 31, p. 51; Gauthier 1987, no. 36, pp. 38–60, ills. 85–110, 118, pls. XIII–XVIII, XXIII.

# 8. Carrand Medallions

**Conques, ca. 1110–30**

### a. *Winged Dragon*
Copper: gilt; champlevé enamel: lapis and lavender blue, turquoise, emerald and meadow green, and white
Diam.: 10 cm (9 cm without the lugs); (3¹⁵⁄₁₆ in. [3⁹⁄₁₆ in. without the lugs])
Mark on reverse: |

### b. *Eagle Attacking a Fish*
Copper: gilt; champlevé enamel: lapis and lavender blue, turquoise, emerald and meadow green, and white
Diam.: 9.9 cm (9 cm without the lugs); (3¹⁵⁄₁₆ in. [3⁹⁄₁₆ in. without the lugs])
Mark on reverse: ∨

### c. *Winged Dragon with Coiled Neck*
Copper: gilt; champlevé enamel: lapis and lavender blue, turquoise, emerald and meadow green, and white
Diam.: 9.7 cm (8.9 cm without the lugs); (3¹³⁄₁₆ in. [3½ in. without the lugs])
Mark on reverse: //

### d. *Quadruped with Scale-Covered Back*
Copper: gilt; champlevé enamel: lapis blue, turquoise, emerald and meadow green, and white
Diam.: 9.8 cm (8.8 cm without the lugs); (3⅞ in. [3⁷⁄₁₆ in. without the lugs])
Mark on reverse: ∨|

Ex colls.: Jean-Baptiste Carrand, Lyons (before 1856), then Louis Carrand, Lyons and Florence. Bequeathed by Louis Carrand to the city of Florence in 1888.

Condition: Wear to the gilding.

Fig. 8a. Alms box, thought to have been made from a coffret. Abbey Church of Saint-Foy, Conques

Museo Nazionale del Bargello, Florence (nos. 623–626 C)
Exhibited in Paris only.

### e. *Eagle Attacking a Fish*
Copper: gilt; champlevé enamel: lapis and lavender blue, turquoise, emerald and meadow green, and white
Diam.: 9 cm (3⁹⁄₁₆ in.)
Mark on reverse: ⟨

### f. *Quadruped*
Copper: gilt; champlevé enamel: lapis and lavender blue, turquoise, emerald and meadow green, and white
Diam.: 9 cm (3⁹⁄₁₆ in.)
Mark on reverse: ∧

### g. *Combat Between Dragon and Dog*
Copper: gilt; champlevé enamel: lapis and lavender blue, turquoise, emerald and meadow green, and white
Diam.: 9 cm (3⁹⁄₁₆ in.)
Mark on reverse: //

### h. *Winged Dragon*
Copper: gilt; champlevé enamel: lapis and lavender blue, turquoise, emerald and meadow green, and white
Diam.: 9.1 cm (3⁹⁄₁₆ in.)
Mark on reverse: ∫

### i. *Combat Between Dragon and Dog*
Copper: gilt; champlevé enamel: lapis and lavender blue, turquoise, emerald and meadow green, and white
Diam.: 9.1 cm (3⁹⁄₁₆ in.)
Mark on reverse: ///

Ex colls.: Jean-Baptiste Carrand, Lyons (before 1856); Michel Boy;

e

f

g

Sigismond Bardac, Paris (in 1900); Georges Hoentschel, Paris (before 1911); J. Pierpont Morgan, London and New York

CONDITION: Slight wear to gilding; loss of enamel on dragon's tail, medallion g.

The Metropolitan Museum of Art, New York; Gift of J. Pierpont Morgan, 1917 (17.190.689–693)

### j. *Winged Dragon*
Copper: gilt; champlevé enamel: lapis and lavender blue, turquoise, emerald and meadow green, and white
Diam.: 9.2 cm (3⅝ in.)
Mark on reverse: ⚌

EX COLLS.: Probably Jean-Baptiste Carrand, Lyons (before 1856), as with the nine others; Victor Gay, Paris (before 1878a; see Darcel 1878). Gift of a group of friends of the Musée du Louvre, 1909

CONDITION: Loss to enamel of wing.

Musée du Louvre, Paris; Département des Objets d'art (OA 6280)

Although preserved in three different collections since the end of the nineteenth century, these medallions are probably the ones—also ten in number—acquired by J.-B. Carrand (1792–1871) and described by Labarte in 1856. Labarte, who felt that they came from a bandolier, assigned a German origin to them. The link with the enamels of the coffret discovered at Conques in 1875 was suggested by Darcel in 1878 in his commentary on the Exposition Universelle, where Victor Gay exhibited the medallion now at the Louvre. Actually, however, Gay had already pointed out that link when he visited Conques in late 1875 or early 1876. This is indicated by an unpublished letter from the head friar of the Premonstratensian community that served Saint-Foy of Conques; the letter, addressed to Monsignor Bourret, bishop of Rodez, was dated January 26, 1878.[1] It was no doubt during this trip that Gay noticed on Abbot Boniface's coffret the two medallions bearing inscriptions, which are portrayed in watercolors in an album preserved in the Louvre (see cat. 7). This letter is also the oldest document to report that the medallions could have come from an alms box placed at the church entrance. Later, this infor-

mation was repeated by several authors, notably Rupin (1890), then Bouillet and Servières (1900). The alms box, probably part of an old coffret divided into sections, is now placed at the entrance of the Saint-Foy chapel (fig. 8a). It bears traces of eight medallions (four circles and four arcs) whose diameter (8.6 cm or 3⅜ in.) corresponds to those from the former Carrand Collection, exclusive of the lugs.[2] It is therefore very possible that the coffret, old but later than the Middle Ages, held these medallions for a period before their removal from Conques.

The great similarity between these medallions and those on the Boniface coffret (cat. 7) is obvious and has never been disputed. The shapes, sizes, and decorations are close enough to justify the theory that they once constituted the decoration of a second coffret commissioned by Boniface—a hypothesis suggested by the plural reference to *scrinia* (coffrets) used in the inscription on the periphery of one medallion on the coffret still in Conques (see cat. 7). The assumption of a link between the two groups is substantiated by various details: thus, the Greek frieze on one medallion of the Conques coffret is analogous to the friezes on the Carrand medallions.

Nevertheless, certain features distinguish the two ensembles. The Carrand medallions form a more homogeneous group than those on the Conques coffret. The Carrand medallions all have the same shape and same size, even taking into account the variation of two to three millimeters inherent in a handcrafted work. They all have the same border, consisting of striations perpendicular to the edge, the same semicircular attachment rings, and the same Greek frieze surrounding the central motif. The colors of the enamels are slightly different from those on the Conques coffret: the darker green and the yellow do not appear here; on the other hand, lavender blue, absent from the preceding series and characteristic of the palette of Limousin enamelers, appears on eight medallions.

While the creatures on the Conques medallions are all winged, the animals—or, rather, the monsters—adorning the Carrand medallions are more diverse. Four of them are

h                    i                    j

winged dragons (a, c, h, j), two are quadrupeds (dogs?) without wings but with heads resembling those of the dragons (d, f). The last four (b, e, g, i) show two animals in combat: a dog and a dragon (g, i) or an eagle and a fish (b, e). This theme of a struggle between two animals—along with the more complex drawing, the entanglement of the bodies and of the animals, and the more luxuriantly deployed rinceaux—distinguishes the Carrand medallions.

These medallions seem remote from possible oriental sources—whether from textiles or other media. The Greek frieze accentuates this distance. This motif is widespread in western Aquitainian art of about 1100—for instance, in the frescoes of Saint-Savin-sur-Gartempe and in the illuminated manuscripts of Saint-Aubin of Angers or Saint-Martial of Limoges.[3] As for the animal themes, they are typical of Aquitainian decoration and are often linked to Carolingian sources. The quadruped whose body curves within the periphery of the disk (d) is very close to the quadrupeds adorning certain initials in Aquitainian illuminations of the tenth and eleventh centuries, for instance, the one inside the letter *E* of folio 23 of the Lectionary of Saint-Martial of Limoges:[4] its source might be found in the Carolingian manuscripts of the School of Tours, such as the Book of Alcuin at the Bibliothèque Municipale in Tours.[5] Likewise the theme of combat between a dragon and a quadruped and especially the theme of an eagle attacking a fish have followed a similar path from the initials in Tours during the ninth century to the Aquitainian manuscripts of the eleventh century. The latter motif can be found in the Works of Saint Jerome (Bibliothèque Universitaire, Ghent)[6] and also in the Lectionary of Saint-Martial of Limoges (fol. 279v) and the Gospels at the Bibliothèque Municipale, Perpignan (end of the eleventh century; fig. 8b),[7] where the arrangement and proportions of the animals show a particular kinship with those on the Conques medallions. Finally, the winged dragon that spews foliage and whose tail ends in a scroll (a, c, i) is one of the most characteristic motifs of Romanesque illuminations of the West. We are familiar with its birth at Mont-Saint-Michel in the early eleventh

century, which renewed the experiments of the eighth and ninth centuries. We also know that it spread through the major scriptoria in Normandy and Anjou during the eleventh and the early twelfth century.[8] Thus, these dragons can be linked to, for instance, the one depicted on folio 13v, volume 2, of Cassiodorus's Commentary on the Psalms, from Saint-Aubin of Angers.[9] One of the best parallels to this juxtaposition of single or confronted animals can be found in manuscripts attributed to the Agen-Moissac region, especially a New Testament (at the BNF, Paris): here, inside a letter L, an animal combat is juxtaposed with a bird having a coiled neck, as on one of the Bargello medallions.[10]

Beyond their diversity the animals on the Carrand medallions are united by the drawing of the heads of dragons and quadrupeds; these heads show an obvious kinship with those of the animals in southern French illuminated manuscripts. One such example is presented by the animals adorning the letter T in a Bible attributed to the Moissac region during the second half of the eleventh century;[11] this

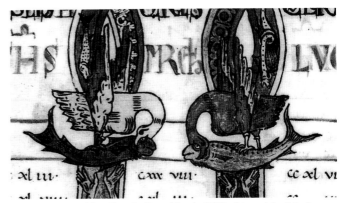

Fig. 8*b*. Canon tables (detail). Gospel book. Bibliothèque Municipale, Perpignan, ms. 1, fol. 13v

confirms that this decoration was rooted in the art of south-western Aquitaine. The complex entanglement of animal bodies and the violence of their combat evoke some of the most famous works of Romanesque art in southwestern France: for example, the abaci of the Moissac monastery or the piers of the old portal of Souillac.

The medallions from the former Carrand Collection, which are linked by technique, shape, and decoration to those on the Conques coffret, were undoubtedly produced at the workshop maintained under Abbot Boniface at the famous abbey in the Rouergue region. They are also linked closely to the art developed in Aquitaine during the late eleventh and the early twelfth century, especially in Quercy and the Limousin. This clear kinship with one of the main trends in the Romanesque art of southwestern France most likely indicates that the Conques artisans were more open to creations from other workshops in the same artistic milieu. Perhaps we should attribute the perceptible change in inspiration and style to the fact that the atelier of Boniface included several artists or groups of artists who had been trained elsewhere.

ET-D

NOTES

1. Rodez, Archives diocésaines, Série des paroisses, Conques (2), liasse B, XIX, XX. Letter discovered by Claire Delmas, curator of antiquities and works of art at Aveyron, to whom I am extremely grateful: "Regarding the enamels of the chasse, or rather the alms box, of Saint-Foy, which at a point unknown to me became the property of an amateur [collector] of antiquities, I must inform your Grace that it was I who was told this secret by Monsieur Gay during his trip to Conques some two years ago. . . . He had promised to take some steps with the antique dealer, his friend, at whose place he had had the occasion to see four enamels similar to those on the chasse of Saint-Foy, and he tried to get him to return them to us. . . ."

2. I wish to thank Jean Taralon and Claire Delmas, who helped me find the coffret and provided me with the dimensions.

3. For example, in the Bible of Saint-Aubin of Angers, Angers, Bibliothèque Municipale, ms. 4, fols. 205v and 206 (see Porcher 1959a), or the Second Bible of Saint-Martial of Limoges, Paris, BNF, ms. lat. 8, Canons, fol. 171 (see Gaborit-Chopin 1969, pp. 177–78 and fig. 103). Incidentally, this motif had already appeared in a drawing by Adémar de Chabannes (d. 1034), Leiden, Bibliotheek, ms. Voss, lat. 8015, fol. 43v; see Gaborit-Chopin 1968, fig. 31.

4. Paris, BNF, ms. lat. 5301, fol. 23v, ca. 1030; see Gaborit-Chopin 1969, pl. 80.

5. Ms. 1742, fol. 52b; see Köhler 1930, vol. 1, pl. 7b.

6. Ms. 102, fols. 1 and 195a; see Köhler 1930, vol. 1, pl. 11, figs. f and g.

7. Ms. 1, fol. 13, Canons; see Porcher 1959b, fig. 26.

8. See, especially, Avril 1983; Rouen 1975, pp. 41–42; Angers 1985, pp. 62–68; Avranches 1991, pp. 37–49.

9. Angers, Bibliothèque Municipale, ms. 44; see Angers 1985, p. 73.

10. Paris, BNF, ms. lat. 254 (fol. 10: letter L); see Porcher 1959c, pl. 21; Avril 1983, p. 187 (first quarter of the twelfth century).

11. Paris, BNF, ms. lat. 7 (fol. 259v); see Gaborit-Chopin 1969, p. 106.

GENERAL BIBLIOGRAPHY: Labarte 1856, p. 61; Labarte 1865, vol. 3, p. 474, and *Album* 1864, pl. CVI (nos. 623c, 624c; here, a and b); Linas 1883, p. 125; Molinier 1887a, p. 474; Rupin 1890, p. 61, figs. 105, 106 (captions transposed here, a and j); Bouillet 1892, p. 74, n. 3; Bouillet and Servières 1900, p. 190; Guibert 1901, p. 13; Hildburgh 1936, pp. 64–66; Gauthier 1950, pp. 27, 68; Gauthier 1972b, p. 78; Garnier 1980, p. 62, n. 66; Gaborit-Chopin 1983b, fig. 282(j), p. 318; Gauthier 1987, nos. 38–47, pp. 61–62, ills. 119–28, pl. XXIV.

a–d BIBLIOGRAPHY: Supino 1898, p. 121, nos. 623–26.

EXHIBITION: Florence 1989, no. 208.

e–i BIBLIOGRAPHY: Pérate 1911, nos. 30–34, pl. XVI; Breck and Rogers 1925, p. 57; W. H. Forsyth 1946, pp. 233, 235 (fig.).

EXHIBITION: Paris 1900, no. 2409.

j BIBLIOGRAPHY: Darcel 1878a, p. 550; Gay 1887–1928, vol. 1, p. 617, engraving; Rupin 1890, p. 61, fig. 106 (captions transposed); Migeon 1909; *Bulletin des Musées de France* no. 5 (1909), p. 69; Marquet de Vasselot 1914, no. 84, pl. IX; Gauthier 1950, pl. 2; Gaborit-Chopin and Taburet 1981, no. 18.

EXHIBITIONS: Limoges 1948, no. 4; Vatican City 1963, no. 2, pl. III.

# 9. Chasse of Bellac

Limoges, ca. 1120–40

Copper: gilt; champlevé enamel: lapis blue, turquoise, solid almond green, white, and pinkish white; 108 glass or hard-stone cabochons, including fifteen intaglios and one cameo; wood core

19.9 x 27.5 x 11.5 cm (7⅞ x 10⅞ x 4½ in.); diam. of the medallions: 5.5–7.8 cm (2⅛–3⅛ in.)

INSCRIBED: SANCTA MARIA MATER D[OMI]NI (Holy Mary, Mother of God); IHESUS

PROVENANCE: First mentioned as in the church of Bellac (Haute-Vienne), 1851. Classified, Monuments Historiques, 1891

CONDITION: The wood core is entirely covered with modern red baize and cannot be examined. The chasse was probably opened from the bottom, which is divided into two parts. Three notches visible on roof of wood core; original crest missing; three medallions missing from reverse (see Roy-Pierrefitte [1851], who mentioned the presence of relics at these locations). One medallion, ornamented with a bird, was published by Rupin in 1890 (fig. 218, p. 145; see fig. 9b, below). At that time, it was in the F.-A. Wasset Collection, Paris, and was identified as coming from the chasse of Bellac. This identification was plausible, since the medallion had the same border as the Paschal Lamb medallion on the roof of the principal face of the chasse. Among the one hundred ten precious stones that originally ornamented the chasse, sixteen intaglios and one cameo were counted by Palustre and Barbier de Montault (1887), Rupin (1890), and in the catalogue of the 1948 exhibition at Limoges. Today, however, only fifteen intaglios have been preserved.

The 1961 restoration consisted principally of disassembly and cleaning; the missing gems were replaced (Archives, Monuments Historiques, dossier Bellac).

Church of Notre-Dame, Bellac

Briefly described by Roy-Pierrefitte in 1851, the Bellac chasse was widely seen for the first time at the retrospective exhibition mounted at Limoges in 1886. Molinier (1886), Guibert (1886), and Rupin (1890) linked the enameled medallions to those of Conques (cats. 7, 8), thereby enabling scholars to restore the chasse to its rightful place in the history of Romanesque enamel work.

The enamels are arranged in an iconographic pattern evoking the Glory of Christ: on the principal face, Christ in Majesty and the Paschal Lamb (on the roof) are surrounded by the four symbols of the evangelists, in accordance with the vision of the Apocalypse. On one gabled end the Virgin appears (the largest medallion), designated as "Mother of God"; on the opposite end, the Paschal Lamb reappears—a major break in iconographic coherence. The three medallions preserved on the reverse are decorated with animals: confronted birds and passant lions. The medallion acquired by Wasset (see above) might be one of the three that are missing from the reverse.

We do not know for which relics the chasse was made. According to Lecler (1920), it contained "relics of Saint Lawrence, Saint Francis of Assisi, and Saint Clare." Yet

according to a description of 1775, the church had no altars dedicated to these saints, two of whom, incidentally, lived a century after the presumed execution date of the work. Abbé Roy-Pierrefitte's description (1851) makes no mention of these relics; instead, he explains that the three voids left by the medallions missing from the reverse had been filled with relics, only one of which was identified by an inscription read as S.P. (for Sanctus Paulus?).

As emphasized by most scholars since the end of the nineteenth century, the kinship between the enamels on the Bellac chasse and the enamels executed at Conques is obvious. The two depictions of the Paschal Lamb are close to the one on the portable altar of Saint Foy (cat. 1), which has been pointed out by Wixom (1967). The medallion with the two confronted birds pecking at the Tree of Life is very close to those with the same subject on the coffret of Abbot Boniface (cat. 7). The winged quadruped with the bird's head, representing the eagle of Saint John, is, for its part, similar to the two creatures of the same type present on the coffret: not only in the silhouette but also in the draftsmanship of the wings and the head, with its elongated eye and long, hooked beak.

As in certain medallions on the Conques coffret and, above all, on those from the former Carrand Collection (cat. 8), the animal motifs are often associated with a rinceau decoration. By the same token these animal depictions are distinguished by their expressive force, together with great precision in the strongly stylized draftsmanship.

Nevertheless, the differences between the Conques and the Bellac enamels should not be overlooked. The palette is less extensive here: yellow is absent, and there is only a single shade of green. Most of the medallions have only two or three colors, and the ability to change color within the champlevé cell—a possibility employed in Conques for coloring beaks and claws—seems to have been lost here. The figures are thinner, and a less expert hand seems to have drawn them. Then again, the Bellac chasse shows the first human figures depicted on champlevé enamel—at least in France and southern Europe. Compared with the figures on the portable altar of Saint Foy or the winged man of Saint Matthew (cats. 1, 6), which were done in the cloisonné technique, these figures are much further away from the examples of the High Middle Ages.

More than in Conques, certain animal depictions here probably justify the hypothesis of a parallel with Islamic textiles and especially with those made in Spain during the eleventh century: above all, the lions on the back can be likened to those covering the inside of the lid of the reliquary of Saint Pelagius at San Isidoro in León.[1]

Fig. 9*a*. The Prophet Nahum. Bible. Bibliothèque Mazarine, Paris, ms. 1, fol. 199v

Fig. 9*b*. Medallion. Formerly Wasset Collection (after Rupin 1890)

Left side

However, the most convincing hypothesis links the Bellac chasse to Limousin illuminations of the early twelfth century, especially with the two Bibles preserved at the town hall of Saint-Yrieix and at the Bibliothèque Mazarine, Paris (ms. 1–2).[2] The simplifying draftsmanship and the dynamic silhouettes attest to an undeniable common spirit, as do certain details such as the surrounding circle transformed into a rinceau or the way each lion's tail terminates with foliage. The figures of Christ and the Virgin offer striking similarities with the figures adorning the initials in these two Bibles. We observe not only the same extreme stylization, the same slightly awkward, demonstrative gestures, which are further accentuated by the inordinate size of the hands. We also notice the same definition of the faces: the hairline, with an inverted V over the forehead, continues as bands framing the cheeks (fig. 9*a*).

Since 1851, when it was first published, the chasse has been preserved in the parish church of Bellac, the ancient chapel in the château of the counts of La Marche. Its presence in the Limousin during the mid-nineteenth century suggests that it has been there since its creation. Thus both the style and the historical data allow us to identify it as the oldest extant work of the Limoges workshops. This attribution, already proposed in the late nineteenth century by Molinier (1886) and Guibert (1888b), is also borne out if we compare the chasse with several other pieces that were unknown at the time of those publications. The major piece is the decorative band from the shaft of a crosier, excavated in 1941 from the site of the ancient abbey of Saint-Augustin-lès-Limoges and formerly preserved at the Musée Municipal (fig. 3).[3] The drawing and coloring of the two stylized birds seem to place them halfway between the enamels of Conques and those of Bellac. Nor should we neglect the evidence of the Frassinoro candlestick (Museo Civico, Modena), which bears the inscription CONSTANTINUS FABER DE LEMOIE (for Lemovice) ME FECIT (Constantine, artisan of Limoges, made me) (fig. 15).[4] It is ornamented with engraved animals, which, although they are slight, can be compared with those on the enameled medallions of the Bellac chasse.

This work is a rare and important example for the history of reliquaries in the form of a chasse. It is an immediate successor to the earliest medieval chasses having this form (for instance, the small reliquary inscribed with the name "Mumma" in Saint-Benoît-sur-Loire).[5] At the same time, it heralds those Limousin creations conceived in a similar manner but made entirely of enameled copper plaques.

It is difficult to assign a precise date for the fabrication of this chasse. Its parallels with the Conques enamels and with Limousin illuminations of the early 1100s place it incontestably in the first half of that century. The most likely date appears to be about 1120–40.

In the light of these observations, it is possible to suggest that this first known example of Limousin enamel work might be the work of an artist trained at Conques—or at least aware of the champlevé enamels originating there. He could have assimilated a certain technical mastery without retaining the wealth of refinements, but his art could also have been forged through contacts with painters active in Limoges during the early decades of the twelfth century.

ET-D

NOTES
1. About 1059 or earlier; see New York 1993, no. 109.
2. About 1100 or first quarter of the twelfth century; Gaborit-Chopin 1969, pp. 129–40, 176, 212–13; Zaluska 1979; Cahn 1982, no. 108 (Bible of Saint-Yrieix).
3. Stolen in 1980. See Delage 1941; Gauthier 1987, no. 59.
4. See Gauthier 1987, no. 52.
5. See Paris 1965, no. 192.

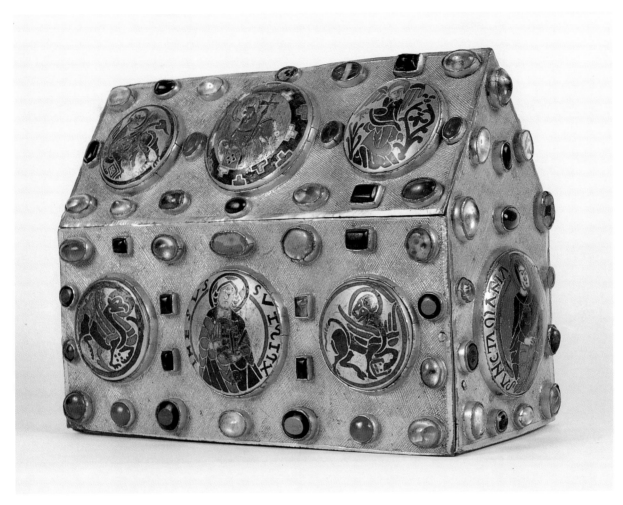

Principal face and right side

Medallions on the reverse of the roof

EXHIBITIONS: Limoges 1886, no. 4, *Album*, pl. I; Paris 1900, no. 2407; Paris 1937, no. 1202; Limoges 1948, no. 3; Barcelona 1961, no. 425; Vatican City 1963, no. 4, pl. II; Paris 1965, no. 356, pl. 52; Cleveland 1967, no. III, I; Prague-Bratislava 1978, no. 25.

BIBLIOGRAPHY: Roy-Pierrefitte 1851, p. 119; Tixier 1857, col. 1258; Molinier 1886, pp. 172–74, with fig.; Guibert and Texier 1886, no. 4, pl. XXIII; Palustre and Barbier de Montault 1887, pp. ix–xii, nos. I, II; Arbellot 1887, pp. 21–27; Guibert 1888b, pp. 201–9; Molinier 1889, p. 21; Rupin 1890, pp. 62, 144–45, pls. I, V, VI; Molinier 1891, p. 132; Guibert 1901, pp. 12–13; Molinier 1901, p. 178; Lecler 1920, pp. 57–58; Demartial 1923, p. 434; Hildburgh 1936, pp. 66–67, 77, 109; Gauthier 1950, pp. 24ff.; Gauthier 1960a, pp. 280–86; Lasko 1972, p. 229; Gauthier 1972b, no. 38, pp. 78–80, 326–27; Gauthier 1976a, pp. 186–90; Favreau et al. 1978, vol. 2, no. 3, p. 91; Gaborit-Chopin 1983b, p. 318, figs. 283, 284; Gauthier 1987, no. 57, pp. 71–74, pls. XXVIII, XXXI–XXXIII.

# 10. Chasse of Champagnat

Limoges, ca. 1150

Copper: engraved and gilt; champlevé enamel: blue-black, medium blue, turquoise, green, red, and white

12.4 × 18.9 × 8.5 cm (4⅞ × 7⁷⁄₁₆ × 3⅜ in.)

INSCRIBED: MARIA/MARCIALIS (Saint Martial)/A/ω (Alpha/omega)

PROVENANCE: Church of Saint-Martial, Champagnat (Creuse)

EX COLLS.: Laforge, Lyons (1868); [art market, Paris, 1891]; Sigismond Bardac, Paris (in 1900); Georges Hoentschel, Paris (after 1900); [Jacques Seligmann, Paris (in January 1912)]; J. Pierpont Morgan, London and New York

CONDITION: Losses to gilding, especially on end plaques; original core lost, modern replacement fabricated at the Metropolitan Museum.

The Metropolitan Museum of Art, New York; Gift of J. Pierpont Morgan, 1917 (17.190.685–87, 695, 710–11)

Christ appears at the center of the principal face of the reliquary, beneath the roof plaque, which depicts the Hand of God flanked by two angels swinging censers. Christ, with cruciform nimbus, holds a book in his left hand and raises his right in blessing; the Greek letters alpha and omega are at either side. At his right is a female saint identified as "Maria." She clasps an unguent jar and a palm frond. To Christ's left is a bearded male saint identified as "Marcialis," who also raises his right hand in blessing. Saint Peter, holding the red-enameled keys to the kingdom of heaven, and another saint, probably Paul, appear on the end panels. At the upper and lower corners of the reverse stand the four symbols of the evangelists, clasping their Gospel texts (or, in

Fig. 10a. *Urna* of Santo Domingo de Silos (detail). Museo Arqueológica Provincial, Burgos

the case of the eagle of John, a scroll) in hooves, talons, claws, or veiled hands. Each of the symbolic figures turns his head to the center of the panels, which have richly worked foliate decoration springing, in the case of the lower panel, from fantastic beasts with human faces.

The chasse is remarkable for the sense of movement that is conveyed through the gestures of Christ and Martial, the forward strides of the censing angels, the strong patterning of the lines of drapery, and the multicolored, sinuous forms of the beasts and the winged man of Saint Matthew. The palette displays a rich range of blues and greens, reminiscent of the chasse of Bellac (cat. 9).

The interlacing hybrid forms on the reverse and the overall palette relate closely to the decoration found on enamels created for the monastery of Santo Domingo de Silos (Burgos), especially to the beasts on the *Urna*[1] (fig. 10a)[1] and the palette of the reliquary casket of Saint Dominic preserved at Burgos.[2] The explanation for the common style is probably to be found in the close political and monastic links between Castile and the Limousin. Plantagenet patronage of the abbey of Grandmont, outside Limoges, began by at least 1167,[3] and Eleanor of England and her husband, Alfonso VIII of Castile, were patrons of Santo Domingo de Silos at the same time.[4] The importance of this patronage seems to be especially significant for Silos: although champlevé enameling on copper was an established tradition in south-central France from the early twelfth century to the fourteenth, there are no known antecedents or succedents of the enamels made for Silos in the twelfth century.

Gauthier[5] identifies "Maria" as the Magdalen, though it is curious that she should be given more prominence than the apostles by being placed at Christ's right side. More likely she represents the Virgin Mary, as on the chasse preserved at Bellac, where she likewise holds a palm as an attribute and is identified by inscription as "Holy Mary, Mother of the Lord."[6]

The representation of Saint Martial next to Christ on the roof of the chasse—in a position of greater importance than that of either Peter, prince of the apostles, or of the figure that probably represents Paul—is a function of both Limousin devotion and the original destination of the chasse. Saint Martial was the first bishop of Limoges, one of seven sent to Gaul in the mid-third century. By the time his vita was recorded by Gregory of Tours in the second half of the sixth century,[7] his relics were being venerated at Limoges.[8] Another vita was written in the ninth century, and by at least the late tenth century the relics of Martial had been dispersed.[9] By the eleventh century, it was

Principal face and right side

believed that Martial had been in the company of the original twelve apostles, baptized by Peter, witness to the Miracle of the Loaves and Fishes, and present at the Last Supper. By decree of a council convened at Limoges in 1031, Martial was declared to be of equal rank with the apostles,[10] a status purportedly confirmed by papal letter.[11] Only this status would entitle him to his assigned place on the Metropolitan chasse.

A drawing made by Auguste Bosvieux, archivist of the Creuse between 1851 and 1864,[12] establishes that this chasse once belonged to the church at Champagnat, located about 100 kilometers (62 miles) northeast of Limoges.

Consecrated in the ninth century, the church was among the earliest of the sanctuaries dedicated to the apostle of the Limousin.[13] The presence of relics of the saint there at that time probably was due to the fact that the church was the property of the counts of La Marche, whose importance in the region would explain the acquisition of Martial's relics.[14] With the chasse of Bellac, then, the chasse of Champagnat stands as one of the earliest preserved Limousin enamels: both trace to a church belonging to the counts of La Marche.

The Champagnat chasse is perhaps a generation later than the Bellac chasse, but it must be nearly contemporaneous with other enameled monuments that also reflect

Reverse and left side

devotion to Limousin bishop-saints, notably the chasse of
Saint Loup of 1158 and that of Saint Alpinien of 1160–74.[15]

BDB

NOTES

1. New York 1993, no. 134.

2. Ibid., no. 132.

3. With the gift of Matilda of England and her son King Henry II after the
consecration of the church at Grandmont. See Souchal 1963, pp. 136–40.

4. Férotin 1897, pp. 103–25.

5. Gauthier 1987, no. 132.

6. The palm appears as an attribute of the Virgin on a miniature of the
eleventh-century Sacramentary of Mont-Saint-Michel at the Pierpont
Morgan Library. See Alexander 1970, pp. 155–57, pl. 42a.

7. *Liber in gloria confessorum,* chap. 27, in Gregory of Tours 1885, pp.
764–65; *History of the Franks* I, chaps. 28 and 30, in Gregory of Tours
1963–65, p. 55.

8. Gauthier 1962b, pp. 205–48.

9. For a summary of the devotion to Saint Martial, see Paris 1993, pp.
94–98.

10. C. Lasteyrie 1901, pp. 74–80.

11. On the letter of apostolicity supposed to have been signed by Pope

John XIX, see Landes 1983.

12. *Notes de voyage* (Limoges, archives départementales, fonds Bosvieux 5F,
discovered by Marvin Ross). See Ross 1939.

13. Aubrun 1981, pp. 328–29.

14. After 1177, the properties of the counts of La Marche were purchased
by Henry II of England; see Souchal 1963, p. 138; Lecler 1907–9, vol.
4(1908), pp. 451–52.

15. Gauthier 1987, no. 98. Created for the priory of Saint-Martial at
Ruffec-en-Berry (Cher).

EXHIBITIONS: Paris 1900, no. 2408, pl. p. 77; New York 1954, no. 8, pl.
VII; New York 1970, no. 140, pp. 136–37.

BIBLIOGRAPHY: Molinier 1891, pp. 132–33; Hildburgh 1936, pp. 57–58,
77, 89, fig. 8b; Ross 1939, pp. 467–77; Lasko 1972, pp. 229–31, fig. 262;
Gaborit-Chopin 1983b, pp. 324–25 (ill.); Gauthier 1987, no. 132, pp.
123–24; Gauthier 1990, p. 386; New York 1993, no. 133, pp. 276–77.

# 11. Three Medallions Decorated with Birds

Silos(?), third quarter of the 12th century

Copper: gilt; champlevé enamel: lapis and lavender blue, emerald and meadow green, and garnet red

a: diam.: 5.9 cm (5⅚₆ in.); b and c: diam.: 5.67 cm (2¼ in.)

EX COLLS.: Augustin Lambert; [Altounian, Paris]; Adolphe Stoclet; Mme Féron-Stoclet (in 1956). Acquired in 1995.

CONDITION: The holes are probably late; wear to the gilding; a few losses to the enamel, especially on the fleuron of medallion a.

Musée du Louvre, Paris; Département des Objets d'art
(OA 11794–11796)

These elegantly drawn birds—each with a blossoming fleuron at the end of a scrolling tail that coils around the curves of a predatory head, a long neck, and a winged body—are among the most seductive inventions of the Romanesque imagination, combining fauna and flora in a single hybrid creation. Birds of similar design decorate the small plaques set into the border of the enameled front panel of the *Urna* of Silos (Burgos, Museo Arqueológico Provincial).[1] They are also seen on other isolated medallions.[2]

Moreover, the fleurons at the ends of these creatures' tails closely parallel those on the reverse of the chasse of Champagnat (cat. 10), compounding the problem of the attribution of the medallions. This problem cannot be dissociated from the ensemble of the *Urna* of Silos, linked by its style and decoration to the Romanesque art of the Plantagenet West[3] and to early Limousin enamelwork. As Boehm has noted, one of the elements of the solution certainly lies in the role played by the patronage of the Plantagenet princes.[4] Were it not for these small plaques, the complexity of the question would be increased by the discovery of medallions—similar to those from the former Stoclet Collection—in England, at two important sites of British illumination, Canterbury and Bury Saint Edmunds (today, respectively, in the Ashmolean Museum, Oxford, and in Moyse's Hall Museum, Bury Saint Edmunds).[5] Were these English medallions products of export, as Marie-Madeleine Gauthier has suggested? Did some serve as models for others? Or were such medallions and small plaques created by artisans working in various places that all benefited from Plantagenet patronage?

The piercings in the three medallions appear modern; they must have been set into a mounting such as those on the small plaques of the *Urna* of Silos and probably, on the comparable medallions at the Kunstgewerbemuseum, Cologne, and at Oxford, which also lack attachment holes. The comparison with the larger and earlier medallions decorating the coffret of Conques (cat. 7) or those, slightly smaller and later, of the coffret in the Metropolitan

Museum (cat. 36) allows us to suppose that these three medallions decorated such an object. The comparable medallion discovered at Bury Saint Edmunds is also pierced with a keyhole, which reinforces this hypothesis.

ET-D

a

b

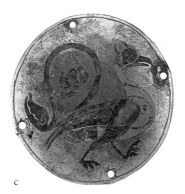

c

NOTES
1. Gauthier 1987, no. 80.
2. Ibid., nos. 84, 85, 87.
3. As, notably, Boehm has emphasized (New York 1993, no. 134).
4. Ibid., and above, cat. 10.
5. See Gauthier 1987, nos. 84, 85.

BIBLIOGRAPHY: Lion-Goldschmidt 1956, p. 170, ill. p. 171; Gauthier 1987, no. 89, p. 94, ill. 300, pl. LXXII; Boehm 1992, p. 157, fig. 12.

# 12. Book Cover: Christ in Majesty

Spain (Silos?) or the Limousin, third quarter of the 12th century

Copper: chased and gilt; cloisonné and champlevé enamel: midnight, saturated, ultramarine, azure, and lavender blue; turquoise, emerald and meadow green, yellow-green, blue-green, red, and white

23.6 x 13.6 cm (9⁵⁄₁₆ x 5⅜ in.)

INSCRIBED: A/ω (Alpha/omega)

EX COLL.: Frédéric Spitzer, Paris (before 1890); sale, Paris, April 17–June 16, 1890, no. 216, pl. VIII; acquired from the Spitzer sale in 1893.

CONDITION: Slight wear to gilding; some losses to enamel.

Musée National du Moyen Âge, Thermes de Cluny, Paris (Cl 13070)

This book cover, exceptional in its quality of workmanship and style, was unknown[1] until its publication in 1890 by Léon Palustre in the catalogue of the Spitzer Collection. Palustre commented on the rarity of the late-twelfth-century work in copper with both cloisonné and champlevé enameling. He recognized that this book cover was distinct, a fact that sheds considerable light on the independence and skill of the Aquitainian enamelers. He attributed it to the Limoges tradition. Since 1936 it has been linked with a plaque having the same format, dimensions, technique, and style, from the former Valencia Collection of Don Juan in Madrid,[2] whose earlier history is also unknown, but which, at one point, passed through the Maison André, Paris, where a plaster image of it is preserved. W. L. Hildburgh argued in favor of a Spanish origin for both plaques, which he associated with the "Silos group"—a reference to the enameled altar frontal at Silos. The plaques have been separated since the dismantling of the evangelical or sacramental liturgical manuscript to which they were originally affixed.

The plaque was first cut from a copper sheet that was three millimeters thick. In the absence of perforations around the periphery, one can assume that it was set in the center of the rectangular field of a book cover, forming the lower plaque, in a frame that was possibly enameled. Using the same metallic support in an exceptional manner, the two traditional techniques of medieval enamel work, cloisonné and champlevé,[3] were employed on one copper support with such virtuosity that at first it is difficult to distinguish the thin cloisons from the fine champlevé cells. A remarkable variety of hues and degrees of opacity characterize the palette, which includes five shades of blue, four of green, turquoise, red, and white. The solid areas of color are outlined with reserved gilt metal. The five enameled figures, with cast appliqué heads in half-relief, are placed against a gold ground that enhances the intensity of the palette. Centered in a mandorla that fills the plaque, Christ in

Majesty sits enthroned on the arc of a rainbow with the Greek letters alpha and omega at either side. (A similar alpha is engraved on the reverse of the Madrid plaque.) Christ raises his right hand in blessing, and in his left, he holds a turquoise book secured with red clasps. The book rests on his left knee, partly obscured by drapery; his feet rest on a turquoise footstool. Over his azure tunic, lined in white and edged in red, the Savior wears a flowing midnight-blue robe with meadow-green orphreys; an emerald green mantle completes the ensemble. Christ's large green halo is stamped with a red trefoil cross and seemingly studded with precious stones, enhancing the applied gilt head, the only part of the figure in relief. The symbols of the evangelists appear in the four corners: at the top left, the winged man of Saint Matthew raises a green book in draped hands; on the right, the eagle of Saint John holds a white scroll in his talons; at the bottom left is the lion of Saint Mark, and at the right, the ox of Saint Luke, turning and raising their heads toward the Savior. Their heads are haloed in green or turquoise, and multicolored wings support their blue, turquoise, and green bodies.

"Sublime" is the only fitting description of this image, a supernatural vision of the Almighty in the Majesty of His Second Coming, according to Marie-Madeleine Gauthier (1987), who sees this piece as the work of a consummate Silos engraver and sculptor of about 1165–70. In the famous abbey of Castile, the central panel of the monumental enameled frontal of the *Urna* of Saint Dominic, preserved at Burgos,[4] probably recently completed at the time, appears to have been a direct inspiration for the Spitzer plaque. Aside from its affinities with the Silos group, however, and due to subtle stylistic differences, Daniéle Gaborit-Chopin, who does not specifically favor either a Spanish or a Limousin origin, suggests, by extending the range of the dating to 1150–70, that the Spitzer plaque be considered as earlier than the *Urna* enamels. This book cover and its Madrid companion remain, in any case, the oldest existing examples of southern champlevé enamels in the category of liturgical articles—book covers in this case—having a format, style of execution, and iconographic scheme that would become the glory of the Limousin workshops at the end of the twelfth and the beginning of the thirteenth century.

GF

NOTES

1. According to M.-M. Gauthier, it is not impossible that one day it will be identified as the "oblong plaque on a gilded ground, Christ giving His Blessing," acquired for 262 francs by Jacob at the Fould sale on June 4, 1860, in Paris, lot 1710. (Research on this point provided by Françoise Arquié-Bruley).

2. Gauthier 1987, no. 133, ill. 369.

3. The simultaneous use of these two techniques is not unique; it is also found on certain early Romanesque enamels (see cats. 3, 54).

4. Gauthier 1987, no. 80, ills. 206, 209, 213, 211.

EXHIBITIONS: Utrecht 1939, no. 139; Paris 1947, no. 2; Limoges 1948, no. 6; Vatican City 1963, no. 13; New York 1970, no. 136.

BIBLIOGRAPHY: Palustre and Molinier 1890, no. 3, pp. 96–97, pl. III; Rupin 1890, p. 150, pl. XXL, fig. 225; Molinier 1891, p. 147, fig. 149; Huici and Juaristi 1929, p. 88, fig. 56; Hildburgh 1936, pp. 90–91, pl. XVII, fig. 23b; Gauthier 1972b, pp. 90–91, nos. 42, 43, fig. pp. 86–87; Gaborit-Chopin 1983b, p. 322, ill. 286; Martín Ansón 1984, pp. 74–75, fig.; Gauthier 1987, no. 134, p. 125, ills. 370, 371; New York 1993, no. 135a, b, pp. 280–81, ill.

# 13. Pendant Medallion

Limoges, ca. 1180–90

Copper: engraved, stamped, and gilt; champlevé enamel: traces of medium and light blue, light green, and red

Diam.: 6.6 cm (2¾ in.)

CONDITION: Losses to gilding; enamel largely lost.

The Metropolitan Museum of Art, New York; Gift of J. Pierpont Morgan, 1917 (17.190.787)

A pair of fantastic birds stands addorsed with gracefully curving necks and long exotic wings filling the outer area of the medallion. Elegant foliate decoration, growing from the birds' elongated, intertwined tails, fills the center. Delicate engraved lines and fine punches decorate the birds' bodies; the gilt ground is hatched, and a beaded border provides the frame.

The coffret of Abbot Boniface (cat. 7) bears witness to the early practice of pairing fantastic birds in medallions. The design of a number of medallions related to the Metropolitan example, especially one in the Kunstgewerbemuseum, Cologne,[1] recalls the designs of Spanish silks, but, in the case of the Metropolitan roundel, the birds' shapes are more elongated, more curvilinear, and the overall effect is more lyrical.

Similarly conceived creatures are frequently found in Limoges work and animate the cover of the Souvigny Bible (cat. 34) and the National Gallery chasse (cat. 22). The similarity of the birds on the Metropolitan medallion to those on a crosier found in a tomb at Silos[2] led Marie-Madeleine Gauthier to link the New York roundel to works found in Spain, but, given their portability, there is no compelling reason to argue that these objects were necessarily fabricated there.

The original use of this pendant medallion is unknown. A similarly decorated roundel, with attachment holes rather than a suspension loop, was excavated at Gimel-les-Cascades (Corrèze) in 1969. That example was found alongside a knight's spur,[3] which suggests that such enameled roundels were used not only to ornament coffrets but also as garniture for a horse, as documented in the fourteenth century.[4]

BDB

NOTES
1. Gauthier 1987, no. 143 bis.
2. Ibid., no. 139.
3. Ibid., no. 141.
4. Gay 1887–1928, vol. 1, pp. 318–19.

BIBLIOGRAPHY: Gauthier 1987, no. 142, p. 141, fig. 463.

# 14. Base of a Candlestick

Limoges, ca. 1180

Copper: engraved, scraped, and gilt; champlevé enamel: lapis and lavender blue, turquoise, medium green, red, and white

H.: 5.4 cm (2⅛ in.); Diam.: 8.5 cm (3⅜ in.)

Ex coll.: Alphonse Kann, Paris (until 1940)

Condition: Considerable loss to gilding; incidental losses to enamel.

Musée du Louvre, Paris; Bequest of Alphonse Kann, 1949 (after restitution) (OA 9485)

The hemispherical base is decorated with a continuous scrolling frieze of the fantastic birds and foliate decoration characteristic of Limousin enameled candlesticks and medallions of the late twelfth century. Each of the three feet is in the form of an animal's paw issuing from a lion's-head mask. The square opening at the top of the domed base originally served to hold a shaft, the means of threading the superimposed elements of the candlestick—the knop(s), bobeche, and pricket—together. The original ensemble was probably similar to such surviving examples as the single candlestick in the Bargello[1] or the one in the Bayerisches Nationalmuseum, Munich,[2] though neither forms a pair with the Louvre example. The type of bird depicted on these candlesticks, like those on the knop of the Cloisters candlesticks (cat. 33), recalls the fantastic winged creatures that decorate the borders of the *Urna* of Saint Dominic at Burgos[3] (fig. 10*a*) and, as a result, it has frequently been argued that such enamels were produced in Spain.[4] It is important to note, however, that creatures of the same or related species also decorate enamel objects excavated as far north as Bury Saint Edmund's and Canterbury[5] and as far east as the Holy Land.[6]

It is more reasonable to view such portable objects in the context of the international taste for "opus lemovicense," which was well established by the late twelfth century, and to consider that at least one Limousin workshop specialized in the production of candlesticks and medallions to meet that taste.

Pairs of Limoges candlesticks are preserved in churches and recorded in ecclesiastical inventories, but their use in a secular context need not be excluded.

BDB

Notes
1. Gauthier 1987, no. 71.
2. Ibid., no. 70 bis.
3. New York 1993, no. 134.
4. For a summary, see Gauthier 1957, pp. 146–67, and Boehm 1991, pp. 149, 162, n. 8.
5. Gauthier 1987, nos. 84, 85.
6. Ibid., no. 70.

Bibliography: Gauthier 1987, no. 88, pp. 93–94, fig. 299.

# 15. Effigy of Geoffrey Plantagenet

Le Mans(?), shortly after 1151

Copper: engraved, chased, and gilt; émail brun; champlevé enamel: lapis and lavender blue, turquoise, dark and medium green, semi-translucent dark green, yellow, golden yellow, pinkish white, and white

63.5 × 33.2 cm (25 × 13⅛ in.)

INSCRIBED: ENSE TVO PRINCEPS PREDONVM TVRBA FVGATVR / ECCLE[S]IIS Q [VE] QVIES PACE VIGENTE DATVR (By your sword, prince, the crowd of pillagers is put to flight / and through your vigilant peace, tranquillity is given to the churches)

PROVENANCE: From cathedral of Saint-Julien at Le Mans, where it was part of tomb of Geoffrey Plantagenet (identified by description in Jean de Marmoutier's *Histoire de Geoffroi le Bel*, written about 1170–80 [see Halphen and Poupardin 1913, p. 224]); probably separated from tomb when the cathedral was sacked in 1562; described several times in sixteenth and eighteenth centuries; placed with its back against the next-to-last pillar in nave; disappeared in 1792 (see Renouard 1811, p. 251) and discovered by the scholar Maulny; rediscovered in 1815 or 1816 and deposited in the Musée de Tessé at Le Mans (mentioned there for the first time in Stothard 1817).

CONDITION: Dented around all the edges. Significant wear in gilding. Losses to enamel, notably in lower right corner, on neck and right shoulder of figure, and on shield. Blade and hilt of sword have lost their revetment. Traces of violent blow on reverse, which pierced plaque at upper part of sword.

Musée de Tessé, Le Mans (Inv. 23-1)

The effigy has been recognized, since the fundamental study by Hucher (1860), as that of Geoffrey Plantagenet (1113–1151), count of Maine and Anjou, whose tomb inside the cathedral of Le Mans is mentioned in many texts. The first of these documents is the charter by which Geoffrey's son, Henry II, king of England, established an endowment for a daily Mass to be said "at the altar in front of my father's tomb."[1] It was in existence before 1158 and was probably drafted in 1155. The earliest version of the *Gestes des comtes d'Anjou*, compiled no doubt by Thomas of Loches (d. 1168), indicates that this tomb bore an epitaph.[2] Lastly, the *Histoire de Geoffroi le Bel* by Jean de Marmoutier, written about 1170–80 and dedicated to Bishop William of Passavant (d. 1187), specifies that the tomb included an image of the count "in gold and gems, in the pose of a prince who seems to want to strike down the pride of the mighty and show mercy to the humble."[3] It is difficult not to see in this description the enameled plaque, now in the Musée de Tessé.

Designated in the inscription by the prestigious title prince, the count of Maine and Anjou, dressed in ceremonial garb—long tunic, robe, mantle lined with vair, and conical helmet. He appears ready for combat, brandishing his sword and protected by a large shield decorated with lions.[4] His energetic gesture, the wide-open eyes, and the youthful face of this prince who died at the age of thirty-seven give the image a striking presence and rare touches of realism.

Geoffrey the Fair, the fifth count of Anjou to bear this first name but the first to be called Plantagenet, was the son of Fulk V the Young and Eremburge of Beaugency, countess of Maine, through whom Maine was joined to Anjou. In 1128 Geoffrey married Matilda, the daughter of King Henry I of England and the widow of the Holy Roman emperor Henry V. When his father became king of Jerusalem through his second marriage, to the daughter of King Baudouin II (1131), Geoffrey was put in charge of the Plantagenet domain. He then set out to conquer Normandy and England, his wife's patrimony, and succeeded in making himself master of Normandy (1135–44). The eldest of his three sons, Henry, acquired control of the duchy of Aquitaine through his marriage in 1152 to Eleanor of Aquitaine; and just two years later, Henry completed his father's mission by becoming king of England. Geoffrey Plantagenet is thus a key figure in medieval dynastic and territorial history, as is suggested by this exceptional effigy in the cathedral that he himself chose to be the site of his sepulchre and where his baptism and marriage had also taken place.

According to Jean de Marmoutier, the tomb was commissioned by Bishop William of Passavant. There is no reason to doubt this, especially because the inscription on the tomb itself makes clear reference to benefits bestowed on the church by the count.[5] It is nonetheless possible that contributions to the conception or financing of the tomb were made either by the count's widow or by his son Henry.

Moreover, it is unlikely that the tomb or the enameled effigy associated with it were conceived or executed long after the death of the count. On the contrary, as H. Landais (1976) has demonstrated, comparisons with works of western France point to an execution immediately following 1152. The tomb of Ulger, the bishop of Angers (r. 1142–53), was decorated in the center of the principal side with a monumental effigy—known through the surveys done for Gaignières—slightly smaller (48 × 30 cm [18⅞ × 11¹³/₁₆ in.]) but otherwise very similar to the Musée de Tessé enamel.[6] Furthermore, as we shall see, no aspect of the decoration or style of the latter compels us to consider a later date.[7] Because the tomb was in place about 1155, at the time of the above-mentioned endowment made by Henry II, the enamel was probably there also. At most, the consecration of the cathedral in 1158 can be taken as the ultimate completion date for the decoration of Geoffrey's mausoleum.

Two questions remain that are more difficult to resolve: where might the enamel have been placed on the tomb, and what are the technical, iconographical, and stylistic sources

of the enamel, not to mention the precise place of its execution? *Plainctes et doléances du chapitre du Mans* (Complaints and grievances of the Le Mans chapter), written after the 1562 sacking, appears to be the only document that describes the tomb: "A monument and sepulchre of dressed stone . . . very old and magnificent, finished in *franc d'espic*, on which there are three very old heads, one of which is marble. . . ."[8] This brief description leads one to imagine a sculpted tomb, perhaps inspired by monuments of antiquity and even reusing sculpted fragments as Gauthier (1987) has suggested—a concept not well suited to the insertion of a central enameled plaque. Descriptions from the seventeenth and eighteenth centuries specify that the enamel was attached to the penultimate pillar in the nave.[9] This placement, postdating the sacking, was no doubt a rearrangement, yet these texts suggest that the enameled plaque might have been originally placed above the tomb itself. This would mean that it is not really a funeral effigy but represents instead the figure in life attending the tomb. This hypothesis is strengthened by a comparison with what we know of the tombs of certain predecessors of Geoffrey the Fair. Indeed, the tombs (no longer extant) of Geoffrey Martel (d. 1060) and his great-grandnephew Geoffrey (d. 1106), the son of Fulk IV, situated in the church of Saint-Nicolas at Angers, were accompanied by "their effigies . . . painted in fresco above their tombs, armed, garbed in mail, with a drawn sword pointing upward in the right hand and a shield in the left, on which, as well as on their helmets . . . and their sleeves, are depicted their arms: Azure Escarbuncle Or a Chief Gules."[10] It is very likely, as Port has asserted, without, however, having seen the frescoes, which have also disappeared, that these were later than the tombs themselves; the shields decorated with an escarbuncle, fictive armorial bearings sometimes ascribed in the thirteenth and fourteenth centuries to people whose armorial bearings were not known or who did not carry them, allow one to think that they were executed in the Gothic age (see cat. 120).[11] The repetition of the armorial bearings on the clothing, on the other hand, suggests a date in the second half of the fourteenth century (oral communication from Michel Pastoureau). These texts nevertheless strengthen the idea that the Musée de Tessé effigy would have been placed above the tomb of Geoffrey the Fair.

These considerations may also be taken into account in an examination of the artistic context of the plaque. De Linas (1885b), Rupin (1890), and Guibert (1901) were the first to underscore the relationship between the Musée de Tessé plaque and Limousin enamelwork,[12] especially the plaques from the main altar of Grandmont (cat. 57). Most of these common characteristics—the arcades surmounted by bell turrets, the two colors of the architecture, and the long-limbed figure with its angular drapery, rounded in an ellipse on the abdomen—were widespread in the Romanesque art of the West from the beginning of the twelfth century.[13] On the other hand, Landais (1976) has emphasized the predominantly "Norman" character of the figure's costume and his stance full of life and energy, which seems the opposite of the disembodied classicism of the art of the Rhine and Meuse regions. In short, the oft-mentioned similarity to works from Anjou or Maine, such as the mid-twelfth-century stained glass devoted to saints Gervais and Protais in the cathedral of Le Mans,[14] confirms that the effigy of Geoffrey Plantagenet is properly placed amid the Romanesque art of western France. These parallels are much more convincing than comparisons that have been suggested with English works, such as the plaque in the British Museum, London, depicting Henry of Blois[15] or the frontispiece of the manuscript of Pliny the Younger now in the Bibliothèque du Mans, the style of which is profoundly different.[16]

Identifying the source of the techniques used to create the work is a more delicate matter. Laboratory study has indicated that the lions adorning the shield and the decorative motifs on the robe were created by applying to the copper émail brun (baked linseed oil) enhanced with gold in the engraved lines; it also indicated the probable use of gold foil under certain enamels, such as the dark green of the lower band of the helmet and the golden yellow of the beard and hair, while traces of gold at the bottom of the now empty cavities of the blade and hilt of the sword suggest that these elements were originally covered with a sheet of gold.[17] This diversity of goldsmithing techniques accompanies a perfect mastery of the art of enameling, manifested in the brilliant and rich range of colors, in the skillful juxtaposition of different shades within a single cell—as many as five gradations in the bell turrets on top of the arcades—and in certain effects that were new at that date, such as the pebbled appearance of the colors on certain cloths and columns. Many authors have supposed that the goldsmith learned his trade from Mosan artists or, at least, was inspired by their works.[18] One must not, however, overestimate the similarities that exist between the Musée de Tessé plaque and Mosan enamels. Thus the borders in which tones shade off all along the cells, often cited as examples, are actually done differently here, the edge obtained being finer and more regular, very much like those in later works of the Grandmont workshop (see cat. 57). The coloration of the faces, using white or flesh-colored enamel, can be observed on two Limousin works close in date to the Musée de Tessé plaque: the chasses of Bellac and Champagnat (cats. 9, 10). These two works—especially the latter—present a range of color completely comparable to that of the Musée de Tessé plaque. While certainly more limited, the Champagnat chasse is similarly dominated by the brilliance of lapis blue and dark green in expanses of solid color, enhanced with turquoise, white, and red. The Musée de Tessé plaque adds to these an intense yellow, which is also present in the medallions of the Conques coffret (cat. 7), as well as the new effects—principally the golden yellow of the beard and the pebbly touches. The diversity of procedures used in making this exceptional enamel indicates that the artist brought to bear a wide range of technical experience, yet the work appears principally nourished by the traditions of Aquitaine and the Loire valley. Anjou or Le Mans seems, therefore, to be the most probable site of its execution. Unique in conception and in the richness of the experiences assimilated for its creation, the enamel is also an important link in the genesis of southern French enamelwork and the understanding of the artistic culture nourishing the artists who, two or three decades later, would have a hand in the prodigious achievements of the Limousin workshops.[19]

ET-D

NOTES

1. "Ad altare illud, quod est ante sepulchrum patris mei," cited in Delisle and Berger 1916, pp. 172–73.
2. "Mausoleum . . . cum epitaphio," *Gesta consulum Andegavorum*, in Halphen and Poupardin 1913, pp. 71–73. For the date and the authors of the redaction of this chronicle, see pp. vii–lxxii.
3. "Effigiati comitis reverenda imago ex auro et lapidibus de center impressa, superbis ruinam, humilibus gratiam distribuere videtur," *Historia Gaufredi ducis Normannorum et comitis Andegavorum*, in Halphen and Poupardin 1913, p. 224. For the date of this drafting, see pp. lxxxiv–lxxxv.
4. For the importance of this shield in the history of heraldry, see Pastoureau, below, p. 39.
5. Nikitine (1981) emphasized that this inscription could just as easily be a direct or indirect allusion to the abrogation by Geoffrey, under the preceding bishop, Hugh of Saint-Calais (r. 1135–43), of the custom according to which the bishop's possessions were seized after his death.
6. See Gauthier 1987, no. 103.
7. The comparison with Limousin enamels has at times caused the Musée de Tessé enamel to be dated later (see Gauthier 1979). But, as de Linas (1885b) has already suggested, the evident traces of a common culture are quite suited to a date some twenty to thirty years earlier.
8. Text formerly kept in the archives of the chapter, cathedral of Le Mans, cited in Ledru 1903, p. 208.
9. Le Corvaisier de Courteilles 1648, p. 445; Le Paige 1777, p. 160. *Plainctes et doléances du chapitre du Mans,* of 1562, had already located the tomb, identified as that of an "English lord," against a pillar "between the two last altars . . . toward the nave" (cited in Ledru 1903, p. 208).
10. "Leurs effigies [. . .] peintes à fresques au dessus de leurs tombeaux, armés, maillés, en main droite l'espee nüe la pointe en haut et en la gauche l'escu, sur lequel comme aussy sur leurs easaques d'armes [. . .]

et sur les manches [. . .] sont leurs armes représentées d'azur au chef de gueules à une escarboucle d'or." Dubuisson-Aubenay, *Mémoires historiques du pays d'Anjou et duché de Beaufort*, manuscript from the first half of the seventeenth century, fol. 83v. Duplessis-Villoutreys Collection, library of the Université catholique de l'Ouest, Angers. This document and this comparison were made known to me by Beatrice de Chancel-Bardelot, whom I warmly thank.
11. "To the left of the main altar . . . there still existed in the seventeenth century a small stone tomb . . . which people said was that of Geoffrey Martel, who was represented by more modern paintings on the wall" (Port, note in Péan de la Tuilerie 1776 and 1869 p. 461). Similar representations of Fulk and Geoffrey Martel were painted in the cloister of the same abbey of Saint-Nicolas (see Bruneau de Tartifume 1632 and 1977, vol. 1, p. 5).
12. Labarte (1856) and Verneilh (1863) have attributed it to the Rhineland.
13. For example, in the Bible de Saint-Aubin d'Angers (Angers, Bibliothèque Municipale, MS. 4 and 5) and Vie de Saint Aubin (Paris, BNF, MS. nouv. acq. lat. 1390). See Porcher 1959a, pp. 179–219, pls. 3–5, and pp. 194–97. See also cat. 76.
14. See Hucher 1860 and Hucher 1878b; Nikitine 1981; for the stained-glass windows, see Grodecki 1977, pp. 60–62, fig. 50, and Granboulan 1994.
15. See Landais 1976; Gaborit-Chopin 1983b. The plaque has recently been attributed to a Mosan artist, probably working in England (Stratford 1993, nos. 1, 2).
16. Comparison suggested in Cordonnier 1961 and often repeated afterward (e.g., Nikitine 1981, Gaborit-Chopin 1983b). For the attribution to England, see Avril 1983, p. 214 ("student of the Master of the Shaftesbury Psalter").
17. See Gauthier 1979; Gaborit-Chopin and Lahanier 1982.
18. See especially Gauthier 1979, p. 130.
19. The work has enjoyed great fame since the beginning of the nineteenth century and has been, notably, the subject of many painted or engraved reproductions, a list of which, drawn up by the authors, appears in Gauthier 1987, pp. 113–15, nos. 108–10.

EXHIBITIONS: Paris 1867, no. 2050; Paris 1900, no. 2515, pl. 81; Paris 1937, no. 1201; Limoges 1948, no. 7; Paris 1957, no. 107; Paris 1964, no. 133; Bordeaux 1971, no. 41; London 1978, no. 1; Paris 1980b, no. 35.

BIBLIOGRAPHY: Trouillart 1643, pp. 106–7; Bernard de Montfaucon, vol. 2 (1730), pp. 71–72, pl. XII; Le Paige 1777, p. 160; Renouard 1811, vol. 1, pp. 251–52; Stothard 1817, pp. 2–3, color pls.; Lenoir 1821, vol. 7, p. 83, pl. 237; Dussieux 1841, p. 57; Planché 1846; A. Du Sommerard 1838–46, p. 190, and *Album*, 10th series, pl. XII; Texier 1851, pp. 144–45; L. Laborde 1852, pp. 36–37; Labarte 1856, pp. 199–206; Hucher 1860; de Verneilh 1863, pp. 27–30; Dugasseau 1864, pp. 14–16; Viollet-le-Duc 1871–72, pp. 216–20, pl. XLI, fig. 22; Port 1874–78, p. 255 (1874); Hucher 1878a; Hucher 1878b; Molinier 1881, pp. 90–95; Linas 1885b, pp. 105–12; Guibert 1888b, p. 206; Molinier 1889, p. 21; Rupin 1890, pp. 89–96, fig. 156, pl. 13; Molinier 1891, pp. 135–40, fig. p. 137; Guibert 1901, pp. 19–20; Le Feuvre and Alexandre 1932, pp. 1–2, no. 1; Gauthier 1950, pp. 28–66; Cordonnier 1961; Gauthier 1967b, pp. 146–47; Lasko 1972, pp. 231–32, pl. 270; Gauthier 1972b, no. 40, 327; Landais 1976, pp. 114–20, pl. 1; Gauthier 1979; Pastoureau 1979, pp. 29–30, fig. 9; Nikitine 1981; Gaborit-Chopin and Lahanier 1982; Gaborit-Chopin 1983b, p. 292, fig. 257; Deyres 1985, pp. 214–19, pl. p. 229; Gauthier 1987, pp. 109–14, no. 108, ills. 333–36, 340–44, pls. LXXXVI, LXXXVII, LXXXIX.

# II

*Opus lemovicense* and the Creation of a European Taste

1160–1190

The extraordinary flowering of Romanesque enamelwork in Limoges during the years 1160–1200 is reflected in the imposing number of surviving examples, inventive in their imagery and decoration. Within this rich body of works, the collections of the Louvre and the Metropolitan Museum not only are of great quality and importance but also are complementary. Hence the various aspects of the group of pieces with vermiculé grounds are evoked, thanks, in particular, to the narrative chasse of Saint Martial of Limoges from the Louvre (cat. 17) and the Christological ones from the Metropolitan Museum (cats. 24, 27, 28), as well as the very beautiful group with secular themes formed by the Louvre medallions (cat. 34), and the coffret and candlesticks from the Metropolitan Museum (cats. 36, 33). Augmenting these are select works from other collections that are particularly important for their use, iconography, or style, as well as for the quality of the metalwork and the richness of the palette, such as the Gimel chasse (cat. 16)—one of the most extraordinary and very earliest of the chasses with vermiculé grounds; the book covers from the Musée de Lyons (cat. 18)—probably the oldest extant pair; the chasse of Saint Valerie from the State Hermitage Museum, Saint Petersburg (cat. 20)—undeniably the most beautiful of those devoted to this Limousin saint by the enamelers of Limoges; and the chasses from Apt and Washington (cats. 21, 22), which are extraordinary as much for their iconography as for the style of their principal faces and the decoration of ornamental scrolls and fantastic birds on their reverse faces. With scenes from the lives of Limousin saints and universal iconography such objects illustrate the local success of Limousin production as well as its broad diffusion outside the area. Moreover, the aristocratic flavor of certain works probably reflects the presence of the Plantagenets in the Limousin and, perhaps,

the role played by their patronage in the creation and diffusion of such masterpieces.

The chasse of Saint Thomas Becket, recently acquired by the Louvre (cat. 39), and the reliquary casket of the True Cross from the treasury of Saint-Sernin of Toulouse (cat. 40)—likewise examples of the expansion of the clientele of Limousin workshops—are also among the earliest examples of a new technical variation—the grounds are enameled while the figures are in reserve and engraved or formed of relief appliqués—used to illustrate new iconographic themes. But their style is still dominated by the animation and stylization associated with Romanesque art and observed on the pieces with vermiculé grounds. Other works, chasses or book covers, for example (cats. 42–44), although less innovative in their iconography, present the same technical and stylistic characteristics, conforming to a canon of Limoges work that was recognized and admired.

The chasse from Mozac (cat. 45), the largest extant Limousin chasse and the best example of the diffusion of Limousin production in the Auvergne, also demonstrates the capacity of Limousin artists to translate particular hagiographical scenes into copper and enamel and confirms their extraordinary narrative sensibility and innovative design. It is also one of the earliest surviving works decorated with appliqués in half relief. The images of Christ in the Louvre closely allied to imposing crosses preserved in Sweden (see figs. 49a, 50a) and the fragments of the ensembles executed for Spain (cat. 50) or Italy (cat. 52) bear witness to the immediate reception given this type of object and the international success of Limousin production before the end of the twelfth century.

BDB and ET-D

*1*

---

*The first patrons and the vermiculé style*

# 16. Chasse of Saint Stephen

Limoges, ca. 1160–70

Copper: engraved, chased, and gilt; champlevé enamel: dark, lapis, medium lavender, and sky blue; turquoise, green, yellow, brick red, translucent wine red, and white; enameled beads; wood core

25.4 x 28.8 x 11.9 cm (10 x 11⅜ x 4¹¹⁄₁₆ in.)

INSCRIBED: IO X H ∽ O Λ (on the phylactery presented by Saint Stephen) for VOX IHESUS (the voice of Jesus)?

PROVENANCE: Originally from the church of Saint-Étienne in Braguse (Corrèze); preserved in the church of Saint-Pardoux, Gimel, probably since the second half of the seventeenth century.[1] Classified, Monuments Historiques, 1891. Stolen in 1991; restored in 1993.

CONDITION: According to de Linas (1883, p. 30): "The Gimel chasse has endured only the loss of the lateral knobs on its crest and a half square centimeter of enamel at the bottom of the robe of the angel on the gable." Today we must also add a loss in the halo of Saint Stephen on the roof, following the theft of 1991.

Church of Saint-Pardoux, Gimel-les-Cascades (Corrèze)

First noted by Poulbrière in 1875, the chasse of Gimel is rightfully thought to be one of the most important vermiculé Limousin works. Nevertheless, it has elicited highly diverse judgments. De Linas (1883) placed it in the first rank of Limousin pieces, and Guibert (1901) considered it a "veritable jewel." On the other hand, it was disparaged by Marquet de Vasselot (1905), who judged its figures as uniquely heavy and uncommonly vulgar.

Indeed, the chasse combines markedly different qualities: a complex iconography, an exceptionally varied palette, and a lively and expressive style. On the principal face, the main episodes of the life of Saint Stephen, patron saint of the church for which it was made, are portrayed as they were related in the Acts of the Apostles (6–7): below, left: Saint Stephen preaching; below, right: Saint Stephen is arrested and led outside the city of Jerusalem. On the roof: Saint Stephen is stoned. The scenes faithfully illustrate certain specific passages in the text; thus, during the sermon, the haloed Christ appears as the bystanders cover their ears. "He then saw the glory of God. . . . Emitting loud shrieks, they covered their ears" (7:55–57). The text explains that during the stoning, "the witnesses placed their garments at the feet of a young man named Saul" (7:58–59); Saul, the only person sitting, is second from the left.

On the lower face of the reverse, four apostles stand under arcades; only Saint Peter, holding his key, can be identified with certainty. On the roof, above the apostles, three half-length angels appear within circles. On the left gable, an angel stands under the keyhole, and on the right gable, an apostle holds a scroll in one hand and a staff surmounted with a cross in the other.

The iconography is very likely related to the relics for which the chasse was made, for it contained a group of relics preserved with the chasse at the church of Saint-Pardoux. Chief among them is, according to tradition, a drop of Saint Stephen's blood contained in a medieval glass vial. Various other fragments, accompanied by authentications, were also kept at the church. The earliest of these authenticating documents, probably written in the thirteenth century, mentions the relics of Saints Peter, Paul, Andrew, Martial, "and many others."[2] The first four saints might be those depicted on the reverse. The figure holding a staff with a cruciform top is probably the apostle Andrew.

The figures all express the same animated style, seen most notably in the dancing postures and demonstrative gestures. Beginning in western and central France about 1100, this style blossomed fully in a group of vermiculé chasses with narrative themes (see cats. 17, 20). The complex Gimel chasse is certainly the most "baroque" example, given its large number of figures (ten in the stoning scene alone) and its skillful execution of strong anecdotal themes and numerous decorative details, including the buildings with bicolored stones that seem protuberant; the doors with hinges, lock plates, and decorative hardware; the marbleized columns; the florets in the halos; the dotted lozenge patterns on the vestments; and the vigorous gestures of the onlookers as they stop their ears or of the executioners hurling stones. Each half-relief appliqué head is distinct; the expressions on the engraved faces vary, yet they all reveal a somewhat crude treatment, which, together with the violent gestures of the agitated bodies, lends the ensemble a "popular" flavor that is the very opposite of the aristocratic tone of other works (see cats. 19, 20, 22), which probably explains the negative opinion of Marquet de Vasselot.[3]

The particular interest in decoration and narrative seen here is also expressed in the highly colorful and varied palette. The range of blues is exceptional. A very pale sky blue enriches the borders of the reverse of the roof and of the rotulus held by the third apostle. The translucent wine red was rarely used (see cat. 20); it was obtained by placing a colorless flux on the copper. The quality and diversity of the metalwork rivals the enameling. The ridges of the cells are unusually thin, and the vermiculé grounds are interrupted by other motifs: imbrications that evoke grassy hillocks or decorative checkerboard patterns. The eyes in the appliqué heads are enameled blue beads.

The Gimel chasse is one of the few enameled works from Limoges in which we can identify the direct or indirect influence of the illuminators who were active about 1100. As recognized by Gauthier (1987, p. 96), the architecture and

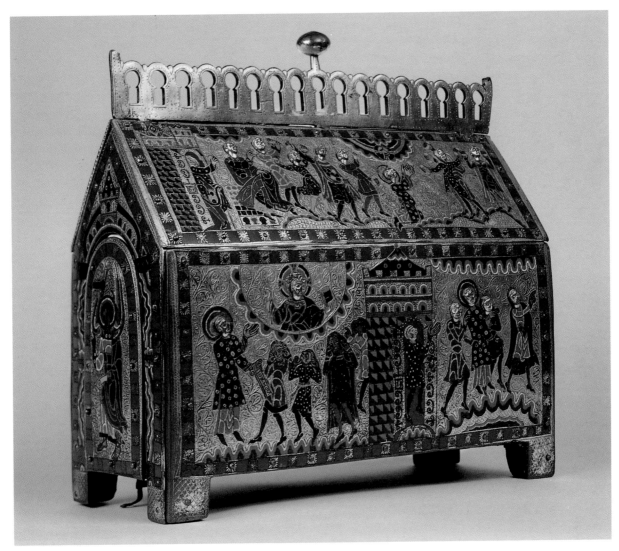

Principal face and left side

Reverse

Fig. 16*a*. The letter S: beheading and entombment of Saint Julian. Lives of the Saints. Bibliothèque Nationale de France, Paris, ms. lat. 5365, fol. 73

especially the composition of the stoning scene (at least the group formed by Saint Stephen and the two executioners at his right) seem to derive from the Sacramentary of Saint-Étienne of Limoges.[4] Those illuminations or other works deriving from them may have inspired other stylistic aspects here, such as the powerful twisting of bodies, the distorted position of heads, or the stylizing of the faces, especially profiles with a three-quarter view of an eye and a protruding nose.

A few manuscripts illuminated in Limoges or western France toward the middle of the twelfth century show how this style, from the beginning of the twelfth century, gradually developed until, at a later date, it directly influenced this chasse. The illustration (fig. 16*a*) of a Limousin manuscript of the lives of the saints[5] contains certain pertinent features: the faces are akin to those engraved on the Gimel chasse; small, animated figures in silhouette wear garments that, as though agitated by the wind, float upward in a bell-like shape, very much like the vestments of the apostles on the reverse of the chasse. A book of legends, probably executed about 1150,[6] includes distorted profiles that are in every way comparable to those in the scenes of the martyrdom of Saint Stephen.

The delight in decorative details that we see here is obvious in other works produced in the Plantagenet west during the third quarter of the twelfth century. The bicolored effects seen in the building stones had already been employed by about 1100, for example, in the Sacramentary of Saint-Étienne of Limoges (fol. 76v). They recur in the Visitation fresco at Tavant and in the funerary plaque of Geoffrey Plantagenet (cat. 15). The doors with decorative fittings recall those in the Life of Saint Peter, a stained-glass window at the cathedral of Poitiers, and the architecture is similar to that of Saint Peter Delivered from His Bonds, stained glass at the cathedral of Le Mans of about 1165–70.[7]

Nevertheless, the style seen here is distinguished from that of earlier or contemporaneous works produced in central and western France by the abstractly stylized drapery, defined by short, curving lines that seem to confine the bodies rather than hint at their shapes, the final manifestation of western Romanesque style.

These observations allow us to view the chasse of Gimel as one of the oldest extant works with vermiculé grounds known to us—without, however, assigning to it a date prior to the years 1160–70,[8] which seems to be in keeping with the comparisons with other Limousin works (cats. 20, 21).

ET-D

NOTES

1. The entire treasury was moved to Gimel. Saint-Étienne, constructed in the twelfth century, was, if we are to believe Poulbrière (1894), "so thoroughly abandoned by 1707 that Mass had not been celebrated there for quite a while, and it was so ruinous by 1732 that the bishop of Limoges placed it under an interdict." On the other hand, according to Abbé Nadaud, as quoted by Aubrun (1981, p. 243), "babies were still baptized there in 1636, in preference to other parishes in the area."

2. "He sunt reliquie s(an)c(t)i Petri, s(an)c(t)i Pauli, s(an)c(t)i Andree, sancti Marcialis, Beate Marie Virginis, Beate Marie Macdalene, s(an)c(t)i Laurentii, s(an)c(t)i Vincentii, s(an)c(t)i Marcelli, s(an)c(t)i Aniani, s(an)c(t)i Pardulphi, s(an)c(t)i Antonini, s(an)c(t)i Taulani [for Taurini?], et multe alie" (Here are the relics of Saint Peter, Saint Paul, Saint Andrew, Saint Martial, the Blessed Virgin Mary, the Blessed Mary Magdalen, Saint Lawrence, Saint Vincent, Saint Marcellus, Saint Aignan, Saint Pardoux, Saint Anthony, Saint Taurin [?], and many more). The other authentications, which list the relics of Saint John, the Virgin's milk, a stone from Calvary, and one from the column of the Flagellation, come from a later time (two of these documents are in Latin, two in French, fifteenth century?).

3. The chasse of Malval (Musée de Guéret; see Gauthier 1987, no. 178), also dedicated to Saint Stephen, but executed with less skill, was part of the same trend.

4. Paris, BNF, ms. lat. 9438, illuminated between 1095 and 1105, fol. 20v; see Gaborit-Chopin 1969, pp. 127, 211.

5. Paris, BNF, ms. lat. 5365, first half of the twelfth century; see Gaborit-Chopin 1969, p. 208.

6. Paris, BNF, ms. lat. 5323; see Burin 1985.

7. Tavant: see Demus 1970, pp. 140–41 (ca. 1150), pl. LVI; stained-glass windows of Poitiers and Le Mans, see Grodecki 1977, fig. 59, and no. 74, p. 287; Granboulan 1994.

8. Not very far from the date proposed by Gauthier in 1987: ca. 1170–75.

EXHIBITIONS: Tulle 1887, no. 12; Paris 1889, no. 455; Paris 1937, no. 1202, pl. CXLIV; Limoges 1948, no. 13; Vatican City 1963, no. 15, pl. XIII.

BIBLIOGRAPHY: Poulbrière 1875, pp. 536–40, pls. II–IV; Linas 1883, pp. 105–7, 126–31; Molinier 1887a, no. 12, pp. 493–94; Guibert 1887, pp. 420–22; Rupin 1890, pp. 381–84; Molinier 1891, pp. 155–57, fig. p. 156; Poulbrière 1894; Guibert 1901, pp. 5–7; Marquet de Vasselot 1905, pp. 232–33, pl. XIV; Forot 1913, p. 187; Fayolle 1921, pp. 337–39; Forot 1924, pp. 247–54; Hildburgh 1936, p. 121; Marquet de Vasselot 1941, p. 129; Gauthier 1950, pp. 30–152, pl. 9; Gauthier 1960a, pp. 286–87, pls. 269–73; Gauthier 1966, pp. 940–50; Taralon 1966, pp. 288–89, ills. pp. 186–88; Gauthier 1967b, p. 152; Fillitz 1969, p. 256, ill. 342b; Landais 1976, p. 126, fig. 3; Favreau et al. 1978, vol. 2 no. 34, p. 45; Leniaud 1983, p. 380; Gauthier 1987, no. 90, pp. 94–97, ills. 301–5, pls. LXXIII–LXXV.

# 17. Chasse of Saint Martial

Limoges, ca. 1165–75

Copper: engraved, chased, stippled, and gilt; champlevé enamel: black, lapis and lavender blue, turquoise, light green, yellow, brick red, and white; wood core

12.5 x 16.5 x 6.7 cm (4¹⁵⁄₁₆ x 6½ x 2⅝ in.)

Ex colls.: Melchior de Saint-Rémy, Villefranche-de-Rouergue (in 1863); Mme A. Cibiel, Villefranche-de-Rouergue (before 1889); Victor Martin Le Roy, Paris (before 1900). Gift of Victor Martin Le Roy, 1914

Condition: Wear to gilding; bottom plank missing; the lower parts of the feet and their copper revetment have been filed.

Musée du Louvre, Paris; Département des Objets d'art (OA 8101)

This chasse is almost certainly the one described at the Archaeological Congress of Rodez in 1863, when it belonged to Melchior de Saint-Rémy. It was exhibited in Paris in 1889, then mentioned by Rupin (1890), by which date it was in the Cibiel Collection at Villefranche-de-Rouergue.[1] After it entered the Martin Le Roy Collection, it became particularly famous because of Marquet de Vasselot's publications (1905, 1906a). He was the first to group together the enamels with vermiculé grounds and to compare this chasse with other examples illustrated with scenes from the life of Christ or the lives of the saints.

Of all the extant chasses from the Limoges workshops, this is the only one dedicated to Saint Martial, who was ranked as an apostle and who served as the first bishop of Limoges.[2] The chasse also provides one of the oldest illustrations of the saint's life. The iconography is sometimes hard to interpret, especially since it does not adhere strictly to the Life of Saint Martial.[3]

On the principal face, Saint Martial, followed by a haloed companion, blesses a kneeling girl who is presented by a veiled woman and a man. This scene has usually been identified as the blessing, or the baptism, of Saint Valerie, accompanied by her mother, Suzanne. Recently, Marie-Madeleine Gauthier (1987) proposed that this scene might rather be a portrayal of the healing and resurrection of the daughter of Arnoul, a rich man; the other two figures would then be the girl's parents. But the figure in the next scene has always been regarded as Stephen, duke of Aquitaine—called Tève the duke—ordering the execution of Valerie. Thus the iconography of this face is more coherent if the first scene is read as the blessing of Saint Valerie.

On the roof, Saint Martial, using the long staff he received from Saint Peter, touches two mitered ecclesiastics. This is probably the resurrection of Martial's first two companions, Alpinian and Austriclinian (Leningrad-Moscow 1980; Gauthier 1987). However, although some texts men-

tion the resurrection of Austriclinian, they never include Alpinian. On the other hand, a later episode does refer to the resurrection of two people: while Saint Martial was preaching in Limoges, Aurelian and Andrew were among the pagans who had taken refuge in the temple of the idols; struck and killed by lightning, they were resuscitated by Saint Martial and converted to Christianity. According to the *Vita prolixior*, Aurelian succeeded Saint Martial as bishop of Limoges and Andrew was placed at the head of the church where Martial was buried[4]—which virtually makes him a predecessor of the abbots of Saint-Martial.[5] Thus the two mitered figures who are being resurrected would most likely be Aurelian and Andrew.

The iconography of the reverse is easier to read and has therefore never given rise to divergent interpretations. The vertical lower part shows the healing of a possessed man laden with chains; the deliverance here is, as often in the Middle Ages, depicted literally: a small demon with a human body emerges from the possessed man's mouth. The saint is next led to his torment, chained and accompanied by an executioner who strikes him with a club. The burial depicted on the roof is very probably that of Saint Martial himself, attended by Aurelian, his successor.[6] On each end an angel holding a censer attends the scene and adds to its solemnity.

Of major importance because of its iconography, dedicated to the first bishop and patron saint of Limoges, this chasse occupies a primary place equally for its stylistic qualities. The liveliness of the gestures and postures and the variety of the color effects combine with the painstaking care employed in the drawing of the engraved faces to differentiate characters and expressions. This is typical of the narrative and decorative experiments found in all Limousin works with vermiculé grounds. Several aspects reveal the quality of the execution: all the reserved copper lines are emphasized with stippling; the contrasting colors are greatly varied in shadings along the lines in reserve and on the dotted pattern sprinkling the vestments and orphreys. The faces are quite individualized: Saint Martial's bears the imprint of a good nature, which sharply distinguishes him from the other male figures. The quality of the two beautiful angels adorning the sides equals, at the very least, that of the other figures.

But then, unlike most works with narrative subjects, but like the chasse from Nantouillet relating the life of Christ (preserved at the Musée de Meaux),[7] the chasse of Saint Martial lacks any depiction of architectural framework. All in all, the affinity of each chasse to the other is very close, especially in the treatment of faces; the face of "Tève the

Principal face and left side

Fig. 17 a. Risen Christ appearing to the Apostles, detail of the chasse of Nantouillet. Musée Municipal, Meaux

duke" on the Louvre chasse is very similar to the one from Nantouillet, that of the resurrected Christ appearing to the apostles (fig. 17 a). Undoubtedly, these two works come from the same hand.

As has been repeatedly emphasized[8]—and here also in regard to the Gimel chasse (cat. 16)—this style obviously draws on the art of the Plantagenet west. However, like those on the Gimel chasse, the enamels here are distinguished by an accentuated stylization of the vestments with their fragmented lines; there is no "illusionist" representation of their actual appearance. The faces have regular contours, calm expressions, and fine features. The only exception is the hirsute, grimacing figure of the possessed man, who conforms to a stereotype widespread in western and central France during the twelfth century. This treatment, perhaps inspired by the art of areas north of the

Loire, is found in central and western France, from about the middle to the end of the twelfth century. Examples can be seen on the frescoes of the Legend of Saint Gilles at Saint-Aignan (Loir-et-Cher),[9] where the two angels flanking the Paschal Lamb are comparable to those on the gables of the chasse of Saint Martial and on the stained glass around the central window of the cathedral of Poitiers—probably a gift from the Plantagenet king Henry II and Eleanor of Aquitaine in about 1165–70,[10] which allows us to assign a date of about 1165–75 for the execution of this chasse.[11]

ET-D

NOTES

1. In an easily explained error, Marquet de Vasselot wrote in the catalogue of the Martin Le Roy Collection (1906a) that the chasse "originally came from Villefranche (Rhône)."

2. An ecclesiastical report of 1775 indicates that the relics of Saint Martial were preserved in the church bearing his name in Limoges as well as in

Reverse and right side

churches at Ahun, Toulx-Sainte-Croix, and Brive (see Lecler 1903, p. 31).

3. On the different texts of this Life, see, most recently, Aubrun 1981, pp. 73–82; Landes and Paupert 1991; Limoges 1995, pp. 65–67; Barrière, above.

4. Chap. XIV. See Birch, n.p., p. 31. "Beatissimus itaque Martialis Aurelianum benedicens ordinavit eum, ac praefecit post suum discessum urbi Lemovicensium. Andream vero prebiterum praefecit ecclesiae qua ipse sepultus est" (The blessed Martial, blessing Aurelian, ordained him and after his departure placed him at the head of the city of Limoges. On the other hand, it was the priest Andrew whom he put at the head of the church in which he himself [Martial] was buried). Most of the ancient episcopal rosters put Aurelian right after Martial. Only one list, in a manuscript belonging to Adémar de Chabannes, inserts the names of Alpinian and Austriclinian between the two others. See Aubrun 1981, pp. 87–91.

5. We know that originally the church of Saint-Pierre-du-Sépulcre was built in the sixth century on the tombs of Saint Martial and his companions and that it was eventually subsumed during the construction of the abbey of Saint-Martial (see, especially, Gauthier, Perrier, and Blanchon 1961).

6. There is really no reason to read this scene as the burial of Saint Valerie or of her mother, Suzanne, by Saint Martial (Gauthier 1987, p. 163), especially since the scenes linked to Saint Valerie are on the other face and Saint Martial is never shown here with a halo.

7. See Gauthier 1987, no. 177.

8. See, particularly, Gauthier 1972b, p. 98; Landais 1976.

9. See Demus 1970: second half of the twelfth century.

10. See Grodecki 1977, pp. 72–73, 286 (no. 73); Granboulan 1994.

11. Here again the suggestion is consistent with others that have been recently made: ca. 1170–80 (Leningrad–Moscow 1980) or ca. 1165–75 (Gauthier 1987).

EXHIBITIONS: Paris 1889, no. 459; Paris 1900, no. 2443; Vatican City 1963, no. 21; Leningrad–Moscow 1980, no. 30.

BIBLIOGRAPHY: Castelnau d'Essenault 1864, p. 157; Rupin 1890, p. 426; Molinier and Marcou 1900, p. 88; Molinier 1901, p. 189; Migeon 1902, p. 19, fig. 2; Marquet de Vasselot 1905, p. 245, pl. XVI; Marquet de Vasselot 1906a, pp. 35–36, no. 22, pl. XV; Leroux 1911, p. 366; Mâle 1922, pp. 195–99, fig. 145; Duthuit 1929, p. 12, fig. 10; Huyghe 1929, pp. 3–4, ill.; Gauthier 1950, pp. 74, 77–78; Gauthier 1955, pp. 54–56; Gauthier 1958b, pp. 73–74; Gauthier 1966, pp. 943, 951; Landais 1976, p. 126, fig. 10; Gauthier 1987, no. 175, pp. 162–65, frontispiece, pl. K, ills. 1, 2, 18, 19.

# 18. Book-Cover Plaques

Limoges, ca. 1170–85

Copper: engraved, chased, stippled, and gilt; champlevé enamel: lapis, medium, and lavender blue; turquoise, green, yellow, brick red, garnet red, and white

21 x 13.5 x .3 cm (8¼ x 5⁵⁄₁₆ x ⅛ in.)

INSCRIBED: CRUCIFIXION: IH[ESU]S / XPS (Christus); CHRIST IN MAJESTY: A/ω/PA[X] E[TE]RNA / EGO SO (for *sum*) (I am eternal Peace)

EX COLLS.: [M. Roche, Lyons, 1884)]; M. de Chabons, Lyons

CONDITION: Wear to gilding; a tear in the metal at each of the four corners of the plaque depicting Christ in Majesty; small losses to enamel, especially at corners.

Musée des Beaux-Arts, Lyons (D. 283–284)

These book-cover plaques are among the oldest preserved examples of such works, which were produced in large numbers by Limoges workshops (see cats. 12, 25, 42, 43). Originally covering the two wood boards that bound a manuscript, they are among the few unseparated pairs to come down to us. The decorative borders are usually distinct from the plaque bearing the central motif, but in this exceptional instance, they are one.

The iconography contrasts the Crucifixion with the vision of Christ in Majesty. These two scenes are closely linked with depictions executed in western and central France during the eleventh and early twelfth century, which, in turn, were derived from Carolingian models. Christ on the cross is flanked by the Virgin and Saint John. Above the arms of the cross are two half-length female figures, one holding a luminous circle, the other a crescent; they personify the sun and the moon. Christ in Majesty, set within a mandorla, is seated on a rainbow, as described in the Apocalypse. The symbols of the evangelists appear at the corners.

The style of these plaques demonstrates the close relationships within the different groups of works with vermiculé grounds and, at the same time, the stylistic diversity known to the Limousin enamelers of the late twelfth century.

The treatment of the engraved and enameled faces is very close to that seen in the narrative group—as defined by Marquet de Vasselot: the calm, regular faces of the chasse of Saint Martial (cat. 17), the chasse from Nantouillet, and the Visitation plaque now in Limoges (cat. 19). The similarity of the female faces in these different works is striking. Each has been engraved and enhanced with fillets of blue enamel. The convention of small circlets seen on those works is also used here—probably in imitation of the highlights in painting that suggest the modeling of hands or cheekbones. The same refinements have also been employed: the reserved metal is all edged in shaded tones; stippling enlivens the rosettes on the frames and the overlapping pattern on the hillock repre-

senting Golgotha. Stippling is also employed with great skill on the muzzles of the animals. The taste for decorative details is expressed particularly in the treatment of the lion of Saint Mark and the ox of Saint Luke: the interplay of colors is exquisite—touches of brick red on hooves and paws and in the open mouths—and the same skill is seen in the complex drawing in the upper part of the loincloth, above and below the cincture.

It is entirely possible that these four works were made in the same workshop. The motif of rosettes set within circles and highlighted by stippling is similar to the pattern seen on the reverse of the chasse in Nantouillet and several others (cats. 24, 27). The stylized V folds of the drapery of the mantle of Christ in Majesty derives from a style dominant in northwestern Europe during the twelfth century; seen about 1150 in manuscript illuminations and in the stained-glass windows at Chartres, then about 1160–80 in stained glass in the west, especially at Poitiers and Le Mans (see cat. 17).

The iconography—which juxtaposes the Crucifixion with Christ in Majesty—the disposition of the drapery according to Romanesque conventions but retaining some link with realism, and the absence of ornamental motifs such as a dotted pattern on orphreys (see cat. 16) tie these plaques to Christological works executed in a severe style (cats. 26, 27).

The extreme curvature of Christ's body, rare in Limoges work, probably reveals an acquaintance with models inspired by Byzantine works. The pronounced slouch recalls the Christ figure of the Crucifixion in the Sacramentary of Clermont-Ferrand,[1] one of the best examples of the diffusion of this type of design in central France at the very end of the twelfth century.

The different comparisons indicate a date of about 1170–85. The two plaques are thus more or less contemporaneous with the famous reference to a "book cover" *de opere Lemovicino* seen in the abbey of Saint-Victor, Paris, about 1169—the first known text documenting the work of Limoges.[2]

ET-D

NOTES

1. Clermont-Ferrand, Bibliothèque Municipale, ms. 63, fol. 58r; see Brisac 1974.

2. See above, Boehm, p. 40.

BIBLIOGRAPHY: Giraud 1887, nos. 123–24; Giraud 1897, nos. 191, 192, engraving p. 307; Marquet de Vasselot 1905, pp. 424–26; Gauthier 1967c, p. 152; Gauthier 1968d, no. 6, p. 283, fig. 1; Gauthier 1987, nos. 101, 102, pp. 102–5, figs. 326, 327, pl. LXXXIII; Briend 1993, pp. 24–25.

The Crucifixion

Christ in Majesty

# 19. The Visitation

Limoges, ca. 1170–80

Copper: engraved, chased, and gilt; champlevé enamel: black, lapis and lavender blue, turquoise, light green, yellow, garnet red, and white

9 x 13.9 cm (3⁹⁄₁₆ x 5½ in.)

Ex colls.: M.-A. Champier, Lyons? (see Gauthier 1987); Rougier (d. 1873), Lyons (sale, Paris, May 3–4, 1904, no. 53); Victor Martin Le Roy, Neuilly-sur-Seine; Jean-Joseph Marquet de Vasselot, Paris; acquired in 1992.

Condition: Wear to gilding, especially at center; small losses to enamel.

Musée Municipale de l'Évêché, Limoges (92.482)

The Visitation, the meeting of the Virgin Mary with her cousin Elizabeth (Luke 1:39–45), is frequently illustrated in Romanesque art and is seen in most cycles depicting the infancy of Christ. But the composition of the scene here is exceptional. Mary and Elizabeth embrace in a manner seemingly Syrian in origin, which was frequently employed in the twelfth century,[1] especially in western and central France: in the frescoes of the Petit-Quevilly (Seine-Maritime), Vicq (Indre), Tavant (Indre-et-Loire), and Rocamadour (Lot), among others.[2] The two women are sometimes accompanied by attendants, as in the Genoels-Elderen ivory diptych of the eighth century (Musée de Bruxelles).[3] Then again, there are no other known depictions of the attendants carrying spindles and distaffs, as they do here. Most likely this iconography, whose direct source is unknown, derives from that of the Annunciation and was inspired by the Apocrypha, chiefly the Protevangelium of James. Indeed, this text describes the two cousins as occupied with spinning prior to their meeting: "And she [Mary]

spins the purple and the scarlet. . . . And Mary, overjoyed, went to Elizabeth . . . and Elizabeth, having heard, threw down her scarlet [and] ran to the door."[4] Spinning was equated with feminine virtue.[5]

The sinuous, long-limbed silhouettes, the broad gestures, and the precise delineation of spindle and distaff place this plaque in the Limousin group of works with vermiculé backgrounds and narrative styles. The conventional treatment of the hands, depicted by engraved arcs around an oval, and, most of all, the full faces with regular, engraved features are comparable to the same elements in the chasse of Saint Martial (cat. 17) and share the same effort toward calm and noble expressions. This tendency is accentuated here by the quieter narrative and the flowing drapery, in which the lines follow the movements of the body, hinting at its forms. The garments, delineated with obvious care, give the figures an aristocratic elegance: long, clinging robes with sleeves revealing a fine white tunic, orphreys and purple-red fastenings, mantles lined with vair. This refinement, which evokes a courtly milieu, links the Visitation to the secular coffret at the British Museum (fig. 19a)[6] and to the chasse of Saint Valerie at Saint Petersburg (cat. 20) and the Wise Virgins of Florence and Vienna (cat. 56). Both in style and in the costume details, these works reveal numerous common traits, as already pointed out by Marie-Madeleine Gauthier. The proposed parallel with the luxury of the court of Eleanor of Aquitaine is probably justified, even though we have no precise knowledge of the fashions. (Eleanor took direct control of her duchy in 1168 and spent long periods there as of that date, especially 1168–73 and 1183–89.) W. L. Hildburgh linked the mantles with their linings of vair to Spanish sources of the first half of the twelfth century;[7] but these garments were frequently seen in France during the second half of that century, especially in the Plantagenet west, notably on the funeral plaque of Geoffrey Plantagenet (cat. 15) and the late-twelfth-century frescoes at Loroux-Bottereau (Loire-Atlantique) and at Saint-Radegonde of Chinon.[8]

The kinship with the other Limoges works having vermiculé backgrounds and with the art of the Aquitainian west allows this plaque, with its seemingly enigmatic iconography, narrative appeal, and highly refined workmanship, to be dated about 1170–80. Devoid of any piercings, the plaque was meant to be set into a larger composition. It probably belonged to a cycle of the life—or at least the infancy—of Christ that was extensive enough to include a Visitation. It may have been part of an altar frontal or, more likely, given its dimensions, on a large chasse such as the thirteenth-century examples at the Metropolitan Museum

Fig. 19a. Coffret with secular scenes. British Museum, London

(cat. 115), the Nationalmuseet, Copenhagen (fig. 115*a*),[9] and the Musée de Cluny, Paris[10]—where, however, the Visitation scene is limited to figures of the Virgin and Elizabeth.

ET-D

NOTES

1. It appears notably on a sixth-century ampulla preserved in Monza. See Mâle 1922, fig. 58; Schiller 1966, vol. 1, pp. 65–67.
2. See Deschamps and Thibout 1963, p. 95, pl. XX; Demus 1970, pp. 139–41, fig. 18, pl. LVI.
3. See Gaborit-Chopin 1978, no. 48, pp. 48, 185.
4. See Michel 1911, pp. 22–23.
5. See Biscoglio 1995, pp. 165–76.
6. Gauthier 1987, no. 174.
7. Hildburgh 1936, p. 108; in the *Libro de los Testamentos* illuminated in Oviedo between 1126 and 1129.
8. Deschamps and Thibout 1963, pp. 68–69; Héron 1965; Trocmé 1966; this motif also appears, albeit more summarily, on the Vannes coffret (see Paris 1965, no. 336, pls. 100, 101).
9. Liebgott 1986, pp. 47–48, ills. 39–42, pp. 49–50.
10. Rupin 1890, pl. XXXIV.

EXHIBITION: Limoges 1992, no. 1.

BIBLIOGRAPHY: Marquet de Vasselot 1905, p. 28; Marquet de Vasselot 1906a, no. 32, pl. XIV; Hildburgh 1936, pp. 107–8; Gauthier 1987, no. 138, ill. 455, pl. CXXVIII; Biscoglio 1995, p. 167, fig. 1, p. 177.

# 20. Chasse of Saint Valerie

Limoges, ca. 1175–85

Copper: engraved, chased, and gilt; champlevé enamel: lapis and lavender blue, turquoise, green, yellow, translucent garnet red, opaque brick red, and white; wood core

23.2 × 28 × 11.5 cm (9⅛ × 11 × 4½ in.)

INSCRIBED: Graffito on the open book held by the second apostle on the reverse: + ARAS / ONMI. Difficult to interpret; Darcel (1874) has suggested *pax hominibus* (peace to all mankind)

EX COLLS.: Bouvier, Amiens; Alexandre Basilewsky (before 1870); acquired in 1884 by Czar Alexander III, along with the entire Basilewsky Collection, for the State Hermitage Museum, Saint Petersburg.

CONDITION: Crest probably modern; modern door on the right end.

State Hermitage Museum, Saint Petersburg (Inv. φ 175).

This chasse was first published in 1874 in the catalogue of the Basilewsky Collection.[1] J.-J. Marquet de Vasselot (1905) linked it to another chasse of Saint Valerie, preserved at the British Museum,[2] because of the face dedicated to this saint, and to the chasses at Apt and Washington (cats. 21, 22), because of the Adoration of the Magi portrayed on the reverse.

In fact, this chasse combines three different themes. On one face, we see the Adoration of the Magi. On the roof, four apostles appear under arcades; they include Saint Peter and perhaps Saint Paul, holding an open book. A fifth apostle is portrayed on the left end (the other end is largely modern). The other face is dedicated to the main episodes

in the martyrdom of Saint Valerie: on the lower part, and then on the roof, Duke Stephen orders the young girl's execution; she is led from the city and decapitated, but then picks up her head with her hands; the executioner, on his way to relate the incident to the duke, is struck by lightning; an angel lifts the lifeless body of the saint, who then, escorted by the angel, carries her own head to Saint Martial.[3]

The portrayal of the various episodes is identical with that on the British Museum chasse (figs. 20a, 22a), revealing, undoubtedly, a common model. However, on the Saint Petersburg chasse picturesque details such as the soldiers' turbans are more numerous or more refined (Valerie's hair ribbons); the style here is also more varied and more sophisticated.

At the same time, this entirely narrative face combines diverse tendencies. The taste for decorative details, observed in the Gimel chasse (cat. 16), is developed further here: the drapery, defined by curving, fragmented lines, except on Saint Valerie's long robe; the expressive faces of the executioners—especially the blunt profiles with a frontal eye; the buildings, which also allow comparison with the scenes of the life of Saint Stephen on the chasse from Gimel. The robe of the saint is similar to those worn by feminine figures on the secular coffret at the British Museum (fig. 20a).[4]

As for the Adoration of the Magi, it is very similar to the same scene on the chasse at the National Gallery, Washington (cat. 22). The only difference, as pointed out by Marquet de Vasselot (1905), lies in the disposition of the three-quarter-length figures and in the liveliness of the Virgin and Child portrayals; this marks a break with the Romanesque tradition.[5] Moreover, all the figures here are slimmer. The apostles on the roof are comparable to those on the Apt chasse, but the Byzantinizing severity is less accentuated here.

Despite the artist's use of diverse tendencies from various sources, the chasse embodies a certain stylistic unity, especially in the elongated figures, the curved lines that define the drapery, and the presence of the same physical types in both scenes. The Virgin's face is not very different from Saint Valerie's, and the youthful, beardless face of Saint Martial's assistant resembles that of the third king; the physiognomies of the first two kings are very close to that of Duke Stephen in the two scenes in which he appears. The elegant—indeed luxurious—costumes also contribute to the stylistic unity. The Magi wear mantles lined with vair, already seen in the funeral effigy of Geoffrey Plantagenêt and the Visitation at the Musée de l'Évêché, Limoges (cats. 15, 19). Valerie wears a shapely, dotted robe; flowing scarves, attached at the wrist, extend the long sleeves. This style,

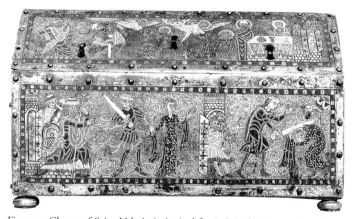

Fig. 20a. Chasse of Saint Valerie (principal face). British Museum, London

Principal face

Reverse

described in courtly poems and chivalric romances,[6] derives from the wide, flared sleeves that had been fashionable both in England and on the Continent since the end of the eleventh century.[7] These "long, attached sleeves" are frequently depicted in the art of western France of the late twelfth and early thirteenth century—for instance, on the frescoes at Saint-Rémy-la-Varenne (Maine-et-Loire).[8]

On the other hand, the artist focused on more remote sources in order to provide an "exotic" aspect for the soldiers, whose garb differs from the apparel usually encountered in twelfth-century portrayals, even in depictions of the Saracen enemy. Thus, the turbans seem to be borrowed from the costumes of Islamic Spain: they are found in Mozarabic illuminations such as the Beatus of Gerona.[9] The imbricated cuirasses stem from military attire of late antiquity. Frequently seen in Byzantine art of the eleventh and twelfth centuries, they are also widespread in Carolingian art, for example, in the Vivian Bible and the illustrations of the Psychomachia at the Bern Bürgerbibliothek.[10] Such armor also occurs extensively in the Bible of Souvigny—one of the principal manuscripts illuminated in central France at the end of the twelfth century.[11]

The artist who created this chasse knew how to combine effects from different sources of inspiration to make a highly original work that blends popular taste with aristocratic refinement and religious feeling in contrasting the majesty of the apostles and the quiet Adoration of the Magi with the narrative intensity of the scenes of Saint Valerie's martyrdom. Although it certainly seems that the narratives of the Martyrdom of Valerie and of the Adoration of the Magi reflect the luxuriant, refined taste developed at the court of Aquitaine under the Plantagenets (see cat. 19), the appealing suggestion that this chasse was commissioned by Eleanor of Aquitaine herself or that it was created for a specific occasion during her long reign in the duchy[12] cannot be demonstrated. Marie-Madeleine Gauthier recalls that during his installation as duke of Aquitaine in 1172, Richard II, the Lion Hearted, was given the ring of Saint Valerie, undeniable evidence of the importance accorded by the Plantagenets to the duchy of Aquitaine and to the cult of the saint at that time in the Limousin. The chasse, like the one in the British Museum, is witness to this fact.[13]

ET-D

NOTES

1. It is this catalogue that indicates that the work came from the Bouvier Collection. However, the entries in the *Exposition Provinciale* catalogue of Amiens in 1860 (no. 1133: "Eighteen chasses and reliquaries") are too vague to allow us to conclude that the chasse was on display. On the other hand, we can affirm that it was not included in the exhibition of the Union Centrale des Arts Décoratifs in 1865 and that it had left the Bouvier Collection before 1866, for it does not appear in Roger's catalogue of that year. The piece certainly entered the Basilewsky Collection before 1870, for it is quite recognizable on two watercolors of the interior of the collector's Parisian townhouse executed in that year (see Krysanovskaïa 1990, pp. 145–47, figs. 2, 3).
2. See Gauthier 1987, no. 91.
3. These scenes are inspired by different texts describing the life of Saint Valerie; the texts were widespread in the Limousin, where she was known first through the Lives of Saint Martial (see cat. 17), then through an independent Life written in the late tenth century at the monastery of Chambon (which, notably, introduced the carrying of the head) and through several sermons and liturgical writings of the eleventh and twelfth centuries. See Gauthier 1955, pp. 36–44.
4. See Gauthier 1987, no. 174.
5. This arrangement can be linked, for example, to that of the Adoration of the Magi at Le Petit-Quevilly (see Jalabert 1954): between 1183 and 1190. For an analysis of the iconography of the Adoration of the Magi on Limousin chasses, see cat. 22.
6. The most famous episode was the one in *Perceval le Gallois*, where the father gives the "Maiden with small sleeves" a "long and wide" sleeve that she can give to the knight (v. 4987ff.).
7. They can be observed in, for example, the embroidery of Bayeux (see Wilson 1985, pl. 50).
8. See Deschamps and Thibout 1963, pp. 59–61 (end of the twelfth century) and pl. XXI, figs. 2, 3. Such sleeves are also depicted on the Vannes coffret and the embroidered burse from Chelles (last quarter of the twelfth century; see Laporte 1988b, pp. 64–69). See also the examples pointed out by Demay (1880a, pp. 92–93, 98) on seals dated between 1170 and 1220.
9. See Williams 1977, p. 99, pl. 30.
10. Paris, BNF, ms. 1, for example, fol. 215v; Bern, Bürgerbibliothek, ms. 264, for example fol. 41r; see Mütherich and Gaehde 1977, ill. v, pl. 22. For antique and Byzantine depictions, see Kolias 1988.
11. For example, fol. 284. See cat. 34.
12. Lapkowskaya 1971; Gauthier 1987.
13. According to Geoffrey de Vigeois (d. 1185), Saint Valerie's relics were preserved mainly in La Souterraine (Creuse), Chambon (Corrèze), and Limoges, where a church was built on the site of her grave in 1160 and consecrated in 1212. Duchesne 1910, p. 110; Gauthier 1987, p. 98; Boehm 1991, pp. 201–2; Paris 1993, pp. 98–99.

EXHIBITIONS: Paris 1878; Moscow 1956, p. 142; Brunswick 1995, no. D100.

BIBLIOGRAPHY: Darcel and Basilewsky 1874, vol. 1, no. 199, vol. 2, pl. XXVI; Darcel 1878b, pp. 250–51 and engraving, p. 251; Linas 1883, pp. 132–36; Linas 1886b, no. 198, p. 251; Rupin 1890, pp. 402–4, fig. 460; Kondakov 1891, no. 18, p. 201; Marquet de Vasselot 1905; Souchal 1963, no. 1, p. 59; Gauthier 1966, pp. 940–51; Lapkowskaya 1971, pls. 13, 14; Gauthier 1972b, no. 48, pp. 94–96, pl. p. 95; Krysanovskaïa 1986, no. 108; Gauthier 1987, no. 94, pp. 100–102, figs. 320–24, pls. LXXX–LXXXIII; Krysanovskaïa 1990, p. 145; *The State Hermitage* 1994, no. 426.

Dr Boshart

# 21. Chasse of Saint Peter

Limoges, ca. 1175–85

Copper: engraved, chased, and gilt; champlevé enamel: lapis, medium, and lavender blue; turquoise, medium green, yellow, brick red, translucent wine red, and white; wood core

19 x 26.5 x 11.5 cm (7½ x 10⅜ x 4½ in.)

PROVENANCE: Possibly from the church of Saint-Peter, Apt. Classified, Monuments Historiques, 1898

CONDITION: Some wear to gilding; small losses to enamel; crest and thirteen nails missing. The original wood core and its opening in the lower plank are preserved, with a lock that possibly dates to the fifteenth century.

Church of Saint-Anne, Apt (Vaucluse)

The Apt chasse was first exhibited and published on the occasion of the retrospective exhibition of French art organized in 1900. J.-J. Marquet de Vasselot (1905), who placed the chasse at the head of his grouping of works with vermiculé backgrounds, linked it to the chasse that was then in the Gambier-Parry Collection and is now in the National Gallery of Art, Washington, D.C. (cat. 22). Denise Jalabert (1954) was the first to show the kinship of foliate decoration on the reverse with the English illuminations of the Winchester Bible and their succedents in western France. Since then, Marie-Madeleine Gauthier[1] has suggested a Byzantine or Byzantinizing influence for the decoration of the principal face.

In effect, two very different types of decoration come together on this chasse. At the center of the principal face the half-length figure of Christ appears above a similar image of Saint Peter, each with an apostle at either side. Trefoil arcades surmount both ensembles. Two other apostles are shown on the ends. Yet, the reverse is decorated with a rinceau pattern ending in large, blossoming fleurons. Still, the two faces, though very different, are equally exceptional in the quality of their execution, as evident in the variety of the palette as in the excellence of the engraving and stippling. Particularly notable are the extensive use of a colorless flux placed against the copper, which creates an effect of translucent wine red on Saint Peter's robe and on that of the apostle on the right gable, and the variety seen in the lively decorative effects achieved by the varied shades of enameling on the columns, capitals, and turrets of the arcades. The preeminence accorded this work by Marquet de Vasselot is fully justified.

The decoration on the reverse of the chasse is part of the late-Romanesque style of western France, described by Gauthier (1967b) as "Plantagenet." There is no doubt that this style originated in England, inspired by Continental works of the eleventh and early twelfth century. Indeed, the palmette motif became one of the most admired decorative features of Limoges enamels.[2] The fleurons lack the "tentacles" (the so-called octopus flowers) found, for instance, on the Souvigny appliqués (cat. 34). However, the kinship with English illuminations seems very direct: the Apt chasse, along with the frescoes of Petit-Quevilly (slightly before 1190), is one of the finest Continental examples of the diffusion of this English decoration,[3] seen, for example, in the Bury Bible (ca. 1135).[4] The parallel of the foliate decoration to that of English ciboria attributed to the 1160s[5] is also very enlightening: although the large fleurons are quite different, the smaller motifs, resting on an engraved, reticulate motif, are quite similar to those on the Apt chasse.

The Byzantinizing aspect of the apostles on the principal face and the gabled ends has been repeatedly emphasized; indeed, it has been suggested that the influence may be Sicilian mosaics of the second half of the twelfth century.[6] The composition is without precedent among Limoges chasses, so it might theoretically have been inspired by Byzantine caskets with figures of saints under arcades bordered with rosettes, as in the Bargello casket.[7] However, the seeming contradiction between the decoration on the two faces of the chasse may not point to different sources of inspiration. In fact, the Limoges artists might have again been moved by the example of English illuminations; the flowering of a "Byzantine" style in England during the last third of the twelfth century has been amply established. The figures of Christ and the apostles on the Apt chasse seem more directly linked to those of the Winchester Bible and related works than to the ones on the Sicilian mosaics. Certain schematic elements, such as the small scalloped curls that frame the foreheads of Saint Peter and the apostle at Christ's left side, correspond to motifs encountered in English illuminations; they are, for example, quite visible on an unfinished page of the Winchester Bible, ca. 1160 (the story of Judith, fol. 331, vol. 2).[8]

The Washington chasse (cat. 22) has a striking kinship with the chasse of Apt: the treatment of the faces of Christ and the apostles on the roof of the principal face and, even more, on the ends, is so similar that it is impossible not to attribute both works to the same artist. The reverses may seem very different, but it is clear that the fleurons terminating the tails of the bird-sirens on the Washington chasse come directly from those on the Apt chasse.

The Apt chasse is one of the finest examples of the extraordinary adaptability shown by the Limoges enamelers. While enriching and diversifying both style and decoration through contact with new artistic developments, they remained faithful to their best technical and decorative

Reverse

traditions. If we accept the dates proposed for the works linked to the chasse, we can place its execution at about 1175–85.

As suggested by Gauthier (1964a and 1987), the preeminence of the apostle Peter within the scheme may be explained by the fact that the chasse was originally made for the church of Saint Peter at Apt. This church, which replaced the old cathedral in the ninth century, was, by the twelfth, a simple parish church. It was deconsecrated in 1464 and then restored by the Black Penitents, who acquired it in 1594.[9] Certainly, the quality of the work might suggest a more prestigious destination: in this regard, it is worth noting that, at the time the chasse was completed, the bishop of the town was Peter of Saint-Paul. Documented between 1162 and 1179, he perhaps died in 1186, and his successor was first cited in 1190. Peter of Saint-Paul was, most notably, responsible for

the solemn translation, in 1179, of the relics of Saint Castor to the new cathedral, which was jointly dedicated to the Virgin and Saint-Castor before being dedicated to Saint Anne.[10]

ET-D

NOTES
1. See, especially, 1958a, p. 363; 1962a, p. 407; 1964a, pp. 78–80.
2. Marquet de Vasselot 1941, p. 142.
3. See also the now-vanished frescoes at Saint-Quiriace de Provins and the Tree of Jesse at the abbot's residence in Moissac (see Jalabert 1954; Deschamps and Thibout 1963, p. 97, pl. XLI).
4. Cambridge, Corpus Christi College, ms. 2, for example, fol. IV or 201V; see Kauffmann 1975, no. 56, pls. 148, 151.
5. See Stratford 1984; London 1984, nos. 278–80.
6. Notably, Gauthier 1964, p. 80a.
7. Twelfth century; see Goldschmidt and Weitzmann 1979, no. 99, p. 56, pl. LIX.

Principal face

Left and right sides

8. See Kauffmann 1975, no. 83, fig. 233.

9. Boze 1820, p. 425.

10. Gams 1873, p. 492; Boze 1820, pp. 119–20.

EXHIBITIONS: Paris 1900, no. 2455; Barcelona 1961, no. 427; Vatican City 1963, no. 14, pl. XI; Paris 1965, no. 644.

BIBLIOGRAPHY: Molinier and Marcou 1900, p. 86; Arnaud d'Agnel 1904, pp. 329–30, pls. XXII–XXIV; Marquet de Vasselot 1905, pp. 24–25, pls. V, VI; Lacrocq 1933; Marquet de Vasselot 1941, pp. 141–42, pl. XXX; Roux 1949, p. 66; Gauthier 1950, pp. 30ff.; Jalabert 1954, p. 25; Gauthier 1958a, p. 363, fig. 7; Gauthier 1962a, p. 404 (ill.), 407; Gauthier 1964a, pp. 78–80; Gauthier 1966, pp. 943–45; Barruol 1967, p. 361; Gauthier 1967b, p. 152; Gauthier 1987, no. 147, pp. 144–45, figs. 506–9, pls. CXXXVIII, CXXXIX; Distelberger et al. 1993, pp. 21–23.

# 22. Chasse with the Adoration of the Magi

Limoges, ca. 1175

Copper: engraved, scraped, stippled, and gilt; champlevé enamel: dark, medium, and light blue; turquoise, dark and light green, yellow, red, and white; wood core, painted red on exterior

18.6 × 26.1 × 11.5 cm (7⅜ × 10¼ × 4½ in.)

Ex colls.: Thomas Gambier Parry, Highnam Court, Gloucestershire, England (by 1862); Hubert Parry (1888–1918); Ernest Gambier-Parry (1918–20); [Durlacher Bros., London (in July 1920)]; Joseph E. Widener, Elkins Park, Pennsylvania

Condition: Some wear to the gilding; minor losses to enamel; crest missing; proper right end panel of core replaced.

National Gallery of Art, Washington, D.C. (1942.9.278 [C-2])

On the lower plaque of the principal face, the Virgin and Child are enthroned under a Romanesque trefoil arch. Above enameled columns, paired towers rise to a roof. Outside this churchlike space, Saint Joseph stands to the right with a flowering rod in his hand as the Magi approach from the left, bearing gifts for the Child.

On the roof panel of the principal face stands the three-quarter-length figure of Christ flanked on each side by two apostles: Saint Peter, holding a key, and an unidentified apostle to his right; the youthful John the Evangelist and an unidentified apostle to his left. Three-quarter-length apostles also appear on the end panels, each holding a book and standing under a trefoil arch surmounted by an array of fanciful turrets. The gilt-copper plaques of the front and side faces are engraved with a rich, overall vermiculé pattern, and the scenes are framed by blue enamel punctuated with reserved florets. The hair of all the figures is enameled in dark blue, and the same deep blue enameling gives definition to fingers and palms and even to the bones and muscles of the figure of John the Evangelist. Red is used for accents such as books, the kings' stockings or shoes, the points of their crowns, the star of Bethlehem, and the flowering rod of Joseph.

Both the upper and lower plaques of the reverse are decorated with fantastic winged creatures enclosed in circles—three on the lower plaque and four on the roof. Each birdlike figure has a foliate tail and the head of a man wearing a hat. The lower left and right roundels are mirror images of each other, but the images on the roof alternate in tandem. The roundels are surrounded by stylized foliate ornament set against a blue field that is framed by a rectangular band of florets reserved against deep blue enamel. All the borders of the chasse are decorated with a delicate vermiculé pattern.

The National Gallery chasse has its closest stylistic parallel in the chasse from Apt (cat. 21), both in overall decorative scheme and in the details of its figural ornament.[1] Comparisons between the figures of Christ and of the apostles on the end panels are especially striking. Particularly remarkable are the similar but distinctive facial features—the peculiar dimples on the apostles' cheeks, the curling mops of hair, the parallel lines encircling the figures' necks, the wrinkled brows, and the long, straggly hairs of the beards. The palette is the same except for the use at Apt of translucent wine red, which is not present on the Washington example.

The disposition of the decorative motifs on the reverse of the Washington chasse is very similar to that found on the chasse of Saint Valerie in the British Museum (fig. 22a), but the draftsmanship of the narrative is different: the Saint Valerie figures are squatter and their brows and eyes are less expressive. However, certain details, such as drapery that terminates in arrowhead shapes, and the drawing of the slippers with reserved lines to indicate the crease across the instep, considered with the similarities of the reverse, suggest that the London and Washington chasses—as well as the Saint Petersburg (cat. 20) and Apt chasses—may be products of the same workshop. The drawing of the apostles and the overall palette are also close to those found on the Metropolitan Museum chasse (cat. 24), but there the overall program is sufficiently distinct to indicate a different workshop.

Gauthier (1987) has attributed the appearance of the golden Christ Child to the enameler's direct experience of monumental works of Byzantine art, probably the mosaics of the Martorana at Palermo or those at Monreale; however, it is not necessary to assume that the golden image of the Divine Child derives from this source.

Left side

Principal face

Right side

123

The mise-en-scène of the Adoration of the Magi corresponds to the one on the reverse of the Saint Petersburg chasse of Saint Valerie (cat. 20), including the rather charming narrative detail of the three horses that have apparently been tethered nearby. (The tree to which their reins are tied appears in the Saint Petersburg scene but is absent in the Washington example.) The iconography of the figure of Joseph follows that of the altar frontal at San Miguel de Excelsis,[2] where Joseph also appears with a flowering rod, emblematic of his divinely ordained position as Mary's spouse. As on that altar frontal, the Three Kings here carry their gifts in vessels that have been precisely drawn, in forms that seem to anticipate later Limousin works. The vessel carried by the first king, for example, suggests the ciborium of Master Alpais (cat. 69); the paten-shaped object carried by the third anticipates the form of gemellions (cats. 125, 126, 131, 132).

Patristic interest in the Magi as witnesses to the coming of God in human form can be traced to the writings and artistic representations of the early Christian period.[3] It is exceptional, however, that they should appear on a reliquary. Whether this chasse was meant to contain relics of the Magi, as their portrayal would seem to suggest, or was designed for another use is not certain.[4] More than twenty chasses of Limoges enamel depict the Journey or the Adoration of the Magi.[5] Yet the relics of the Three Kings, discovered near Milan in 1158 and translated to Cologne Cathedral in 1164,[6] were not widely distributed before the fourteenth century, when church and altar dedications to the Three Kings became widespread.[7] There is no record of the monks of Grandmont returning from Cologne with relics of the Magi, their success in procuring other relics notwithstanding. Nor does the feast of the translation of the Kings' relics, celebrated in Cologne on July 24, appear in any Limousin calendar.[8] A chasse with images of the Magi preserved in the parish church at Beaulieu (Corrèze) contains a thirteenth-century "etiquette," or official listing, of its contents,[9] specifying relics of the Cross of Jesus and of Peter, Paul, and/or Saint Hilary, with no mention of the Magi.[10]

The presentation made by the Three Kings was a common ecclesiastical metaphor for the appropriate homage rendered to God, regardless of earthly status. In the twelfth-century liturgy of Limoges, the gifts of the Magi were specifically linked to the Offertory of the Mass for the Feast of the Epiphany. During the Mass, in a reenactment of the trip to Bethlehem, three church officials dressed as kings in elaborate costumes of silk, wearing crowns and carrying liturgical vessels, approached the lectern for the reading of the Gospel and then advanced to the high altar. The "stage directions" for this theatrical performance specify: "Et vadunt ad offerendam, relinquentes ibi sua jocalia" (And they go to make offering, leaving their jewels there). It may be that Limoges reliquaries depicting the Magi were first intended for use at the altar in this context.[12]

BDB

NOTES

1. The comparison was first noted by Marquet de Vasselot (1906, p. 12) and has appeared in all subsequent literature.
2. Gauthier 1987, no. 135.
3. See Cologne 1982, especially the essays by Ernst Dassmann and Johannes Deckers.
4. There is nothing surprising about the presence of Saint Peter at the right hand of Christ on the roof plaque; his preeminence among the apostles cannot be considered grounds for suggesting that the chasse was made for his relics, as stated in the 1993 National Gallery catalogue. See Distelberger et al., 1993, no. 1.
5. Gauthier 1972b, p. 94.
6. See Schiller 1971, vol. 1, pp. 106, 110.
7. Werner Schäfke, in Cologne 1982, p. 74.
8. Dom Becquet, O.S.B., communication with author.
9. It should be noted that the date of the authentication, often discernible on paleographic grounds, is unpublished. The mention of the authentication appears in Rupin (1890, p. 353).
10. Rupin 1890, p. 353.
11. Young 1933, vol. 2, pp. 32–37.
12. Marie-Madeleine Gauthier has suggested that the participation of the Plantagenet princes in this ceremony in 1173 would be an appropriate occasion for the creation of a chasse such as the one in the National Gallery, which, on stylistic grounds, is closely linked to other works associated with Plantagenet patronage. Gauthier 1972b, p. 94 (without citation of the primary source).

Fig. 22a. Chasse of Saint Valerie (reverse). British Museum, London

EXHIBITION: London 1862, no. 1072.

BIBLIOGRAPHY: Rupin 1890, p. 424; Marquet de Vasselot 1906, pp. 11–13, pl. 2; Widener 1935, p. 29, no. 345; Christensen 1952, pp. 10–15; Souchal 1963, p. 48, fig. 6, pp. 49–60, n. 1; Gauthier 1967b, p. 152; Gauthier 1966, no. C-55; Gauthier 1972b, pp. 98–99; Gauthier 1987, no. 148; Distelberger et al., 1993, pp. 19–24.

Reverse

# 23. Crosier with Palmette Flower

Limoges, ca. 1175–85

Copper: engraved, chased, and gilt; nielloed silver; champlevé enamel: blue-black, lapis and lavender blue, turquoise, medium and light green, yellow, brick red, and white

H.: 21.5 cm (8½ in.); diam. of volute: 7.2 cm (2¹³⁄₁₆ in.)

Ex colls.: Acquired by the Musée Sainte-Croix, Poitiers, before 1842. It might have come from the Dom Mazet Collection, which the museum bought in 1818 and which included some Limoges enamels. During the French Revolution, Dom Mazet had taken part in opening tombs, notably at the abbey of Saint-Cyprien (information from M. Rérolle).

Condition: Wear to gilding, especially around shaft and knop; losses to enamel of knop; slight oxidation in some areas; deterioration of base of shaft caused by burial.

Musée de la Ville de Poitiers et de la Société des Antiquaires de l'Ouest (Musée Sainte-Croix) (890-256)

The theme of the flower blossoming in the center of a volute originated with a group of crosiers that are among the most beguiling works produced by Limousin metalsmiths. The Poitiers crosier, although the smallest in the group, holds a primary position because of the quality of its craftsmanship and because it is very likely the oldest preserved example.

This choice of motif was certainly inspired by the story of Aaron's flowering rod (Num. 17:6–8), but the decoration itself is closely tied to a style that was widespread in the second half of the twelfth century (see cats. 21, 34). Other goldsmiths, especially those in the Anglo-Norman domain, had previously used the motif in decorating the volutes of crosiers. Thus, a blossoming flower adorns the crosier of an abbess of Saint-Amand, Rouen—a piece assigned to the third quarter of the twelfth century;[1] the same motif is also found on the English crosiers from Saint David's Cathedral in Wales and Whithorn, Scotland.[2]

The elegance of the foliate motifs and the perfection of their orientation within the scrolling volutes—like the brilliant colors of the enamels—undeniably place these works among the finest achievements of the Limousin workshops. Moreover, they enjoyed great esteem, for some thirty enameled crosiers of this type are preserved today (see cats. 77, 81, 107).[3]

The Poitiers crosier differs in several ways from other examples. Smaller in size, it is the product of an original fabricating process, and its nielloed silver decoration is unique in the Limoges oeuvre. The volute was not constructed by joining two pieces of metal, each the mirror image of the other, along a vertical seam. Rather, it consists of one gradually tapering shaft. X rays have revealed that this shaft is solid; furthermore, the nielloed silver is made of thin bands that encircle the gilt-copper shaft. The shaft is not composed of alternating silver and gilt-copper cylinders, as was assumed by Marquet de Vasselot (1941).

The regular pattern of nielloed silver bands adorned with fine vermiculé rinceaux is also found on another crosier at Saint David's Cathedral.[4] This is additional proof—if any be needed—of the links that existed between the Limousin workshops and those of Plantagenet England during the final decades of the twelfth century. Nielloed bands, with somewhat more subdued vermiculé decoration, also appear on the volute of the crosier found at Silos in the tomb of Abbot Jean, who died in 1198.[5] These three examples clearly show how the "Plantagenet domain" could have been the source of this type of work and style of decoration.

A large flower blooms at the center of the volute, whose outer curve supports the tips of the five slender and supple petals. This motif places the crosier, along with the Souvigny appliqués (cat. 34), in the group of Limousin works decorated with so-called octopus palmettes—a motif that is recognized as having an English origin. The affinity with the Souvigny fleurons is again emphasized through the important use of engraving and chasing to bring out in bas-relief the details of the foliate ornament of the volute and the flower: small stems and leaves with beaded bands, fillets, buds, and stippled motifs.

The large enameled flowers on the knop are similar to those on the Apt chasse (cat. 21), whereas the smaller ones, in the form of a half corolla with pointed tips, recall the small enameled appliqués of the same shape on the reverse of the Ambazac chasse (cat. 55).

It is probably in the Winchester Psalter and the works deriving directly from it[6] that we find the octopus palmettes closest to those on the Poitiers crosier, as seen in the drawing of the slender petals and in the stippled motifs separating them. Small, rounded leaves with serrated edges, their form defined by beaded strips, alternate with the nielloed silver bands in ornamenting the volute. These leaves also resemble ones in the Winchester Psalter. However, Continental sources should not be neglected: from the mid-twelfth century, Cistercian illuminations, most notably, offer numerous examples of palmettes with tentacled tips coiling around the encasing scroll. Certain initials in the City of God from the Cistercian monastery of Bussière (Côte-d'Or) are among the finest examples of this motif, accompanied, moreover, by diminutive foliage with beaded strips, which recalls the foliate decoration of the Poitiers crosier.[7]

The crosier of the cathedral of Carpentras (Vaucluse)[8] offers another, less spectacular, example of a crosier with octopus palmettes that may be related to English sources (fig. 23a): the closest parallels can be found in the Canterbury Psalter.[9] The fleurons ornamenting these Limousin crosiers are distinguished by clearer, less jagged draftsmanship, a development of the Limoges artists themselves or emanating from their knowledge of intermediary works that familiarized them with English illuminations.

ET-D

NOTES

1. Rouen, Musée des Antiquités de la Seine-Maritime; see Rouen-Caen 1979, no. 275.
2. Saint David's Cathedral and National Gallery of Scotland, Edinburgh; see London 1984, nos. 270, 285.
3. Marquet de Vasselot 1941, nos. 1–34c; the thirty-seven examples listed by Marquet de Vasselot are joined by the one discovered in 1953 in the cathedral of Poznań; see Gauthier 1956, p. 284; ibid., 1987, p. 152.
4. See London 1984, no. 268 (attributed to the mid-twelfth century, but could be slightly later).
5. See Gauthier 1987, no. 139.
6. Ca. 1150, British Library, London, ms. Cotton Nero CIV, for instance, fol. 9, Tree of Jesse; see Kauffmann 1975, no. 78; London 1984, no. 61.
7. Dijon, Bibliothèque Municipale, ms. 759; see Zaluska 1991, pl. xxxv, fig. 8 (third quarter of the twelfth century).
8. Marquet de Vasselot 1941, no. 9; Gauthier 1987, no. 288.
9. Paris, BNF, ms. lat. 8846, ca. 1180–90; see Morgan 1982, no. 1, and London 1984, no. 73.

Fig. 23*a*. Crosier. Musée d'Art Sacré, Carpentras

EXHIBITIONS: Paris 1867, no. 2080; Paris 1878; Paris 1889, no. 544; Tours 1890; Limoges 1948, no. 77; Vatican City 1963, no. 108, pl. xxxiii; Paris 1968b, no. 382; New York 1970, no. 158.

BIBLIOGRAPHY: Texier 1842, p. 247; Texier 1857, col. 572, fig. 3, col. 1485; Brouillet 1885, vol. 2 no. 3897; Rohault de Fleury 1883–89, vol. 8, p. 105, pl. DCCIII; Rupin 1890, p. 79, fig. 146; Palustre 1891, p. 71, pl. xxxvi; Barbier de Montault 1894, pp. 532–34, fig. 5, p. 533; Molinier 1901, p. 195, fig. 192, p. 192; Dez 1934, pp. 10–11; Marquet de Vasselot 1941, no. 8, pp. 185–86, pl. II; Gauthier 1950, p. 34; Gauthier 1958a, p. 366, pl. III; Gauthier 1967b, p. 152; Gaborit-Chopin 1983b, p. 286, ill. 383, p. 375; Gauthier 1987, no. 158, pp. 151–52, ill. 534, pl. CXLIX, ill. 558, pl. CLIX.

# 24. Chasse: The Crucifixion and Christ in Majesty

Limoges, ca. 1180–90

Copper: engraved, chiseled, stippled, and gilt; champlevé enamel: dark, medium, and light blue; turquoise, dark and light green, yellow, red, and white; wood core, painted red on exterior

26.2 × 30.2 × 11.6 cm (10⁵⁄₁₆ × 11⅞ × 4⁹⁄₁₆ in.)

INSCRIBED: IH[ESU]S / XPS (Christus)

EX COLLS.: William Beckford, Fonthill Abbey, Wiltshire (sold 1823); Anne, countess of Newburgh (d. 1861); Hon. Robert Curzon, M.P. (d. 1863); Robert Curzon (14th Baron Zouche, in 1870), Parham, near Pulborough, Sussex (1863–73); Robert Nathaniel Curzon (15th Baron Zouche, 1873–1911); Georges Hoentschel, Paris (by 1911); [Jacques Seligmann, Paris (in January 1912)]; J. Pierpont Morgan, London and New York

CONDITION: Minor wear to gilding at edges; side lock panel replaced; two crystals missing from crest.

The Metropolitan Museum of Art, New York; Gift of J. Pierpont Morgan, 1917 (17.190.514)

Centered on the front panel of the vermiculé chasse is the image of the crucified Christ in a knee-length loincloth, his torso gilt against a deep blue cross ornamented with reserved rinceaux. Beneath the arms of the cross stand his mother, her hands clasped to her breast, and Saint John the Evangelist, who raises his right hand to his head, which, like those of all the figures on the front of the chasse, has been separately modeled and applied. Above the cross, enclosed in circles, are half-length figures that represent the sun and the moon. Standing figures of apostles are at either side of the central field of the Crucifixion, two at the left and two at the right. Each pair stands under a double arcade that is supported by enameled columns patterned to resemble porphyry. Fanciful towers rise from the springing of each arch. As witnesses to the Crucifixion, the apostles turn toward the scene, raising a hand or touching the chest. Each holds a book or a scroll, and each enameled halo is distinct.

At the center of the roof plaque is the figure of the enthroned Christ in Majesty, posed in a mandorla and surrounded by symbols of the four evangelists. Three apostles appear at either side, framed, as on the lower plaque, by elaborately enameled columns and arches. A border of half circles with inscribed trefoils frames both the upper and lower plaques.

The reverse of the chasse has an overall pattern of quatrefoils enclosed in circles, a motif frequently found on vermiculé chasses (see cats. 27–29).

The engraved and gilt image of the cross on the proper left end replaces the original lock plate, which, judging from the usual Limoges repertory, depicted the standing figure of Saint Peter holding the keys to the kingdom of heaven. On the proper right end stands a figure holding a scroll, set

beneath an arcade and tower, probably intended as Saint Paul. The image of Saint Paul has some of the same defining attributes that appear on the ends of the Apt and Washington chasses (cats. 21, 22): the individually articulated enameled lines of hair and beard; the heavy eyebrows and paired lines around the eyes; the lines across the hands and on the arm. The plaques are not the work of the same artist, but the creator of the New York chasse was surely aware of the work of the Apt and Washington reliquaries' enameler. The apparent dialogue between the two allows for the proposed dating of the Metropolitan chasse.

The image of the crucified Christ finds a close parallel in the book-cover plaque in the Louvre (cat. 25) and also with a cross in Guéret.[1] Marie-Madeleine Gauthier is probably correct in assigning them to the same workshop, though the figures of the Metropolitan chasse are not so elongated and their faces not so carefully worked as those on the Louvre plaque. The chasse that most closely resembles the Metropolitan example is one that was once in the Meredith Collection and is now in a private collection in Canada.[2] So similar are the arrangements of figures that the collection histories of the two were often confused in early literature. In fact, however, the style of the figures of the Meredith end panels is remarkably distinct from that of the Metropolitan chasse and must represent a different hand. The range of the workshop defined by Gauthier as "du Queyroix" seems rather too wide, allowing, for example, another Metropolitan chasse (cat. 28) to have been produced by the same atelier only five years later than the present example. Furthermore, it seems inconceivable that the Crucifixion plaque in the British Museum[3] should be the product of the same workshop. In the absence of reliable evidence, however, it is not possible to account fully for the common threads and distinctions seen among the enamels.

The Metropolitan chasse was once said to be from the Royal Abbey of Saint-Denis, a provenance that cannot be supported by documentation. In fact, only one Limoges chasse, a comparatively modest example, is recorded at Saint-Denis (cat. 112). Given the fact that the present chasse is recorded in England before 1823, it is also possible that it was preserved there since the Middle Ages, despite the destruction of objects during the Reformation[4] inaugurated by Henry VIII.                                        BDB

NOTES
1. Gauthier 1987, no. 204.
2. Ibid., nos. 208–211.
3. Ibid., no. 201.
4. On this phenomenon, see Caudron 1976 and Caudron 1977.

Principal face

Reverse

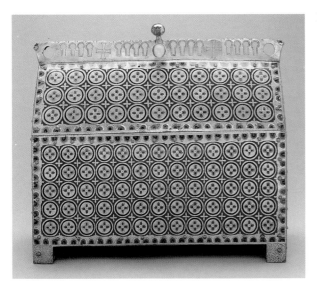

EXHIBITIONS: London 1862, no. 1070; London 1897, p. 8, no. 32, pl. x.

BIBLIOGRAPHY: Rutter 1823, chap. II, heading, p. 7; Marquet de Vasselot 1906, pp. 21–22, pl. v; Pératé 1911, no. 49, pls. XXIV, XXV; Conway 1915, pl. XX, facing p. 156; Rubinstein 1917, pp. 304–6; Breck and Rogers 1929, p. 102, fig. 52; Gardner 1954, p. 47; Seligmann 1961, p. 74; Gauthier 1972b, no. 49, ill. p. 97; Gauthier 1976, no. C38, fig. 12; Caudron 1976, pp. 137–68; Gauthier 1987, no. 205, pp. 181–82, ill. 21 pl. N and pp. 636–38, pl. CLXXXV.

# 25. Book-Cover Plaque: Crucifixion

Limoges, ca. 1180–90

Copper: engraved, chased, and gilt; champlevé enamel: blue-black, lapis and lavender blue, turquoise, green, translucent garnet red, and white

23.5 x 14.1 cm (9¼ x 5⁹⁄₁₆ in.)

INSCRIBED: IH[ESU]S / XPS (Christus)

EX COLLS.: Octave Homberg (acquired in 1913; sale, Paris, June 3–5, 1931, no. 126, pl. LVI). Acquired through the Raoul Duseigneur bequest at that sale.

CONDITION: Wear to gilding on corners and edges; some losses to enamel (medallions surrounding the sun and the moon, mantle of the Virgin); lower right corner broken.

Musée du Louvre, Paris; Département des Objets d'art (OA 8205).

Fig. 25a. Book-cover plaque: Christ in Majesty. Kunstgewerbemuseum, Berlin

The plaque belongs to the small group of book covers with vermiculé backgrounds assembled by J.-J. Marquet de Vasselot in 1932, including, aside from the two plaques at the Lyons museum—the only ones that have come down to us as a pair—the Louvre Crucifixion and another preserved at the Musée Municipal at Nevers, and two images of Christ in Majesty (British Museum, London, and Kunstgewerbemuseum, Berlin [fig. 25a]).[1] Carefully executed, all these plaques portray figures, both severe and elegant, with expressive faces. Their iconography also reveals numerous similarities.

In 1932, Marquet de Vasselot published the plaque, which had recently been acquired by the Louvre. At that time, he emphasized that apart from their almost equivalent dimensions (23.4 x 13.9 cm [9¼ x 5½ in.] for the Berlin plaque), this Crucifixion had certain analogies with the Berlin Christ in Majesty (fig. 25a). But he pointed out that they did not seem "striking enough to allow the assumption that the pieces were once part of the same binding." He did, however, feel that "the two works might have been created at the same atelier."

Today, we can reiterate these conclusions. The affinity seen in the postures and the drapery of the figures, as well as the similarities in certain details—for instance, the drawing of the bare feet with prominent bones—certainly justifies an attribution to the same workshop. The difference of four millimeters in width is not in itself significant, but nothing immediately suggests that the two plaques are a pair. We must add that the known history of these pieces also discourages such an inference, since the Berlin plaque supposedly comes from the treasury of the abbey of Saint-Maximin in Trier, whereas the Louvre plaque was "found" in Spain early in this century.[2] It is necessary to note, however, that this information cannot be verified today.

The two binding plaques are undeniably related to certain works preserved in Spain—notably the altar frontal of San Miguel in Excelsis, Pamplona,[3] and above all to the group of vermiculé chasses classified by Marquet de Vasselot as being in the "severe" style. Chief among these is a chasse at the Metropolitan Museum (cat. 24), the principal face of which is decorated with a Crucifixion that is remarkably similar to the Louvre book cover.

ET-D

NOTES
1. See Gauthier 1987, nos. 153, 223, 231.
2. Marquet de Vasselot 1932, p. 3: "I first saw M. Homberg's plaque when
   he acquired it in 1913. He got it from a foreign antique dealer who had
   found it in Spain." The only certainty about the Berlin plaque is that it
   was previously in the Görres Collection (1776–1848; D. Kötzche, written
   communication, 1994).
3. Gauthier 1987, no. 135.

EXHIBITIONS: London 1932, no. 580e; Paris 1934, no. 374; Vatican City
1963, no. 24.

BIBLIOGRAPHY: Demartial 1932; Marquet de Vasselot 1932, pp. 1–12;
Marquet de Vasselot 1941, p. 129, pl. XXIX; Thoby 1959, p. 146; Gauthier
1968d, p. 283; Gauthier 1987, no. 203, pp. 180–81, ill. 634, pl. CLXXXIV.

# 26. Cross

Limoges, ca. 1180

Copper: engraved, scraped, stippled, and gilt; appliqué: engraved, chased, and gilt; champlevé enamel: dark and medium blue, turquoise, dark and light green, red, and white

37.7 × 28.6 cm (14⅞ × 11¼ in.)

INSCRIBED: IH[ESU]S/XPS (Christus) (S reversed in each instance)

EX COLLS.: Gaillard de La Dionnerie, Poitiers (sold, November 30, 1903, no. 206); Georges Hoentschel, Paris (by 1906); [Jacques Seligmann, Paris (in January 1912)]; J. Pierpont Morgan, London and New York

CONDITION: Extensive wear to gilding; three enamel plaques missing from reverse.

The Metropolitan Museum of Art, New York; Gift of J. Pierpont Morgan, 1917 (17.190.773)

On this slender, double-sided cross with stepped terminals, the image of Jesus is set against a vermiculé ground enameled in deep blue. His figure is in reserve and gilded; his head has been separately carved and applied to the copper plate of the cross. He wears a long loincloth of deep blue, and his feet rest on a blue-and-white dotted suppedaneum. The half-length figures of the Virgin and Saint John, their heads also applied, appear on the stepped terminals at left and right. In the upright terminal above the identifying inscription are half-length images symbolizing the sun and the moon; below Christ's feet is the standing image of a male saint, his face damaged by having been pierced, probably during a later mounting of the cross.

Exceptionally, the reverse of the cross is decorated with an overall pattern of vermiculé without enameling on the shaft and crossarm, and with three inset enameled plaques with figures in reserve at the top and at either side. A circular medallion set into the crossing and two rectangular plaques from the lower half of the shaft are missing.

The cross is remarkable for its rarity and for its fine quality, which is particularly noticeable in the carving of the head and torso of Christ, in the delicate stippling of the inscription, and in the skill of the enameled vermiculé work. It has traditionally been compared with a cross in Barcelona,[1] which may be of slightly later date. Most recently, Marie-Madeleine Gauthier has compared the open scrolling form of the vermiculé with the *Urna* of Silos, attributing the cross to a Hispano-Limousin master. It is not necessary, however, to look to Spain to explain the style of this cross, which was first recorded in a French collection. Though its condition has been somewhat compromised, the cross can be compared with the book-cover plaque of the Crucifixion in the Louvre (cat. 25), with its stepped cross and similarly defined head and shoulders of Christ, or,

rather more generically, with the central plaque of the Metropolitan chasse (cat. 24). The form of the suppedaneum is remarkably similar to that of conventional Limousin book covers, including the use of engraved lines to suggest pages and reserved metal to indicate the closing strap.

The construction and decoration of the reverse of the cross are without known parallel. The inset plaques of saints (at least some of whom are female) holding either a tall cross or a book—one at the viewer's right holds a disk—have no recognized iconographic antecedents. It is to be presumed that these figures are worshiping a now-lost central image of Christ in Majesty, or, perhaps more likely, given the circular form, of the Lamb of God. The sketchy drawing of the figures might seem to suggest that these enameled plaques are later in date; yet there is no technical evidence to indicate that this is the case. It should be noted, moreover, that the pattern of dots aligned in triangular patterns, found on the garments of the saints on the terminals of the front face, is similar to the pattern of white dots on blue enamel often found in late-twelfth-century Limoges work (see cats. 18, 55a and b).

Gauthier suggested that a relic may have been set into the copper beneath the missing central medallion. However, the shallow cavity was certainly created as a means of receiving the enameled disk, and there is no inscription or other evidence to confirm that a relic was contained here as well. The additional cutting away of metal tab shapes where each terminal plaque was inset suggests that other elements were meant to protrude from the body of the cross. If these cuttings were only on the horizontal crossarm, they might have served to hold the supporting tabs of pendant decorative elements, such as the letters alpha and omega. In this instance, however, there are corresponding channels in the upright shaft as well, and their function is unknown.

BDB

NOTE
1. First by Ross (1933a); illustrated in Thoby (1953, no. 22, pl. XIII).

BIBLIOGRAPHY: Marquet de Vasselot 1906, p. 39; Pérat é 1911, no. 51, pl. xxvii; Hildburgh 1936, p. 86, n. 5; Ross 1933a; Thoby 1953, no. 21, pl. XIII; Gauthier 1987, no. 76, ills. 214, 215 pl. LIII and p. 293 pl. LXXI.

Reverse

# 27. Chasse: Christ in Majesty, the Lamb of God

Limoges, ca. 1180–90

Copper: engraved, scraped, stippled, and gilt; champlevé enamel: blue-black; dark, medium, and light blue; turquoise, dark and light green; yellow, red, translucent wine red, and white; wood core, painted red on exterior

22 × 23.5 × 10.7 cm (8⅝ × 9¼ × 4¼ in.)

INSCRIBED: A/ω (Alpha/omega, twice, surrounding Christ and the Lamb)

EX COLLS.: Frédéric Spitzer, Paris (sold, Paris, April 6–June 17, 1893, no. 231, pl. VII–3); Georges Hoentschel, Paris (sold 1911); [Jacques Seligmann, Paris (in January 1912)]; J. Pierpont Morgan, London and New York

CONDITION: Slight wear to gilding along edges; some loss to enamel near lock; crest repaired at each end.

The Metropolitan Museum of Art, New York; Gift of J. Pierpont Morgan, 1917 (17.190.523)

Christ in Majesty, framed within a mandorla and surrounded by the symbols of the evangelists, sits on the arc of a rainbow at the center of the principal face. His right hand is raised in blessing and his left holds a book; his feet rest on a polka-dot footstool. On either side of Christ stand two male saints, each holding a book. Perhaps representing the four evangelists, they are barefoot beneath arches that spring from fancifully enameled columns surmounted by small windowed towers. The gilt ground against which all the figures are set is richly engraved with a rhythmic pattern of scrolling vermiculé vines.

On the upper plaque of the principal face the Lamb of God is in front of a cross within a circle, clasping a book with his right hoof. As in the image of Christ directly below, a cruciform nimbus encircles his head and the Greek letters alpha and omega appear on either side. Four three-quarter-length angels are paired on either side of the Lamb in positions corresponding to those of the standing saints below. All the figures, except those of the symbols of Matthew and John, hold red-enameled books.

Saint Peter, holding the keys to the kingdom of heaven, appears on the proper right end of the chasse, where the hinged door could be opened by means of a key inserted into the original lock opening. Peter's depiction is an explicit reference to his having been given the keys to the kingdom of heaven by Jesus and also a kind of visual pun referring to the need for a key to gain access to the sacred contents of the box. Peter's standing figure, also set against a vermiculé ground, is framed by a trefoil arch and paired enamel columns from whose capitals spring a fantastic array of towers reminiscent of Romanesque architecture. On the proper left end of the chasse stands a beardless saint holding

a book and, like Peter, set between the columns of a fantastic cutaway church. The chasse has retained its original crest, though repaired at each end, and is set with three rock-crystal cabochons.

The figure style is remarkable for its strong delineation of features by means of dark blue enamel and for such details as the lines that crease the skin of the figures' necks, feet, and wrists, all rendered with enameling. The drawing of this example is less mannered and individual than that seen on the Apt and Washington examples (cats. 21, 22), yet it is clearly related to that signature style, which is also articulated on the Munich chasse.[1] The graphic quality imparted by this use of dark blue enamel seems to invite comparisons with contemporaneous stained glass or illuminated manuscripts. Yet precise comparisons of these motifs are elusive, testament to the inventive hand of the master of this enameling workshop. The approximate chronology of the chasses can be suggested by comparison with glazing of about 1180[2] at Angers where, for example, the craggy features of Peter in the Death of the Virgin recall the Peter of the Munich chasse. The New York chasse, like the one in Apt, employs a seldom-found translucent red as well as an opaque red.

Marquet de Vasselot noted that the iconography of the Metropolitan chasse is "almost identical" to that of an example preserved in the parish church of Zell-an-der-Moselle,[3] but the quality of both the drawing and the enamel is distinct—as Marie-Madeleine Gauthier clearly recognized by assigning them to different workshops—and far superior on the Metropolitan example.

Gauthier has suggested that the decoration of the reverse—an overall pattern of quatrefoils enclosed in roundels—was meant to suggest the silken texture and the design of a funeral pall. However, the pattern does not conform to any medieval silks preserved in France. Moreover, the same pattern occurs in other Limousin objects where the context is distinct—as a single line on the borders of the pair of book covers in Lyons (cat. 18) and as the "stained-glass windows" on the chasse of Ambazac (cat. 55).

BDB

NOTES
1. Gauthier 1987, no. 151; Munich 1992, no. 5, pp. 106–9. The chasse comes from Schloss Tüssling, where it is recorded in 1688.
2. Grodecki 1977, p. 83, fig. 64, pp. 276–77, no. 34.
3. Gauthier 1987, no. 229.

BIBLIOGRAPHY: Rupin 1890, p. 423; Palustre and Molinier 1890, no. 20, p. 103, pl. VIII; Marquet de Vasselot 1906, p. 24; Pérathé 1911, no. 48, pl. XXIII; W. H. Forsyth 1946, p. 238; Gauthier 1966, no. C39; Gauthier 1987, no. 150, ills. 514–16 pl. CXLI and 520 pl. CXLIII.

Principal face and left side

Reverse and right side

135

# 28. Chasse: The Crucifixion and Christ in Majesty

Limoges, ca. 1190

Copper: engraved, stippled, and gilt; champlevé enamel: dark, medium, and light blue; turquoise, green, yellow, opaque and translucent red, and white; wood core, painted red on exterior

26 × 29 × 11.5 cm (10¼ × 11⅜ × 4½ in.)

INSCRIBED: IH[ESU]S / XPS (Christus)

EX COLLS.: Alexis Joseph Febvre, Paris (sold April 17–20, 1882, no. 151, p. 74); Frédéric Spitzer, before 1890–93 (sold, Paris, April 6–June 17, 1893, no. 229, pl. VII–3); Picard; Michel Boy, Paris (sold, Paris, 1905, Galerie Georges Petit, no. 145); [L. Daguerre, Paris]; George and Florence Blumenthal, New York (before 1926)

CONDITION: Some wear to gilding; losses to enamel, especially at edges; crest added before 1890, removed at the Metropolitan Museum.

The Metropolitan Museum of Art, New York; Gift of George Blumenthal, 1941 (41.100.155)

On the lower plaque of the principal face, set against an elaborate scrolling ground, is the Crucifixion with the Virgin Mary and Saint John standing at either side under the arms of the cross and half-length angels, probably symbolizing the sun and the moon, above. To the left and the right, two over-lifesize angels stand outside the central scene, framed under the arch of a vaulted interior space that is decorated with fancifully enameled columns and brickwork. The angel at the left turns toward Christ, gesturing with an outstretched right arm. The angel at the right turns his foot outward and holds his hand against his chest.

On the roof plaque above the Crucifixion is the image of Christ in Majesty enthroned within a mandorla, flanked by the symbols of the evangelists. Four apostles are paired at either side of Christ, each standing beneath an arch and framed by enameled columns. They turn toward Christ. At the left, one holds a book, the other a scroll; those at the right raise their arms in gestures of exclamation. On each end is another apostle standing beneath an arch surmounted by a windowed tower. One apostle holds a book, the other a scroll.

The heads on the principal face of this example, unlike those of the chasse from the Morgan Collection (cat. 27), have been separately worked and applied. Only the heads on the ends are engraved and enhanced with red enamel. Though some of the same conventions are used on both chasses—the horizontal lines crossing the feet and defining the edges of the necks—here the graphic line is weaker, betraying a less accomplished hand.

The reverse of the chasse is decorated with an overall repeating pattern of quatrefoils inscribed in circles, common to many vermiculé chasses. However, only one other known vermiculé chasse depicts oversize standing angels flanking

the Crucifixion;[1] its similarity to the present example was first noted by Marquet de Vasselot.[2] Though its enamel has been lost, the style of the engraved copper is legible, and Gauthier[3] correctly attributes it to the same workshop, but there is no evidence to confirm the suggestion that this atelier was located near the church of Saint-Pierre-du-Queyroix in Limoges.

Neither the iconography nor the collection history of this chasse provides any hint of its original destination. Indeed, its subject matter would have been appropriate for use in any church. Nor is there sufficient collection history to suggest its provenance. Alexis Febvre, the first recorded owner, was trained as a gilder in Brussels but was established as a dealer in Paris by about 1850.

BDB

NOTES
1. In The Metropolitan Museum of Art, New York; Gift of George Blumenthal, 1941 (41.100.161).
2. Marquet de Vasselot 1906, pp. 22–23.
3. Gauthier 1987, no. 220.

EXHIBITION: New York 1943.

BIBLIOGRAPHY: L'Art (1882), p. 257; Palustre and Molinier 1890, no. 18, pp. 102–3; Rubinstein-Bloch 1926, pl. XI; Marquet de Vasselot 1906, pp. 22–23; Gauthier 1966, no. C42; Gauthier 1987, no. 220, figs. 668–70 pl. CXCVI.

Reverse and left side

Principal face and right side

# 29. Chasse: Christ in Majesty

Limoges, ca. 1185–95

Copper: engraved, chased, and gilt; champlevé enamel: blue-black, lapis and lavender blue, turquoise, green, yellow, red, and white; wood core with traces of red paint

14.8 x 16 x 7.8 cm (5¹³⁄₁₆ x 6⁵⁄₁₆ x 3¹⁄₁₆ in.)

Ex coll.: Acquired from M. Bourgeois, Cologne, 1903

Condition: Considerable wear to gilding; relief of heads worn; some losses to enamel; crest and bottom plank of core missing; vestiges of original hinge mechanism survive on vertical boards of core.

Musée du Louvre, Paris; Département des Objets d'art (OA 5892)

This small chasse has drawn but scant attention, no doubt because of the wear to the principal face. Yet the engraving and the enamelwork are of very high quality, reflecting the refinement typical of the most beautiful late-twelfth-century Limousin works with vermiculé backgrounds, especially the stippling around the enameled rosette design on the reverse and the rich and contrasting palette whose colors border the reserved lines of the metal and highlight the vestments with their polka-dot patterning. Sometimes three or four colors are juxtaposed in a single cell to produce an effect of "scales" on the tips of the angels' wings.

This work is one of a group of small chasses distinguished by the same decorative elegance but also by an array of common characteristics, iconographic and stylistic.[1] Each chasse has identical decoration, with angels and apostles surrounding the central figure of Christ, who is usually, as here, depicted in majesty. The reverse is always ornamented with a regular pattern of rosettes inscribed in circles. These works were executed in one or several ateliers, not for one precise destination but for a varied and expanding clientele. On all the chasses, stylized drapery with short, curving lines clothes the figures. (This drapery style was called "torique," or "curving," by Marie-Madeleine Gauthier.[2]) The faces are always in the form of half-relief appliqués, also strongly stylized.

The portrayal, on the principal face, of Christ in Majesty flanked by two apostles places the Louvre chasse among those preserved at Auxerre, Lyons, Saint Petersburg, and Moûtiers, and in the Keir Collection, on deposit at the Nelson-Atkins Museum of Art, Kansas City. But in most of these examples a Paschal Lamb is depicted at the center of the roof; here the roof is ornamented by three angels inscribed within circles—similar to the scheme on the reverse of the Gimel chasse (cat. 16). The scalloped pattern outlining the central angel is close to that found on the mandorla surrounding Christ on the book-cover plaque at Lyons (cat. 18). The highly refined decoration on the reverse, where the petals of the rosette are, alternately, enameled or stippled, is also seen only on the chasses of Auxerre, Lyons, and Zell-an-der-Mosel.[3]

The principal aspects of their style and decoration place these small chasses—especially the Louvre's piece—in immediate succession to the large chasses with vermiculé backgrounds and Christological or narrative decoration. The stylized drapery recaptures, in simplified form, the drapery frequently found on the works with narrative themes (cats. 16, 17). The scaly effects on the angels' wings can also be seen on the Metropolitan Museum chasse (cat. 27) and on the chasse of Saint Valerie at the British Museum, London;[4] the marbled effects on the columns also appear on another Metropolitan chasse (cat. 28). The pattern of small crosses on the borders, alternating on a light or dark background, appears on several chasses with vermiculé backgrounds—for example, the one from Nantouillet and the British Museum coffret ornamented with secular scenes (fig. 19a).[5] The rosettes engraved on the roof, between angels in circles, are, for their part, analogous to the ones on the Ambazac chasse (cat. 55).[6]

ET-D

Notes
1. Gauthier 1987, nos. 224–34.
2. Ibid.
3. Indeed, it was not included in Gauthier 1987, while the other comparable examples that were cited did figure there.
4. Gauthier 1987, no. 91 (on the small sides).
5. Ibid., nos. 174, 177.
6. They are also found on the Auxerre chasse and on the chasses from the former Burns Collections (Gauthier 1987, nos. 226, 230), belonging to the same group of small chasses with vermiculé backgrounds.

Bibliography: Migeon 1904a, p. 197, fig. p. 198; Marquet de Vasselot 1905, p. 243; Marquet de Vasselot 1914, no. 61, pl. XII.

Reverse and right side

Principal face and left side

# 30. Chasse: Crucifixion and Christ in Majesty

Limoges, ca. 1185–1200

Copper: engraved, chased, and gilt; champlevé enamel: lapis, light, and lavender blue; turquoise, dark and light green, yellow, red, and white; wood core

17.7 × 17.4 × 10.1 cm (7 × 6⅞ × 4 in.)

Inscribed: ih[esu]s / xps (Christus)

Ex coll.: Edme Durand. Acquired in 1825

Condition: Crest missing; early lock attached on back.

Musée du Louvre, Paris; Département des Objets d'art (MR 2648)

This chasse from the former Durand Collection belongs to a small group of works that mark a turning point in the evolution of the Limousin enamelers' practices. In style and iconography, these pieces are the immediate succedents of the late-twelfth-century chasses with vermiculé backgrounds and Christological themes (see cats. 24, 27), although the two figures of Christ here are accompanied by a total of sixteen depictions of apostles. But here the enameling procedure has been reversed: the engraved figures are in reserve, set on an enamel ground. Moreover, on a chasse from Chamalières (Puy-de-Dôme), preserved at the cathedral of Clermont-Ferrand (figs. 30a, 30b),[1] both techniques appear: the figures on the front are in reserve and those on the reverse and sides are enameled. Another chasse—from the Soltykoff, then later the Spitzer Collection, and now in the Toledo (Ohio) Museum of Art (fig. 31a)—differs only very slightly from the Louvre example.[2] On the Paris and Toledo chasses, as on the front of the Chamalières piece, the engraved lines that suggest the drapery clearly reproduce the effects of the champlevé technique (see cat. 31).

The faces engraved on the reverses of the Louvre and Toledo chasses present the austere expressions, strong styli-zation, and elegant individuality to be found—also engraved but with enameled features—on the reverse of the Chamalières chasse; this is a simplified version of the "severe" style of the large chasses with vermiculé backgrounds and Christological themes (see cats. 24, 27). The heads fixed to the front of the Louvre chasse are also close to those on other small chasses (see cat. 29) and are distinguished by the rare precision of the engraving that individualizes each face. On the Toledo chasse, by contrast, the appliqué heads are less differentiated and closer to classical heads. Finally, these chasses present a highly original palette, which gives the greens a much larger role than was usual.

These works thus acquaint us with an independent and innovative artist. Several other works can be attributed to the same workshop or to an immediate successor, notably a plaque at the Nationalmuseet, Copenhagen,[3] depicting Christ in Majesty and four apostles, which is very close to the Toledo piece, the chasse from the former Dzialynska Collection, the National Museum of Warsaw, but preserved at the Museum of Poznań,[4] and a chasse at the Metropolitan Museum (cat. 31).

J.-R. Gaborit (1976a) drew attention to the exact correspondence between the Louvre chasse and the "coffret of Saint Brandan" described in the 1666 inventory of the Grandmont Treasury (no. LXXV), the most precise of the Grandmont inventories.[5] Since the iconography of the Paris and Toledo chasses is virtually identical, we must assume that one or the other could have come from the Grandmont Treasury—unless both were copied from a Grandmont model.

ET-D

Figs. 30a, 30b. Chasse of Chamalières (principal face and reverse). Cathedral Treasury, Clermont-Ferrand

Principal face and left side

Reverse and right side

NOTES
1. Gauthier 1987, no. 222.
2. Ibid., no. 307. Soltykoff Collection, Paris sale, April 8, 1861, no. 140.
3. See Liebgott 1986, p. 46, fig. 37.
4. Molinier 1903, no. 142.
5. Texier 1857, cols. 877–78: "A casket made as a chasse, in gilt copper, enameled on the outside, wooden on the inside, with the image of the Crucifix and that of the Virgin and Saint John and two other apostles on its sides; above, a Savior who is in the midst of the four evangelists, and two other saints; on the back, four figures as apostles below, four others above, and two on each side." The object was "reserved" for Abbé Sicelier (ibid., col. 903) and could thus escape being melted down; Gaborit 1976a, p. 246, n. 63.

EXHIBITIONS: Limoges 1886, no. 36; Vatican City 1963, no. 87; Paris 1979, no. 208.

BIBLIOGRAPHY: L. Laborde 1852, nos. 71–76; Courajod 1888, p. 52; Rupin 1890, p. 421; Darcel 1891, nos. 94–101; Marquet de Vasselot 1914, no. 62; Fontaine 1931; Gaborit 1976a, n. 63, p. 246; Gauthier 1987, no. 306, ills. 839, 839bis, 841, 842, pls. CCLII, CCLIII.

# 31. Chasse with Christ in Majesty and Apostles

Limoges, ca. 1190–1200

Copper: engraved, chiseled, stippled, and gilt; champlevé enamel: blue-black, dark, medium, and light blue; turquoise, green, yellow, red, translucent red, and white

16.2 × 16.3 × 9 cm (6⅜ × 6⁷⁄₁₆ × 3½ in.)

INSCRIBED: A/ω (Alpha/omega)

EX COLLS.: Georges Hoentschel, Paris (before 1911); [Jacques Seligmann, Paris (in January 1912)]; J. Pierpont Morgan, London and New York

CONDITION: Wear to gilding at edges; some losses to enamel, especially on lower plaque of principal face and on rear roof plaque; hinges and lock replaced; iron bracing added on interior.

The Metropolitan Museum of Art, New York; Gift of J. Pierpont Morgan, 1917 (17.190.513)

Christ in Majesty appears at the center of the lower plaque of the principal face. Seated on a pillow set on the arc of a rainbow and framed within a mandorla, he raises his right hand in blessing; in his left, he holds a book. The keyhole of the lock is cut into his left side. Heads of the symbols of the evangelists appear around the mandorla. A male saint, probably an apostle, stands at either side of Christ, framed within a rounded arch that springs from capitals. The one at the left holds a book. Christ appears again at the center of the roof plaque, set against a field of green enamel. He is framed under an arch, as are the single figures to either side of him. Again he gestures in blessing and holds a book. The saint on his left also holds a book, whereas the one on his right clasps a large disk to his chest. It is marked with a cross, perhaps signifying the Host. The saints' garments are decorated with fine stippling set in a triangular pattern. The heads of Christ and his apostles on the principal face are separately worked and applied. A bearded male saint appears under an arch on the proper left end of the chasse; a youthful and beardless one appears on the proper right end, each standing against a field of green. The reverse of the chasse is decorated on the roof with large enameled flowers enclosed in roundels; the lower panel has small quatrefoils and stars set against a diaper ground. The feet of the chasse are engraved with either a vermiculé or a diaper pattern.

The chasse is inventive both in its design and its construction. It is distinguished by a rare and beautiful green enamel ground and by the lively figure of a lion on the proper right gable of the lid. (The opposite gable, somewhat more predictably, depicts a grouping of towers surmounted by a cross.) Both saints on the end panels and the lion on the gable are partly in reserve and partly enameled.

The Metropolitan chasse finds its closest parallels in examples preserved in the Louvre (cat. 30) and in the

Toledo Museum of Art (fig. 31a),[1] especially in its unusual palette and in the disposition of the figures on the sides. Like these examples, the Metropolitan Museum chasse was clearly created by an enameler who was aware of decorative conventions used in the production of vermiculé chasses. The figures of the side plaques are characterized by paired engraved lines that define the neck and the muscles of the arms, following the example of vermiculé work, but they are no longer enameled, and they have lost definition and distinction: the lines of the apostle's beard mix indiscriminately with the lines of his neck muscles. The floral pattern of the reverse of the Metropolitan chasse is unique among the three, surely based on the pattern of quatrefoils in circles found on vermiculé chasses. The disposition of the vermiculé pattern on both front supporting feet but on only one of the reverse is particularly surprising. The varying pattern of the framing enamel of each of the plaques is also remarkable.

The Metropolitan chasse is probably from the same workshop as the Paris and Toledo examples, but they are not the work of the same hand. The products of this atelier are few, but not, consequently, exclusively local: one chasse is preserved in the Puy-de-Dôme, another in Denmark (see cat. 30).

As Gaborit noted, the chasse at the Louvre (and in fact the one in Toledo) conforms to the description of a "coffre" of Saint Brandan listed in the inventory of the abbey of Grandmont.[2] This parallel offers further testament to the range of Limoges work acquired for Grandmont.

BDB

NOTES
1. The Toledo chasse has a later Mosan-style crest, probably added when it was in the Spitzer Collection. See cat. 30.
2. Gaborit 1976a, p. 246, n. 63.

EXHIBITION: Oklahoma City 1985, no. 40, pp. 144–45.

BIBLIOGRAPHY: Pérate 1911, no. 52, pls. XXVIII, XXIX; Gauthier 1966, pp. 944, n. 18, 949.

Principal face and left side

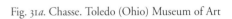
Fig. 31a. Chasse. Toledo (Ohio) Museum of Art

Reverse

# 32. Glove Plaques

Limoges, ca. 1180–90

Copper: engraved and gilt; champlevé enamel: lapis and lavender blue, turquoise, green, yellow, red, pinkish white, and white

Diam.: 4.2 cm (1⅝ in.)

INSCRIBED: + AGNUS DEI QUI TOLLIT P[ECCATA MUNDI] (Lamb of God, who takes away the sins of the world) + DEXTERA DOMINI (The right hand of the Lord)

PROVENANCE: Unearthed from a tomb, along with a thirteenth-century crosier depicting the Annunciation and the Visitation, in 1872 during work under the choir of the cathedral of Cahors. Classified, Monuments Historiques, 1901

CONDITION: Some wear to gilding; deterioration of the enamel, caused by burial.

Treasury of the Cathedral, Cahors (Lot)

The two circular plaques, one depicting the Paschal Lamb, the other, the Hand of God Blessing, were discovered in a tomb beside a crosier. They are pierced with sixteen regularly spaced holes, which allowed them to be sewn to fabric; undoubtedly, the plaques adorned liturgical gloves. According to Xavier Barbier de Montault, one of the first to write about them (1876), the plaques were still sewn to the tops of the gloves when they were found.

Few objects of this type have come down to us, and these medallions from the treasury of the cathedral of Cahors are the only preserved ones that were executed in the workshops of Limoges. Thus, despite their small size and decorative restraint, they are documents of major importance. Rupin linked them to an isolated plaque that depicted the Paschal Lamb, then owned by Ferdinand de Lasteyrie, all trace of which has since been lost.[1]

The identity of the bishop buried with these plaques is not known. Pons d'Antejac (r. 1235–36) and Anthony of Luzech (r. 1497–1508) have been suggested, but without proof.[2] Be that as it may, it appears that these gloves were reused for funerary purposes. They certainly predate the thirteenth century: as correctly pointed out by Marie-Madeleine Gauthier, the concentric curves employed in the modeling of the Paschal Lamb link these plaques to the group of vermiculé works executed at the Limousin ateliers during the last three decades of the twelfth century. This Lamb is very close to the one on the roof of a chasse at the Metropolitan Museum (cat. 27), which allows us to attribute a date of about 1180–90 to the two glove plaques.

ET-D

NOTES
1. Rupin 1890, fig. 641; see also Gauthier 1987, no. 231bis.
2. Costa 1993; Rupin 1890; Gauthier 1987.

EXHIBITIONS: Paris 1889, no. 441; Paris 1965, no. 530.

BIBLIOGRAPHY: Malinowski 1873, pp. 49–50; Barbier de Montault 1876, p. 785, fig.; Rohault de Fleury 1883–89, vol. 8 p. 195; Rupin 1890, pp. 567–69, figs. 639, 640; Enlart 1916, p. 384; Calmon 1931, p. 489; Beaulieu 1968, nos. 48, 144; Favreau et al. 1984, vol. 9, no. 5, figs. 62, 63, pl. XXIX; Gauthier 1987, no. 231, pp. 195–96, ills. 699, 700, pl. CCVI; Costa 1993, p. 81, fig. 4, p. 82.

*2*

*Secular ornament and imagination*

# 33. Pair of Candlesticks

Limoges, ca. 1180–90

Copper: engraved, scraped, stippled, and gilt; champlevé enamel: dark, medium, and light blue; green, yellow, red, and white

H.: 23.7 cm (9⅜ in.)

EX COLLS.: Thomas Barrett, Lee Priory, Kent (sold August 11–15, 1834); [Horatio Rodd, London, 1834]; George Weare Braikenridge, Bristol (1834–d. 1856); W. Jerdone Braikenridge (sold Bristol, February 27, 1908, lot 51); Mrs. Otto H. Kahn, New York (sold 1936); [Joseph Brummer, New York]

CONDITION: Considerable loss to gilding; worn engraving; some loss to enamel, especially on the knops and bobeches; prickets replaced.

The Metropolitan Museum of Art, New York; The Cloisters Collection, 1947 (47.101.37, 38)

Each candlestick stands on a truncated pyramidal base. On each face of the pyramid is a striding warrior armed with shield and club, inscribed in a green-enameled circle and flanked by lions set against a blue ground dotted with white enamel. Three short feet extend from the base of the pyramid. Each is in the form of an animal's paw issuing from a lion's-head mask. Threaded onto each shaft are two cylinders, with vermiculé decoration, and a knop with fantastic birds of blue, red, and green enamel. The original disk-shaped bobeches survive; the prickets, although dating to the early twentieth century, reflect a twelfth-century form.

Fig. 33a. Candlestick. Musée National du Moyen Âge, Thermes de Cluny, Paris

The stylized birds on the knop and the vermiculé sleeves of the shafts are common to enamels dated to the second half of the twelfth century from the Limousin and from Silos in northern Spain. Two decorative elements of the Cloisters candlesticks, however, appear to be uniquely Limousin hallmarks: the use of red enamel to fill the engraved lines of the lions and the use of white "snowflakes" on a blue ground.[1] Many ornamental elements seen on the candlesticks occur on a wide range of other objects presented here. The Cloisters candlesticks are closely related to a single stick in the Bayerisches Nationalmuseum,[2] to another single example in the Glencairn Museum, Academy of the New Church, Bryn Athyn, Pennsylvania, and to one recently acquired by the Musée de Cluny (fig. 33a), which suggests that the workshop responsible for their production specialized in the fabrication of candlesticks.[3]

The creation of enameled candlesticks at Limoges, beginning in the twelfth century, was coincident with their increased use on the altar itself, either singly, as seen on the altar at Mozac (cat. 45), or, more commonly, in pairs. Pope Innocent III (r. 1198–1216) referred to paired candlesticks set at the corners of the altar in *De sacro altaris mysterio*, apparently indicating what had become standard practice. The chronicles of the abbey of Saint-Martial of Limoges record a pair of enameled candlesticks in 1208.[4] They were also commonly exported. A pair of Limoges candlesticks was among the objects that Cardinal Guala Bicchieri presented to the abbey of Sant' Andrea, Vercelli, in 1226 (see cat. 88),[5] and a thirteenth-century inventory of Saint Paul's, London, includes a pair of candlesticks of Limoges work.[6] The collection history of the Cloisters candlesticks suggests that they were in England before the French Revolution and, consequently, may have been exported during the Middle Ages.

BDB

NOTES
1. See Boehm 1992, p. 150.
2. Ibid., p. 155, figs. 10a–c.
3. Ibid., pp. 155–57, figs. 11a,b, p. 158.
4. Duplès-Agier 1874, p. 73.
5. Castronovo 1992, p. 166, n. 2.
6. "Item duo candelabra de cuprea de opera lemovicensi" [Two candlesticks of copper of Limoges work]; Lehmann-Brockhaus 1956, vol. 2, p. 185, no. 2892.

EXHIBITIONS: London 1857b; London 1862, nos. 1109–10; London 1875, nos. 875–77; London 1897, nos. 20–21, pl. VII.

BIBLIOGRAPHY: London 1857a, p. 26; *Transactions* (1878–79), pp. 36–38; Rupin 1890, p. 525; Boehm 1992.

# 34. Appliqués of the Souvigny Bible

Limoges(?), ca. 1175–85

a, b. *Two Medallions Decorated with Rosettes*

Copper: chased and gilt; champlevé enamel: lapis and lavender blue, dark and almond green, yellow, and white

Diam.: 4.4 cm (1¾ in.)

c, d. *Two Rectangular Plaques with Vermiculé Decoration*

Brass (alloy identical to that of plaques and bosses e to o)

5.6 x 11.8 cm (2³⁄₁₆ × 4⅝ in.)

e, f, g. *Three Rectangular Plaques of Openwork Decorated with Palmette Flowers*

Brass (alloy of 86–90 percent copper, 5–7 percent zinc, and 4–6 percent tin with many impurities, principally lead, iron, and antimony), engraved, chased, plaque g reusing a plaque with vermiculé decoration

e. 5.6 × 20.2 cm (2¼ × 8 in.)

f. 5.6 × 20.1 cm (2¼ × 7⅞ in.)

g. 5.6 × 12.3 cm (2¼ × 4⅞ in.)

h. *A Circular, Openwork Boss with a Decoration of Winged Dragons and Palmette Flowers*

Fig. 34a. Appliqués mounted in 1832–33 on the binding of the Souvigny Bible, before their removal in 1980

Brass (alloy identical with preceding) probably cast; engraved and chased

Diam.: 12 cm (4¾ in.)

i to o. *Seven Openwork Bosses.* i to n: *Decoration of dragons with tails ending in scrolls of palmette flowers;* o: *Siren-dragon with tail ending in a scroll of palmette flowers.*

Brass (alloy identical with preceding), probably cast, engraved, and chased

Height and width: 4.9–5.1 cm (1⅞–2 in.)

PROVENANCE: From priory of Saint-Maïeul in Souvigny (Allier), where they decorated the binding of a Bible (see inventory of March 23, 1795: "Numerous copper plates very artistically engraved and festooned with openwork," quoted in Conny, *Description de la Bible de Souvigny*, Sept. 6, 1856, Paris, BNF, ms. nouv. acq. fr. 22328, fol. 155). The Bible itself is probably the one given to Souvigny by the sacristan Bernard; see *Livre des Anniversaires* prepared under the prior Dom Geoffrey Chollet (1424–54), Moulins, Bibliothèque Municipale, ms. 13, fol. 60, quoted in Cahn 1980, p. 12.

Bible and binding were brought to Moulins in 1783 (attic of Collège des Jésuites, then deposited at Bibliothèque de Moulins, created in 1832).

Appliqué *n* was perhaps detached in 1832 or 1833 and later would have passed into the hands of the mayor of Moulins, who supposedly owned four grotesque masks (see "Chronique" 1884, p. 291); J.-B. Moretti Collection, Moulins (1884 and 1893); A. Figdor Collection, Vienna (before 1901; see Bertrand 1903; sale cat., Berlin, Sept. 20, 1930, no. 447); Robert von Hirsch Collection, Frankfurt, then Basel (sale, Sotheby's, London, June 22, 1978, no. 238). Acquired at the von Hirsch sale by the city of Moulins and deposited at the Bibliothèque Municipale.

CONDITION: Bible and binding were restored in 1832–33 (fig. 34a). At that time the "old pigskin" binding was replaced with velvet (see D'Avout 1847). Fourteen pieces of goldsmiths' work were added in 1856 (see Conny, BNF, ms. nouv. acq. fr. 22328, fol. 156). The fourteen appliqués were detached in 1980 and placed on an independent support.

The numerous small holes scattered across the plaques and openwork bosses undoubtedly once contained enamel beads.

The Musée d'Art at Moulins owns the molds of twelve appliqués (Gauthier 1987, no. 146). Eleven of them correspond to the unenameled elements earlier attached to the binding; the twelfth appliqué, broken, with floral openwork decoration, points to a sixteenth appliqué, now lost, which was certainly part of the same set.

Bibliothèque Municipale, Moulins

The appliqués of the Souvigny Bible were linked by Marie-Madeleine Gauthier to the work done at Limoges.[1] Within that body of work, however, they occupy a singular place.

The two enameled medallions decorated with rosettes (a, b), numerous stippled highlights, and a refined palette belong indisputably to Limousin production of the final

a

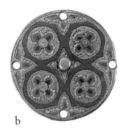

b

decades of the twelfth century and are related to the decorations on the back of the framing strips of certain works with vermiculé backgrounds (cats. 18, 24). The vermiculé decoration of the two rectangular plaques (c, d) and of the corners of the six bosses (i–l, n, o) enhances this parallel. On the other hand, these works are distinguished by the use of a three-part alloy (copper, zinc, tin)[2] which does not seem to have been gilded and indicates, probably in the case of the bosses, that they were formed by casting.

It is certainly possible to imagine an association between the two sets, one made at Limoges, the other imported—an assumption that would seem to be confirmed by the reuse of a plaque with vermiculé background to make one that is pierced and ornamented with foliage (g). Nonetheless, rectangular plaques and bosses have all been made of the same metal, of a composition similar to brass. This suggests, on the contrary, that the plaques with engraved decoration have been deliberately executed in the same material as those obtained by casting to make them match. Pure copper has only been used for the two enameled medallions, following typical Limousin technologies.

c

Furthermore, the decoration of these appliqués is dominated by "palmette flowers," which, because certain long, slender petals join the coiled stem, are called "octopus palmettes." They are, as M.-M. Gauthier (1967) has pointed out, one of the chief characteristics of the "Plantagenet style." These palmettes are comparable to those on the chasse of Apt (cat. 21) and the ornamental facing of Silos and San Miguel de Excelsis.[3] They are distinguished mainly by their design, more detached from examples of English illumination and different from derivations encountered north of the Loire Valley. The stylization of this design gives their contours a less jagged and more compact appearance that especially evokes the manuscript painting created in central France at the end of the twelfth century. While the artists who illuminated the Souvigny Bible had little interest in this ornamental motif, the Clermont-Ferrand Bible—specifically the second volume,[4] which is all that remains—

d

i

j

k

g

has a huge and varied number of initials decorated with pal-mette flowers in which one can easily find the principal species depicted in the Souvigny appliqués and showing the same simplification as the English models (fig. 34*b*). Their contour is often bounded by the body of a winged dragon similar to those on the Moulins bosses.

One of the openwork bosses (o) is adorned not with a dragon but with a creature having a female head and breasts, usually identified as a siren even though the lower part of her body is more like a serpent than a bird or a fish. This figure, with her supplely modeled bare torso and small, round, sweet face surrounded by short curly hair, is set apart from the world of Romanesque monsters such as the drag-ons that coil around the other bosses. She is much more closely related to the small nude figures enlaced in the scrolls of certain capitals of early Gothic sculpture—for example, those on the northern portal of Saint-Étienne in Bourges (ca. 1160) and in the Musée Saint-Remi in Reims (ca. 1170).[5]

It remains to identify the original destination of these appliqués. Gauthier[6] has emphasized that, unlike numerous Gospel books, no known medieval Bible has a cover of pre-cious metal. We must add that the themes and techniques of the decorations on these appliqués are not in keeping with such a use; instead, they evoke that of secular coffrets produced at about the same time by Limousin workshops (see cat. 36). Hence, it is quite probable that these appliqués—the only survivors of an important set—origi-nally adorned a coffret and were later remounted on the binding of the Souvigny Bible.

It is nevertheless difficult to explain why the Limousin workshops abandoned a technique and a type of decoration that they had so perfectly mastered.

ET-D

Fig. 34*b*. The letter V. Bible, Bibliothèque Municipal, Clermont-Ferrand, ms. 1, p. 8

l

m

n

o

e

f

h

NOTES

1. Gauthier 1981 and 1987.
2. This type of alloy seems to characterize a certain number of bronzes of the twelfth century; see Oddy, La Niece, and Stratford 1986, p. 13.
3. Gauthier 1987, nos. 80, 135. M.-M. Gauthier (1967b, no. 38, pp. 152–53) had already suggested attributing the appliqués of Souvigny and the frontal of San Miguel de Excelsis to the same hand, but this seems far-fetched.
4. Bibliothèque Municipale, ms. 1; see Cahn 1982, no. 63 (Central France, ca. 1170–80).
5. See Sauerländer 1972, nos. 35, 55.
6. Gauthier 1981, p. 146, and Gauthier 1987.

EXHIBITIONS: Paris 1878; Paris 1937, no. 741; Paris 1954, no. 331; Paris 1958, no. 116.

BIBLIOGRAPHY: Allier 1832, p. 2; Allier 1833, p. 1 (vignette reproducing three bosses and two plaques); Ripoud 1840, p. 90, pl. 1; D'Avout 1847; Fanjoux 1847, pp. 367–68; Viollet-le-Duc 1871–72, vol. 2, pp. 191–92, fig. 12, p. 191 (appliqué *k*); "Chronique," in *Revue bourbonnaise* 1 (1884), p. 291; Bertrand 1893, pp. 264–67; Bertrand 1903, p. 340; Gauthier 1967b, pp. 152, n. 38; Cahn 1967, pp. 376–88; Gauthier 1968d, p. 283, no. 1, 2; Cahn 1980, p. 12; Cahn 1981, pp. 155–59; Gauthier 1981; Cahn 1982, no. 76, pp. 273–74 (English ed.); Gauthier 1987, no. 145, pp. 142–43, 466–94 ill., pls. CXXXII–CXXXVI.

# 35. Medallions and Straps from a Coffret

Limoges, ca. 1180–90

Copper: engraved, punched, and gilt; champlevé enamel: blue-black, medium and light blue, light green, yellow, garnet and brick red, and white

Medallions: Diam.: 6.9–7 cm (2¹¹⁄₁₆–2¾ in.); 6 Straps: H.: 11.2–11.7 cm (4⅜–4⅝ in.); W.: 3.4–4.4 cm (1⅜–1¾ in.); 2 Straps: H.: 10.5 and 10.7 cm (4⅛ and 4¼ in.); W.: 3.2 and 3.95 cm (1¼ and 1⁹⁄₁₆ in.)

CONDITION: Some losses to gilding; some degradation of enamel surfaces.

EX COLLS.: Comte de Mailly (before 1825); Marquis de Mailly-Nesle, La Roche-Mailly (Sarthe) (in 1880)

Musée du Louvre, Paris; Département des Objets d'art (OA 10889-1–19)

The six historiated quatrefoil medallions represent scenes of hunting. Two show a man on the back of a lion, wielding sword and shield, one the mirror image of the other. One represents a man wrestling a lion, pulling its jaws apart with his bare hands. Another presents a centaur who has just let spring his bow, and the arrow pierces the neck of a stag in a fifth medallion as predators sink their teeth into its flesh. In the sixth medallion, the only one in which the figure is not inscribed in a circle, a man riding a dappled horse carries a shield and an ax. Five more quatrefoil medallions are decorated with foliate ornament. In addition, there are eight decorative straps, two of which are later in date, arguably of the thirteenth century.[1]

By their subject matter and their form, these elements are recognizably from a leathern coffret similar to the Morgan coffret, preserved in the Metropolitan Museum (cat. 36). When first published in an engraving of 1839,[2] however, the Louvre pieces were already isolated elements; their exact number was not recorded until 1880. Nor is it now possible to reconstruct their original arrangement or to determine how many enamels may be missing from the ensemble. The Morgan coffret is decorated with thirty applied medallions, albeit of smaller size. A coffret in the British Museum,[3] which is entirely enameled but which has the semblance of having applied roundels on three faces, has four narrative medallions on the top face and two on each of the end panels. The medallions on the ends seem to be paired by subject. One side shows a continuous narrative— a battle between a hybrid creature and a lion, each in a roundel. This arrangement is similar to the relationship between the Louvre plaques with the centaur armed with a bow and the stag struck by his arrow. The second side of the London coffret shows the same man in a single pose, but in two distinct guises: first playing a string instrument while a maiden dances and then wielding a sword (instead of a bow) to kill a lion. This kind of compositional pairing

recalls the two Paris plaques of the armed man astride a lion. In all probability, the plaque with the man on horseback came from a different face of the coffret than the others, since that plaque alone has a decorative quatrefoil within the picture field and no inner circle in which its figure is inscribed.

The stylistic and technical comparisons between the London and Paris medallions are striking. They have the same palette and use deep red enamel to accent the reserved areas; each includes a dappled horse and a figure with enameled hair streaming out as the wind sweeps through it. The floral medallions recall the quatrefoils applied to the chasse of Ambazac (cat. 55), both in form (especially OA 10889-7) and in palette, but the design of the Louvre quatrefoils overall is at once looser and more intricate.

The early collection history of these medallions in the Sarthe region around Le Mans and their stylistic similarities to works associated with aristocratic, possibly royal, taste, such as the British Museum coffret, may suggest their original provenance in this Plantagenet domain.

Two other quatrefoils bear some similarity to the Louvre series. One is the medallion reused on a cross preserved in Saint Petersburg;[4] the second was applied to a cross acquired by the church at Bagneux in the nineteenth century.[5] In each case, the foliage is constrained within a more rigid geometric configuration; these probably are medallions from distinct ensembles.

BDB

NOTES
1. See Durand 1985, p. 40.
2. The research of Françoise Arquié-Bruley has established that the engraving was made before 1825, though the publication did not appear until 1839. See Arquié-Bruley 1983, pp. 145, n. 55; 154.
3. Gauthier 1987, no. 174, figs. 563–64, 566–68.
4. Ibid., no. 195.
5. Paris 1965, no. 453; Thoby 1953, no. 38, pp. 112–13. Quatrefoils such as these were also produced in the nineteenth century—four quatrefoils preserved in the Fundación Lazao, Madrid, were fabricated to decorate the roof of the Germeau chasse.

EXHIBITIONS: Le Mans 1880, nos. 1201–3; Le Mans 1980, vol. I, p. 104, Ema 5–7; Paris 1985, no. 8, pp. 37–41.

BIBLIOGRAPHY: Willemin 1839, vol. I, p. 67, pl. 110; *Revue du Louvre* 1981, p. 288; Gaborit-Chopin 1983b, p. 327, fig. 293; Arquié-Bruley 1983b, pp. 154–55; Gauthier 1987, no. 189, ills. 598–616, 618, pls. CLXXVI, CLXXVII, CLXXIX.

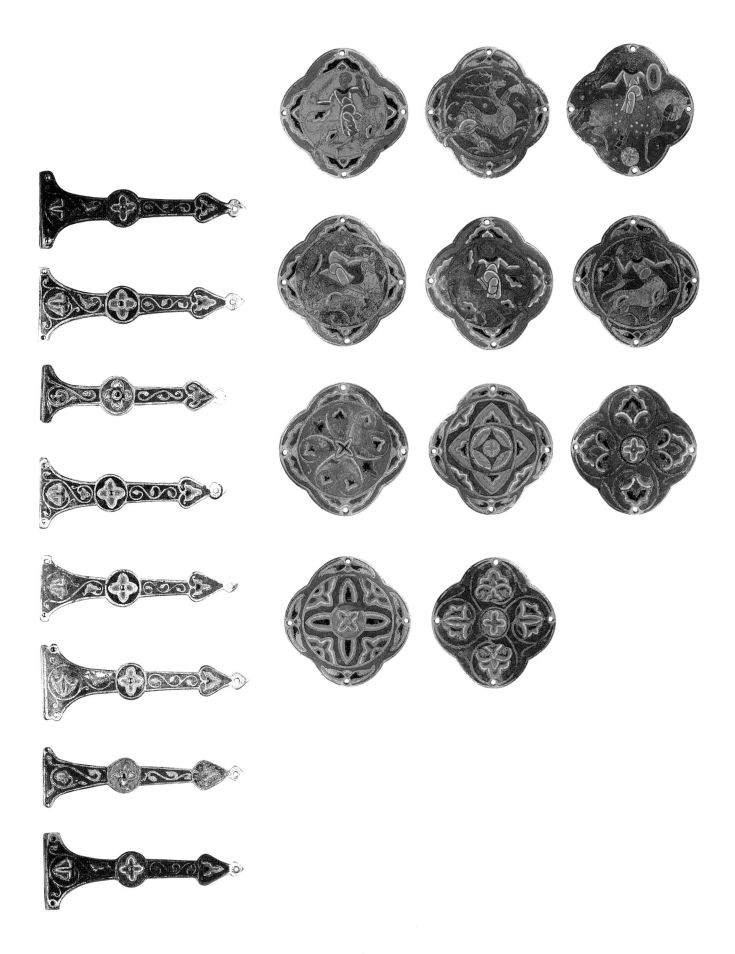

# 36. Coffret

Limoges, ca. 1190

Copper: engraved, stippled, and gilt; champlevé enamel: dark, medium, and light blue; green, yellow, red, and white; wood core with applied parchment, gesso; traces of red paint

18 × 46 × 22 cm (7⅟₁₆ × 18⅛ × 8⅝ in.)

CONDITION: Parchment and wood core consolidated and enamels cleaned, 1985; handle and lock plate missing; bottom wood panel replaced.

The Metropolitan Museum of Art, New York; Gift of J. Pierpont Morgan, 1917 (17.190.511)

Thirty enameled roundels decorate the rectangular wood coffret, each secured by four rivets. On the ends a roundel is set at each corner and at the center. The roundels are disposed in the same way across the top and front panels. An enameled strap with foliate decoration is fastened to the center of the lid; a similar band once extended down the center front. Gilt-copper strapwork decorated with vermiculé patterning is set above and below the centered roundels and on the side edges. The lock plate from the center front is missing, but it may have resembled the openwork medallion on the Glencairn Museum coffret (cat. 38). A pair of winged, lizardlike creatures are hinged together to form the hasp of the lock. The keyhole of an additional lock is set into the strapwork of the front panel at the proper left, and the lock mechanism is still in place inside. Traces of red paint on the inside lip of the coffret suggest that its original color was the same as that found on the undersides of Limoges chasses. The box is sturdy; the leather stretched over the wood core and the studs encircling the medallions serve both to decorate and to protect. The leather provides a warm contrast to the finely executed, colorful roundels with their richly inventive, energetic designs.

Most of the roundels depict fantastic animals; three portray armed men, and three have floral designs. Of the floral medallions, one is markedly smaller but apparently not a replacement, since no stud holes from a larger medallion are perceptible. In fact, small, button-size medallions of this type appear on a coffret preserved in the Museo Arqueológico Nacional, Madrid.[1] The medallions are unique, except for one pair; yet several are clearly subtle variations of the same design. As Marie-Madeleine Gauthier observed, eleven show striding lions or griffins facing to the left, a rear paw raised, the tail lifted and splayed out in a decorative fleuron. Six others present dragonlike creatures similarly disposed in their medallions. The roundel representing an attack on a hart is a simplified version of the subject presented on the Louvre quatrefoil (cat. 35). The placement of the medallions on the coffret seems to be random, and no program is implied either by the range or by the number of subjects. This aspect is particularly important to note, since the Morgan coffret retains its original core and disposition of its decorative elements.

Although its subject matter provokes the suggestion that the Morgan coffret was intended for use by a layperson, it is

The sides

Principal face and right side

Medallion on lid: Dog attacking a stag

Medallion on lid: Fantastic bird

well documented that such boxes were also used in an ecclesiastical context (see cats. 7, 88). An inventory of the property of Saint Paul's, London, taken in 1295, includes a "black coffret which is said to have belonged to Bishop Gilbert, containing many enameled roundels, in which are contained many relics, sealed by the seal of the dean."[2] (Gilbert Foliot of London [r. 1163–87] was a supporter of Henry II against Thomas Becket.) The coffret at Saint Paul's was roughly contemporaneous with the present example.

The inside of the Morgan coffret bears a label numbered 1311, but it has not been possible to associate this with any collection or exhibition history.  BDB

NOTES

1. I am grateful to Charles T. Little for information concerning this object.
2. Gauthier 1987, no. 97, p. 103: "Coffra nigra, quae dicitur fuisse Gilberti episcopi continens multas rotellas aymallatas, in qua reponuntur multae reliquae, modo sigillita sigillo decani"; Lehmann-Brockhaus 1956, vol. 2, pp. 158–59, no. 2902.

EXHIBITION: New York 1970, no. 160, pp. 154–55.

BIBLIOGRAPHY: Breck and Rogers 1925, p. 62; Gauthier 1987, no. 294, pp. 230–32, ills. 43, 793–826.

# 37. Candlestick

Limoges, late 12th century

Copper: engraved, scraped, stippled, and gilt; champlevé enamel: dark and medium blue, green, yellow, red, and white

H.: 40 cm (15¾ in.)

Ex colls.: Georges Hoentschel, Paris (by 1911); [Jacques Seligmann, Paris (in January 1912)]; J. Pierpont Morgan, London and New York

Condition: Considerable losses to gilding; minor losses to enamel; engraving and appliqués worn.

The Metropolitan Museum of Art, New York; Gift of J. Pierpont Morgan, 1917 (17.190.345)

On each of the three faces of the pyramidal base appears a hybrid figure, with the appliqué head of a man who holds a shield and a club in his outstretched arms. The creature has the trunk and short forelegs of a lizard, with a scaly, reptilian back and a yellow and green underbelly. Its curling, attenuated tail terminates in an enameled fleuron. The figure is enclosed within the arc of a circle that is enameled with a cloudlike design. In the lower corners of the triangular faces are enameled flowers. Each of the feet of the candlestick is engraved with a lion's-head mask (now largely effaced) and a lion's paw. Threaded onto the shaft are four gilt-copper cylinders and three knops; the upper and lower knops are enameled in blue with reserved foliate design; the center, with enameled fleurons. The crownlike bobeche is enameled with a flamelike design; the original pricket remains.

The earliest candlestick associated with Limoges, preserved in the Museo Civico, Modena (fig. 15), stands over five feet tall and bears the inscription "Constantinus faber de lemoei civitatis me fecit" (Constantine, artisan of the city of Limoges, made me). Although the style of its engraving recalls the fantastic animals found in early-twelfth-century works from Conques and the Limousin, its size makes it anomalous among surviving Limoges candlesticks. Typical, rather, are candlesticks of more modest size, appropriate for use on the altar. Their decoration with foliate ornaments and fantastic animals, sometimes locked in combat with men, is remarkably similar to the repertory used on Limousin enameled roundels by the last decades of the twelfth century (see cat. 36). The appliqué heads on the base of this candlestick and the simpler foliate decoration argue for a slightly later date than the Cloisters candlesticks (cat. 33).

A single candlestick appears on one altar of the chasse of Mozac (cat. 45), but the evidence of inventories from both the thirteenth and fourteenth centuries suggests that they were more commonly made in pairs.[1] The Avignon inventory of Pope Innocent VI (r. 1352–62), a native of the

Fig. 37a. Candlestick. Kunstgewerbemuseum, Berlin

156

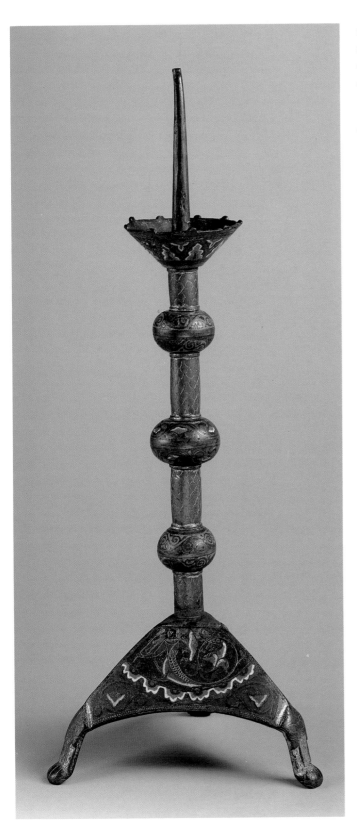

diocese of Limoges, includes a profusion of Limoges candlesticks, cited, with one exception, as pairs.[1] One pair was part of the furnishing for a chapel, offered by a Limousin priest along with vestments, a chalice, a paten, and a cross.[2]

A candlestick of the same size, with identical decoration, is preserved in the Kunstgewerbemuseum, Berlin (fig. 37*a*). Tracing to the nineteenth-century collection formed at Schloss Glienicke, it is probably the mate of the Metropolitan example.[3] Both, in turn, are particularly close to a pair from the Church of Saint Thomas the Martyr, Bristol.[4]

BDB

NOTES
1. For the Avignon references, see Hoberg 1944, pp. 132, 191, 341, 348, 426: p. 132 (1353: "15 candelabra de cupro de opere Lemovicano"); p. 191 (1353: "4 candelabra de cupro de opere Lemovicano"); p. 341 (1355: "12 candelabra de cupro de opere Lemovicano"); p. 348 (1357: "2 candelabra de cupro de opere Lemovicensi"); p. 426 (1369: "6 candelabra de opere Lemovicensi cum punctis").
2. "De cupro de opere Lemovicano," Hoberg 1944, p. 346.
3. Inv. 17.102. See Zuchold 1993, no. 980; Brunswick 1995, no. D103.
4. Gauthier 1987, no. 186, pp. 172–73, ill. 593 pl. CLXXIII.

BIBLIOGRAPHY: Pérate 1911, no. 72, pl. XLVI; Gauthier 1987, no. 185, ill. 593 pl. CLXXIII; Boehm 1992, pp. 154–55, fig. 9.

# 38. Coffret

Limoges, ca. 1200–1210

Copper: pierced, repoussé, engraved, chased, and gilt; champlevé enamel: dark and medium blue, green, yellow, red, and white; parchment over wood core

19 × 36 × 20 cm (7½ × 14¼ × 7⅞ in.)

Ex colls.: Charles Stein, Paris (in 1878); Eugen Felix, Leipzig (sold October 25, 1886, Cologne, no. 360, ill.); J. Campe, Hamburg (in 1900); R. Heilbronner, Paris (sold 1921, Galerie Georges Petit, lot 39, ill. p. 12); Raymond Pitcairn, Bryn Athyn, Pennsylvania

Condition: Overall wear to gilding; some losses to enamel, with later fill; some copper corrosion, especially on the enamels of the reverse; loss of definition on central lock plate through wear; handle replaced.

The Glencairn Museum, Bryn Athyn, Pennsylvania (12.EN.89)

Lid

Fig. 38a. Medallion. Musée National du Moyen Âge, Thermes de Cluny, Paris

Nine medallions decorate this deep, oblong leathern coffret. The largest, set at the center front, serves as the lock plate. This pierced copper medallion represents a man striding through vinelike, foliate ornament; a delicate vermiculé pattern decorates the perimeter. Paired hasps in the form of winged lizards (*serpenses*) are hinged at the edge of the lid and secured to the lock plate through paired openings set at either side of the man's figure. Each of the remaining openwork medallions—one on either side of the lock plate, one on each end, two on the lid, and two on the back—depicts a fantastic creature inscribed within an enameled ring of leafy ornament. By playing on the circular form of the roundels, the artist created a sense of turning motion in the design of the fantastic beasts, with their twisting postures, and the swirling floral motifs. Pierced copper strips protect the edges of the box; the corner braces of the lid are decorated with enameled flowers.

The leather and the roundels have apparently been rearranged, since shadows having the same diameter as sev-

Principal face

Right side

Reverse

eral of the roundels are perceptible on the leather of the proper right end at the bottom right corner, in two places on the lower back, and above and to the right of a medallion on the proper left side. A single medallion in the Musée de Cluny having the same diameter and similar design may have originally come from this ensemble (no. 945*b*, Fonds Du Sommerard; fig. 38a).

The composition of the central lock medallion, with its striding figure enmeshed in foliate vines, recalls the tradition of Limousin Romanesque manuscript illumination, as seen, for instance, in the Bible of Saint-Yrieix.[1] A date of about 1200–1210 is suggested by the classicized figure on the lock plate, the applied heads, the persistence of vermiculé decoration on the central medallion, and by the form and palette of the foliation.

Inside the lid of the coffret are two ink stamps on paper, one marked "Firenze," suggesting an earlier collection history in Italy, like the coffrets of Cardinal Bicchieri (see cats. 88, 89). A stamp on the bottom bears the number 110.

BDB

NOTE
1. Gaborit-Chopin 1969, fig. 202.

EXHIBITIONS: Paris 1878; Paris 1880; Paris 1900, no. 2588; New York 1987, no. 15.

BIBLIOGRAPHY: Darcel 1878b, vol. 2, p. 253, ill. pp. 261–62; Giraud 1881, pl. VI; Marquet de Vasselot 1941, p. 139, n. 1; Mallé 1950 and 1951, p. 81; Verlet 1950, p. 7.

# 3

*Innovation and diversification*

# 39. Chasse of Saint Thomas Becket

Limoges, ca. 1180–90

Copper: engraved, chased, and gilt; champlevé enamel: black, lapis and lavender blue, turquoise, green, yellow, red, and white; wood core

20.6 × 19.8 × 8.8 cm (8⅛ × 7¾ × 3½ in.)

Ex colls.: Doire, Évreux (in 1885); Victor Martin Le Roy, Paris (1900); Jean-Joseph Marquet de Vasselot, Paris. Acquired as payment of estate taxes, 1991.

Condition: Wood core and closing mechanism are old.

Musée du Louvre, Paris; Département des Objets d'art (OA 11333)

The murder of the archbishop Thomas Becket in 1170 at the cathedral of Canterbury by knights in the service of King Henry II is one of the best-known events in English history. In the years following Becket's canonization in 1173, it was also one of the most frequently depicted episodes. The wide and rapid spread of his cult, especially in western France,[1] the recent union of England and Aquitaine under Plantagenet rule, and perhaps the special role played by certain English sovereigns or prelates contributed to the considerable success of Limoges in the creation of chasses of Thomas Becket.[2] The most recent inventory lists fifty-two extant examples, which constitute the most important group of Limoges chasses dedicated to a saint.[3]

The chasse acquired by the Louvre in 1991 is one of four that are distinguished by their large size, the exceptional care devoted to their execution, and their style—still fully Romanesque. The others are now in British collections: one

in the Ashmolean Museum, Oxford (fig. 39a); the second belonging to the British Rail Superannuation Fund, deposited in the British Museum (fig. 39b); the third at the London Society of Antiquaries.[4] The principal face of each chasse depicts the saint's martyrdom at the altar, the chief variations being the number of assailants (two here) and the posture of the bishop, turned toward the altar (as in the Louvre chasse) or toward the assassins. Represented on the roof is either the saint's entombment (here) or his soul rising toward heaven—except for the example belonging to the British Rail Superannuation Fund, which juxtaposes the two episodes. The two small sides most often show an apostle: on the Louvre chasse, the figure with a receding hairline, who holds a book, is probably Saint Paul, and the one holding a cross, Saint Andrew. The treatment of the back is more varied; the chasse in the Louvre and the one in the Ashmolean Museum are both adorned with a field of crosslets.

The animated figures in dancelike attitudes, their expressive gestures, the strong stylization of the lines sometimes schematized to the extreme—like the long curve of the winding sheet under the saint's body, the large arcs suggesting the modeling of the legs, and the circles indicating the knees—all are connected to the traditions of Romanesque art in western France as embodied in the group of narrative works with vermiculé backgrounds (see cats. 16–20). On the other hand, reserving and engraving the figures and supplying them with "classical" heads in half relief mark a change in the development of Limoges enamelwork.

Fig. 39a. Chasse of Saint Thomas Becket. Ashmolean Museum, Oxford

Fig. 39b. Chasse of Saint Thomas Becket. British Museum, London, on loan from British Rail Superannuation Fund

Principal face and left side

Reverse and right side

Certain peculiarities found only in this small group of chasses dedicated to Thomas Becket should be highlighted. Thus, the first assailant's robe is open in front to the waist, exposing a large flap of the garment underneath—a detail found, apparently, only in England, for example on the verso of folio 69 of a psalter (ca. 1170) in the Royal Library, Copenhagen.[5] Likewise, the vertical lines on the second assailant's robe, terminating in an arc, are found in English illuminations of about 1170 to 1190, notably in the Canterbury Psalter in the Bibliothèque Nationale de France, Paris.[6]

The care taken in the execution of the Louvre chasse is obvious in the brightness and variety of the colors and especially in the quality of the engraving, where the depth of the main lines contrasts with the refinement of numerous details, such as the drawing of the cult objects on the altar, the engraved orphreys on the garments, the checkered motif of the sainted bishop's alb, and the stippling that emphasizes the reserved metal and decorates the framing bands.

The style of this chasse and the three others in its small group identifies them as having been made long before the solemn elevation of the saint's body in 1220, a fact that others have already recognized.[7] Further established is that the relics had been distributed immediately after his canonization in 1173 and during the following years; thus it was that the bishop of Poitiers went to Canterbury in 1174 and brought back several relics of the saint.[8]

The use of stylistic conventions and decorative elements found in the group of works with vermiculé backgrounds (like the stylized modeling of the legs of the figures and the finely stippled highlights on the reserved and gilt parts), as well as the stylistic kinship with other works of the late twelfth century (the resemblance, for example, of Saint Paul's face to the engraved faces on the chasse in the former Durand Collection [cat. 30]) all suggest that this small group of the oldest Limoges chasses of Thomas Becket were executed before 1200, probably about 1180–90. This hypothesis is substantiated by a comparison with similar works—for instance, the reliquary casket of the True Cross in the treasury of Saint-Sernin at Toulouse (cat. 40), although its more "relaxed" style implies a slightly later date.

The fact that three of the four early chasses of Thomas Becket are now located in England gives rise to the assumption that they were made primarily for an English clientele. However, their present locations appear to be mainly the result of a precocious interest in these works on the part of British connoisseurs. While the chasse on deposit at the British Museum was already in England at the beginning of the eighteenth century,[9] the chasse at the Ashmolean Museum[10] has been known only since the end of the nineteenth century, and the one given to the London Society of Antiquaries in 1801 by William Hamilton had been bought in Naples and very likely was already in southern Italy by the end of the Middle Ages.[11]

As to the chasse acquired by the Louvre in 1991, it corresponds precisely to the description of one exhibited in Rouen in 1884 that was part of the Doire Collection in Évreux.[12] Hence it is possible that the Louvre chasse came from Normandy, where devotion to this martyred English saint reached its apogee.[13]

ET-D

NOTES
1. Foreville 1976.
2. Ibid., pp. 360–36.
3. Gauthier 1975b, p. 247; Caudron 1993, p. 56.
4. See Borenius 1932, pp. 84ff.; Caudron 1975, pp. 237–39, pl. v, figs. 2, 4, pl. vii, fig. 2; Caudron 1993, pp. 68, 69, 73.
5. Ms. Thott 143-2; see Kauffmann 1975, no. 96, fig. 274. These observations could support the frequently stated hypothesis that Limoges artists were inspired by an English model (see Gauthier 1975b, p. 254; Foreville 1976, pp. 363–64).
6. Ms. lat. 8846; see Morgan 1984, no. 1, e.g., folio 21, fig. 6.
7. Gauthier 1975b; Caudron 1975; London 1984, no. 292; Gaborit-Chopin 1991.
8. Foreville 1976, p. 355.
9. At that time it was the property of a family at Saint Neots. Next it passed into various British collections, then into the Kofler-Truninger Collection (sale, Sotheby's, Dec. 13, 1979, no. 6). See Borenius 1932, p. 86, pl. xxxiv; London 1984, no. 292; Caudron 1993, p. 68.
10. Caudron 1975, p. 239, pl. v, fig. 2, and 1993, p. 73.
11. Caudron 1975, p. 237, pl. v, fig. 4, and 1993, p. 69; Thompson 1980.
12. See Darcel 1884, p. 71, and especially Linas 1885b, p. 52, whose description leaves little doubt: "It is distinguished both by the beauty of its workmanship and by its remarkable state of preservation. . . . Up above, the divine hand . . . behind the prelate, two murderers. On the roof is depicted the entombment of the deceased; a bishop, assisted by an acolyte carrying out the ritual, proceeds to the ceremony. At each side a standing saint. On the back, a tapestry of florets."
13. See Foreville 1975.

EXHIBITIONS: Paris 1900, no. 2443; Paris 1995, no. 8.

BIBLIOGRAPHY: Marquet de Vasselot 1906, no. 26, pl. xix; Borenius 1932, p. 89; Gaborit-Chopin 1991, p. 9, fig.; Caudron 1993, p. 77.

# 40. Reliquary of the True Cross

Limoges, 1178–98

Copper: engraved, scraped, stippled, and gilt; champlevé enamel; dark, medium, and light blue; turquoise, dark and light green, yellow, red, and white; wood core

12.9 × 29.2 × 14 cm (5⅟₁₆ × 11½ × 5½ in.)

Inscribed (on the left side): s[ancta] [h]ele/na//ivd/as (Saint Helen, Judas [Cyriacus]); on the back: [h]ier[usa]l[e]m//abbas de io/saphat de//crv/ce dat//oremvs (on banner)//raimvndo botarde/lli//hi[c] intr/at/ma/re (Jerusalem; Abbot of [Notre Dame of] Josaphat gives [a fragment] of the Cross to Raymond Botardelli. He takes [the cross across] the sea.); right side: hi[c] da(t)//abbati pon/cio ([Raymond Botardelli] gives [it] to Abbot Pons [of Saint-Sernin, Toulouse].); front: ca/nonici cv[m] abbate//offer[un]t//crvcem//satvrni/no to/losa (The canons and abbot offer the cross to Saint-Sernin, Toulouse.) A/ω (Alpha/omega)

Provenance: First recorded, 1836; classified, Monuments Historiques, 1892

Condition: Wear to gilding; some loss to enamel, especially near hinges on rear roof panel; later lock, studs, supports, and base.

Treasury of Saint-Sernin, Toulouse (Haute-Garonne)

The oblong chasse presents Christological scenes on the lid and, on the base, the story of the Finding of the True Cross and of the translation of a relic of the cross from the Holy Land to Saint-Sernin, Toulouse, the monastic church for which this reliquary was made. On the hinged side of the roof plaque, the Three Holy Women arrive at the tomb of Jesus on Easter Sunday, following the account in the Gospel of Mark (16:1–8). An angel on the far side of the empty sarcophagus holds a scepter and gestures toward them. The tomb is enameled with diagonal stripes of green and yellow to suggest that it is made of hard stone. The blue-and-white cloth that had covered the body is now draped over the tomb. One of the soldiers meant to guard the site still sleeps; a second near the angel awakens, and the third, next to the Holy Women, bows deeply. On the opposite face of the lid, the half-length nimbed figure of Christ appears in glory amid the clouds, holding a book in his left hand and raising his right in blessing. The Greek letters alpha and omega are at either side of his head and angels surround him. The Virgin Mary appears at his right under an arcade, carrying a lily; a striding angel behind her carries a crown. To Christ's left, Saint John stands within an arcade; a second angel advances behind him, also carrying a crown. On the proper left end of the roof, two angels converse, one gesturing heavenward; on the proper right end, the angel Gabriel announces the forthcoming birth of Christ to the Virgin.

The story of the Finding of the True Cross is depicted on the proper left end of the chasse. Saint Helen, mother of the emperor Constantine, commands Judas Cyriacus to begin digging with a pickax on Golgotha, the place of the skull, where Jesus was crucified.[1] Along the back panel, Abbot Jean of Josaphat, standing outside the walls labeled "Jerusalem," gives a double-armed cross to Raymond

Principal face

Botardelli. Botardelli, holding the cross, then turns and, as an oarsman steadies it, steps into a boat with a raised sail. On the right end, Raymond Botardelli presents the cross to Abbot Pons of Saint-Sernin. In the culminating scene on the lower front, the canons of Saint-Sernin stand while Raymond Botardelli kneels with the cross before the abbot, receiving his blessing. The doors of the church of Saint-Sernin are flung wide to receive the offering.

The Toulouse reliquary is exceptional for its elaborate narrative, the quality of its execution, and the inscriptions that allow the reconstruction of the historical circumstances underlying its creation. The figures named on the reliquary can be identified as follows: Raymond Botardelli is twice mentioned as a scribe in the cartulary of Saint-Sernin, Toulouse, first in 1162 and then again in 1164.[2] Pons de Montpezat was abbot of Saint-Sernin from 1176 to 1198(?).[3] During Pons's abbacy, the abbot of Notre-Dame-de-

Josaphat, the monastery built on the site of the tomb of the Virgin, was Jean, first mentioned in the cartulary of the Holy Sepulchre in Jerusalem in 1178 and replaced by 1195.[4]

The small Saint-Sernin chasse, consequently, dates to the final decades of the twelfth century, contemporaneous with the great Mozac chasse of 1185–97 (cat. 45), and like it an important commission for a thriving monastic community. Like the Mozac chasse, the Saint-Sernin reliquary is highly inventive and appealing in its narrative. Both display bursts of colorful enameled decoration in the field behind the figures, with floral patterns enclosed in variously sized and colored roundels. The figures' heads are delicately worked and applied.

The elongated dimensions of the reliquary are not typical of Limoges work; they may have been determined by those of the intended contents—a relic of the True Cross, perhaps encased in a small cross of the type illustrated on the lower plaques of the chasse, as Gauthier (1983) suggested. The inventory of the treasury of Saint-Sernin taken in 1246 mentions two relics of the True Cross in "small crosses of silver" but does not mention an enameled chasse.[5] Though its absence from the inventory provoked some scholars to date the chasse after 1246,[6] its decoration can be seen only in the context of ornament initiated in the late twelfth century and current only through the first decade of the thirteenth.

BDB

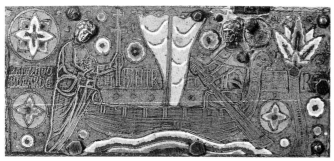

Raymond Botardelli embarks with the relic (detail)

Left side

NOTES
1. The account is recorded in *Golden Legend* 1969, pp. 272–74.
2. Douais 1888, pp. 163, 167.
3. The first publications of the chasse by Du Mège and Texier indicate that he died in 1183, the date given in *Gallia christiana*; Douais, Rupin, and subsequent literature give the date as 1198.
4. Given the incomplete dates associated with the career of Raymond Botardelli, the possibility that the abbot of Notre-Dame-de-Josaphat represented on the chasse is not Jean cannot be excluded. Douais cites two additional candidates: Pierre, mentioned in cartularies of 1170 and 1176; and Amelius, mentioned in 1195. It is even possible that Botardelli obtained the relic from an earlier abbot of the 1160s. See Douais 1888, p. 166.
5. Inventory published by Douais (1904); see also Douais 1888, p. 166; Gauthier 1972b, p. 337.
6. Douais 1888, p. 169; Rupin 1890, p. 389; Gauthier 1950, revised in subsequent publications to reflect the date of acquisition of the relic; Frolow 1961, p. 435, no. 546.

EXHIBITIONS: Vatican City 1963, no. 59, pl. XXIII; Paris 1965, no. 498, pl. 61; Cologne 1985, no. H15; Toulouse 1989, no. 410.

BIBLIOGRAPHY: Du Mège 1836–37, pp. 321–24, pl. VII; Texier 1857, cols. 1346–49; Douais 1888, pp. 161–69; Rupin 1890, pp. 386–89, figs. 442–45; Auriol and Rey 1930, pp. 320–24; Aubert 1933, pp. 63–67; Gauthier 1950, *passim*, pls. 26–28; Frolow 1961, p. 144, nos. 369, 546; Frolow 1965, pp. 22 n. 1, 25, 160, 170, fig. 2; Gauthier 1972b, pp. 111–12, no. 62, ill. 62; Gauthier 1982, pp. 57, 61, pl. 43; Gauthier 1983, pp. 86–88, no. 46; Durliat 1986, pp. 139–41.

Reverse

Right side

167

# 41. Chasse of the Holy Innocents

Limoges, ca. 1190–1210

Copper: engraved, chased, and gilt; champlevé enamel: lapis and lavender blue, turquoise, green, yellow, red, and white; wood core

18.8 × 21.7 × 9 cm (7⅜ × 8½ × 3½ in.)

PROVENANCE: Probably from a church in Monflanquin (Lot-et-Garonne; see Forestié 1885)

EX COLLS.: Lagravère, Montauban (in 1866 and 1885); de Boissière-Lagravère, Montauban (in 1890); Henri Daguerre, Paris; Julien Chappée, Le Mans (in 1933 and 1947); Claude Vaudecrane, Le Mans. Acquired in 1971.

CONDITION: Some losses to enamel; gilding partly worn; wood core covered with later paint; crest missing. Modern hinges and rivets.

Musée du Louvre, Paris; Département des Objets d'art (OA 10406)

Chasses dedicated to the Infancy of Christ are rare in Limoges—aside from the Adoration of the Magi, which was a product of renewed devotion to the Three Kings at the end of the twelfth century (see cats. 20, 22). Hence the Monflanquin chasse, which shows on its principal face two episodes from the Infancy cycle—the Presentation in the Temple (Luke 2:22–38) on the roof and the Massacre of the Innocents (Matt. 2:16) on the case—presents an iconography that is uncommon in Limoges enamelwork. The back and the small sides are decorated with half-length angels inscribed in circles, by contrast a frequent motif in Limoges. This, however, is one of the motif's earliest uses on three of the four faces of a chasse.

This chasse belongs to the group of works with reserved and engraved figures that embodied one of the final blossomings of late Romanesque art according to formulas spe-

cific to Limousin enamelers (see cats. 39, 40). Various decorative motifs employed during the final third of the twelfth century are still used here, such as the stippled highlights and, on the back, the frieze of crosslets on an alternately blue and red background (see cat. 29). The large fleurons adorning the upper part of the gables come directly from those found on certain works with vermiculé grounds (see cat. 21).

The figures have preserved the dancelike stances and demonstrative gestures of Romanesque art of western France. The faces offer fine examples of classical, or "severe," heads (made separately and attached by means of small rivets), also reflecting Romanesque stylization, which appears, as well, in the engraved faces of the angels on the back and on the end panels. The representation of the garments, however, departs further from Romanesque conventions than that in most earlier works executed in the final quarter of the twelfth century. Here the more supple lines suggest lighter fabrics, allowing the shape of the body and the movement of the legs to be more clearly perceived, especially in the scene of the Massacre of the Innocents. This more "naturalistic" striving is served by sensitive and finely finished engraving. The figure of Herod is a very beautiful example of such skillful use of lines to define form, while engraved orphreys suggest the richness of the costume and a fine checkerwork indicates the facing of the robe. This royal figure is perhaps related to those found at the end of the twelfth century or at the beginning of the thirteenth, for example, the emperors Julian and Mauritius (ca. 1175–80) in the stained-glass windows of Canterbury Cathedral[1] and in the stained-glass window depicting Lot (dated by Grodecki about 1210–15) in the cathedral at Poitiers.[2]

The Presentation in the Temple takes place in an atmosphere that is peaceful, almost static—the very opposite of the violence of the Massacre of the Innocents. It is characterized by an interest in the precise rendering of details that are at the same time naturalistic and poetic—like the basket of doves carried by Joseph.

These observations lead to a proposed date for the chasse of about 1190 to 1210. According to Forestié, the work "belonged to a church in Monflanquin (Lot-et-Garonne)."[3] Much later (1933), it was published as coming from Saint-Antonin (Aveyron), an assertion that does not seem justified. However, the fortress of Monflanquin was not built by Alphonse of Poitiers until 1256,[4] so the chasse could not have been originally intended for him. Still, it could have been given to the church at the time of its founding.

ET-D

Left and right sides

Principal face

Reverse

NOTES
1. See London 1984, no. 92, fig. 92c.
2. See Grodecki and Brisac 1984, p. 56, fig. 45.
3. Forestié 1885, p. 184.
4. Lavedan and Hugueney 1969, pp. 18, 30.

EXHIBITIONS: Nantes 1933, no. 325; Paris 1935, no. 156; Paris 1937, no. 1206; Mexico City 1993, p. 50, fig. 16.

BIBLIOGRAPHY: Pottier 1866, p. 329; Forestié 1885, p. 184; Rupin 1890, pp. 356, 358; Gauthier 1950, p. 39; Gaborit-Chopin 1972.

# 42. Book-Cover Plaque: Christ in Majesty

Limoges, ca. 1190–1210

Copper: engraved, chased, and gilt; champlevé enamel: lapis, medium, and lavender blue; turquoise, dark and light green, yellow, red, and white

22.5 × 11 cm (8⅞ × 4⁵⁄₁₆ in.)

INSCRIBED: A / ω (Alpha/omega), mark on back: Λ

EX COLL.: Pierre Révoil. Acquired in 1828.

CONDITION: Wear to the gilding on figure of Christ.

Musée du Louvre, Paris; Département des Objets d'art (MRR 248)

Exhibited in Paris only.

The book-cover plaque from the former Révoil Collection—the only book cover in the collections of the Louvre to depict Christ in Majesty—once had as a companion piece a plaque of the same size showing the Crucifixion. Both must have been attached to wood press boards and equipped with framing strips to cover the two binding boards according to a formula illustrated by rare fully preserved examples (see cat. 18 and figs. 12a, b) and also by countless detached plaques depicting one scene or the other.[1] Those of Christ in Majesty, however, occur less often: Marie-Madeleine Gauthier has listed forty-six examples done between 1170 and 1230, whereas Crucifixions are twice as numerous. As she has suggested, certain bindings may have had only one enameled plaque, depicting the Crucifixion.[2]

This work is part of a group of book-cover plaques that follow the late Romanesque style of Limoges. The figure of Christ is either reserved and engraved (see cat. 43) or, as here, repoussé in bas-relief and applied. Romanesque traditions are manifest especially in the dynamic attitudes, powerful expressions, and vigorous stylization of the animals symbolizing the evangelists, which evoke the monumental sculpture of western France.

The majestic attitude of Christ and the draping of his mantle belong to an iconography encountered throughout the twelfth century. The orphrey indicating the kneecap of Christ's right leg had previously appeared in Limoges illuminations of the late eleventh and early twelfth centuries, such as those in the *Sacramentary of Saint-Étienne de Limoges* and the *Rule of Saint Benedict* from Saint-Martial.[3] The figures of Christ in certain frescoes of the late twelfth century—for instance, the one in Saint-Jacques-des-Guérets (Loir-et-Cher)[4]—demonstrate the longevity of these traditions in western and central France.

This attachment to an old iconographic and stylistic type accentuates the difficulty of arranging chronologically the diverse Limousin representations of Christ in Majesty. Moreover, Christ's face and especially the face of the angel of Saint Matthew place the Louvre plaque in the major group of works using applied "classical" heads, made at the end of the twelfth century and in the opening decades of the thirteenth. The small lozenge-shaped motifs imitating gems, which are scattered along the orphreys, and the fine stippling that highlights the reserved portions of the friezes and rosettes are characteristic of works from the late twelfth and early thirteenth centuries; this justifies the dating of the Louvre plaque to about 1190–1210.

ET-D

NOTES
1. Gauthier 1967c, p. 152, n. 6, and p. 155, n. 11, lists drawn up by the author.
2. Ibid., p. 155.
3. See Paris, BNF, ms. lat. 5243, fol. 45v, and ms. lat. 9438, fol. 58v; Gaborit-Chopin 1969, pp. 127f., figs. 162, 171.
4. Demus 1970, pp. 144–45, pl. LXII (ca. 1200).

BIBLIOGRAPHY: L. Laborde 1852, no. 35; Rupin 1890, p. 318, fig. 381; Darcel 1891, no. 83; Havard 1896, p. 292; Marquet de Vasselot 1914, no. 91; Gauthier 1950, p. 44; Gauthier 1967c, p. 155 and fig. 62; Gauthier 1968d, no. 15, p. 284.

# 43. Book-Cover Plaque: Christ in Majesty

Limoges, ca. 1200

Copper: engraved, chased, scraped, stippled, and gilt; champlevé
enamel: dark, medium, and light blue; turquoise, dark and light green,
yellow, red, translucent red, translucent rose, and white

23.7 × 11.8 cm (9⅜ × 4⅝ in.)

INSCRIBED: A/ω

CONDITION: Incidental losses to gilding; incidental losses to enamel
under Christ's right arm and at center top of plaque.

The Metropolitan Museum of Art, New York; Gift of J. Pierpont
Morgan, 1917 (17.190.783)

Exhibited in New York only.

The gilt figure of Christ, enthroned on an enameled cushion, sits before a field of blue punctuated with stars. He
raises his right hand in blessing. His head is separately
worked and applied to the plaque, as are the heads of the
symbols of the evangelists that surround him, following the
convention of Limoges book covers. Noteworthy features of
this particular plaque are the refined carving of the reserved
metal, especially on the drapery of Christ and the body of
the eagle of Saint John; the delicate pointillé work used to
enliven the surface; the finely worked appliqué heads; the
refined enamel and rich palette, including the use of a
translucent rose color. These features suggest a date at the
end of the twelfth century.

It is unusual that this plaque, like the Christ in Majesty
in the Musée de Cluny (cat. 12), should originally have had
no holes for attachment to a wood backing. Rather, it has
only a single hole at the center top, which was apparently
added later, damaging the surrounding enamel.

There is evidence that Limousin plaques representing
Christ in Majesty and the Crucifixion were used in other
contexts.[1] A 1482 inventory of the priory of Aureil, just
south of Limoges, indicates that two square plaques depicting the Crucifixion and Christ in Majesty, referred to as
"retables," were to be found on the altars of the Virgin and
of Saint Gaucher.[2] Of necessity, these had either attachment
holes, like the plaques incorporated into altar frontals at
Novgorod[3] and in Icelandic churches,[4] or framing metal
bands, which, in the case of the Metropolitan plaque, would
have obscured the delicate border decoration.

Gauthier assigned to the same atelier the Morgan plaque
and four others, including the upper and lower plaques of a
book cover recorded in the collection of Abbé Fauvel in 1723
by Dom Bernard de Montfaucon.[5] The points of comparison, notably the vigorously rendered, sinewy symbols of the
lion of Saint Mark and the ox of Saint Luke, and the rich
mandorlas surrounding Christ appear to be hallmarks of the
ateliers that produced hundreds of surviving book covers
and do not in this case argue sufficiently for a common

workshop. Neither common palette nor style argues for
Gauthier's association of the plaque with crosses in
Barcelona and in the Keir Collection.[6]

Before entering the collection of the Metropolitan
Museum, the plaque was numbered on the reverse by stick-
ers marked "15" and "1285," but it has not been possible to
establish either the exhibition or collection history of the
plaque prior to its gift to the Metropolitan Museum by
J. Pierpont Morgan.                                              BDB

NOTES

1. For a more complete range of possibilities, some of which, by virtue of
   their dimensions or appearance, do not apply in this instance, see
   Gauthier 1968d, pp. 274–75. A plaque representing the Crucifixion is
   said to have been mounted on a coffin in Saint Knud's Cathedral,
   Odense, until the mid-nineteenth century. See Liebgott 1986, p. 57, fig. 51.
2. Gauthier 1968d, p. 276; Decanter 1960, pp. 383–92.
3. Rupin 1890, p. 322 (when there were still six plaques, of which only four
   now survive); Gauthier 1987, nos. 163, 164.
4. Liebgott 1986, p. 59, figs. 52, 53.
5. See Gauthier 1987, no. 302, ill. 834.
6. Ibid., nos. 304, 305.

BIBLIOGRAPHY: Gauthier 1987, no. 299, p. 233; ill. 832, pl. CCXLVII.

# 44. Chasse: Crucifixion, Christ in Majesty

Limoges, ca. 1190–1200

Copper: engraved, chased, and gilt; champlevé enamel: black, lapis, medium and lavender blue; turquoise, green, yellow, brick red, garnet red, and white; wood core

23.2 × 21.2 × 9.2 cm (9⅛ × 8⁷⁄₁₆ × 3⅝ in.)

Ex colls.: Adolphe de Théis (in 1867; sale, Paris, May 13, 1874, no. 154); Eugène Fichel, Paris; L. Goldschmidt; Charles Séguin, Paris. Bequest of Charles Séguin, 1908

Condition: Wood core and closing device are old.

Musée du Louvre, Paris; Département des Objets d'art (OA 6183)

Exhibited in Paris only.

Except for its inclusion in the exhibition of 1867 and its mention in Marquet de Vasselot's catalogue (1914), this chasse formerly in the Théis Collection is virtually unpublished. Yet it is a fine example of the reliquaries produced by Limousin workshops at their full maturity and employing an iconography sufficiently general to suit the needs of a widening clientele.

This iconography consists of the Crucifixion and Christ in Majesty surrounded by apostles, here eight, on the two plaques of the principal face and on the ends (the back is strewn simply with rosettes). Already widespread among chasses with vermiculé backgrounds (cat. 24), these images appeared very often during the late twelfth and early thirteenth centuries on works of varying size and quality. Several other chasses are similar in dimensions, iconography, and style. The closest, probably, is the one in the treasury of Klosterneuburg, except that its backgrounds are covered with scrolls of small trefoil florets.[1]

The style and decoration of the chasse are connected in large part to the Romanesque works of Limoges. The two depictions of Christ are in half relief whereas the other figures are reserved and engraved, but they all have heads that recall the classical style. The back is decorated with an even scattering of rosettes, which are distinguished from the ones adorning vermiculé chasses by their larger sizes and more circular shapes. The stippled highlights have also evolved slightly: the dots are finer and they form small notched wheels around the enameled rosettes. The cabochons on the crest are embellished with coiled-up scrolls similar to those on the crowns of the two Christs in half relief (cats. 49, 50) and the small spandrels surrounding the summit and the base of the Alpais ciborium (cat. 69). Also, next to the rosettes strewn across all the backgrounds, we notice scrolls ending in trefoil florets like those, for example, on the chasse from Mozac (cat. 45).

The style of the figures likewise carries on Romanesque traditions. The mantle of Christ in Majesty is distinguished by strongly stylized, firm, angular lines and by the sashlike draping across the abdomen. Drapery arranged in this way recalls many Romanesque depictions, such as the one on the southern apse of the church of Saint-Gilles in Montoire (Loir-et-Cher, mid-twelfth century).[2] It is still found in the early thirteenth century—for instance, in Lavaudieu (Haute-Loire)[3]—and on Limousin figures of Christ of the late twelfth and early thirteenth centuries—for example, on a book-cover plaque in the treasury of the cathedral of Lyons.[4] The garments of the apostles, the Virgin, and Saint John are likewise defined by an intense linear abstraction that masks the actual volume of the bodies and, where the lines are interrupted, by ideal forms that have no tie to reality, like ellipses with one pointed end, placed at the level of the waist or around the bent arms of the Virgin and Saint John. As for the faces in half relief, they are among the finest examples of so-called classical heads; here the severe expressions are softened by subtle engraving and individualized by small details, such as the wrinkle across the forehead and crown of the head of the apostle at Saint John's left. Faces of the same high quality and diversity, some of them identical to those found on this chasse, can be seen in the Nationalmuseet, Copenhagen, on a chasse with the same iconography, from the royal treasury.[5]

Only the figure of Christ on the cross escapes this linear abstraction; instead, it suggests a sinuous movement of the body, perceptible even through the lightweight fabric of the perizonium—the sole early Gothic touch on this otherwise thoroughly Romanesque work. The supple treatment of this figure is comparable to that in works executed during the final fifteen years of the twelfth century, such as the figures of Christ from Grandmont and the one on the aforementioned chasse from Mozac.

This style, recognizable mainly by the elliptical shapes, allows us to identify with certainty the production of one particular workshop. The style recurs on several book covers and book-cover plaques, which can be grouped around the Crucifixion in the former Feltrinelli Collection in Geneva[6] and also the tabernacle that came from the château of Tüssling and is now in the Bayerisches Nationalmuseum in Munich.[7] The last (fig. 70a) is among the oldest extant example of this type of object. Thus it seems possible to identify a single hand that produced objects of different types but all of extraordinary quality. Although output of the workshop would have extended over a relatively long period, the date for the Louvre's chasse is suggested to be about 1190 to 1200.

ET-D

Principal face and left side

Reverse and right side

NOTES
1. Rupin 1890, p. 336, fig. 402; Fillitz and Pippal 1987, no. 98, pp. 356–57; for this iconography, see also Rupin 1890, pp. 421–23.
2. Demus 1970, fig. 133 and pp. 141–42.
3. Ibid., fig. 149 and p. 147.
4. Gauthier 1968d, no. 18 and fig. 3.
5. See Liebgott 1986, figs. 32, 33.
6. Gauthier 1972b, nos. 64, 65.
7. Munich 1992, no. 6 (entry by L. Seelig).

EXHIBITION: Paris 1867, no. 2068.

BIBLIOGRAPHY: Marquet de Vasselot 1914, no. 63.

# 45. Chasse of Saint Calminius

Limoges, after 1181, probably ca. 1185–97

Copper: engraved, chased, and gilt; appliqué figures: repoussé, engraved, chased, and gilt; champlevé enamel: midnight, lapis, and lavender blue; turquoise, dark and light green, yellow, red, and white; midnight and lapis blue, turquoise, yellow, and white enameled beads; wood core

45 × 81 × 24 cm (17¾ × 31⅞ × 9½ in.)

INSCRIBED:
*Face*
Case:
Center: MARIA (Mary) / HIOANNES (John)
Left: MATEVS (Matthew) / TOMAS (Thomas) / PETRUS (Peter)
Right: ANDREA[S] (Andrew) / IACOBVS (James) / PAVLUS (Paul)

*Roof*
Center: A / ω (Alpha/omega). On phylacteries of eagle and lion: S[ANCTUS] IOANNES : E[VANGELISTUS] (Saint John the Evangelist)
S[ANCTUS] MARCVS (Saint Mark)
Left: MACIAS (Matthias) / SIMON (Simon) / IVDAS (Judas)
Right: FILIPVS (Philip) / BARTOLOMEV[S] (Bartholomew) / IACOBVS (James)

*Reverse*
Case:
Left: S[ANCTUS] : CALMINIVS : CO[N]STRVIT : VNAM : ABBA[T]IAM : IN / PODIENSI : EP[ISCOP]ATV : IN[H]ONORE[M] : S[AN]C[T]I : THEOFREDI : MARTIRIS. (Saint Calminius builds an abbey in the diocese of Puy in honor of Saint Theofrede the Martyr) NAMADIA / CALMINIVS
Center: S[ANCTUS] : CALMINIVS : SENATOR ROMAN[US] CO[N]STRVIT : S[E]C[VN]D[A]M: / ABBA[T]IAM : IN LEMOVICENSI EP[ISCOP]ATV : NO[M]I[N]E : THVELLAM (Saint Calminius, Roman senator, builds a second abbey, named Tulle, in the diocese of Limoges) CALMINIVS / NAMADIA
Right: S[ANCTUS] : CALMINIVS : CO[N]STRVIT TERCIA[M] : ABBA[T]IAM : NOMINE : MAVZIACVM : IN ARVERNENSI EP[ISCOP]ATV : in [H]ONORE / S[ANCT]I : CAPRASII M[A]R[TIR]IS : ET : S[AN]C[T]I PETRI : QV[A]M OF[F]ER[T] : EISDEM S[AN]C[TI]S (Saint Calminius builds a third abbey, named Mozac, in the diocese of Auvergne in honor of Saint Caprasius the martyr and Saint Peter—[an abbey] that he offers to these saints). NAMADIA / CALMINVS / S[ANCTUS] CAPRASIVS : MARTIR: DEI. X [CHRISTI] (Saint Caprasius the martyr of God Christ)

*Roof*
Left: HIC : ANIMA : AB ANGEL[IS] PORTATVR / HIC SEPELIT[UR] S[ANCTUS] : CALMINIUS CONFESOR XPI [CHRISTI] IN MO[NASTERIO] (Here the soul is borne by the angels. Here Saint Calminius, confessor of Christ, is buried in the monastery.)
Center: BEATA NAMADIA SEPELIT[UR] [H]IC HIC / IN MONASTERIO MAVZIACO AB ANGEL[IS] DVCITVR (Here the blessed Namadia is buried. Here she is led by the angels to the monastery of Mozac)
Right: PETRVS ABBAS MAVZIACVVS FECIT CAPSAM PRECIO (Peter, abbot of Mozac, has a chasse made for its price) PETRVS ABBAS M[AUZIACUM] (Peter, abbot of Mozac)

Graffito on book carried by an assistant: DE / US / I / S [EST] CARI / A / N [CARITAS?] (God is charity[?])
Right gable: MARIA (Mary)
Left gable: S[ANCTUS] AUSTREMONIVS (Saint Austremonius)

PROVENANCE: First mentioned in church at Mozac by Dom J. Mabillon in 1707. Inspection of relics on October 24, 1839: the chasse contained an authentication dated 1705, various bones, and two heads (Gomot 1872). Classified, Monuments Historiques, 1901. Stolen, then returned anonymously, 1907 (Leniaud 1983).

CONDITION: Appliqués and enameled plaques in very good condition. Partial wear to the gilding. Most of the gilt-copper framework is modern: the long horizontal band separating case from roof on the principal face, all the friezes on the back (except the four vertical bands decorated with circles), and all the friezes on the small sides and on the feet. On the vertical friezes of the principal face, the enameled cabochons adorned with a yellow cross on a blue background are undoubtedly restorations.

The appliqué of the ox, the symbol of Saint Luke, is missing from the left side of Christ on the roof of the principal face. If one is to believe the engraving published in 1829 by Nodier, Taylor, and de Cailleux (which shows the principal face but is imprecise as to style and is in reverse), the ox was still intact and the framing friezes contained a total of forty-five medallions. An engraving of the back, published by Du Sommerard in 1846, indicates that all the framing friezes decorated with circles within a diamond-shaped grid were in place at that time. The reproduction of the principal face in Rupin (1890), however, reveals the same condition that exists today. Reproductions published by Mallay in 1838 show only the enameled panels of the back; lacking an overall view, they are of no help in precisely dating these removals or "restorations."

The crest, disproportionate and poorly executed in comparison with the general quality of the work, might be a restoration; yet it was already in place in 1829.

The lock, with its long metallic shank, used to be on the back (see photographs in Archives des Monuments Historiques) but has been reinstalled under the chasse.

Church of Saint-Peter, Mozac (Puy-de-Dôme)

In the history of Limousin enamelwork, the Mozac chasse occupies a major position because of the originality of its iconography, its style and technique, and its crucial place in the chronology. It was also one of the first works to be emphasized in the scholarly literature, for it was mentioned in 1707 by Dom Mabillon and reproduced in 1829 by the authors of *Voyages pittoresques et romantiques dans l'ancienne France* (Picturesque and Romantic Travels in Old France).

This work is unique chiefly because of its dimensions, as it is the largest Limousin chasse to have come down to us. The iconography is also exceptional: on the principal face, the twelve apostles, sheltered by arcades, are arranged on both sides of superimposed depictions of the Crucifixion and Christ in Majesty surrounded by the four symbols of the evangelists, according to a well-established formula in Limousin art. On the back, however, an unusual and complex iconography is accompanied by inscriptions, an exceptional development in the enamelwork of Limoges: the six large panels are dedicated to the story of Saint Calminius, a

Principal face

Reverse

seventh-century Roman senator who eventually became duke of Aquitaine and count of Auvergne, and his wife, Namadia.[1] The panels narrate the successive construction under their direction of three monasteries—at Le Puy, Tulle, and Mozac. The roof depicts the burial of each of these two saints and the elevation of their souls to heaven; it then shows, at an altar, a priest with a halo, designated as Abbot Peter, flanked by two other people. Saint Austremonius, the first bishop of Clermont (third century), appears on one end, while the figure of the Virgin holding the Child is on the opposite end.

The final scene on the roof, accompanied by the inscription *Petrus abbas Mauziacus fecit capsam precio*, has, since its first publication, aroused question and debate.[2] Most scholars agree that this figure is the client and not the artist, with the more abbreviated *fecit* substituted for *fecit fieri*, subordinating the role of the artist to that of the patron. They are divided, however, on which abbot commissioned the piece: Peter II (mentioned in 1165), Peter III, of Marsac (mentioned from 1168 to 1181), or one of their thirteenth-century namesakes—Peter IV (mentioned in 1245) or Peter V (mentioned from 1252 to 1267).[3] Marie-Madeleine Gauthier was apparently the first to observe, correctly, that because the abbot has a halo, this depiction could not possibly have been executed during his lifetime.[4] On the contrary, we must deduce that this Abbot Peter decided to have the chasse made and perhaps gathered the necessary funds but that the work was executed or, at least, completed after his death, probably under his successor. The abbot depicted here could thus readily have been Peter III; in that case, the chasse would have to be dated after 1181 and would have

been executed under William of Bromont, who was mentioned in 1195 and 1197. This analysis agrees perfectly with another given, according to which the relics of Saint Calminius were translated in 1172 to the church of Laguenne (Corrèze).[5]

Furthermore, the dating of this chasse from about 1185 to 1195 corresponds entirely to what we can discern about the evolution of Limousin enamelwork. This piece, exceptional in all regards, combines the various technical and decorative procedures used by Limousin artists during the late twelfth and the early thirteenth century—a period in which their production reached its apogee. The figures are in half-relief repoussé or reserved and engraved. Backgrounds strewn with rosettes alternate with others covered by floral rinceaux. The design of these florets and the roses decorating the backgrounds of the mandorlas with the Virgin and Saint Austremonius are slightly simplified versions of motifs already encountered on the Ambazac chasse (cat. 55). Also noteworthy is the presence of friezes with pseudo-Kufic inscriptions, frequently found in Limoges work of about 1200 and the early thirteenth century (see cats. 51, 69, 70). The artist has also obviously played with the diversity of these decorative motifs, which are emphasized by stippling, as in the best works of that period.

Many other aspects of decoration and style link the Mozac chasse to Romanesque Limoges work during the years from 1160 to 1190. Thus the scenes engraved on the back are surrounded by friezes of small crosses on grounds alternately red and turquoise or dark blue and lavender—identical with the friezes on numerous works with vermiculé backgrounds. The design of the buildings, with turrets whose doors are defined by meticulously detailed ironwork, pierced at the top with very closely spaced semicircular openings, is very similar to that of buildings depicted on the chasse of Saint Valerie at the State Hermitage Museum, Saint Petersburg (cat. 20). The bell turrets with concave or convex roofs surmounting the turrets at the back or the arcades of the front can be likened to those on the effigy of Geoffrey Plantagenêt (cat. 15) and the plaques of the main altar of Grandmont (cat. 57). The animated, dancelike attitudes of the workers building the three monasteries, as well as the stylization of their drapery, are comparable to those of figures on "narrative" chasses with vermiculé backgrounds and older chasses with reserved figures and "classic" heads in relief (see cats. 16–19, 39–41). The engraved faces of Saint Calminius and Saint Namadia directly follow the style of certain works with vermiculé backgrounds, such as the book-cover plaques from Lyons (cat. 18), on which can also be seen the folds ending in a T that marks certain drapery. The clothing of other figures reserved on copper or in half-relief repoussé are suggested by fine juxtaposed lines, which, although characteristic of a trend widespread in northwestern Europe during the final

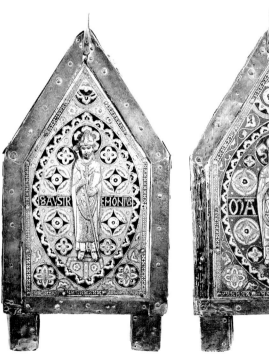

Left and right sides

176

third of the twelfth century, reached Limousin workshops belatedly and had limited influence (see cat. 4). Nonetheless, this treatment is used with suppleness and harmony to suggest the forms and movements of the four very beautiful figures of assistants holding the sudarium in the two burial scenes. In other cases, the artist was unable to avoid a certain clumsiness, especially in the drawing of the sleeve covering Saint Calminius's outstretched arm as he supervises the final two scenes of construction.

The faces of the figures in half relief on the obverse seem inspired by entirely different examples: these massive, almost sculptural heads with wide, strong features, big round eyes, and puffy lower lips are quite distinct from the ones usually depicted by Limousin artists of the late twelfth century. They could, however, recall the sculpture of Mozac—for instance, the capital of the Resurrection. The figures of the Virgin and Saint Austremonius on the gables seem marked by this sensitivity to local sculpture.

On the other hand, the Christ on the cross eludes this attachment to Romanesque traditions: the soft modeling of the torso, as well as the suppleness of the contour and of the lines of the perizonium, reveal the artist's interest in a more humane and naturalistic depiction—an interest shared by the maker or makers of the Grandmontain Christs (see cats. 62, 65) of about 1185 to 1195.

Open and even eclectic, the Master of Mozac showed himself to be particularly gifted and inventive in enamelwork, in which he probably surpassed his contemporaries. His extremely broad palette contains several notable colors—two greens and four blues, including a rare midnight blue. There are also many gradations that encompass four colors. Most striking of all are the intensity and quality of the enamel and these unusual effects: on the ends, the red strokes scattered across the turquoise ground on which the Virgin and Saint Austremonius are standing; on the back, the alternation of red and turquoise in the openings in the architecture of the monastery of Tulle; and, in the next scene, the contrast between the superb, slightly bluish white material of the mortar used by the builders of the abbey of Mozac and the very dense midnight blue of the vat that contains it.

It should be noted that the plaque depicting the figure of Saint Austremonius has a slightly different coloration from the others: the background blue is lighter. Does this indicate an execution stretching over some length of time, or is it due to the collaboration of several enamelers? Or can one perhaps push the hypothesis still further and speculate that this plaque might have been the last element made, at the time of the elevation of the relics of Saint Austremonius in 1197?[6]

ET-D

NOTES

1. In the seventeenth century, historians were divided as to whether or not Saint Calminius of Auvergne was the hermit of the same name honored in the Limousin. See especially Thomas d'Aquin 1646; Collin 1672, pp. 338–49; Bonaventure de Saint-Amable 1676–85, pp. 41–44 (1683).

2. *Precio* suggests, as noted by Gauthier (1972b), that Abbot Peter left money for the realization of the chasse.

3. The *Gallia christiana* (1720), followed by certain authors of the nineteenth century (e.g., Bouillet 1846), attributed its execution to Peter V.

Two workers preparing mortar for the construction of the monastery of Mozac (detail)

Mallay (1838), followed by Texier (1842), remarked that the style pointed more to an execution under Peter II or Peter III. More recently, the attribution to Peter III was vigorously supported in Souchal (1963). For the various abbatial dates, see *Gallia christiana*, vol. 2, col. 353, and Gomot 1872, pp. 241–42.

4. Gauthier 1972b, p. 108.

5. Father Thomas d'Aquin (1646, p. 294) was the first to mention the discovery of a parchment in the Laguenne chasse (today in the Musée Dobrée, Nantes). The parchment bore the following inscription: *Hic est corpus Beati Calminii confessoris, quod fuit inuentum in tumulo ipsius, qui est infra Ecclesiam Aquinae, iuxta magnum altare, anno ab incarnatione Domini nostri Iesu Christi 1172* (This is the body of the Blessed Calminius the Confessor, which was found in his tomb, which is under the church of Laguenne near the large altar, in the year 1172 after the incarnation of our Lord Jesus Christ).

6. See Labbé 1657, vol. 2, pp. 505–7, and *Chronique de Mozac*, quoted in Gomot 1872, pp. 49, 256.

EXHIBITIONS: Paris 1867, no. 2063; Vatican City 1963, no. 54, pl. XIX; Paris 1965, no. 445, pl. 59; New York 1970, no. 135.

BIBLIOGRAPHY: Mabillon 1707, p. 852; *Gallia christiana*, vol. 2, col. 351; Nodier, Taylor, and de Cailleux 1829, Auvergne, vol. 1, pp. 8, 14, 15, engraving p. 15; Mallay 1838, pp. 24–26, pls. XX–XXII, XXIX–XXXI; Texier 1842, p. 138; A. Bouillet 1846, pp. 223–26; Du Sommerard 1838–46, vol. 5, pp. 190–91, and *Album*, 10th ser., pl. XIII; Caumont 1850, p. 160; Texier 1851, pp. 146–47; Labarte 1856, pp. 204–5; Texier 1857, cols. 308–12, 1187–89; Gomot 1872, pp. 209–18; Rohault de Fleury 1883–89, vol. 1, p. 81; Gondelon 1886, pp. 13–17; Rupin 1890, pp. 100–107; Molinier 1891, p. 176, fig. p. 174; *L'Art pour tous* no. 1039 (1903), p. 4253; Moschetti 1920, pp. 44–46; Fayolle 1924, pp. 436, 438; Ross 1933b, pp. 272–78, ill.; Marquet de Vasselot 1941, pp. 119ff., pl. XXVIII; Gauthier 1950, pp. 26ff., pls. XXXVIII, XXXIX; Erdmann 1953, no. 57, p. 484; Craplet 1962, pp. 239–49, pls. 114–18; Souchal 1963, pp. 307–18; Gauthier 1972b, no. 57, pp. 106–10, 333–35; Leniaud 1983, p. 378.

# 46. Saint Peter in Glory

Limoges, ca. 1185–1200

Copper (plaque): chased, engraved, stippled, and gilt; (appliqué): repoussé, engraved, chased, scraped, stippled, and gilt; champlevé enamel: medium and light blue, turquoise, green, yellow, red, and white; dark blue and turquoise glass beads

24.7 × 14.3 cm (9¾ × 5⅝ in.)

Ex colls.: Abbé Jacques-Rémi Texier, diocese of Limoges (in 1842); Muret de Pagnac, Saint-Junien? (see Ardant 1855); [Frédéric Spitzer, Paris (before 1890; sale, Paris, April 16–June 17, 1893, no. 255, pl. VIII)]; [Gaspard, Jean, and Stephen Bourgeois (sale, Cologne, 1904, no. 386, ill.)]; Georges Hoentschel, Paris (by 1892); [Jacques Seligmann, Paris (in January 1912)]; J. Pierpont Morgan, London and New York

Condition: Some loss to gilding on highest areas of relief and on edge of enameled back plate; losses in dark blue enamel along borders and on a number of florets, which have been filled with colored mastic, especially along the right edge (partially restored); four rivets for attachment, heads of which are not engraved, may be replacements; no further restorations; beads inset in eyes and on cover of small book are original.

The Metropolitan Museum of Art, New York; Gift of J. Pierpont Morgan, 1917 (17.190.779)

The appliqué figure, worked in low relief, sits on a bolster supported by an integral arc of heaven. The traditional attributes of Saint Peter include the tonsure framed by short, tight curls and a short beard and mustache, here each carefully and exquisitely engraved. The saint's book, with a handsomely decorated cover, is held by the left hand, slight-

ly to the side. Grasped in the right hand, two keys, symbolic of Peter's authority (Matt. 16:19), extend from the knee to the upper chest in a manner that prominently proclaims the saint's identity.

Curved and almost straight lines delineate the crisp folds of the drapery and emphasize the swelling of the figural masses beneath. The richness of this drapery is enhanced by the subtle irregularities of the engraved and modeled lines and by the cross-hatching, circles, and beading of the borders.

The enameled, vesica-shaped back plate is beveled at the edges and pierced by ten holes provided for attachment to a larger work. The undulating enameled border is outlined by a plain gilt line in reserve; the dark blue background is partially restored. The inner gilt field of the plaque is engraved with small circles, quatrefoils, and near-horizontal lines and dashes. Dominating this inner field is an imposing and colorful series of enameled circles, ovals, and lozenges, each decorated internally with multicolored rosettes, circles, and quatrefoils.

The style of the figure, although Romanesque in inspiration, may be designated early transitional, as with the following works (cats. 47, 48). Also, the decorative motifs in enamel, although not identical, are similar to those on the book-cover plaque (cat. 48) and to those on some of the enamel plaques on the large chasse of Saint Calminius at Mozac (cat. 45), both datable from about 1185 to 1200. A less dramatic continuation of such motifs can be seen in the background enamels on the vertical face of the large, late chasse of Saint Fausta in the Musée de Cluny, Paris.[1]

Writing in 1886, Guibert and Tixier stated that this plaque "probably came from Grandmont," but it is not clear whether their attribution was based on historical documentation or was merely inspired by the fine quality of the piece. The argument for its provenance in the Limousin itself is bolstered by its having been in Limousin collections. However, another plaque representing an apostle and having the same dimensions and details of its enamel and copper decoration entered the Musée de Poitiers before 1842, a collection that seems to have been primarily formed of objects coming from Poitou (fig. 46a).[2] A third appliqué figure with the same dimensions and probably representing Saint John the Evangelist, its enamel base plate missing, is preserved in the Kestner Museum, Hanover (fig. 46b),[3] and a fourth was on the art market in 1933 (fig. 13).[4] Given the dimensions, known to be common to three of them, and the similarity of their decoration and figural type, it seems most likely that all four plaques once formed part of an altar frontal.[5]

WDW

Fig. 46*b*. Apostle. Kestner Museum, Hanover

Fig. 46a. Apostle. Musée Sainte-Croix, Poitiers

NOTES

1. Gauthier 1972b, pp. 188–89, 373 ill., no. 132.

2. Inv. no. 892.258; Texier 1843, p. 247, no. 6; Gauthier 1950, p. 45. Some of the museum's collection belonged to a local priest. We are grateful to M. Rérolle, conservateur, Musée de Poitiers, for this information. The link between the Poitiers and Metropolitan Museum plaques was first noted by Texier.

3. 1904, no. 322; H.: 17.27 cm (6¾ in.; feet missing); Stuttmann 1966, pp. 60, 164, ill. 60. The Hanover plaque comes from the Collection of Friedrich Georg Hermann Culemann (1811–1886), a collector of medieval manuscripts and goldsmithwork. See Ernst and Heusinger 1963, p. xi.

4. Offered to the Fogg Art Museum, Cambridge, Mass., in 1933 by Arnold Seligmann, Rey & Co. Information preserved in the Stohlman Archives, Princeton University.

5. A copy of the Metropolitan Museum Saint Peter, presumably made while it resided in the collection of Frédéric Spitzer, may be seen in the large collection of modern copies of Limoges enamels in the Museo Lazaro Galdiano in Madrid. An engraving of a plaque of Saint Peter, said to have been found in excavations at Trier in the nineteenth century, is reproduced in Weerth 1857, pl. LXI; 1868, vol. 3, p. 97. Although virtually identical in design to the Metropolitan plaque, the example in Trier is smaller and presents a number of stylistic anomalies, notably the misunderstanding of the form of the footstool and of the saint's drapery. There are no attachment holes, and the base plate seems thicker and more three-dimensional. It may be that the Trier plaque was based on the engraving in Texier 1843. It does not appear in subsequent publications concerning Trier.

BIBLIOGRAPHY: Texier 1843, p. 25, pl. III; Ardant 1855, p. 58; Guibert and Tixier 1886, p. 59, pl. XXXII; Palustre and Molinier 1890, p. 112, no. 47; Rupin 1890, p. 432; Pératé 1911, no. 60, pl. XXXVI.

# 47. Saint Peter

Limoges, ca. 1185–1200

Copper (plaque): engraved, stippled, and gilt; (figure): repoussé, engraved, chased, scraped, and gilt; champlevé enamel: medium and light blue, turquoise, green, yellow, red, and white; blue-black glass inset eyes

24.8 × 9.5 cm (9¾ × 3¾ in.)

EX COLLS.: Georges Hoentschel, Paris (by 1911); [Jacques Seligmann, Paris (in January 1912)]; J. Pierpont Morgan, London and New York

CONDITION: Some losses to gilding on highest areas of reliefs; small losses to enamel on back plate; repair to enamel at proper right of head.

The Metropolitan Museum of Art, New York; Gift of J. Pierpont Morgan, 1917 (17.190.812)

The standing figure, modeled in low relief, has been expertly engraved, chased, stippled, and gilt; the eyes are glass bead insets. The figure is held in place by two domed copper rivets, both of which may be replacements of the original ones. The saint's collar is engraved with rinceaux closely related to the standard vermiculé pattern. The right sleeve ends with a cuff engraved with a series of alternating horizontal and vertical ovals.

The figure shows all the traditional attributes of Saint Peter,[1] including the tonsure, ringed by tight curls, and the short, cropped beard and mustache, which in this piece are finely engraved with a series of parallel curves. The other attributes are the book (here, held tightly against the chest) and the keys to the kingdom of heaven (grasped firmly in the right hand).

The saint's feet rest on a "pillow" that has been engraved with three points resembling arrowheads. Parabolic and nearly straight lines create an elegant pattern that describes both the long, flowing drapery folds and the underlying tubular masses of the figure. The hem of the saint's tunic is beaded.

The arched, enameled, and gilt back plate is framed by an undulating border of enameled half circles set against a narrow gilt ribbon and a band of gilt metal engraved and stippled with straight lines and dots. The outer edge of the plaque is beveled. Reserved gilt rinceaux in each field explode into colorful, enameled flowers. Horizontal bands of turquoise enamel set with engraved, stippled, and gilt horizontal lozenges are positioned behind the shoulders and knees of the saint. The dark blue halo is enriched by a radiating series of five colorful lobes alternating with red enamel rays that are enlivened by gilt dots in reserve.

The style, while Romanesque in inspiration, may be designated as early transitional or late twelfth century by reason of the rhythmic, linear portions of the drapery, the suggestion of underlying figural mass, and the hint of transitional classicism in the handsome and well-delineated head that recalls some of the bearded heads on several of the large Mosan-Rhenish shrines of the late twelfth century, such as the Heribert Shrine of about 1160–70 and the shrine of the Three Kings of about 1181–91, both in Cologne.[2] The best comparisons with works by Limousin artists include the chasse of Saint Calminius in Mozac (cat. 45) and the dismounted enamel plaques preserved in the cathedral at Orense (cat. 51).[3] A similar attribution may be entertained for the present work on the basis of the style and engraving of the relief as well as the decorative motifs and the colors of the enameled backplate.

The Museum's plaque probably once served as one of a series of individual plaques representing apostles set under arcades and flanking the central image of Christ on the vertical face or sloping roof of a large, now lost Limoges shrine, in a program comparable to that of the chasse of Saint Viance in Saint-Viance (Corrèze) (cat. 118). Certainly by the third quarter of the twelfth century, this convention of using appliqué plaques and figurative elements for the sides of such chasses was well established. Between 1160 and 1174, a chasse of Saint Alpinien was made for the priory church of Saint-Martial at Ruffec-en-Berry. It was described in the eighteenth century as having "plusieurs testes en relief et nombre de figures en émaillé appliqués sur les dites lames de cuivre" [several heads in relief and a number of enameled figures applied to the sheets of copper].[4]

WDW

NOTES
1. Carr 1978.
2. Cologne and Brussels 1972, pp. 277 (ill.), 316 (ill.).
3. Ross 1933b, figs. 1, 2, 4; Hildburgh 1936, pl. XXIII, figs. 29a, 29b.
4. Gauthier 1987, p. 103.

BIBLIOGRAPHY: Pérâté 1911, no. 59, pl. XXXVI.

# 48. Book-Cover Plaque: Christ in Majesty

Limoges, ca. 1185–1210

Copper (plaque): engraved, scraped, stippled, and gilt; (appliqués): repoussé, engraved, chased, and gilt; champlevé enamel: blue-black, dark, medium, and light blue; green, yellow, red, brownish red, and white; blue-black glass inset eyes

27.5 × 14.2 cm (10⅞ × 5⅝ in.)

INSCRIBED: A/ω (Alpha/omega)

EX COLLS.: Georges Hoentschel, Paris (by 1911); [Jacques Seligmann, Paris (in January 1912)]; J. Pierpont Morgan, London and New York

CONDITION: Some loss of gilding on highest areas of reliefs; enamel partially lost on three corners; no restorations.

The Metropolitan Museum of Art, New York; Gift of J. Pierpont Morgan, 1917 (17.190.757)

The seated Christ in the center and the symbols of the evangelists in the corners of this plaque were formed in relief and are secured to the enameled backplate by two of the three original rivets. These rivets were engraved and gilded in place to match the surrounding surfaces of each relief. A lozenge pattern has been engraved on Christ's robe at his neck, chest, and collar. Many of the lines of the drapery folds have been enriched by the chatterlike effect that resulted from the rocking motion of the engraver's hand. Christ's crown, cut from a separate piece of copper and engraved and gilded, curves to fit his head. Christ's right hand is raised in blessing; his mantle-covered left hand holds a book; his feet rest firmly on a three-legged stool (engraved on the backplate). The reliefs of the evangelists' symbols are engraved and stippled with various patterns and textures appropriate to each symbol: the feathered imbrications of the eagle, the tightly parallel lines and dots of the mane and chest of the lion, the burnished grooves of the ribs of the ox, and the curvilinear drapery patterns of the angel. Christ's eyes, like those of the symbols, are insets of glass beads.

The rectangular plaque backing these reliefs is fully enameled with an overall pattern of circles, circles with interior quatrefoils, and circles with rosettes or crosses. The enameled cruciform nimbus of Christ and the reserved Greek letters alpha and omega are part of this overall decorative pattern; varied but repeated colors enrich the total effect. Dominating this scheme is the almond-shaped frame and an arc, both engraved with a vermiculé pattern, which offset the central figure of Christ. A series of undecorated ovals, the centers of which lack gilding, punctuate the frame and the arc. These portions of the frame decoration may have once been covered by bezels holding crystal or glass cabochons, as suggested by the pair of holes within each oval and by comparison with the mandorla frames of the frontals at Burgos (ca. 1150–70)[1] and at Pamplona

(ca. 1175–80).[2] This motif does not appear on other published book covers. A second arc with an undulating pattern of enamel appears outside the vesical frame. All the gilt edges of the various patterns, including the alpha and omega, are enlivened by a series of minute, consecutive punches. The flat, gilt outer edge of the plaque is engraved with a series of hatchings.

The style of the appliqué figures is late Romanesque of the early transitional manner associated with the years before and somewhat after 1200. The frontal, static, and monumental Majestas Domini of Italo-Byzantine fifth to sixth-century painting and mosaics,[3] of Carolingian court art,[4] of early Romanesque manuscript painting in northern Europe,[5] and of the central tympanum of the west facade of Chartres Cathedral[6] may each be said to be part of the ancestry of the imposing Christ in Majesty. In the rendering of the masses, the triangular configuration of the lower legs, and the abstract delineation of the taut drapery folds, the inheritance of the image becomes simplified. The angel of Matthew, however, suggests a sense of animation because of the raised leg, the arm pulled across the chest, the upraised hand, and the three-quarter view of the whole figure. The relatively fluid delineation of the angel's drapery and his attenuated, outstretched wings (the upper one is broken) heightens the sensation of movement, which distantly recalls earlier Carolingian Reimois work, such as the figures in the ninth-century Utrecht Psalter, and parallels some Mosan objects by Nicholas of Verdun or made under his influence, such as Nicholas's enameled ambo plaque of the Annunciation at Klosterneuburg, with its twisting figure of Gabriel. In fact, the symbol of Matthew on the book cover might serve equally well as the angel Gabriel of an Annunciation scene. Such animation is found in contemporaneous Limoges work, such as the base of the Alpais ciborium (cat. 69).

Except for the angel of Matthew, the style of the symbols as well as of Christ, together with the enameled background plaque, recalls the style of the panel on the roof of the chasse of Saint Calminius in Mozac (cat. 45), usually dated between 1185 and 1197.[7] The Metropolitan Museum book cover may also be compared with the "Alfonso" plaque in the series of dismounted enamels preserved in Orense Cathedral,[8] on the pilgrimage road to Santiago de Compostela in northern Spain (fig. 51a). The stance and some of the draperies of the supplicating figure of "Alfonso" appear to be enlargements of the figural concept of the angel symbol. Such comparisons confirm a date of about 1185—at the earliest—to about 1210 for the Metropolitan Museum's plaque.

The iconography of the Metropolitan's piece is characteristic of many other Limoges book-cover plaques, although here the reliefs are higher. The quality of the present work is also extremely fine, both in the metalworking and the enameling. Complete covers with their framing borders, like the plaque from Lyons (cat. 18), suggest the original context for the present work. The twelve holes along the outer edge indicate where additional rivets would have held the plaque to the wood support of the bookbinding. The significance of the single line drawn across, near the upper edge of the reverse, is unknown.

WDW

NOTES

1. Hildburgh 1936, pl. v, fig. 6c; Boehm, in New York 1993, no. 134.
2. Gauthier 1987, no. 135, pl. D, fig. 7; pl. CXVI.
3. Catacomb of Commodilla, Rome; apse mosaic, S. Pudenziana, Rome; apse mosaic, San Vitale, Ravenna; east end, south wall mosaic, S. Apollinare Nuovo, Ravenna; Kitzinger 1977, figs. 80, 106, 156.
4. Godescalc Evangelistary, 781–83, Paris, BNF, nouv. acq. lat. 1203, fol. 3r; Mütherich and Gaehde 1976, pp. 32–33, pl. 1; Codex Aureus of Saint Emmeram, about 870, Munich, Staatsbibliothek. MS lat. 14000; Swarzenski 1954, pl. 10.
5. Bible for the abbey of Stavelot, 1093–97, London, British Library Add. MS 28107, fol. 136; Cahn 1982, pp. 129 (ill. 83, colorplate), 265.
6. Sauerländer 1972, pl. 5.
7. Hildburgh 1936, p. 119; Souchal 1963, p. 309, fig. 23; Paris 1965, no. 445, pp. 245–46, pl. 59.
8. Hildburgh 1936, p. 119, pl. XXIV, figs. 29b, 30.

BIBLIOGRAPHY: Pératé 1911, no. 56, pl. XXXIII; Hildburgh 1936, p. 119, pl. XXIV, fig. 30; Marquet de Vasselot 1941, p. 144.

# 49. Appliqué Christ

Limoges, ca. 1190–1200

Copper: repoussé, engraved, chased, and gilt; champlevé enamel: lapis and lavender blue, emerald and light green, yellow, red, and white; glass cabochons and enameled beads

27.7 × 19.2 cm (10⅞ × 7½ in.)

Ex colls.: Victor Martin Le Roy, Paris (Marquet de Vasselot [1906a] erroneously identifies it with the one published by Grimouard de Saint-Laurent in 1869). Gift of Victor Martin Le Roy, 1914

Condition: Right hand and both feet missing. Most of cabochons on orphreys have been lost; only two glass cabochons and fifteen turquoise-enameled beads remain on robe. The two blue-enameled beads of the eyes and the four cabochons on the crown, however, have been preserved.

Musée du Louvre, Paris; Département des Objets d'art (OA 8102)

This Christ, the exceptional quality of which has been underscored whenever it has been published, is truly one of the finest examples of a Limousin figure of Christ in which the Savior is represented in glory, alive, eyes wide open, crowned, and clad in superimposed robes, the top robe enhanced by orphreys richly encrusted with gems. This depiction harks back to the oldest Christian traditions: images of the Crucifixion, which, as deftly summed up by Grabar (1968), aimed at showing Christ's glory and none of the reality of his death. The earliest example of this iconography of Christ, dressed in a long robe with sleeves, is the one on a silver plate from the sixth or seventh century, found in Perm, Syria, and now in the State Hermitage Museum, Saint Petersburg.[1] As emphasized by several scholars, the origin of these depictions must be sought in the vision of the Apocalypse (1: 12–13): "I saw . . . one like to the Son of man, clothed with a garment down to the feet, and girt about the paps with a golden girdle."[2]

The Christ in the Musée du Louvre is very similar to those affixed to the center of the front face of twin crosses (fig. 49a) from Nävelsjö, Sweden, now in the Statens Historiska Museum, Stockholm, and to the figure from the former Bouvier Collection, now in the State Hermitage Museum, Saint Petersburg. The latter is probably the one published in 1869 by Grimouard de Saint-Laurent.[3] These figures, incontestably from the same workshop, constitute the oldest group of appliqués of this type, intended to decorate the center of a large cross. Other examples, of smaller size and much inferior quality, testify to the prolonged success of this iconographic formula.[4]

The two Nävelsjö crosses give a good idea of the kind of cross to which the Louvre Christ must have been affixed. Large in size (H. 70 cm [27½ in.]), they are covered with plaques of stamped copper and strewn with cabochons. There are appliqués in half-relief on the principal face and enameled medallions on the back. As pointed out by B. M. Andersson (1980), these works were undoubtedly inspired by the large crosses of gem-encrusted goldsmith's work known throughout the West since the fifth century.

Countless features link the Louvre Christ and its peers to Limousin production during the late twelfth century.

The rigid frontality is still completely Romanesque. The engraved decoration of the orphreys ties them to the group of works with vermiculé backgrounds made during the final third of the century. The style, full, with a perfectly oval face surrounded by a beard and by fine hair with soft curls, especially suggests a connection with the enameled figures of Christ on crosses executed for the Order of Grandmont from about 1185 to 1200 (see cat. 62). The rosettes, one each on the backs of the Nävelsjö crosses, resemble those on the Ambazac chasse (cat. 55). As noted by M.-M. Gauthier (1972) and B. M. Andersson, these crosses are similar to the works of the very early thirteenth century—for instance, the chasse of Saint-Marcel (cat. 75). The Christ in the Louvre, the one in Saint Petersburg, and the two in Stockholm could therefore have been executed during the final decade of the twelfth century or the very beginning of the thirteenth. Comparison with the enameled appliqués in the Confessio of Saint Peter in the Vatican, put in place under Pope Innocent III (r. 1198–1216), which seem later, confirms this date.[5]

According to Andersson (who sees the three superimposed robes as evoking the tunic, the alb, and the dalmatic worn by deacons and bishops), the diffusion in Sweden of this iconographic type of Christ—"sovereign, priest, and king"—might have been in response to a specific demand by Scandinavian clergy, as the result of a local tradition of putting cloth robes on the figures of Christ on choir crosses. However, this tradition was not limited to Scandinavia. A passage in the *Chroniques de Saint-Martial de Limoges*, published by H. Duplès-Agier, who dates the passage to about 1216 to 1218, alludes to a cloth tunic that was made for the crucifix placed on the altar of the Holy Savior at Saint-Martial. Another, older passage shows us that this Christ was covered with silver: therefore, it must have been a Christ sculpted in relief.[6] T. Sauvel, who published these references in 1951, had earlier linked them to the Louvre Christ coming from the Martin Le Roy Collection, the garments of which—sharply distinguished by different colors at the bottom and at the ends of the sleeves—undoubtedly mimic cloth tunics.[7] This iconographic form of Christ in glory harmonized especially well with the dedication of this altar to the Savior.

We must therefore conclude that Limoges, like other places in southern France and other regions of western Europe, from Germany in the north to Catalonia, was one of the centers in which the vision of Christ, at once crucified yet in glory, came back into favor at the end of the twelfth century. The most poignant testimony to this revival is, no doubt, to be found in the restoration of the *Volto Santo* in Lucca and the rapid spread of devotion to it.

ET-D

NOTES

1. Grabar 1968, pp. 131–32; Schiller 1966–68, vol. 2, pp. 98ff.
2. Barbier de Montault 1898b; Durliat 1989, p. 73.
3. For the Stockholm crosses (H. of Christ: 32 cm [12⅝ in.]), see Gauthier 1972b, no. 71, pp. 115–17, 340; B. M. Andersson 1980, pp. 18–22. For the Saint Petersburg Christ (H. 30.5 cm [12 in.]), see Grimouard de Saint-Laurent 1869, pl. p. 357; Lapkowskaya 1971, pl. 16.
4. Examples are in Oxford, designed for Montfaucon (see Caudron 1976, pp. 140ff.); the Musée de Cluny, Paris (Cl. 959); the Victoria and Albert Museum, London (834.1891); Walters Art Gallery, Baltimore (44–108); and the Museum of Fine Arts, Boston (49.472) (see Swarzenski and Netzer 1986, no. 23, which provides a complete list, including exemplars

Fig. 49a.
Cross from Nävelsjö.
Nationalmuseum,
Stockholm

known through reproductions). One formerly preserved in the Kunstgewerbemuseum in Berlin has disappeared (Zuchold 1993, no. 84).
5. See Gauthier 1968a.
6. Duplès-Agier 1874, pp. 298–99: "pallium . . . ad faciendam tunicam vultus preciosi qui gloriosus eminet super altare sancti Salvatoris" (a veil . . . to make the tunic of the precious figure, which, glorious, surmounts the altar of the Holy Savior); p. 285 (Rigaud): "Iste deargentavit vultum sancti Salvatoris jubente Ludovico rege" (This one covers with silver the figure of the Holy Savior by order of King Louis).
7. Sauvel 1951, p. 104. This author saw the term *vultus* for this Christ as an allusion to the *Volto Santo* at Lucca—a hypothesis that seems at the very least far-fetched.

EXHIBITION: Vatican City 1963, no. 30, pl. XVIII.

BIBLIOGRAPHY: Marquet de Vasselot 1906a, pp. 37–38, no. 23, pl. XVI; Duthuit 1929, fig. 15; Huyghe 1929, p. 4, ill.; Sauvel 1951; Thoby 1953, p. 11; Caudron 1976, p. 145; B. M. Andersson 1980, p. 21, fig. 30; Swarzenski and Netzer 1986, no. 23, pp. 80–81.

# 50. Appliqué Christ

Limoges, ca. 1195–1210

Copper: repoussé, engraved, chased, and gilt; champlevé enamel: lapis and lavender blue, red, and white; glass cabochons and enameled beads

29.5 × 18.8 cm (11⅝ × 7⅜ in.)

Ex coll.: Charles Mège, Paris. Bequest of Elisabeth Mège, 1958

Condition: Gilt very worn, especially on legs and torso. Arms broken; hands missing; top of crown broken. Four cabochons missing from belt, one from crown.

Musée du Louvre, Paris; Département des Objets d'art (OA 9956)

This figure presents a vision of the crucified Christ as supreme and in glory. It differs from the preceding only in his garment, a simple loincloth, and retains, on the garment and around the neck, luxurious orphreys encrusted with gems. The orphrey at the neckline—unrealistic because it is not applied to fabric—can only be explained as an imitation of figures of Christ dressed in a tunic (cat. 49). The strict frontality and massiveness of the figure and the rigid, purely rectilinear treatment of the loincloth are still entirely Romanesque in style.

Several figures of Christ with the same features are comparable to the one in the Louvre. Only a single exemplar, from the church of Ukna, Sweden, and now located in the Statens Historiska Museum, Stockholm, is still attached to its cross, which is decorated with sheets of stamped copper and additional appliqués (fig. 50a).[1] However, works of this type with similar dimensions were also executed for churches in central France. Before the Revolution, the church in Mozac had a cross 84 cm (33⅟16 in.) in height—hence, even larger than the ones from Ukna and Nävelsjö in Sweden. The Mozac crosier was described in 1846, when it belonged to "M. Thévenot, a former major in Clermont-Ferrand," as "remarkably rich and brilliant."[2]

The Walters Art Gallery in Baltimore and the Toledo (Ohio) Museum of Art have figures of Christ detached from their support like the one in the Louvre.[3] They show the same fine quality but are less vigorous. Moreover, the figure at Toledo has undergone several restorations. Despite the absence of an orphrey around the neck, the Christ on a cross from a church in the department of Mayenne and now preserved in the Musée Dobrée in Nantes can, by its style and quality, be linked to this group.[4] Especially close to the Christ from Ukna, the Musée Dobrée figure is attached to a large cross decorated with stamped plaques highlighted by cabochons, very much like examples found in Sweden. Numerous crosses of the first half of the thirteenth century follow suit, presenting an appliqué Christ at the center of their principal face. While clearly very similar in conception to the Louvre and related examples, they are less sumptuously adorned, often of inferior quality, and reflect a gradual softening of the figure and its garment.[5]

The Christ in the Louvre is one of the largest, undoubtedly the oldest, and certainly the finest in the small group of these appliqués still in Romanesque style. It is, notably, bigger and more highly finished than the one on the Ukna cross. The sculpting and engraving, as well as most of the decorative elements, link it to the preceding one in this catalogue. Only a slight difference can be detected in the engraved design of the scrolls on the orphreys in which the vermiculation has been replaced by a simpler and more relaxed motif that became widespread during a later phase in the evolution of Limousin enamelwork. Slightly larger, it must have come from a monumental cross, of dimensions equivalent to those of the crosses of Mozac, Nävelsjö, and Ukna.

ET-D

Fig. 50a. Cross from Ukna. Nationalmuseum, Stockholm

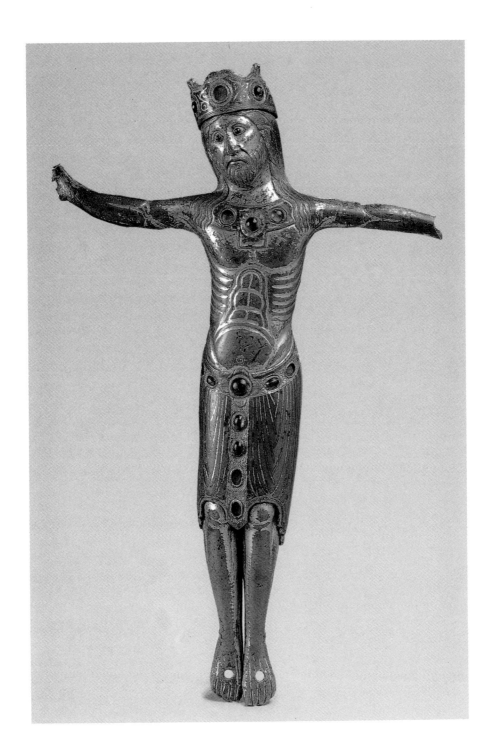

NOTES

1. Inv. 11504, H. 66.4 cm (26⅛ in.); see B. M. Andersson 1980, pp. 18–22, fig. 32. As in the preceding case, we note that the central Christ is the most polished and most sumptuously decorated piece.

2. Bouillet 1846, pp. 325–26, pl. 31, fig. 4. According to Craplet 1962, p. 215, it "ended up in a museum in England." The engraving that accompanies the reproduction is too summary to permit a judgment of the style and date. A slightly later figure (ca. 1210–30), in a private collection, measures 35.6 cm (14 in.). See Gauthier 1992, pp. 99–100, fig. 6d.

3. Walters Art Gallery (44.228); see B. M. Andersson 1980, n. 98, p. 33, fig. 34. Toledo Museum of Art (inv. 69.295); ex coll. M. Boy (sale, Paris, May 15–24, 1905, no. 151); see *Chronique des Arts* 1970, no. 238, p. 49.

4. Height of cross 85 cm (33½ in.), of Christ 27.2 cm (10¾ in.); inv.

896–1–24. See Thoby 1953, no. 32, pl. XIX; Freiburg 1972, no. 80. The Christ on another cross at the Musée Dobrée, from the collections of Bastard d'Estang and Thoby (Thoby 1953, no. 28, pl. XV, and Gauthier 1987, no. 272, pp. 218–19, fig. 747, pl. CCXXVIII), attached to a cross that was not its original support, belongs in the "wake" of that first group of enameled appliqué Christs.

5. Thoby 1953, nos. 28–45. See also the Christ cited in n. 2 above, of great quality and exceptional dimensions.

EXHIBITION: Paris 1967, no. 235.

BIBLIOGRAPHY: Landais 1961b, p. 12.

# 51. Saint John the Evangelist

Limoges, ca. 1174–1213 (probably by 1188)

Copper (plaque): engraved and gilt; (appliqués): repoussé, engraved, chased, scraped, and gilt; champlevé enamel: dark, medium, and light blue; turquoise, medium green, yellow, red, and white

32 x 13 cm (12½ x 5⅛ in.)

INSCRIBED: s[ANCTUS] IOKAN[N]ES (Saint John)

PROVENANCE: Cathedral of Saint Martín, Orense (Galicia), Spain (first described by Villa-amil y Castro in 1898; see Villa-amil y Castro 1907)

CONDITION: Slight wear to gilding; some loss to enamel at proper left between turquoise bands; plaque curved slightly at bottom.

Private collection, New York

The gilt-copper image of Saint John is identified by the inscription that appears at shoulder level on the enameled plaque to which the repoussé figure is attached. He clasps a book in his left hand and raises his right to his chest, his index and fourth fingers extended. His downturned feet rest on a half circle of enamel. Two horizontal turquoise-enameled bands with reserved pseudo-Kufic inscriptions appear behind him, and he is surrounded by simple patterns of foliate vine ornament.

The plaque certainly belonged originally to the ensemble of fifty-three enameled elements preserved in the cathedral of Orense. Chief among these are sixteen gilt-copper relief figures of apostles and saints, each set against an enameled backplate whose scheme of decoration and inscription correspond point for point with the Saint John plaque. The combination of standing apostles and saints and symbols of the evangelists set on triangular plaques that clearly once surrounded an image of Christ in Majesty suggests that these enamels originally formed an altar frontal in the cathedral,[1] as their present installation there indicates. Recent literature has suggested an alternate to this traditional position: that the elements were incorporated on a large chasse like the one preserved at Saint-Viance (cat. 118).[2]

Critical to either theory is a review of the surviving figural ensemble. The apostles of the series (now incomplete) are identified by inscriptions as James (two plaques, for James the Great and James the Less), Bartholomew, Simon, Thaddeus, Thomas, Andrew, Philip, and Peter. In addition to Paul, traditionally numbered with the apostles, Saint Martial—included among the apostles in the Limousin—appears.[3] The Virgin Mary is also represented,[4] as is the evangelist Saint Mark.

Other plaques in the series represent Saint Vincent, a fourth-century bishop of Saragossa, and a saint identified as TIRIS or CIRIS. Associated by Ross with Saint Cyr (Cyricus; Cyrus), martyred with his mother as a boy of three, he more logically represents Saint Quirico, a seventh-century bishop of Toledo.[5] This would make a more consistent program, wherein all saints other than apostles, evangelists, and the Virgin are bishops. The last and most important is Saint

Martin, a fourth-century bishop of Tours, with a donor kneeling at his feet (fig. 51a). The inscription on the plaque of Saint Martin provides the means for identifying the patron responsible for these enamels. Though early authors offered several interpretations, the most convincing is that proposed by Marvin Ross: "S. Alfonso Arer D." (Suo Alfonso Areri dedit) refers to a gift by Alfonso, bishop of Orense between 1174 and 1213.[6] It is recorded that Bishop Alfonso obtained relics of Saint Martin for the consecration of the altar dedicated to him in the new cathedral on July 4, 1188.[7]

It is difficult to conceive of a chasse sufficient in size to incorporate all the existing elements of this enameled ensemble. Nor is there evidence that the relics of Martin obtained for Orense from Tours, where his body remained enshrined, were significant enough to warrant a chasse of tombal proportions, in the tradition of the *Urna* of Silos. It is also unnecessary to posit the existence of a chasse, since the individual elements can logically be seen within the context of altar decoration. The four triangular plaques bearing the symbols of the evangelists, each with a side cut away to form an arc, could have accommodated a scalloped mandorla similar to the one surrounding the figure of the Virgin and Child in Majesty at San Miguel in Excelsis, Pamplona—dated by Gauthier to 1175–80,[8] roughly contemporaneous with the Orense pieces.

The format of the Saint John plaque, the disposition of its foliate ornament, and the gently looping drapery of the figure invite comparison with the roughly contemporaneous plaque of Saint Peter (cat. 47), which, however, is not so finely worked as the Orense plaque. The chasse of Mozac (cat. 45), also sharing the same general characteristics as well as the combination of Latin and pseudo-Kufic inscription bands, is similarly inventive in the design of the enameling set under the feet of the applied figures of its apostles.

The series now at Orense lacks—in addition to the standing image of the apostle-evangelist Saint John shown here—the evangelist Luke, the apostle-evangelist Matthew, and the apostle Matthias. The symbol of Saint John the Evangelist as an eagle is also missing, along with more than a dozen enameled arches and columns, crenellated tower finials, lenticular and semicircular plaques of angels, and floral border plaques. It is presumed that the missing plaques were removed from the cathedral during the Napoleonic era; the plaque of Saint John does not appear in any of the early publications of the Orense ensemble.[9]

BDB

Fig. 51a. Bishop Alfonso at the feet of Saint Martin. Orense Cathedral

NOTES

1. Ross 1933b, p. 273.
2. Lorenzo 1989, pp. 61–69. The hypothesis is accepted in Santiago de Compostel, 1990.
3. See cat. 58 and Ross 1933b, p. 278.
4. The color illustration of this plaque in Madrid 1981, vol. 1, fig. 65, p. 199, suggests that it may be reenameled or is a replacement.
5. See *Bibliotheca Sanctorum* 1961–70, vol. 10, col. 1324.
6. Ross 1933b, pp. 273–76. The problematic part of the inscription was the word "ARERI," thought by Villa-amil y Castro and Leguina to stand for *aurifaber* (goldsmith) and by Juaristi to refer to Aredius (Saint Yrieix). Ross logically connected it to the *sedes Auriensis* (see of Orense).
7. *España sagrada* 1763, vol. 17, pp. 95–100 (on Alfonso I).
8. Gauthier 1987, no. 135.
9. See Villa-amil y Castro 1907, pp. 226–27; Leguina 1909, pp. 147–51; Bertaux 1910, pp. 329–30, fig. 108.

BIBLIOGRAPHY: Unpublished.

# 52. Virgin and Child

Limoges, ca. 1200

Copper: formed, repoussé, engraved, chased, scraped, and gilt; hands cast; champlevé enamel: dark, medium, and light blue; turquoise, green, yellow, red, and white; glass cabochons; wood core

H.: 36 cm (14⅛ in.)

Ex colls.: Marqués de Castrillo, Madrid (in 1892); J. Pierpont Morgan, London and New York

Condition: Losses to gilding overall; Virgin's left hand, Child's left foot, and cabochons from footstool and hemline lost.

The Metropolitan Museum of Art, New York; Gift of J. Pierpont Morgan, 1917 (17.190.125)

Fig. 52a. Virgin and Child. Old Cathedral, Salamanca

The Virgin Mary, sitting erect on an enameled and jeweled throne, balances the Christ Child on her lap. In her right hand she once grasped an object, probably a lily. The Virgin's left hand had rested on the knee of her son, who sits equally erect, raising his right hand in blessing and holding an enameled book in his left. The crowns, though ill fitting, appear to be original, as the metal is engraved with a pattern that also appears on the neck and hem of the Virgin's gown. Glass cabochons, some backed with silk, decorate the robes of the Mother and Child in imitation of gemstones. The Virgin's slippered feet rest on a footstool set with additional cabochons. The sides of the throne are sheathed with sheets of copper stamped with rosettes inscribed in circles and studded with cabochons set around an applied lozenge decorated with an enameled flower.

The Metropolitan Virgin belongs to a tradition of small-scale copper-and-enamel sculptures of the enthroned Virgin and Child that share a common decorative vocabulary.[1] Many were preserved into the nineteenth century in French and Spanish churches; a number are now found in museum collections as well. The existence of such sculpture in Spanish collections and the presence of a celebrated example in the Old Cathedral, Salamanca (fig. 52a), underlie the traditional attribution to Spain. Since goldsmiths are recorded at Salamanca from the twelfth century, the suggestion that such images were made there is not unreasonable.[2]

However, the decorative vocabulary is that of Limoges enamels of the late twelfth and early thirteenth century. The type of stylized enamel flower set in a lozenge and fixed here to the sides of the throne appears in Limousin work as early as the chasse of Ambazac of about 1180–90 (cat. 55). Glass cabochons of the type found on the Metropolitan Virgin and Child commonly decorate liturgical objects made in the Limousin, such as the knops of crosiers or the arms of processional crosses (cats. 49, 50). This indicates that the artists who produced them were, at the very least, schooled in Limoges; in all likelihood, these images were produced there and sent to Spain.

Most of the surviving Limousin images of the Virgin and Child are made from sheets of copper hammered in relief to form the front and back; they are joined mechanically along the sides like two halves of a mold. This example, however, is composed of a number of sheets of copper worked over a carved wood core; the hands and head were made separately and then inserted. In this regard, the Virgin and Child corresponds to the established technique used for earlier monumental sculpture in precious metal seen in preserved examples like the Virgin of Orcival[3] and prevalent through the twelfth century. This technical distinction and the rigid

190

Reverse

sculptural frontality of the figures argue for a late-twelfth-century date for the Metropolitan work. Although the Metropolitan Virgin and Child has been thought to derive from the one at Salamanca,[4] it seems more likely that the Salamanca example, with its relatively freer drapery and more decorative cabochons, is slightly later.

Posed in this way, the Virgin and Child corresponds to a type found throughout France and northern Europe by the Romanesque period and known as the *Throne of Wisdom* (*sedes sapientiae*). According to this pervasive metaphor, the Child symbolizes the word of God and Mary, the throne of the biblical King Solomon.[5] But the presence of a now-missing hinged door at the back of this small-scale sculpture may indicate a more specific context for this image. As Gauthier has suggested, the Metropolitan Virgin and Child may have been intended as a container for the Host. Since the consecrated bread of the Eucharist was believed to be the body of Christ, it was appropriately secured in an image of the Mother of God, whose body once nurtured the infant Jesus.[6] Support for this hypothesis is provided by a copper-and-enamel Virgin and Child preserved in the Episcopal Palace, Palencia,[7] which still has a door decorated with an enameled roundel of the Hand of God in blessing, a motif commonly found on Eucharistic doves, like the one formerly at Laguenne.[8]

BDB

NOTES
1. In the Metropolitan Museum Collection, for example, compare 25.120.435 and 17.190.333.
2. Hildburgh 1955, pp. 138–39.
3. Barral i Altet 1983, pp. 348–49, figs. 304, 305.
4. See, most recently, Boehm in New York 1993, no. 137.
5. I. H. Forsyth 1972, pp. 22–30.
6. See Gauthier 1968b, pp. 87–88, pl. IV, no. 3, and Bynum 1986b, p. 81.
7. Hildburgh 1955, pl. XLIXa.
8. Rohault de Fleury 1883–89, vol. 5, pl. CCLXXVI.

EXHIBITION: Madrid 1893, no. 224, salle XVIII, pl. CXVI.

BIBLIOGRAPHY: Hildburgh 1955, pp. 139, pls. LVIIa–c, 142, 144–45; Cleveland 1967, no. IV-24, p. 162; Gauthier 1968b, pp. 72–81, pl. II; New York 1993, no. 138, pp. 283–84.

# 53. Saint Barontus

Limousin workshop, active in Italy (Rome or Pistoia?), ca. 1190–1210

Copper: repoussé, chased, engraved, scraped, and gilt; champlevé enamel: medium and light blue, turquoise, intense light green, and white; applied bezel with blue glass cabochon

47.5 × 12.9 × 3.7 cm (18¾ × 5⅛ × 1½ in.)

INSCRIBED: S[ANCTUS] BARONTUS

EX COLLS.: Baron A. Schickler de Pourtalès, Paris (in 1900); Arnold Seligmann, Rey & Co., New York (in 1940)

CONDITION: Much loss of gilding, especially on highest areas of relief, face and hands; losses of enamel throughout, especially in halo, rear edge of proper left arm, and folds of tunic below knees; nose slightly crushed; copper partially torn around nostrils and back edge next to right cheek; double-sheathed bezel in gilt copper mounted on center of chest may not be original; glass cabochon and copper screw forced into back of bezel appear to be modern.

Allen Memorial Art Museum, Oberlin College, Oberlin, Ohio (46.44)

A thick sheet of copper was hammered and shaped into the relatively high relief of this large standing figure. The work has been expertly chased, engraved, enameled, and gilt, a point that is not clear in most photographs and published illustrations of the piece. The folds of the drapery are rendered in a series of nearly parallel, elliptical yet clearly tapered divisions in reserved gilt copper. The sleeveless dalmatic, enameled in medium blue, is bordered with a narrow band of white enamel. The long-sleeved tunic with light blue collar and cuffs is enameled in green around the curve of the collar, almost to the rear edge. The lower hem of the tunic is engraved with undulating, crisscrossed lines that form a horizontal series of linked ovals. On three sides of the book, its leaves are suggested by engraved parallel lines. Two gilt clasps and several gilt oval petals, each engraved in the reserved metal, interrupt the intense turquoise panel of the book cover. Similarly, the engraved cross in gilt reserve contrasts with the medium blue of the dalmatic. The slippered feet of the saint rest on a steep slope of gilt-copper ground that is entirely without enamel.[1]

The head, whose magnificence can be fully appreciated only when seen in the original, is worked and engraved in a style and technique partially related to such Limoges works as the plaques with Saint Peter and Saint James (cats. 46, 47, 58b). The head, projecting forward in a comparable way, also rests on a very broad neck. Despite the wear throughout, many of the engraved lines, still bearing a residue of gilding, are similar, particularly the irregular edges (or chatter) resulting from the rocking motion of the engraver's hand, as seen in the curving lines of the beard, mustache, and the hair that frames the forehead and temples. (Behind the hair there is a series of short, parallel hatchings.) The

features—almond-shaped eyes with engraved iris and pupils and high-placed small ears—are clearly different from those of the Limoges plaques cited above.

The halo, cut from a separate sheet of copper, appears to be original. Its palette is consistent with that of the figure, as Parkhurst noted. The brilliant green enamel in the narrow outer border and the medium blue in the background of the circular scalloped band are identical with enamels on the figure. Moreover, the incisions of the scalloped pattern are similar to those of the petals on the book cover. The curve of the halo, just inside the narrow green border, is inscribed: S BAθRONTVS. (A rivet was positioned between the A and the R.)

The halo and the figure both have holes for attachment to a larger work. The halo has three holes spaced along the gilded band bearing the inscription and one near the center of the ungilded section that lies behind the head of the saint, which can be seen only when the work is disassembled. The figure was originally held in place by two rivets, one between the feet and the other centered in the chest. The circular hole in the chest and its fully gilded surround are revealed only when the bezel and cabochon are removed. At the back of the relief this hole has been cut to allow for a countersunk square, which would assure a secure fit.

The inscription set in the halo naming Saint Barontus is key to the reconstruction of the presumed original context of the Oberlin figure. Saint Barontus was a seventh-century hermit who lived with several companions—one of whom was known as Desiderius—in the hills outside Pistoia in Tuscany. After his death he was interred in an oratory near his cell. The *Atti* of the church of Pistoia, written in the eleventh/twelfth century, conflates this legend with the life of Saint Barontus, a nobleman of Berry thought to have traveled to Italy.[2] The tomb of Barontus of Pistoia became the focus of local devotion; by the eleventh century Restaldo, bishop of Pistoia (r. 1012–23), established a Benedictine monastery there; it was consecrated in 1018, at which time the saint's relics were translated.[3] Moreover, in 1107 a pilgrim's hospital of significant size and architectural importance was built at the same site and placed under special protection.[4]

Sabatino Ferrali and Marie-Madeleine Gauthier have theorized that the Oberlin Saint Barontus comes from the church at Pistoia, which, according to the description given in an earlier publication by Mazzanti, contained "several graceful, valuable, and very old figures . . . in relief from the background and panels which were affixed to the reredos with nails."[5] According to this theory, the other figures would have represented the hermit companions of

Barontus. In addition, Ferrali recounts a tale popular in Pistoia, which claims that an "American" bought these same figures from the church. This story may be based on Baron Schickler's purchase of the Saint Barontus figure between 1875 and 1895 by private sale. Although this evidence is compelling, it is regrettable that it cannot be confirmed by earlier documentation concerning the church.

An important point of reference for the probable dating of the Barontus figure and support for an Italian provenance are provided by the enameled copper appliqué reliefs of Christ in Majesty, Peter, Paul, and three unidentified saints in the Museo Sacro at the Vatican[6] that undoubtedly came from the large enameled and arched rectangular closing panel that originally covered, yet gave access to, the niche of the Confessio of Saint Peter in Saint Peter's basilica in Rome. This panel was made for Innocent III (r. 1198–1216), apparently coincident with the Twelfth Ecumenical Council of Lateran IV, in 1215. The most telling feature that invites comparison is the classicistic treatment of the drapery folds. Notable are the elliptical divisions for the enamel in reserved gilt metal, often tapered, a rare feature held in common[7] that suggests that these works are contemporaneous.[8] Although the standing saints in the Vatican series are considerably smaller (22–26 cm), the enthroned Christ is of nearly comparative size (41 cm). Thus, following the suggestion of Gauthier,[9] both the Vatican series and the Oberlin relief may be assigned to Limousin workshops active in Italy. A date of about 1200—or certainly before 1215—seems reasonable, considering the stylistic elements they hold in common. The comparison is limited, however, because the figures have been almost completely reenameled and physiognomies are distinct. The location of these workshops is unknown; it could have been Rome or somewhere in Tuscany, possibly Pistoia, but there could also have been itinerant workshops that moved from commission to commission.

WDW

NOTES

1. Barontus's attire, which conforms neither to that of a hermit nor to a standard liturgical costume, may represent a regional variant, as observed by M.-M. Gauthier (1972a, p. 288).
2. This is the legend summarized by Parkhurst (1952, p. 94).
3. See Ferrali 1964, p. 57.
4. Beani 1912, p. 131.
5. "Delle graziose pregevoli, antichissime figurine . . . rilevate nella parte anteriore e piane dalla parte per cui erano affisse con chiodi al gradino" (Ferrali 1964, p. 61). Ferrali's work is indebted to Mazzanti 1920. Gauthier (1972a, pp. 287–89) sees it as a châsse-sarcophage. Ferrali (1964, p. 59) and an eighteenth-century author whose manuscript is preserved at the Vatican indicate that the tomb of Barontus included a stone fenestella (small gate) at the main altar but offer no evidence concerning the incorporation of enamelwork in or around this tomb. See Galletti Cod. Vatic. lat 8053. We are grateful to Dr. Massimo Ceresa of the Biblioteca Apostolica Vaticana for his consultation of this reference.
6. Gauthier 1964, pp. 43–61; Gauthier 1968a, pp. 237–46, figs. 4, 5, 9; Wixom 1982, no. 3, illus. The statement in the Oberlin files, reported by Parkhurst, that the Saint Barontus is "said to come from the abbey of Grandmont (Haute-Vienne), parish of Saint Silvester" appears to be completely without foundation and apparently represents the equivalent of attributing a Gothic sculpture to Chartres or Saint-Denis.
7. This characteristic is quite different from the rather flaccid, untapered divisions seen in such later enameled plaques as the one of Saint William of Bourges (cat. 102).
8. Gauthier (1972a, p. 289) suggested "2e quart du XIIIe siècle" for the Oberlin figure, which appears to be far too late, considering the style of both the head and the drapery.
9. Gauthier (1972a, pp. 286, 288, figs. 4, 9) cites the several characteristics that distinguish the Vatican and Oberlin figures from works produced in Limoges proper.

EXHIBITIONS: Paris 1900, no. 2411, p. 292; Paris 1913, no. 228, p. 120; Boston 1940, no. 234; London 1962, no. 9, ill.; Minneapolis 1966.

BIBLIOGRAPHY: Parkhurst 1952, pp. 92–96, figs. 1–3; Ferrali 1964, pp. 57–70, ill., p. 51; Gauthier 1972a, pp. 287–88, fig. 9, pl. CXVII.

# III

The Abbey of
Grandmont

12th–13th century

That the monastery of Grandmont is still somewhat renowned today, even though its buildings have been dismantled, is due essentially to the objects once preserved in its treasury. Neither the spirituality of the order, however unusual at least at the beginning, nor the architecture of the Grandmontain monasteries, which was quite modest, has commanded much attention. On the other hand, all books on medieval goldsmithwork, even the most general, mention at least one piece from this treasury. The possibility that a goldsmithery or an enameling workshop existed at Grandmont itself is no longer considered likely; but this "substantial monastic client"[1] of the workshops of Limoges, or in more precise terms, "this very important initial recipient of major works conceived and executed in urban workshops run by secular masters established at Limoges"[2] would have had a significant influence on the development of *opus lemovicense.*

A paradoxical situation: to be influential a client must have sufficient funds to maintain a regular flow of commissions and must be in a position to dictate an iconographic and artistic program that, if not totally new, is at least sufficiently structured to be able ultimately to serve as both model and reference. The order of Grandmont, however, professed poverty, and its entire history is marked by financial crises that grew out of its meager economic foundations. Furthermore, nothing in the most authentic of Grandmontain tradition, issuing from the Rule and the Sentences of its founder, Stephen of Muret, foreordained a role for his disciples in artistic matters, especially in the realm of the precious arts: more opposed to luxury even than the Cistercians, more sober in their liturgy than the reformed canons, the Grandmontines in their solitude practiced a sort of communal eremitism, without parish ministry (*cura animarum*) or preaching. Even though the *Regula Sancti Stephani,* drawn up about twenty years after the death of the founder in 1124, contains no prescription relating to liturgical furnishings, it is clear that to conform to the spirit of this text the Grandmontines should have limited themselves, in this sphere as in others, to bare essentials. Since ordinary worshipers could only occasionally enter the churches of the order, the richness of decorations, of sacred vessels, and of all that contributes to embellishing the liturgy could find no justification in the need to spark the imagination of the Christian people, a necessity admitted even by Saint Bernard.[3] Furthermore, all material matters in the Grandmontain cells, particularly purchases and the use of alms, were in the hands of a lay brother (the *curiosus* or *dispensator*); the priests in the order, who were few in number, did not have the liberty to acquire objects of great luxury for the holy service.

Later than the *Regula,* the *Institutio ad castigationem,*[4] the first custumal (*coutumier*) of the order of Grandmont, gives precise instructions of an architectural nature and prescribes the use of the *vouta plana* (that is, an unembellished barrel vault without reinforcing arches or the transverse rib) for the covering of the churches of the order. More than the prescription itself, it is the reason invoked that is worth noting: all ornament is similar to a useless word for which the monks would have to account on Judgment Day. All painting and sculpture were likewise banished from Grandmontain buildings. Liturgical equipment was not specifically discussed; we do not know if the prohibition against using luxurious fabrics extended to liturgical vestments. The use of enameled objects, which were luxurious in the use of color and decoration—if not intrinsically so in the materials of copper and glass—was thus antithetical to the Grandmontain spirit.

A closer reading of the history of the order, however, requires a modification of this assessment. Impecuniousness was indeed chronic among the Grandmontines: the few hectares of land—always obtained in fact through charity—upon which the cells were constructed were inadequate to support a community, however few its members, but it seems that the cells received enough gifts to compensate for this scarcity. The motherhouse of the order, established on a relatively large but quite infertile piece of land, benefited from the largesse not only of the local lords but also of the Plantagenets. The late historiography of the order no doubt had a tendency to exaggerate the importance of the ties between this Franco-English dynasty and Grandmont, but objective accounts do attest to the importance of the gifts received.[5]

Papal interventions, which became more frequent at the end of the twelfth century and during the course of the thirteenth century,[6] had the effect of gradually "mitigating" the Rule, notably in all that concerned collective poverty, one of the distinctive traits of the order. The prohibition against preserving written testimonies of the donations was broken or overturned in 1170. At the beginning of the thirteenth century, the regularization of possessions *extra metas*—that is, not part of the initial endowment of the cells—and the possibility of collecting tithes were allowed. And it appears that after the eclipse of the Plantagenets, the order found new protectors in the Blois-Champagne family. One can advance the hypothesis that dividends from the assay of the mint at Limoges and the foundation of an altar candle by the count of Nevers were a source of regular and considerable revenue for the motherhouse of the order.[7]

Without going into the details of the financial history of the order, which is perhaps impossible to reconstruct, one may recognize that Grandmont undoubtedly experienced two periods of relative pecuniary ease. The first stretches from the years 1160–70 to about 1200 and corresponds to the period of the Plantagenet donations. The crisis that struck the order between 1185 and 1189 and pitted clerics (*clercs*) and lay brothers (*convers*) against each other momentarily interrupted these liberalities but did not stop them. The years 1218–24 seem marked by a financial crisis whose origins lay rather, it appears, in bad management than in aggravated penury. Apparently it was overcome, and if the thirteenth century seems, on the whole, to coincide with an impoverishment of the cells to the profit of the motherhouse of the order, this phenomenon can in part be explained by the policy of prestige and embellishment carried out by the prior of Grandmont to the detriment of the rest of the order.

It is significant that the stages in this evolution of Grandmont on the temporal plane coincided with profound changes in the intellectual and ecclesiastical character of the order. The Grandmontines had long been reticent with regard to the cult of relics. The custumal, in its first version (before 1170), prescribed only that the brothers bow down each time they passed before the Cross (*ante crucem*),[8] and the body of Stephen of Muret himself seems to have been honored decently but discreetly. Without going into the complex detail of the translations of which it was the object, one can note that nothing was done to encourage his public cult.[9]

By the end of the twelfth century, however, Grandmont's reputation was based in part on the relics it held.[10] The gift, in 1174, by Amaury, the king of Jerusalem, of a fragment of the True Cross may have played a decisive role. But the initiative of the sixth prior, William of Treignac (r. 1171–88), who in 1181 sent four monks (two clerics and two lay brothers) to Cologne to procure relics was a determining factor. He justified his actions by the fact "that in the churches where bodies of saints have been buried in peace, not only the monks but also the secular clerics are more assiduous in their divine office, come there more willingly, and pray there more devoutly."[11] The translation of these relics (seven whole bodies of companions of Saint Ursula, two heads, and various bones) from Cologne to Grandmont was followed by a distribution of the relics among the Grandmontain cells. Therefore it is hardly surprising that the new draft of the custumal, no doubt about 1190, specifies that henceforth the brothers should bow down "before the Cross, altar, and holy relics" (*ante crucem, altare, sacrosanctas reliquias*).[12]

It is likely that this introduction of the cult of relics was deeply felt at the heart of the order, particularly among the lay brothers, as one of the "novelties" that led to the revolt by lay brothers (*barbati*) and to the deposing and the death in exile of William of Treignac. In obtaining the canonization—and thus the elevation on the altar—of the founder of the order in 1189, the new prior, Gérard Itier, maintained the cult of relics in the order, but by offering for veneration by the brothers the relics of their own spiritual father, not just those of martyrs from faraway Cologne.

Under these circumstances, it is natural that in the usual evolutionary process other relics joined the core of those of the sainted founder and the followers of Saint Ursula from Cologne. Detailed analysis of inventories allows us to suggest that some relics were of Roman origin, although we cannot know if their presence is related to the many appeals the Grandmontines made to the Curia during the thirteenth century. Others attest to the ties to the Holy Land. But the majority of relics evoke mostly other Limousin monastic establishments (Saint-Martial of Limoges, Chambon-sur-Voueize, Saint-Léonard de Noblat, Arnac, Saint-Junien, Salagnac, Tulle, Brive, Bénévent-l'Abbaye), with a few notable absences (Aureil, Lesterps). The most important donation of relics—and the most surprising, given that it came from a lay lord—was that made by Thibaut VI of Champagne in 1269, of the bodies of Saint Panaphreta (also a follower of Saint Ursula) and Saint Macarius, one of the followers of Saint Maurice of Agaune. This pairing of the virgin martyrs of Cologne and the martyrs of Agaune is not without precedent.[13]

To "house" these relics, the monks of Grandmont naturally called upon the workshops of the closest town, Limoges. Extant objects and inventory listings demonstrate that when the nature of the relic permitted it, or when tradition required it (in the case of reliquaries containing parts of the True Cross), goldsmiths working in silver and gilt silver were summoned, although gold seems to have been totally absent from the treasury.[14] And in this respect, what remains of the treasury of Grandmont attests to the permanence of work in precious metals at Limoges, no doubt since a very early date, and then paralleling the development of *opus lemovicense.*

It was this enamelwork that the Grandmontines requested for all the larger works, works for the most part now destroyed but which seem to have been made in at least three successive stages. The first is situated between 1181 and 1185 and corresponds to the execution of chasses destined for the relics brought from Cologne; there were four, and, although there is no certain vestige of them, according to the descriptions in the inventories they seem to have had enameled figures with, at least in some cases, heads in relief.[15]

The second phase is tied to the canonization of Stephen of Muret in 1189. Although it is thus chronologically very close to the first phase, there was most probably an abrupt break between 1185 and 1189. During this second period, presumably, the chasse of the new saint (the chasse of Ambazac [cat. 55], if one accepts the hypotheses developed elsewhere)[16] and the plaques of the Life of Christ and the Life of Saint Stephen were made and placed on the retable of the main altar.

The third phase, whose beginning can be put about 1220–25 but which might have extended into the second quarter of the thirteenth century, is marked by the execution of the antependium of the main altar with, surrounding Christ in Glory, the company of the apostles, including Saint Martial (cat. 58), as well as the Great Cross of the altar for the office of matins (of which the deacon of Billanges [cat. 59] is most probably a part).

Should one assume that the creation of a new chasse for Saint Stephen of Muret, decorated with scenes from his life, belongs to the same phase of the embellishment of the main altar and probably of the whole sanctuary of Grandmont? If we accept that the appliqué in the Musée Jacquemart-André in Chaalis (cat. 61) comes from this chasse, we must note that its stylistic differences from the figures on the antependium are obvious, although without necessarily implying a true chronological break. Was the "chasse with figures of the apostles" perhaps the missing link between the two works? But its date may have been considerably later, perhaps contemporaneous with the gift of the relics of Saint Macarius and Saint Panaphreta (1269). It is moreover about this time that the only document attesting to a direct link between Grandmont and the workshops of Limoges is dated: in 1267 the prior Gui Archer notified Thibaut VI of Champagne-Navarre that the tomb of his father was executed by John Châtelas(?), burgher of Limoges.[17]

One thus sees that the monks of Grandmont were indeed "good clients" and faithful clients of the Limousin enamelers. The absence or the rarity in the treasury, such as we know it, of the most characteristic objects of the production then current in the workshops of Limoges is all the more remarkable:

no gemellions or cruets, only one processional cross (with an iconography, moreover, of an unknown type),[18] two small chasses, and several reliquary bases, more often engraved than enameled.[19] In any case, the descriptions in inventories, compounded by the absence of any graphic sources, barely allow us to reconstruct the appearance of some objects. If the angel now in Saint-Sulpice-les-Feuilles (cat. 54) had not been preserved, would we be able to tell from the simple description in entry XLVI of the inventory of 1666 that it was one of the most seductive and precious examples in all of Limousin production?

This exceptional object was considered, in 1575, to have been brought to Grandmont from the house of Balezis, a very modest establishment of the order of Grandmont whose history remains obscure and which was not considered a true cell, and still less a priory, after the reform of the order in 1319. Whether it had been deposited there temporarily or whether it was part of the belongings of this Grandmont annex, this angel obviously raises the question of the liturgical objects (one does not dare speak of treasury) used and preserved by the monks of Grandmont in their some 150 establishments. The chasse of Villemaur, originally from the house at Macheret in Champagne, is probably the only example identified as a receptacle used at the time of the distribution of the relics of the Virgins of Cologne. We are better informed about the crosses (whose use was probably both processional and stationary) used in the Grandmont cells and of which there remain a considerable number, several of which form a coherent stylistic series.[20] We should add to this series the cross in the Musée d'Arts Decoratifs, Saumur, and with it the ensemble of the Treasure of Cherves (cats. 98–100), which is quite representative, by the quality of certain objects and the materials of which they are made, of what could have been the liturgical equipment of a church of medium importance, more likely monastic than parochial. The site of the discovery suggests even more that it came from the Grandmont cell of Gandory, rather than from the powerful abbey of la Couronne, as Msgr. Barbier de Montault suggested.[21]

The relative frequency, again confirmed by recent finds,[22] of objects of Limousin enameling with a Grandmontain provenance is explained relatively easily: in a completely centralized order where, until the fourteenth century, everything—regulatory or disciplinary decisions, recruitment, administration—was referred to the motherhouse, it is likely that indispensable liturgical objects were sent from Grandmont to the dependents of the order. However, it is risky to suppose that the network of Grandmontain cells facilitated dispersion of Limoges enamels across Europe, a diffusion that, furthermore, reached areas such as Scandinavia and Italy, where the Grandmontain order never took root. We have nevertheless seen that the prior of Grandmont did intervene as intermediary, and although it is unlikely that the monks themselves took part in commercial activity, one cannot totally preclude the possibility that the laymen who, in the thirteenth century, took over for the lay brothers in managing the material and financial affairs of the cells[23] might have, occasionally, intervened in certain transactions.

JEAN-RENÉ GABORIT

NOTES

1. Gauthier 1960a, p. 288.

2. Gauthier 1992, p. 93.

3. There is a large body of literature on the position of Saint Bernard. But the best analyses are undoubtedly those of Dimier (1964, pp. 82–90); and Pressouyre (1990, pp. 59–67).

4. For the complete texts of the Order of Grandmont, see Becquet 1968; for the *Institutio*, see pp. 516ff.

5. Gifts by Henry I are probably only legend. Geoffrey of Vigeois (see Labbé 1657, vol. 1, p. 387), who was generally well informed, mentioned the bequest of 30,000 gold sous by the empress Mathilda, augmented by an equivalent gift from Henry II. The gift of lead for the church roof is confirmed by the *Great Pips Roll*: Richard the Lion-Hearted seems to have carried out the bequest of his father (see Souchal 1964b, p. 130).

6. Becquet 1960, pp. 310–13.

7. Gaborit 1976a, no. 3, pp. 244–45. The gift of assay from the Limoges mint by Hugh of Lusignan was made explicitly for the fabrication of liturgical objects.

8. See Becquet 1968, p. 516, statute I.

9. The Grandmontain historiography makes much of the position taken by the prior Peter of Limoges (r. 1224–37), who solemnly entreated his predecessor, Stephen of Muret, to cease performing miracles so that the peace of his disciples would not be troubled by pilgrims (see *Vita Stephani*, in Becquet 1968, p. 129). In the time of the fourth prior, Stephen of Liciac, the relics of Stephen of Muret seem to have been withdrawn from their tomb in front of the altar in the chapel to be "hidden" (*sepulchrum . . . conditum*) in the cloister, possibly because of the start of work on the new church; they were returned to the church in the time of the prior Peter Bernard (r. 1163–70). At the time of the *revelatio* of 1190, it was

10. Geoffrey of Vigeois (see Labbé 1657, vol. 2, p. 286).

11. See Becquet 1968, p. 251, nos. 9–13.

12. Ibid., p. 526.

13. See Gaborit 1976a, p. 236, n. 20.

14. However, the texts are ambiguous. The Eucharistic dove that Henry the Younger seized is generally considered to have been made of gold; Geoffrey of Vigeois speaks of the *Columba aurea*; the martyrology of Grandmont designates the reliquary of the True Cross sent by King Amaury as a *phylacterium aureum*, whereas it was of gilt silver.

15. See Gaborit 1976a, pp. 238–39. I have suggested that the two appliqués representing the Wise Virgins, one of which is now in Florence (Museo Nazionale del Bargello), the other in Vienna (Österreichisches Museum für angewandte Kunst), might be the remains of one of the chasses of the companions of Saint Ursula, who are often compared with the Wise Virgins of the Gospel, from which the iconography might have been borrowed.

16. Ibid., pp. 239–61 and 245.

17. Martène 1717, vol. 1, cols. 1123–24.

18. The principal face, with Christ between the Virgin and Saint John and the image of Saint Peter on the upper part, conforms to a well-known schema, but the decoration on the reverse, with Saint Peter crucified, does not seem to be found on any other extant example (see Texier 1857, col. 843, art. XVI).

19. The quite unusual iconography of one of the small chasses (the coffret of Saint Brandan) allows one to think that it is either from one of those now in the Louvre (MR 2618) or from the one now in the Toledo Museum of Art (fig. 31a). Like the Eucharistic coffret formerly in the Musée Municipal de l'Évêché, Limoges, the coffret of Saint Brandan was not included in the distribution of the relics and reliquaries in 1790.

20. The stylistic coherence between this group and the plaques in the Musée de Cluny from the main altar (see Souchal 1967, pp. 21–71) is the only argument in favor of the existence, perhaps brief, of a workshop specifically for Grandmont and established at the motherhouse of the order. (See Guibert 1888a, pp. 45–99.) Nonetheless Souchal has demonstrated (1964b, p. 141) how doubtful is the authenticity of the inscription F. REGINALDUS ME FECIT, cited only by Abbé Texier.

21. Barbier de Montault 1897a and b; Québec–Montréal 1972, no. 29.

22. Gauthier 1992, p. 96 and n. 35 (fragment found on a non-specified Grandmontain site in Champagne), and p. 104, n. 61 (enamels found in the excavation of the priory of Pinel).

23. This idea was briefly studied in a law thesis (Farnier 1913).

specified clearly that the remains were still in a buried tomb, but the location is not precisely known.

*1*

## *The treasury of Grandmont*

# 54. Angel Surmounted by a Reliquary

Limoges(?), ca. 1120–40 (statuette) and 13th century (base and mounting of reliquary)

Solid copper: engraved, chased, and gilt; copper wings engraved on the back; champlevé and cloisonné enamel: lapis and medium blue, turquoise, almond green, red, and white; lapis blue enameled beads; rock crystal

23.2 × 9.8 cm (9⅛ × 3⅞ in.)

PROVENANCE: The treasury of the abbey of Grandmont: mentioned in inventories 1496–1790; inventory of 1575 specifies that object "came long ago" from cell of Grandmont monks at Balezis (Haute-Vienne); see Texier 1857, col. 868; for 1496 inventory, no. 51, see Guibert 1888a, p. 88; for 1666 inventory, no. 47, see Texier 1857, col. 868; for May 1790 inventory, no. 46, see Texier 1842, p. 336. In December 1790, placed in the church of Saint-Sulpice-les-Feuilles (*Procès-verbal de distribution des reliques*, no. 20; see Ardant 1848, p. 72). Classified, Monuments Historiques, 1891.

CONDITION: In excellent condition except for crack near feet; crude reattachment done later, using a rod placed at the back and fastened with two thick rivets, the heads of which are visible on the obverse at the bottom of the robe and at the knees. Gilding of base is worn. Opening at top of reliquary suggests that it was originally surmounted by a cross.

Church of Saint-Sulpice, Saint-Sulpice-les-Feuilles (Haute-Vienne)

The object is a combination of pieces of different date: the statuette of an angel, on the one hand; the base on which it stands and the crystal reliquary in a mounting of gilt copper, on the other hand. The inventory of 1575—the oldest to mention the base—also indicates the provenance of the object: "a small base of gilt copper on which there is an angel of gilt copper, enameled, once belonging to the Balezis chapel." It is very likely, however, inasmuch as this provenance does not appear in earlier inventories, that Balezis was only a temporary depository, a supposition supported by the modesty of this priory, mentioned for the first time in 1234 and apparently abandoned by the monks by the end of the thirteenth century and in ruins a century later.[1] Only the inventory of 1666 clearly describes the three distinct parts and, for the first time, mentions the relic: "An angel in relief, of gilt copper, enameled, on a square base and having on its head a little crystal decorated with gilt copper, on which there is this note, 'de sancto Juniano' [of Saint Junian] and in which we have found a finger bone, folded into a violet cloth with this note, 'sancti Juniani confessoris' [of Saint Junian, confessor]." In May 1790, it was a relic of the True Cross that had been found in the reliquary.[2] Many nineteenth-century writers identified it as Saint Leonard's finger.[3]

The assemblage, which made it possible for the statuette to be used as an independent reliquary, no doubt took place in the thirteenth century: the dentilated friezes summarily suggesting leaves on the base and on the mounting of the crystal reliquary, as well as the circles engraved on the top of the mounting, appear frequently in goldsmiths' work of the time and are seen, for example, on the gilt-copper mounting of the altar cruet from Milhaguet.[4] The feet of the Saint-Sulpice angel are one with a small metal plate, which is soldered to the top of the base—an attachment completed, no doubt at a later date, by the rod placed on the reverse.

Identifying the original use of the angel statuette is one of the difficult problems posed by this work. Indeed, it is neither an appliqué figure nor one truly in the round. The very thin statuette is smooth on the reverse and slightly concave, while the reverse of the wings has an engraved decoration repeating the motifs of the obverse. The most convincing

Profile view

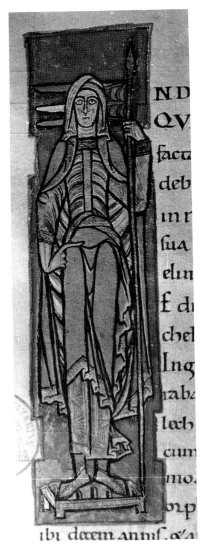

Fig. 54a. Ruth. Bible. Bibliothèque Mazarine, Paris, ms. 1, fol. 82

hypothesis seems to be that it was once a corner figure attached to the foot of a cross or reliquary.[5] The iconography may possibly confirm the hypothesis, as this book-bearing angel might be a symbol of the evangelist Matthew.

The attribution of the work to the Limousin has scarcely been contested since the publications by Molinier (1886) and Guibert (1888b); nonetheless, its singularity should be acknowledged. Indeed, the statuette is exceptional both in the profound originality of the concept and the quality of its realization. It is striking, first of all, for the rigidity and strict frontality of the pose, rare in western Romanesque art; for the graphic complexity of the drapery, treated with perfectly mastered rigor; and for the striking presence of the face with large triangular eyes, high cheekbones, and a small pinched mouth.

Some of these characteristics are common to the art of western and central France. Thus the frontal disposition of the figure, the juxtaposition of long vertical falls and abrupt horizontal slashes in the drapery, the large ellipsis drawn on the abdomen, as well as the rich orphreys and the sleeves flaring at the ends, easily find parallels among the painted figures in these regions at the beginning of the twelfth century. In this milieu, it is among the illuminations of the *Sacramentaire de Saint-Étienne de Limoges* (Sacramentary of Saint Stephen of Limoges), about 1100, and the manuscripts derived from it that one finds the most convincing comparisons, such as the figure of Ruth in folio 82 of the first volume of the Bible (fig. 54a), from the beginning of the twelfth century, now in the Bibliothèque Mazarine, Paris, which shares traits such as the definition of the face, especially the long, thin, straight nose and the small mouth with pinched lips.[6] Other aspects might suggest the sculpture of the southwest, notably the south portal of the abbey of Saint-Pierre at Moissac (ca. 1120–40), where the sculptor delighted in a similar graphic complexity of drapery and where the angels beside Christ on the tympanum, including the one representing Matthew, have comparable features, including high cheekbones and prominent chin, and the same hairstyle of short stiff locks ending in curls.[7]

The extreme rigidity of the statuette's outline, the calligraphic precision—highly decorative but totally unrealistic—of the drawing of the drapery and the hands, and the strange beauty of the face, with large eyes animated by enamel beads, are nonetheless unique. Certain bronzes originating in the west of France, generally dated after the middle of the twelfth century, may perhaps compare with it,[8] while other, slightly later works executed in the central regions, such as the bust of Saint Baudime in Saint-Nectaire, seem to prolong the very particular aesthetic of this face.[9] The angel does not seem to have been made by casting but cut and wrought from a single ingot of copper (the wings are not separately attached) and engraved and chased with special dexterity.

206

The enameled ornamentation of the wings confirms the association of this angel with the art of central and south-western Aquitaine. Indeed, the technique, which combines champlevé with cloisonné on the small circles on the horizontal band, is directly related to the practices of the enamelers working at Conques and may be compared with effects observed in a few works attributed to the Limousin shortly after the middle of the century, such as the Christ in Majesty from the former Spitzer Collection (cat. 12), a comparison already proposed by Molinier (1891) and Gauthier (1960). It should be especially noted that the range of color is identical to that on the chasse of Bellac (cat. 9), where we find in particular the same range of blues and the same almond green. Enamel beads animate the pupils, as on some of the earliest works known to have been executed by Limousin workshops, from about 1160 to 1170 (see cat. 16).

Comparison with paintings and sculpture of the years 1110 to 1140, the object's sculptural power, and the astonishing "personality" of the figure place this statuette among the greatest masterpieces of Romanesque art in Aquitaine before the middle of the twelfth century.

Although it cannot be determined whether or not this statuette came to Grandmont as soon as it was made, the angel certainly is both contemporary with the beginning of the order and the oldest extant work from its treasury.[10]

ET-D

Detail

NOTES
1. Texier 1857, col. 868, no. 48. On Balezis, see Guibert 1877a and 1877b.
2. Texier 1857, col. 868, no. 47; Texier 1842, p. 336.
3. Texier 1855b, p. 290.
4. See Paris 1965, no. 368, pl. 73 (stolen in 1981 from the Musée Municipal de l'Évêché, Limoges). For examples outside the Limousin, like the mounting of the reliquary of Saint Riquier, see Hahnloser and Brugger-Koch 1985, no. 310, pls. 258, 406. The same authors attribute a date of the third quarter of the twelfth century to the reliquary of Saint-Sulpice-les-Feuilles and its mounting, which seems, at least for the mounting, too early.
5. Suggested in Guibert 1888b, p. 197; taken up again in Gauthier 1960a and Gaborit-Chopin 1983b.
6. See Gaborit-Chopin 1969, fig. 173, p. 101.
7. See Barral i Altet 1983, pl. 91; Vergnolle 1984, p. 244, fig. 332, p. 245.
8. For example, the Christ from Soudan (Loire-Atlantique), in the Musée des Beaux-Arts, Angers. See Gaborit-Chopin 1983b, p. 294.
9. Middle of the twelfth century (perhaps between 1146 and 1178); see Paris 1965, no. 447, pl. 79; Boehm 1991, pp. 285–94, figs. 8a–f.
10. Many copies or pastiches were executed in the nineteenth century, such as the one in the Martin d'Arcy Art Gallery of Loyola University, Chicago (see Chicago 1970, no. 2) and the one from the former Adolphe de Rothschild Collection, now in the Musée du Louvre, Paris (OA 5963; see Marquet de Vasselot 1914, no. 105).

EXHIBITIONS: Paris 1867, no. 1955; Limoges 1886, no. 3; Limoges 1948, no. 11, pl. III, fig. 33; Paris 1965, no. 374, pl. 70; Cleveland 1967, no. III, 33, ill.; Cologne 1985, no. H 50, pl. 146.

BIBLIOGRAPHY: Ardant 1848, p. 72; Texier 1855b, pp. 290–91, 293, engraving on p. 285; Texier 1857, col. 128, pp. 901–2, 1259; Didron 1859, pp. 41–42, fig. 32, p. 41; Viollet-le Duc 1871–72, p. 216; Molinier 1881, pp. 102–3; Linas 1883, p. 73; Molinier 1886, p. 171, fig. p. 167; Guibert and Tixier 1886, p. 53, pl. XXIV; Palustre and Barbier de Montault 1887, pl. VII; Guibert 1888b, pp. 195–97; Molinier 1889, p. 22; Guibert 1889, p. 478; Molinier 1891, p. 147; Rupin 1890, p. 477, fig. 527; L'Art pour tous 33 (1894), no. 812, figs. 7487–90, p. 3348; Guibert 1901, p. 14; Drouault 1905, p. 248; Demartial 1923, p. 434, fig. p. 435; Perrier 1938, p. 52; Gauthier 1950, pp. 28, 74; Gauthier 1960a, pp. 288–89, pl. 10; Souchal 1963, pp. 234–35; Taralon 1966, p. 290, pl. pp. 194–95; Gaborit-Chopin and Taburet 1981, no. 22; Gaborit-Chopin 1983b, p. 219, pls. pp. 260, 321; Hahnloser and Brugger-Koch 1985, no. 506, pl. 407; Gauthier 1987, no. 74, p. 83, pls. XXXVIII, XLIII; Boehm 1991, p. 141, fig. 39.

# 55. Chasse from Ambazac

Limoges, between 1180 and 1190

Copper: repoussé, stamped, engraved, chased, and gilt; gilt-copper filigree; champlevé enamel: lapis, medium, and lavender blue; dark and
light green, garnet red, translucent wine red, and white; rock crystal
and colored glass cabochons; some precious stones (amethyst, chal-
cedony), beads, and terracotta; wood core

58.6 × 79 × 26.2 cm (23⅛ × 31⅛ × 10⁵⁄₁₆ in.)

PROVENANCE: From the abbey of Grandmont; placed on the main
altar and therefore not described in the oldest inventories, which documented only the objects in the treasury; mentioned by Pardoux de La
Garde before 1591 (see Souchal 1963, p. 51) and by Levesque in 1662 (p.
78); no. 2 in the inventory of 1666, when it held relics of Saint
Macarius (ms. 1, Sem. 8, fol. 123v; see Texier 1857, cols. 827–28);
removed before 1771, when decoration from main altar was relegated
to a chapel, and perhaps placed in "treasury of relics" (*Description* of
Martial de Lépine; see Guibert 1877a, p. 333; Souchal 1964b, p. 135, n.
2); no. 30 in inventory of May 1790 (see Texier 1842, p. 333, and
Souchal 1964b, p. 135).

In 1790 placed in the church at Ambazac, at which time it held
relics of Stephen of Muret, no. 2 in *Procès-verbal de distribution des
reliques*; see Ardant 1848, p. 69, and Souchal 1964b, pp. 134–35.

Classified, Monuments Historiques, 1881. Stolen in 1907; recovered
in London the same year and returned to Ambazac (see Leniaud 1983).

CONDITION: Bottom of wood core missing. Restoration by L.
Toulouse in 1966, "consolidated, cleaned . . . refastened the plaques
with nailing and tightened up" (*Archives des Monuments historiques*,
Ambazac dossier, note from Nov. 24, 1966). Some glass cabochons are
modern.

Church of Saint-Anthony, Ambazac (Haute-Vienne)

Left and right sides

When this chasse was deposited in Ambazac in 1790, it contained "a large leg-bone" of Saint Stephen of Muret.[1] In the
sixteenth and seventeenth centuries, it had housed various
other relics, the most important being those of Saint
Macarius, given to the abbey of Grandmont by Thibaud
VI, count of Champagne, in 1269.[2] Guibert was the first to
point out the discrepancy between the late date of the
arrival of the relics of Saint Macarius and the "characteristics
of the style," which seemed to him "to indicate an earlier
time" and led him to suggest that, before receiving the body
of Saint Macarius, the Ambazac chasse could have contained the relics of another saint.[3] The transfer of relics was
in fact a frequent phenomenon, illustrated elsewhere by the
different inventories at Grandmont. Two hypotheses have
developed subsequently: Souchal (1964b) suggested that the
chasse was made to contain relics of the companions of
Saint Ursula brought from Cologne (the site of their martyrdom) in 1181; Gaborit (1976a) considered it to be the first
chasse devoted to Stephen of Muret (the founding saint of
Grandmont), made for the elevation of the body following
his canonization in 1189.

Both hypotheses are solidly supported. Souchal emphasized that the inventory of 1666 mentioned, together with
the relics of Saint Macarius, those of martyrs of the Theban
Legion, brought back from Cologne in 1181. The relics
donated in 1269 would thus have joined the relics from 1181
in the chasse. One might add that, because the voyage to
Cologne was extremely brief and the number of relics
brought back considerable,[4] the swift execution of a large
chasse might have seemed the best solution for holding,
provisionally at least, all or some of the relics. Grandmont
seems to have kept only four chasses for the seven bodies
from Cologne.

Gaborit, on the other hand, based his argument on three
main points. In spite of the total absence of figurative ornamentation—a highly unusual characteristic—the chasse
corresponds to a very elaborate program that, with "both
naive and sophisticated" symbolism, suggests the "tomb" of
the saint placed beneath the jeweled cross representing
Christ in Glory and the dove evoking the Holy Ghost. Only
the seventh prior of Grandmont, Gérard Itier (1188–97)
"seems to have been capable of carrying out this type of
compromise between the eremitic tradition of Grandmont
and liturgical splendor." Moreover, the chasse bears signs of
"brilliant but hasty" execution, which might be explained by
the brief time available to Itier (elected in September 1188)
before the elevation of the body of Stephen of Muret on
August 30, 1189. These first two points are not beyond criti-
cism: we know very little about the spirituality of the prede-

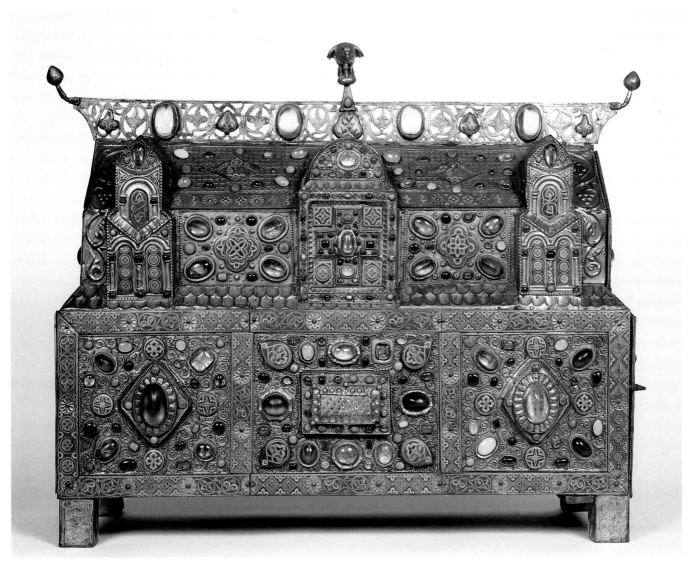

Principal face

cessors of Gérard Itier, especially that of William of Treignac (r. 1170–87; see cat. 60), who was clearly endowed with a strong personality and initiated the voyage to Cologne. Furthermore, the second point, as we have just seen, could support either hypothesis. There is little doubt, however, that, if one accepts the identification of the appliqué from Chaalis (cat. 61) with one of the reliefs on the chasse that housed the relics of Stephen of Muret from 1666 to 1790, the latter must have been executed about 1250–1270. The chasse that had previously held these same relics thus became available for reuse, and it seems logical that it was used in 1269 to house the recently received relics of Saint Macarius, for it is unlikely that an object of this import would have been left for long without relics.

The question of the original destination of the Ambazac chasse thus remains difficult to resolve—and few writers have expressed themselves on this point.[5] If one takes the most positive elements of both arguments, it is conceivable that the chasse found today in Ambazac originally housed the Cologne relics; then, as it was available after the transferal of these relics into several other chasses executed after 1181, it could have received the body of Stephen of Muret, elevated in 1189. The absence of figurative decoration would have favored the subsequent reuse of the chasse. One might, however, object that these are somewhat vain conjectures, especially as both hypotheses allow the Ambazac chasse to be dated fairly precisely to the decade 1180 to 1190—in 1181 in the first case, and between 1188 and 1189 in the second.[6]

All the writers on the subject agree on the quality and originality of the ornamentation of the chasse. An image of the celestial Jerusalem[7] or an evocation of the abbey church at Grandmont,[8] it is without a doubt, as J.-R. Gaborit (1976a) has well demonstrated, a sumptuous tomb placed under the protection of Christ and the Holy Ghost. The form is, furthermore, intermediate between that called "of the sarcophagus," to which the lower part is related, and that called "architectural," evoked by the upper part. The execution, described in minute detail by Gauthier (1987), makes use of all the resources of the Limousin goldsmiths. A unique example, it shows as much accomplishment in the actual goldsmithing as in the enameled ornamentation. The filigree, made by tightly twisting smooth, thin wire, is a splendid example of the flowering of this technique in Limousin goldsmiths' work of the thirteenth century. The crystal cabochons, accompanied by many precious stones, form, because of their number and size, the most important ensemble on a Limousin work. On the principal face and the ends, these gems, along with small enamel plaques, constitute the main decoration. On the back, on the other hand, there is a pattern of intertwining foliage in low relief, enhanced by many small enameled plaques. Gauthier (1960) suggested that such ornamentation was inspired by "Byzantine or Georgian goldsmiths' work of the eleventh and twelfth centuries." There is no way of knowing what particular works of this sort influenced the goldsmiths working for Grandmont at the end of the twelfth century. In particular, we do not know whether the reliquary of gilt silver bearing a Greek inscription of the name *Alexis Doukas*, which enshrined a fragment of the True Cross, sent by King Amaury of Jerusalem in 1174, had vegetal ornamentation;[9] but it is true that the ornamental foliage with large florets on the chasse invites comparison with Byzantine works of the twelfth century, such as the Khakouli triptych.[10] It is no less certain that the design of this foliage and its florets in half relief also have parallels in French illumination and goldsmiths' work of the third quarter of the twelfth century. They are found, among other places, in a

Bible in the Bibliothèque Nationale de France (ms. lat. 116), attributed to the region around Chartres about the middle of the twelfth century,[11] and in the Commentary on the Pentateuch (Burgundy?, third quarter of the twelfth century) by Bruno de Sigena in the library in Dijon.[12] As Gaborit-Chopin (1992) has noted, the florets can be compared to those decorating the back of the gilt-silver cross from La Roche-Foulques (Maine-et-Loire), kept in the Musée des Beaux-Arts, Angers, since 1851 and prefiguring those used on the backs of many crosses of Limousin goldsmiths' work in the first half of the thirteenth century, notably the one now in Eymoutiers (Haute-Vienne). The signs of rapid execution, noted by Gaborit, are evident in details of the ornamentation in half-relief repoussé, essentially several small bosses not retouched on the upper part of the principal face and perhaps the egg shapes on the upper and lower points of the large enameled plaques on the back.

The enameled ornamentation, on the other hand, is executed with all the perfection and refinement of the most successful Limousin workmanship of the last decades of the twelfth century and, in addition, presents a formal repertory of great variety. Made up entirely of separate plaques applied to the gilt-copper facing and using two principal motifs—rosettes and florets—it takes the form of framing friezes or borders on the front face and gable; rosettes in the shape of quatrefoils, lozenges, or squares on all four sides; small plaques inserted in the arches of the "towers" on the principal facade, perhaps to suggest stained-glass windows; crosses ornamenting the upper part of each end panel; isolated appliqués in circular or ovoid shapes on the face; cutout florets on the back; and, lastly, small half-relief bosses, also in the shape of florets, sprinkled over the crest. The figurative ornamentation is limited to two plaques with busts of angels on the roof of the principal face. Their style is close to that of the works with vermiculé backgrounds, but the heads are made in half-relief appliqué (see cats. 39, 57). The variety and brilliance of the colors, the deftness in the execution of the color gradations, and the importance and subtlety of the engraved and chased work animating the backgrounds and outlining the curled petals rank these enamels among the greatest masterpieces of Limousin enamelwork. The decoration of rosettes inscribed in circles spread across several friezes is similar to that which surrounds the book covers in the Musée des Beaux-Arts, Lyons (cat. 18). It "carpets" the back of many chasses with vermiculé grounds (see cats. 24, 27–29) and adorns the two enameled appliqués on the Bible of Souvigny (cat. 34). The rosettes have been justly compared to those in the Musée du Louvre that decorated a secular coffret (cat. 35);[13] on the other hand, the small plaques inserted in the arcades of the towers on the roof and particularly the appliqués cut in the shape of florets and entirely enameled on the back are origi-

Fig. 55*a*. Canon tables (detail). Bible. Bibliothèque Municipale, Clermont-Ferrand, ms. 1, p. 272

Reverse

Detail of the reverse

Detail of the reverse

nal inventions, which seem to have remained unique.

The design of the small florets with trefoil tops outlined by a wavy line in the center of the front face and in a lozenge at the back is similar to that of the "palmette-flowers" on the back of the chasse from Apt (cat. 21) and especially the one from Washington, D.C. (cat. 22). The half-corollas with curved edges are comparable to those on the knop of the crosier at Poitiers (cat. 23). The small rosettes formed by a square placed on its tip, with concave edges, and inscribed in a circle (the large appliqué on the back left), prefigure those on the chasses of Mozac (cat. 45) and Saint-Sernin of Toulouse (cat. 40). Other motifs are original in their color combination—for example, the use of a single range of blue-red-white in certain florets on the back—and especially their cutout shapes and the play between the enameled surfaces and the reserved parts enhanced with fine stippled lines suggesting the folds of the petals that sometimes give them an asymmetrical design. These florets can be compared in particular to those, also in the shape of corollas or trefoil buds with simplified contours, situated at the top of the pages of the canons of the Clermont-Ferrand Bible (fig. 55a). Although the place of execution of this manuscript is not precisely known, it illustrates in an exemplary fashion the synthesis between local traditions and innovations inspired by outside contributions—in this case, especially English—that occurred in the artistic milieus in central France during the last decades of the twelfth century.[14]

The only chasse originating for certain in the Grandmont Treasury to come down to us intact, the Ambazac chasse was one of its most precious ornaments. Exceptional for the decoration combining goldsmiths' work and enamels, its importance is increased for us by the precision of its dating, between 1180 and 1190.

ET-D

NOTES

1. See report published in Ardant 1848, p. 69, and more completely in Souchal 1964b, pp. 133–35, n. 4.
2. Pardoux de La Garde: "Another chasse of enameled, gilt copper, well-made, garnished with small gems, in which lies the body of Saint Macarius" (Limoges, Archives of the Haute-Vienne, ms. 1, Sem. 8d, fol. 123v).

Inventory of 1666 (no. 2): "The chasse of Saint Macarius . . . of the same material and shape, but smaller than the preceding, embellished with numerous large gems and without figures."

Inventory of May 1790 (no. 30): "Another chasse, of medium size, having roughly two pieces lengthwise, also of wood and ridged, covered with strips of enameled and gilt copper, with some figures and other ornaments, crystals and small gems, many of which are missing; its door, which is at one end, closing with a key."

These different descriptions have been published many times, most recently in Souchal 1963, p. 51, and 1964b, pp. 134–35, and in Gauthier 1987, p. 154.
3. Guibert 1888b, p. 210.
4. The abundant bibliography on this subject (see especially Guibert 1888a and 1888b and Lecler 1907–9) has been perfectly summarized in Gaborit 1976a, p. 237: "Having departed at the beginning of Lent in 1181, the four messengers returned on April 28, 1181, with seven whole bodies, two heads, and other minor relics."
5. Gauthier (1987, p. 154) at first chose the identification suggested by Souchal and dated the chasse about 1183, then (1992) opted for the identification suggested by Gaborit but retained the date 1182.
6. Souchal (1964b) suggested that it could have just as easily been executed under the "intruding" Stephen, interim prior between 1185 and 1188, during a period of political unrest in the governance of the abbey, but, according to his hypothesis, a date about 1181–82 seems much more likely.
7. See the often cited Texier 1857 and, in particular, Gauthier 1960a and Paris 1993.
8. Texier 1842; also Rupin 1890, pp. 142–44.
9. See Lecler 1907–9, pp. 444–51 (1908); Frolow 1961, no. 319, pp. 320–21.
10. National Museum of Fine Arts in Georgia, Tbilisi; see Javakhishvili and Abrahamishvili 1986, pp. 167–75.
11. See Avril 1983, p. 191.
12. See Zaluska 1991, no. 97.
13. Paris 1985, no. 8 (entry by J. Durand).
14. Regarding the manuscript, see Cahn 1980, p. 13, and above, cat. 34.

EXHIBITIONS: Limoges 1886, III, no. 13; Paris 1900, no. 2412; Paris 1937, no. 1194; Limoges 1948, no. 22, fig. 35; Paris 1965, no. 353.

BIBLIOGRAPHY: Texier 1842, pp. 189–91; Texier 1842a, p. 144; Texier 1843a, pp. 411–12; Texier 1857, cols. 98–100, fig. 1, cols. 1475–76; Linas 1883, pp. 145–49; L'Art pour tous 23, no. 576 (June 15, 1884), fig. 5014, p. 2403; Molinier 1886, p. 171; Guibert and Tixier 1886, pp. 56–57, pl. XXVI; Palustre and Barbier de Montault 1887, pls. IV–VI; Gay 1887–1928, vol. 1, p. 341; Guibert 1888a and b; Guibert 1889, pp. 471–72; Rupin 1890, pp. 137–44, figs. 210–17; Molinier 1891, pp. 172–76; Lecler 1907–9, pp. 565–79 (1908); Braun 1940, pp. 164, 173, fig. 80, pl. XXVI; Gauthier 1950, pp. 35, 64, 154, pl. 19; Gauthier 1960a, pp. 289–90, figs. 5–7, p. 248; Souchal 1964b, pp. 132–37, fig. 43; Gauthier 1967b, p. 153; Gauthier 1972b, no. 53, pp. 102–3; Gaborit 1976a, pp. 239–41, 245, figs. 3–6; Leniaud 1983; Gaborit-Chopin 1983b, p. 327, fig. 291; Gauthier 1987, no. 159, pp. 152–55, figs. 530–32, 535–47, 550–52, pls. CXLVIII, CL–CLIII, CLV; Gauthier 1992, p. 94, pl. 1; Gaborit-Chopin 1992, pp. 421–24, fig. 13.

# 56. Two Plaques: Wise Virgins

a. Limoges, ca. 1180

Copper: engraved, scraped, stippled, and gilt; champlevé enamel: dark, medium, and light blue; green, yellow, red, and white; blue glass beads

24.9 × 9.25 cm (9¾ × 3⅝ in.)

Ex colls.: Khevenhüller, Venice (sold 1804); Austrian Imperial Collections, Vienna (until 1891)

Condition: Gilding almost entirely lost, except along proper right edge; enamel loss and copper corrosion at proper right bottom.

Österreichisches Museum für angewandte Kunst, Vienna (Em247)

b. Limoges, ca. 1180

Copper: engraved, scraped, stippled, and gilt; champlevé enamel: dark, medium, and light blue; turquoise, light green, yellow, red, and white

24.9 × 9.3 cm (9¾ × 3⅝ in.)

Ex colls.: J.-B. Carrand, Lyons (before 1871)?; Louis-Claude Carrand, Lyons and Florence (d. 1888)

Condition: Overall wear to gilding.

Museo del Bargello, Florence (632c)

Exhibited in Paris only.

Each plaque portrays a standing maiden, dressed in luxurious costume. Slipped over a white bodice that is visible at the neck—and the sleeve of the Bargello figure—is a green-and-yellow gown with a decorative band at the wrist and at one thigh. Over the gown is a mantle of blue-and-white dotted fabric, probably meant to be silk, lined, as can be seen on the Bargello plaque, with vair and faced with polka-dot silk of a contrasting turquoise. Each of the veiled heads is separately worked and applied to the plaque. The maidens are nimbed. As each raises her left hand, she holds a burning lamp in her right. With this attribute, they are recognizable, not as ladies of the court, but as two of the Five Wise Virgins, following the parable recounted in the Gospel of Matthew (25:1–13). In this metaphor of the Elect and the Damned, the Wise Virgins conserve the oil of their lamps in readiness for the Bridegroom (Christ), whereas the Foolish Virgins squander their reserves.

The parable of the Wise and Foolish Virgins is frequently represented in Romanesque and Gothic sculpture and in manuscript illumination. In fact, its relative popularity in the representational arts in central and southern France has been explained by the importance of the story in a medieval liturgical play preserved in a manuscript from the abbey of Saint-Martial in Limoges.[1] The Bargello and Vienna plaques are, however, the only known depictions of this subject in enamel. Marie-Madeleine Gauthier has suggested that the enamels were part of a series of the Five Wise and Five Foolish Virgins from an altar frontal; certainly their shape

a　　　　　　　　　　　　　　　　　　b

and size are consistent with plaques from surviving altar frontals, including the one of San Miguel de Excelsis (Museo de Navarra, Pamplona), with which she appropriately compares the Virgins.[2] However, the choice of this parable as the subject for an altar frontal would be entirely exceptional in medieval art. In such a program, the Virgins, Wise and Foolish, would presumably flank a central image of Christ. But such a scheme is inconceivable in the case of the Bargello and Vienna plaques, since their steadfast gaze to their right-hand side presupposes that they would be placed at the left hand of Christ. The sinister side must necessarily be occupied by the Foolish Virgins, representing the Damned.

Alternately, Jean-René Gaborit proposed that the two plaques with the Wise Virgins were part of the decoration of one of the chasses of the Companions of Saint Ursula at the abbey of Grandmont,[3] where a long-standing tradition of associating the Wise Virgins with virgin-martyr saints[4] was extended specifically to the Virgins of Cologne. The twelfth-century chronicler Geoffrey de Vigeois, listing the relics venerated in the Limousin, mentioned the Cologne Virgins as "cinq vierges sages rapportés de Cologne" (Five Wise Virgins brought back from Cologne).[5] Prior to this, the same association was made in the *Itinerarium fratrum*, the account of the voyage of the Grandmont hermit monks to Cologne in 1181.[6] The presence of halos on the enameled Wise Virgins may reflect this conflation. As Gaborit noted, however, the descriptions of the chasses for relics of the Cologne Virgins at Grandmont are not sufficiently precise to confirm his suggestion that the Vienna and Florence plaques were part of one of the Cologne reliquaries. Chasse VI, containing the relics acquired at the monastery of the Virgins, was described in the eighteenth century as having heads in relief.[7] The roof of Chasse V had three figures of virgins and others of bishops or abbots.

When the Vienna plaque was sold at the Khevenhüller[8] auction in 1804, it was part of a lot of "6 Bronzen 'griechisch' christlicher Zeit byzant. Periode mit verschiedenen Farben geschmolzen in schönem eleganten Behältnis" (six bronzes of the "Greek" Christian era, Byzantine period, enameled with various colors in pretty, elegant attire).[9] It is reasonable to presume that the Bargello plaque—later acquired by Carrand, who is known to have purchased works of art in Italy—was at that time part of the same group. The plaque in Florence is marked, or numbered, on the reverse, perhaps indicative of its being part of a series. The fact that there were six plaques in the Khevenhüller sale—one more than the number of the Wise Virgins—could argue either for an ensemble, already incomplete in 1804, of both Wise and Foolish Virgins or, as seems more likely, for a grouping of Wise Virgins with Ursula's companions.

In the first volume of the *Émaux méridionaux,* Gauthier suggested further that the plaque of the Visitation (cat. 19)

may have come from the same ensemble. Although the figures in that scene are similarly and elegantly attired, the style of the Visitation plaque is distinct, the figures being somewhat less elongated and slightly less accomplished in drawing. The feet and hands are outlined in dark enamel, whereas they are simply reserved in the Vienna and Florence plaques. The style of the Vienna and Florence plaques does not recur in other enamels. The plaques have an elongated elegance in common with the ones from the San Miguel altar, and the applied head of the Virgin Mary from that ensemble provides a further point of comparison. The courtly attire of the Wise Virgins links them to works associated with Plantagenet patronage, notably the coffret in the British Museum (fig. 19a).[10] Both the exceptionally fine quality of the plaques and their links to Plantagenet taste serve to bolster the circumstantial historic and documentary evidence in favor of associating the Wise Virgins with the reliquaries of the abbey of Grandmont.

BDB

NOTES

1. Mâle 1978, pp. 151–53; Collins 1972, pp. 189–95. For the text of the play, see Young 1933, vol. 2, pp. 362–69.
2. Gauthier 1972b, p. 92; Gauthier 1987, no. 135.
3. Gaborit 1976a, pp. 238, 239, n. 28.
4. Heyne 1922.
5. Gaborit 1976a, p. 238, n. 29.
6. Ibid., p. 238, n. 29.
7. Ibid., p. 237, nn. 21, 30.
8. It is not known how long the plaque was in the Khevenhüller family collection. It should be noted, however, that members of this Austrian noble family traveled in France as early as the sixteenth century. Barthelmae Khevenhüller traveled from the south of France to Paris, where he described the treasury of Saint-Denis in 1550, an account that predates the looting by Huguenots in 1567. See Czerwenka 1867, p. 141.
9. I am grateful to Dr. Elisabeth Schmuttermeier, curator, Department of Metalwork, Österreichisches Museum für angewandte Kunst, for this research.
10. Gauthier 1987, no. 174.

EXHIBITIONS: a: Zurich 1946, no. 29, pl. xv; Krems-an-der-Donau 1964, no. 153, pp. 191–92; b: Florence 1989, no. 211.

BIBLIOGRAPHY: Sacken and Kenner 1866, p. 483 (a); Supino 1898, no. 632, p. 122 (b); Marquet de Vasselot 1906, p. 33, pl. vii; Falke 1931, p. 282 (a); Hildburgh 1936, p. 107 (b); Gauthier 1963a, p. 55; Gauthier 1972b, nos. 45–46; Gaborit 1976a, pp. 238–39; Gauthier 1987, nos. 136, 137, pp. 138–39, ills. 453, 454, pl. cxxviii.

# 57. Plaques from Main Altar of Grandmont

Limoges, last quarter of the 12th century (before 1190)

Copper: engraved, chased, and gilt; champlevé enamel: black, saturated blue, deep ultramarine blue, azure, turquoise; medium, emerald, and meadow green; yellow, vermilion, pinkish white, and white

26 × 18 cm (10¼ × 7⅛ in.)

INSCRIBED (plaque a): NIGOLASERT : PARLAAMNE TEVEDEMVRET (The lord Hugh Lacerta speaks with the lord Stephen of Muret)

PROVENANCE: From the decoration of the main altar of Grandmont

EX COLL.: Baron Salomon Roger until the sale in 1841[1] in which the two plaques were bought by Alexandre Du Sommerard.[2] In Musée de Cluny since 1843

CONDITION: Some losses to enamel in lower part of both plaques (bases of columns, feet of the Three Kings, mounds of earth). Gilding preserved on the whole but traces of wear in upper part (a and b) and in a large area between the Three Kings and the Virgin.

Musée National du Moyen Âge, Paris; Thermes de Cluny
(Cl 956 a and b)

The Grandmontain origin of these two plaques, suggested by the imagery and especially by the inscription on plaque a, was established as early as 1842 by Abbé Texier following their publication in color in *Les Arts au Moyen Âge* by Alexandre Du Sommerard in which, because of their format, they were regarded as book covers and were presented without provenance. The Limousin scholar Texier was already proposing that they be considered fragments of the main altar of the famous abbey. The construction of the priory church, the motherhouse of the order of Grandmont, near Limoges, was largely completed in 1166. Slightly more than twenty years later, it was endowed with a precious decoration of goldsmiths' work and enamels celebrating the canonization, in 1189, of its founder, Stephen of Muret (d. 1124). His relics, raised at that time, were integrated somehow into the main altar; but the structure is difficult to imagine in spite of the old descriptions compiled from the sixteenth century onward.[3] Jean-René Gaborit, who proposed the restoration of its arrangement, structure, and decoration, has demonstrated that the altar underwent several modifications between 1181 and 1230. Today these two objects are believed to be the only surviving elements of a series of plaques on the retable of the main altar that combined Gospel scenes with scenes from the life of the saintly founder of the Grandmont order. The singular fate of these two enamels, considered the most famous of those surviving from the oeuvre of Limoges, remained unknown from 1791, the date of the destruction of the altar of Grandmont, until 1841, when they appeared in Paris at the Roger sale. We know that by February 1791, a Limousin smith and smelter,

M. Coutaud, had acquired from the Grandmont abbey all its enameled and wrought "old copper," including the altar, to be hammered into pieces so that the metal could be salvaged.[4] It is likely that a witness, perhaps Abbé Legros[5] (who obtained from Coutaud the transcription of inscriptions on the objects destined for destruction),[6] succeeded in saving these two remarkable fragments.

The two plates were cut from a sheet of thick copper (2.5 mm [1/10th in.]) and riveted to a wood core, as indicated by the perforations along the edge. The rich palette comprises twelve colors.[7] Plaque a is dominated by blues and turquoise, while b has different tones of blue and green in nearly equal dominance; on a and b luminous touches of yellow and red make the details stand out vividly. The enamels are placed sometimes in large areas of single colors or in incrustations, and sometimes in spaces picked out with colors or clouds of two or three tones. On both plaques there are incrustations of blue enamel outlining the stones in the walls and red enamel emphasizing the tiles of the roofs. Similarly, incrustations of red enamel set off the hair, mustaches, and beards of the Three Kings and Saint Stephen; only Hugh Lacerta's beard and hair are encrusted with white enamel. The flesh of the faces and hands of all the figures is treated in pinkish white, and their eyes and eyebrows are outlined in black enamel. Two techniques used in Limousin workshops at the time are combined here: figures and motifs in polychrome enamel on a plain gold ground and small heads in half relief on reserved and chased bodies, as shown on plaque b in the appliqué treatment of the Infant Jesus.

The first plaque represents Saint Stephen of Muret and Hugh Lacerta, identified by an inscription in the Limousin dialect of the Romanesque period engraved on the intrados of the arch in the form of a suite of thirty-two symbols divided into three compact groups of ten, nine, and eleven letters: NIGOLASERT : PARLAAMNE TEVEDEMVRET. Despite its relative clarity, this inscription has baffled epigraphists for more than a hundred years. It should be read according to the definitive interpretation given by Jacques Bousquet before 1962: N IGO LASERT : PARLA AM N ETEVE DE MVRET, which is to say, "The lord Hugo Lacert speaks with the lord Stephen of Muret." This decisive reading allowed Geneviève Souchal to identify the scene as a conversation between Stephen of Muret and his secular disciple Hugh Lacerta, or Hugo Lacert—the former depicted at the right as a priest with a tonsure and a halo; the latter to the left as a Grandmont hermit leaning on a staff. Until that time, people thought that the hermit was Stephen of Muret, conversing with Saint Nicholas, bishop of Myra, whose relics were

a. Stephen of Muret
and Hugh Lacerta

preserved in the Middle Ages at Bari in southern Italy. This scene illustrates a posthumous episode involving the sainted founder of Grandmont. According to the texts,[8] after his death Stephen appeared in a vision to his faithful companion Hugh Lacerta, undoubtedly quite old by then, as suggested here by his white beard hair. The second plaque represents the Adoration of the Magi, a traditional gospel scene honoring the Virgin Mary, Mother of God, to whom the church of Grandmont was dedicated. These two representations reflect, according to M.-M. Gauthier, the iconological system underlying the undoubtedly complex design of the altar.

GF

NOTES

1. Salomon Roger sale, Paris, December 20, 1841, no. 209.
2. As demonstrated in Arquié-Bruley 1981.
3. *Les Antiquités de Grandmont*, manuscript of Grandmont's Pardoux de La Garde, written in the sixteenth century (before 1591), Archives de la Haute-Vienne (ms. 1, Sem. 81).
4. See letter of the municipality of Saint-Sylvestre (the commune in which the abbey of Grandmont was built) of February 17, 1791, Archives Nationales (H. 1200).
5. Martial Legros, "priest of the collegiate church of Saint-Martial in Limoges," an impressive scholar, and secretary to the abbot Sicelier, was charged in 1789 with distributing the Grandmont reliquaries among the churches of the diocese.
6. Texier 1842; published separately, 1843, p. 267.
7. Observation of the two plaques through a binocular loupe, when they were cleaned in May 1995 at the Musée de Cluny by M.-E. Meyohas, restorer of art objects, revealed certain information especially concerning the enamels (treatment, opacity, color): they are all opaque except the

b. Adoration of
the Magi

emerald green, which is distinctly translucent, and the pinkish white, which is half opaque, half translucent; the vermilion is composed of grains stuck together; the color that looks like black to the naked eye— seen in the shoes of the Virgin and those of Saint Stephen and Hugh Lacerta—is in reality a dark garnet red (perhaps obtained from manganese). The condition of the gilding, very worn, is due to the thinness of the original layer of gold, thinner here, according to the restorer, than on most of the enamels that pass through her hands.

8. Becquet 1958, pp. 9–36.

EXHIBITIONS: Utrecht 1939, no. 138, p. 46; Limoges 1948, nos. 8, 8a; Vatican City 1963, nos. 11–12; Cleveland 1967, no. III-30 (a); New York 1970, no. 141; Moscow–Leningrad 1980, no. 31.

BIBLIOGRAPHY: Texier 1842, published separately 1843, pp. 72–76, 149–50; A. Du Sommerard 1838–46, vol. 4 (1843), pp. 67–70; vol. 5 (1846), pp. 184–85, pls. XXXVIII–1, 2; Texier 1846; L. Laborde 1853, pp. 36–38; Ardant 1855, p. 79; Labarte 1856, pp. 199–204; Texier 1857, cols. 893, 903–4; Labarte 1865, vol. 3, pp. 465–66, 677–81, 698; Guibert 1877b; E. Du Sommerard 1883, no. 4492 (a), no. 4493 (b); Linas 1884a, 341–43; Linas 1884b, pp. 164, 166–67; Guibert 1889; Rupin 1890, pp. 96–99, pl. XIV, figs. 163–64, pp. 384–85; Molinier 1891, pp. 144–47, ill.; Lecler 1907–9, pp. 67–70 (1908); Marquet de Vasselot 1941, pp. 149–59; Souchal 1962, pp. 339–57, figs. 1–4; Souchal 1963, pp. 41–64, 123–50, 219–35; Souchal 1964b, pp. 7–35; Souchal 1967, pp. 21–26, figs. 1–2; Gauthier 1972b, no. 52, pp. 98, ill., 100–101, 331; Gaborit 1976a, p. 231; Landais 1976, pp. 122–24, fig. 5; Gaborit-Chopin and Taburet 1981, no. 29; Gaborit-Chopin 1983b, pp. 327, 290 ill.; Gauthier 1987, no. 247, pp. 204–7, 716–18, ill.; Hutchison 1989, pp. 97–102; Martin and Walker 1990, pp. 1–12; Gauthier 1992, pp. 94–95, ill.

# 58. Apostles of Grandmont

Limousin workshop, ca. 1231

Copper (plaques): engraved and gilt; (appliqués): repoussé, chased, engraved, scraped, and gilt; champlevé enamel: dark, medium, and light blue; turquoise, dark green, yellow, red, and white; dark blue glass inset eyes; dark blue and turquoise glass beads inset in collars, cuffs, hems, and book covers

Provenance: Probably from the frontal of the high altar, destroyed in 1790, of the abbey church at Grandmont, near Limoges[1]

Condition: Losses to gilding on highest surfaces of each relief figure, especially the foreheads, noses, cheeks, hands, and knees; enamels damaged near attachment holes (below the shoulder levels) along edges of backplates; Saint James (b) has further damage in hair, thin cracks in neck area, while there is a larger, opened crack and loss of metal at the side of the proper left shoulder. There is some reenameling on Saint Thomas (e). The original glass beads appear to be in place except for single beads missing from the collars of Saint Matthew (a) and Saint James (b).

### a. *Saint Matthew*

29 x 14 cm (11 7/16 x 5 1/2 in.)
Inscribed: s[anctus] mateus; engraved mark on reverse of backplate: N
Ex coll.: Edme Durand (in 1825)
Musée du Louvre, Paris. Acquired in 1825 (MR 2650).

### b. *Saint James the Great*

29.8 x 14.5 (11 3/4 x 5 3/4 in.)
Inscribed: s[anctus] iacob; engraved mark on reverse of backplate: ◊
Ex colls.: M. Astaix, Limoges (by 1883); Adolf Hommel, Zurich (sold August 18, 1909, lot 996); [Hamburger Frères, Paris, 1909]; J. Pierpont Morgan, London and New York (in 1909)
The Metropolitan Museum of Art, New York; Gift of J. Pierpont Morgan, 1917 (17.190.123)

### c. *Saint Philip*

29.6 x 14 cm (11 5/8 x 5 1/2 in.)
Inscribed: s[anctus] philipp'; reverse of backplate has not been examined.
Ex colls.: Prince Peter Soltykoff, Paris (before 1852; see L. Laborde 1852; sale, Drouot, Paris 1861, lot 159); Morland; A. P. Basilewsky, Paris (before 1870). State Hermitage Museum, Saint Petersburg (Ф194) Acquired in 1884.

### d. *Saint Paul*

29.8 x 14.7 cm (11 3/4 x 5 3/4 in.)
Inscribed: s[anctus] paulus; (on banderole) si secundum carnem vixeritis moriemini (If you live according to the flesh, you will die. Romans 8:13); engraved mark on reverse of backplate:
Ex colls.: Albert Germeau (before 1852; acquired "at Limoges about 1837" [see L. Laborde 1852; Linas 1884]; sold May 1868, lot 55); Eugène (d. 1886) and Auguste Dutuit, Rouen and Paris (until 1902)
Musée du Petit Palais, Paris (OD 1239[1])

### e. *Saint Thomas*

29.7 x 14.5 cm (11 11/16 x 5 11/16 in.)
Inscribed: s[anctus] th[o]mas; no marks on reverse of backplate.
Ex colls.: Albert Germeau (before 1852; acquired "at Limoges about 1837" [see L. Laborde 1852; de Linas 1884a and 1884b]; sold May 1868, lot 55); Eugène (d. 1886) and Auguste Dutuit, Rouen and Paris (until 1902)
Musée du Petit Palais, Paris (OD 1239[2])

### f. *Saint Martial* (not in the exhibition)

30 x 14 cm (11 13/16 x 5 1/2 in.)
Inscribed: s[anctus] marcialis; engraved mark on reverse of the backplate: ▽
Ex colls.: Louis Fould, Paris (in 1861); Louis-Claude Carrand, Lyons and Florence (d. 1888)
Museo Nazionale del Bargello, Florence (Inv. 649)

The overall form, details of technique and decoration, and measurements of these imposing, large appliqué seated figures are strikingly similar. Each figure, secured to the enameled backplate by two large rivets, is worked in repoussé, direct chasing, and engraving. The elongated proportions of the bodies, heads, and noses, together with the small mouths, the close-set dark blue glass eyes, the irregularly engraved lines of the beards (Philip lacks a beard), and other physical details, are instantly recognized as common to the group. The backs of the heads, each of which projects forward, are closed with a separate sheet of gilt copper. The engraved, parallel lines of dashes defining the hair are virtually identical. The classicizing character of the heavy yet fluid folds of the mantles is perhaps the most dramatic shared feature. Details of the draperies, such as the engraved vair design along some of the borders, recur among them. The fluted folds of the mantles continue in softer relief on the tunics beneath, the only exception being Saint Martial tunic's (f), with centered crescents in a diaper pattern. The collars and hems of the tunics have a shared yet varied vocabulary of beading and/or engraving. Except for Saint Paul, who holds a sword and an inscribed banderole, each figure grasps a book whose engraved cover is studded with glass beads. An engraved band behind the shoulders of each figure bears a large enameled inscription identifying the saint represented. The sloping plinths and the figures were worked together from a single piece of copper. Although the angle of the slope and the visible surfaces of the plinths vary, their engraved motifs are limited mostly to paired and facing acanthus leaves or to undulating, foliated vines. The separately worked halos, each riveted to the backplate, are saucer-shaped and embellished with narrow, radiating petals. The arched backplates have closely related yet varied motifs of engraved rinceaux and enameled flowers executed with a common palette against a blue enamel background. The enameled bolsters and the architectural details of the thrones correspond but vary slightly. The evidence of the common technique, style, and decorative vocabulary clearly supports the concept that this series represents the product of one workshop and also that these six appliqué figures were once part of a single monument, probably a large arcaded altar frontal.

Placed against their original, brilliantly enameled backplates, these appliqué figures of impressive size, powerful form, and classicizing mien express an unforgettable sense of monumentality and ennobled spirit. Although Marquet de Vasselot dated the series to the mid-thirteenth century on the basis of comparison with Limoges tomb figures, the styl-

Nicholas (1210–15) attributed to one of the radiating chapels in the cathedral of Saints Gervais and Protais in Soissons.[3] The appliqué figures, as well as the comparative works in stone and glass, are examples of the "1200 style," which had special focus and momentum in northeastern France and in the valleys of the Meuse and Rhine Rivers. That this style was pervasive in the Limousin is clear not only in the case of the present series but also in the smaller appliqué figures produced in the region. An especially powerful example is

a

b

istic expression evidenced in the apostle figures is related more to some of the early Gothic sculptures of the great cathedrals, such as the Coronation of the Virgin group of the central tympanum of the north transept of Chartres Cathedral (ca. 1205–10).[2] The appliqué figures with their classicizing drapery may also be compared with stained glass, such as the panels with episodes from the life of Saint

the Virgin and Child appliqué in the Cleveland Museum of Art (fig. 8), which was probably produced in the same workshop.[4]

The plaque of Saint James (b) was the first to be linked with the abbey of Grandmont. When it was exhibited at Limoges in 1886,[5] Guibert recognized that the Saint Matthew plaque (a) could be placed in the same context.

219

Noting their exceptionally high relief, he was reminded of the sixteenth-century description of the altar at Grandmont by a Grandmontain brother, Pardoux de La Garde (d. 1591): "Between these four beautiful pillars stands the high altar, which retable and frontal are made of copper gilded and enameled, with figures of the Old and New Testaments, the thirteen apostles and other saints, the whole thing raised in relief and enriched with small gemstones, the whole magnificently worked and excellent, as precious or even more precious than if it were made of silver."[6] Guibert's hypothesis found favor with subsequent authors. Rupin concurred with Guibert, noting that the high altar had been sold as scrap copper to one M. Coutaud in Limoges in 1790 and suggesting that the plaques had been saved in the same manner posited for the deacon of Les Billanges (cat. 59).

For Marquet de Vasselot, however, the previously cited arguments for associating the apostle plaques with Grandmont were too tenuous, especially given the documentary evidence of other altar frontals at Saint-Martial, Bourganeuf, and Saint-Pierre-du-Sépulcre in Limoges.[7] Indeed, the plaque with Saint Peter (cat. 47) was associated by Guibert with Grandmont at the same time that he linked the Saint James specifically to Pardoux de La Garde's description. But Marquet de Vasselot then identified the plaque representing Saint Martial as an integral part of the same group of plaques. He considered the presence of Martial, who was given apostolic status at the Council of Limoges of 1034 and who was surely the "thirteenth apostle" mentioned by Pardoux de La Garde, to be the pivotal point of evidence for associating the ensemble with the altar of Grandmont.[8]

In fact, the presence of Martial is not definitive evidence of

c

d

the provenance of the ensemble from Grandmont, since he appears among the plaques at Orense, for example (see cat. 51), and may well have been included elsewhere. The appearance of Martial with the attribute of a loaf of bread is distinctively Limousin (though not specifically Grandmontain). As Gaborit noted, it is a reference to the belief that Martial's apostolicity included his presence at the Miracle of the Loaves and Fishes (see Barrière, above p. 22). It was the Grandmontain authors in particular, however, who favored the text from the Epistle to

ness.[10] Furthermore, Gaborit introduced into evidence gifts made to Grandmont in the thirteenth century, notably one from Hugues de Lusignan, count of La Marche, of liturgical vessels.[11] Guillaume de Nevers's appropriation for an altar candle in 1231 may provide the date of the second campaign for the decoration of the high altar and for the apostles.[12]

The documentary evidence from which an association with Grandmont can be inferred is further strengthened by the exceptional number of small glass beads that decorate

e

f

the Romans that is inscribed on Saint Paul's banderole.[9]

The only real objection to the Grandmontain origin of the apostle plaques has been posed by Souchal. But her reasoning that the group could not come from Grandmont, since the abbey was too impoverished in the thirteenth century to afford such a commission, is clearly untenable, as the thirteenth-century figure of the deacon (cat. 59) bears witness.

the drapery of each apostle—thirty-seven on the figures of Saint Martial and of Saint Paul—conforming to Pardoux de La Garde's specific mention of the numerous "petites pierreries" on the high-relief figures. There is, moreover, a remarkable coincidence of drapery pattern between that of the figure of the deacon from the Grandmont Treasury and the robe worn by the Bargello Saint Martial (f).[13]

In 1855, Maurice Ardant described another plaque that could belong to this, or a closely allied, ensemble: "At the cemetery of 'Louyas' at Limoges, there is, set on the tomb of Madame Chabrol, born de Mondion, a figure of Saint Peter made of copper, gilt and repoussé, in very high relief, set against an enameled and arched backplate, the colors of which are very bright."[14]

The renown of this series provoked the creation of a number of nineteenth-century copies in enamel of the Saint Matthew, the Saint James, and the Saint Martial, and a copy in ivory of the Saint Matthew.[15]

BDB/WDW

NOTES
1. Rupin 1890, p. 428.
2. Sauerländer 1972, pls. 78, 79.
3. *Scenes from the Life of Saint Nicholas*, pot-metal glass, 54.6 x 39.3 cm each (21½ x 15½ in.), attributed to Soissons, 1210–15. The Metropolitan Museum of Art, New York; The Cloisters Collection, 1980 (1980.263.2–3). Grodecki and Brisac 1985, pp. 40–43 (figs. 27, 28), 260, 261, no. 84 (bibl.).
4. *Enthroned Virgin and Child*, gilt copper and glass beads, 21.6 x 10.5 cm (8½ x 4⅛ in.); Limoges, ca. 1220. The Cleveland Museum of Art, Purchase from the J. H. Wade Fund (62.29). Cleveland 1967, pp. 148–51 (ill.), 362, 63, no. IV-17 (bibl.); Gaborit 1976a, p. 243.
5. Marbouty 1886, p. 85.
6. "Entre ces quatre excellentz pilliers est ledict grand autel, et tant le contretable que le davant d'ycelluy est de cuyvre doré esmailhé et y sont les hystoires du vieux et nouveau testament, les treze apostres et aultres sainctz, le tout eslevé en bosse et enrichi de petite piererie; le tout fort bien ouvré et excellent, aultant ou plus riche que si le tout estoyt d'argent." Cited in Guibert 1888a; Souchal 1963, p. 51; Limoges, Archives départementales de la Haute-Vienne, ms. I, Sem. 81, fol. 122v.
7. Marquet de Vasselot 1941, p. 152.
8. Marquet de Vasselot 1941, pp. 153–55. The classic study of the apostolicity of Saint Martial is that by Msgr. Duchesne (1892, pp. 289–330). See, most recently, Landes and Paupert 1991 and Barrière, above p. 22. Only Brisset (1949, pp. 100–102) has dissented from the identification of Martial as the thirteenth apostle, considering that the reference could equally be to Saint Paul. However, whereas Paul commonly appears with the apostles, not only in this series but in the one at Orense, it was the distinctive honor of the Limousin that Martial was considered to be the thirteenth apostle.
9. Gaborit 1976a, p. 244.
10. The relationship of the deacon to the apostles and to a Virgin and Child in Cleveland was manifest in the exhibition in Cleveland (1967, Section IV, nos. 16–18).
11. Gaborit 1976a, p. 244.
12. Ibid., p. 246, n. 61.
13. See Cleveland 1967, nos. IV-16–18.
14. "Au cimetière de rouyas à Limoges, on a placé sur la tombe de madame Chabrol, née de Mondion, un saint Pierre en cuivre doré et repoussé, en bosse très saillante, adossé à un émail cintré et incrusté dont les couleurs sont très brillantes." Ardant 1855, p. 170. The association was suggested by E. Delahaye. No enamel is now to be found on the tomb. We are grateful to Thierry Zimmer and Paul Robinne for verifying this information.
15. Madrid, Lazaro Collection. Copies noted by Marquet de Vasselot (1941, p. 152, n. 6). Photograph of ivory sent to the Medieval Department of the Metropolitan Museum in the 1980s.

EXHIBITIONS (the letters a through f in brackets refer to the plaques listed above): Limoges 1886, no. 32, pl. XXXI [b]; Vatican City 1963, nos. 90–92 [a, d (pl. XXX), e]; Cleveland 1967, pp. 148–51, 362–63, no. IV-16 ill. [d]; Paris 1968b, pp. 252–54, no. 393 [a–f].

BIBLIOGRAPHY (the letters a through f in brackets refer to the plaques listed above): L. Laborde 1852, no. 70, p. 76 [a, c, d, e]; Chabouillet 1861, no. 1703 [f]; Darcel 1865b, p. 440 [d, e]; Darcel and Basilewsky 1874, no. 216 [c]; Guibert 1877c; Linas 1884b, pp. 164–67; *BSAHL* 1885 [b]; Garnier 1886, pp. 433–34, fig. 88 [d]; Linas 1886a, no. 4, p. 21 [c]; Marbouty 1886b; Palustre and Barbier de Montault 1887, no. XII, pl. XII [b]; Courajod 1888, no. 115; Guibert 1888a, pp. 17–19 [a, b]; Rupin 1890, pp. 427–28 [a, b (pl. XXXVIII, fig. 478), d, e]; Darcel 1891, no. 120 [a]; Kondakov 1891, p. 201 [c]; Supino 1898, no. 649, p. 125 [f]; Marquet de Vasselot 1914, no. 93 [a]; Breck and Rogers 1925, p. 101, fig. 55; Lapauze 1925, no. 1296 [d, e]; Marquet de Vasselot 1941, pp. 120, 149–58 [a, b, c, d (pl. XXXIV), e, f]; W. H. Forsyth 1946, p. 239 ill. [b]; Gauthier 1950, pp. 67, 157, pl. 44 [d]; Souchal 1963, pp. 126–30, 146 [a, b, c, d, e (fig. 12), f]; Souchal 1964a, pp. 75–76 [d (fig. 7), e]; Gauthier 1967b, pp. 139–55; Lapkowskaya 1971, p. 20 (ill.), no. 23 (color ill.) [c]; Gauthier 1972b, pp. 184–86, 371–72, ill. 129 (color) [f]; Gaborit 1976a, pp. 243–44, figs. 8–15 [a–f]; Florence 1989, pp. 432–33, no. 212 (ill.) [f]; Paris 1993, p. 41 (color ill.) [a].

# 59. Figure of a Deacon

Limoges, ca. 1220–30; later base, 13th(?) century

H.: 50 cm; H. of figure: 33 cm (19¹¹⁄₁₆ in.; 13 in.)

Copper: repoussé, engraved, chased, scraped, and gilt; champlevé enamel: medium blue; dark blue glass cabochons

PROVENANCE: The abbey of Grandmont. The figure of the deacon was probably placed at the lower terminal of the large cross on the matutinal altar (see Pardoux de La Garde, ms. 1, Sem. 81, before 1591, fol. 124; transcribed in Souchal 1963, p. 53); Abbé Legros, handwritten note, May 10, 1790, published by Texier (1857, col. 904); "the entire facing of this cross" was purchased by "M. Coutaud, metal founder of Limoges," in 1789 (Abbé Legros, ibid.). First mentioned as in the church at Les Billanges in 1853. Classified, Monuments Historiques, 1891.

CONDITION: Base reenameled; ensemble restored by M. Toulouse in 1965.

Parish Church of the Nativity of Saint John the Baptist, Les Billanges (Haute-Vienne)

The tonsured deacon wears a dalmatic with crescent moons set within a diaper pattern and decorated at the wrists and neck with cabochons. Beneath the dalmatic he wears a long stole and an alb; on his feet are richly embroidered shoes. In his hands he holds a pillow engraved with quatrefoils; on it rests a reliquary for the True Cross in the form of a rectangular box pierced with a double-armed cross.[1] A rock-crystal cover, now missing, protected a relic of the True Cross. The figure is hammered and worked from a single sheet of copper, except for the head, which has been rounded by the attachment of a second piece along a vertical seam.

The face of the deacon, with high, prominent cheekbones, small eyes, and large forehead, is remarkably similar to that of the appliqué figure of the Virgin in the Cleveland Museum of Art (fig. 8),[2] as Wixom first noted in assigning both figures to the same workshop as the apostles of Grandmont (cat. 58). The pattern of the deacon's dalmatic is identical to the one on the figure of Saint Martial from Grandmont preserved in Florence (cat. 58f). The presentation and vesting of the figure irresistibly suggest an analogy with the trumeau image of Saint Stephen of about 1200 at the cathedral of Sens;[3] stylistic similarities are, however, utterly lacking.

Because of the proximity of Les Billanges to Grandmont and the figure's reputed provenance from that abbey, Texier identified it as Stephen of Muret, the founder of the Grandmont order, who remained a deacon until his death as a sign of humility, following the example of his namesake, Stephen the protomartyr. Texier also recognized that the relief figure and its base did not originally belong together, suggesting that the figure would once have been applied to an altar or a chasse. Gauthier has followed the hypotheses put forth by Texier.[4] Souchal, following Abbé Brisset, ascribed the figure to the chasse of Stephen of Muret.[5] This identification necessarily affects the dating of the deacon. If it is a figure representing Saint Stephen of Muret from his chasse, it must be dated either close to 1189 or about 1260 (see cat. 61),[6] both of which seem inconsistent with its style, especially given its relationship to the apostles of Grandmont.

Jean-René Gaborit was the first author to examine systematically the rather richly embroidered hypotheses concerning the identification of the deacon figure and its original use; his arguments for dismissing them are compelling. First, the figure is exceptionally large for a chasse. More important, there are no known or recorded examples of a figure on a chasse, or even an altar, that bears an actual relic of the True Cross. Scholars who identify the figure as Stephen of Muret further posit that the relic he holds is the one sent from Jerusalem by Amalric I, king of Jerusalem, in 1174. It is difficult to fabricate the circumstances that would cause Stephen of Muret to be shown holding a relic that was sent to his abbey some fifty years after his death, as Frolow noted;[7] nor is it necessary, given the numerous relics of the True Cross recorded at Grandmont, to infer that the relic held by the figure of a deacon be identified with the one sent by Amalric.[8]

Gaborit suggested rather that the figure be identified with one of the appliqués on the "Croix de l'Autel de Prime," recorded by Texier in 1857. The cross, "neuf pieds de longueur et la croisée quatre pieds" (nine feet in length and the crosspiece four feet) had a figure of the crowned Christ, a figure of Saint Peter below, and still another figure below that, with an inscription indicating that the cross contained wood of the Cross and other relics.[9] The considerable size of the cross would allow for the inclusion of a figure like the Billanges deacon. Gaborit correctly notes that if the figure below Peter had been Adam, following conventional iconography, he would have been easily identified. At the same time, if the figure was the Billanges deacon, it is curious that the fact that he held a relic is not noted. A small enameled figure in the Walters Art Gallery, Baltimore (fig. 59a),[10] representing a deacon holding a True Cross reliquary—much like the figure of the deacon of Les Billanges—probably comes from a cross terminal, suggesting that such figures were used to decorate Limousin crosses, as Gaborit suggests in this instance.

In 1789, this cross was in the possession of a M. Coutaud, a metal founder in Limoges who had purchased the old scrap copper from Grandmont. By 1857, Texier did not know what had become of the cross, though he noted that some of its relics had been given to the sisters of the Visitation at Limoges.

Fig. 59*a*. Appliqué from a cross: Deacon. Walters
Art Gallery, Baltimore

The marriage of the foot and the base did not occur at
the abbey of Grandmont, for this object is not described in
any of the inventories; nor was the church at Les Billanges
given this piece at the distribution of the holdings of the
treasury of Grandmont among churches of the diocese of
Limoges.[12] Might one suggest that the heavy reenameling of
the base[13] and the marriage of the disparate elements were
accomplished after the purchase by M. Coutaud, incorporating elements he had acquired—a cross base and the
unidentified figure from the large cross (this unique element
preserved because it was a reliquary)—to create a new reliquary for the church of Les Billanges?

BDB

Palustre and Barbier de Montault (1887) suggested that
the base was the one described as supporting a double-
armed gilt-silver cross in the inventory of the abbey of
Grandmont of 1666: "sur un pied rond, de cuivre doré, qui
a trois figures de serpents dessus et est porté sur trois autres
petits" (on a round foot of copper gilt, which has three figures of serpents above and which is supported by three
other small ones). Gauthier, like earlier authors, noted that
the deacon figure was affixed "later" to a cross base, without
suggesting the date at which this happened or the probable
date of the base. Gaborit, too, was convinced of this identification of the cross base with the inventory reference,
though he noted that the base had been "heavily
reenameled." The base is similar in form to other examples,
including one preserved at Obazine,[11] but the carving of the
copper, the foliate ornamentation, and the deep, not fully
filled troughs of enamel are unexpected in Limoges work.

NOTES

1. The format of the reliquary can be found in surviving thirteenth-
   century reliquaries such as the one in The Cleveland Museum of Art
   (Inv. 52.89), published with related examples by Verdier (1982, pp.
   94–110).
2. Inv. 62.29; Cleveland 1967, pp. 148–51.
3. Sauerländer 1972, fig. 62.
4. Gauthier 1950, p. 49; Gauthier 1992, p. 99.
5. Souchal 1963, p. 62.
6. Paris 1965, no. 357, p. 194.
7. Gaborit 1976a, p. 242.
8. For a listing of the True Cross relics at the abbey of Grandmont, see
   Frolow 1965, pp. 586–87.
9. The inscriptions are recorded by Texier (1857, col. 904). A second listing of relics was inscribed on a small box at the foot of the cross. Texier
   identified the lettering as twelfth or thirteenth century.
10. Inv. 44.18. See Verdier 1965, pp. 1–3.
11. Rupin 1890, p. 307, fig. 378. The two are illustrated together in the
    exhibition catalogue (Paris 1993, p. 235) and in the album from the
    1886 Limoges exhibition, pl. xv.
12. The church at Les Billanges did receive a silver arm reliquary of Saint
    Félicien at the time of that distribution, however. *Procès-verbal de dis-
    tribution des reliques*, December 1790; see Ardant 1848, art. 8, p. 70.
13. The restoration may include not only reenameling but also the replace-
    ment of other elements, such as the serpents.

EXHIBITIONS: Paris 1867, no. 1966; Limoges 1886, pl. XXIV; Limoges
1948, no. III; Paris 1965, no. 357, p. 194; Cleveland 1967, no. IV-18, pp.
148–51; Paris 1993, no. 87, pp. 233–35.

BIBLIOGRAPHY: Texier 1853, pp. 324–25, ill.; Texier 1857, cols. 891–93,
1494, fig. 3; Didron 1859, p. 28; Palustre and Barbier de Montaut 1887,
pl. XXIX; Guibert 1888b, p. 217; Rupin 1890, pp. 475–77, figs. 525, 526;
Frolow 1941, p. 233; Brisset 1949, pp. 101–2; Gauthier 1950, p. 49, 157, pl.
46; Frolow 1961, no. 629, p. 471; Souchal 1963, pp. 62–64; Verdier 1965,
no. 8, pp. 1–3; Gaborit 1976a, pp. 242–43; Gauthier 1992, p. 99, fig. 2.

# 60. Effigy of William of Treignac

Limoges, ca. 1235–50

Copper: engraved and gilt; champlevé enamel: lapis and lavender blue, turquoise, green, yellow, red, and white; lapis-blue enameled bead

14.4 × 7.9 cm (5⅝ × 3⅛ in.)

INSCRIBED: GUI / LELMUS / VI P / RIOR / GRAN / DI MON / TIS (William, sixth prior of Grandmont)

PROVENANCE: From the treasury of Grandmont Abbey, where it was affixed to a "coffret" cited in inventories from 1496 to 1790 (no. 65 in 1496 inventory; see Guibert 1888a, p. 90; no. LXXVI in the one of 1666; see Texier 1857, col. 878; no. 19 in the one of May 1790; see Texier 1842, pp. 329–30). In 1790 the coffret was placed in the church at Razès (Haute-Vienne, no. 7 in *Procès-verbal de distribution des reliques* [Proceedings of the distribution of the relics]: "the large chasse on which was Saint Stephen of Muret . . . as well as a wooden casket decorated with gilded copper strips, only on the front"; see Ardant 1848, p. 70), then confiscated on Jan. 31, 1794 (see Lecler 1891, p. 174).

Plaque alone: detached before 1887; Bordes Collection (in 1887 and 1891; see Guibert 1877c). Acquired at Munich (Hess) in 1891.

CONDITION: A few areas missing in enamel (lower part of left column, beneath feet, and in halo). One enamel bead broken.

Uméleckoprùmyslové Muzeum, Prague (3959)

William of Treignac is a prominent figure in the history of Grandmont. From a noble Limousin family, he was described by his contemporaries as a simple, righteous man.[1] It was during his priorate, from 1170 to 1188, that Grandmont received the relic of the True Cross sent by King Amaury of Jerusalem (1174) and that four brothers were sent to Cologne (1181) to bring back relics of the companions of Saint Ursula.[2] In 1185, a revolt of the *convers*, or lay brothers, against the *clercs* over the interpretation of the rules regarding the subservient status of the brothers within the order forced William of Treignac to resign. In 1187, deposed by the *convers* in a violent coup, he left Grandmont for Rome. He died in 1188, either in Rome or while returning home.

This plaque decorated the front of an object described as a coffret in no. LXXVI of the inventory of the treasury of Grandmont drawn up in 1666: "Another wooden coffret, larger than the preceding, embellished with gilt copper, on the front of which are four images in relief on as many plaques of enameled, gilt copper, upon the first of which is written *Guilelmus, prior Grandimontis*, and the same on the second; on the third, *Sanctus Stephanus protomartyr*, on the fourth, *Bernardus de sancto Eligio*. There is also engraved on the base of the cover, over the heads of these four figures, *Hic sunt sanctorum sacrosancta memoria quorum sit consolamen nobis orantibus. Amen.* We found this coffret full of rotted cloth, which no doubt once served to wrap the holy relics."[3] The report of the meeting of March 27, 1888, of the Société Archéologique et Historique du Limousin mentions the discovery "at the house of someone living on Saint Leonard Street" of "a box of relics from Grandmont dating from the thirteenth century," stripped of the "enamels that decorated it" but preserving the inscription "*Hic sunt sanctorum sacrosancta memoria. . . .*"[4] The coffret was thus still intact at that date; moreover, the inscription was published by Rupin in 1890;[5] but since then all trace of it has disappeared.

Whereas Souchal (1964a), like nineteenth-century authors, considered this "coffret" to be the chasse for the relics of William of Treignac, Gauthier (1992) has suggested that it might be only an assemblage, a "commemorative portrait" of William once "linked perhaps to a funerary monument." Yet, many chasses in the inventory of 1666 are designated by the term *coffre* and the four figures with identifying inscriptions over them seem to have formed a homogeneous ensemble. Furthermore, the priors of Grandmont were buried under simple stone markers.[6]

Admittedly, it would be surprising that a chasse should be executed to preserve the remains of the sixth prior of Grandmont, who was never canonized but is nonetheless shown with a halo. This might be explained by the return to Grandmont of the body of the prior.

Dom Jean Becquet and Geneviève Souchal have suggested convincingly that the figure depicted on the fourth plaque of the chasse, identified by the inscription *Bernardus de sancto Eligio*, might be Bernard de La Roche, a brother at Grandmont and disciple of William of Treignac, who, having visited the tomb of his mentor, died in Rome and was buried at his side.[7] This would explain why the remains of the prior and the brother, brought back from Rome at an unknown date, were housed in the same chasse.

According to Souchal (1964a) and Gauthier (1992), the chasse would have been executed at the beginning of the thirteenth century under the authority of Adhémar of Friac (r. 1199–1216) or early in the administration of the prior Caturcin (r. 1216–28). This hypothesis is predicated on the belief that later financial difficulties would have prevented artistic commissions,[8] a weak argument as Gaborit has shown.[9] Nor, was the chasse of William of Treignac necessarily luxurious. None of these descriptions mentions the presence of "gemstones" or other elements that might enhance its value. We know only that the principal face had four half-relief appliqué figures on enameled plaques. According to the inventory drawn up in May 1790, the "wooden coffret" was "embellished with gilt copper on the front only,"[10] but we do not know if the object looked this way originally because no preceding inventory described the back.

On the plaque now in Prague there is a surprising contrast between the enamelwork, of modest quality, and the very carefully executed gilt-copper relief, which is striking in its precise representation of the figure. The priestly attire— alb, chasuble, and maniple—is precisely rendered; the austere face with visible bone structure, high cheekbones, and cleft chin is meticulously fashioned and contrasts with the regular roundness of most of the faces of repoussé copper executed by Limousin artists. It seems difficult to accept the dating of this figure to the beginning of the thirteenth century. Indeed, the treatment of the face makes it much closer to the reliefs executed toward mid-century: the way the hair falls in a continuous mass surrounding the temples and forehead, the hollowing out of the eyes, and the drawing of the small mouth with pinched lips call to mind the appliqués of the Passion (cat. 119), the Ascension in the Louvre (cat. 116), and especially the apostles on the chasse of Saint Geoffrey of Châlard (fig. 118a).[11]

The decoration of roses strewn over the background is certainly archaizing at this date, but other works show that this type of decoration had not been abandoned by the first half of the thirteenth century—for example, the large chasse of Saint Fausta in the Musée de Cluny, Paris, executed in the years 1220 to 1240,[12] and the chasse of Saint Hippolytus from the treasury of Saint-Denis (cat. 112). Their rosette motifs are similar to those on this plaque, although the design and the separation between the different colors are less precise here.

If the plaque really did come from a chasse preserving the remains of William of Treignac, the iconography of this chasse could, moreover, confirm a dating of about 1235–50. The 1666 description specifies that, among the four figures, two bore the inscription *Guilelmus prior*. It is quite unlikely that one would represent the same person twice, side by side. The second prior named William could only be William Dongres, the thirteenth prior of Grandmont (r. 1245–48), who possibly commissioned the work.[13]

ET-D

NOTES

1. See Lecler 1907–9, pp. 444–78(1908), and Becquet 1960 and 1963, which correct a number of errors. Accordingly, two letters attributed to William, including the often mentioned epistle to the English king Henry II reproaching him for the murder of Thomas Becket, would be forgeries (Becquet 1963a, p. 65).

2. Gaborit 1976a, pp. 237–38, has underscored how much this quest for relics, with the explicit goal of attracting the faithful, corresponded to a new style of "politics"; for more information, see Becquet 1960, pp. 291–307.

3. Du Boys 1855, p. 68; Texier 1857, col. 69. The inscriptions can be translated, "William, prior of Grandmont," "Saint Stephen protomartyr," "Bernard of Saint Eloi," "Here lie the sacred [remains] of the saints whose memory is a consolation for us who pray."

4. *BSAHL* 37(1890), pp. 442–43.

5. Rupin 1890, fig. 204, p. 134 (transcribed by L. Guibert).

6. Levesque 1662, p. 401, who mentions the discovery of a number of tombs, including the one of William of Treignac in 1639.

7. Souchal 1964a, p. 73, citing an unpublished remark by Dom Jean Becquet. See also Becquet 1968, pp. 232–33.

8. Ibid., p. 75.

9. Gaborit 1976a, pp. 243–44.

10. Texier 1842, pp. 329–30 ("the identification is made possible by the mention in this inventory of the inscription *Hic sunt sanctorum . . .*"). A note specifies, furthermore, that this reliquary was together with chasse no. 29, which is the chasse of Saint Stephen of Muret (cat. 55).

11. Paris 1993, no. 79.

12. Gauthier 1972b, no. 132, pp. 188, 373, pl. 189.

13. Dom Jean Becquet has graciously pointed out that it was William Dongres who succeeded in relegating the *convers* to a lower status, "which would certainly make him a logical artistic sponsor of poor William of Treignac," by placing the remains of this prior and his disciple in a single chasse. (Dom Jean Becquet to the author, April 6, 1995.)

EXHIBITIONS: Tulle 1887, no. 72; Prosinec-Úmor 1986–87, no. 22, ill.

BIBLIOGRAPHY: Guibert 1877c, pp. 430–31; Guibert 1887 and 1888, p. 610; Molinier and Rupin 1887, pp. 258–60, ill.; Rupin 1890, pp. 133–34, fig. 203, pp. 533–34; Decoux-Lagoutte 1891, p. 273; Marquet de Vasselot 1905, p. 428, n. 3; Graham and Clapham 1926, pp. 167–68, pl. xxx, fig. 1; Marquet de Vasselot 1941, p. 113; Becquet 1960, p. 307, n. 73; Souchal 1964a; Špaček 1971, no. 10, pp. 179–80, fig. 4; Gauthier 1992, p. 97, fig. 1.

# 61. Miracle of the Child from Ambazac

Limoges, ca. 1250–70

Copper: repoussé, engraved, chased, and gilt; medium and light blue enameled beads

33 × 26 cm (13 × 10¼ in.)

PROVENANCE: Probably from chasse of Saint Stephen of Muret at the abbey of Grandmont, where the chasse was placed on main altar and therefore not described in the oldest inventories, which documented only objects in the treasury; no. 1 in the inventory of 1666 (see Texier 1857, cols. 826–27); no. 29 in that of May 1970 (see Texier 1842, p. 332).

    Attributed in 1790 to the church at Razès (Haute-Vienne, no. 7 in *Procès-verbal de distribution des reliques* [Proceedings from the distribution of the relics; see Ardant 1848, p. 10) and confiscated on April 31, 1794 (see Lecler 1891, p. 174 and above, cat. 59).

EX COLLS.: Appliqué: Frédéric Spitzer (see Palustre and Molinier 1890, no. 254, pl. VI; sale, Paris, April 16–June 17, 1893, no. 254); acquired by Mme Nélie Jacquemart-André at this sale. Bequeathed by Mme Nélie Jacquemart-André, with the rest of her collection, to the Institut de France, 1912.

CONDITION: Gilding worn on highest parts of relief.
Musée Jacquemart-André, Chaalis (1020)

This relief has been variously identified—as the Presentation in the Temple (Molinier 1890), "a birth" (Rückert 1959), and the resurrection of the son of Sarepta's widow (3 Kings 17: 17–24) (Otavsky 1973). The most convincing interpretation seems to be J.-R. Gaborit's suggestion that it is an appliqué from the chasse of Saint Stephen of Muret, placed on the main altar of the abbey of Grandmont and depicting an episode from his life—the miracle of the child of Ambazac.[1] Indeed, many arguments support this interpretation. First of all, the inventories specify that the chasse was decorated with scenes from the life of Saint Stephen of Muret depicted "in relief." The inventory of 1666 describes it as follows: "There is, on the main altar of the church, the chasse of our blessed father Saint Stephen, which is of enameled and gilt copper on the outside and wood inside, decorated all over with a great many small figures in relief, set with gems and curiously worked."[2] Furthermore, a marginal note by an eighteenth-century commentator alongside the description of the altar in *Antiquitiés de Grandmont* by Pardoux de La Garde seems to refer specifically to this scene. Pardoux's comment that the chasse also contained the "body of a Holy child" is annotated thus: "Some say it is the body of the child from Ambazac who saw the soul of Saint Stephen rise up to heaven, and that is probable, because the child died after the vision."[3] This scene with a somewhat enigmatic iconography seems, moreover, to correspond to the story of the life of Saint Stephen of Muret: at the moment of the saint's death, the dying child called to his mother that he saw the soul of the saint rising up to heaven surrounded by angels. He repeated this assertion to the brothers who came running when his mother summoned them.[4] The woman in a headdress, leaning over the child in the center, would thus be his mother, and the figure dressed in a robe and hooded mantle either a brother come to confirm the miracle or, as suggested by de Chancel (1991), the father of the child. Hermits or members of hermetic orders are usually represented in such garments by Limousin artists in the thirteenth century—for example, Saint Savinian on the back of the chasse of Saint Viance (cat. 118).

The accidents of history or, more likely, the care of a devout person or art lover would thus appear to have been responsible for the removal of a significant fragment of each of the two chasses that were initially sent to the church at Razès in 1790 at the dissolution of the abbey of Grandmont, and then confiscated by order of the authorities of the French Revolution during the Reign of Terror in 1794 (see cat. 60).[5]

The inventories describe the chasse of Saint Stephen as one of the most sumptuous of those preserved at Grandmont, where it occupied the place of honor at the center of the main altar. It must have been one of the largest ever made by the Limousin workshops because the dimensions given in the inventory of May 1790, and confirmed by the registers of the commune at Razès published by Lecler, are those of a chasse of more than a meter (107–110 cm [42⅛–43¼ in.]) in length, about 90 centimeters (35⅜ in.) in height, and 34 centimeters (13⅜ in.) in depth.[6] The ridged ("donkey-backed") roof described in the May 1790 inventory must have been comparable to that of the Ambazac chasse, described in the same terms. The register of Razès mentions gems "of little worth," whereas the May 1790 inventory describes a chasse "extremely enriched and ornamented with precious stones, crystals . . ." We must then imagine a work embellished with crystal cabochons and colored glass, such as the one from Ambazac (cat. 55) but principally decorated with appliqués in half relief depicting scenes from the life of Stephen of Muret.[7] The inventories of Grandmont confirm, furthermore, that the two chasses were not dissimilar; the 1666 inventory specifies that the chasse of Saint Macarius, now at Ambazac, was of the same "material and shape but smaller" than the one of Stephen of Muret.[8]

Only the style of the relief presently at Chaalis suggests an approximate date of execution for the chasse. Rückert (1959) and, later, Otavsky (1973) compared it to the group formed by the appliqués of the Passion, those of the Germeau chasse, and the Christ from Amiens (cats. 119–21).

The style clearly embodies the same general characteristics, notably the cutting of the drapery in deep furrows with the breaks indicated by "spoon-shaped" openings at the ends. But one notices here a more accentuated cutting, especially in the vertical falls of the drapery, which have a tubular look at the bottom. As for the faces, they are somewhat pared down, and the scene as a whole is treated in a more anecdotal manner, the mother of the child wearing the hint of a smile, perhaps at her young son's assertion that he would join Stephen of Muret in heaven. The high quality of the workmanship combining monumentality in the scene and subtlety in the details has not been sufficiently emphasized. The softness of the feminine faces and the lighter blue enamel of the child's eyes should also be noted.

It is very likely that the chasse to which this relief originally belonged was made slightly after midcentury by an artist trained among those who executed the large ensemble devoted to the Passion of Christ destined for an important Limousin building.[9]

The fabrication of a new chasse for the relics of Saint Stephen of Muret about 1260 might be confirmed by two historical facts: in 1256 the bishop of Limoges, Aymeric of La Serre, instituted the celebration of the feast of Saint Stephen of Muret in his diocese—a decision that could have led to the execution of a more sumptuous and especially more narrative chasse than the preceding one; in 1269 Count Thibault of Champagne repeated the order to pay ten livres annually for the maintenance of candles for the altar—a possible indication that works enriching the altar had been completed.[10]

ET-D

NOTES

1. Gaborit 1976a, pp. 241–42.
2. ". . . il y a sur le grand autel de l'église, la châsse de notre bienheureux père Saint Étienne, qui est de cuivre doré et émaillé par dehors, et de bois par dedans, ornée de toutes parts de grand nombre de petites figures en bosse, garnie de pierreries et curieusement travaillée." See Texier 1857, col. 827.
3. See Souchal 1963, p. 51. The text of Pardoux de La Garde (before 1591), from the Bibliothèque du Séminaire de Limoges and kept in the Bibliothèque des Archives Départementales de la Haute-Vienne (ms. I Sem. 81), has been published many times, the most recent and most precise edition being the one in Souchal 1963, pp. 50–53, based on a study by J. Decanter.
4. Becquet 1968, p. 125.
5. It is not clear how much credence we should give the mention made among the *Comptes rendus des séances* (Records of the meeting) of the Société Archéologique et Historique du Limousin: "M. Royer indicates that a reliquary from the church at Razès, from the abbey of Grandmont, has just been sold to a dealer in Paris for a relatively high price" (*BSAHL* 55 [1906], p. 861). In fact, the two chasses deposited in Razès seem to have been dismantled long before this date inasmuch as the Chaalis appliqué, from the chasse of Saint Stephen, and the plaque in Prague, from the second chasse (see cat. 60), are both mentioned as individual objects by 1890. On the other hand, the assertion that the Chaalis appliqué would have been part of the Londesborough Collection (Gaborit 1976a, p. 241 n. 40, based on information given by M.-M. Gauthier) would seem to be mistaken, for this object does not appear in the work devoted to the collection by F. W. Fairholt (1857).
6. Texier 1842, p. 332; Lecler 1891, p. 174.
7. The dimensions of the Chaalis relief lead one to think that it could have come either from the roof, or, more likely, from the body, because each of these two parts could bear three or four appliqués of this type, depending on the space left for gems and other decorative elements.
8. Texier 1857, col. 827; see also cat. 55.
9. This chasse could thus be the one that inspired the representation, often cited and reproduced, of the leaf of the *Speculum Grandimontis*, now in the Archives Départementales de la Haute-Vienne, showing the elevation of the body of Stephen of Muret. The style of the illumination suggests that it was made in the third quarter of the thirteenth century. Regarding this illumination, see, most recently, Paris 1993, no. 35.
10. See the similar conclusions drawn from the gift of a luminary by Count William of Nevers in 1231 (cat. 58). For historical references, see Lecler 1907–9, esp. pp. 461–62, 465–66(1908), which publish these texts.

BIBLIOGRAPHY: Palustre and Molinier 1890, no. 254, pl. VI; Rückert 1959, pp. 7, 14 n. 23; Otavsky 1973, pp. 59, 65, fig. 19; Gaborit 1976a, pp. 241–43, fig. 7; Chancel 1991, p. 38, fig. 17; Bautier 1992, pp. 18–19, ill. p. 16.

# 2

## *The Grandmontain style*

# 62. Central Plaque of a Cross

Limoges, ca. 1185–95

Copper: engraved, stippled, and gilt; champlevé enamel: dark and light blue, translucent dark and opaque medium green, yellow, translucent and opaque red, rose, and white

37.2 × 30 cm (14⅝ × 11¹³⁄₁₆ in.)

INSCRIBED: IH[ESU]S/XPS (Christ)

EX COLLS.: [Alfred André, Paris (sold 1910)]; J. Pierpont Morgan, London and New York

CONDITION: Some losses to gilding overall; cut at top.

The Metropolitan Museum of Art, New York; Gift of J. Pierpont Morgan, 1917 (17.190.409a)

The enameled figure of Christ, his torso starkly white, hangs against a gilt cross. A knee-length loincloth is knotted at his proper left hip; his head is bowed, and his red hair falls against his shoulders. His beard, too, is red; his face, a delicate rose color. The curving lines of his ribs and the stylized musculature of his arms and legs are defined by reserved copper.

Although its provenance is unknown, the cross is closely allied in style, palette, and execution to works created especially for the abbey of Grandmont, its dependencies, and religious communities with patrons in common. The defining characteristic of these works is the rose-colored enamel used for faces, as on the plaque depicting Hugh Lacerta and Stephen of Muret, from Grandmont itself, now preserved in the Musée de Cluny (cat. 57a). The Metropolitan cross is one of a small number distinguishable by the pink faces, chalk-white torsos with fine lines in reserve that articulate ribs, chest bones, nipples, and the muscles of the arms and abdomen and by a knee-length, deep blue loincloth. The cross in the Louvre (cat. 65), excavated at Cherves, was apparently part of the liturgical furnishings of a Grandmontain community. The example at Bordeaux, which, like the Metropolitan and Cleveland crosses,[1] portrays Christ against a green-enameled cross, was made for the abbey of Sauve-Majeure. Since Henry II and Eleanor of Aquitaine were patrons of both Grandmont and Sauve-Majeure, Gauthier's argument that this style represents Plantagenet taste is compelling.[2] Only the example in the Metropolitan Museum, however, clearly shows the green cross as tree trunks, displaying the stubby marks of sawed-off crossed branches.

The palette and drawing of the Grandmont group of enamels represent a tradition distinct from that of contemporaneous vermiculé works. Souchal suggested Byzantine influences for certain iconographic motifs, such as situating the crucifix itself inside the boundaries of a cross form. Yet, as she herself noted, that design was already manifest in glazing at Poitiers by the mid-twelfth century, as were other design elements found on the enamel, such as the articulation of the ribs and the sinews of the arms,[3] which obviates the need to cite Byzantine sources in accounting for anatomical style.[4] Something of the artistic heritage of the Grandmont group can perhaps be seen in the pinkish palette of the faces on the Bellac chasse (cat. 9) and the plaque of Geoffroy Plantagenet (cat. 15), which, like the plaques in the Musée de Cluny (cat. 57), show the figure framed under the enameled arch of a fictive church building. The deep turquoise found on a number of the terminal plaques of these crosses was first seen on the chasse of Champagnat (cat. 10).

When the plaque was acquired by J. Pierpont Morgan, it was associated with two terminal plaques, one representing the eagle of Saint John the Evangelist and the other, the winged man of Saint Matthew (figs. 62a, b). However, the drawing of these figures, the palette, and even the pattern of the border decoration are distinct from that of the central plaque, making it clear that the terminal plaques cannot be from the same original ensemble.[5] Nor can their style be recognized on any other surviving Grandmontain cross elements. Nevertheless, the composition of the enamels is consistent with twelfth-century Limousin examples.

J.-P. Suau has posited that the Metropolitan cross is to be identified as the one once in the collection of Louis-Adolphe Loisel, a Norman collector of the second half of the nineteenth century.[6] That cross, however, was described in 1889 as still having the terminal plaque with the figure of Saint John. It seems unusual that the Saint John plaque should have been dissociated from the cross and two alternate plaques substituted before the sale to Morgan in 1910.[7] The Morgan cross might also be the one mentioned in the same 1889 publication of the Loisel cross as being in Brives[8] and said by Palustre in his 1890 catalogue of the Spitzer Collection to be related to the Spitzer cross now in Cleveland (fig. 63a).[9] Though he specifies in the 1890 publication that the cross was at the church of Saint-Martin, it is peculiar that it would not have been exhibited or at least mentioned in the exhibition at Tulle in 1887 if the cross were in Brives in the Corrèze. Could it rather have been at Brives in the Indre?[10]

BDB

Figs. 62*a*, 62*b*. Terminal plaques of a cross. The Metropolitan Museum of Art, New York (17.190.409b,c)

NOTES
  1. Gauthier 1987, no. 255.
  2. Ibid., no. 261.
  3. See Aubert et al. 1958, fig. 1; Grodecki 1977, fig. 58.
  4. Souchal 1967, p. 42.
  5. This was recognized by Souchal (1967).
  6. Suau 1982, 1983.
  7. This problem was noted by Suau (1982, p. 70, n. 57).
  8. Mentioned in *Congrès archéologique de France* 1890, p. 101.
  9. Palustre and Molinier 1890, p. 85, n. 1.
  10. Cottineau 1935–39, vol. 1, p. 507.

EXHIBITION: New York 1970, no. 156.

BIBLIOGRAPHY: Breck and Rogers 1929, p. 101, fig. 54; Thoby 1953, no. 16, pp. 989–99, pl. XI; Gauthier 1967b, p. 148; Souchal 1967, p. 34, fig. 8; Gauthier 1972c, p. 625, pl. 469-2; Gauthier 1972b, no. 51; Hospital 1979, pp. 29–30, fig. 5; Gauthier 1987, no. 245, pp. 203–4, ill. 721, pl. CCXVII.

# 63. Two Plaques from a Cross

## a. *The Ox of Saint Luke*
Limoges, ca. 1185–95

Copper: engraved, stippled, and gilt; champlevé enamel: dark, medium, and light blue; deep turquoise, translucent dark and opaque medium green, yellow, translucent wine and opaque red, and white

8.75 × 9.2 cm (3⁷⁄₁₆ × 3⅝ in.)

Ex colls.: Abbé Jacques-Rémi Texier, diocese of Limoges (see Rupin 1890); Rémi and Hubert Texier, Limoges; Georges Chalandon, Paris (in 1913); Guennol Collection (before 1951)

Condition: Incidental wear to gilding.

The Metropolitan Museum of Art, New York; lent by Cynthia Hannah Moore (L51.10.2)

a

## b. *The Eagle of Saint John*
Limoges, ca. 1185–95

Copper: engraved, stippled, and gilt; champlevé enamel: dark and light blue, deep turquoise, translucent dark and medium green, yellow, red, and white

10.15 × 8.7 cm (4 × 3⁷⁄₁₆ in.)

Ex colls.: Georges Hoentschel, Paris (by 1911); [Jacques Seligmann, Paris (in January 1912)]; J. Pierpont Morgan, London and New York

Condition: Incidental wear to gilding.

The Metropolitan Museum of Art, New York; Gift of J. Pierpont Morgan, 1917 (17.190.771)

The eagle, symbol of Saint John the Evangelist, stands clasping a scroll in his talons, his head turned back, his proper left wing raised over his head. The entire figure is outlined with fine stippling, and the wings are accented by dark blue. His talons are red, but his body is rendered in muted tones of light blue with white dots. A leaf-shaped reserved area defines the upper part of the leg muscle. Its vertical orientation clearly indicates that this plaque is from the top of a cross.

The ox, symbol of Saint Luke, strides forward as he clasps an enameled book between his front hooves. Straining the sinewy muscles of his neck, he turns and raises his head to look up. Stylized, almost leaflike areas of enamel and reserved gilt suggest powerful shoulders and haunches; regular reserved, stippled lines define the ribs. The tail curls down, falling between the legs, and the entire figure is outlined by fine stippling.

Like the image of the eagle of Saint John, the figure of the ox is distinguished by foliate ornament used to define shoulder and haunch. The leaflike shape is realized by reserved metal stippled to resemble the veins of a leaf's underside, which suggests that it has folded over against its more colorful enameled side. In seeking to account for this unusual design, Souchal suggested the possibility of Byzantine influence, without, however, pointing to specific comparisons.[1] It seems unnecessary to look so far afield for a source when the better explanation may be creative adaptation of motifs: a similar system is employed for leafwork decoration on the chasse of Ambazac (cat. 55). The form of the book, with its applied bosses and heart-shaped clasp, recalls the one held by Stephen of Muret on the Musée de Cluny plaque (cat. 57a). The deep turquoise enamel of the ox's face finds a parallel on the orphrey of Saint Stephen. In details of both technique and palette, then, the plaques are closely allied to works made for the abbey of Grandmont and its dependencies.

The composition of the plaque of John is almost identical to one last recorded in the Martin Le Roy Collection.[2] The composition of the plaque of Luke is virtually the mirror image of the one from the cross preserved at the Musée de Cluny,[3] though the New York example is remarkable for its finer condition. Such similarities typify the crosses of the Grandmont group.

Since they represent symbols of the evangelists, these plaques must have come from the reverse of a cross, where, following standard convention, they would have framed the central image of Christ in Majesty.[4] Both the Luke and John plaques have been incised on the reverse with two parallel lines, indications that they were intended for the same ensemble.[5] Furthermore, based on their dimensions and stippling, their common border decoration, and the placement of the holes for attachment to a wood support, these two plaques are demonstrably from the reverse of the cross associated with the Grandmont workshop preserved in Cleveland (fig. 63a).[6] The image of the winged man of Saint Matthew last recorded in the Martin Le Roy Collection,[7] on which the similarity of enameling on the wings is particularly noticeable, is also from the same ensemble. Gauthier has also posited that the almond-shaped plaque with the image of Christ blessing from the Martin Le Roy Collection comes from the same cross.[8]

BDB

NOTES
1. Souchal 1967, p. 45.
2. Gauthier 1987, no. 252.
3. Ibid., no. 254a.
4. See Thoby 1953; de Linas 1885 and 1886, pp. 453–78, especially 453–65.
5. See Stohlman 1934.
6. See Souchal 1967, pp. 30, 51.
7. Gauthier 1987, no. 258.
8. Ibid., no. 256.

EXHIBITION (b): Paris 1913, no. 190, p. 100.

BIBLIOGRAPHY (a): Pératé 1911, no. 41, pl. XI; Gauthier 1967b, p. 148; Souchal 1967, fig. 18, pp. 50, 52–53; Gauthier 1987, no. 257, pp. 210–12, ill. 730 pl. CCXXII. (b): Rupin 1890, p. 249, fig. 312; Souchal 1967, fig. 19, pp. 50, 53; Rubin 1975, pp. 183–90; Gauthier 1987, no. 259, pp. 210–12, ill. 733 pl. CCXXII.

b

Fig. 63a. Cross. Cleveland Museum of Art

# 64. Plaque from a Cross: The Winged Man of Saint Matthew

Limoges, ca. 1185–95

Copper: engraved, stippled, and gilt; champlevé enamel: dark and light blue, deep turquoise, translucent dark and opaque medium green, yellow, red, rose, and white

12.6 × 8.55 cm (5 × 3⅜ in.)

Ex colls.: Georges Hoentschel, Paris (by 1911); [Jacques Seligmann, Paris (in January 1912)]; J. Pierpont Morgan, London and New York

Condition: Incidental losses to gilding.

The Metropolitan Museum of Art, New York; Gift of J. Pierpont Morgan, 1917 (17.190.772)

The torso of the winged man, symbol of Saint Matthew, appears above a band of clouds; he clasps a book in his right hand, which is veiled by his dark blue mantle. His wings rise above his nimbed head, crossing near their tips. The entire figure is outlined by fine stippling, which also serves to define the reserved areas of his wings and the bodice of his robe.

The dimensions, border decoration, and pattern of attachment holes of this plaque are distinct from those of the other plaques included here (cat. 63a, 63b) but allow this one to be associated with the central plaque of a cross in the Musée de Cluny and related terminal plaques also in the Cluny and in Baltimore, and one last recorded in the Martin Le Roy Collection.[1] Like the plaque of Saint John in Baltimore and the one of Peter in the Musée de Cluny, the Metropolitan plaque is distinguished by stippling in a diaper pattern to define the bodice of the saint's garments. The drawing of the face, with an oval loop outlining the earlobe, a full, rounded nose, and downward turn of the cleft of the chin, recalls especially the face of the Virgin on the Adoration of the Magi plaque from the abbey of Grandmont (cat. 57b).

A similar composition can be found on other plaques associated with the Grandmont workshop, notably one last recorded in the Martin Le Roy Collection,[2] and another, apparently slightly less accomplished, in Moscow.[3]

The plaques in the Musée de Cluny are said to be from the Grandmontain priory of Mathons. Indeed, an inventory of the "petit séminaire" of Troyes, where the holdings of the priory of Mathons had passed by the time of an inventory of May 19, 1770, includes "six grands chandeliers et une croix de cuivre pour le maître autel" (six large candlesticks and a cross of copper for the high altar), as well as three less important crosses and a number of candlesticks large and small.[4] The first recording of the putative provenance of the Cluny cross elements, however, was made only sometime between 1926 and 1947, by M. de Montrémy, then curator of the museum, and so must remain hypothetical. Indeed, the identification may be further complicated by the presence of a plaque of Saint Peter from a similar cross in the treasury of the cathedral at Troyes, recorded before 1861.[5]          BDB

Notes
1. See, most recently, Gauthier 1987, nos. 249–54.
2. Ibid., no. 258.
3. Ibid., no. 265b. First recorded in the Texier Collection in 1856.
4. Souchal 1967, pp. 37–38, n. 1.
5. Gauthier 1987, no. 260, pp. 212–13, ill. 736 pl. ccxxvi.

Bibliography: Pérate 1911, no. 42, pl. xix; Souchal 1967, pp. 47–48, fig. 16; Gauthier 1967b, p. 148; Gauthier 1987, no. 253, pp. 208–10, ill. 726 pl. ccxxi.

# 65. Central Plaque of a Cross

Limoges, ca. 1185–95

Copper: engraved, stippled, and gilt; champlevé enamel: blue-black, dark and medium blue, turquoise, medium and light green, yellow, red, and white

34 × 21 cm (13⅜ × 8¼ in.)

INSCRIBED: IH[ESU]S/XPS (Christus)

PROVENANCE: Treasure of Cherves (Charente). Excavated at Château-Chesnel, near Plumejeau, December 11, 1896 (see also cats. 98–100).

EX COLLS.: Comte Ferdinand de Roffignac, Angoulême (in 1898); Félix Doistau, Paris (in 1900)

CONDITION: Wear to gilding in area of inscription; loss to enamel, especially on abdomen, loincloth, and lower right leg of Christ.

Musée du Louvre, Paris; don Doistau, 1919 (OA 7284)

The figure of the crucified Christ, enameled white and wearing a deep blue loincloth edged in turquoise and yellow, is set against a gilt-copper cross. His arms are stretched wide, and his feet rest against a turquoise suppedaneum. The nails of crucifixion in each hand and foot are suggested by circlets of reserved copper. A turquoise and dark blue cruciform nimbus frames his head and his shoulder-length red hair and beard.

The Louvre cross is clearly bound by palette and style to the Metropolitan Museum cross and others associated with the enamels produced for the abbey of Grandmont. (See discussion in cat. 62.) The Louvre example is closest to the central plaque of a cross from the Biais Collection in Angoulême, preserved at the Ackland Art Museum, Chapel Hill, North Carolina,[1] especially in such details as the use of scrolling vine decoration set against the reserved field of the cross, the definition of Christ's halo, the sagging stomach muscles divided into four distinct areas, or the turquoise suppedaneum that supports Christ's rather broad feet with their widely splayed toes.

The Louvre cross was excavated at Cherves, along with a number of other liturgical objects. It seems that some of the contents of the treasure were separated from it prior to Barbier de Montault's publication,[2] but no other elements from this cross are recorded. The plaques from either side would have had half-length figures of the Virgin and Saint John,[3] with the hands of angels appearing from the sky at either side: the angels' censers can be seen on the Louvre cross just under the hands of Christ. The treasure unearthed at Cherves may have belonged to the Grandmontain community at Gandory in the Charente;[4] it was buried for safekeeping at an unknown date.

BDB

NOTES

1. Gauthier 1987, no. 270; large areas of the Ackland cross have, however, been reenameled.
2. See, for example, Geneviève François's entries below (cats. 99, 100).
3. The appearance of the Saint John plaque is suggested by an example in the Walters Art Gallery, Baltimore. Gauthier 1987, no. 251.
4. The community at Gandory was suggested first by Gaborit in Québec–Montréal 1972, no. 29, for the cross of Saumur, acquired with the Lair Collection in 1919, which, according to the donor, belonged to the Treasure

of Cherves. The reasons for this attribution are, however, unknown. The cross does not appear in the publication of the Treasure in 1897.

EXHIBITIONS: Paris 1900, no. 2417; Leningrad–Moscow 1980, no. 32.

BIBLIOGRAPHY: Barbier de Montault 1897b, pp. 61–76, pl. VII; Barbier de Montault 1898c, pp. 217–19, 223; Barbier de Montault 1898a, p. 166, fig. 1, pp. 165–75; Marquet de Vasselot 1921, pl. 45; Migeon 1922, p. 134; Gauthier 1950, pp. 30, 59, 152, pl. 6; Thoby 1953, pp. 21, 22, 94, no. 9, pl. IV; Gauthier 1967b, p. 148, n. 27; Souchal 1967, pp. 37, 59–62, fig. 28, p. 60; Gauthier 1987, no. 269, p. 217, ill. 743 pl. CCXXV; Taburet-Delahaye 1994, pp. 19–20.

# 66. Plaque with Christ Blessing

Limoges, ca. 1190–1200

Copper: engraved, stippled, and gilt; champlevé enamel: dark, medium, and light blue; mauve, turquoise, green, yellow, red, and white

17 × 12.1 cm (6¹¹⁄₁₆ × 4¾ in.)

EX COLLS.: Georges Hoentschel, Paris (by 1911); [Jacques Seligmann, Paris (in January 1912)]; J. Pierpont Morgan, London and New York

Christ, seated on a mauve-colored pillow that is placed on a rainbow, raises his right hand in blessing as he holds a turquoise book aloft in his veiled left hand. He rests his feet on a triangular footstool. The reserved copper ground against which he sits is engraved and stippled with a pattern of lozenges enclosing small circles. His hair and beard are red; his mantle, deep blue. He wears a robe of green and yellow, with decorative bands of turquoise. These decorative bands on Christ's robe, reminiscent of a Byzantine imperial *loros*, follow a tradition already established in Limousin manuscript illumination.[1]

Like the upper terminal of a cross (cat. 67), this enamel is clearly related to works that were produced for the abbey of Grandmont. This is particularly noticeable in the use of enamel in combination with reserved copper to define the hair, in the slightly mottled tone of the face that distinguishes it from the white of other flesh, in the red enamel eyes (seen, for example, on the Grandmont workshop plaque with Saint Matthew, cat. 64), and in the use of stippling and engraving to highlight areas of reserved metal. However, the overall palette is distinct from that of the Grandmont workshop: there is only one green, and there is a highly unusual mauve tone used for Christ's pillow and in his nimbus. Whereas the figures of the Grandmont workshop are outlined by stippling, this figure's outline is unbroken.

Almond-shaped plaques were used in a variety of contexts, but given the size of this example, which is rather too large for a cross and too small for an altar frontal, it is most likely that it was used on a chasse.

BDB

CONDITION: Wear to gilding, especially at proper left side; minimal loss to enamel of nimbus and border at top.

The Metropolitan Museum of Art, New York; Gift of J. Pierpont Morgan, 1917 (17.190.759)

NOTE

1. Gaborit-Chopin 1969, figs. 162, 171, for example.

EXHIBITION: New York 1970, no. 150.

BIBLIOGRAPHY: Pérat é 1911, no. 46, pl. XXI; Souchal 1967, pp. 56–57, fig. 26; Gauthier 1987, no. 267, p. 216, ill. 738 pl. CCXXIV.

# 67. Plaque from a Cross

Limoges, ca. 1190–1200

Copper: engraved, stippled, and gilt; champlevé enamel: dark and light blue, translucent dark and opaque light green, yellow, red, and white

9.23 × 8.7 cm (3⅝ × 3⁷⁄₁₆ in.)

Ex colls.: Sigismond Bardac, Paris (by 1911); Georges Hoentschel, Paris; [Jacques Seligmann, Paris (in January 1912)]; J. Pierpont Morgan, London and New York

Condition: Gilding almost entirely lost; some loss to enamel, especially in border at upper and lower left.

The Metropolitan Museum of Art, New York; Gift of J. Pierpont Morgan, 1917 (17.190.761)

The nimbed, three-quarter-length figure of a man, posed above a stylized cloud on a T-shaped plaque, glances down as he clasps between veiled hands a disk inscribed with a cross. The white enamel of his face contrasts sharply with the dark blue of his mantle and the deep red of his hair, eyes, and mouth.

The orientation and T-shape of the plaque indicate that it comes from the upper terminal of a cross. Since symbols of the evangelists conventionally occupied the reverse, this figure must have been set on the front, above the crucified Christ, in a place usually reserved for one or more angels. The sacrificial and sacramental nature of the Crucifixion is emphasized by the representation of the Host—the body of Christ—held between the hands of the saintly figure. To further the metaphor, the lateral arms of the cross from which it came probably depicted censing angels, as on the Louvre cross (cat. 68), rather than the Virgin Mary and Saint John as witnesses to the Crucifixion.

The style of drawing, with the wide, rounded face and reserved lines to define features, the highlighting of reserved lines with stippling, and the enameling of the hair suggest that this plaque is contemporaneous with works produced for the abbey of Grandmont. However, the juxtaposition of the colors in strong contrast to one another creates a starker aspect than that found on the Grandmont group of enamels. As Gauthier recognized, the Metropolitan plaque appears to be from the same workshop as the Louvre cross. It is, however, certainly distinct from one of the crosses in the Walters Art Gallery, Baltimore, with which she associated it,[1] and not so closely allied to either the Milan cross or a second in the Walters.[2]

BDB

Notes

1. Gauthier 1987, no. 197. Certain aspects of this work seem rather awkward in comparison with other Limousin crosses of the twelfth century.
2. Ibid., nos. 196, 198.

Bibliography: Pératé 1911, no. 45, pl. xx; Souchal 1967, pp. 56–57; Gauthier 1987, no. 199.

# 68. Central Plaque of a Cross

Limoges, ca. 1190–1200

Copper: engraved and previously gilt; champlevé enamel: black, lapis and lavender blue, turquoise, green, yellow, red, and white

43.5 × 29.8 cm (17⅛ × 11¾ in.)

INSCRIBED: IH[ESU]s / XPS (Christus)

EX COLLS.: P. Soltykoff? (sale, Paris, April 8–May 1, 1861, no. 97); Baron Jean-Charles Davillier, Paris. Bequest of Baron Jean-Charles Davillier, 1883.

CONDITION: Gilding has almost entirely disappeared.

Musée du Louvre, Paris; Département des Objets d'art (OA 2956)

The plaque of a cross from the former Davillier Collection belongs to a small group of Limousin works that are distinguished especially by the particularly harsh style in which they depict Christ. An exaggerated hollowing out of the main features confers on the face an expression of unusual severity. It is intensified by the thickness and density of the beard and hair, both enameled in blue. This small group, identified by Souchal, then by Gauthier,[1] comprises chiefly the plaque at the Musée du Louvre, the one at The Metropolitan Museum of Art that depicts a young saint (cat. 67), and another central plaque of a cross at the Walters Art Gallery, Baltimore. The Christ placed at the center of the principal face of a large cross at the Museo Poldi Pezzoli, Milan, shows a similar austerity in the treatment of the face but differs in the vermiculé background and other aspects of the style.

The Louvre plaque is set apart by the figure symbolizing the moon, above Christ. On the other hand, the censers seen here that would have been held by angels depicted on separate plaques attached at the ends of the arms constitute a recurring theme on this type of plaque (see cat. 65) and are found also on the one in Baltimore.

The Christ of the Louvre is the one that most clearly demonstrates the "severe style." The eyes, half closed by blue-ringed lids and protected by thick dark brows, are sunken into the hollows of their sockets, emphasized by two thin lines joining the sides of the nose. A long line, probably indicating a deep wrinkle, cuts across the forehead—a detail missing from similar Christs, as are the spurts of blood flowing from the nail holes that pierce the feet, which accentuate the dramatic intensity. The severity of the figure and the hardness of the features are, however, heightened by the nearly complete loss of the gilding.

Other aspects link the Louvre plaque to its fellows among the works of the "Grandmont group" (see cats. 62, 65), especially the supple treatment of the loincloth in which the echo of "early Gothic art" can be perceived and the refinement of numerous details, such as the orphrey and the belt, which are both dotted with red. The figure of the moon at the top is very close to figures of angels or young saints adorning those Grandmont crosses.

The style of the Louvre cross can thus be explained as either a more extreme interpretation of the models used by the artists working at Grandmont or as a preference for other works of Byzantine derivation. Indeed, a similar "severe" tendency can be observed in the illuminations and stained-glass windows made in the Auvergne about 1200. The most famous and most significant examples are the Sacramentary in the library at Clermont-Ferrand and certain stained-glass windows in the cathedral there.[2] These models would explain not only Christ's facial expression but also the presence of particular details, such as the thin lines marking the sides of the nose, probably to simulate a cast shadow. These parallels, like the kinship with the works of the Grandmont group, justify dating the Christ of the former Davillier Collection to the final decade of the twelfth century.

ET-D

NOTES
1. Souchal 1967; Gauthier 1987, nos. 196–200.
2. Brisac 1974; Grodecki 1977, pp. 193–94, no. 52, ill. 166.

EXHIBITIONS: Paris 1865, no. 607; Brussels 1958, no. 52; Mexico City 1993, p. 48, fig. 15.

BIBLIOGRAPHY: Courajod and Molinier 1885, no. 270; Linas 1885 and 1886, p. 456, pl. XVIII; Rupin 1890, pp. 276–77, fig. 335; Barbier de Montault 1898c, pp. 222–23, fig. 4 p. 222; Marquet de Vasselot 1914, no. 72, pl. XIV; Thoby 1953, pp. 15ff., no. 4, pl. III; Souchal 1967, pp. 54ff.; Gauthier 1972b, pp. 179ff., no. 125c, ill. p. 179; Gauthier 1987, no. 200, pp. 178–79, figs. 631–32, pl. CLXXXII.

# IV

## Limoges in Transition

1190–1230

The first three or four decades of the thirteenth century are probably the period when the production of the Limousin workshops was the most abundant and varied. Indeed, it is difficult to evoke the full richness of Limoges work from this time. Nonetheless, the joining of the collections of the Louvre and the Metropolitan Museum again constitutes the most complete reference group possible, composed as it is of examples of very fine quality of the principal types of objects created, from a time when, concurrent with the success of Limousin production, in particular for the furnishing of ecclesiastical objects, artists diversified their production. The variety of the forms of Eucharistic reserves—tabernacles, doves, or pyxes (cats. 70, 74, 98, 105, 106)—offers but one example. Rarer objects, such as the chrismatories (cats. 71–73), the belt buckle from Vienna (cat. 90), or the stunning knight from Gourdon (cat. 93)—which might have decorated a secular work or a funerary monument—complete the demonstration. The exceptional loan of the coffret of Cardinal Bicchieri (cat. 88), from before 1227, allows for a better illustration of the evolution of secular decoration, especially since the collections of the Louvre preserve exceptionally fine isolated medallions, probably from another of the coffrets executed for this same cardinal (cat. 91).

The principal stylistic tendencies adopted during this period by the Limousin workshops, divided between an embrace of forms associated with early Gothic art and an attachment to their own traditions, are represented by some of the most significant pieces, notably the ciborium of Master Alpais (cat. 69), the tabernacle of Cherves (cat. 98), whose eloquent pathos rivals that of the finest Gothic sculpture anywhere, and the reliquary of Saint Francis of Assisi (cat. 101).

The two works from the thirteenth century bearing an inscription of the name of a person who is most likely the creator, accompanied by a geographic indication, "from Limoges," or "in Limoges," are presented here: the ciborium of Master Alpais and the cross of Garnerius (cat. 103). These two very different pieces are important documents that, together with the abundant text references to *opus lemovicense* found after about 1168, attest to the establishment of the principal workshops within the city itself. Although the provenance of these two objects is not certain, the crosiers unearthed at Luxeuil (Haute-Saône) in eastern France (cat. 77), Montmajour, in Provence (cat. 80), Preuilly (Seine-et-Marne), near Bourges (cat. 82), and Nieul-sur-l'Autise (cat. 107), combined with examples dispersed over all of Europe, illustrate the broad diffusion of Limousin production. The abiding importance of the European monastic clientele is seen not only in the works destined for the abbey of Grandmont (cats. 57, 58) and its dependencies, such as Gandory (Charente), from which the Treasure of Cherves (cats. 98–100) probably came, but also in those belonging to Cardinal Bicchieri or given by him to the Victorine community he established in Sant'Andrea in Vercelli (cat. 88) and the first commissions for the new mendicant orders (cat. 101).

BDB and ET-D

*1*

---

*Master Alpais and the*
*classic Limousin style*

# 69. Ciborium of Master Alpais

Limoges, ca. 1200

Copper: stamped, engraved, chased, and gilt; champlevé enamel: lapis, medium, and lavender blue; turquoise, dark and light green, yellow, red, and white; glass cabochons and enameled beads

30.1 x 16.8 cm (11⅞ × 6⅝ in.)

INSCRIBED (inside cup): + MAGI[S]TER : G : ALPAIS : ME FECIT : LEMOVICARUM (Master Alpais from Limoges made me)

PROVENANCE: Possibly from the abbey of Montmajour (Bouches-du-Rhône)

EX COLL.: Pierre Révoil. Acquired in 1828.

CONDITION: Two cabochons missing: one on upper cup, one on fruit-shaped knop. Rim of upper cup fitting inside lower one is broken; only a small part remains. Some losses to enamel, principally in lower frieze of angels on upper cup. Inside, gilding is very worn in center of lower cup, whereas it is perfectly preserved on the upper cup, which proves that the object was used over a long time. The excellent preservation of the external surface, on the other hand, hardly seems compatible with a very lengthy burial.

Musée du Louvre, Paris; Département des Objets d'art (MR R 98)

The ciborium of Master Alpais is one of the most famous creations of the Limousin enamelers—rightly so, to be sure, but perhaps not for the best reasons. Indeed, publications have commented on the inscription and argued about the function of the object, but few have sufficiently stressed the technical perfection and exemplary style, which are at the peak of work by Limousin artists and at the crossroads of Romanesque and Gothic art.

Most writers about the ciborium have maintained that it came from the abbey of Montmajour. The inventory drawn up in 1828 when the Révoil Collection entered the Musée du Louvre, however, is very guarded on this point: "M. Révoil thinks that, based on several facts, this vessel was made for the Benedictine monks of Montmajour, near Arles." According to de Laborde (1852), this assumption is based solely on a parallel with the crosier "that [Révoil] knew to have come from the tomb of an abbot of Montmajour" (cat. 80). Only the crosier, however, is certain to have belonged to the Arlesian connoisseur J.-D. Véran. The two objects, therefore, could not have been brought together until a later phase of their history. Given the lack of other documents, the provenance of Montmajour for this ciborium cannot be verified.

The object has almost always been considered a ciborium, although several writers have noted that the shape and the dimensions are more those of a drinking cup (see Barbier de Montault 1881 and Molinier 1891). This double cup resting on a short foot in the shape of a truncated cone belongs to a group of pieces of similar form and dimensions, all made at the end of the twelfth century and the begin-

ning of the thirteenth. The most famous are the English enameled cups in the British Museum and the Victoria and Albert Museum, as well as the silver ones in the treasuries of the churches at Sens and Saint-Maurice d'Agaune.[1] These works bear witness to the appearance at this time of a new means of reserving the Eucharist—a receptacle frequently designated by the French term *coupe* (cup or chalice).[2] As Skubiszewski (1965) has shown, its shape derives from that of secular drinking cups (the transition was perhaps assured by the practice of bringing wine to the Mass in vessels that the texts call *scyphi*) and seems to have been inspired by Oriental goldsmiths' work in which one finds vessels with the same profile and the same lozenge-shaped pattern, for example, on a cup in the archaeological museum in Tehran.[3]

The Eucharistic use, suggested by the oldest inventory references to the "coupe" in the treasury of Sens[4] and by the iconography of the English pieces, is, in the case of the cup bearing the name Alpais, clearly designated by the half-relief figures of four angels bearing consecrated hosts on the knop at the top of the vessel. The hand of God engraved inside the upper cup can also be found on the cover of many Eucharistic doves (see cat. 105).

The inscription engraved at the center of the lower cup has inspired the greatest comment. Although the origin of the name Alpais gave rise to bitter argument among authors of the second half of the nineteenth century,[5] they, like their successors, were in agreement that it was the "signature" of the artist,[6] the first example in Limousin enamelwork of the maker claiming his work—a rare occurrence. The person being designated as "Master" (*Magister*) and his name followed by his place of activity indicate that the name probably belongs to the artist and not the patron. De Linas (1883) and Rupin (1890) were the first to publish references in the consular registries of Limoges noted by Guibert, particularly the one in 1216, to two brothers, J. and W. Alpais, property owners in common in the quarter near the church of Saint-Pierre-du-Queyroix, and another, a few years later, about 1230 to 1240, to a "lot that belonged to G. Alpais." This G. Alpais also appears several times in the *Rentes de la confrérie des suaires* (Revenues of the Confraternity of the Shrouds).[7] Rupin raised the possibility that this person might be the enameler of the Louvre's chalice—which seems all the more likely because the name Alpais was apparently uncommon in Limoges.[8]

The iconography of the object has received little comment. Engraved on the inside, at the center of the upper cup, is the hand of God in benediction; at the center of the lower cup, a half-length angel is surrounded by an inscription. On the outside, busts of figures emerging from clouds

247

Angel blessing, surrounded by an inscription, engraved inside the lower cup (detail)

Detail

are inscribed within a lozenge-shaped network. On each cup, two friezes of eight angels frame eight male figures, all but one bearing a book. Most often the sixteen male figures have been considered to be the twelve apostles and four prophets,[9] although prophets hold a scroll more often than a book. It is not unusual, however, to encounter more than twelve apostles in Limousin iconography (see cat. 30). One of the figures, the one without a book, is also the only one surrounded by enameled florets, no doubt to distinguish him from the others. Since their publication by Darcel (1854), the busts of angels have been most often associated with the identification of the Eucharist as "the bread of the angels" (*panis angelorum*) in the Lauda Sion drawn up by Saint Thomas Aquinas for the celebration of Corpus Christi in 1264. Although the ciborium is certainly much earlier than this date, the reference to *panis angelorum*, an expression borrowed from Saint Augustine, recurs frequently in the writings of medieval theologians.[10]

The ornamentation that has elicited the greatest amount of scholarly comment is the frieze of pseudo-Kufic inscriptions that runs around the lip of the lower cup. These were, moreover, the basis of one of the very first publications of the object, by de Longpérier (1842), who, like most of the later writers, saw this as proof of contact between Limoges and Spain. While such contact is undeniable, it is equally true that this decoration, often used in Limoges work (see cats. 45, 70, 77), has been long established in Aquitaine, where it could be found before 1034 in the manuscripts of Adémar of Chabannes and before 1072 in the Apocalypse of Saint Severus.[11]

Likewise, it is the decoration of manuscripts of the late eleventh and early twelfth centuries that is evoked by the alternating colored rectangles and engraved motifs (circles or broken rods) on the bands defining the lozenges on the coupe. This decorative approach recalls that of the uprights on many ornamental capital letters and columns of canon tables, for example, in the second Bible of Saint Martial and the Bible of Saint-Yrieix.[12]

The exceptional historical interest of the work should not overshadow its incomparable artistic qualities, which display, as Viollet-le-Duc (1871) noted, the full range of techniques used by Limousin artists at the height of their perfection.

The ciborium of Master Alpais is thus, incontestably, the most beautiful of the rare objects created fully in the round that have come down to us from the Limousin workshops. The perfection of its form and proportions, the beauty of the hammerwork in creating, on the obverse and reverse, the lozenge pattern within which the half-length figures are inscribed—all are without equal in Limousin production. The subtle play of color is coupled with the quality of the metalwork: the ground, alternating lapis blue and medium blue, is enlivened with red touches sprinkled among the foliage, while graduated tones of green and yellow illuminate the clouds from which the figures emerge. The champlevé enamel and the engraving are handled with a very precise dexterity, to indicate, for example, the jagged contour of the leaves. The figures on the two cups are engraved and furnished with appliqué heads, the angels on the knop on top are in half-relief repoussé, and the small figures and animals on the foot are done in repoussé and pierced. All display extremely high-quality workmanship.

Corresponding to this difference in techniques, there is also a difference in style: the engraved figures on the two cups are among the finest expressions of the last phase of Romanesque art in the oeuvre of Limoges, while the figures and birds inscribed in the foliage on the foot, as well as the very beautiful engraved angel giving his blessing at the bottom of the lower cup, are striking examples of the Limousin artists' acceptance of the full forms and search for naturalness typical of the "1200 style"—a major, early testimony to the southern diffusion of this trend, which rivals the best northern examples in its subtlety and ease. It is revealing to observe how subtly the artist has varied his style: most of the heads on both cups display the stylization characteristic of "classical heads" like those found on the chasse of

Ambazac and many other works from the late twelfth and early thirteenth centuries (cats. 55, 39–44), but they are distinguished by a softening of contour and expression. This tendency is found, accentuated, on the full and peaceful faces in high-relief repoussé of the figures on the foot and the angels on the knop and, even more, on the figure of the angel engraved inside the lower cup, to whom this roundness lends a youthful, almost childish air. Furthermore, while the drapery on the engraved busts is still related to linear Romanesque tendencies, the supple curves drawn in certain details—such as the mantle held under the book of one of the young blessing figures, on the lower cup, and that of the angel engraved on the bottom of this same cup—serve as a prelude to the deeply furrowed folds of the robes of the figures enclosed in the scrolls on the foot. These figures, the most innovative by reason of their Gothic manner, nonetheless preserve the dynamic postures and vehement gestures that Limoges work shared with the Romanesque art of western France.

It is difficult to identify the precise sources of the stylistic mutation of which this object offers striking evidence. A similar combination of forms borrowed directly from early Gothic art with those undeniably surviving from Romanesque art (notably the vehement gestures) can be found, however, in the art of central and western France between 1200 and 1215. Examples are in the stained-glass windows of the cathedral at Poitiers and the church of the Trinity at Vendôme, where the sweet, round faces of the angels swinging censers near the Virgin are not unrelated to those engraved or worked in repoussé on the ciborium of Master Alpais.[13] It is with such works, more than with the creations of the Île-de-France or more northern regions, that Master Alpais shares a real commonality of style. Such parallels also allow us to confirm the date of about 1200 suggested by the place of the work in the evolution of Limousin production and by the few texts that might indicate the presence of its maker in Limoges.

It should be noted that there are several copies or imitations of this famous work, which was illustrated many times, beginning in the mid-nineteenth century. The Victoria and Albert Museum has two of them, acquired in 1880 and 1913, respectively.[14]

ET-D

NOTES

1. See London 1984, nos. 278–80, 309; Stratford 1984; Skubiszewski 1965; Paris 1965, no. 817, pl. 102.

2. The term *ciborium*, in fact, does not appear before the fourteenth century and remains rare until the sixteenth century (see Braun 1932, pp. 283–84).

3. Skubiszewski 1965, pp. 38–39, ill. Some western cups, probably for secular use, have the same form as ciboria; the best examples are those in the Treasure of Dune (Stockholm Museum; see London 1984, nos. 304–6).

4. The first inventory, dated 1446, mentions it as follows: "Item une coppe diargent doré à porter Corpus Domini" (Item a cup of gilt silver to carry the Body of the Lord) (Sens, Archives Départementales de l'Yonne). Later texts mention the same use. I heartily thank Mme Glattauer, who thoroughly reviewed the Sens archives and was kind enough to share with me the unpublished documents that she found.

5. A. Du Sommerard (1838–46, vol 3) considered the name to be of Greek origin. Others, Ardant (1855) the first among them, countered that it was found in the Limousin.

6. With the exception of Hildburgh (1936), who tried to overlook all the elements that might contradict his thesis of a Spanish origin for champlevé enamel—and contrary to the interpretation proposed for the inscription on the chasse of Mozac (cat. 45), in which the subject of *fecit* is the abbot of Mozac.

7. *Registres du consulat de Limoges*, extracts noted by Guibert and published several times (Linas 1883; Rupin 1890); edited as a whole by Chabaneau in 1895: see fol. 89r, no. 253, p. 102: *J. Alpais ha sobre la frairescha W. Alpais, so fair MCCXVI* (J. Alpais has in his fraternity, W. Alpais, his brother, 1216); and fol. 87r, no. 230, p. 96: *lo cuminals de lemotges a x. sl. redens en claus qui fo G. Alpais* (the community of Limoges owns ten sous of income from a land that first belonged to G. Alpais). The *Rentes de la Confrérie des suaires*, published by Chabaneau as an appendix, mentions also *Eu Solar G. Alpais deu gras deu cairoi*, p. 256, and *VI d. red. per don G. Alpais*, p. 258 (six deniers of income granted by G. Alpais), which proves both the wealth and the generosity of this person.

8. Rupin 1890, p. 122.

9. See Darcel 1854 and many later publications. Abbé Barraud (1858) suggested, on the other hand, that they were priests and clerics.

10. Father P.-M. Gy to the writer; I am deeply grateful for his precise details. See also Madec 1975.

11. Leyden, Bibliotheek der Rijksuniversiteit Leiden, ms. Voss. lat. 8. 15, fol. 210v; see Gaborit-Chopin 1968, fig. 36, p. 213; Paris, BNF, ms. lat. 8878, from fol. 1; see Erdmann 1953, pp. 468ff.

12. Paris, BNF, ms. lat. 8, fol. 136v; town hall of Saint-Yrieix, fol. 75; see Gaborit-Chopin 1969, figs. 199, 220.

13. Grodecki and Brisac 1984, pp. 55–57, fig. 62, p. 74.

14. Information from M. Campbell, whom I thank warmly.

EXHIBITIONS: London 1932, no. 572f.; Limoges 1948, no. 85; Vatican City 1963, no. 105; Paris 1968b, no. 385; Paris 1970, no. 67; Leningrad–Moscow 1980, no. 33.

BIBLIOGRAPHY: Dussieux 1841, p. 49; A. Du Sommerard 1838–46, vol. 3 (1841), p. 219, and *Atlas*, chap. 14, pl. III, fig. 3; Longpérier 1842, pp. 154–55; Texier 1842, p. 155; L. Laborde 1852, no. 31; Darcel 1854; Ardant 1855, pp. 82–83; Labarte 1856, pp. 57, 211; Texier 1857, cols. 85–96; Barraud 1858, pp. 573–75, 596–98, fig. p. 575; Didron 1859, p. 167, fig. 131; Labarte 1865, vol. 3, pp. 692–93; Viollet-le-Duc 1871–72, pp. 222–24, pl. XIX; Barbier de Montault 1881, pp. 153–54; Linas 1883, p. 56–58; Guibert 1885, p. 67; Linas 1885a, pp. 112–13; Molinier 1885, pp. 7–8; Corblet 1885b, p. 325; Longpérier 1842; Rohault de Fleury 1883–89, vol. 5(1887), pp. 84–85; Rupin 1890, pp. 119–22, pl. XVI, figs. 183–86; Darcel 1891, no. 125; Molinier 1891; Havard 1896, p. 286, pl. XVI; Molinier 1901, p. 191, ill.; Rosenberg 1907, p. 19; Marquet de Vasselot 1914, no. 67; Benoît 1927; Burger 1930, p. 110; Braun 1932, pp. 314–17; Ross 1934, pp. 88–89; Clouzot 1934, pp. 56–57, ill.; Hildburgh 1936, pp. 15–17; Marquet de Vasselot 1941, pp. 121, 141, 160–61, pl. XXVIII; Evans 1948, pl. 49; Boinet 1948, p. 81, ill.; Gauthier 1950, pp. 25ff., frontispiece; Delage 1950; Marquet de Vasselot 1952, p. 73; Morel 1952; Thoby 1953, pp. 65–73; Erdmann 1953, no. 67, p. 486; Gauthier 1958b, p. 69; Skubiszewski 1965, pp. 17–18, fig. 9; Gauthier 1967b, p. 144, n. 16; Gauthier 1972, no. 63, pp. 109, 112; Gauthier 1976a, pp. 177, 185, 191; Landais 1976, pp. 128–30, fig. 12; Gaborit-Chopin 1983b, pp. 328–30, fig. 294.

# 70. Two Tabernacles

a. Limoges, ca. 1200–1210

Copper (plaques): engraved, scraped, stippled, and gilt; (appliqués): repoussé, chased, engraved, scraped, and gilt; champlevé enamel: dark, medium, and light blue; turquoise, dark and light green, yellow, red, and white

34.5 x 15.2 cm (13⅝ x 6 in.)

INSCRIBED: ih[esu]s xps (Christus)

EX COLL.: Antonin Personnaz, Bayonne and Paris (before 1889)

CONDITION: Some loss to enamel; serpents and rock crystal on roof added in nineteenth century; core replaced.

Musée du Louvre, Paris; Bequest of Antonin Personnaz, 1936 (OA 8984)

b. Limoges, ca. 1200–1210

Copper (plaques): engraved, scraped, stippled, and gilt; (appliqués): repoussé, chased, engraved, scraped, and gilt; champlevé enamel: blue-black, dark, medium, and light blue; turquoise, dark and medium green, yellow, red, and white; wood core, painted red

36.2 x 15.5 cm (14¼ x 6⅛ in.)

INSCRIBED: ih[esu]s xps (Christus)

PROVENANCE: Found in Bethlehem Square, Prague, in the nineteenth century, on the site of the Romanesque church of Saints Philip and James. (Information kindly received from Dr. Dana Stehlíková, Úměleckoprùmyslové Muzeum, Prague.)

EX COLLS.: J. Neuberg; Baron Adalbert von Lanna, Prague (purchased 1885, sold, Berlin, Nov. 9–16, 1909, no. 87, taf. 1, 2); George Blumenthal, New York

CONDITION: Roof finials missing from top and corners; some repair to enamel of Saint Peter plaque; core partly modern.

The Metropolitan Museum of Art, New York; Gift of George Blumenthal, 1941 (41.100.184)

Both tabernacles present the same subjects: the Crucifixion is on the principal face. Set immediately above is the figure of Christ Blessing, standing on a cloud and flanked by

Fig. 70a. Tabernacle from Schloss Tüssling. Bayerisches Nationalmuseum, Munich

angels, in what is probably an abbreviated representation of the Ascension. The Holy Women at Christ's tomb appear on the proper left face; the Virgin (as Ecclesia, personification of the Church) is enthroned on the proper right face, as is Saint Peter on the door face at the back. Angels in roundels occupy the three other roof plaques. The heads of the figures on the principal faces are applied, as is the entire body of Christ on both tabernacles. Each of the figures on the other faces of the Metropolitan tabernacle is reserved; the heads of the figures on the Louvre example are applied on each face. There are further subtle variations of design and execution between the two objects; they are not counterparts. The Louvre tabernacle has the arms of censing angels on the Crucifixion plaque and a pseudo-Kufic inscription band. The Metropolitan tabernacle has a pseudo-Kufic band on the plaque with the image of the Virgin enthroned. The shape of the roof plaques, with the bottom corners of the triangles cut away, suggests that finials were intended at each corner as well as at the top. However, the original format of these is unknown. When it was in the Lanna Collection, the Metropolitan tabernacle had finials of later date composed of four superimposed, flattened spheres. These were removed prior to its acquisition by the Museum.

Overall, the quality of the engraving and of the appliqués on the Louvre example is superior to those of the Metropolitan's. Both are part of a small group that is remarkably similar in size, composition, and technique. Noteworthy among these are examples in the Bayerisches Nationalmuseum, Munich (fig. 70a), recorded at Schloss Tüssling in Bavaria in 1688;[1] from La Voûte-Chilhac (Haute-Loire) now in the Bargello, Florence (fig. 70b);[2] elements in the treasury of Saint-Riquier, near Abbeville (Somme) in the diocese of Amiens;[3] and at San Sepolcro, Barletta,[4] as well as two complete examples of lesser quality in the Metropolitan and the Louvre.[5] All are close—but not identical—to the examples under consideration here. The production techniques seen in these works are similar to those of other liturgical objects created at about the same time, such as book-cover plaques and containers for holy oils. They have a common iconographic scheme, yet the workmanship is unique.

The surviving tabernacles have diverse provenances; clearly, they were in use not only in France but also in Italy, Bavaria, Bohemia, and Spain.[6] However, none has an established history sufficient to provide a basis for dating. Marquet de Vasselot compared the scene of the Holy Women at the tomb with the same scene illustrated in the Psalter of Saint Louis and Blanche of Castile[7] and in the

a. The Virgin

a. The Crucifixion

a. Holy Women at the Tomb

a. Saint Peter

b. Saint Peter

Fig. 70*b*. Tabernacle. Church of
La Voûte-Chilhac, Haute-Loire (after
Aymard and Malègue 1857, pl. 14)

Bible Moralisée,[8] suggesting, consequently, a date after 1240 for the tabernacles. These manuscript comparisons demonstrate, rather, conventions for the representation of the Easter story that cannot be used to date the tabernacles. A number of compositional aspects of the scene were already established locally in the illumination of the same scene in the Sacramentary of the cathedral of Saint-Étienne, Limoges, of about 1100:[9] the placement within an arcade surmounted by paired towers and the attire and grouping of the Three Holy Women opposite the gesturing angel who hovers above the sarcophagus. The approximate chronology of the tabernacles can, moreover, be established by comparison with works such as the True Cross reliquary in Toulouse (cat. 40), which shares a similar palette and decoration with quatrefoils inscribed in circles. The representation of the Three Holy Women at the tomb of Jesus on Easter Sunday can be compared with the same scene on the True Cross reliquary, particularly in details such as the enameling of the sarcophagus. The rich carving of the engraved surfaces, the abundant stippling, and the varied palettes suggest a date not much later than that of the Toulouse reliquary and contemporaneous with the Saint-Viance chrismatory (cat. 71). The Bargello and Louvre tabernacles are arguably the earliest, followed by the Metropolitan Museum example; the one in Munich is distinct.

On the tabernacles, the narrative of the visit to the tomb is compressed, and the figures of the soldiers are omitted altogether. Two of the women carry unguent jars, which they intended to use to anoint the body of Christ, a narrative detail in the Gospel of Mark (16:1) that was commonly included in medieval theatrical presentations of the Passion.[10] The metaphoric association of the tomb with the altar of a church is emphasized by the sanctuary lamp suspended over it and perhaps by the columnar supports for the sarcophagus. The use of these enameled containers as tabernacles is inferred both from their subject matter—with its emphasis on the sacrifice of Christ and his bodily resurrection—and their shape.[11] In this context, it should be noted that both text and images of the Bible Moralisée link the scene of the Holy Women at the tomb as recounted in Mark (16:1) with the faithful receiving Communion.

A plaster of a Crucifixion plaque and a roof plaque from a tabernacle preserved at the André workshop, Paris, seem to indicate that at least one tabernacle of this type was restored there in the nineteenth century. The format of the Crucifixion plaque is distinct, however, from both the Metropolitan and Louvre examples.

BDB

NOTES

1. Munich 1992, no. 6, pp. 110–15.

2. While still in the church of La Voûte-Chilhac, the tabernacle was published by Aymard and Malégue (1857, pl. 14). It probably belonged to the Cluniac priory, La Voûte, founded in 1025. See Cottineau 1935–39,

252

b. The Crucifixion

b. The Virgin

b. Holy Women at the Tomb

vol. 2, col. 3422.

3. Photo in the departmental archives, Département des Objets d'art, Musée du Louvre.

4. Braun 1924, p. 624 and pl. 344.

5. MMA 41.100.185; Louvre OA 8103.

6. The Louvre tabernacle has a putative Spanish provenance; Personnaz said he bought it from a dealer who had been traveling in Spain.

7. Martin 1909, pl. XXXI.

8. A. Laborde 1913, vol. 3, pl. 530.

9. Paris, BNF, ms. lat. 9438; illustrated in color in Paris 1993, p. 233.

10. See, for example, Young 1933, vol. 1, p. 350. In the Gospel of Mark, the three women are Mary Magdalen, Mary the mother of James, and Salome.

11. Marquet de Vasselot (1938, p. 130, n. 1) rejected the suggestion that the Munich example be considered a tabernacle; Braun 1924, pp. 624–25, pl. 344.

EXHIBITIONS (a): Paris 1889, no. 463; Vatican City 1963, no. 51, pl. XXI; (b): New York 1943; New York 1970, no. 159; Tokyo and Kyoto 1972, no. 49; Moscow and Leningrad 1990, no. 42.

BIBLIOGRAPHY (b): Rubinstein-Bloch 1926, vol. 3, pl. XIV; Marquet de Vasselot 1938, no. 2; Gauthier 1972a, fig. 3 (with correct label information under fig. 7); Špaček 1971, no. 17; Gauthier 1972b, no. 66; Gauthier 1978c, p. 29, fig. 11; (a): Rupin 1890, p. 424; Marquet de Vasselot 1938; Fontaine 1937, pp. 101–3; Erdmann 1953, p. 482, no. 47; Foucart-Borville 1990, p. 362, fig. 9.

# 71. Chrismatory

Limoges, ca. 1200

Copper: engraved, chased, and gilt; champlevé enamel: midnight, lapis, and lavender blue; turquoise, dark and light green, yellow, red, and white

14.4 × 11.8 × 9.2 cm (5⅝ × 4⅝ × 3⅝ in.)

PROVENANCE: Mentioned in 1880 as being in church at Saint-Viance. Classified, Monuments Historiques, 1891.

CONDITION: Wear to gilding.

Church of Saint-Viance, Saint-Viance (Corrèze)

It is exceptional that this small box still has within it the gilt-copper plaque pierced with three circles to hold small receptacles of chrism, oil for catechumens, and oil for exorcists.

Anointing with holy oils during baptism is mentioned by the first Christian writers, notably Tertullian in the second century and Hippolytus of Rome in the early third century.[1] Blessed by the bishop on Holy Thursday, the oils were also used for other sacraments during the Middle Ages—chiefly the confirmation and ordination of priests, as well as the consecration of churches. Vessels containing holy oils were among the gifts given by Pope Innocent I (r. 401–17) to the new Roman basilica dedicated to Saint Gervase and Saint Protase: *Vasum ad oleum crismae argenteum. . . . Vas alium ad oleum exorcidiatum.*[2] It seems that for a long time holy oils were kept thus in separate vials (known in Latin as *ampullae*, *phyala*, or *vasa*), usually made of silver like those left by Bishop William III (d. 1248) to the cathedral of Notre-Dame, Paris.[3] The first mentions of small boxes to

keep these vials together appear in the late twelfth and early thirteenth centuries—for example, in the inventory of Saint Paul's Cathedral, London, dated 1295, among the gifts of Bishop Gilbert (r. 1163–87): *Crismatorium Gilberti episcopi, interius ligneum est coopertum exterius foliis argenteis cum ymaginibus elevatis.*[4] A more explicit mention was made among the gifts presented by Bishop Stephen Tempier (r. 1268–79) to the cathedral of Notre-Dame in Paris: *Parvum scrinium cum vase argenti, in quo reponitur crisma.*[5] However, very few individual vials or boxes to contain them have down to us.

The little group of small chrismatories made by Limousin workshops—M. C. Ross (1942) mentioned about a dozen, to which were added a few preserved in public and private collections[6]—is therefore all the more interesting. The example at Saint-Viance takes top priority: its excellent condition—which includes the interior metal plate, pierced by three circles, that attests to its usage—and its very fine quality rival the most beautiful pieces of Limousin enamel-work. It is adorned on four sides with the busts of eight apostles emerging from clouds inscribed in medallions and on the roof with four busts of angels with outspread wings. The very wide range of colors includes, notably, two greens and four blues. The many shadings within a single cell, the dots strewn across the clouds and the florets, the precise engraving, and the stippled highlights all reveal the artist's dexterity and the care given to the execution.

Rupin (1881) noted similarities of style and iconography between this box and the ciborium of Master Alpais (cat. 69). The association of apostles emerging from clouds and busts of angels, to the exclusion of any other figurative decoration, is sufficiently rare for this iconographic kinship to be emphasized. The drapery of the garments is likewise very similar. But, above all, the appliqué heads in half relief give these two works the "family resemblance" correctly pointed out by Rupin. They share a softening of the contours and a sharpness in the final touches to the engraving that indicate that they were made at about the same time, perhaps in the same workshop.

ET-D

The box open, showing
its compartments

NOTES
1. See Hippolytus of Rome, chap. 21, pp. 80–89; see also Cabrol 1925, cols. 2779–90.
2. "A vessel of silver for the chrism . . . another vessel for the oil of exorcism," quoted in Duchesne 1955 and 1981, vol. 1, p. 220. For chrismatories, see Helleputte 1884 and Schnütgen 1884.
3. *"Phyalas argenteas in quibus reponitur crisma, et etiam alia duo vasa chrismate"* (silver vials for storing chrism, and two other vessels with chrism), quoted in Guérard 1850, p. 38.

4. "A chrismatory of Bishop Gilbert, the interior made of wood, the exterior covered with silver foil with images in relief," quoted in Lehmann-Brockhaus 1956, vol. 2, no. 2734, p. 125.

5. "A small casket with a silver vessel in which chrism is stored," quoted in Guérard 1850, p. 178.

6. In addition to those at the Musée du Louvre in Paris and The Metropolitan Museum of Art in New York, Ross (1942) mentions ones at the Musée des Beaux-Arts, Rouen; the Musée of Guéret (Creuse); the Victoria and Albert Museum, London; the Walters Art Gallery, Baltimore; and the Rütschi and Dzialynska Collections (sale cat., Lucerne, Sept. 5, 1931, no. 16, and Molinier 1903, no. 141, respectively). An example was stolen in 1981 from the Musée Municipal de l'Évêché, Limoges. Chrismatories are also preserved in the Vatican Museum and the State Hermitage Museum in Saint Petersburg (see Malbork 1994, no. 17), and various others have been sold publicly or are in private collections. (See *Sculptures et objets* 1993, no. 5, p. 40.) Since most of these objects are now without any interior device, their use as boxes for holy oils—suggested by their form and decoration—can only be inferred, not confirmed.

EXHIBITIONS: Tulle 1887, no. 68; Paris 1965, no. 408, pl. 55.

BIBLIOGRAPHY: Rupin 1881a, pp. 247–51; Rupin 1881b, ill.; Molinier 1887a, no. 68, pp. 531–32; Rupin 1890, pp. 444–45, fig. 493; Forot 1913; Ross 1942, pp. 341–42; Pérol 1948, pp. 90–91.

# 72. Chrismatory

Limoges, ca. 1200–1220

Copper: engraved, chased, and gilt; champlevé enamel: lapis and
lavender blue, turquoise, light and dark green, red, and white

9.9 (8.6 without feet) × 14.2 × 9.6 cm (3⅞ [3⅜] × 5⅝ × 3¾ in.)

Ex coll.: [Brimo de Laroussilhe, Paris]; J. Pierpont Morgan,
London and New York

Condition: Feet, hinges, and lock are modern. Plaque on bottom
perhaps also replaced. Crowning element missing. A few marks on
enamel, mostly on principal face: at lower left and at top near center
on body, at lower left on roof.

The Metropolitan Museum of Art, New York; Gift of J. Pierpont
Morgan, 1917 (17.190.853)

This small box is made of copper plaques assembled with
rivets—four plaques for the body, with just one for the roof.
There is no wood core. The underside of the roof is entirely
gilded; only the upper part of the rectangular base was gilded, suggesting that the interior originally had a pierced plate
similar to the one on the chrismatory from Saint-Viance
(cat. 71). M. C. Ross (1942), who compared the two objects
in great detail, concluded that they obviously came from the
same workshop.[1]

The enameled ornamentation here is in the form of busts
of angels emerging from clouds inscribed in medallions on
alternating lavender blue and turquoise backgrounds except
on the roof, where all the grounds are turquoise but the
medallions are alternating squares and circles. Four angels
each carry a book, two a crown. The squares and circles
stand out on a ground of lapis blue enamel sprinkled with
rosettes. The variety of colors, the graduated hues of the
clouds and rosettes, and the considerable refining of the
decorative motifs of the parts in reserve by engraving and
stippled chasing bear witness to the care expended on the
execution of the object. The characteristics of this
decoration—the engraving of the busts of angels and especially their "classical" appliqué heads—place this small box
among the works of essentially Romanesque style executed
in Limoges at the very end of the twelfth century and the
first decades of the thirteenth. The rapid stylization of the
busts of angels through the use of vertical lines suggests the
early thirteenth century as the more likely date of execution.
The simplification noticeable in the iconography as well as
in the engraving justifies a date slightly later than that of the
chrismatory from Saint-Viance, mentioned above.

ET-D

Note
1. Ross 1942, pp. 342–44.

# 73. Chrismatory

Limoges, ca. 1210–20

Copper: engraved, chased, and gilt; champlevé enamel: lapis and lavender blue, turquoise, green, yellow, red, and white

12.7 × 10.5 × 6.8 cm (5 × 4⅛ × 2⅝ in.)

PROVENANCE: From the church of Neuville (Corrèze), where it was mentioned in 1887. Classified, Monuments Historiques, 1891. Sold by the commune of Neuville in 1897 (erroneously cited in several later publications as still located in Neuville).

EX COLL.: The Marquise Marie Arconati-Visconti (1900). Gift of the Marquise Arconati-Visconti, 1916.

CONDITION: Modern crest.

Musée du Louvre, Paris; Département des Objets d'art (OA 6935)

The chrismatory that comes from Neuville varies in several respects from the one preserved in Saint-Viance (cat. 71). The form is slightly different, the roof with four pitches being replaced by one with two pitches, which gives the work the same shape as a chasse. The box is made, however, only of enameled copper plaques, with no wood core. The interior device for holding the three vials of oil has disappeared except for two parallel shafts on the smaller sides that must have anchored a plaque like the one on the box in Saint-Viance.

The decoration here is more sober, made up solely of busts of angels—a favorite theme of Limousin enamelers for decorating liturgical objects (see cats. 69, 74). Its treatment here is of excellent quality; indeed, M. C. Ross regarded this object as the most beautiful of all those that have come down to us.[1] The six half-length angels emerging from clouds—two on the principal face, two on the roof, and one on each of the small sides—are inscribed in circular medallions except the two on the front face, which are placed in mandorlas. The backgrounds are strewn with rosettes; on the back, they are inscribed in a lozenge grid. The vegetal decoration appears only on the small sides of the roof, each side adorned with a huge trefoil, which, although more sketchy, resembles those in the same place on the chasse of the Holy Innocents from Monflanquin (cat. 41).

The angels on the sides are engraved, and the four on the front have, in addition, "classical" heads in half relief. Two are crowned, a peculiarity that may point to rather hasty execution in which the goldsmith used available elements without giving much thought to how well they fit into the design.

The engraving of the busts confirms the attribution of this object to Limousin production of the early thirteenth century, when Romanesque traditions still flourished, hardly diminished by contacts with the nascent Gothic art (see cat. 69).

ET-D

NOTE
1. Ross 1942, p. 342.

EXHIBITIONS: Tulle 1887, no. 69; Paris 1889, no. 553; Paris 1900, no. 2586; Vatican City 1963, no. 68.

BIBLIOGRAPHY: Guibert 1877c, p. 417; Molinier 1887a, no. 69, p. 532; Rupin 1890, pp. 445–46, fig. 494; Poulbrière 1894–99, vol. 2(1899), p. 362; Forot 1913; Marquet de Vasselot 1917, no. 92, pl. XXXVIII; Migeon 1918, p. 12; Ross 1942, pp. 341–43; Pérol 1948, p. 87.

# 74. Pyx

Limoges, ca. 1200–1210

Copper: engraved and gilt; champlevé enamel: midnight, lapis, and lavender blue; dark and light green, yellow, brick red, and white

H.: 9.3 cm (3⅝ in.); (without cross: 6.1 cm [2⅜ in.]); diam.: 6.4 cm (2½ in.)

Ex colls.: Charles de Meixmoron, Nancy (1895; see old label inside object); Claudius Côte, Lyons. Bequest of Madame Claudius Côte, 1960

Condition: Small cross on top of object undoubtedly a later addition.

Musée du Louvre, Paris; Département des Objets d'art (OA 10027)

Inventory references and examples preserved to this day attest to the fact that pyxes were among the objects most frequently made by Limousin workshops. Although the Latin word *pyxis*, which comes from the Greek and denotes vessels of wood, is used in medieval inventories to refer to various kinds of boxes,[1] the term was commonly associated with the boxes in which the consecrated wafers were kept for the sacrament of the Eucharist. It is likely that, as early as the ninth century, pyxes containing the body of Christ (*cum corpore Domini*) were placed on the altar and used for the Communion of the sick.[2] The first mention of pyxes of Limoges work used in the Eucharist is by William of Salisbury, about 1220.[3] The instructions given by Bishop William of Blois to the council of Worcester in 1229 concerning the liturgical objects needed by a church called for "two pyxes, one in silver, ivory, or Limoges work, or any other pyx to hold the consecrated wafers, to be furnished with a key and kept well protected."[4] About a dozen years later, when Bishop Walter of Worcester presided over a synod, an allusion was made to "two pyxes . . . of Limoges work in which the Eucharistic wafers are preserved."[5] It was recommended that there be two pyxes, one for the wafers before their consecration, the other for those distributed after the celebration of the sacrament.[6]

None of these texts describes the appearance or size of the pyxes. Nonetheless, the cylindrical bodies and conical covers common to the lids of Limoges examples that survive seem to be the traditional forms for Eucharistic use and recall the mention by Gregory of Tour of a deacon carrying a round vessel containing Eucharistic wafers.[7] Furthermore, the decoration of some pyxes reflects this function.

The body of the pyx was made from a strip of copper, hammered to form a circular band, then soldered with a vertical seam, and finally affixed to a flat, circular base. The cover, made from a single piece of hammered metal often capped by a separately worked element, was attached to the back of the body by a hinge. The interior of the pyx was usually gilded.

The pyx from the former Côte Collection is among the oldest and most beautiful examples. It is furnished on the inside with a cupule of gilt copper (see cat. 139). Moreover, it is part of the largest group of pyxes adorned with busts of angels separated by rinceau decoration. Bust-length and inscribed in medallions, the six angels here are placed three on the body of the pyx and three on the roof. Emerging from clouds, they are entirely engraved on the cylindrical body, smaller and furnished with appliqué heads on the roof. The wide range of colors includes in particular two greens and three blues. The midnight blue recurs on a certain number of works executed at the very end of the twelfth century and the beginning of the thirteenth (cats. 45, 70, 71, 75). The systematic use of shadings in the clouds and the florets confirms the quality of this work and a date of about 1200. The entirely engraved angels decorating the cylindrical body of the object were meticulously executed. Their round faces with tiny features surrounded by curls look like an engraved transcription of the "classical" appliqué heads with softened expressions, about 1200—of which the best example is undoubtedly the ciborium of Master Alpais (cat. 69). It is no exaggeration to say that the angels on the Louvre pyx are among the loveliest ever produced by Limousin artists.

BDB and ET-D

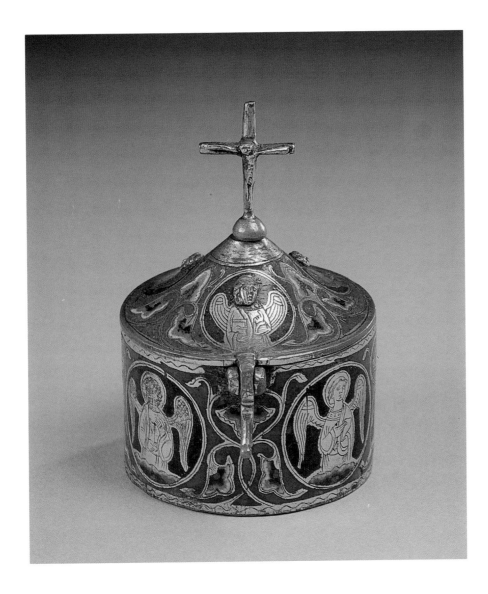

NOTES

1. See, for example, the inventories of the popes at Avignon in which the word *pyxis* is used to describe vessels containing chrism, a finger of Saint Barbara, relics, and the blood of Christ. Hoberg 1944, pp. 248, 96, 58, 96, 127, 157.

2. "*Super altare nihil ponatur nisi: capsae, et reliquiae, et quatuor Evangelia, et pixis cum corpore Domini ad viaticum infirmis. . . .*" This reference appears in the *Decretals* of Pope Leo IV (r. 847–855); see *PL*, 115, col. 677.

3. "*Pixis dependens supra altare cum Eucharistia, de opere Lemovicensi*" (Pyxes hanging above the altar, containing the Eucharist, of Limoges work), seen on a visit to Hurst (Berkshire) by William of Wanda, bishop of Salisbury beginning in 1210; see Lehmann-Brockhaus 1955–60, vol. 1, no. 2170, p. 577.

4. "*Due pixides, una argentea vel eburnea bel de opere Lemovitico vel alia ydonea in qua hostie reserventur et sub fideli custodia clavi adhibita conserventur,*" in *Concils and Synods*, vol. 2, p. 171.

5. "*Due pixides . . . de opere lemovicino, in quo hostiae conserventur.*" See Du Cange 1845–50, vol. 4, p. 119. The Latin text is variously given. See also Rohault de Fleury 1883–89, vol. 5, p. 90.

6. See Rupin 1890, p. 206, nos. 3, 4. On pyxes in general, see Barraud 1858; Corblet 1885a, pp. 379–90; Rohault de Fleury 1883–89, vol. 5, pp. 89–95; Rupin 1890, pp. 201–14; Braun 1932, p. 338; King 1965, pp. 112–20; Rubin 1991, pp. 44–47.

7. "*Acceptaque turre, diaconus, in qua mysterium Domini corporis habetur, . . . ingressusque templum, ut eam altari superponeret. . . .*" Gregory of Tours, *Glory of the Martyrs* 1988, no. 85, p. 110.

EXHIBITIONS: Vatican City 1963, no. 130; Mexico City 1993, p. 82, fig. 30.

BIBLIOGRAPHY: Landais 1961a, p. 132.

# 75. Chasse of Saint-Marcel

Limoges, ca. 1200–1215

Copper (plaques): engraved, scraped, stippled, and gilt; (appliqués): repoussé, engraved, chased, scraped, and gilt; (heads): stamped, engraved, and gilt; champlevé enamel: dark, medium, and light blue; turquoise, dark and light green, yellow, red, and white; glass and rock-crystal cabochons; wood core

42.4 (with crest; 36.4 without) x 43.3 x 17.3 cm (16¹¹⁄₁₆ with crest [14⅜ without] x 17⁵⁄₁₆ x 6¹³⁄₁₆ in.)

PROVENANCE: Church of Saint-Marcel; recorded during the French Revolution in an inventory of August 6, 1790 ("une châsse contenant les reliques de saint marcel et saint anastase en cuivre");[1] hidden subsequently, with other objects from the treasury; returned 1802. Classified, Monuments Historiques, 1897

CONDITION: Restored between 1867 and 1898; cleaned 1958 by M. Toulouse; one enameled band lost from proper right side; almost complete loss of enamel to lock plate with roundel of angel on proper left side; two finials replaced on ends of crest; wood core replaced; wear to gilding; some copper corrosion.

Church of Saint-Marcel, Saint-Marcel (Indre)

On the principal face, figures in gilt copper with glass eyes are applied to enameled quatrefoil plaques which, in turn, are applied to the sheets of gilt copper that cover the wood core of this large chasse. Glass cabochons in floret mounts punctuate the surface of the gilt copper, and enameled strips decorated with a repeating pattern of quatrefoils enclosed in circles frame the edges of the chasse and the individual figures. At the center of the lower face, the crowned Virgin holds the Child Jesus on her lap and a lily in her right hand. She is flanked by two seated apostles holding books. On the roof, Christ in Majesty appears, holding a book and raising his right hand in blessing. Like the Virgin, he is flanked by apostles. Each of the figures of the principal face sits on a pillow reserved in gilt copper. On the reverse are six more figures, each seated on the arc of a rainbow with his feet on a cloud. Each figure is reserved in gilt copper against a field of enamel decorated with circles; only the heads are applied. On each of the two ends are two superimposed roundels of three-quarter-length angels in gilt copper. Their heads are separately worked and applied, and each appears above a cloud and holds a book. The upper roundel on the proper left end is cut in half to allow the opening of the hinged door that gives access to the relics. The body of the angel is pierced by a keyhole, which is nonfunctional, since the core has been replaced. The chasse retains its original crest set with rock-crystal cabochons alternating with small enameled towers on the principal face and enameled quatrefoils on the reverse. Three of the ball-shaped finials remain in place. A large iron loop, probably intended to secure the chasse to an altar, is affixed to the bottom.

The chasse is recorded as containing relics of Saint Marcel, patron of the church, and of his companion

Anastasius. According to his legend, Saint Marcel was a Roman of the time of Emperor Marcus Aurelius. Traveling to Gaul in the company of Anastasius to meet their brothers who had preceded them, they came to the Roman city of Argentomagus. Here they cured the blind, deaf, mute, and lame child of an old widow, thereupon converting her to Christianity. Summoned before the Roman magistrate, Marcel was subjected to various forms of torture, all of which were ineffectual, until finally he was decapitated.

The vita of Saint Marcel was not recorded until 1657, though the presence of his relics was first noted after 1005.[2] In the twelfth century, a Benedictine priory dedicated to Saint Marcel was established at Argenton, a dependency of Saint-Gildas-de-Châteauroux.[3] Still, there is little precise information concerning the veneration of Saint Marcel during the Middle Ages.[4] Nor is the decoration of the chasse specific to the saint's legend. It is not surprising, however, that nearby Limoges should have provided the reliquary for the patron saint of Saint-Marcel, just north of the Limousin.

The chasse of Saint-Marcel is among the most technically accomplished and well-preserved Limoges chasses, remarkable for the range of its palette and its sculptural appliqué figures. Great textural richness is achieved by the various uses of the copper and the subtle variations of the designs in enamel. This variety provides links between this chasse and other Limoges works such as book covers, crosiers, and isolated appliqué figures. The elongated figures of the seated apostles with small appliqué heads are reminiscent of their standing counterparts in the Walters Art Gallery, Baltimore.[5] Several of the appliqué heads of the angels on the end panels and the apostles on the reverse seem to have been struck from the same die. By the end of the twelfth century, this technique was already in use at Limoges workshops on works such as the Toulouse True Cross reliquary (cat. 40). This does not diminish the quality of the chasse (even though in several instances the appliqué head is slightly smaller than the reserved area allotted for it). The chasse of Saint-Marcel represents a high point of Limoges enameling, comparable in the quality of its engraving, its appliqué work, and its palette to the Eucharistic coffret from the treasury of Grandmont, from the collection of the Musée Municipal de Limoges,[6] often attributed to Master Alpais, with which it must be roughly contemporaneous. The Eucharistic coffret similarly presents apostles seated on tufted pillows and poised on the arc of a rainbow. At Limoges, the copper is more deeply incised to define drapery and figure, but whether this is a function of a slightly earlier date is not clear.

BDB

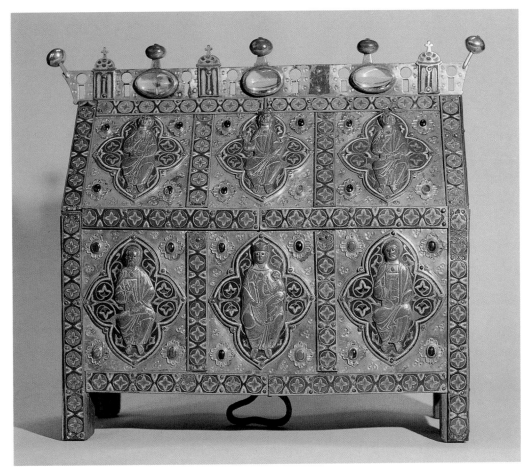

Principal face

Reverse

Left side

NOTES

1. Archives de l'Indre, Q452; cited in Trotignon 1991, p. 43.
2. Trotignon 1991, p. 15.
3. Cottineau 1935–39, vol. 2, col. 2779.
4. Jean, duke of Berry, visited Saint-Marcel to see the relics in 1375. See Trotignon 1991, p. 17.
5. See New York 1970, no. 161, p. 155.
6. No. 267. Stolen in 1981 and not recovered. See New York 1970, no. 149, pp. 143–44.

EXHIBITIONS: Limoges 1948, no. 80, p. 54, pl. VI, fig. 37; Vatican City 1963, no. 99; Paris 1965, no. 202, pp. 104–5.

BIBLIOGRAPHY: Rupin 1890, pp. 339–40, fig. 407; Gauthier 1950, pl. 43, p. 157; Trotignon 1991, pp. 8–23, figs. 3–7.

# 76. Chasse: Burial of a Saint

Limoges, ca. 1200–1210

Copper: engraved, chased, and gilt; champlevé enamel: midnight, lapis, and lavender blue; turquoise, green, yellow, brick red, garnet red, and white; wood core

15.5 × 12.4 × 9.2 cm (6⅛ × 4⅞ × 3⅝ in.)

Ex coll.: Alexandre-Charles Sauvageot, Paris. Gift of Alexandre-Charles Sauvageot, 1856.

Condition: Enamel and gilding very well preserved. Wood core and closing apparatus are old. Finials at ends of crest are undoubtedly modern.

Musée du Louvre, Paris; Département des Objets d'art (OA 940)

This lovely chasse from the former Sauvageot Collection is a work set apart by its somewhat enigmatic iconography and further distinguished by rare chromatic harmonies and singular decoration. The principal face seems to represent the burial of a saint attended by two seated figures with halos, while an angel appears, probably to look for the soul of the deceased. On the roof, two other haloed figures bear a stretcher covered with rich cloth, no doubt used to transport the body before its burial; they are going away after having laid it in the tomb. Rupin (1890) and Darcel (1891) identified the principal scene as the Death of the Virgin, a reasonable hypothesis because the two attendants have halos—which would make them both apostles—but nonetheless surprising inasmuch as the iconography is very different from what is familiar to us in Limoges work (see cat. 78). It is also possible that the scene represents not a burial but rather the elevation of the body of a saint, in which case the two figures shown on the roof would be about to place the body on the richly adorned stretcher. The chasse would thus depict the translation of the relics of the saint. This interpretation might be supported by the parallel example of the frequently reproduced illumination from the *Speculum Grandimontis* (Mirror of Grandmont), now in the Archives Départementales de la Haute-Vienne,[1] in which the elevation of the relics of Saint Stephen of Muret is represented by two bishops bearing the chasse of the saint on a stretcher above the tomb. One might think of the relics of the virgins of Cologne brought to Grandmont in 1181 and later dispersed.[2] The composition of the two scenes seems to have no equal among extant Limousin chasses.

The originality of the work is also evident in the attention paid to the representation of objects that set the scene, especially the stretcher covered with luxurious fabric highlighted with gold motifs. The workmanship, furthermore, is of very high quality. Stippled lines set off the reserved copper framing the principal elements of each scene: the body of the deceased, the tomb, and the cloth covering the

stretcher. The richness of the palette is remarkable—especially its range, notably the garnet red found on a small number of works with vermiculé backgrounds and the midnight blue used by Limousin enamelers in the late twelfth and early thirteenth centuries; the variety of graduated tones on the tomb; and the imaginative plays of yellow on blue and red on turquoise that embellish the fabric of the stretcher.

The back and the two small sides are, on the other hand, of a rare sobriety, totally without images, and simply covered with regular rows of little crosses on a midnight blue ground except on the roof, which has a background of midnight blue and turquoise lozenges.

The drapery and the appliqué heads are characteristic of the "classical" period of the oeuvre of Limoges. The preference for curved lines and the hollowing out of deep folds, particularly on the robes of the two seated figures, here mark the opening of the style to "early Gothic" tendencies. These characteristics, in addition to the remarks made regarding the range of colors, place the execution of the chasse between 1200 and 1210.

ET-D

Notes
1. See, most recently, Paris 1993, no. 85.
2. See Gaborit 1976a, p. 239, and in the present catalogue, p. 199, above.

Exhibition: Vatican City 1963, no. 62.

Bibliography: Sauzay 1861, no. 1109; Lièvre and Sauzay 1863, pl. xxxii; Rupin 1890, p. 372, fig. 432; Darcel 1891, nos. 112–19; Marquet de Vasselot 1914, no. 65.

Reverse

262

Principal face

# 77. Crosier of Aimon of Mollain

Limoges, ca. 1200

Copper: stamped, engraved, and gilt; champlevé enamel: lapis and lavender blue, turquoise, green, yellow, red, and white; openwork copper knop, engraved and gilt

28.2 × 10.9 (11⅛ × 4¼ in.); diam. of knop: 6.55 cm (2½ in.)

PROVENANCE: Discovered in March 1862 approximately three feet below flagstone paving of the chapel of Saint-Nicholas in the abbey church of Saint-Peter of Luxeuil (Haute-Saône) on the site of the tomb of Abbot Aimon of Mollain (see *Registre des délibérations du Conseil municipal de la ville de Luxeuil*, December 3, 1864, and note 6, below). Acquired from the city of Luxeuil, 1864.

CONDITION: Copper and enamel damaged while buried.

Musée du Louvre, Paris; Département des Objets d'art (OA 2023)

The crosier discovered in Luxeuil in 1862 belongs to the most common type of crosiers "with a palmette flower" or "with a large stylized flower"—to use the expression of Marquet de Vasselot (1941). This type is defined by the shape of the flower, composed of three main petals, supple and slender, their tips resting on the outermost curve of the volute and, between them, two much shorter, triangular petals.[1]

This motif obviously derives from the palmette with five petals, of which the Poitiers crosier (cat. 23) is the finest and earliest example. The appearance of this motif in Limousin enamelwork must be dated to about 1200. Indeed, some palmette flowers of similar design can be found in the second volume of the Lyons Bible, which was probably illuminated during the very last years of the twelfth century, about 1190 to 1200.[2]

Other elements justify this date, notably the presence along the volute of a small crest, which was first seen in the twelfth century and developed gradually. Crosiers of this type must have been produced during a relatively long period, however, as indicated by the many extant examples and the diversity of their decorations. Within the group, the Luxeuil crosier is distinguished by several original and refined details that provide a very special charm. Thus, the flower has at its heart a trilobed palmette, which is also found on the small crosiers from the church of Saint-Colombe at Sens, the abbey of Toussaint at Angers (now in the Musée des Beaux-Arts), and the treasury of the church at Monte Cassino;[3] whereas most of the other examples have at the center a motif that is more summary, indeed geometric. While the volute of these crosiers is frequently adorned with a scroll of small trefoil flowers, the one from Luxeuil offers a variant in which scrolls alternate with little horizontal bands of motifs derived from pseudo-Kufic inscriptions.[4]

Likewise, countless crosiers in this group have a cylindrical knop adorned with busts of angels with either engraved or appliqué heads; the knop of the Luxeuil crosier, however, is openwork—a unique example because openwork is ordinarily used for designs of sirens and other fantastic creatures, as on the crosiers of the church of Saint-Maurice at Agaune and the treasury of the cathedral at Troyes.[5]

Finally, on the shaft, the scrolls outline lozenge-shaped compartments in which are inscribed busts of angels that are very beautifully engraved—an arrangement which, though rare, is comparable to the one on the shaft of the crosier of Saint-Maurice of Agaune.

These different decorative aspects place the Luxeuil crosier at the head of the main group of crosiers with palmette flowers and consequently date it to about 1200.

The church at Luxeuil, a celebrated and long-established Benedictine abbey, burned down in 1201 and was reconstructed. The new building was completed about 1330 and dedicated in 1340. The abbots were buried in the apse or the choir—except for Aimon of Mollain (r. 1364–82): two chronicles of the seventeenth and eighteenth centuries mention his sepulchre in the chapel of Saint-Nicholas.[6] The identification of the tomb where this crosier was found seems therefore quite secure. What we have apparently is yet another example of a Limousin crosier being used in the fourteenth century for funerary purposes.[7]

ET-D

NOTES

1. See Marquet de Vasselot 1941, nos. 10, 11, 19, 21, 28.
2. See Cahn 1980; Cahn 1982, no. 75.
3. Marquet de Vasselot 1941, nos. 22, 25, 26.
4. This alternation recurs on several crosiers with a quadrangular volute—for instance, that of Saint Maurice of Agaune; see ibid., no. 11.
5. Ibid., nos. 11, 14; see also nos. 7, 12, 13, 17, 29, 30. For crosiers having a knop decorated with engraved busts of angels, see ibid., nos. 3, 15, 18, 19, 22, 24–26, 28.
6. Dom Placide de Villiers, *Eductum e tenebris Luxovium, seu Chronicon Luxoviense*, 1684; Vesoul, Archives Départementales de la Haute-Saône, IJ 502, fol. 295: . . . *ejus tumuli in Sancti Nicolai oratorio epigrapha* (the inscription on his tomb in the oratory of Saint-Nicholas). Dom Grappin, *Histoire de l'abbaie roialle de Luxeu*, eighteenth century, Besançon, Bibliothèques Municipales, coll. de l'Académie, no. 32. fol. 218: "The abbot of Molans died in Luxeuil on April 20, 1382. He was buried in the chapel of Saint-Nicholas."
7. A similar example is the crosier with a "large flower" discovered in the Troyes Cathedral, in the tomb of Bishop Peter II of Arcis (r. 1378–95), a contemporary of Aimon of Mollain. See Marquet de Vasselot 1941, no. 14, pp. 16–21.

EXHIBITION: Berlin 1989, no. 4/47, fig. 664.

BIBLIOGRAPHY: Nicard 1864; *L'Art pour tous*, no. 593 (Feb. 28, 1864), p. 2472, figs. 5149, 5150; Beauséjour 1891, pp. 35–37, ill.; Darcel 1891, no. 939; Watts 1924, p. 12, pl. v; Marquet de Vasselot 1941, pp. 48–49, nos. 21, 74, pl. XIV; Erdmann 1953, p. 485, no. 62; Dabrowska 1995 (forthcoming), no. 53.

# 78. Death of the Virgin

Limoges, ca. 1200

Copper: engraved, chased, and gilt; champlevé enamel: lapis and lavender blue, dark and medium green, yellow, red, and white

26.2 × 20.2 × .6 cm (10⁵⁄₁₆ × 8 × ¼ in.)

INSCRIBED: : REGINA MUNDI DE TERRIS ET DE (Queen of the world, of the earth, and of . . .) Graffito on the book of one of the apostles: S[ANCTUS] IOA / N[N]ES A[POSTULUS] (Saint John the apostle)

EX COLL.: Pierre Révoil, Paris. Acquired in 1828.

CONDITION: Loss to the gilding and corrosion of the surface of the copper, especially in the central area (robe of the Virgin) and the bottom (the robes of the apostles); denting, probably from the blow of a tool, on the Virgin's mantle. A few missing areas on the enamel, principally in the upper corners.

Musée du Louvre, Paris; Département des Objets d'art (MRR 243)

The Virgin, reposing on a bed covered with swagged drapery, is surrounded by the twelve apostles, standing in the background and forming a horizontal frieze. Two hold candles and two carry books. L. de Laborde (1852) suggested a reading for the graffito engraved on one of the books: *Sanctus Ioannes apostolus*, a convincing interpretation, considering the youthful appearance of the apostle carrying the book, the only one who is unbearded, which would designate him as Saint John. L. de Laborde was also the first to comment on the abrupt interruption of the incomplete inscription, which suggests that another plaque of the same dimensions was originally placed next to this one. The second plaque would have represented the Assumption of the Virgin or her Coronation or Glorification, suggested by her designation as "Queen . . . of the earth" in the inscription. The missing part would have evoked her reign as Queen of Heaven.[1]

Gauthier (1950, 1972b, 1981) compared this plaque with two others preserved in the Fitzwilliam Museum in Cambridge and in the Keir Collection (deposited at the Nelson-Atkins Museum of Art, Kansas City; fig. 78a). Each is pentagonal in form and depicts two crowned saints, seated side by side, bearing palm branches; these works are in fact similar in decoration and style to the Louvre plaque. Gauthier concludes that the three plaques came from the same object, perhaps a chasse with a transept such as those from Pisa and Agrigento.[2] Although the stylistic similarities of the figures and of the rinceau decoration with fleurons are undeniable and sufficiently close to justify an attribution to the same atelier, it seems, on the other hand, difficult to accept that the three plaques are from the same object. Their iconography hardly encourages this assumption, even though it is highly probable, as we have seen, that the plaque from the Louvre was part of a more important ensemble devoted to the Virgin. The dimensions are similar to plaques from the main altar at Grandmont (cat. 57). The condition of the plaques does not justify this hypothesis either; indeed the plaque from the Louvre has traces of corrosion and even a dent, probably from the blow of a tool, which leads one to believe that it had perhaps been buried, whereas the two others are in perfect condition. Lastly, the very particular style of these works was also found on a fourth plaque, in the form of a mandorla, depicting Saint James seated against a ground of rosettes, not of rinceaux, formerly in the Klosterhof Collection in Glienicke Castle, and later deposited in the Kunstgewerbemuseum in Berlin (fig. 78b).[3] Thus it seems more plausible that these plaques are evidence of the activity of a highly individualized atelier, with the Death of the Virgin being its undeniable masterpiece. These works show the fine, supple folds of early Gothic art and manifest, in the calm attitudes and internalized expressions, a "classical" search for idealization uncommon in Limousin work as well as a strong characterization of the faces. They are distinguished by their massive and elongated structure, their long, pronounced noses—slightly upturned—their small, oval eyes lightly engraved on the surface of the faces, and the hair and beards coiffed in regular masses of finely combed locks.

This style is not, however, entirely isolated. A number of features make it similar, notably, to that of the group of works executed for Grandmont in the 1190s. As proof of this, one need only compare the face of the Virgin on the Louvre plaque with her face in the Adoration of the Magi from the main altar, now in the Musée de Cluny (cat. 57b). The faces are engraved here but enameled on the Grandmontain works. The more complete adoption of the supple style of early Gothic art reveals, in the Master of the Death of the Virgin, a greater receptivity to influences from more northern regions and, perhaps, a slightly later date.

The decoration of burgeoning fleurons enclosed within large rinceaux belongs to the series of vegetal "fabrics" on the chasses from Apt (cat. 21) and Ambazac (cat. 55), although the more summary design here reveals, yet again, a certain chronological distance, which justifies the date of about 1200 proposed for the Louvre plaque.

ET-D

Fig. 78*b*. Saint James. Formerly
Kunstgewerbemuseum, Berlin

Fig. 78*a*. Two
saints. Keir
Collection,
Nelson-Atkins
Museum of Art,
Kansas City

NOTES

1. A parallel example of the Death and the Coronation of the Virgin is seen
   on the two tympanums of the south portal of the transept in the cathedral of Strasbourg (ca. 1230); see Sauerländer 1972, pls. 130, 131.
2. Gauthier 1972b, nos. 60, 61, pp. 110–11, 335–36, figs. pp. 104, 105.
3. Zuchold 1993, no. 94, p. 121, fig. 18.

EXHIBITIONS: Vatican City 1963, pl. 40; Cleveland 1967, no. IV, 2.

BIBLIOGRAPHY: L. Laborde 1852, no. 41; Rupin 1890, pp. 148–49, 372,
fig. 224, p. 149; Darcel 1891, no. 82; Havard 1896, p. 292; Marquet de
Vasselot 1914, no. 92; Gauthier 1950, pp. 49, 63, 74, 156, 157, pl.; Lasko
1964, p. 473; Gauthier 1972b, no. 125c, pp. 179–80, 370, fig. p. 179;
Gauthier and François 1981, p. 17.

# 79. Three Figures

Limoges, ca. 1200

Copper: repoussé, chased, engraved, scraped, stippled, and gilt; blue-black glass beads

17.6 x 9.1 x 1.5 cm (6⅞ x 3½ x ⅝ in.)

Ex colls.: Aimé Desmottes, Lille and Paris (sale, March 19–23, 1900, no. 37, ill.); [Brimo de Laroussilhe, Paris]; J. Pierpont Morgan, London and New York

Condition: Gilding worn along some ridges; right foot of foremost figure missing; no restorations.

The Metropolitan Museum of Art, New York; Gift of J. Pierpont Morgan, 1917 (17.190.790)

The three male figures represented on this relief are so tightly placed that only the bust of the central figure is visible. The mantles, knotted over the right shoulders, and the long tunics of the two outer figures are so exquisitely formed and tooled that both the decorative patterns of the drapery folds and the underlying masses of the shoulders, torsos, and legs are clear.

Puffy caps of hair, engraved with multiple parallel dotted lines, surmount the almost fully rounded heads. The beardless central head is attached, by a rivet through the forehead, to a flattened portion of the sheet that forms the rest of the composition. The collar of the central figure and the lower hem of the other beardless figure's robe were crosshatched and stippled with a circular tool. The diagonal lower hem of the bearded figure's mantle is engraved with a row of lozenges alternating with small stippled circles.

The style may be described as transitional, datable to about 1200. The reliefs are shallower and less massive than those seen in three previous entries (cats. 46, 47, 48), and the pleating of the drapery folds dramatically extends the damp-fold style particularly evident in Byzantine and Western manuscript illustrations.[1] More specifically, the figures relate to other works created in central and western France about 1200, especially the stained-glass windows at the cathedral of Poitiers dedicated to Noah and Isaac.[2] The supplicating figure of Alfonso in the Orense series (fig. 51a),[3] with his round head and richly engraved, looping drapery, provides one of the closest parallels in Limoges metalwork.

The clustered figures, bowing toward the left, seem to be witnesses. The upraised hands of one figure imply astonishment if not anguish or lament, whereas the compression of the figures into a single group emphasizes their common reaction. The bowed heads may suggest sorrow and pathos as the figures gaze down upon the larger scene. Their pose, compression, and gaze relate to the apostles grouped around the recumbent Mary in the Death of the Virgin plaque in the Louvre (cat. 78). It is also possible that they come from a scene of the entombment of a saint or of Christ, or from a miracle scene such as the Raising of Lazarus. Although it is only a fragment of a larger composition, this relief has remarkable rhythmic unity. The three rivets used for attachment to the original setting are missing.

WDW

NOTES
1. Wixom 1969, figs. 1–3, and color cover: single leaves from the Phanar Gospels with Saint Matthew and Saint Luke, Byzantine, 1057–63, The Cleveland Museum of Art (42.1512 and 42.1511); Saint Matthew, Pantheon Bible, Rome, about 1100, Vatican Library (lat. 12958); fragmentary leaf from a Bible showing Saint Luke, Burgundy, abbey of Cluny, about 1100, The Cleveland Museum of Art (68.190).
2. Grodecki 1977, pp. 55–56, figs. 44, 46.
3. Hildburgh 1936, p. 119, pl. XXIV, fig. 29b.

BIBLIOGRAPHY: Unpublished, except for the Desmottes sale catalogue cited above.

# 80. Crosier of Bertrand of Malsang

Limoges, ca. 1210–20

Copper: stamped, engraved, and gilt; champlevé enamel: lapis blue, almond green, and brick red; dark and turquoise blue enameled beads

30.2 × 11.6 cm (11⅞ × 5⅝ in.)

INSCRIBED (on both sides, at the top of the volute): AVE MARIA : GR[ATI]A : PLENA (Hail Mary, full of grace)

PROVENANCE: Discovered in August 1799 in the tomb of Abbot Bertrand of Malsang, in the chapel of Notre-Dame-la-Blanche, the abbey church of Montmajour (Bouches-du-Rhône).

EX COLLS.: Jacques-Didier Véran, Arles (in 1817; see *Antiques du cabinet de Mr Véran, notaire à Arles, en 1817*, Arles, Médiathèque, ms. 777, no. 84, fols. 25, 135); Pierre Révoil. Acquired in 1828.

CONDITION: Slight wear to the gilding; small losses to the enamel of the inscription (on both sides); volute slightly dented on the top.

Musée du Louvre, Paris; Département des Objets d'art (MR R 810)

Exhibited in Paris only.

As Benoît (1927) has noted—without, however, giving a precise indication of his source—this crosier was found in the abbey church of Montmajour in August 1799, in the chapel of Notre-Dame-la-Blanche, which was built by Abbot Bertrand of Malsang (r. 1298–1316) to serve as his tomb. Early documentary evidence and the sculpted armorial bearings on the tomb recess attest to the identity of the deceased person buried in the chapel.[1] A sketch of the crosier and a brief mention of its discovery are preserved in the inventory of the collection of Jacques-Didier Véran (1764–1848), a notary of Arles who owned the work in 1817 (fig. 80a). Véran, whose collection was made up mostly of antiques, was the nephew of Pierre Véran (1744–1819), the founder of the Musée Archéologique.[2] Shortly thereafter, the crosier passed into the collection of Pierre Révoil, as indicated by another document, unfortunately not dated, also preserved in the Médiathèque of Arles (ms. 777, fol. 251v). The crosier, entombed in 1316, was evidently reused for this funerary purpose, since the work obviously predates the late thirteenth century.

The Annunciation is frequently depicted on the volutes of Limousin crosiers. Marquet de Vasselot (1941)—and, more recently, Schwarzwälder (1978)—ranked the Montmajour crosier as a primary example in his classification system of the "first group" of crosiers decorated with an Annunciation, and it is distinguished by the spirited stance of the angel, bending his knees while reaching out to the Virgin, and also by the association of the Annunciation theme with that of the serpent.

The style of the two figures is indeed characteristic of the period in which Limousin work is still largely marked by Romanesque traditions and shows only limited openness to Gothic forms. The dynamic movement and the strong gesture of the angel contrast with the immobility of the Virgin, her hands open in an attitude of prayer, as seen, for example, in the fresco at Tavant (Indre-et-Loire).[3] The sensitivity to Gothic tendencies is seen in the gentle, peaceful expressions of the rounded faces. A certain concern with naturalism is also expressed in the sinuous design of the bough ending in a floret.

Marie-Madeleine Gauthier has suggested a comparison of this Annunciation with one on a stained-glass window of the cathedral of Lyons, probably executed about 1215–20,[4] especially in terms of its dynamism and the vigorous gesture of the angel. This angel might also be compared with the angel in the Sacrifice of Abraham in the stained-glass window depicting the Story of Isaac on the north wall of the nave of the cathedral of Poitiers.[5]

Other details confirm an early-thirteenth-century date. The rosettes with enameled beads at their centers, which decorate the pierced knop, recall those on the chasse of Saint Marcel (cat. 75), and the crest, which is very thin and relatively underdeveloped, is comparable to the crest on the crosier from Luxeuil (cat. 77).

As Marquet de Vasselot (1941) has already noted, the crosier of Bertrand of Malsang is close to the one in the treasury of the cathedral of Bamberg, which differs only in a few details, principally in the presence of an arch suggesting the structure within which the scene takes place. Furthermore, several other crosiers strikingly similar to the one from Montmajour are also preserved, mostly in private collections, which, in the words of Marquet de Vasselot, "may surprise us."[6] We can only repeat here his statement that only the example in the Louvre has been known since the early nineteenth century; its precise origin is now well established. Early publications, which included engraved illustrations, notably by Barraud and Martin (1856)—published as an independent study by the restorer Poussielgue-Rusand (see cat. 7)—and then Cahier (1867), might explain the fabrication of copies.

ET-D

NOTES
1. Benoît 1927; Benoît 1928, pp. 41–42; Mognetti 1979, p. 201.
2. For these two individuals, see Benoît 1951, pp. 47-48.
3. See Demus 1970, pp. 140–41 and pl. LVI (ca. 1150).
4. Gauthier 1975a, p. 170; see Grodecki and Brisac 1984, fig. 76.
5. See Grodecki and Brisac 1984, pp. 55–56 and fig. 46 (ca. 1210–15).
6. See, for example, the copy sold in Paris, Palais Galliera, November 29, 1969 (not numbered or illustrated).

EXHIBITION: Paris 1979, no. 212.

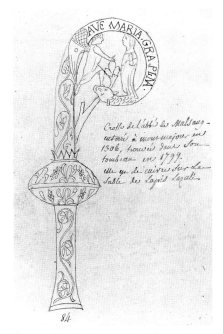

Fig. 80a. Crosier discovered at Montmajour in 1799. Inventory of the J.-D. Véran Collection, 1817, Médiathèque, Arles, ms. 777, fol. 25

BIBLIOGRAPHY: L. Laborde 1852, no. 32; Trichaud 1854, p. 15; Barraud and Martin 1856, pp. 86-88, fig. 105; Cahier 1867, vol. 2, p. 523 fig.; Marin de Carranrais 1877, no. 5, p. 73; Courajod 1886, p. 19 and no. 382, p. 59; Rohault de Fleury 1883–89, vol. 8, p. 103; Rupin 1890, p. 562, fig. 634; Darcel 1891, no. 122; Molinier 1891, pp. 170–71, fig. p. 170; Marquet de Vasselot 1914, no. 75; Benoît 1927; Marquet de Vasselot 1941, no. 65, pp. 232–35, pl. XIII; Reuterswärd 1970; Reuterswärd 1975; Schwarzwälder 1978, p. 206 and fig. 2, p. 209.

# 81. Crosier with Serpent Devouring a Flower

Limoges, ca. 1200–1220

Copper: formed, engraved, chased, scraped, stippled, and gilt; champlevé enamel: medium and light blue, light green, yellow, red, and white; glass cabochons

H.: 23.2 cm (9⅛ in.)

EX COLLS.: Georges Hoentschel, Paris (by 1911); [Jacques Seligmann, Paris (in January 1912)]; J. Pierpont Morgan, London and New York

CONDITION: Wear to gilding; some loss to enamel on volute; shaft probably added (before 1911).

The Metropolitan Museum of Art, New York; Gift of J. Pierpont Morgan, 1917 (17.190.833)

Exhibited in New York only.

The volute of the crosier is in the form of a serpent grasping an enameled flower in its mouth. Horizontal strips of enamel define the volute, and rounded crockets decorate its spine. The knop, in the form of a flattened sphere, is decorated with cabochons set within gilt fleurons. On the shaft in reserve is a band of half-length angels whose halos are enhanced by a bright outer circle; they are framed at top and bottom by a repeating pattern of enameled triangles.

Like the crosier from Nieul-sur-l'Autise (cat. 107), the Metropolitan crosier is part of a larger group characterized by a large fleuron set at the center of the volute. Both show the flower emerging from the mouth of a serpent, an allusion to the rod of Moses and Aaron that, in the presence of Pharaoh, miraculously turned into a serpent at the command of God and to the flowering rod of Aaron, symbol of his election to the priesthood by God (Exod. 7:9–12; Num. 17:6–8).

Pératé believed that the knop was added later, and Marquet de Vasselot questioned whether disparate elements were used to compose the ensemble. However, the decoration of both the knop and the volute is consistent with a date at the beginning of the thirteenth century. (For the dates of knops with fleurons, see above, cats. 80 and 81). Similar gilt-copper bosses with glass cabochons appear on a serpent's-head crosier in Estella[1] and at the horizontal terminals of the processional cross of about 1195–1200 at Nantes[2] and are not, therefore, disconsonant in this context. Nor are images of angels in reserve unexpected in combination with early crosiers with large enameled leaf decoration, appearing, for example, on the crosier from Luxeuil (cat. 77) and on a crosier preserved at Saint-Maurice-d'Agaune.[3]

On the other hand, the metal shaft is highly unusual in Limoges work. An example in Tours has gilt-copper sheeting over a wood core;[4] but it is distinct from the three other crosiers with shafts similar to this example, all of which passed through the Hoentschel Collection.[5] The original shafts of Limoges crosiers seem, almost exclusively, to have been made of wood and to have had metal terminals, which were sometimes enameled; the wood shafts have fractured and decayed, either through prolonged use or through burial in a bishop's tomb.[6]

BDB

NOTES
1. Marquet de Vasselot 1941, no. 180, pl. XXIII.
2. Gauthier 1987, no. 272, fig. 747.
3. Marquet de Vasselot 1941, no. 11, pl. III.
4. Ibid., no. 74.
5. The Metropolitan Museum of Art, New York; Gift of J. Pierpont Morgan, 1917 (17.190.834–836). The other shafts have been removed.
5. See Marquet de Vasselot 1941, p. 12. See the illustrations in Brandt 1976, especially fig. 10.

BIBLIOGRAPHY: Pératé 1911, no. 68, pl. XLII; Breck and Rogers 1929, p. 105; Marquet de Vasselot 1941, no. 33, pp. 52, 209–10.

# 82. Crosier of John of Chanlay

Limoges, ca. 1200–1215

Copper: gilt; champlevé enamel: lapis and lavender blue, turquoise, and, only on the knop, green, yellow, and red

23.6 × 10.8 cm (9⁵⁄₁₆ × 4¼ in.); diam. of knop: 5.5 cm (2³⁄₁₆ in.)

PROVENANCE: Found in February 1856 at the abbey of Notre-Dame, Preuilly (Seine-et-Marne), in the tomb of John of Chanlay, bishop of Le Mans (d. 1291). Acquired from Claude Vaudecrane in 1971.

EX COLLS.: G. Husson in Preuilly, then Montereau (Seine-et-Marne) (in 1857); Julien Chappée, Le Mans (in 1930 and 1941); Claude Vaudecrane, (before 1971).

CONDITION: Some wear and traces of corrosion, chiefly on the knop (gilding largely lost; hole).

Musée du Louvre, Paris; Département des Objets d'art (OA 10407)

In 1856 the crosier—together with two glove plaques with enamels in the later Gothic technique of *émaux de plique*[1]— was unearthed at the left of the high altar of the abbey church of Preuilly (Seine-et-Marne), where notes and drawings made for Roger de Gaignières and Dom Ythier (17th–18th c.),[2] identified the tomb as that of John of Chanlay, bishop of Le Mans (r. 1277–91).[3] The abbots of Preuilly had always been given modest burials, and only one other bishop, Gauthier, was interred in this choir, but he had died in 1235—far too early for his tomb to contain the glove plaques. Hence the identification of the tomb appears certain.

Limousin crosiers seldom depict a lone serpent. As noted by Marquet de Vasselot, this image may have been inspired by the story of the transformation of Aaron's rod into a serpent (Exod. 7: 9–12)—an episode illustrated more thoroughly by images of the serpent devouring other serpents,[4] or by the story of the brazen serpent (Num. 21:6–9). This form recalls above all that of the crosier of Abbot John of Silos (d. 1198), which is certainly earlier, but there is also a basis for comparison with enameled crosiers that are not from the Limousin, such as the one preserved at the Museo Nazionale del Bargello, Florence, identified with Willelmus; it apparently comes from Saint-Père, Chartres.[5] Certain gilt-bronze crosiers from the late twelfth century can also be mentioned, such as one unearthed at the abbey of Chaalis (Oise).[6]

The rinceau pattern with small trefoil florets on the volute, the half-length angels adorning the knop, and the rosette motif on the shaft are all typical of Limoges works, especially of the crosiers with palmette flowers at Saint-Colombe of Sens and the one of Peter of Charny (d. 1274) in the treasury of the cathedral of Sens.

ET-D

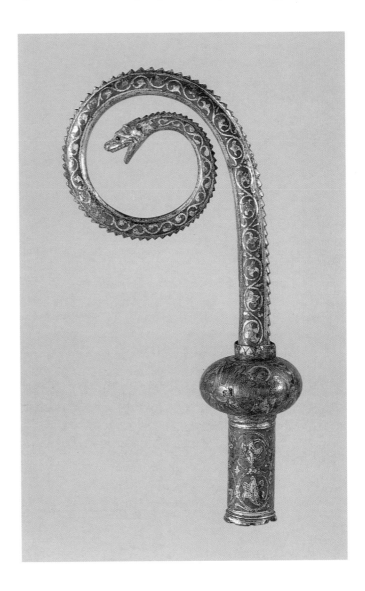

NOTES

1. Louvre OA 3437–3438, acquired 1893; see Enlart 1927–28, pp. 87–89.
2. Ythier, *Histoire ecclésiastique de Provins* (ms., eighteenth century); Provins, Bibliothèque Municipale, ms. 138, fol. 117; de Gaignières, *Dessins d'Archéologie*, Paris, BNF, Cabinet des Estampes, Pe 11a, fol. 47; see Adhémar 1974, p. 83, fig. 428.
3. *Gallia christiana*, vol. 14, cols. 403–4; Piolin 1856, pp. 438–53; Grésy 1857. Certain Le Mans sources place John of Chanlay's death in 1294, contradicting the inscription on his tombstone, which was copied for de Gaignières (Paris, BNF, Cabinet des Estampes, Pe 11a, fol. 47). This apparent discrepancy was due to the long vacancy that preceded the nomination of his successor in 1294.
4. The crosier from Tiron, now preserved at the Musée de Chartres; Marquet de Vasselot 1941, no. 1873.
5. Campbell 1979; Florence 1989, no. 210.
6. Longpérier-Grimoard 1866; Ménégaux 1986, no. 19.
7. Marquet de Vasselot 1941, nos. 20, 26.

EXHIBITION: Le Mans 1980, no. Ema 3.

BIBLIOGRAPHY: Piolin 1856, p. 51, n. 1; Grésy 1857, p. 375; Bastard 1857, pp. 407, 530–31, fig. p. 536; Rohault de Fleury 1883–89, vol. 8, p. 105; Rupin 1890, p. 79, fig. 147; Chappée 1896, p. 7; Rohan-Chabot, 1930, pp. 331–32, fig. 332; Marquet de Vasselot 1941, pp. 19, 97, and no. 189; Vinson 1971, p. 76, fig. 77; Gaborit-Chopin 1972.

# 83. Reliquary

Limoges, ca. 1200–1220

Copper: engraved, scraped, stippled, and gilt; champlevé enamel: blue-black, two medium blues and one light blue, turquoise, light green, yellow, red, and white

11.4 x 5.7 x 7.3 cm (4½ x 2¼ x 2⅞ in.)

INSCRIBED (on the lid): s[ANCTUS]IACOB[US] (Saint James); IH[E]S[US]

EX COLL.: Michel Boy, Paris (sold May 24, 1905, no. 164)

CONDITION: Later keyhole and pierced holes; hinges replaced.

The Metropolitan Museum of Art, New York; Bequest of Benjamin Altman, 1913 (14.40.703)

Exhibited in New York only.

The lower plaque of the principal face depicts two three-quarter-length angels, reserved in gilt copper, with an enameled fleuron set between them, under the lock plate. On the roof two three-quarter-length male saints pose above clouds on either side of the central figure of Christ in a half circle, holding a book in his left hand and raising his right in blessing. At the lower back of the chasse is a half-length figure of an angel, also enclosed within a half circle. At the center of the roof is the Lamb of God within a full circle. At either side, an angel descends from a cloud, reaching with outstretched arms toward the Lamb. The small sides of the chasse are decorated with enameled fleurons and gently scrolling ornament reminiscent of ivy. The interior of the box is gilded and divided into two unequal compartments by means of a vertical wall. The box stands on four small

round feet of wrought gilt copper tenoned through the base and hammered over.

Because of the division of the interior into compartments, the box was catalogued in the Boy sale as a "boîte aux saintes huiles" (a chrismatory) for transporting holy oils. Typically, however, chrismatories were made to hold either one container of oil or three, not two.[1] It is also unclear how the inscriptions naming Saint James and Jesus would be appropriate to this use. The unequal divisions of the interior do seem, however, to correspond in length to the inscriptions on the outside roof, perhaps indicating that the names refer not to two of the three figures represented but rather to relics contained within the box—of Saint James and secondary relics of Christ, perhaps of the Passion. Such an interior division is found in the late-twelfth-century silver-and-niello reliquary of Saint Thomas Becket in the Metropolitan Museum,[2] on a small box from the collection of the Musée Municipal de l'Évêché, Limoges,[3] and in the Museum of Fine Arts, Boston.[4]

The construction of the box within the context of Limoges work is unusual. It represents an early attempt to create a container without a supporting wood core. The roof is created from a single sheet of copper, scored and bent; the lid of the box in Limoges is also made from a single sheet of copper. The front, bottom, and back of the box in the Metropolitan are formed from a single, shaped sheet; the sides are secured by means of broadly spaced finger joins, secured with solder.

The chasse combines figures in reserve on an enamel ground and enamel accents for flowers, clouds, and halos. The form of the Lamb of God, with its forward stride and its sinews defined by paired lines, and the rather sketchy engraving of the faces resemble the style of a chasse at the Louvre (cat. 30) or another chasse in the Metropolitan (cat. 31). The ivy vines with their stippled accenting also occur on the Louvre chasse, and the surprisingly heavy zigzag engraving appears on the Metropolitan example. Such comparisons suggest that the Altman box was made soon after these chasses, at the beginning of the thirteenth century. The wide range of the palette and the elegant form of the fleurons on the sides also argue for this dating.

BDB

The reliquary open

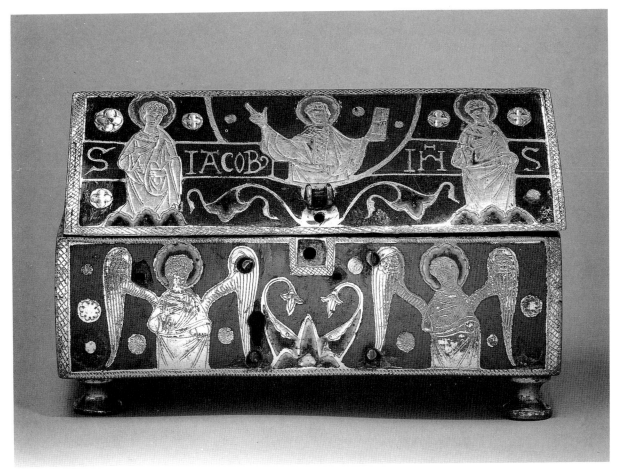

Principal face

Reverse

NOTES

1. See Helleputte 1884, pp. 146–53; Schnütgen 1884, pp. 454–62; Didron 1859, p. 189; Rupin 1890, pp. 443–46; Durandus 1906, pp. 124–41.
2. No. 17.190.520. The reliquary has long been considered an English work of the twelfth century or, most recently, as the product of a Rhenish artist resident in England. See London 1984, no. 302, p. 282. It is interesting to note, however, that it was in the collection of Albert Germeau, prefect of the Haute-Vienne, raising the question of whether it is a work made within the Plantagenet orbit in France. The presence of niello work on the crosier of Poitiers (cat. 23) makes this possibility even more intriguing.
3. See Gauthier 1968c.
4. Swarzenski and Netzer 1986, no. 27, pp. 88–89.

BIBLIOGRAPHY: Unpublished.

# 84. Chasse of Saint Thomas Becket

Limoges, ca. 1220–30

Copper: engraved and gilt; champlevé enamel: lapis and lavender blue, turquoise, green, yellow, brick red, garnet red, and white; wood core

17.5 × 13.2 × 6.3 cm (6⅞ × 5¼ × 2½ in.)

INSCRIBED: (on the bottom of the wood core, in black ink): 27 janvier 1881. M. Noët de La Fortemaison (Remsey)

EX COLLS.: Noët de La Fortemaison, 1881 (see the inscription on the wood core); Édouard Corroyer. Gift of Édouard Corroyer, with lifetime interest, 1905 (entered the museum in 1923).

CONDITION: Wood core has been greatly modified. The plank now on the front face is hollowed out at the center; the corresponding plank of the reverse is missing; what is left is covered with a thick coat of modern orange paint, except underneath, where one can read the inscription in ink. The pierced plaque on the back is modern.

Musée du Louvre, Paris; Département des Objets d'art (OA 7745)

Exhibited in Paris only.

Smaller than the chasse from the former Martin Le Roy Collection (cat. 39) and more modest in quality—despite the excellence of the execution—the chasse from the former Corroyer Collection is an example of the many chasses of Thomas Becket made at a slightly later date, but clearly based on the iconography and style of earlier works. A few details or nuances differentiate the many examples of this group.

The martyrdom of the saint before the altar in 1170, carried out by two executioners, is depicted on the body of the chasse, whereas on the roof the haloed saint appears surrounded by two angels, evoking the ascension of his soul to heaven. This scene, which is not depicted on the chasse from the former Martin Le Roy Collection, is placed next to the burial of the saint on the chasse from Saint Neots, now on deposit at the British Museum, and replaces the entombment on quite a few chasses of this "second group," such as the one from Vigean (Cantal; stolen in 1981), for example, or the one originating in Auriat (Creuse), formerly in the collection of Abbé Texier, now in the Liverpool Museum.[1] Two apostles are depicted on the gable ends. On the reverse, only the roof plaque is old, but it is probable that the other plaque was covered, like the one replacing it, with a row of small crosses on a dark blue ground.

The conventions of western Romanesque art seen on the oldest chasses of this saint are also marked in the style here, notably in the liveliness of the postures and gestures, the oversized hands, and the linear stylization of the drapery; one also finds the use of conventions more particular to the later group, such as the vertical lines ending in semicircles on the folds of the robe of the figure praying on the roof. But the evolution of the style is noticeable in a certain sim-

plification of the engraving and ornamentation; the stippled accents are far less numerous and more summary, the range of colors is reduced, the gradations are less skillful, and the lines indicating the folds of the clothing are more shallow and more quickly drawn. This evolution is noticeable as well in the style—in the softening of the lines, which suggests, particularly in the scene of the martyrdom, finer fabrics and more natural draping, signaling the opening up of the Limousin ateliers to early Gothic tendencies.

The "classic" appliqué heads are still close to the head on the chasse from the former Martin Le Roy Collection, as well as to the heads on many works from the end of the twelfth and the beginning of the thirteenth century. It seems logical to consider that this "second generation" of chasses dedicated to Thomas Becket—much more numerous than the first—owes its birth to the solemn translation of the saint's body in 1220, which certainly made possible the distribution of new relics.[2] These works thus bear witness to the persistence of Romanesque traditions and the practices of typical ateliers at the end of the twelfth century until the years 1220–30, only timidly modified by the new tendencies of early Gothic. By this time, these works were made not only for export but also for a local clientele, as proved by the examples cited above.[3]

ET-D

Reverse

276

Principal face

NOTES

1. See Foreville 1976, pl. I, figs. 3, 4; Caudron 1993, p. 66, fig. p. 65; p. 80;
   Texier 1842, pl. IV.
2. See Caudron 1975; Foreville 1976.
3. For the cult of Thomas Becket in the Limousin, see also Becquet 1975.

BIBLIOGRAPHY: Borenius 1932, p. 89; Boinet 1948, p. 77; Gauthier 1950,
pp. 38ff., pl. 29; Gauthier 1955, p. 68; Caudron 1993, p. 77.

# 85. Chasse of Saint Valerie

Limoges, ca. 1225–35

Copper: engraved and gilt; champlevé enamel: lapis and lavender blue, turquoise, green, yellow, brick red, garnet red, and white; wood core

16.5 × 13.6 × 9.1 cm (6½ × 5⅜ × 3⅝ in.)

Ex coll.: Édouard Corroyer. Gift of Édouard Corroyer, with lifetime interest, 1905 (entered the museum in 1923).

Condition: Badly damaged enamel, replaced by paint fill on most of the face and almost entirely in the lozenge pattern on the reverse; gilding very worn and engraving on the heads worn. The wood core is undoubtedly original.

Musée du Louvre, Paris; Département des Objets d'art (OA 7744)

Exhibited in Paris only.

As Marquet de Vasselot has already noted, when this chasse entered the Louvre (1923), it was mainly its iconographic interest that distinguished it. Indeed, the work is of modest quality, and it has also suffered a great deal; large parts of the gilding and enameling have disappeared, with the enameling replaced by modern additions. But the chasse is an example of Limousin production intended, principally, for a local clientele. Valerie ranks among the great Limousin saints, listed by Geoffrey de Vigeois at the end of the twelfth century and in all calendars of Limousin saints.[1]

Indeed, the work belongs to a group of small chasses of similar dimensions, iconography, and style, and of generally poor quality. Many examples, such as those from Meilhac (Haute-Vienne), Masseret (Corrèze), or Salins (Cantal), are still preserved in the Limousin.[2] Thus it is certain that these reliquaries are evidence of the continued devotion to the cult of Saint Valerie in this region (see cat. 20).

As on other small chasses, the two principal episodes of the martyrdom of the saint, as described in the *Vita* written at the end of the tenth century, are depicted in succession on the roof and then on the lower part of the principal face. On the roof, Valerie is decapitated in the presence of yet another figure who seems to be praying—no doubt the executioner "struck down by the angel of God." Below, she carries her head to Saint Martial, who stands before an altar while the executioner wields a sword behind her. On the gable ends two standing apostles are depicted; the reverse, adorned with a pattern of lozenges, is very similar to that of the Meilhac chasse, whereas the iconography is especially close to that of the reliquary in the church of Masseret, or to that of the work in the Museo Nazionale della Ceramica (Duca di Martina) in Naples.[3]

The style is still marked by the traditions of late Romanesque art, notably in the dancing postures and demonstrative gestures. On the other hand, the drapery, rather summarily engraved, shows the influence of early Gothic art in its more supple lines. The faces represent a lesser rendition, both in size and in quality, of the "classical" heads that were widespread in the decades before and after 1200.

Thus the chasse can be placed among those works that continued the traditions of the late twelfth and early thirteenth century—undoubtedly at a relatively late date, given the distinct evolution seen in the appliqué heads and the skillful simplification of the style and decoration. These elements allow us to accept Marie-Madeleine Gauthier's dating to the second quarter of the thirteenth century, while perhaps suggesting here a more precise date of about 1225–35.

ET-D

Notes
1. See Decanter in Paris 1993, pp. 98–99; Dom Jean Becquet, oral communication. On Geoffrey de Vigeois, see Barrière, above, p. 221.
2. See Gauthier 1955, pp. 50–54; Rupin 1890, pp. 404–5, figs. 461, 462.
3. Naples 1981, nos. 1, 3.

Exhibitions: Paris 1900, no. 2458; Vatican City 1963, no. 76.

Bibliography: Marquet de Vasselot 1923, p. 150; *Le Curieux* 1923, p. 530; Gauthier 1950, p. 38; Gauthier 1955, pp. 52–54.

# 86. Book-Cover Plaque: Christ in Majesty

Limoges, ca. 1210–20

Copper: engraved, chased, scraped, stippled, and gilt; champlevé enamel: dark and light blue, turquoise, green, yellow, red, and white

26 x 16 cm (10¼ x 6¼ in.)

INSCRIBED: A / ω (Alpha / omega)

CONDITION: Some enamel lost in border strips, which have been cut down; wood board replaced.

The Metropolitan Museum of Art, New York; Gift of J. Pierpont Morgan, 1917 (17.190.798)

Exhibited in New York only.

Within a richly decorated mandorla framed with undulating "clouds" of color and filled with enameled fleurons, the enthroned Christ rests his feet on an enameled footstool as he raises his right hand in blessing and balances an opened book on his left knee. The separately worked and applied head of the reserved figure is framed by a cruciform nimbus and the Greek letters alpha and omega. Around the mandorla are poised the symbols of the four evangelists in reserved copper, with applied heads riveted to the backplate.

Although the framing elements with foliate ornamentation have been cut, the overall format suggests the traditional style of all but the earliest Limousin book covers, wherein the central figurative plaque is surrounded by decorative bands that both frame and protect the central image. Missing in this instance is an intermediary beveling sheathed with gilt-metal strips, the absence of which, perhaps due to damage, provoked the cutting of the framing bands.[1]

A book cover with a similar border is recorded in the collection of Guillaume Libri in 1864. The palette of that border is, however, distinct, as are details of the central Christ in Majesty plaque.[2]

The similarity in size and composition of many Limoges book covers argues in favor of Gauthier's belief that these works were the specialty of a single workshop with a quasi-industrial production, which made applied heads *en serie*.[3] Most bear, nonetheless, a distinct fingerprint. This plaque is notable for the quality of the applied heads of the evangelists' symbols, for the palette—especially the juxtaposition of green and red—and for the engraving of the framing bands. By contrast, the figure of Christ is characterized by uneasy proportions, unexceptional engraving, and an ill-conceived rainbow on which he sits.

Although most Limousin book covers survive as isolated plaques, the image of Christ in Majesty was usually paired with a plaque of the Crucifixion, forming the lower and upper covers of a Gospel book,[4] Missal,[5] or Psalter.[6]

BDB

## NOTES

1. Extrapolating from the distance between the brads, Peter Champe (Metropolitan Museum, technical report, autumn 1994) suggests that the strips were cut by about 3 cm (1½ in.).
2. *Monuments . . . Guillaume Libri* 1864, pl. VII.
3. Gauthier 1968d, p. 277.
4. Rupin 1890, p. 314; Gauthier 1968d, p. 273.
5. Schapiro 1954, p. 331, on the pairing of a Crucifixion and a Christ in Majesty in the context of a Missal.
6. Rupin 1890, p. 317, fig. 383.

BIBLIOGRAPHY: Unpublished.

# 87. Book Cover: Crucifixion

Limoges, ca. 1210–30

Copper: engraved, chased, and gilt; champlevé enamel: midnight and lapis blue, turquoise, green, yellow, red, garnet red, and white; filigree border of gilt copper and glass cabochons; wood core

31.7 × 15.5 cm (12½ × 6⅛ in.); central plaque: 24.2 x 11.8 cm (9½ x 4⅝ in.)

INSCRIBED: IH[ESU]S / XPS (Christus); on the wood of the reverse, in black ink: *ff aᵒ / 3004 / mdff* (?)

PROVENANCE: Bequest of Jean-Charles Seguin, Paris, 1908

CONDITION: Signs of wear to gilding; stamped band of frame, decorated with rosettes inscribed in a lozenge pattern, might be modern.

Musée du Louvre, Paris; Département des Objets d'art (OA 6173)

Exhibited in Paris only.

At one time this plaque served as one face of a manuscript cover. It is composed of a wood plank that supports an enameled plaque, which is centered in a shallow recess. The plaque is surrounded by two borders: the first was beveled to meet the level of the second, which is decorated with fili-

gree and scattered cabochons. The opposite face would have depicted Christ in Majesty at its center, as seen in the rare preserved pairs, such as the covers of a manuscript from the Stiftbibliothek of Saint Gall[1] (see cats. 18, 113).

In France, many medieval manuscripts have composite covers that include, in the center, a Limousin plaque depicting a Crucifixion. Such is the case with the Gospels of Saint Ludger (Musée Condé, Chantilly)[2] and the manuscript containing the Gospels of Luke and John, now in the Bibliothèque Nationale, Paris,[3] which both have Gothic borders; or again a gospel book from the Bibliothèque Sainte-Geneviève,[4] with a modern frame.

The composition of this scene is similar to that of many preserved book covers depicting the Crucifixion. This style and decoration are encountered in classic Limousin production of the late twelfth and early thirteenth century. The singularity of this book cover lies in the filigree border that surrounds it.[5] The filigrees coil into regularly spaced volutes that end in rosettes of granulations embellished with small clusters of thorns and crosses. These stylistic elements are closely comparable to those of the filigree on the crosses of Gorre (Haute-Vienne) and Rouvres (Côte-d'Or), which come from the abbey of Grandmont and the Grandmontain priory of Époisses, and can be dated to about 1240, based on the style of the Christ at the center of the Rouvres cross.[6]

The style of the enameled plaque leads us to attribute to it a slightly earlier date than to the border, about 1210–20, so it is possible that the border was added later. But we cannot exclude the likelihood that the enameled plaque and its filigree border were executed as an ensemble about 1220–30, or that the plaque and its border came from separate ateliers active at the same time, over a period of ten to twenty years.

ET-D

NOTES
1. *Règle de saint Benoît et lettres de saint Jérôme*, 9th century, Saint-Gall, Stiftbibliothek, ms. 216; see Gauthier 1968d, p. 273; Steenbock 1965, no. 122.
2. Gauthier 1968d, no. 16, p. 273.
3. Ibid., no. 26; Laffitte 1991, p. 25 and pl. 45.
4. Bernard 1965, p. 31; Gauthier 1968d, no. 25; Étaix and Vregille 1970; there is debate about the provenance of this manuscript, whose cover once held a relic of Saint Ferreolus, martyr of Besançon.
5. Two other book covers with similar filigree borders are known. On the one from the Walters Art Gallery, Baltimore, the filigree seems modern. On the other, depicting Christ in Majesty, which once belonged to the Homberg Collection and then was put up for sale in 1984 (Monaco, Sotheby's, June 26, 1984, no. 921), the filigree could not be examined, but the design seems slightly different (see Taburet-Delahaye 1990, p. 49 and n. 19).
6. See Taburet-Delahaye 1990.

BIBLIOGRAPHY: Marquet de Vasselot 1914, no. 88; Gauthier 1968d, no. 29; Taburet-Delahaye 1990, p. 49.

# 2

*The emergence of the
Gothic style in Limoges*

# 88. Coffret of Cardinal Bicchieri

Limoges, before 1227

Copper: pierced, repoussé, chased, engraved, scraped, and gilt; champlevé enamel: medium blue, turquoise, green, red, and white; wood core

34 x 82 x 39 cm (13⅜ x 32¼ x 15⅜ in.)

PROVENANCE: Cardinal Guala Bicchieri (d. 1227); Sant' Andrea, Vercelli; immured at Sant' Andrea after 1611; discovered by Carlo Emmanuele Arborio Mella, architect / restorer of Sant' Andrea, 1823; presented to him by Cardinal Grimaldi.

CONDITION: Modern wood core.

Private collection

Exhibited in Paris only.

On the principal face of the coffret are seven openwork roundels with fantastic creatures ringed by bands of enamel and surrounded by small domed studs, each engraved with a star pattern. An eighth medallion has been replaced by a gilt-copper disk. Two more, each with an arc cut from its circumference, once framed the central lock plate, which has paired openwork creatures in combat with clubs and shields. Ornamental strapwork and corner pieces further enliven the front as well as the sides and lid. Each side has four enameled medallions with chivalric figures in gilt copper as well as strapwork and corner reinforcements. No medallions are preserved on the lid except for a small enameled shield with a coat of arms, discovered at the same time as this coffret but clearly from a different ensemble. The lid retains its decorative strapwork as well as a pair of hasps for the lock in the form of lizardlike beasts from whose mouths issue smaller reptilian creatures, from whose mouths in turn issue the small-headed lizards that fit into the paired keyholes of the lock plate. Strips of stamped gilt copper serve to reinforce all the edges of the coffret. Like the coffret from the Morgan Collection, the reverse is undecorated. The original core, subsequently replaced because of its poor condition, was covered with dark red parchment, like the Longpont and Morgan coffrets (cats. 133, 136).

Two roundels in the Museo Civico, Turin, correspond in their central design, border decoration, and dimensions to the type found on the Bicchieri coffret.[1] They entered the museum's collection in 1868 from an unknown source, so it is possible that they originally decorated the coffret but became separated from it.

Among the few surviving Limoges coffrets, only the one preserved at Aachen (Schatzkammer Cathedral Treasury) is as imposing in size as the Bicchieri example (fig. 9). Particularly noteworthy in the two coffrets are the similarities in composition and execution of the lock plates and the medallions depicting falconers on horseback.[2]

When the present coffret was discovered by C. E. Arborio Mella, it contained human remains and an authentication on parchment written in 1611 by the abbot of Sant' Andrea, Pietro Francesco Maletto, testifying that these were the bones of Cardinal Guala Bicchieri.[3] This use of a Limoges coffret was paralleled in the coffret used for the bones of Blessed Amadeus IX of Savoy, elements of which are preserved in a private collection[4] and in France for relics of Saint Louis and of Blessed John of Montmirail (cats. 123, 133). In this instance, it seems that the coffret so adapted was one that Bicchieri himself had owned. In the inventory of his possessions taken at his death in 1227, the following listings appear: "Item in uno scrineo operis lemovicensis sunt VII cuppe cum cooperculis suis laborate et deaurate . . . item in alio scrineo eiusdem operis sunt nonaginta tres marche et VII unce et dimidium in vasis pertinentibus ad hospitale, et cuppa aurea que ponderat duas marchas et duas uncias . . . item in uno scrineo operis lemovicensis sunt ea quae sunt Capelle: turibulum argenteum, duo calices aurei et calix argenteus et crux aurea cum pede argenteo, ampulla argenthea pro vino et ampulla cristallina pro aqua, IX anuli aurei in rubizis . . . et sex anulis in saphiris et tredecim anuli parvi et IV anuli pastorales et crux aurea ad pectus et una imago argenthea cum reliquis et X coclearia aurea et IV adamantini et XIV anuli inter jacintos et saphiros et tres magni anuli cum tribus magnis saphiris et optimis . . . ."[5] It is reasonable to assume that the coffret found by Count Arborio Mella corresponds to one of these, whereas the medallions incorporated into the choir at Biella (cat. 91) constitute additional pieces from the cardinal's collection.

Principal face

The sides

Their profane decoration notwithstanding, these containers were used by the cardinal for storing his liturgical vessels; they were not adapted for ecclesiastical use at a later date. Nor, apparently, is the example of Cardinal Bicchieri's use of such coffrets unique. The inventory of Clement VI at Avignon of 1342–43 includes "Item in coffro rubeo de opere Lemovicensi signato per . . . sunt 4 bacilia argenti, quorum 2 sunt ad figuras et alii albi, 1 thuribulum cum navicula, ponderis 22 m. 6 u. 3 q."[6]

Bicchieri was an avid traveler, collector, and patron whose career provided ample opportunity for him to acquire the Limoges coffrets recorded in the inventory. From 1208 until 1209 he was a legate to France and visited Limoges itself. In 1219 he laid the first stone of the new church of Sant' Andrea at Vercelli,[7] bringing Thomas Gallus and three other canons from Saint-Victor in Paris,[8] the celebrated Augustinian foundation known to have possessed Limoges work in the 1160s.

BDB

283

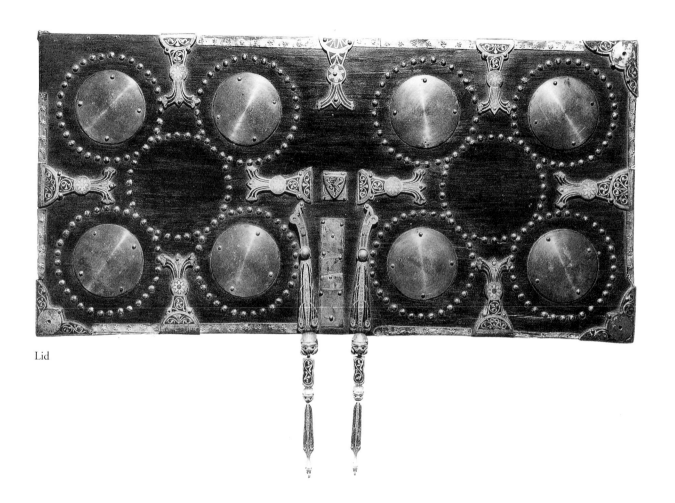

Lid

NOTES

1. See Castronovo 1992, p. 175, no. 35.

2. Vaivre 1974, pp. 100–101, figs. 4, 5.

3. When the coffret was presented by Bishop Grimaldi to Count Arborio Mella, the bones were transferred to a new container.

4. See Mallé 1950 and 1951, figs. 42–45.

5. Transcribed in Castronovo 1992, p. 169: ". . . in a coffret of Limoges work, seven cups with lids elaborated and gilded; in another coffer, of the same work, ninety-three marks and seven and a half ounces of vessels belonging to the hospital and one golden cup weighing two marks and two ounces . . . in another coffret of Limoges work, objects which belong to the Chapel: one silver tower-shaped ampulla for the wine and another one in crystal for the water, nine gold and ruby rings, six rings with sapphires, thirteen small rings and four pastoral rings . . . one gold pectoral cross, one silver 'image' with relics, ten golden spoons, four dia-

mond (or steel?) [rings] and fourteen rings with topaz(?) and sapphires, and three big rings with three big and beautiful sapphires."

6. Hoberg 1944, p. 91. "The same, in a red coffret of Limoges work, marked by [. . .] are 4 silver basins, of which 2 have figures and the others are white, a censer and its navette, weighing 22 marks, 6 ounces, 3 quarters."

7. Castronovo 1992, p. 167.

8. Pastè and Arborio Mella 1907, pp. 43–45.

BIBLIOGRAPHY: Arborio Mella 1856; Arborio Mella 1883, pp. 256–62; Pastè and Arborio Mella 1907, p. 488; Mallé 1951, pp. 80–94; Gauthier 1972b, p. 191; Castronovo 1992, pp. 166–76; Vaivre 1974, pp. 100–101, figs. 4, 5.

Scenes of hunting (knight with falcon, capture of a hare) and of combat (details)

# 89. Medallion

Limoges, ca. 1210

Copper: pierced, repoussé, chased, engraved, scraped, stamped, and gilt; champlevé enamel: medium blue and red

Diam.: 8.3 cm (3¼ in.)

Ex colls.: Georges Hoentschel, Paris (by 1911); [Jacques Seligmann, Paris (in January 1912)]; J. Pierpont Morgan, London and New York

Condition: Gilding worn; incidental losses to enamel.

The Metropolitan Museum of Art, New York; Gift of J. Pierpont Morgan, 1917 (17.190.794)

At the center of the medallion, a standing, naked man throttles a pair of *serpenses*, one with his right hand and the other with his left. As they turn to look at him, their tails entwine each other and the man's legs. The openwork figures have been raised from the copper, then tooled and chased. The head of the man was worked separately and then riveted to the body.

In medieval bestiaries, the creatures depicted on the medallion conform to the classification *serpens*, which included snake- and lizardlike beasts. According to these texts, one of their chief characteristics was their fear of naked men.[1] The portrayal here may be as much a reference to that quality as a representation of a more general theme: the struggle between good and evil.

Now isolated, the roundel was once secured to a coffret like the ones in the Glencairn Museum Collection (cat. 38) and in a private collection (cat. 88).

BDB

Note
1. George and Yapp 1991, p. 194.

Exhibition: South Hadley, Mass., 1977, no. 12.

Bibliography: Pératé 1911, no. 74, pl. xlviii.

# 90. Medallion

Limoges, ca. 1210–20

Copper: pierced, repoussé, engraved, chased, scraped, and gilt; champlevé enamel: medium and light blue, green, yellow, red, and white

Diam.: 8 cm (3⅛ in.)

Ex colls.: [S. Goldschmidt, Paris (in 1919)]; Alexandre Eugène Joseph Chompret, Paris

Condition: Some wear and loss to gilding on openwork figures and at border.

Musée du Louvre, Paris; don Chompret (OA 9352)

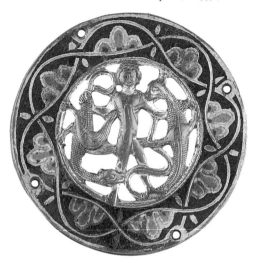

At the center of the medallion, a man, standing with his left foot forward, holds two fantastic reptiles: in his right hand, a winged one by the tail, and in his left, a horned one, which turns to bite his shoulder. The openwork figures have been raised from the copper, then tooled and chased. The enameled border is patterned with intertwining quatrefoils of reserved metal enclosing enameled floral ornament.

In medieval bestiaries, the creatures depicted on the medallion, like those on the Metropolitan medallion (cat. 89), fall within the classification *serpens*, which encompassed snake- and lizardlike beasts. Such imagery became part of the standard repertory of coffret medallions, which, as in the case of the Bicchieri coffret (cat. 88), were sometimes used in combination with chivalric or amatory scenes.

BDB

Exhibition: Paris 1945, no. 257.

Bibliography: Marquet de Vasselot 1941, pp. 138 n. 2, 140; Verlet 1950, p. 6.

# 91. Belt Buckle

Limoges, 1200–1220

Copper: pierced, repoussé, engraved, chased, scraped, stippled, and gilt; champlevé enamel: dark, medium, and light blue, green, yellow, and white

6 x 8.1 cm (2⅜ x 3³⁄₁₆ in.)

Ex colls.: Victor Gay, Paris (manuscript inventory, December 1, 1908, no. II, 44; sold, Paris, March 23–26, 1909, no. 57); Albert Figdor, Vienna; acquired by the Kunsthistorisches Museum, 1930; deposited in the Museum für angewandte Kunst, 1940

Condition: Losses to gilding throughout; surfaces worn; losses to enamel on buckle.

Österreichisches Museum für angewandte Kunst, Vienna (F 285)

Notes
1. Rupin 1890, fig. 660. Other buckles illustrated in Fingerlin 1971 may also be Limousin, especially figs. 11 and 14, illustrated on p. 37. Bilimoff (1977) indicates a total of thirty-nine Limousin examples, but many of these are fragmentary or badly damaged.
2. Liebgott 1986, p. 78, figs. 67, 68.

Bibliography: D'Allemagne 1928, vol. 1, pl. xxxv, fig. 15; Weixlgartner 1932, no. 16, p. 27; Fingerlin 1971, no. 540, pp. 472–73, fig. 559, p. 474.

The buckle consists of two principal elements: the loop and tongue through which the belt passed and a rectangular decorative plaque with an openwork medallion that was fixed to the leather. At the center of the plaque stands a fantastic creature with the body of a winged lizard and the head of a man, surrounded by rays of enamel in a sunburst pattern enlivened by the stippling of the copper plate. Enameled trefoils fill the corners. The loop of the buckle, with a thin copper tongue at its center, is decorated with scrolling foliate ornament whose enamel is almost entirely lost. The hasp of the buckle was secured to the decorated plate by folding the copper of the rectangular tab back on itself.

The portrayal of a fantastic creature enclosed in a roundel is part of a decorative vocabulary recognizable in a range of Limoges enamels but especially in openwork medallions which, like belt buckles, were often secured to a leather mount. This openwork decoration recalls the medallions of the Glencairn Museum coffret (cat. 38). The buckle is also technically allied to openwork medallions with secular decoration. The head of the creature on the buckle was separately worked and then applied to the openwork body, like the head of a man throttling a pair of *serpenses* on a medallion in the Metropolitan Museum (cat. 89). Such technical and compositional similarities may suggest that the Vienna buckle was created in a workshop that specialized in objects for secular use.

Similar buckles have been found in excavations in France, including one found at Avesne-le-Comte (Pas-de-Calais) and preserved at Arras until 1915,[1] but also at Roskilde and nearby at Flengemarken, Denmark,[2] where the taste for Limoges work in the Middle Ages is well documented.

BDB

# 92. Fourteen Medallions

### a. *Two Medallions*
Limoges, before 1227

Copper: pierced, repoussé, chased, engraved, scraped, stamped, and gilt; champlevé enamel: medium blue

Diam.: 8.6 cm (3⅜ in.)

The Metropolitan Museum of Art, New York; Gift of J. Pierpont Morgan, 1917 (17.190.795,796)

### b. *Twelve Medallions*
Limoges, before 1227

Copper: pierced, repoussé, chased, engraved, scraped, stamped, and gilt; champlevé enamel: medium and lavender blue, turquoise

Diam.: 8.5–9 cm (3⅜–3⁹⁄₁₆ in.)

Musée du Louvre, Paris; don Larcade (OA 9468–9479)

This series of roundels presents an imaginative array of fantastic creatures—winged apes, centaurs, serpents, and dragons—locked in battle, sometimes with armed men. They are intertwined as they turn within the circular confines of a framing band, which is enameled predominantly in blue with a relatively simple ornamental pattern reserved in gilt

a

copper. The roundels' relationship to thirteenth-century medallions in the church of San Sebastiano at Biella (Piedmont) has long been recognized. In fact, many of them have the same design found in the group, originally numbering twenty-nine,[1] that was installed in the choir there in 1546 (fig. 92*a*).

Earlier literature has attributed this coincidence to the use of the same mold.[2] However, the Metropolitan and Louvre roundels were not produced by casting. Rather, they are distinguished by an accomplished repoussé technique, typical of Limoges work in the thirteenth century. (See "Techniques and Materials in Limoges Enamels," above p. 48.) The copper was hammered from the back and chased from the front so that it swells and dips to delineate the anatomy of the beasts and men; it was then engraved to define lines of foliage and punched repeatedly to create the lizardlike bellies of serpents and dragons. Conversely, the roundels now at Biella were created by casting, evident in the blurred detail on the scales of dragons and serpents and the hard quality of the engraved straight lines that articulate stems and foliage. In addition, the palette of the roundels at Biella is manifestly different from that found on thirteenth-century Limousin work; this is especially noticeable in the brilliance of the blue. The same color appears on a number of casts from the same series now in various collections, including that of the Art Museum, Princeton, New Jersey.[3] Technical analyses of the enamels in the Princeton roundels suggest that the unusual color can be attributed to the use of arsenic in the enamel, a nineteenth-century practice.[4] There can be no question that the celebrated group at Biella represents nineteenth-century replacements of the original series.

The historical circumstances of this substitution can only be partially inferred. It is important to note that copies

b

Fig. 92a. Medallions. Church of San Sebastiano, Biella, Piedmont

already existed by 1868, for in that year two entered the collection of the Museo Civico, Turin, acquired from an unknown source. Although repeatedly published as originals in this century,[5] they were clearly recognized during the nineteenth century as copies of Limoges work; a photo in the Stohlman Archives, Princeton, New Jersey, shows them in the museum in Turin with a label inscribed in a nineteenth-century hand: "Imitazione di Limogia." This may suggest that the campaign of restoration and replacement at Biella was undertaken, at least in part, as an exercise in technique, in keeping with the renewal of the craft tradition during the nineteenth century. It may be that such a campaign coincided with the nineteenth-century restoration of the church's facade and that an exchange of originals for what were deemed fine copies aided the financing of the architectural campaign.

It is peculiar, however, that the series of eighteen roundels now at Biella includes only thirteen distinct designs, since the original series must have comprised twenty-nine unique roundels. Ten can be identified (seven in the Louvre, one in the Metropolitan, and two in the Keir Collection, deposited at the Nelson-Atkins Museum of Art, Kansas City) by matching them to existing casts at Biella.[6] Comparing the borders, dimensions, and collection histories of other dispersed roundels allows that number to be increased to twenty-three. These include one roundel in the Metropolitan (17.190.795) and one in the Musée du Louvre (OA 9479), not represented at Biella but having the same diameter and border design as others and tracing to the Hoentschel and Larcade Collections, plus another from the Larcade Collection (OA 9471) that is not represented at Biella but exists as a plaster and in copies. The borders of two roundels at Turin (26 and 27/S) conform to Biella types and establish their provenance.[7] Similarly, the borders of Turin medallions 24 and 25/S, 28 and 29/S correspond to examples in the Louvre (OA 9473 and 9478), suggesting their common history. It is probable that two remaining in the Louvre (OA 9469) and Turin (23/S) also trace to Biella. A medallion in the Musée de Cluny (945a) and one in the Musée de Picardie, Amiens (MP 992-4-14), with borders common to examples from Biella, may bring the number to twenty-five. Two original designs are known only from copies now at Biella.[8]

The medallions at Turin entered the collection at the same time as the copies. These, however, do not exist in copies, suggesting that the group was divided after removal from Biella, with some originals remaining in the collection at Turin, and the ones chosen to be cast sent to Paris. The copies were produced in Paris by Alfred André; plaster casts of the central designs survive at the André workshop.[9] Apart from those that remained in Turin, the originals were divided among nineteenth-century Paris collections, eventually passing to museums in the present century.

Measurement of the bull's-eyes in the backs of the choir stalls and on each face of the rectangular base of the lectern, where the roundels were placed at Biella, indicates that they were meant to accommodate medallions ranging in size from 8.2 to 9.7 centimeters, with the largest examples reserved predominantly for the lectern. That the original disposition of the roundels has not been maintained is clearly indicated by the fact that some roundels are significantly smaller than the bull's-eyes into which they have been set, with earlier stud holes visible outside the circumference of the medallion. Given this range, it is reasonable to suggest that the larger roundels, which also entered the Museo Civico, Turin, in 1868, likewise came from the Biella choir stalls.

The church at Biella was built in 1546 as a dependency of the religious community at Sant' Andrea in nearby Vercelli.

Indeed, the contract for the choir stalls, given to Gerolamo de Mellis da Vespolate, specified that they should conform to the type at Sant' Andrea.[10] Although the contract does not mention the enameled decoration, the historical circumstances of the use of these thirteenth-century elements for the new choir can nonetheless be convincingly reconstructed. Sant' Andrea, although part of the Order of Canons of the Lateran by the sixteenth century, was founded by Cardinal Guala Bicchieri as a community of Canons Regular, dependent on the Order of Saint-Victor in Paris. He called its first abbot, Thomas Gallus, from Paris in 1219,[11] and in 1224, Bicchieri presented two "cophini de opere lemovicensi" (coffrets of Limoges work) to Gallus, the preeminent Victorine scholar of his day.[12] In addition, at the time of the cardinal's death at Vercelli in 1227, three "scrini" (coffrets) were in his possession. Thus, there were at least five Limoges coffrets in the church of Sant' Andrea, Vercelli, by 1227. Given the dependency of Saint' Andrea on Paris and its alliance to an order that demonstrated a predilection for Limoges enamels (see Boehm, above p. 40), five must be the minimum number of coffrets at Vercelli in the thirteenth century.

One of these is surely the coffret discovered by the architect Carlo Emmanuele Arborio Mella during renovations of Sant' Andrea in 1823 (cat. 88), containing bones that, according to a sixteenth-century authentication, were those of Guala Bicchieri. Given its size (34 x 82 x 39 cm), it probably corresponds to one of the two coffrets designated as "cophini" in 1224. The roundels adapted for use in the sixteenth-century choir stalls at Biella surely came from one or more of the other coffrets preserved in the motherhouse at Vercelli. The disparity of size and subject matter among them does not preclude their being from a single coffret, since unlike types and sizes are incorporated on the Bicchieri and Saint Louis coffrets (cats. 88, 123). However, the sharply divergent style of one roundel type installed at Biella (the location of the original is at present unknown) suggests that it at least came from a different coffret.

BDB

NOTES

1. The number is sometimes given as thirty-one (Sciolla 1980, p. 72–73), but a count of the wood "bull's-eye" mounts for the roundels indicates clearly that the original ensemble consisted of twenty-three enamels in the stalls and six in the lectern.
2. Castronovo 1992.
3. Wadsworth Atheneum, Hartford, Conn. (1966.528–30); The Art Museum, Princeton, New Jersey (30.170, 50.14); Walters Art Gallery, Baltimore, M. 67, inventoried as a pastiche; M. Lazaro Collection, Madrid; Kiev, Museum of Western and Oriental Art (145–146 bk); Rotterdam, Binnenvaartmuseum; Casablanca, private collection; Saint Petersburg F2295 (see Malbork 1994, no. 24, p. 30); sale, Sotheby's, London, December 9, 1993, p. 13, lot 22. Additional roundels are found in the archives of Foto Mas, Barcelona, though it is possible that one or more of these are to be identified with examples now in museum collections or sales catalogues.
4. See Biron, Dandridge, and Wypyski, above. It is also unusual to find that metal backplates have been attached to the roundels, as in the Wadsworth Atheneum examples, and that, apparently, plaster has been set behind the openwork figures for support.
5. Mallé 1950 and 1951; Gauthier 1972b; Castronovo 1992.
6. For the two roundels from the Keir Collection, see Schnitzler, Bloch, and Ratton 1965, p. 27, E60–61; Gauthier and François 1981, nos. 29–30, p. 25.
7. For the Turin medallions, see Mallé 1950 and 1951; Mallé 1969, pp. 81–90; Petrassi 1982, nos. 167–74.
8. The original of the one showing intertwined serpents may have been sold with Alfred André's own collection at the Hôtel Drouot, Paris, April 23–24, 1920, no. 169.
9. At least one of the designs preserved, representing a bird resembling a swallow flying into the mouth of a dragon, seems to have been of André's own invention. This is consistent with the range of workshop practices known to have existed, from direct copies to simple restoration to new compositions inspired by medieval examples. See Distelberger et al. 1993, pp. 282–87.
10. Torrione 1947, pp. 21–27.
11. Pastè and Arborio Mella 1907, p. 44.
12. See, most recently, Castronovo 1992, p. 166 n. 2; Mallé 1969, pp. 84–85.

BIBLIOGRAPHY (for the Biella series): Rupin 1890, p. 442; Roccavilla 1905, pp. 44–47; Mallé 1950 and 1951; Mallé 1969, figs. 21–33; Sciolla 1980, pp. 72–73; Castronovo 1992, pl. 47; (for the Morgan pieces): Yokohama 1989, no. 5, p. 63; (for the Larcade series): Verlet 1950; (for the Keir Collection pieces): Schnitzler, Bloch, and Ratton 1965, p. 27, E60–61; Gauthier and François 1981, nos. 29, 30, p. 25; (for the Turin pieces): Mallé 1950 and 1951, Mallé 1969; Petrassi 1982, nos. 167–74.

# 93. Equestrian Plaque

Limoges, ca. 1220

Copper (plaque): engraved, stippled, and gilt; (appliqué): repoussé, engraved, chased, scraped, stippled, and gilt; champlevé enamel: dark, medium, and light blue; dark translucent and light green, yellow, red and white; blue-black glass beads

24.8 x 13 cm (9¾ x 5⅛ in.)

PROVENANCE: Found at Saint-Germain, near Gourdon (Lot), before 1890 (see Rupin 1890, p. 419).

EX COLLS.: Louis Greil (before 1890; d. 1901), Cahors; Comte Chandon de Briailles, Épernay; J. Pierpont Morgan, London and New York

CONDITION: Substantial loss to gilding; left eye of figure missing glass bead; part of blue enamel of backplate missing; no restorations.

The Metropolitan Museum of Art, New York; Gift of J. Pierpont Morgan, 1917 (17.190.854)

The appliqué figure of a horse ridden by a beardless youth is attached to the vesica-shaped backplate by three domed rivets: at the lower center of the man's neck, on the left rear leg of the horse, and at the terminus of the sinuous groundline near the raised right front hoof. The finely engraved parallel lines of the man's wavy hair and the curved lines of the horse's mane and tail are offset by the firm and dramatic delineation of the folds of the rider's cape and tunic and by the intricate and minute tooling of the saddle and the harness. Both the man and the horse are notable for their smooth, rounded masses, particularly noticeable in the haunch and chest of the horse. Glass beads were inset for the eyes of the rider as well as for the one depicted eye of the horse. Lacking any attributes beyond his rounded, beardless face, the rider is unidentifiable.

The edge of the deeply beveled backplate (up to .7 cm in depth) is engraved with zigzag cross-hatching. There are seven visible holes for attachment to a larger ensemble. (The hole beneath the horse's tail was probably never used.) A large gilt circle in reserve, possibly unique, appears in the lowest portion of the plaque. This circle and an inner one are stippled with wavy lines of dots as are the veins of the four leaves that frame the enameled star that surrounds the inner circle. This motif is close to those used on other Limousin works, notably the chasse of Mozac (cat. 45) and the two tabernacles (cat. 70). This punched decoration is repeated on all the reserved elements of the enameled plaque, which is richly enhanced by a series of enameled circles, rosettes, and stars (or pointed lozenge patterns).

The ornamental richness of this enameling is in fitting contrast to the gentle movement of the pacing horse and the turning pose of the figure of the rider (frontal head and torso vs. profile right leg). The ensemble suggests the elegiac poetry notable in the appliqué relief fragment of the Annunciation to the Shepherds recently acquired from the Martin Le Roy Collection by the Louvre (cat. 94).

Because the rider lacks a crown, he cannot represent one of the Magi riding to Bethlehem, a frequent subject on chasses of the Limousin. Nor can he represent the equestrian Saint Martin, since he does not hold the sword, which, according to legend, Saint Martin used to divide his cloak to aid a beggar. This plaque may have been part of an unidentified hagiographic cycle, as in the story of Saint Fausta depicted on the smaller chasse of that saint in the Musée de Cluny.[1] However, Limousin hagiography, even in well-developed legends like that of Saint Valerie, offers no role for a horseman who is neither royal nor holy.

It is possible that this figure comes from an unknown secular context. Equestrian figures appear on coffret medallions beginning in the twelfth century (see cat. 35), where careful attention is similarly given to the presentation of the horseman, his steed, and details of his harness. In a secular context it is tempting to suggest a relationship to the seigneurs de Salviac, near Gourdon, where this plaque was found in the nineteenth century.[2]

WDW

NOTES

1. Gauthier 1950, pp. 50–51, 78–79, 158, pl. 53: "dernier tiers du XIIIe siècle."

2. I am grateful to Elisabeth Taburet-Delahaye for this suggestion. It has not yet been possible to confirm this through archival research. Nevertheless it is interesting to note that in the mid-thirteenth century (1261), Guillaume de Gourdon, seigneur de Salviac, founded a Cistercian abbey at Gourdon. See Cottineau 1935–39, vol. 1, col. 1310.

BIBLIOGRAPHY: Rupin 1890, p. 429.

# 94. Annunciation to the Shepherds

Limoges, ca. 1210–25

Copper: repoussé, engraved, chased, and gilt; dark blue enameled beads

13.8 × 7.9 cm (5⅜ × 3⅛ in.)

Ex COLLS.: Jacques-Rémi Texier, diocese of Limoges (1813–59); Victor Martin Le Roy (before 1900); Jean-Joseph Marquet de Vasselot. Acquired in 1991.

Musée du Louvre, Paris; Département des Objets d'art (OA 11349)

Only a part of the Annunciation to the Shepherds scene is depicted here: two figures direct their gaze toward the angel who has come to announce the birth of Christ; the angel would have been portrayed on an adjoining appliqué. But the harmonious composition creates a completely independent tableau, one with great poetic charm.

The scene is based on a brief passage from the Gospel of Saint Luke (2:8–9), frequently employed in medieval art, probably because it lent itself to bucolic representations. However, like scenes from the Infancy of Christ (with the exception of the Adoration of the Magi) it is quite rare in Limousin work. It appears that this relief is the oldest surviving example, but the scene is also depicted on the few Limousin chasses dedicated to the Infancy of Christ, such as the ones in the Nationalmuseet, Copenhagen, the Musée de Cluny, and the church of Sarrancolin (Hautes-Pyrénées).[1] All three are of later date and lack the elegiac quality that distinguishes this relief. All of them depict a musician placed among the shepherds: a horn player, as we see here, on the chasse in Copenhagen; a flutist on the one in the Musée de Cluny; and a piper on the one in Sarrancolin. Millet has attributed this detail, found since the sixth century, to the influence of pastoral scenes from antiquity or to alexandrines, but also to the influence of liturgical texts comparing the song of angels to that of shepherds and, perhaps, to a play on the words from the Greek text of the Gospel Ἀγραυλοῦντες, "passing the night in the fields," since Αὐλοῦντες means to play the flute.[2]

The light and airy composition, the serene pastures and gestures, the round faces with gentle, peaceful expressions—all achieved by skillful repoussé work, finished with particularly subtle engraving and chasing—allow a glimpse of the renewal brought about by the emergence of Gothic art. The supple lines and the attention given to naturalistic details confirm this tendency. The work is thus one of the best examples of the renewal of Limousin style through contact with the influences of early Gothic art. Moreover, these two shepherds are not unrelated to the figures on the stained-glass windows executed for the cathedrals of Chartres and Bourges in the years 1200–1225.

Within Limousin production, a commonality of style and spirit links this Annunciation to the Shepherds to the reliefs on the tabernacle of Cherves (cat. 98) and to a few isolated appliqués, such as the Wise Man of the Kestner-Gesellschaft in Hanover.[3] The relief thus takes its place among the works executed in Limoges at the beginning of the thirteenth century, and it is among the first that can be considered truly Gothic.

The relief is first recorded in the collection of the scholar Jacques-Rémi Texier, curé of Auriat, and then superior of the seminary of Dorat, who was born and died in Limoges. He created his collection in the Limousin, so it is quite probable that the Annunciation to the Shepherds came from an ensemble intended for a local building: from a large chasse or, more probably, an altar.

ET-D

NOTES

1. See Liebgott 1986, pp. 47–48, figs. 39, 40; Rupin 1890, pp. 349–50, pl. XXXIV (inv. Cl. 1896, former Debret Collection); Paris 1965, no. 473, pl. 56.
2. Millet 1916, p. 132.
3. Inv. no. 474 (H.: 14.11 cm [5 15/16 in.]); see Stuttmann 1966, no. 85.

EXHIBITIONS: Paris 1900, no. 2523; Paris 1995, no. 9.

BIBLIOGRAPHY: Marquet de Vasselot 1906a, no. 38; Taburet-Delahaye 1992; *Musiques au Louvre* 1994, p. 52.

# 95. Christ in Majesty

Limoges, ca. 1220–35

Copper: repoussé, engraved, chased, and gilt; dark blue, turquoise, and white enameled beads; colored glass beads

26.3 × 13.3 cm (10⅜ × 5¼ in.)

Ex coll.: Pierre Révoil. Acquired in 1828.

Condition: Slight wear to the gilding (mostly on the neck, knees, and book).

Musée du Louvre, Paris; Département des Objets d'art (MR R 305)

The crowned Christ is seated, blessing with his right hand and holding a book in his left. The figure, whose artistic merit is evident, was an appliqué on an enameled backplate and once formed the center of a composition devoted to a vision of the Last Judgment. Or it was at the center of a group of apostles surrounding Christ, as on the main altar of Grandmont, whose figures are of similar dimensions (cat. 58).

The appliqué can be placed among those works that demonstrate the new direction taken by one or more Limousin workshops at the beginning of the thirteenth century. Vigorous working of the copper in repoussé produced figures with sculptural qualities unequaled in earlier works—enamels were reserved for ornamentation of the backplate. For the execution of the figures, the artists turned to early Gothic models. Because of its flexibility, copper lent itself extremely well to the expression of this style, which favored fluid lines and soft curves.

This style is evident in the long lines of the silhouette, in the folds of the robe covering the body—whose forms are nevertheless masked by the play of slender folds and sinuous engraved lines—in the complex curves created by the flow of the robe to the ground, and in the large, bearded face with a mustache of finely engraved curls. This treatment recalls, for example, the recumbent figures of Fontevrault (shortly after 1220?), where one can observe the same play of lines and folds animating the drapery covering the bodies.[1]

The Christ from the former Révoil Collection is comparable to the relief at the center of the right-hand hinged door of the tabernacle of Cherves (cat. 98), even in the details of the orphrey and the decoration of the book. It can also be compared with the very beautiful angel with an undulating silhouette (fig. 7), in the Musée de Cluny, which might very well, with the other three symbols of the evangelists preserved in that museum,[2] have been associated with it. The evangelists seated at lecterns, from the former Spitzer and Desmottes[3] Collections, offer a more intensified version of this style, in the fullness of the drapery and the complexity of the play of curves. The strong stylistic linkage and the high quality of workmanship seen in these reliefs allow us to consider them as the work of a single atelier, similar to the one that executed the Grandmont apostles at about the same date. This Christ is distinguished from the apostles by several nuances in the overall aspect of the slighter silhouette and the markedly less plastic treatment.

ET-D

Notes
1. Sauerländer 1972, pp. 128–29, pl. 142.
2. Cl. 988b–e.
3. Palustre and Molinier 1890, no. 43; Desmottes Sale, Paris, March 19–23, 1900, nos. 35, 36.

Bibliography: Darcel 1891, no. 715; Marquet de Vasselot 1914, no. 107; Marquet de Vasselot 1941, p. 145 and pl. XXXII.

# 96. Appliqué Figure

Limoges, ca. 1220–40

Copper: repoussé, chased, engraved, scraped, stippled, and gilt; blue-black and turquoise glass beads

16.8 x 4.9 x 2.3 cm (6⅝ x 1¹⁵⁄₁₆ x ⅞ in.)

Ex coll.: J. Pierpont Morgan, London and New York

Condition: Some corrosion and loss of gilding, especially on top of head, shoulders, and base beneath feet; one turquoise bead missing from collar.

The Metropolitan Museum of Art, New York; Gift of J. Pierpont Morgan, 1917 (17.190.789)

This appliqué figure with bowed head, holding a rotulus, may represent an apostle or, possibly, an Old Testament prophet. His right arm is bent, and the raised hand and index finger are extended across the chest as though he is witnessing a sacred scene. The mantle, which envelops the left shoulder and waist, hangs below the right knee. The undergarment, or tunic, which touches the ground below the ankles, is completed by a decorated collar set with turquoise beads (two of the original three beads are still intact). The sloping ground is engraved and stippled with dotted horizontal lines and foliate patterns centered on a palmette of three leaves that appears between the feet. The distinctively narrow head is notable for its high relief, engraved curly beard, and parallel dotted lines in the cap of hair. The drapery folds are barely shaped, and their form depends almost entirely on a series of irregular, roughly edged lines. The lines of the mantle that appear over the right thigh are arbitrary rather than rhythmic or descriptive.

The relief was once secured to an enameled base plate by two rivets, which are still in place on the chest and at the bottom of the tunic. Their originality is confirmed by the engraved lines and gilding that are shared with the rest of the figure.

The appliqué figures of Limousin reliquaries, such as the ones on the large chasse of Saint Fausta, preserved in the Musée de Cluny,[1] Paris, and dated by Gauthier to the third decade of the thirteenth century, suggest the original function of this figure and provide a basis for dating. The principal vertical face of this chasse shows standing apostles, each holding a book and bowing his head in the direction of the central scene of the Crucifixion.

WDW

Note

1. Inv. no. 2826; Souchal 1963, fig. 17, p. 226; Gauthier 1972b, pp. 188, 373, no. 132, ill.

Bibliography: Unpublished.

# 97. Crosier: Christ and the Virgin

Limoges, ca. 1220–40

Copper: repoussé, engraved, chased, and gilt; glass cabochons; midnight-blue enameled beads

30.5 × 12.5 cm (12 × 4⅞ in.)

INSCRIBED: IH[ESU]S / x (Christus)

EX COLL.: Pierre Révoil. Acquired in 1828.

CONDITION: Wear to gilding, especially on the shaft; serpent's eyes have lost their enameled beads.

Musée du Louvre, Paris; Département des Objets d'art (MR R 809)

Exhibited in Paris only.

The Virgin and Child
(detail of the reverse)

Side of the crosier with Christ in Majesty

This work belongs to a group of crosiers depicting the addorsed figures of Christ and the Virgin in half-relief at the center of the volute, applied to an engraved copper plaque, which here is in the shape of an almond.

The theme is one that is frequently seen among Limousin crosiers: Marquet de Vasselot (1941) noted twenty-five, making this the largest group after those crosiers decorated with the Annunciation and Saint Michael slaying the dragon. To this number we should add the crosiers discovered since, such as the one found in the church of Saint-Benedict near Poitiers.[1]

This crosier presents an unusual peculiarity: it was conceived almost totally without enamel, which is limited to the blue beads in the eyes, a characteristic found in eight examples of the same type known to Marquet de Vasselot. Two of these were found in thirteenth-century tombs whose identification seems certain: the enameled crosier discovered at Orléans in 1889 in the tomb designated, according to an inscription, for Bishop William of Boesses (r. 1238–58), and the one, made of gilt copper like the Louvre's example, discovered in Angers in 1892 in the tomb of Bishop Michael of Villoiseau (r. 1240–61).[2] Even if they were reused for funerary purposes, these examples offer a *terminus ante quem* of about 1258–61 for dating the group. If the crosiers actually belonged to these two bishops—a hypothesis which should not be rejected—we must then date their execution to the years 1235–40.

Such a date corresponds to the date that can be inferred from an examination of the style and decoration of the Louvre example, which is very similar to that of the crosier from Angers. Indeed, the style of the two figures places them, like the preceding works (cats. 94–96), within the group of Limousin reliefs that adopted the supple drapery and full faces of early Gothic art, but the summary quality here betrays a more rapid execution. The finely engraved foliage—sometimes called "fern leaves"—on the shaft and the volute of the side with the Virgin is not rare in the second quarter of the thirteenth century, and it ornaments, notably, the tabernacle of Cherves (cat. 98), while the engraved background of rosettes in reserve is comparable to that on the reliquary of Saint Francis of Assisi (cat. 101).

ET-D

NOTES
1. Maupeou 1971.
2. Marquet de Vasselot 1941, nos. 36, 52.

BIBLIOGRAPHY: Naudet 1837, pl. XI; Barraud and Martin 1856, p. 89 and fig. 108, p. 88; *L'Art pour tous* (1878), no. 17, fig. 3761, p. 1712; Rohault de Fleury 1883–89, vol. 8, p. 103; Darcel 1891, no. 719; Marquet de Vasselot 1914, no. 78; Marquet de Vasselot 1941, no. 53, pl. IX.

# 98. Tabernacle of Cherves

Limoges, ca. 1220–30

Copper (plaques): engraved, scraped, stippled, and gilt; (appliqués): repoussé, chased, engraved, scraped, and gilt; champlevé enamel: medium blue, turquoise, medium green, yellow, red, and white

78 x 46 x 25 cm (closed) (30¾ x 18⅛ x 9⅞ in.)

PROVENANCE: Treasure of Cherves (Charente). Excavated at Château-Chesnel, near Plumejeau, December 11, 1896 (see also cats 65, 99, 100).

EX COLLS.: Comte Ferdinand de Roffignac, Angoulême; [Lowengard, London (sold 1902)]; J. Pierpont Morgan, London and New York

CONDITION: Loss to gilding and corrosion of copper and enamel due to burial; plaque on inner right gable missing (replaced by copy, on paper or parchment, of opposite scene of the Ascension); head of Virgin on proper right door lost after 1948 and before 1956; appliqué angel lost after 1948 and by 1964, probably before 1956; wood support fabricated after discovery in 1896.

The Metropolitan Museum of Art, New York; Gift of J. Pierpont Morgan, 1917 (17.190.735)

Standing on short legs, the tabernacle is in the form of a gabled cupboard with hinged doors. Gilded repoussé figures are applied to copper plates decorated with enameled foliate ornament. On the outside of the proper left door is the figure of Christ in Majesty, enthroned in a mandorla and surrounded by symbols of the evangelists. Opposite him on the proper right door is the Virgin with the Infant Jesus on her lap. She is framed within a mandorla and surrounded by four angels. Above them on the roof are two full-length angels, each holding a censer. Across the front runs a band of gilt copper inscribed with a decorative pattern derived from Kufic script, apparently based on the Arabic word *yemen*.[1] At the center of the open tabernacle, against its back wall, are appliqué figures representing the Descent from the Cross. Joseph of Arimathea takes the torso of the dead Christ in his arms as Nicodemus uses pliers and a hammer to remove the nails that still hold Christ's feet to the green-enameled cross. The Virgin takes her son's hands in hers and gently pulls them to her cheek; Saint John looks on from the opposite side, his head resting in his hand. Above the arms of the cross, two half-length angels hold emblems of the sun and moon. The Hand of God appears at the top of the cross; another figure of an angel once stood over it.[2]

On the insides of the doors are openwork medallions recounting the events that followed the Crucifixion, reading from lower left to upper right. The first is the Descent into Limbo, a nonscriptural image of Jesus leading souls by the hand out of the mouth of Hell, which is seen as the gaping mouth of a dragonlike beast. Set above it is the scene of the Holy Women arriving at the tomb of Jesus on Easter

Sunday. Following the account in the Gospel of Mark (16:1), they bear jars of unguent to anoint the body and are greeted by a man, seen here as winged, who informs them that Jesus has risen. In the almond-shaped medallion above, Mary Magdalen meets the risen Christ in the garden (Mark 16:9; John 20:14–18), where he backs away and advises her not to

The tabernacle closed

touch him yet. At the lower right, the apostles on the road outside the walls of Emmaus (Luke 24:13–35) are greeted by Jesus, attired as a pilgrim; in the roundel above, they dine with him at Emmaus and realize who he is when he breaks bread with them. In the oval at the upper right, Saint Thomas (Doubting Thomas) touches the wound in Jesus' side and is convinced of his Resurrection (John 20:24–29).

The interior side panels of the tabernacle have large lozenges with engraved figurative scenes framed at the corners by triangular enamel plaques, each depicting an angel in a roundel. At the lower left is the Entombment of Christ; at the upper right is the Ascension. At the lower right, Christ emerges from his tomb, with angels at either side. The base of the cupboard is covered with sheets of gilt copper depicting angels in roundels.

The tabernacle of Cherves is remarkable for its iconographic sophistication and for the dialogue established compositionally and visually between thematically related scenes. On the insides of the doors, Jesus guides souls out of the mouth of Hell at the lower left; at the lower right, he guides the apostles on the journey to Emmaus. On the center left roundel, the Holy Women seek Jesus' body and find it gone; on the center right roundel, Jesus offers his body to the apostles in the sacrament of bread and wine. At the upper left, he tells the Magdalen it is too soon to touch him; at the upper right, he invites Thomas to touch his wound. In the inside lozenge at the left, Jesus is lowered into his tomb; at the right, he rises from it. At the upper left, he leaves his apostles and rises to heaven; at the upper right, the Holy Spirit descends from heaven on the apostles in a representation of Pentecost.

The Descent from the Cross is both elegant and full of pathos, a masterpiece of Gothic relief sculpture. As such, it has rightly served as a point of comparison with works in other media, notably the ivory Descent from the Cross in the Louvre.[3] A number of gilt-copper relief sculptures produced in the Limousin but now isolated from their original contexts can be compared with those on the Cherves tabernacle. Notable among these is the Descent from the Cross preserved in the Abegg-Stiftung, Bern (fig. 121a), first recorded in 1870.[4] Most of these reliefs are presumed to come from altar frontals (see cat. 61).

The enameled ground of the Cherves tabernacle, with its strong concentric circles of reserved gilt copper enclosing full fleurons, seems to anticipate the enameled plate of the tomb effigy of John of France of after 1248 (cat. 146).

The identification of this enameled cupboard as a tabernacle for the consecrated Host has not been confirmed: the Church of the Middle Ages had no universal custom for the reservation of the Eucharistic bread or regulations requiring a tabernacle. Nor is there a wealth of comparative medieval examples. Only one other Limoges tabernacle of this type is known; it was acquired by the cathedral of Chartres in the

Interior panel, left side

The tabernacle open

Fig. 98*a*. Tabernacle. Chartres Cathedral

nineteenth century, and its earlier history is not known (fig. 98*a*).[5] The supposition that the enameled cupboard from Cherves is a Eucharistic tabernacle is based on its resemblance in form to later tabernacles, its subject matter, and even the gilt-copper base plate which would allow an enclosed pyx to slide easily in and out.[6]

Soon after its discovery in 1896, the tabernacle was presented to the Société archéologique de la Charente by Maurice d'Hauteville, curator at Angoulême and son-in-law of Ferdinand de Roffignac, on whose property it was unearthed. He suggested that the treasure could have come from the Benedictine monastery of Fontdouce, founded in 1117.[7] More recently it has been supposed that the treasure at Cherves comes from the Grandmontain foundation at Gandory, of which, unfortunately, there are no remains.[8]

Since its discovery, the tabernacle of Cherves has been recognized as a masterpiece of Limoges work in the Gothic period. Part of a larger treasure, from which several other pieces are included here (cats. 65, 99, 100), it was exhibited successively at Poitiers, Brive, and Limoges, and then at the Musée de Cluny before being sent to Great Britain.[9]

BDB

NOTES

1. Apparently found on the rim of pottery vessels of the twelfth and thirteenth centuries; Medieval Department file note, Mr. Wilkinson to Mr. Forsyth, undated.
2. The earliest records of the excavation at Château-Chesnel indicate that this angel was damaged by the shovel of one of the workmen, but that this was of little concern since the angel appeared to be a cast, not a repoussé, figure.
3. Gaborit-Chopin 1988.
4. *L'Art pour tous* (1870), no. 248.
5. A number of its plaques were reproduced in the nineteenth century and dispersed among prominent collectors. In that process, some elements of the tabernacle itself seem to have been replaced.
6. Gauthier (1978c, p. 39) illustrated a mock-up of the tabernacle enclosing a pyx. Although it need not have been in the form of an image of the Virgin and Child, her reconstruction is plausible.
7. Barbier de Montault 1897a and b; Cottineau 1935–39, vol. 1, col. 1179.
8. See Gaborit in Québec–Montréal 1972, no. 29, and introduction (above).
9. I am grateful to Martine Larigauderie, whose research and dissertation, *Prieurés grandmontains en Charente*, led her to find local press coverage on the treasure, preserved in the Archives Départementales de la Charente.

EXHIBITIONS: Poitiers, Brive, Limoges, Paris (Musée de Cluny), Glasgow, 1897 (per Barbier de Montault 1897a and b; see also note 9 above).

BIBLIOGRAPHY: Barbier de Montault 1897a and b; Guibert 1901, p. 13; Breck and Rogers 1929, p. 104, fig. 50; Souchal 1964b, p. 133; Gauthier 1972b, no. 130, p. 372; Gauthier 1978c, pp. 23–42; Otavsky 1973, p. 57, fig. 12; Gaborit 1979, p. 179, fig. 27; Gaborit 1988, p. 37, fig. 21.

# 99. Bowl from a Censer

Limoges, ca. 1200–1210

Copper: pierced, chased, and gilt

H.: 5.5 cm (2⅛ in.); diam.: 10.5–10.9 cm (4⅛–4⁵⁄₁₆ in.)

PROVENANCE: Trasure of Cherves (Charente). Excavated at Château-Chesnel, near Plumejeau, December 11, 1896, (see also cats. 65, 98, 100). Between 1898 and 1901, in the possession of a merchant named Brauer. Gift of Mme Goldschmidt-Prizbram, Brussels, to the Musée de Cluny, 1901.

CONDITION: External gilding well preserved; a palmette removed from the base of a gadroon. Cleaned in 1995.

Musée National du Moyen Âge, Thermes de Cluny, Paris (Cl 14026)

Msgr. Xavier Barbier de Montault suggested that the liturgical pieces found at Cherves might be the remains of a hidden treasury from one of the abbeys near Cherves, such as Fontdouce, Bassac, or Chastres; more recently, Jean-René Gaborit and Martine Larigauderie have proposed a more likely provenance, the former Grand-montain dependency of Gandory, which was located in the same district as the present-day commune of Cherves (see cat. 98).

Barbier de Montault published the Cherves treasury upon its discovery; he saw this object as a *canistrum*, or container for the Eucharist, which would have been suspended by thin chains above the altar. According to his main argument, an openwork receptacle such as this could not have held incense. However, evidence suggests that this was the lower part of a censer that has lost an inner sheath or a small cup used to contain the aromatic resin. This hemispheric bowl, composed of eight petal-shaped, openwork sections, was probably fabricated by casting; its truncated conical base and the four rings designed for the thin suspension chains are fastened by rivets. The ornamentation of this small receptacle in the form of an eight-petaled chalice was determined by its radiating composition. Two motifs whose outlines stand out distinctly against the openwork ground—the figure of an angel and a cluster of rinceaux, each repeated four times—alternate in eight adjoining lobes. The angels, with wings spread and faces turned in three-quarter profile, hold books in their left hands and arise to mid-leg from the corollas of flowers chased in low relief; the only difference among them is in the gestures of their right hands, some pointing, some touching the book. In the heart-shaped or round scrolling of the rinceaux, trilobed palmettes bloom, each one drawn differently. Within the open spaces of the border, trefoil buds are inscribed. A frieze of engraved scallops decorates the upper border.

Of all the surviving Limoges censers, only the one from Cherves has an openwork bowl. The treatment and the quality of the workmanship of certain details—particularly the small, classicalizing appliqué heads and the variety of chased motifs—link the censer to Limousin works of the years 1200–1210. As far as we know, only the cover of another censer preserved in the Liverpool Museum[1] exhibits similar structural and stylistic characteristics: the same pierced dome with eight gadroons in low relief, the same chased and openwork motifs within the gadroons, the same appliqué heads on the figures of the angels, and, above all, the same dimensions. Nonetheless, certain slight differences in the ornamentation and its handling distinguish the cover from the Cherves incense bowl, such as the inlaid enamel on the angels' wings and on other nonpierced parts, and the fabulous birds with human heads and scrolling tails that alternate with the angels. The complementary parts of the two censers are linked by their execution and ornamentation to a small series of Limousin crosiers from the beginning of the thirteenth century,[2] which exhibit, in their volutes or their knops, analogous openwork and chasing, as well as a similar ornamentation of angels, rinceaux, and fabulous creatures.[3]

GF

NOTES
1. Liverpool Museum (Inv. M. 23); see Gatty 1883, p. 28, no. 88.
2. See the research by M. M. Gauthier in the forthcoming *L'Ecole de Limoges*, vol. 2 of the *Émaux méridionaux* (Gauthier 1987).
3. Marquet de Vasselot 1941, no. 11 (Saint-Maurice of Agaune); no. 12 (Pincot Collection); no. 13 (David-Weill Collection); no. 14 (cathedral of Troyes); no. 33 (Metropolitan Museum, 25.120.443); no. 113 (cathedral of Trier); and finally, a crosier excavated at the cathedral of Cahors in 1980.

EXHIBITION: Angoulême, Poitiers, Brive, Limoges, and Paris (Musée de Cluny), 1897 (according to Barbier de Montault 1897).

BIBLIOGRAPHY: Barbier de Montault 1897b, pp. 33–40, pl. II; Gauthier 1978c, pp. 23, 39, n. 1, 2; Larigauderie-Beigeaud 1994, p. 42.

# 100. Suspension Crown from a Eucharistic Dove

Limoges, second quarter of the 13th century

Copper: stamped and gilt; champlevé enamel: ultramarine blue and turquoise

H.: 6.3 cm (2½ in.); diam.: 26.7 cm (10½ in.)

PROVENANCE: Treasure of Cherves (Charente). Excavated at Château-Chesnel, near Plumejeau, December 11, 1896 (see also cats. 65, 98, 99). Between 1898 and 1901, in the possession of a merchant named Brauer. Gift of Mme Goldschmidt-Prizbram, Brussels, to the Musée de Cluny, 1901.

CONDITION: Stamped facing of revetment largely lost; modern chains. Cleaned in 1995.

Musée National du Moyen Âge, Thermes de Cluny, Paris (Cl 14027)

In 1897 Msgr. Xavier Barbier de Montault, lacking any references to similar pieces, believed this crown to be part of a hanging lamp whose lighting ensemble of glass had been lost. He compared it with the monumental crowns of oil lamps used since the early Middle Ages to illuminate medieval churches. Gauthier sees it rather as the suspension crown for the platform of a Eucharistic dove and compares it with the crown once preserved in the church of Laguenne (Corrèze), reconstructed by Ernest Rupin:[1] the crown was suspended from four thin chains that were joined at the top by a large ring attached to a cord; four other thin chains were hung from the bottom of the crown and held up the round tray upon which the dove rested. The dove was veiled by a cloth canopy, sewn to the crown through holes pierced all along the rim, which could be raised or lowered by a system of counterweights attached to the cord. The presence of such a crown and its platform is confirmed by a description from the inventory of the treasury at Grandmont in 1611: "A receptacle in the form of a gilt-silver dove in which the Blessed Sacrament rests; the dove sits on a gilt-silver coupe, and over the receptacle there is a crown of crimson satin with small turrets and four thin silver chains."[2]

The crown from Cherves is made from a single band of copper three centimeters (1⅛ in.) high and eighty centimeters (31½ in.) in circumference, shaped from a single sheet of copper two millimeters thick, whose overlapping ends are fastened with two rivets. The lower part of the border is pierced with twenty-four holes, from which a veil was suspended. A thin, stamped-metal revetment sheet covers the outside of the band; its upper edge is folded over the inner side. Eight almost identical architectural appliqués are affixed, each by three rivets, at regular intervals around the circumference; these appliqués are in the form of turrets surmounted with crosses, decorated with two types of pseudo-heraldic motifs on an enameled ground of medium blue: on six a turquoise chevron is placed on a turquoise band; the other two have single turquoise bands. Only three of the eight stamped motifs decorating the rectangular compartments between these turrets are still intact: each depicts a half-length angel surrounded by rosettes in a beaded frame. On the inside of the crown, the four rings holding the thin suspension chains are still intact.

The use of the dove, symbol of the Holy Ghost, as a Eucharistic receptacle above the altar is confirmed as early as the fourth century;[3] in the eighth century these Eucharistic suspensions appear to have replaced the bejeweled votive crowns that one also finds hanging from canopies in churches. The combination, in thirteenth-century Limoges, of enameled Eucharistic doves with crowns might well be a vestige of this ancient custom;[4] some fifty doves are now counted in the *Corpus des émaux*, proving, in any case, the use of some of them as enameled Eucharistic receptacles at the beginning of the thirteenth century. They have, however, with the exception of the one from Laguenne, all lost their suspension apparatus. Only one other Limousin crown of this type is still preserved today: that is the one traditionally known as the "crown of Roger II," in the treasury of Saint Nicholas of Bari in the Puglia region of southern Italy. It was long, and erroneously, thought to be the crown from the coronation of that king in 1130.[5]

GF

NOTES
1. This dove, stolen in 1907, has never been recovered; see Rupin 1890, pp. 225–26, 232, figs. 291, 298.
2. Texier 1857, col. 872.
3. Cabrol and Leclercq 1907–53, vol. 3 (1914), cols. 2231–34.
4. A bibliography on this subject was recently compiled by Foucart-Borville (1987, pp. 267–89).
5. Rome 1994, no. 85.

EXHIBITION: Angoulême, Poitiers, Brive, Limoges, and Paris (Musée de Cluny), 1897 (according to Barbier de Montault 1897a and b).

BIBLIOGRAPHY: Barbier de Montault 1897b, pp. 41–46, pl. III-1; Gauthier 1978c, pp. 23, 39, n. 1, 2; Larigauderie-Beigeaud 1994, p. 42.

# 3

## Animation and tradition

# 101. Reliquary of Saint Francis of Assisi

Limoges, shortly after 1228

Copper: punched, engraved, chased, and gilt; champlevé enamel: black, lapis and lavender blue, almond and light green, yellow, brick red, and white; turquoise enameled beads; rock crystal and glass cabochons; wood core

36.2 × 20.6 cm (14¼ × 8⅛ in.); diam. of base: 15.2 cm (6 in.)

Provenance: Said to be from a church in Majorca.

Ex colls.: Charles Ducatel (sale, Paris, April 21, 1890, no. 2, pl. 1); Michel Boy. Acquired in 1899.

Condition: Eight glass cabochons are missing from the back.

Musée du Louvre, Paris; Département des Objets d'art (OA 4083)

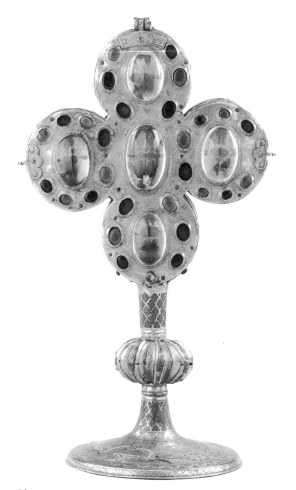

Obverse

This quatrefoil reliquary is made up of a thick wood core with a double revetment of copper sheeting on the obverse. The first gilt-copper plaque, with five openings in the form of a Greek cross, is attached by rivets atop the grooves in the wood for the relics. A second plaque, mounted on hinges, is composed of two gilt-copper sheets and is decorated with five large cabochons of rock crystal, each encircled by smaller cabochons of colored glass to form rosettes, which stand out from the guilloche-patterned background. The reverse, by contrast, has a single, thick plaque containing an enameled depiction of Saint Francis receiving the stigmata. The base also combines goldsmiths' work and champlevé enamels: the elegant ribbed knop is adorned with bands of turquoise-enameled beads, while the shaft is enameled with an imbricated pattern and the socle with confronted peacocks drinking from a large, footed cup.

This type of footed quatrefoil reliquary, previously unknown in Limoges work, had in fact been known since the twelfth century in the Holy Roman Empire and northern France; this is confirmed, for example, by the reliquary of Saint Henry (Hildesheim, ca. 1170) now in the Louvre[1] and the reliquary of Saints Sixtus and Sinicius (early twelfth century) in the treasury of the cathedral of Reims.[2] As seen here, the chief modifications from these earlier works include the elimination of supports under the circular base and the introduction of the ribbed form of the knop, common in the thirteenth century. The structure of the quatrefoil reliquary, superimposing two plaques on the principal face, and the decoration of crystal cabochons evoke works executed in northern France as well as in the Rhine-Meuse region during the first third of the thirteenth century.[3]

The iconography of the enameled plaque is entirely new. Certainly executed after 1226, the year of Saint Francis's death, and more likely after 1228, the year of his canonization—as he is shown with a halo—this is one of the earliest representations of the saint. The only earlier work is the largely repainted fresco in the hermitage at Subiaco, where Francis is shown standing and without a halo.[4] On the other hand, the plaque should be compared with the very similar example from a reliquary of the same type (fig. 101a) transferred by the Louvre to the Musée de Cluny.[5] Both plaques represent the oldest extant depictions of the miracle of the stigmata.[6]

Matrod (1906) has shown how the depictions of the saint on the Louvre reliquary and its "twin" plaque in the Musée de Cluny come closest to the original account of the miracle, as recorded by Thomas of Celano at the behest of Gregory IX in 1229: "Two years before his death, while sojourning in a hermitage known as La Verna, Francis had a

The reliquary open

Reverse: Saint Francis
Receiving the Stigmata

Fig. 101*a*. Plaque from a reliquary of Saint Francis. Musée National du Moyen Âge, Thermes de Cluny, Paris

Fig. 101*b*. Base of a reliquary of Saint Francis (detail). Museo Nazionale del Bargello, Florence

vision: above his head he saw a man with six wings like seraphim, nailed to a cross with his hands extended and his feet joined. . . . As . . . the novelty of this vision weighed painfully on his heart, the marks of the nails began to appear on his hands and feet, just as he had seen them on the man he had perceived above him."[7]

Although this form and this iconography are new in Limoges work, the style and the enameling remain archaic. Flesh and faces were seldom rendered in enamel, and then primarily in the twelfth century, in the group of works executed for the abbey of Grandmont from about 1190 to 1210 (see cats. 57, 62, 65). The face of the seraph here and, to a lesser extent, that of the saint reveal some kinship to the latter group. The juxtaposition of graduated colors to represent the ornamental patterns of peacocks' tails and the leaves of trees recalls the technique frequently employed on chasses with vermiculé backgrounds to decorate the wings of angels (see cats. 24, 27, 29).

Stohlman (1950) had assembled a number of works displaying not only the same technique but also the same type of decoration: rosettes or stars in reserve on a copper background. However, the variance in the style and quality of these objects suggests that they originated in several different workshops, over several decades. This stylistic and chronological diversity is confirmed by other works that have since been published, such as the chalice discovered in the abbey of Rusper (Sussex) and now in the British Museum.[8] These last works have been attributed to artists from Limoges who were active in different countries, including Spain (Hildburgh 1936), Italy (Salet 1958), and even England (the Rusper chalice; see London 1984). However, as Gauthier repeatedly has stressed,[9] numerous aspects of these objects point to Limoges as their place of manufacture. The very diversity of the places where these items are conserved or where they were found argues for a common origin in Limoges. The connections presented by their technique and decoration to the works made in Limoges in the late twelfth or early thirteenth century would likewise support that argument—as would the above-mentioned borrowings of form and construction widespread among northern goldsmiths. Various details confirm the inclusion of the reliquary of Saint Francis among those objects produced in Limoges, such as the reticulate ornamentation of the shaft, which is found on the volutes of numerous crosiers (cats. 107–11), and the confronted peacocks on the base, which are seen, as noted by Marquet de Vasselot, on several gemellions that were probably executed in about the mid-thirteenth century.[10] Even the vegetal decoration, which is undeniably one of the most original and engaging features of the design of this reliquary, recalls the friezes with "fern-leaf" motifs that were widespread in Limoges during the thirteenth century (see cat. 98).

It is equally true that the fabrication of reliquaries dedicated to Saint Francis of Assisi prompted Limoges artists to devise original creations. The undoubtedly deliberate choice of iconography, like that of the present work, and of techniques of production and decoration then considered "old-fashioned" in Limoges could be explained only by the necessity of meeting the precise demands of a patron, who may also have influenced the ultimate form and structure of the object. It has been suggested[11] that Saint Anthony of Padua, who was in the Limousin in 1226–27, may have commissioned the work; his interest in nature, and especially in trees, may have led to the abundance of vegetal ornamentation here, which was unusual on works made in Limoges.

In any event, it is certain that the Franciscans had established themselves at an early date in Limoges: the chronicle of Peter Coral, abbot of Saint-Martin of Limoges, relates that they first settled there in 1223.[12] Thus, it is not surprising that relics arrived in the city shortly after 1228–29, giving rise to the creation of reliquaries whose relatively inexpensive materials and figural decoration corresponded perfectly with the ideals of the first Franciscans. Moreover, it appears that these reliquaries were not executed for a specifically Limousin clientele: the plaque in the Musée de Cluny was bought in Rodez in 1851, and the reliquary acquired by the Louvre in 1899 is said to have come from Majorca, according to data provided by Molinier on the occasion of the purchase[13]—information, however, that still has not been verified.[14]

The base formerly in the Carrand Collection and now in the Museo Nazionale del Bargello, Florence (Inv. 653 C), which is very similar to the base of the Louvre reliquary, may have belonged to the reliquary whose reverse side is in the Musée de Cluny; at any rate, it certainly comes from an object of this type. That base is adorned with peacocks inscribed within medallions comparable to those on the base of the Louvre reliquary, except for one shown presenting a donor who kneels before Saint Francis (fig. 101*b*). The style of the engraving, which is of very fine quality, is close to that on the back of the chasse of Saint-Viance (cat. 118) and to that on the large chasse of Saint Fausta in the Musée de Cluny.[15] This confirms a dating slightly after 1228–30 for this small group of works that would leave behind an artistic legacy in Limoges (see cat. 143).

ET-D

NOTES

1. Inv. OA 49; Hildesheim 1993, no. IX–II.
2. Paris 1965, no. 133, pl. 96.
3. For example, the reliquary from Termonde, Belgium, for the construction, and a quatrefoil reliquary of unknown provenance for the cabochon decoration, both in the Musée de Cluny, Paris. Taburet-Delahaye 1989, nos. 7 (reliquary of Termonde) and 8 (reliquary).
4. *Francesco d'Assisi* 1982, no. 81, p. 107.
5. Inv. OA 84. Acquired in 1851 from the "directeur de la maitrise" (probably the choir director) of Rodez, Monsieur l'abbé Maymard; see Cologne 1985, no. 67b.
6. The first Italian example occurs in the altarpiece by Bonaventura Berlinghieri, dated 1235, in the Museo Civico, Pescia. *Francesco d'Assisi* 1982, nos. 8–20, pp. 92, 116–17; for depictions of Saint Francis and the stigmata in Italian painting, see Scarpellini, ibid., pp. 91–126.
7. Matrod 1906, p. 516, which also supplies the original Latin text, after Rosedale 1904.
8. Gauthier 1972a, no. 73; London 1984, no. 293.
9. Notably, Gauthier 1978a, p. 282.
10. Marquet de Vasselot 1952, p. 21 and nos. 41, 59–61.
11. Matrod 1906, pp. 17-18, and then by Gauthier, especially 1983, p. 138.
12. Becquet 1990, p. 41; Roy-Pierrefitte 1864, pp. 154–55.
13. Louvre, Archives de la Direction des Musées de France, committee minutes of May 4, 1899. The information does not appear in the catalogue of the 1890 Ducatel sale, but perhaps Molinier was privy to special information. While the object may have reached Majorca at a later date, it is nevertheless noteworthy that the conquest of the island by King James I of Aragon took place in 1229–31. The conquest is mentioned in the *Chronicle* of Peter Coral, immediately after the establishment of Franciscans and Dominicans in Limoges (see Becquet 1990, p. 41). The reliquary could, then, have contained relics sent to the churches founded at that time. The object still contains relics with written authentications, some of which are thirteenth century. They are identified as relics of soil from Nazareth, relics of Saint Mauritius and the Theban Legion, Saint Poncius, Saint Cesarius, Saint Severus, Saint Ruffina, Saints Chrysanthos and Daria, and Saint Placidus. However, no relic of Saint Francis is mentioned. Either the authentication has come off, or the relic itself has been removed.
14. I am very grateful to Juan Domenge y Mesquida, who graciously consulted published and unpublished inventories of churches in Majorca.
15. Gauthier 1972b, no. 132.

EXHIBITIONS: Vatican City 1963, no. 27, pl. XV; Paris 1968a, no. 166; Paris 1970, no. 89; Cologne 1985, no. H 66.

BIBLIOGRAPHY: Rupin 1890, pp. 493–94, figs. 546, 547; Matrod 1906; Marquet de Vasselot 1914, no. 110; Hildburgh 1936, pp. 127–28; Stohlman 1950, p. 330; Gauthier 1950, pp. 32, 64, 74; Marquet de Vasselot 1952, pp. 21, 30; Gauthier 1972b, no. 128, p. 283, ills. 184, 371; Gauthier 1972a, p. 292 and pl. CXX; Gauthier 1978a, pp. 281–82, fig. 29; Gauthier 1982, fig. 20, a, b; Gauthier 1983, pp. 138–39, fig. 82 p. 138.

# 102. Saint William of Bourges

Limoges, ca. 1218–25

Copper: engraved and gilt: champlevé enamel: dark blue, turquoise, medium green, yellow, red, and white

20.3 x 17.4 cm (8 x 6⅞ in.)

INSCRIBED: s[ANCTUS] GUILELM[US]; on the reverse: [1758]: EGO P.[etrus] LE LIEVRE CLERICUS / —UEUS JOANNIS E. R. GAS / SELIN FILIUS CO[G]NATOS / ORO NE HUNC PANCEUM AB AN / NIS PLUS CENTUM IN DOMO NOSTRA / SERVATUM AD ALIOS NOBIS ET NOMINE / ET SANG[UINE] / NISI JUNCTOS RELINQUA[NT]. (I, Pierre Lelièvre, priest, son of Gasselin, ask my relatives not to give up this panel [?], which has been preserved in our house for more than a hundred years, except to persons who are allied to us by name and blood.)

EX COLLS.: P. Lelièvre (in 1758); J. Pierpont Morgan, London and New York

CONDITION: Gilding almost entirely lost; losses to enamel and wear to engraving.

The Metropolitan Museum of Art, New York; Gift of J. Pierpont Morgan, 1917 (17.190.694)

A bearded bishop-saint stands at the center of the plaque beneath a trefoil arch, vested in a white alb, red dalmatic, and blue chasuble; a green maniple is at his wrist, and a pall, or superhumeral, around his neck. On his head he wears a small white-and-yellow miter of the type used in the first half of the thirteenth century. In his left hand he holds a crosier with a green shaft and a volute terminating in a yellow fleur-de-lis. He raises his right hand in blessing to the tonsured figure of a monk wearing a white alb and a green cape and cowl. A deacon stands on the bishop's left. Holding both a book and the processional cross of the bishop, he wears a white alb, yellow amice, and blue dalmatic. A pair of half-length angels descend from above the arch. An enameled border of repeating loops is pierced at each corner and at the center of each side for attachment. A scrolling, foliate pattern engraved into the copper ground enlivens the surface behind the figures, while a horizontal band at knee-height is inscribed s GUILELM, identifying the bishop-saint as William.

Since the back of the plaque bears an inscription in Latin indicating that it belonged to a priest named Lelièvre in 1758,[1] it is reasonable to assume that the episcopal saint represented was a French bishop named William, of which there were four. Guillaume de Saint-Lazare can be excluded, since he had no monastic associations and consequently would not be shown in the company of a monk. Guillaume de Saint-Brieuc did not die until 1231 and was canonized in 1247, later than the style of this plaque would suggest. Guillaume Tempier, bishop of Poitiers, died in 1197 and was buried at the monastic church of Saint-Cyprian behind the principal altar. Unfortunately, nothing is known about the

decoration of his sepulchre before the body was translated in the sixteenth century; the church itself was destroyed in 1793.[2] Among French bishop-saints, the most venerated was Guillaume de Bourges; both the biography of the saint and the history of his cult support the association of this plaque with William of Bourges, as Ross—and Gauthier—has proposed.[3]

Son of the count of Nevers, William of Bourges became a canon, first at Soissons and then at Paris. He subsequently retired to the hermitic monastery of Grandmont in the diocese of Limoges. Some years later, he again changed orders, becoming a Cistercian. He was, in succession, abbot of Pontigny, of Fontaine-Jean, and finally of Chaalis until 1199. Because of his links to Grandmont, he was involved in the resolution of disputes there on behalf of the Church hierarchy. Named archbishop of Bourges in 1205, he died in 1209, while preparing to leave on the Albigensian Crusade, and was hastily interred.[4] The first funds for maintaining candles before his tomb date to 1213.[5] William of Bourges was canonized in 1218; soon thereafter, Pierre of Chaalis compiled a history of his life.[6]

An altar dedicated to Saint William was located behind the high altar of the cathedral at Bourges; it was surely built soon after his canonization, for the pope's letter of canonization required that the body of the bishop be placed in a proper and fitting part of the cathedral.[7] In 1222, Robert de Courtenay, grandnephew of William and seigneur of Mehun-sur-Yèvre, provided funds for a perpetual candle before the body of William. The countess of Nevers, William's grandniece, did the same in 1225. She was the patron of the altar on which the chasse of the saint was placed, followed, before 1257, by the gift of a stained-glass window representing her uncle.[8] The chasse containing the relics of Saint William was not described until the sixteenth century, when it was noted as being of precious metal. It is possible that it replaced an earlier, thirteenth-century shrine. But it may be more reasonable to suggest that the Metropolitan Museum's plaque representing Saint William formed part of the revetment of the altar given by the Countess of Nevers and made at Limoges or by a Limousin enameler for Bourges.

The diocese of Bourges maintained close links with Limoges, evidenced, for example, in the thirteenth-century testimony of Jean and Sebrand, bishops of Limoges, that it was their custom to consult the archbishop of Bourges, rather than the see of Bordeaux, on Church matters.[9] The square format and size of the Saint William plaque, with regularly spaced attachment holes around its perimeter, recall the plaques of the Adoration of the Magi and the meeting of Hugo Lacerta and Stephen of Muret from the

altar at Grandmont (cat. 57a). The style of the figures on the plaque, with their simple lines of drapery, broad fields of enamel, and distinctive yellow, seems to anticipate the quatrefoil reliquary of Saint Francis (cat. 101) of after 1228, consistent with a date soon after the saint's consecration in 1218 and the maintenance of his cult with votive candles by 1222–25.

Although it does not include the description of an enameled altar frontal, the history of the cathedral indicates that there was a wealth of copper in the church in an era before such assets were systematically inventoried, that much of it was destroyed before the end of the sixteenth century, and that efforts were made to save works of art at that time.

The cathedral of Bourges and its contents suffered heavy damage during and following the Protestant occupation of the city in 1562. The relics of William were burned and "tossed to the winds."[10] The rood screen and altar decorations were destroyed. The fragments of reliquaries, chasses, and vases taken by Gabriel de Lorges, count of Montgomery, weighed 651 marcs.[11] But precautionary measures had been taken and much was saved. In May of that year, payment was made to Grand Jehan Legier, "orfèvre de la dicte église . . . pour ses peines et vacation d'avoir dessemblé les grandes reliques de la dicte église, et les avoir mises en pièces afin de plus aisement les transporter." Additional payments were made for "fosses" (coffrets) in which to hide the "reliques." These were then buried in the garden of the house of Sieur Thomas Nostin.[12] In addition, two silver plates were immured in a window of the choir. The following month, an order for the confiscation of Church property required "de faire la recherche des reliquaires cuyvres et fers, pour être vendus par les echevins assistés des huit deputés du bureau."[13] After the departure of the Calvinists, the canons of the cathedral were obliged to sell much of what had been hidden, including the chasse of William and the "table de l'autel," which was melted to pay rent due to

the king on November 26, 1563.[14] By the eighteenth century, the altar at the site of Saint William's had become the marble altar of "féries ou anniversaires" (feasts or anniversaries).[15] Bearing in mind that the inscription written by the priest Lelièvre in 1758 indicates that the plaque had been in his family for more than a century, it is entirely plausible that it would have been removed from the cathedral of Bourges, the center of William's cult, for safekeeping during the Wars of Religion.

BDB

NOTES

1. I am grateful to Catherine Gros of the École des Chartes for this transcription.
2. I am grateful to Sigrid Goldiner for this information on the diocese of Poitiers and the church of Saint-Cyprian; see Favreau 1988.
3. Ross 1932b; Gauthier 1992, pp. 100–101, fig. 3. Ross correctly related the style of the plaque to another of a bishop-saint in the Fitzwilliam Museum, Cambridge (M28a, b).
4. The location of the first tomb has been variously understood. See Branner 1989, p. 53; Gauchery and Grossouvre 1966, p. 29.
5. Archives départementales du Cher, série G, 8G351.
6. See *Bibliotheca Sanctorum* 1961–70, vol. 7, cols. 459–60. The first vita was published in *Analecta Bollandiana* 3 (1884), pp. 271–361.
7. *PL*, 2, col. 766. The high altar was consecrated in 1256, but is believed to have been in place by 1214. See Branner 1989, p. 19.
8. "Les Reliques de Saint Guillaume à la cathédrale de Bourges," typescript, p. 4. I am grateful to J.-Y. Ribault, Directeur des Archives départementales du Cher, for a copy of this typescript.
9. Archives départementales du Cher, série G, pp. 140–41.
10. Clément and Guitard 1900, p. 80.
11. Giradot and Durand 1849, p. 192; Boinet 1929, p. 20.
12. Giradot 1859, p. 244. "Goldsmith of the cathedral . . . for his work. He disassembled the great relics of the church, took them apart so they could be carried more easily."
13. Giradot and Durand 1849, p. 192. "To search for reliquaries, copper and iron, to be sold by the magistrates, assisted by eight deputies of the office."
14. Giradot 1859, p. 246.
15. Boinet 1929, p. 23.

BIBLIOGRAPHY: Ross 1932b, Gauthier 1992, pp. 100–101, fig. 3.

# 103. The Cross of Garnerius

Limoges, ca. 1220–30

Copper: engraved and gilt; champlevé enamel: black, medium blue, almond green, pink, red, and white

44.5 × 21.2 cm (17½ × 8⅜ in.)

INSCRIBED: IOHANNIS : GARN / ERIVS : LE / MOVICEN / SIS : ME FE / SIS : FRAT / RIS : MEI (John Garnerius of Limoges made me . . . of my brother) IH[ESU]S / XPS (Christus)

PROVENANCE: Said to come from the Benedictine abbey of Savigny-le-Vieux (Manche)

EX COLLS.: Private collection, department of the Manche (in 1884); Victor Gay, Paris (before 1887; Louvre, Archives, handwritten inventory, December 1, 1908, no. II, 1). Gift of a group of friends of Victor Gay, 1909

CONDITION: Much in-filling with colored mastic; wear to the gilding.

Musée du Louvre, Paris; Département des Objects d'art (OA 6278)

The cross of Garnerius shares with the Alpais ciborium (cat. 69) the rare and exceptionally interesting feature of bearing an inscription that in all probability indicates the name of the maker.[1] Such exact information is only found in Limoges on a later work by one of its enamelers—the reliquary bust of Saint Ferreolus, made by Aymeric Chrétien in 1346 (cat. 157).

The text of the present inscription, IOHANNIS GARNERIUS LEMOVICENSIS ME FESIS [for *fecit*] FRATRIS MEI, is more developed than that on the Alpais ciborium. The last words, which have led to much speculation on the part of most scholars,[2] do not form a comprehensible text: the inscription thus is probably either incorrect (*fratris* for *fratribus*) or missing something, but it nevertheless records that the cross was made by Garnerius for or in memory of his "brother"—fraternal brother or member of the same monastic community.

The exceptional singularity that the inscription gives this work justifies its renown in spite of its poor condition, rarely encountered in other works. The enameling, in fact, is almost totally lost on the body and the drapery of Christ, and has been replaced by colored mastic. The poor condition does not, however, entirely mask the stylistic qualities of the work. The Christ is clearly distinguished from other late-twelfth-century[3] and early-thirteenth-century examples in its enduring respect for the artistic traditions of the preceding epoch. This is one of the first representations of Christ on the cross in Limoges that can be considered as fully Gothic, in the position of the body and the expression of the suffering Christ, whose arms and legs are neatly bowed; in the face, which is turned sharply toward the shoulder; and in the drapery of the long loincloth, with its gracefully curving folds. The Christ could, for example, be compared with the Christ on the cross in the stained-glass window of the Passion in Chartres Cathedral, where one notes the same overall position of the figure and the same supple quality of the loincloth, one end of which falls below Christ's right knee.[4]

Likewise, the representation of Adam rising from his tomb, evoking the Redemption of mankind made possible by the death of Christ on the cross, is original here: while on most other Limousin crosses of the late twelfth and the thirteenth century the small, naked figure is simply depicted in half length, with arms raised. Here he is shown almost in his entirety to better suggest the movement propelling him from the tomb.

According to de Lasteyrie, the first to publish this work (1884), it belonged to a private collector in the department of the Manche, who obtained it from the former abbey of Savigny-le-Vieux—information often repeated later although never confirmed by Victor Gay, who owned the plaque by 1887.

The Benedictine abbey of Savigny-le-Vieux, dedicated to Saint Barbara, was suppressed in 1790, and most of its possessions were sold in 1791. We know that its last curé, Edme Bourdoin de La Perrière, left the abbey in 1792, taking with him "the holy vessels and ornaments," while other objects were entrusted to "the assuredly faithful hands of those who would guard them preciously and later give them back to the church."[5] It is therefore possible that part of a cross was preserved separately. However, the archives of Savigny, deposited in the Departmental Archives of the Manche, were destroyed in 1944, making verification of the assertion concerning the origin of the Garnerius cross virtually impossible.

ET-D

NOTES

1. It is well known that the words *me fecit* could, in the Middle Ages, be used for *me fecit fieri*; such is the case, for example, on the chasse from Mozac (cat. 45). In the cases of the Alpais ciborium and this cross, however, the indication of origin, *lemovicensis* (from Limoges), makes more natural sense if we are dealing with an artist than if the inscription designated the patron.
2. See, notably, R. Lasteyrie 1884; Linas 1885 and 1886; Molinier 1885 and 1891; Rupin 1890.
3. Notably, as Souchal well noted (1967, p. 36), those decorating Grandmont crosses.
4. Ca. 1210–20; see Manhes Deremble and Deremble 1993, no. 37, p. 353.
5. Lemasson 1886, pp. 89–90.

EXHIBITIONS: Paris 1934, no. 372; Vatican City 1963, no. 6.

BIBLIOGRAPHY: R. Lasteyrie 1884; Linas 1885 and 1886, pp. 457–58, pl. XIX, no. 1; Molinier 1885, pp. 33–34; Gay 1887–1928, vol. 1, p. 618; Guibert 1888b, pp. 208, 230; Molinier 1889, pp. 21–22, fig. p. 23; Rupin 1890, pp. 108–9, fig. 176; Molinier 1891, pp. 151–52; Molinier 1901, p. 185; Migeon 1909, pp. 424–25, fig. p. 423; *Bulletin des Musées de France* (1909), p. 69; Marquet de Vasselot 1914, no. 71; Ross 1934; Hildburgh 1936, p. 16; Gauthier 1950, pp. 30, 34; Thoby 1953, pp. 15ff., no. 2, p. 90, pl. 1; Gauthier 1957b, p. 156; Souchal 1967; Gauthier 1967, p. 144; Lasko 1972, fig. 269 (with an error in the caption); Gauthier 1972a, pp. 276–77; Favreau et al. 1978, vol. 2, p. 74 (listed as lost).

# 104. Cross of Bonneval

Limoges, ca. 1225–35

Copper: engraved, stamped, and gilt; champlevé enamel: blue-black; saturated cobalt, ultramarine, and azure blue; turquoise, dark and light green, yellow, red, pinkish white, and opaque white; wood core

61 × 35.2 cm (24 × 13⅞ in.) with knop and handle; without, 43.6 × 35.2 cm (17⅛ × 13⅞ in.)

INSCRIBED: XPS (Christus) / IH[ESU]S. The letters A and ω (Alpha and omega) are enameled on the reverse.

EX COLLS.: M. Amable Frayssinous de Saint-Côme (Aveyron) (in 1863); given (ca. 1875) to the former Cistercian abbey of Bonneval, commune of Espalion (Aveyron), then restored as a convent for Trappist nuns; Heugel (in 1904). Acquired in 1977.[1]

CONDITION: Good. Lower part of the shaft, with two enameled plaques, is missing.[2] A few losses to the enamel. Restoration (19th century?) of part of the stamped revetment riveted to the edge, which modified the wood core.

Musée National du Moyen Âge, Paris; Thermes de Cluny (Cl 22888)

Obverse

Although the cross was not documented with certainty until after 1875, it was probably made originally for the Cistercian abbey of Bonneval, founded in 1161 in the diocese of Rodez, then an ecclesiastical province of early Aquitaine. In fact, as Marie-Madeleine Gauthier demonstrated very convincingly in 1978, its last owner, Amable Frayssinous of Saint-Côme, was a collateral descendant of the last abbot of Bonneval, Jean-Aymard Frayssinous. Before leaving the Rouergue in 1789 to find refuge in Paris, the abbot must have placed the cross in a safe place, and it was probably kept by his family until its surreptitious return to the Trappist convent.

This processional cross, endowed with its original knop and tubular finial, is decorated entirely on both sides with precious enameled plates. It is considered to be one of the most beautiful and unusual pieces of southern enamelwork produced in the first decades of the thirteenth century. Nonetheless, the disappearance of two support plaques that formed the base of the shaft on both sides deprived it of its original height, decidedly modifying its remarkable proportions as a Latin cross and its iconographic scheme: the obverse has probably lost the figure of Saint Peter, and the reverse, certainly, that of the winged man, symbol of Saint Matthew. One can nevertheless estimate the original height of this cross, without the knop and the handle, at some sixty-one centimeters. The nine copper plates—four on the obverse and five on the reverse—that cover the wood core were engraved, enameled, and gilded and then attached with rivets; narrow bands of revetment, cut from a thin sheet of copper and decorated with repoussé motifs of stars, cover the edges. Although varied in size and shape, the enameled facings nevertheless present the same type of exe-cution—rarely found on both sides of a cross: enameled figures on a reserved and gilt ground, in this case, strewn with simply engraved ornaments. Practiced in Limoges until the end of the twelfth century,[3] this particular treatment of copper—evocatively termed "flat painting" in the nineteenth century—reappears here as an archaism.

The blues and turquoise dominate the rich palette of eleven colors. They are placed, as are the green and pinkish white, in solid areas at times accented by bright red or

315

white. These enamels, contrary to the practices current at the time, only have graduated shades or fields of two, three, or four colors in rare spots: in the clouds of the angels and the mounds of earth beneath Adam's tomb. Also unique is the singular shape of this cross, which, in keeping with the Limousin custom of the time, combines a mandorla at the center of the crossbar with curvilinear supports. But in quite an unusual way, it presents a convex oval form at each end of the crossbar. According to Gauthier, the Bonneval cross is one of the first Limousin crosses to assume such an unusual shape; she finds its genesis in Carolingian art, but also in the twelfth-century art of Catalonia and Byzantine Italy.

Following traditional Limoges iconography, the Crucifixion is placed on one side and Christ in Majesty on the other. However, very unusual figures and motifs have been added. On the obverse, at the ends of the arms, the usual figures of the Virgin and Saint John, here within cartouches, are accompanied by Ecclesia with Sun on the left and Synagogue with Moon on the right; both are portrayed with feminine features and bear their traditional attributes, the former a chalice, the latter a banner with a broken staff. On the supports of the reverse, three small, enigmatic figures, each carrying a scroll, accompany the symbols of the evangelists, superbly rendered in cartouches: the youthful features of these small, haloed half-length figures prevent their being identified as prophets.[4] On the Bonneval cross, there are also certain peculiarities in the images of Christ: on the obverse, the figure on the cross is neither living nor crowned; on the reverse, as Savior of the World, he sits enthroned between the Greek letters alpha and omega and displays—atypically—the stigmata on his hands and feet.[5] Other unusual details, such as the postures of the angels, led Gauthier to see this cross in relation to the liturgy of the Mass of Pope Innocent III, who had summoned Limousin artists to Rome in 1215.[6] On the other hand, the ornamentation in reserve on a guilloché ground, which decorates all the gilt surfaces of the cross—the eight-pointed stars and the quatrefoils—comes from the same decorative repertory that characterizes a series of enamels known as the "Star Group," for their patterned ground,[7] and which are contemporaneous with our cross. These works[8] present the same stylistic characteristics and the same type of execution; they are dated with certainty to the years 1225–30, using the reliquary of Saint Francis of Assisi (d. 1228) in the Louvre as a reference (cat. 101).

Because of these iconographic, stylistic, and historical factors, whether or not the Bonneval cross and the "Star Group" belong to the body of Limousin works—as Gauthier suggested in her important 1978 article—is still being debated.[9] Were these enamels executed in Limoges or in central Italy by Limousin enamelers, or are they the work of Italian artists influenced by Limousin models?

GF

NOTES

1. The cross was in the catalogue of the Sotheby sale in London (December 15, 1977; no. 22, pp. 12–13, color pl., fig.), but it was not sold.
2. According to de Castelnau's notes (Castelnau d'Essenault 1864 [1863], p. 156) in 1863,: "its condition would be perfect if one of the copper plaques at one of the two points of intersection had not been stolen"; the median plaque of the reverse with Christ in Majesty was thus stolen before that date, then found, it appears, as the cross today bears witness.
3. See in particular a group of crosses published in Gauthier 1987, cats. 248–70.
4. As Erlande-Brandenburg suggested in 1980 (Paris 1980a, no. 17).
5. A mandorla-shaped plaque, contemporaneous with the cross, also bears this unusual image (Victoria and Albert Museum, inv. M. 210-1956).
6. See Gauthier 1968a, pp. 237–46.
7. A term given by Stohlman (1950, pp. 327–30).
8. In particular: the chasse in Copenhagen (Nationalmuseet; see Liebgott 1986, figs. 39–42; below, fig. 115a); reliquary of Saint Francis of Assisi (cat. 101); plaque from a reliquary (Louvre OA 84, deposited in the Musée de Cluny); an icon (Mount Sinai, monastery of Saint Catherine); a cross in the Odiot Collection (sale, Paris, 1889, no. 45).
9. This group will be published in volume 3 of *Émaux méridionaux,* in preparation under the direction of Marie-Madeleine Gauthier.

EXHIBITIONS: Rodez 1863 (see Gauthier 1987, p. 287); Paris 1980a, no. 17.

BIBLIOGRAPHY: Castelnau-d'Essenault 1864 [1863], pp. 151–62; Thoby 1953, pp. 13, 15, no. 17, pp. 99–100, pl. XI; Gauthier 1978a, pp. 267–85, figs. 1–6; Gauthier 1978b, pp. 53–56.

# 105. Eucharistic Dove

Limoges, ca. 1215–35

Copper: formed, engraved, chiseled, scraped, stippled, and gilt; champlevé enamel: dark and medium blue, red, yellow, turquoise, medium green, and white; blue-black glass inset eyes

19 x 20.5 cm (7½ x 8 1/16 in.)

INSCRIBED: IH[ESU]S

EX COLLS.: Aimé Desmottes, Paris (sale, Paris, March 19–23, 1900, no. 5, ill.); Georges Hoentschel, Paris (by 1911); [Jacques Seligmann, Paris (in January 1912)]; J. Pierpont Morgan, London and New York

CONDITION: Overall wear to gilding; loss and subsequent repair of enamel on proper left wing; modern base and suspension plate.

The Metropolitan Museum of Art, New York; Gift of J. Pierpont Morgan, 1917 (17.190.344)

This standing dove is formed of six sheets of worked copper; the two halves are mechanically joined along a vertical seam. Attached by means of rivets, the enameled wings lie flat against the body and come to rest on the gently dipping enameled tail feathers. A hinged, tear-shaped copper plate is set over a cavity in the center back and secured by a domed pin. The underside of the plate is engraved with the monogram of Jesus inscribed in a disk. The dove perches on a modern, circular base plate that has four projecting arms for the attachment of suspension chains.

The first mention of a vessel in the form of a dove used in a Christian context dates to the beginning of the third century.[1] In a specifically Eucharistic context, the figure of a gilt dove was part of the silver ensemble for the Host presented by Pope Innocent I (r. 401–417) to the basilica of Saints Gervase and Protase.[2] The use of Eucharistic doves is also mentioned in late-fifth-century France: Saint Perpetuus, bishop of Tours, gave a "columbam argenteam ad repositorium."[3] This text alone argues against King's supposition that the use of Eucharistic doves in the West derived from a Byzantine tradition introduced in the eleventh century.[4] Although not all citations are specific as to their use, the frequent references to gold or silver doves in France—at Saint-Denis, Cluny, Auxerre, Dijon, Angers, and Poitiers[5]—suggest that they were part of the standard liturgical furnishings for those churches and communities that could afford them. The *Ordo* of Cluny of 1093 instructs that the containers for the Body of Christ be filled afresh each Sunday, and that, along with two other pyxes, the deacon should refill the dove hanging over the altar during the celebration of the Peace.[6]

Despite the fact that it contained the Body of Christ, the golden dove at the Limousin abbey of Grandmont was looted by Henry II's son and namesake in 1183;[7] indeed, the value of such doves probably accounts for their being known almost uniquely from textual evidence.[8] On the other hand, doves of Limoges work fashioned from copper survive in large numbers. Rich with gilding, their overall surface is engraved in a pattern that, although stylized, nevertheless suggests the nervous rustling of layers of feathers. The beak is ridged; the joint of the upper leg is marked and the talons are jointed; the wings are worked in brilliantly colored enamel. The tear-shaped door in the center back covers a small opening that reveals a cavity similar to the inside of a pyx.

Though modern, the suspension of the Metropolitan dove is inspired by a medieval convention found, for example, on doves preserved at Amiens,[9] Copenhagen,[10] and formerly at Laguenne.[11] The form of these base plates usually resembles a gemellion or imitates a crenellated enclosure.[12] The only other instances of the exact form of suspension plate found on the Metropolitan dove occur on examples that either passed through the Spitzer Collection or are of doubtful authenticity, suggesting that these elements represent nineteenth-century design.[13] On a practical level, suspension may have been advantageous in protecting the farinaceous contents of Eucharistic doves from mice or rodents. At the same time, suspension enhanced the dove's metaphoric association with the Holy Spirit. The doves were sometimes covered with cloth while suspended, as recorded at Le Monastier in the Auvergne and in Milan.[14]

Although the provenance of a number of Limousin doves is known,[15] none is associated with a date. Limoges doves represent a highly accomplished use of enamel: as many as three colors are aligned in each cell of the horizontal feathers and as many as four in each of the numerous small lozenges at the broad end of the wing. This richly variegated and carefully laid palette is in keeping with enamel work in Limoges through the early decades of the thirteenth century. At the same time, the fully sculptural aspect of the doves—formed from two halves, each of worked copper riveted together—links them to trends that emerged in Limoges during the course of the thirteenth century. This confluence suggests a date early in the second quarter of the thirteenth century.

BDB

NOTES

1. See Cabrol and Leclercq 1907–53, vol. 3 (1914), pt. 2, col. 2231.
2. Duchesne 1981, vol. 1, XLII, p. 220: "turrem argenteam cum patenam et columbam deauratam, pens. lib. XXX." See *Book of Pontiffs* 1989, p. 31, for translation.
3. Cited in Texier 1857, col. 453, and by Gay 1887–1928, vol. 1, p. 415. *Gallia christiana*, vol. 11 (1759), col. 2.
4. King 1965, p. 44.
5. Ibid., p. 92. For a general discussion of Eucharistic doves, see

Schnütgen 1887; Rohault de Fleury 1883–89, vol. 5, pp. 77–83; Texier 1857, cols. 453–55; Barraud 1858, pp. 567–72, 633; Rupin 1890, pp. 223–34; Raible 1908, pp. 131–52; Gauthier 1973.

6. *Ordo Cluniacensis*, pars 1, c. 35: "Omni die Dominica corpus Domini mutatur . . . et . . . recens confectum a Diacono in pixide aurea reponitur . . . praedictam autem pixidem . . . Diaconus de columba iugiter pendente super altare . . . abstrahit . . . missaque finita in eodem loco reponit." Hergott 1726, p. 226.

7. The event was recorded by Geoffrey de Vigeois; see Labbé 1657.

8. An important exception is the sixth- or seventh-century silver dove from the Attarouthi treasure preserved in the Metropolitan Museum (1986.3.15). See "Medieval Art and The Cloisters," in *Annual Report 1989–1990*, p. 28.

9. See, most recently, Gauthier 1973, with earlier literature.

10. Liebgott 1986, pp. 53–55.

11. Rupin 1890, fig. 291, p. 225; Forot 1907.

12. Luchs, in Distelberger et al. 1993, pp. 36–40.

13. This type of base plate is described as early as 1857 in the Londesborough Collection. See Fairholt 1857, pl. XXXVII, fig. 3, and in the sale catalogue of M—. [Alexandre Basilewsky], December 13–15, 1864, p. 7, no. 22.

14. See the discussion in Raible 1908, p. 142.

15. See Gauthier 1973.

BIBLIOGRAPHY: Breck and Rogers 1929, p. 105; W. H. Forsyth 1946, p. 237, ill.

# 106. Eucharistic Dove

Limoges, ca. 1215–35

Copper: stamped, engraved, and gilt; champlevé enamel: midnight, lapis, and lavender blue, turquoise, dark and light green, yellow, red, and white; lapis blue enameled beads

14.7 × 19.8 × 6.9 cm (5¾ × 7¾ × 2¾ in.)

Ex colls.: Charles Ducatel (sale, Paris, April 21–26, 1890, no. 3); Victor Martin Le Roy (before 1900). Gift of Victor Martin Le Roy, 1914.

In spite of several restorations, the dove from the former Ducatel and Martin Le Roy Collections is a fine example of the evocative Eucharistic receptacles favored by Limousin enamelers. The plumage, especially, is highly finished, exhibiting many gradations of hue in a varied palette, including three blues and two greens.

It is singularly difficult to date these objects precisely because they lack strong stylistic elements. The earliest examples, such as the one in the Musée de Picardie, Amiens, from the abbey of Raincheval,[1] present different decorative elements that justify a dating of about 1200, notably the use of the vermiculé motif in the decorative bands. The dove from the Louvre, like the one from the Metropolitan Museum (cat. 105), takes its place in the "second generation" group, in which one notes a varied palette and the systematic use of gradations of hue but also a certain simplification of the decoration— the absence of gems and the separation of the plumage on the wings by a single decorative band—which are probably indications of a slightly later date. The dove from Notre-Dame at Erfurt, which later became part of the Robert von Hirsch Collection and is still privately owned,[2] also exhibits these characteristics in the arrangement of the plumage and the engraved decoration of the middle band, suggesting cabochons.

Some believe that the position of the feet, folded under the body, was designed to represent the bird in flight,[3] but since these feet are of brass and not of copper, they are probably a later addition. Therefore, one cannot draw any conclusions about the original positioning of the dove.

ET-D

Condition: Chains, inner cupule (of silvered iron), and, very probably, the brass feet, soldered to the underside of the body, are later additions; beak and feet have been resoldered. Significant wear to gilding, especially on the wings; some losses to the enamel, especially on the left wing and the tip of the tail.

Musée du Louvre, Paris; Département des Objets d'art (OA 8104)

Notes
1. Gauthier 1973, p. 177 and fig. 5, p. 176.
2. Ibid., p. 179 and fig. 11, p. 181.
3. Ibid., p. 177.

Exhibition: Paris 1900, no. 2556.

Bibliography: Rupin 1890, p. 230, fig. 296; Molinier and Marcou 1900, p. 230, fig. 296; Molinier 1900, p. 354; Molinier 1901, p. 191, fig. p. 190; Migeon 1902, p. 20; Marquet de Vasselot 1906a, no. 29, p. 49, pl. XXI; Duthuit 1929, fig. 17; Huyghe 1929, p. 4, with fig.; Gauthier 1973, pp. 177–78, fig. 10, p. 180.

# 107. Crosier with Flower Palmette

Limoges, ca. 1210–20

Copper: stamped, engraved, chased, and gilt; champlevé enamel: lapis blue, turquoise, green, yellow, and red. Nine lapis blue and turquoise enameled beads

30.6 × 14.5 cm (12 × 5¾ in.)

Provenance: Discovered in 1868 in a gallery of the cloister of Nieul-sur-l'Autise (Vendée)

Ex colls.: A. Martineau; de Pongerville (1892) in Nieul; Victor Martin Le Roy (acquired ca. 1896; see Marquet de Vasselot 1941). Gift of Victor Martin Le Roy, 1914

Musée du Louvre, Paris; Département des Objets d'art (OA 8105)

This crosier, found in the cloister of Nieul-sur-l'Autise, belongs to the group defined by the "flower-palmette" blossom at the center of the volute (cats. 23, 77). However, it also shows several characteristics that seem to mark the final stage in the evolution of the type. The fleuron is composed of three long petals separated only by a slight swelling of its central part. This simplification of the motif already appears on several crosiers among those with quadrangular volutes and openwork knops.[1] Here, moreover, the fleuron does not rest on foliage but on the serpent's head in which the volute terminates, emphasizing the allusion to Aaron's flowering rod by evoking its transformation into a serpent before the Pharaoh.[2]

A crosier in the Musée Masséna in Nice and a restored crosier in the Museo Nazionale del Bargello, Florence,[3] combine this same simplified fleuron with a serpent's head. On the present crosier, the evocation of the animal is underscored by the reticulate decoration of the volute and the stylized rendering of the scales that cover the serpent's body.[4] The execution is extremely fine: the blue enamel is punctuated with dots of red enamel or gilt copper in reserve against the enameled surface. On the shaft, the large trefoil fleurons are in reserve and are chased but not enameled. The Nieul crosier is distinguished above all by the knop with its openwork pattern of entwined serpents—unique among flower-palmette crosiers but frequently occurring on crosiers with figural decoration executed during the first half of the thirteenth century, which depict such subjects as Saint Michael and the Dragon or the Annunciation (cats. 109–11). It is thus quite probable, Marquet de Vasselot (1941) proposed, that this crosier is one of the latest examples with a large flower, suggesting a date of about 1210–20.

The provenance of the crosier is certain because it was published in the same year as its discovery. On the other hand, little information has come down to us about the tombs in the cloister of the abbey: we know only (Fillon 1868) that abbots were buried on the north side in the thirteenth century. Two other crosiers, each of which depicts the Annunciation, were recovered in the course of the same excavation.[5]

ET-D

Notes
1. Like the crosier found at Verdun in 1922, which then passed into different private collections (Marquet de Vasselot 1941, no. 13).
2. Episode of the flowering rod: Num. 17: 1–8; staff changed into a serpent: Exod. 4: 1–3.
3. Marquet de Vasselot 1941, nos. 29, 30.
4. On the example in Nice this motif has been even more stylized to form a checkerboard pattern.
5. Marquet de Vasselot 1941, nos. 73, pl. iv (Octave Homberg Collection, Paris; Gilbert Collection), 95 (Munich, Bayerisches Nationalmuseum).

Exhibition: Mexico City 1993, p. 78, fig. 28.

Bibliography: Fillon 1868; Berthelé and Drochon 1892, photo p. 6; Migeon 1902, p. 20; Marquet de Vasselot 1906a, no. 34; Duthuit 1929, fig. 16; Huyghe 1929, p. 4 with fig.; Marquet de Vasselot 1941, no. 31.

# 108. Crosier: Lion and Serpent

Limoges, ca. 1220–30

Copper: stamped, engraved, chased, and gilt; champlevé enamel: medium blue and turquoise; (on the shaft): red, green, yellow, and white; medium blue enameled beads

28.5 × 13.3 cm (11¼ × 5¼ in.)

EX COLL.: Félix Doistau, Paris (in 1900). Gift of Félix Doistau, 1919

CONDITION: Wear and some losses to the enamel; eight enameled beads missing; wyverns attached to the shaft are modern.

Musée du Louvre, Paris; Département des Objets d'art (OA 7287)

Exhibited in Paris only.

This crosier from the former Doistau Collection belongs to a group of some ten known examples[1] in which the volute portrays a lion and a serpent in combat: the volute terminates with the mouth of the serpent, which grips the tail of the lion in the center, head turned toward the attack.

The reticulated motif of the volute, suggesting the scales that cover the serpent's body, the decoration of coiled serpents on the knop, and the rinceaux with flowers on the shaft are commonly found on crosiers of the thirteenth century (cats. 107, 109–11).

The iconography of these crosiers has inspired various interpretations; the lion and the serpent have been endowed with both benevolent and malevolent symbolism. De Bastard, followed recently by Dereux and François,[2] suggested seeing the lion as symbolic of the tribe of Judah, prefiguring Christ. Marquet de Vasselot, on the other hand—objecting that this interpretation is not at all compatible with the role of the serpent, which bites the lion's tail—preferred to see the serpent, as on many other crosiers on which it figures, as an allusion to Aaron's rod (see cat. 82) and suggested that the lion here could represent the devil, referring to the First Epistle of Saint Peter: "the devil, roaring like a lion . . ." (5:8). It is possible that the Limousin artists sought to portray the struggle between good and evil in these crosiers; but, above all, they created a new variation on one of their favorite themes: combat between two or more animals, using as a point of departure the motif of the serpent, essential in giving these crosiers a biblical theme, through the allusion to the episodes of Aaron's rod and the brazen serpent (cat. 82).

In this respect the Louvre crosier and others like it should be compared with the medallions representing the same subject on secular coffrets. This parallel affords a glimpse of how Limousin artists were able to create objects for both religious and secular use; its also helps to date these crosiers precisely. Indeed, the only crosier of this group found in a tomb that can be identified with any certainty is the one from the treasury of the cathedral of Troyes, found, with several other objects, in the sepulchre of Bishop Hervée (r. 1207–23). The identification of the tomb seems certain, and, moreover, it has never been contested, since Hervée was the only bishop of Troyes buried in the Lady Chapel, where this sepulchre was located. The publication of Arnaud followed immediately after the discovery, at which he was present.[3]

Marquet de Vasselot, however, judged it "difficult to accept that this crosier belonged to a prelate who died in 1223."[4] The parallel with coffret medallions depicting the same theme, and in particular the similarities observed in the treatment of the lions, allow us to remove any doubt. We recall that the coffrets belonging to Cardinal Bicchieri are mentioned in an inventory of 1227 (see cat. 88). Hervée's crosier, made before 1223, is certainly one of the earliest in this group: the graduated hues of the reticulated ornamentation of the volute and the rounded forms of the lion still suggest Romanesque works of the end of the twelfth century. The Louvre's crosier, which is slightly later, can be dated to about 1220–30.

ET-D

NOTES
1. Dereux and François (1992, pp. 195–96, 199) counted eleven examples, to which are added two mentioned in former collections but not presently localized. However, this number probably includes some pastiches.
2. Bastard 1857, pp. 433–34; Dereux and François 1992.
3. Arnaud 1844, pp. 7ff.
4. Marquet de Vasselot 1941, no. 182 and pp. 18–19.

EXHIBITIONS: Paris 1900, no. 2438; Vatican City 1963, no. 119.

BIBLIOGRAPHY: Marquet de Vasselot 1941, no. 181; Dereux and François 1992, p. 194 and fig. 6c.

# 109. Crosier: Saint Michael Slaying the Dragon

Limoges, ca. 1220–30

Copper: formed, engraved, chased, scraped, stippled, and gilt; champlevé enamel: medium blue, turquoise, green (on the replaced lower shaft), yellow, red, and white

34 x 13 cm (13⅜ x 5⅛ in.)

Ex colls.: Aimé Desmottes, Lille and Paris (in 1882; sale, Paris, March 19–23, 1900, no. 19, ill.); Baron Léopold Goldschmidt, Paris (d. 1904); Michael Friedsam, New York

Condition: Wear to gilding; some losses to enamel; crest above knop added (after 1882, before 1927); shaft and serpents below knop replaced (perhaps after 1882).

The Metropolitan Museum of Art, New York; The Friedsam Collection, Bequest of Michael Friedsam, 1931 (32.100.289)

Exhibited in New York only.

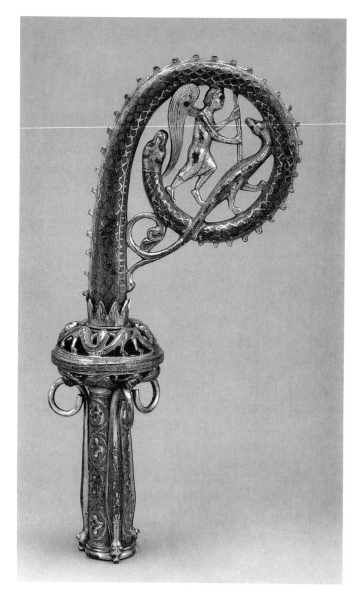

Within the curving volute of the crosier, the gilt figure of Saint Michael steps forward and plunges the point of his lance into the back of the dragon. The volute, composed of two pieces, is enameled with blue lozenges and decorated along its outer edge with crockets. It has a dragon's head at its terminus, and the tail of the wounded dragon crosses over the volute and terminates in leafy ornament against the shaft. The top and bottom halves of the openwork knop are decorated with intertwined reptiles.

The archangel overcoming the devil in the form of a dragon is the most common subject found among preserved enameled crosiers from Limoges.[1] Saint Michael commonly assumes the form seen in this example and only rarely appears in a slightly different guise.[2] A crosier of Saint Michael preserved in the Musée d'Évreux was discovered in the tomb of Bishop Jean II de la Cour d'Aubergenville, who died in 1256,[3] providing a *terminus ante quem*. In some instances, Limoges crosiers were used in a burial context well after their manufacture: a crosier with Saint Michael was found at the cathedral of Notre-Dame, Paris, in the tomb of a bishop who died in the mid-fifteenth century.[4] The dragons and their disposition within a circular composition are natural outgrowths of the Limousin production of medallions with fantastic creatures, as Marquet de Vasselot noted.[5]

With their inventive decoration, crosiers were much admired in the nineteenth century;[6] an example with Saint Michael slaying the dragon is among the drawings of three crosiers executed by Eugène Delacroix.[7]

BDB

Notes

1. See the publications by Bilimoff (1979, p. 37) and Marquet de Vasselot (1941, pp. 85–91, nos. 122–67). Another example from the Metropolitan's collection published by Marquet de Vasselot, no. 163, is a nineteenth-century copy.
2. See Schrader 1971, pp. 42–45.
3. Marquet de Vasselot 1941, no. 158, pp. 288–89, pl. XIX.
4. Ibid., no. 166, pp. 293–94.
5. Ibid., pp. 130–32.
6. One example was successively in the collections of Abbé Fauvel, Bernard de Montfaucon, and Alexandre Lenoir; see Marquet de Vasselot (1941, no. 150). Crosiers were cited in the 1882 Brussels exhibition catalogue as the most beautiful of Limoges liturgical objects.
7. See Marquet de Vasselot (1936, pp. 141–43, ill). The crosier is preserved in the Detroit Institute of Arts. See, most recently, Barnet (1988, frontispiece and figs. 6–8).

Exhibitions: Brussels 1882, p. 24, fig. 11, p. 10; Paris 1884, p. 225, no. 14; New York 1927, no. 86.

Bibliography: Reusens 1886, vol. 2, pp. 491–92, fig. 717; Marquet de Vasselot 1941, no. 149, pp. 284–85; Bilimoff 1979, p. 48, n. 25; Schrader 1986, p. 44.

# 110. Crosier: Saint Michael Slaying the Dragon

Limoges, ca. 1225–35

Copper: stamped, engraved, and gilt; champlevé enamel: lapis blue; lapis blue enameled beads

31.5 × 12.2 cm (12⅜ × 4¾ in.)

Ex coll.: Félix Doistau, Paris. Gift of Félix Doistau, 1919

Condition: Two wyverns missing; two enameled beads missing: one on Saint Michael and one on the dragon; gilding partially worn.

Musée du Louvre, Paris; Département des Objets d'art (OA 7286)

Exhibited in Paris only.

This crosier, like that in the Metropolitan Museum (cat. 109), belongs to the most important group of crosiers depicting the combat of Saint Michael with the dragon, in which the scrolling of the volute terminates with the head of a serpent, while another small animal with two feet and a tapering body is lodged in the volute.[1] As on almost all the other examples, the saint thrusts the staff of his cross vertically into the animal's body. The volute is covered by a reticulated motif suggesting scales on a serpent's body, and the openwork knop is decorated with intertwined dragons. In this uncommonly coherent group, which has only slight variants, the Louvre crosier is among the examples distinguished by a certain simplification, in the treatment of the figure of the saint as well as in the ornamentation of the serpent.

The features of the face are quite schematic, and the slender silhouette, whose drapery is still linked to early Gothic art, is somewhat flattened. An example with the same characteristics, though with a treatment that is at once simpler and more "plastic," preserved in the Museo di Capodimonte in Naples,[2] has a ribbed knop similar to the one on the reliquary of Saint Francis of Assisi (cat. 101), which leads us to suggest dating them to about 1230–40. The Louvre crosier might be slightly earlier.

ET-D

Notes

1. Bilimoff (1979, p. 46) counted about thirty crosiers of this type, in an ensemble of fifty-eight, which makes the subject, as Marquet de Vasselot (1941, p. 85) has noted, the one most frequently represented on Limousin crosiers.
2. Marquet de Vasselot 1941, no. 128, pl. xvii; Naples 1981, no. 1, p. 5.

Bibliography: Marquet de Vasselot 1941, no. 126, p. 272, pl. xviii.

# III. Crosier: The Annunciation

Limoges, second quarter of the 13th century

Copper: formed, engraved, chased, scraped, stippled, and gilt; champlevé enamel: medium blue

32 x 13.5 cm (12⅝ x 5⅕₆ in.)

EX COLLS.: Count Grégoire Stroganoff (1829–1910), Rome; Georges Hoentschel, Paris (by 1911); [Jacques Seligmann, Paris (in January 1912)]; J. Pierpont Morgan, London and New York

CONDITION: Wear to gilding and enamel; applied lizards lost from lower shaft.

The Metropolitan Museum of Art, New York; Gift of J. Pierpont Morgan, 1917 (17.190.835)

Exhibited in New York only.

Enclosed within the volute of the crosier, the angel Gabriel, holding a stylized lily, greets the Virgin Mary, who has risen from her high-backed chair. She holds a book in her right hand and raises her left in greeting. The volute is enameled with blue lozenges and decorated with crockets along its spine. A crest of leaves separates the upper volute from the intertwined, lizardlike creatures of the knop. The lower shaft is decorated with scrolling, foliate ornament to which small lizards were once affixed.

The depiction of the Annunciation is consistent with the second of three types identified by Marquet de Vasselot in his corpus on Limousin crosiers.[1] Although Marquet de Vasselot made comparisons of the composition with representations in early Gothic stained glass, this ubiquitous imagery should be seen as a Gothic convention common to several media. The Metropolitan example is similar to two examples excavated in Germany subsequent to Marquet de Vasselot's publication: one at Trier and the other at Bremen. The example at Bremen was found in the grave of Archbishop Gerhards II (r. 1219–?1258);[2] curiously, however, the example at Trier is considered to be from the grave of Bishop Arnold I (r. 1169–83),[3] but it is inconceivable that the swirling Gothic forms of the leaf ornament or the realization of the appliqués could have been created within these twelfth-century dates. Rather, as Gauthier indicated for the Bremen example,[4] the crosiers should be attributed to the second quarter of the thirteenth century, when the emphasis in Limoges work was on gilt appliqué figures complemented by restrained enamel decoration.

BDB

NOTES
1. Marquet de Vasselot 1941, pp. 76–79, nos. 64–102e.
2. Brandt 1976, p. 26 and figs. 21, 22.
3. Trier 1984, pp. 121–22, no. 56.
4. Gauthier 1975a, pp. 165–80.

BIBLIOGRAPHY: Rohault de Fleury 1883–89, vol. 8, p. 103, pl. 653; *Les Arts* (1904), p. 37; Pératé 1911, no. 70, pl. XLIV; Breck and Rogers 1925, p. 105; Marquet de Vasselot 1941, no. 72, p. 239.

# 112. Chasse of Saint Hippolytus

Limoges, second quarter of the 13th century

Copper: repoussé, engraved, and gilt; champlevé enamel: medium and lavender blue, turquoise, green, yellow, red, and white; glass cabochons; glass plate; lapis blue enameled beads; wood core

15.3 × 21.6 × 9.9 cm (6 × 8½ × 3⅞ in.)

PROVENANCE: Originally from the treasury of Saint-Denis; possibly mentioned in the Tabernacle of Martyrs' Remains in 1534. Found in the chapel of Saint-Hippolytus in 1634 (no. 318 in the inventory taken that year); mentioned in 1714, in the fifth cabinet of the treasury; deposited at the Museum, December 5, 1793.

CONDITION: Crest is missing. On the face with small appliqué figures, a rectangular hole was cut in the center of the roof and then closed with glass, eliminating two appliqués, traces of which remain visible above and below the opening. The numerous holes pierced in the central part of the roof of this face lead us to surmise that another element, perhaps a plaque bearing a crystal cabochon, was once affixed there. Two joined friezes border the lower part of the two plaques of this face; their location is marked by an ungilded band in reserve at the bottom of each plaque.

Missing enamel on the roof-plate, on the left; two missing areas on the edges of the roof decorated with engraved figures, on the other side. Wear to the gilding.

Musée du Louvre, Paris; Département des Objets d'art (MR 343)

The two faces of this chasse are dissimilar. The reverse is decorated with ten engraved figures within arcades, in reserve on an enameled ground of rosettes. They seem to be depictions of the apostles, with the probable exception of the two figures at the left on the roof, who perhaps represent the Annunciation, or, at the very least, an encounter between two people. On the principal face, glass cabochons mounted in broad flat settings alternate with appliqué figures in half relief with enameled bodies; both repetitive and summary, these figures belong to a type that was quite widespread in the thirteenth century and generally designated as *poupées* (dolls). The guilloche-patterned copper background sets off the halos of the figures and the decoration of rinceaux with trefoil fleurons in reserve—a design that is likewise greatly simplified. Each of the end panels is ornamented with an engraved angel—full length on one side and half length on the door on the other end—raised on a ground of enameled rosettes.

The date of the work is difficult to determine precisely; chasses with *poupées* have rarely been dated with certainty. The graduated tones of the enameled rosettes and the style of the engraved figures invite comparisons with works of the early thirteenth century. Some chasses with *poupée* figures, such as the one in the Keir Collection, deposited at the Nelson-Atkins Museum of Art, Kansas City,[1] where the figures are placed on a vermiculé ground, would tend to suggest, moreover, that this technique of manufacture first

appeared in Limoges in the late twelfth century. The stylized *poupée* figures no doubt derive directly from the appliquéd figures in low relief, with enameled bodies, produced in Limoges at the close of the twelfth century and in the first part of the thirteenth century (cats. 49, 50). The oval settings of some of the cabochons, like the design of rinceaux with small trefoil fleurons, here confirms the comparison with the two book covers in the Musée de Cluny (cat. 113). The engraved background strewn with rosettes and stars recalls those of the reliquary of Saint Francis of Assisi (cat. 101) and related works.

There are other chasses—often more worn—that are decorated on one face with similar figures but are ornamented on the other face with busts of angels engraved in a rather summary style that instead invites comparison with works of the second and third quarters of the thirteenth century. Such is the case, notably, with the examples at Nexon (Haute-Vienne), Chamberet (Corrèze), and Saint-Merd-Lapleau (Corrèze), or even of the chasse in the Museo Nazionale "Duca di Martina," Naples.[2] Several successive workshops therefore produced this type of chasse during the course of the thirteenth century; rather summary in execution, such chasses were made either for a less well-to-do clientele in the Limousin—as the extant examples in the region demonstrate—or for export. The chasse in the treasury of Saint-Denis, one of the most highly finished, may be dated—on the basis of the style of the engraved figures

Principal face

Reverse

and the characteristics of its ornamentation—to the second quarter of the thirteenth century.

Gaborit-Chopin (1991) has shown that this chasse corresponds exactly to the description of the one that in 1634 was at Saint-Denis in the chapel of Saint-Hippolytus, which at the time contained a relic "of the most holy saint Hippolytus." The chasse came, perhaps, from the Tabernacle of the Martyrs' Remains, destroyed in 1626, which, indeed, housed two "coffrets of gilded and enameled copper."[3] The present object is an example of a work of modest quality, made without a precise commission, which, during the random transfer of relics, perhaps found its way into one of the most prestigious treasuries of medieval Europe, where it is known to have held a number of different relics, among them the pallium of Stephen II in 1714 and 1793.

ET-D

NOTES
1. From the Kofler-Truniger Collection, Lucerne; see Gauthier and François 1981, no. 17, p. 20 and pl. 15.
2. See Jouhaud 1949; Paris 1965, no. 369; Naples 1981, no. 1, 7.
3. Paris 1991, no. 42.

EXHIBITIONS: Vatican City 1963, no. 78; Paris 1991, no. 42.

BIBLIOGRAPHY: L. Laborde 1852, nos. 42–47; Rupin 1890, p. 330 and fig. 385, p. xxxi; Darcel 1891, nos. 102–11; Marquet de Vasselot 1914, no. 64; Conway 1915, pp. 117, 155; Gauthier 1976b; Montesquiou-Fezensac and Gaborit-Chopin 1973, vol. 1, p. 277; vol. 2, no. 29; vol. 3, pp. 122–23, pl. 107.

# 113. Two Book Covers and Two Appliqués

Limoges, first quarter of the 13th century

## a. *Christ in Majesty with Symbols of the Evangelists* (upper plaque)

Copper: gilt; champlevé enamel: dark and lavender blue, turquoise, green, yellow, and white; modern cabochons in the center and on the border, with some reuse of old mounts; wood core covered with gesso and glazed red with minium

30 × 19.8 (11 ¹³⁄₁₆ × 7 ¾ in.)

Inscribed (on the scroll held by the eagle of Saint John; it must be turned to read it): s[anctus] i[ohannes] ac[c] ep[i]t (Saint John takes . . .)

## b. *Crucifixion* (lower plaque)

Copper: gilt; champlevé enamel: dark and lavender blue, turquoise, green, yellow, brick red, and white; modern cabochons of glass paste: green, blue, and garnet red; wood core covered with gesso and glazed red with minium

30 × 19.5 cm (11¹³⁄₁₆ × 7⅝ in.)

Inscribed (on the reverse, on the wood core, in ink, very faded, in cursive letters): *Evangelistarium solempne [conven]tus f[ra]trum gross[eti]* (Solemn evangelistary of the monastery of the brothers of Grosseto)

Ex colls.: Alexandre Du Sommerard (before 1842). Formerly exhibited in the chapel of the Hôtel de Cluny.

Condition: (a) Modern cabochons. The work has undergone several transformations. The appliqués of the lion of Saint Mark and the ox of Saint Luke are missing. The copper plaque forming the ground, probably original, has exactly the same stamped motif (circles, quatrefoils) as the Crucifixion, but two small, smooth modern plaques have been added to join the upper and lower plates. The central part of this plate was entirely dismantled at some point, probably at the time the symbols of Saint Mark and Saint Luke were removed; later the plaque was the object of a clumsy reassembly: it would have been regilded, modern cabochons attached, and the appliqués (angel, eagle, Christ, mandorla) reattached last, which explains the overlapping of the appliqués on certain cabochons. The angels surrounding the cross were reversed: they were put in their correct positions on the occasion of the cleaning of the work for the exhibition. (b) Slight chips in the enamel.

Musée National du Moyen Âge, Paris; Thermes de Cluny
(Cl 971 a and b)

These two book covers were in the collection of Alexandre Du Sommerard (died 1842), who is said to have bought them in Italy from another collector. The inscription on the reverse of the Crucifixion plaque may confirm this provenance. Indeed, based on the principle that a place name should follow the expression *conventus fratrum*, there are few names beginning with "gross" other than Grosseto, a Tuscan city. If the book cover belonged to a monastery in Grosseto, it is not surprising that it was purchased in Italy,

but this is merely a hypothesis based on the reconstruction of a very incomplete inscription.[1]

The two plaques were published in *Les Arts au Moyen Âge*, where Du Sommerard showed the book cover in its entirety, the two plaques joined by an enameled spine. This spine never figured in the inventory of Du Sommerard's collection, and it is unlikely that he acquired the book cover in this condition; it seems more probable that he filled out the illustration with an imaginary plaque to make the book cover appear complete.

In *Les Arts au Moyen Âge*, the collector assigned the cover a prestigious provenance: "Cover of an evangelistary given by Saint Louis." The inscription on the reverse of the lower plate does not, however, confirm Du Sommerard's fanciful allegations.

This pair of plaques nevertheless occupies a special place among enameled book covers: it is quite unusual to find plaques such as these with their original wood supports intact. The highly original features of these plaques consequently might shed new light on the iconography of the covers of Gospel books. Traditionally, the Crucifixion was believed to be on the upper plaque with Christ in Majesty on the lower, following the temporal sequence of events, the Crucifixion necessarily preceding the Second Coming. Here, on the other hand, the order seems reversed, with the Crucifixion on the lower plaque, as clearly attested by the extant traces of the binding cord. Among the six book covers indexed in the first volume of *Émaux méridionaux* (Gauthier 1987) there is only one pair of plaques whose order is apparent, since the cover still contains its original manuscript: the Missal of Saint Rufus, from Avignon.[2] As in the present pair of plaques, Christ in Majesty is on the upper plaque, and the Crucifixion on the lower.[3] It is not clear whether this is the result of an error in the assembly of the two plaques or merely that the rule indicating that the Crucifixion should precede Christ in Majesty was not absolute.[4]

Although the Crucifixion plaque is extremely well preserved, that of the Majesty shows considerable wear: the lion and the ox, symbols of Saint Mark and Saint Luke, which were in the lower part of the central scene, are missing, and modern cabochons now fill the empty spaces. The plaque came to the Du Sommerard Collection in this condition. Research undertaken for this exhibition has confirmed that the two missing symbols are not lost: they now belong to The Metropolitan Museum of Art.[5]

As for the eagle of Saint John, it is notable as well as unusual that the letters engraved on the scroll form a complete phrase. Indeed, if we assume that the eagle's claws on the scroll cover some of the letters, the inscription can be reconstructed thus: s[ANCTUS] I[OHANNES] AC[C]EP[I]T—a reference to the Apocalypse (10: 9–10), in which the angel returns the book to Saint John and orders him to eat it: "*Dixit mihi: Accipe librum, et devora illum . . .*" (He has said to me: take this book and devour it . . .)

These two plaques are characteristic of Limousin production at the turn of the twelfth century, when certain elements of Byzantine iconographic tradition are still present in Limoges enamels: the two angels carrying books on either side of the cross in the Crucifixion scene have replaced the Sun and the Moon, and the figures of the Virgin and Saint John are seen in frontal view, the Virgin holding her wrist; in the Majesty plaque Christ holds the book in his draped hand. The Christ of the Crucifixion, crowned and triumphant, is, however, an innovation of the early thirteenth century,[6] as is the crown of the Christ in Majesty, which, in the same period, supplanted the cruciform halo. The chasse of Saint Dulcida in Chamberet offers several interesting points of comparison. The Christ in Majesty on the front side of this chasse is very close indeed to the Christ on the present book cover: his general posture, gesture of benediction, manner of holding the book, and crown are similar; the copper plaques of the chasse also bear the same stamped decoration of quatrefoil leaves. The rapid execution of the ornamental decoration and the absence of certain details such as Christ's throne, the footstool on which he would rest his feet, and the hillocks upon which the Virgin and Saint John would stand indicate that the workshop that produced this book cover already had embarked on a phase of "industrial" production, which suggests a date between 1200 and 1220 for these plaques.[7]

TWO APPLIQUÉS: SYMBOLS OF THE EVANGELISTS

Limoges, first quarter of the 13th century

c. *The Ox of Saint Luke*

d. *The Lion of Saint Mark*

Copper: gilt; champlevé enamel: dark and lavender blue, turquoise, green, yellow, and white

9.5 × 7.3 cm (3¾ × 2⅞ in.)

EX COLLS.: Rev. Walter Sneyd, Keele, Staffordshire (by 1855); Michael Friedsam, New York

CONDITION: Slight wear to gilding; flattening of raised wing ox, with loss to enamel.

The Metropolitan Museum of Art, New York; The Friedsam Collection, Bequest of Michael Friedsam, 1931 (32.100.290, 32.100.291)

These two appliqués, which were given to the Metropolitan Museum in 1931 by Michael Friedsam, were linked as early as 1935, by Ross, to the plaque of a book cover in the Musée de Cluny, which was missing these elements (the upper plate with Christ in Majesty, presented above). The hypothesis, however, was never pursued.

The combined study of these objects within the framework of this exhibition allowed for a reexamination and confirmation of this hypothesis. The outlines of the Metropolitan Museum appliqués correspond perfectly to the shapes of the missing pieces on the book cover, which are still visible despite the cabochons placed there at a later date. The colors of the enamel, some of which were relatively rarely used, such as the pale yellow or the turquoise flecked with pinkish white, support the hypothesis of the supposed origin of the appliqués, for the same colors are on both plaques. Stylistically, as well, the parts are connected.

The two evangelists' symbols already were missing when the book cover entered Du Sommerard's collection in the nineteenth century. By 1855 at the latest they were in the collection of Rev. Walter Sneyd, who presented them to the Archaeological Institute on February 3 of that year. Sneyd, a well-known English bibliophile, traveled widely in Italy, attended numerous auctions in England and France, and was active in circles of neo-Gothic collectors, especially that of Prosper Mérimée. Photos of these appliqués appear in undated early photographs of Sneyd's collection at Keele, which are preserved at the Society of Antiquaries, London.[8]

Highly finished in execution and displaying a range of refined colors, the symbols of the evangelists are the elements on which the enameler concentrated his talents. A tetramorph framing Christ in Majesty is a recurring motif on enameled book covers, but the symbols of the evangelists were rarely executed in half relief and enameled. Only a few works may be compared with these: the appliqués on the Mozac chasse (cat. 45), those on a Spanish cross in the Museu Arqueológico-Artistico Episcopal in Vich,[9] and the fragments from a retable in the cathedral of Orense. The dates assigned to these works by various scholars have ranged from 1180 to 1220. Based on the stylistic analysis of the elements on the book covers, we favor a dating between 1200 and 1220.[10]

EA

NOTES
1. I wish to thank Marc Smith, professor at the École Nationale des Chartes, for his scholarly help in deciphering these inscriptions.
2. Now in the treasury of the cathedral of Tortosa (Catalonia). See Gauthier 1987, no. 287 a, b, pl. CCXXXVIII, ills. 784, 785.
3. An enameled book cover from the Österreichisches Nationalmuseum in Vienna (EM 430, 431) is arranged in the same order (figs. 12a, b).
4. Among the book covers studied by Steenbock (1965), the order

a. Upper plaque: Christ in Majesty

b. Lower plaque: The Crucifixion

d. Lion of Saint Mark

c. Ox of Saint Luke

Crucifixion/Majesty is more common but not the only one possible. The liturgical use of manuscripts contained in the book covers—which generally have disappeared today—helps, no doubt, to understand these variations.

5. See *infra*.

6. See Gauthier 1968d, p. 279.

7. These two plaques will be published in the second volume of *Émaux méridionaux, L'École de Limoges, 1190–1216*, under the direction of M.-M. Gauthier (forthcoming).

8. Portfolio, "Early Medieval Prints and Drawings," fol. 75ff. I am grateful to Neil Stratford for verifying this collection history at the Society of Antiquaries and for the information about Walter Sneyd that he and Barbara Boehm provided. See Munby 1972, pp. 107–19.

9. See Thoby 1953, p. 161 and pl. XVII.

10. Ross (1935) and Souchal (1963, pp. 219–35) lean toward an early dating, at the end of the twelfth century. Gauthier, on the other hand, dates these enamels to about the first quarter of the thirteenth century (*L'École de Limoges*, forthcoming).

EXHIBITIONS (a, b): Paris 1934 (Cl 971a); Limoges 1948 (Cl 971a, b); Vatican City 1963, no. 129 (Cl 971a); Paris 1979, p. 103, no. 234 (Cl 971b); (c, d): New York 1970, no. 139; South Hadley, Mass., 1977, no. 10; New York 1978, no. 8.

# 114. Chasse of Saint Ursula

Limoges, ca. 1235–45

Copper (plaques): engraved and gilt; (appliqués): repoussé, chased, engraved, scraped, and gilt; champlevé enamel: medium and light blue, turquoise, green, red, and white; clear glass cabochons, some with silk backing; semiprecious stones, including malachite, coral, smoky quartz, and lapis lazuli; wood core; interior painted red with white cross

42 x 43 x 16.8 cm (16½ x 16¹⁵⁄₁₆ x 6⅝ in.)

Principal face

INSCRIBED: COLONIA (Cologne)

EX COLLS.: Mme Camille Lelong, Paris (sale, Paris, December 8–10, 1902, no. 76, ill. opp. p. 28); M. La Rouchefoucauld, Paris; [Brimo de Laroussilhe, Paris]; [Simon Seligmann, Paris (ca. 1909–12)]; [Georges Seligmann, New York]

CONDITION: Original crest, finials missing; lock system missing; upper left plaque of reverse, left angel, and possibly upper right appliqué figures of transept replaced.

Private collection, New York

Exhibited in New York only.

Enameled figures applied to the principal face of this chasse portray the martyrdom of Saint Ursula and four of her companions. On the projecting transept, Saint Ursula is killed by an arrow shot from the bow of one of the Huns, whose king she had declined to marry; at either side another vir-gin-martyr is executed by sword-wielding Huns wearing short tunics; on the roof above, a third is killed by an ax, the fourth by a sword. The figures are affixed to sheets of gilt copper engraved with a star pattern, which cover the wood core of the chasse. The surface is decorated with glass cabochons, some of them backed by silk, and numerous hard stones, including malachite and lapis lazuli.

The reverse of the chasse is dominated by the enameled baptismal scene that decorates the transept. A naked man and woman, each crowned, stand in water, their hands clasped in prayer, while a bishop, assisted by a deacon, baptizes them with water flowing from a large ewer poised over their heads. A second tonsured attendant is partly obscured by the figure of the bishop. The scene is enclosed within a trefoil arch springing from a column at either side; the Hand of God descends from the top of the arch. The double baptism does not conform to the story of Ursula as recorded in *The Golden Legend*, but the crowns worn by the two figures suggest that they are Ursula and her devoted fiancé, Etherius, who, as a suitor of the Christian princess, was obliged by her to embark on a three-year study of her religion and to raise a fleet of ships for her to make a pilgrimage to Rome (via Cologne and Basel) in the company of eleven thousand virgins. Episodes from the journey of this remarkable armada are disposed in three of the four scenes flanking the transept. At the top are mirror images of the virgins in two sailing vessels, just at the moment that the sails are allowed to luff and the maidens are helped ashore at an unknown stop. At the lower left, two shiploads of virgin pilgrims are on the waters. Red enamel highlights the sails.

At the lower right, the first ship docks at Cologne, its city walls identified by the inscription COLONIA on the fortifications. As the virgins approach, an executioner grabs one by the hair, pulling her from the boat to behead her, just like her hapless companions whose decapitated bodies have fallen into the flowing waters of the Rhine below. A second man looks on from the ramparts of the city, while the Hand of God descends from a cloud above. On the triangular plaques of the roof are figures of angels holding books and swinging censers. Saint Peter appears on the lock plate end; Saint Paul appears on the opposite end.

The repetitive nature of the enameled scenes on the reverse probably provoked the dismissal of the entire reverse as being modern in the listing of the chasse in the 1902 Lelong sale. The upper left plaque, distinguished by weaker engraving and by the grayish rather than the bluish white enamel of the water, appears to be a replacement for a lost original; it is based on the plaque on the right that it mirrors. Still, a repetitive sequence of boat scenes conforms to certain iconographic conventions evidenced in gilt-copper bowls of the twelfth century that represent the Ursula legend: the arrival at a port (either Cologne or Basel en route to Rome, or Cologne on the return), and the disposition of the women in crescent-shaped vessels riding on choppy waters.[1] The joint baptism of Ursula and her fiancé, Etherius, conforms neither to the written legend of Ursula nor to a pictorial tradition of her legend; rather, it seems compositionally derived from depictions of the Baptism of Christ (see cat. 122).

The chasse of Saint Ursula, like the chasse from the Jonas Collection (cat. 115), belongs to a small group of highly decorative chasses whose wood cores are covered with thin sheets of gilt copper that has been stamped and pierced, decorated with applied enameled figures, and then punctuated with glass cabochons. The chasse of Saint Ursula, along with the chasse of the Magi in Copenhagen (fig. 115a), is among the finest.

The narrative scenes on the reverse recall the style of the quatrefoils on the reverse of the chasse of Saint Viance (cat. 118), especially in the sketchy rendering of facial features and hair, the summary drapery, and the undulating enameled stripes meant to indicate water. The appliqué figures of the martyrdom resemble the image of the martyrdom of Saint Catherine from the church of Solignac.[2] The composition of the baptismal scene, especially of the water pouring forth from an inverted ewer, recalls the Boston relief (cat. 122), whereas the engraving of the bishop's and the deacon's vestments recalls the liturgical garments on the quatrefoils of the entombment of Saint Viance. The engraving of the figures of Saints Peter and Paul recalls that on the interior of the plaques of the tabernacle of Cherves (cat. 98). These several comparisons allow the suggestion of a date in the second quarter of the thirteenth century.

The provenance of the chasse has traditionally been given as the church at Lezoux (Puy-de-Dôme), presumably Notre-Dame, a Cluniac priory of Thiers,[3] though it has not been possible to confirm this history. Since relics of the virgin companions of Saint Ursula were largely preserved in the Rhineland, the possibility that this chasse was made for export cannot be excluded. However, important relics of the Cologne maidens came to France via the mission of the monks of Grandmont to Cologne in 1181, which resulted in the acquisition of the remains of five virgins. An additional relic of one of Ursula's holy companions was presented to King Louis IX about 1250. Descriptions from the abbey of Grandmont itself give tantalizingly vague descriptions of the reliquaries themselves. At the time of the suppression of the abbey in 1790, one chasse was given to the parish of Saint-Priest-Palus (Creuse).[4] In addition to statuette images of two of the virgins and of the donors, it included a six-part narrative representation of the legend of Saint Ursula; the present example has small enameled "statuettes" on the front and five scenes on the reverse. It is recorded that Solignac had an enameled thirteenth-century chasse in poor condition but containing relics of six of the companions of Saint Ursula.[5] Additional evidence for the meting out of Ursuline relics in the Limousin is offered by the head-reliquaries of Limoges work of the late thirteenth century (cats. 153, 154).

BDB

Reverse

NOTES
1. Beard 1929, pp. 83–90.
2. Rupin 1890, fig. 437, Arminjon et al. 1995, p. 79.
3. Cottineau 1935–39, vol. 1, col. 1600. The Cistercian priory at Lezoux was not founded until 1656.
4. Pardiac 1860, pp. 369–70.
5. Ibid., p. 370.

EXHIBITIONS: New York 1968, no. 165; New York 1974.

# 115. Chasse: The Life of Christ

Limoges, ca. 1235–45

Copper (plaques): engraved, stamped, and gilt; (appliqués): repoussé, engraved, chased, scraped, and gilt; champlevé enamel: dark and medium blue, turquoise, light green, red, and white; glass cabochons; wood core

34 x 34.3 x 13.5 cm (13⅜ x 13½ x 5⁵⁄₁₆ in.)

INSCRIBED: IH[ESU]S

Reverse

IN COLLS.: Frédéric Spitzer, Paris (sale, Paris, April 6–June 17, 1893, no. 237, pl. VII); [Seligmann and Co., N.Y., 1908]; Thomas Fortune Ryan, New York (sold November 23–25, 1933, no. 403); Harriet H. Jonas, New York

CONDITION: Appliqué figure of Christ and some striped border strips replaced by Alfred André, Paris, probably while the chasse was in the Spitzer Collection.

The Metropolitan Museum of Art, New York; Bequest of Harriet H. Jonas, 1974 (1974.228.1)

The chasse is in the form of an oblong casket standing on cubical feet, with pitched roof, gabled ends, and a projecting transept. The wood core has been covered with gilt-copper sheets that are pierced and studded with glass cabochons and decorated with engraved quatrefoils. The principal face is decorated with appliqué figures enameled in medium blue and turquoise with accents of white, light green, and red. The figure of Christ appears on the projecting transept at the center of the principal face, his hands

and feet nailed to a green-enameled cross bearing the titulus IHS. To his right is the Virgin and another of the Marys; at his left, Saint John and the centurion who recognized Jesus as the son of God.[1] On the roof above, flanking the transept, appear standing, winged figures, symbolic of the sun and the moon, and half-length figures of angels. Saint Peter appears on the gabled end panel with the keyhole and lock mechanism; seen holding the keys, the figure is reserved in gilt copper against a blue enamel ground. Saint Paul stands on the opposite end. Four enameled quatrefoils with figures in reserve decorate the roof and lower panels of the reverse. Reading clockwise from upper left, they present scenes from the Infancy of Christ: the Annunciation, the Visitation, the Nativity, and the Annunciation to the Shepherds. The standing figure of Christ displaying his wounds appears on the transept panel of the back. The crest is pierced with a pattern of alternating quatrefoils and superimposed canted squares and is surmounted by glass cabochons. The wood core is original and is gessoed and painted on the underside. As in many chasses with preserved original cores, the inside of this one has the remains of white paint; a red cross appears on the interior of the left gable.

The Metropolitan chasse should be grouped with a number of other reliquaries with enameled figures and cabochons that are datable to about the mid-thirteenth century, notably the chasse of Saint Ursula (cat. 114), one of the Adoration of the Magi in Saint Petersburg, and another in Copenhagen (fig. 115a),[2] which was first documented in the royal collection there in 1696.

The enamels of the reverse, with their reserved figures in gilt copper against blue grounds, recall similar medallions with anecdotal narrative scenes on the chasse of Saint Viance (cat. 118), though the example at Saint-Viance, with its stockier figures and scrolling enameled fleurons within the quatrefoils, is probably earlier than the Metropolitan chasse.[3] Reserved figures on an enameled ground are also found on the reverse of the coffret of Saint Louis (cat. 123), dated 1234–37 on heraldic evidence. The Saint Louis chasse is notable, too, for its use of opaque, deep red enamel, also found on the New York, Copenhagen, and Saint Petersburg chasses.

The punctuation of the principal face with glass cabochons is typical of a number of Limoges crosses and quatrefoil reliquaries dating to the second and third quarters of the thirteenth century, notably the cross of Bonneval (cat. 104) and the reliquary of Saint Francis (cat. 101), which is dated after 1228, the year of his canonization.

Plaster casts of the appliqué figures are preserved at

Maison André, Paris, where the Jonas chasse was apparently restored when it was part of the Spitzer Collection.[4] The majority of the elements now in place appear to be original: the metal is worked, not cast, and displays wear consistent with a medieval date. The figure of Christ, however, is a replacement, betrayed by the more regular lines of the drapery, the bright turquoise of the loincloth, the thinner metal, and the lack of wear. The figure corresponds to a positive plaster cast at André marked on the reverse: "Chasse no. 1, Christ refait" and perhaps based on a second plaster labeled "Christ moulé sur une chasse ancienne."

BDB

NOTES
1. Following the text of the gospels of Matthew (27:54), Mark (15:39), and Luke (23:47).
2. Liebgott 1986, figs. 39–42, pp. 49–50.
3. By contrast, the chasse at Sarrancolin, with its elongated, sinuous, more Gothic figures, probably dates to the late thirteenth century. See Paris 1965, no. 473, pl. 56.
4. For a discussion of André's work for Spitzer, see Distelberger et al. 1993. We are grateful to M. Michel André for his generous welcome to his studio and for access to his ancestors' collection of plaster models and casts.

BIBLIOGRAPHY: Palustre and Molinier 1890, vol, 1, no. 26; Rupin 1890, p. 421.

Principal face

Fig. 115a. Chasse: Adoration of the Magi (lower register). Nationalmuseet, Copenhagen

# V

## The Last
## Flowering

1240–1320

The final chapter of the history of Limoges enameling in the Middle Ages is the least known, and the objects are widely dispersed. Consequently, we have endeavored to bring together the most significant works here. In the evolution of this period, the development of relief sculpture in gilt copper is of capital importance (cats. 118–22). A number of works that are dated—or datable—are critical in establishing a chronology from the mid-thirteenth to the early fourteenth century. Among these, the coffret of Saint Louis (cat. 123), created between 1234 and 1237 is highly significant because of its rich array of armorial bearings. On the other hand, the presence of heraldic ornament on other objects, while visually appealing, may be purely decorative. Some works can be approximately dated by their religious imagery, such as the reliquary from Brive (cat. 142) on which Saint Clare of Assisi, canonized in 1255, appears. This reliquary likewise bears witness to the continuing taste for Limoges work among the mendicant orders. The works that bear a date range from the relatively simple dedicatory plaque from an altar of 1267 (cat. 141) to the appealing funerary image of 1307 representing

Guy de Meyos kneeling before the sainted French king, Louis IX (cat. 152), to the Virgin and Child from La Sauvetat of 1319—whose startling enameled eyes, richly enameled slippers, and well-preserved gilt surface were more fully revealed during cleaning for this exhibition (cat. 156)—and the commanding head reliquary of the bishop-saint Ferreolus of Limoges, signed by the artist Aymeric Chrétien and dated 1346 (cat. 157).

Funerary ensembles were among the most important commissions undertaken by the enamelers at Limoges in the thirteenth and fourteenth centuries. Though now largely destroyed, their importance and beauty can still be gleaned from the images of two children of Louis IX and the one of Blanche of Champagne (cats. 146, 151). The richness of such ensembles is tantalizingly suggested by watercolors of other tombs made before the French Revolution (cats. 147–49). These monuments of Limoges work are so little known and yet of such consequence that a special chapter is devoted to them (chapter VI).

BDB and ET-D

338

# The Use of Heraldry in Limousin Enamels

Heraldry and enameling have ancient, close, and complex ties, going back to the emergence, diffusion, and composition of the first armorial bearings between the mid-twelfth and the early fourteenth century.

For a long time, heraldists have considered the most ancient extant arms to be those found on the funerary plaque (cat. 15) of Geoffrey Plantagenet (1113–1151), count of Anjou and duke of Normandy. Geoffrey appears as a knight in full military dress, bearing an immense shield emblazoned with *Azure eight lionls rampant Or*.[1] This enameled copper plaque, made at the request of his widow, Mathilda, in the years following Geoffrey's death, is the sole visual evidence of his arms that has come down to us. An imprint of his equestrian seal, appended to a document dated 1149, bears no trace of it.[2] By this date, other important figures—dukes and counts in possession of lands in different parts of western Europe—had for one or two decades placed their coats of arms on their seals.[3] Geoffrey did not do this. Even though a chronicler, compiling a history of the counts of Anjou during the years 1170 to 1175, tells us that, when Geoffrey was made a knight in 1127, he received from his father-in-law, King Henry I of England, a buckler decorated with lions,[4] it is not at all sure that our count of Anjou used armorial bearings until the very end of his life.

The arms of Geoffrey Plantagenet are therefore not the oldest known armorial bearings, but they are probably the oldest for which we have evidence of the original colors. Thus it is an enamel that first shows us arms in color—a generation before they appear in illumination and mural painting, two generations before they appear in stained glass, and three generations before they appear in textiles. This is important because heraldry is first and foremost a system of color. There are many armorial bearings without figures[5] but there are none without color, although a large number of them, from the medieval period, are now known only through monochrome documents (seals, coins, and sculpted arms that have lost their polychromy).

This strong tie between the art of enameling and the heraldic system is underscored by the use, since at least the

fifteenth century, of *émaux* (enamels) as a generic term to designate colors in the language of French blazonry. During the course of time, in fact, the language of blazonry has become progressively detached from the literary language of the twelfth and thirteenth centuries that had witnessed its birth. Heraldry developed a specific vocabulary—for example, *gueules* (in English heraldry, *gules*) for red, *sable* for black, *sinople* (English, *vert*) for green, and so forth. The etymology of these words has nothing to do with armorial bearings, and their original meanings may therefore have been quite different from those assigned in blazonry,[6] wherein they gradually became so specialized as to have an exclusively heraldic meaning. In the fifteenth century—and perhaps as early as the end of the fourteenth[7]—French heralds, considering that each color was designated in the language of blazonry by a word different from ordinary words, chose the word *émaux* specifically to mean the ensemble of heraldic colors. This lexical peculiarity is preserved in all the manuals and treatises of French blazonry printed from the fifteenth to the twentieth century. The *émaux* of French blazonry are the seven tinctures, or heraldic colors, of which armorial bearings could be composed: white or silver (*argent*), yellow or gold (*or*), red (*gules*), blue (*azure*), black (*sable*), green (*vert*), and purple (*purpure*). The use of this unusual and somewhat precious term in this context is itself a rich historical document.

The link between *émaux* and the colors of heraldry is not, however, limited to lexical matters. It also concerns the distribution of these colors on coats of arms. An exceedingly powerful constraint, which existed from the beginning and allowed no exceptions, prohibited the use of color in just any fashion. It divided the tinctures of blazonry into two groups: the metals comprised silver and gold, represented as white and yellow; the second, the five colors. When creating a coat of arms, it is not possible to overlay or juxtapose two tinctures belonging to the same group. Let us take, for example, a coat of arms with the figure of a lion. If the background is white, the lion could be red, blue, black, green, or purple, but it could not be yellow, because yellow and white belong to the same group. Conversely, if the background is red, the lion could be white or yellow but not blue, black, green, or purple. This simple rule, affecting only the first state of any given coat of arms and not its subsequent modifications (which are often unknown), is related to the question of visibility (the first coats of arms were made to be seen from a distance on the battlefield and in tournaments); to symbolical considerations (according to medieval sensibility, yellow was often thought of as a half-white or under-white); and to technical problems. Separating two colors belonging to the same group, enclosing the colors in a cleanly outlined perimeter, articulating the distributions according to rigorous and recurring principles—all this makes one think of certain techniques and customs of enamelwork, both cloisonné and champlevé. There is no doubt that future scholarship on the origins of this rule concerning the association of colors in the earliest coats of arms—a rule about which much has been written without solving its mysteries[8]—will have to look in this direction.

Let us return to the plaque of enameled copper that is (as explained above) the oldest remaining monument depicting armorial bearings in color. It begins a relatively abundant series of objects and works of art made with enamel techniques that, until very late in the fourteenth century, somewhat widely in western Europe but principally in France and England, depicted armorial bearings in enduring colors so that today we have a source enabling us to recognize a growing number of them. From the middle of the thirteenth century, many other documents and monuments have transmitted to us the representation of coats of arms in color: mural paintings, illuminated manuscripts, stained-glass windows, embroidered and woven cloth, polychrome sculpture, and especially painted armorial bearings, which became more and more numerous as the decades passed. But for the period preceding the years 1230 to 1250, no books of heraldry have survived, the stained-glass windows adorned with armorial bearings have almost all been destroyed, and the armorial bearings depicted on cloth—once by far the most numerous—no longer exist. Very rare indeed are the armorial bearings in color that have come down to us from the first century of Western heraldry. In the meager corpus that they form, it is probable that the enamels provide the most important source—the most important but certainly not the easiest to study, because the presence of armorial bearings on enamels, whether found before or after the middle of the thirteenth century, poses complex problems of reading, identification, and interpretation. These problems, which only the heraldist can solve, are still unrecognized or misunderstood by enamel specialists. Confronting these questions involves reviewing and reconsidering a great many datings, localizations, and attributions proposed by our predecessors.

These predecessors have almost always indicated that the arms found on works in enamel, notably on coffrets, small boxes, crosiers, crosses, cups, and gemellions belonging to Limousin production, were true armorial bearings and that this presence responded to precise intentions. Identifying these armorial bearings, situating them in space and time, disentangling the network of kinship and vassalage, researching the commissions, gifts, homages, alliances, marriages, and historical circumstances explaining their presence on an object—all this seemed legitimate and appropriate to endow these enameled objects with a civil status. Such a procedure, however, has meaning only if the armorial bearings found there are indeed authentic, belonging to real people who are precisely identifiable and tied, directly or indirectly, to the production or possession of the objects on which the arms are depicted. This is far from being always the case, particularly in the thirteenth century.

Indeed it would be a mistake to believe that, each time one encounters the arms of France, England, Castile, Aragon, or Navarre on one of these objects, one should search for an attribution at the heart of these royal houses. Such prestigious arms, on enamels as well as other objects and works of art, quickly became ornamental motifs and do not always require a precise attribution. To try to attribute them to a patron or recipient belonging to these dynasties often leads down the wrong track. This fact, which for a long time was unknown to art historians, is better understood today. We now know, for example, that whenever we see the lilies of France combined with the castles of Castile on a thirteenth-century object or monument, we need not systematically think of Blanche of Castile or Alphonse of Poitiers, as had been done for so long. The combination of the arms of France and Castile is an ornamental motif that took root perhaps as early as the 1230s and lasted until the first decades of the fourteenth century (when the combined arms of Évreux and Navarre began to play the same role).

We are less certain, on the other hand, that what is true for royal armorial bearings is also true for those of families situated farther down the social scale, such as the armorial bearings of dukes (Burgundy, Brittany), counts (Toulouse, Champagne, Anjou, Dreux, Bar), viscounts (Turenne), and also simple lords (Lusignan, Rohan, Courtenay, not to mention the many younger branches of the house of Chatillon). In this regard, the case of the arms of the lords of Coucy, *Barry Vair and Gules*, is typical: as early as the beginning of the thirteenth century, they functioned as a stereotype, used here and there notably in Parisian and northern French illumination,[9] to give the idea of armorial bearings.

It is likewise less commonly known that armorial bearings belonging to real people are mixed with those of imaginary people (Prester John, King Arthur, the knights of the Round Table) and historical figures who lived before the appearance of armorial bearings (David, Alexander, Charlemagne). The arms of Tristan or Lancelot, for example, could well appear on a gemellion decorated with a courtly scene, while those of David or Alexander might be integrated into a group of armorial bearings. This is a constant fact in heraldry and medieval iconography.

Identification and attribution are still further complicated by the fact that the representation of these stereotypical armorial bearings often takes liberties with the exactitude and rigor of heraldic composition. Certain elements are missing, others poorly interpreted, the colors reversed, the animals distorted, the components badly arranged, the figures reduced in number to three or one. This makes it difficult to identify such arms correctly and to recognize that the escutcheon seen on a certain gemellion is, in spite of its differences, the same as that on some other gemellion, small box, object, or monument. Here again the case of the Coucy arms is illustrative: one finds many variants of *Barry Vair and Gules*, the gules becoming white, the vair losing its shape or represented in the wrong colors (sometimes Vairy, Vair and Or!), or each bar of Vair being horozontally divided—into Vair and another color. Anything is possible, and only the trained eye of the heraldist—accustomed always to giving priority to structure over form in the reading of medieval images—knows how to distinguish what is and what is not significant among these imprecisions and deformations.

Added to the difficulties of reading and interpretation are those arising from the poor state of preservation of certain objects (notably the disappearance of certain colors) and the clumsy restorations or fraudulent transformations that have been made from one epoch to another.

The armorial bearings encountered on enamels have a further characteristic that often encourages us to see them not as armorial bearings needing to be precisely attributed but as simple decorative arms. It is the very particular frequency with which certain figures and colors recur in the composition. As they appear in the "armorial" (book of heraldry) of Limoges enamels (many hundreds of escutcheons from the end of the twelfth century to the beginning of the fourteenth), the frequency of these elements has very little relation to what one finds on other documents, notably seals (for those related to the figures) and armorial bearings (for those related to the figures and the colors). The most striking example is that of animal figures. In the armorial bearings of the thirteenth century, the proportion of "animal" arms—which is to say, those having one or more animals as principal or secondary figures—in relation to the totality of armorial bearings is about thirty to thirty-five percent.[10] The figured arms on enamels make up clearly less than ten percent; and those only on gemellions, probably less than five percent. This abnormal proportion underscores the selective and artificial character of such a group of armorial bearings. Conversely, the frequency of some partitions (barry, barruly, bendy, cotised, paly, and especially chevrony) is much higher than in true armorial bearings: twenty or twenty-five percent instead of three or five percent!

Similar divergences, more serious still, exist for the colors:

azure, which infrequently appears in early armorial bearings, is strongly overrepresented here (in more than half of them!). Sable (black), on the contrary, is very rare in enamels, whereas in any assortment of thirteenth-century armorial bearings it is found in over twenty percent. As for vert (green), it is never found on enamels, but on "true" armorial bearings its frequency of appearance, although small, is in general five percent.[11] Such divergences are not the fruit of chance: they underscore the way in which certain armorial bearings were chosen and placed for purely decorative ends: to balance the distribution of figures and colors, join red with blue, contrast horizontal and vertical partitions, form sequences, repeat the same armorial bearing many times on the same object, place arms with straight or wavy bars—that is to say, those suggesting water—beneath a gemellion, and so forth.

Enameled objects, however, do not have a monopoly on this selection of armorial bearings for ornamental purposes. One finds the same thing from the late twelfth century until the middle of the fourteenth century, on coffrets and small boxes of wood covered with painted parchment; on alms purses, pouches, and cushions;[12] on various embroidered or woven objects such as those on English embroidery (*opus anglicanum*), and even on stained-glass windows and painted murals. The method was the same: choosing armorial bearings, prestigious if possible, with figures and colors that allowed for a harmonious distribution; giving preference to geometrical figures over animals, which were too contorted by heraldic style; eliminating brisures, marks of difference for younger members of a family, which were hard to see and difficult to represent; avoiding sable, which cuts too much into the other colors, and favoring azure, which forms a pair with gules just as or pairs with argent. Four colors instead of six, a select repertory of figures, few or no brisures, great liberties taken with the customs of blazonry, variations on alternated or repeated motifs: we are far from the polychromy of true heraldry.

This said, even though many of the identifications and attributions must be revised and too many objects and enameled works have, from earlier heraldic study, received a civil status or provenance that is not justified, still one must not go too far in the other direction. Not everything is ornamental; not all the armorial bearings represented on enamels are decorative. Some can and should be identified and attributed with precision, notably those represented in large numbers on tombs (such as the effigies of Alix of Thouars and Yolanda of Brittany [cat. 149]), on chests (such as the one said to have belonged to Saint Louis (cat. 124) and that at Aachen (fig. 9), and on most of the objects made before the year 1230 or 1240. It seems to me, in fact, that it is after this time that the stock of armorial bearings tends to become fixed and that, on objects made in series, there begin to multiply instances of imitation, of reuse, of deterioration into hackneyed models, and of sequences at the center of which the decorative aspect becomes more important than the meaning of the symbols. The case of the gemellions of the second half of the thirteenth century is typical in this respect, for these provoked many enamel specialists into extravagant and pointless heraldic and genealogical acrobatics, which led to their endowing these objects with a provenance and civil status.

This is often a vain exercise, as Alain-Charles Dionnet recently reminded us in regard to a Limousin gemellion of the 1270s.[13] It is further complicated, as he emphasized and as stated above, by the great liberties that the engraver-enamelers sometimes took with the rules of blazonry and the representation of heraldic figures. Lastly, it is methodologically dangerous because the armorial bearings depicted on the enamels are almost always very simple, even too simple: *Azure a Lion rampant Or, Barry Argent and Gules, Bendy, or Paly Or and Gules,* and so forth. For many of them, several identifications are possible; and, naturally, any researcher has a tendency to retain only the identification that accords with the demonstration he or she is undertaking and to leave aside all the contradictory ones, a lamentable method of analysis in any field. The fragility of such methods—and the classification, typologies, stylistic analyses, and historical syntheses that result—is well known. Let us hope that this presentation will make us more aware of our lacunae and insufficiencies, and encourage us to question most of these certitudes.

MICHEL PASTOUREAU

NOTES

1. The gold has practically disappeared, and the figures placed on the azure field appear today more or less red. Only four lions are visible; but, because the shield is very concave, we must imagine four others on the half that is not visible. The correct heraldic description should therefore speak of eight lions or, more strictly, of a semy (scattered field) of lions (i.e., a field scattered with small lions), but not six lions, as in the past. On this subject, see the most recent discussions of Roger Harmignies (1983, pp. 55–63).

2. Demay 1880b, no. 20. This treats the only remaining imprint of one of Geoffrey's seals.

3. One can find the list in Galbreath (1942, pp. 26–27), completed by Wagner (1956, pp. 13–17).

4. According to John, monk of Marmoutier, "Clipeus leunculos ymaginarios habens collo ejus suspenditur"; see Marmoutier 1913, in Halphen and Poupardin 1913, *p.* 179.

5. Pastoureau 1979 and 1993, p. 100.

6. The word *sinople,* for example, which in the literary texts of the twelfth and thirteenth centuries designates the color red, comes to signify green in the language of blazonry at the end of the fourteenth century.

7. A close study of this new lexical use proper to armorial heralds, the only professionals of heraldic language in the late Middle Ages, would be welcome. I wish to thank my student Claire Boudreau, our best specialist in French treatises on heraldry of the fifteenth and sixteenth centuries, for the information that she gave me on this subject.

8. Pastoureau 1979 and 1993, pp. 108–10.

9. Thus in the famous Psalter of Queen Ingeburge (ms. 1965, Musée Condé, Chantilly), one can see on the verso of folio 28, in the scene of the Holy Women at the tomb, a clumsily emblazoned escutcheon of the Coucy arms, held by one of the sleeping soldiers. I hope soon to devote a study to this important image, which, coupled with the captivity of Ingeburge, might help to date and localize precisely this copy of the psalter.

10. Pastoureau 1982, pp. 105–16.

11. Pastoureau 1979 and 1993, pp. 116–21.

12. Faustino Menendez Pidal recently presented, in a colloquium devoted to medieval armorial bearings, a cushion embroidered in England in the mid-thirteenth century and decorated with thirty-six emblazoned escutcheons of which not one includes an animal as a heraldic figure. He rightly attributes a purely decorative function to these escutcheons and stresses how vain it would be to try to identify them for the purpose of dating the cushion and attributing it to a precise recipient. See his study "Armoriaux et Décors Brodés au Milieu du XIIIe Siècle," to be published in *Cahiers du Léopard d'Or* 8 (1996).

13. Dionnet 1994, pp. 94–112.

# 1

## *The taste for sculpture*

# 116. The Virgin and Six Apostles

Limoges, ca. 1230–40

Copper: repoussé, engraved, chased, and gilt; lapis blue enameled beads

27.2 × 17.7 cm (10¾ × 7 in.)

Ex colls.: Carlo Micheli (1809–1895), Paris (Musée Mayer van den Bergh, handwritten catalogue, 1898, no. 239); Victor Martin Le Roy, Paris (before 1900); Jean-Joseph Marquet de Vasselot, Paris. Acquired in 1992.

Condition: Some wear and scratches on the gilding; small fissures in the metal.

Musée du Louvre, Paris; Département des Objets d'art (OA 11352)

This group, composed of the Virgin and six apostles, is certainly from a depiction of one of the New Testament episodes that followed the Resurrection of Christ, narrated in the Acts of the Apostles: the Ascension (1:6–11), the Meeting in the Upper Room (1:13–14), or the Pentecost (2:1–4). The second of these episodes was seldom illustrated in the thirteenth century. We know of no portrayals of the Pentecost from the same period in which the figures are not seated, so the appliqué very probably comes from a composition dedicated to the Ascension, which was accompanied by the figure of Christ rising to heaven in a cloud and by another group of apostles as a counterpart. It is not surprising that the figures do not clearly gaze heavenward. During the thirteenth century such figures were often oriented frontally, as here, or toward the center of the composition. A number of illuminated examples can be cited, such as those in a collection of polyphonic music at the Biblioteca Medicea Laurenziana, Florence, the Douce Psalter at the Bodleian Library, Oxford, and the Missal of Saint-Nicaise of Reims, just limiting ourselves to French works of the second quarter of the thirteenth century.[1] In Limoges work, the closest parallel, slightly earlier, is undoubtedly the Ascension engraved on the interior of the tabernacle of Cherves (cat. 98).

Among the appliqués executed in half relief by Limousin workshops in the thirteenth century, this group is distinguished by a rare sobriety and monumentality. The draperies are enlivened by barely curving vertical or oblique lines, indicating a departure from early Gothic conventions. The round faces, with full cheeks and regular features, are dominated by large eyes widened by enameled beads denoting the pupils. This style directly precedes that of appliqué groups portraying the Passion (cat. 119) and the figures of the Germeau chasse (fig. 120*a*), which present the same overall characteristics and are ornamented with similar details, especially in the decoration of the orphreys. But on this Ascension the stereotypes typical of these appliqués do not appear, chiefly the treatment of the mustache and beard as a continuous projection and the hollowing of the drapery folds into grooves terminating in spoon shapes. Furthermore, the mouths here are smaller and pinched. The sober expressions, devoid of any schematizing, reveal a quest for individualized faces. Although the dimensions are comparable—probably corresponding to a widespread Limousin canon—this Ascension seems to have been created by a hand different from that of the appliqués cited above.[2] But actually, it is one of the rare examples of the efforts made by Limousin artists to create an independent and unconventional monumental style. This rare style is found on other appliqués, for instance, a deacon, probably Saint Stephen, preserved at the Musée de Cluny, which also presents the same embroidered orphreys.[3]

ET-D

Notes

1. Florence, Biblioteca Medicea Laurenziana, ms., Plut. 29.1, fol. 346, see Branner 1977, fig. 222 (Paris, after 1236); Oxford, Bodleian Library, ms. Douce 50, fol. xxiiir, see ibid., fig. 301 (Paris, ca. 1240); Reims, Bibliothèque Municipale, ms. 230, fol. 246v, see ibid., fig. 294 (Paris or Reims, ca. 1250).
2. Contrary to a frequently made suggestion, when this relief was known only by its reproduction in Marquet de Vasselot (1906a); see Otavsky 1973, p. 65; Cleveland 1967, p. 132.
3. Inv. Cl. 14 133 (acquired in 1902, Lelong sale).

Exhibitions: Paris 1900, no. 2524; Paris 1995, no. 10.

Bibliography: Marquet de Vasselot 1906a, no. 37, pl. xxvii; Coo 1965, no. 239, p. 359; Cleveland 1967, p. 132; Otavsky 1973, p. 44 and no. 35, pp. 58, 65; Swarzenski and Netzer 1986, p. 100, n. 12.

# 117. Angel Carrying a Crown

Limoges, ca. 1230–50

Copper: repoussé, engraved, chased, and gilt

10.8 × 9.3 cm (4¼ × 3⅝ in.)

Ex coll.: Harry G. Friedman, New York

Condition: Wear to the gilding; left wing of the angel is missing; right wing, with tip missing, has been reattached.

The Metropolitan Museum of Art, New York; Gift of Harry G. Friedman, 1958 (58.110)

This small relief, so captivating in the youthful charm of the face and the quality of the workmanship, has nonetheless been largely ignored.

The theme of the angel emerging from a cloud to intervene in a human story or a scene from the life of Christ is common in the art of the thirteenth century, but it is not easy to determine the composition into which this figure, bearing an enormous crown, might fit. The most natural hypothesis suggests that it is a figure from the Coronation of the Virgin. It was not, in fact, uncommon in the thirteenth century for such compositions to be surmounted by a representation of one or two angels carrying a crown—for example, in the Coronation in the Glorification of the Virgin window at the cathedral of Chartres,[1] on the tympanum of the left portal of the west facade of Nôtre-Dame in Paris,[2] and in the upper part of the left side of the ivory diptych in the Vatican Museum, belonging to the "Soissons diptychs."[3] An angel bearing a crown appears also in many representations of the Virgin in Glory, surrounded by angels and sometimes by saints, notably on leaves or tabernacles of ivory, like the one in the Toledo (Ohio) Museum of Art.[4] But the same iconographic theme is sometimes found in other scenes: the Crucifixion—for example, in the window of the Passion in the cathedral of Bourges[5]—or the Last Judgment—as in the tympanum of the central portal of the south transept of the cathedral of Chartres.[6]

The round face with full cheeks and the robe entirely covered in a lozenge-shaped grid decorated with small circles place this angel within the immediate circle of the large appliqués or group of appliqués executed in Limoges shortly before the middle of the thirteenth century (cats. 119–22). Though the motif decorating the robe is not identical to any on these reliefs, it is nonetheless comparable to the checked pattern strewn with three little marks that decorates the robe of the Virgin in the Deposition in the Abegg Stiftung, Bern (fig. 121a) and the one on a Holy Woman in the Entombment in the Minneapolis Institute of Art (cat. 119). It is perhaps still closer to the orphrey on the neckline of one of the two Holy Women, probably the Virgin, on another group preserved in Bern.[7]

The massive structure of the face, with full cheeks and rounded chin, as well as the hair of fine and even strands curled toward the outside above the neck, justify its comparison with works executed in the Île-de-France about 1220 to 1230, such as the angels on the right side of the lintel of the central portal of Notre-Dame in Paris.[8] However, the arrangement of the robe on the upper part of the body and the similarities with the above-mentioned appliqués suggest a slightly later date.

ET-D

Notes

1. About 1200 to 1210; see Grodecki and Brisac 1984, pp. 62–64; Manhes-Deremble and Deremble 1993, no. 42, p. 361.
2. About 1210 to 1220; see Sauerländer 1972, fig. 153.
3. About 1210 to 1215; see Gaborit-Chopin 1978, no. 214.
4. About 1280 to 1290; former Spitzer and Baboin Collections; see Randall 1993, no. 32, pl. 6.
5. See Verrier 1942, pl. XVI.
6. About 1210 to 1215; see Sauerländer 1972, fig. 108.
7. For the appliqués of Bern, see cats. 120, 121.
8. Deposited at the Musée de Cluny; see Sauerländer 1972, fig. 147.

# 118. Chasse of Saint-Viance

Limoges, ca. 1230–50

Copper (plaques): engraved and gilt; (appliqués): repoussé, chased, engraved, scraped, and gilt; champlevé enamel: medium blue, gray-blue, turquoise, dark and light green, yellow, red, and white; glass cabochons; wood core

57.8 (with crest but excluding finials) x 80.6 x 25 cm (22¾ x 31¾ x 9⅞ in.)

INSCRIBED: SAINSMACNSA.

PROVENANCE: Church of Saint-Viance; first recorded 1530,[1] first published 1669;[2] stolen 1908; recovered in part before 1913 and restored; deposited at the Musée de Cluny, Paris, until about 1939. Classified, Monuments Historiques, June 1891

CONDITION: Recorded as being in poor condition in 1842; two angels above the Virgin and enameled base plates of two evangelist symbols lost after 1859 and before 1891; restored just before 1859; eleven trefoil plaques with angels replaced before 1891; restored in Paris, 1913, by M. André, who replaced lost elements with casts; cleaned and restored by M. Toulouse, Paris, 1983, who added gilt-copper strips at the middle and at the base; door missing from proper left end (lost after 1859; replaced in 1984 by a thin copper plaque); mandorlas and some arches, towers, and columns replaced; original crest; modern core.

Parish Church of Saint-Viance, Saint-Viance (Corrèze)

A diaper pattern and small, punched concentric circles decorate the thin sheets of gilt copper that cover the wood core of this pitched-roof chasse. On the base of the principal face, the repoussé and appliqué figure of the Virgin is enthroned against an almond-shaped enameled plaque. She is framed by figures of angels; those at the bottom are repoussé relief figures set against triangular enameled plaques; similar figures at the top are replacements. At either side two applied figures of male saints, probably apostles, stand against an enameled base plate decorated with foliate rinceaux. In a mandorla on the roof the figure of Christ in Majesty is surrounded by symbols of the evangelists. Corresponding to the program below, two male saints stand at either side of Christ, each set under an arcade. Glass cabochons set within strips of gilt copper frame the vertical edges of the roof; enameled strips run beneath the crest and down the proper right side of the base. The enameled plaque from the end with the door is missing; applied to an enameled plate on the opposite end is a repoussé image of Saint Paul holding a scroll. On the front face and roof, seven columns, two arches, and two enameled towers are original. Five of the original triangular plaques with angels survive.

The reverse of the chasse has six enameled quatrefoils affixed to sheets of stamped gilt copper; the position of the plaques was indicated by scoring on the gilt-copper sheets. On the roof are the Flagellation of Christ, the Crucifixion with the Virgin and Ecclesia on the left and Saint John and

Synagogue on the right, and the Holy Women at the tomb of Christ. On the body of the reverse are medallions portraying the life of the Merovingian hermit-saint known as Viance. The illustrations conform to his legend, first published in 1669 on the basis of an earlier manuscript.[3] At dawn on a Sunday morning, an angel appeared to a priest named Savinian to inform him that his friend Viance had died and to instruct him to remove the body from Rouffiac, where it lay, to the place where they were to meet. After celebrating Mass, Savinian went to recover the body from the bishop of Rouffiac, Rustique, and the duke, Barontus. In the central quatrefoil, the body of Viance is carried, not on an elegant litter but on a simple wagon accompanied by Savinian and a single witness, representative of the crowd that followed the body. The procession began with a pair of

Entombment of Saint Viance (detail of the reverse)

347

The chasse of Saint Viance has traditionally been dated to the middle of the thirteenth century. Ferdinand de Lasteyrie suggested this dating based on the armor of the soldiers at the Sepulchre of Christ, which he characterized as being of "the time of Saint Louis." In fact, however, the flat-topped helmet, or pot helm, worn by one of the soldiers is typical of those from the late twelfth century.[6] The arms on the shield of one of the soldiers have been identified as Sable an Eagle displayed Or, corresponding to those of the de Lasteyrie, seigneurs of the upper and lower Limousin. It has further been proposed that the chasse was the gift of Pierre de Lasteyrie, who went on the Sixth Crusade in 1248. In the first instance, it seems highly unusual that the donor of a reliquary would wish to be identified with one of the Roman soldiers sent to guard the tomb of Jesus, but, in any event, the arms present a golden fleur-de-lis on a dark green field, not an eagle on a black ground (as Rupin's engraving would suggest), and are probably fictive. Nevertheless, according to archives preserved at the church from the time of de Lasteyrie's publication, there was a head-reliquary of Saint Viance that had been given by the marquis du Saillant, the title given, in the seventeenth century, to one branch of the family descended from Pierre de Lasteyrie, raising the question whether the same family may have been responsible for the gift of the enameled chasse of Saint Viance. If so, it was not necessarily Pierre de Lasteyrie. Giraud, of the same family, died in 1355, and Guy de Lasteyrie is mentioned in a record of 1372.[7] At the same time, however, it is equally possible that the chasse was commissioned as the repository for the relics of the patron saint by the Benedictine community of monks established, since 1048, at Saint-Viance as a priory of Uzerche.[8]

Even though the specific historical circumstances surrounding the creation of the chasse of Saint Viance cannot be reconstructed from present evidence, a date close to the middle of the thirteenth century seems correct on stylistic grounds. The rather loose, curving line of the drawing bears comparison with the engraved gilt-copper plaques of the Cherves tabernacle (cat. 98), especially the figure of Christ risen from his tomb, or the angels on the copper plate that lies flat, who have curly hair like that of the angel greeting Savinian. The chattering line that defines the drapery and the hands and feet characterizes both works. On the Cherves tabernacle, the foliate enameled rinceaux are significantly more prominent and more expertly inscribed in circles. The difference is at least partly attributable to scale and probably to a slightly earlier date.

The appliqué figures on the Saint Viance chasse are also the successors of those on the Cherves tabernacle, as a juxtaposition of the enthroned figures of Christ surrounded by symbols of the evangelists or the Pilgrims on the road to Emmaus with the standing apostle plaques suggests. The iconographic formulation of the Flagellation is identical to

Transporting the body of Saint Viance (detail of the reverse)

oxen drawing the cart, but when the party stopped for a meal, a bear devoured one of the oxen. Savinian, undaunted, pursued the bear, commanding the beast in the name of God and of Viance to allow itself to be yoked to the cart in the place of the ox. Upon arriving at the place of his meeting with the angel, Savinian and the bishop took the body of Viance into a church, which had been built next to the Vesère River, at the order of Barontus. The third medallion shows Savinian and an assistant lowering the body into its sarcophagus, recorded as being on the north side of the church near the water's edge, as the bishop, assisted by a deacon, presides.

The inscription, enameled in red over the scene of Saint Viance's entombment, has been variously interpreted. It reads SAINSMACNSA and, as such, is unintelligible. It has been suggested, therefore, that it might be an acronym: S[ub] A[nno] I[ncarnationis] N[ostri] S[alvatoris] M[illesimo] A[volca] C[urta] N[omen] S[ancti] A[ccepit], or SA[vinianus] I[n] S[epulchru]M A[volcae] C[urtae] N[ostrum] S[anctum] A[ttulit].[4] Alternately, and perhaps more likely, it may represent an error in the transcription of SAINT VIANCSA.[5]

Principal face

Reverse

that of the appliqué figures in the Musée de Cluny (cat. 119c), including the representation of the column with a decorated shaft and carved capital, the bloused drapery of the soldiers who beat Christ, the binding of Christ's hands, and the placement of the soldiers' hands as they hold their whips.

The palette of the Saint Viance enamels is exceptional, both in its tones—including gray-blue, stark white, and dark forest green—and in its combinations. This is not surprising for the second quarter of the thirteenth century, when new color schemes were introduced.[9] (See cats. 101, 102, 104: the quatrefoil reliquary of Saint Francis, the cross of Bonneval, and the plaque of William of Bourges.) Not coincidentally, technical evidence suggests that new "recipes" were being employed in the manufacture of glass (see "Techniques and Materials in Limoges Enamels," above, p. 48).

The plaque representing Saint Peter formerly in the collection of the Musée Municipal de l'Évêché of Limoges[10] suggests the original appearance of the missing end panel of the chasse of Saint Viance, but the enameling of the flowers and the palette are distinct. A number of isolated quatrefoils are related in style to those found on this chasse. Among them are a scene of the Annunciation to the Virgin in the Musé de Sainte-Croix, Poitiers, and a Presentation in the Temple published in the Hoentschel catalogue (no. 66, which does not form part of the gift of the Morgan Collection to the Metropolitan). Finally, the appliqué figures are remarkably similar to those of the chasse made for the relics of Saint Geoffrey for the monastic community at La Chalard (fig. 118a),[11] and surely contemporaneous with the chasse of Saint Viance.

BDB

Fig. 118a. Chasse of Saint Geoffrey (detail). Church of Le Chalard (Haute-Vienne)

NOTES

1. In an account of a procession with relics at Brive; transcribed by Rupin 1890, p. 414.
2. "Les Reliques du Saint sont renfermées dans une châsse des plus remarquables, sur laquelle on publiera sous peu une notice." M. Jeoufre, curé de Saint-Viance, in *La Vie miraculeuse* 1859 (p. 46).
3. *La Vie miraculeuse de S. Vincentian*, 1859.
4. Transcribed also by Rupin (1890, p. 412). "During the millennium year of the Incarnation of our Savior, Avolce-Courte received the name of saint." "Savinian places in the tomb of Avolce-Courte [the remains of] our saint." The interpretation depends on the use of an earlier place name for Saint-Viance.
5. Poulbrière 1894, p. 230.
6. Oral communication, Robert Theo Margelony. See also, Edge and Paddock 1988, p. 53.
7. See Jougla de Morenas 1939, vol. 4, p. 421; *Titled Nobility of Europe* 1914; *Armorial Général* 1926–54, vol. 2, p. 29. I am grateful to Robert Theo Margelony for this research.
8. See Cottineau 1935–39, vol. 2, cols. 2914–15.
9. See the discussion in Gauthier 1972c.
10. Gauthier 1950, pl. 52. Stolen in 1981.
11. See Gauthier 1950, pl. 54; Paris 1993, no. 79.

EXHIBITIONS: Tulle 1887; Limoges 1948, pl. IX, fig. II; Paris 1968b, no. 394, pl. 104.

BIBLIOGRAPHY: *La Vie miraculeuse* 1859, p. 46; Texier 1843, pp. 119–20; Texier 1857, col. 1261; F. Lasteyrie 1859; Roux 1889; Rupin 1890, pp. 405–10, figs. 463–70; Poulbrière 1894, pp. 229–32; Forot 1913, pp. 193–94; Gauthier 1950, pls. XLVIII–L; Gauthier 1972c, pp. 621–33, fig. 471, 8; Gauthier 1972b, pp. 373–74, no. 133; Favreau et al. 1978, vol. 2, no. 40.

# 119. Appliqué Reliefs of the Passion of Christ

Limoges, ca. 1240–50

Copper: repoussé, chased, engraved, scraped, and gilt; dark blue and turquoise enameled beads

## a. *The Last Supper*

33 × 30 cm (13 × 11¹³⁄₁₆ in.)

Ex colls.: Alexandre Du Sommerard, Paris (after 1838, before 1842)

Musée National du Moyen Âge, Paris; Thermes de Cluny (Cl 973)

## b. *The Betrayal of Christ*

34.9 × 27 cm (13¾ × 10⅝ in.)

Ex coll.: Acquired in 1923.

Walters Art Gallery, Baltimore (53.10)

## c. *The Flagellation of Christ*

*Christ*: 28.5 × 6.7 cm (11¼ × 2⅝ in.); *Flagellants*: 24.5 × 7.5 cm (9⅝ × 2¹⁵⁄₁₆ in.)

Ex colls.: Debruge-Duménil? (sale, Paris, December 9, 1839, no. 29 [a group of three figures and one figure of a saint on a Gothic column in enameled gilt copper, Byzantine workmanship]); Alexandre Du Sommerard, Paris (ca. 1838, before 1842)

Musée National du Moyen Âge, Paris; Thermes de Cluny (Cl 942)

## d. *The Entombment of Christ*

29.2 × 27.9 cm (11½ × 11 in.)

Ex colls.: Countess Isabelle Dzialynska, Paris (in 1865 and 1867); Prince Witold Czartoryski, Goluchow Castle (Poznań, Poland, in 1903); Adam-Louis Czartoryski; Kenneth Clark, London; [Seymour Rey, in 1956]; [Wildenstein and Co., New York, 1956].

Institute of Arts, Minneapolis (58.8)

Exhibited in New York only.

The four relief sculptures in gilt copper represent scenes of the final day of the life of Jesus, beginning with the Last Supper.[1] Christ, surrounded by his apostles, appears behind the banquet table. Of the twelve, only six are fully realized as relief figures; a second row of faces appears at intervals between the shoulders of those in the front row, according to isocephalic medieval principles. Jesus raises his right hand in blessing and cradles the young Saint John the Evangelist beneath his left arm. Isolated on the opposite side of the table, Judas, smaller in scale to suggest his proximity, steals a fish from the table. This representation follows an iconographic convention established by the twelfth century and reflects a tradition of medieval mystery plays in which the disaffected apostle is physically separated from the others.[2]

The second relief depicts the scene in which Judas betrays Jesus with a kiss.[3] As in the appliqué of the Last Supper, Judas is again removed from the other apostles, appearing alone to the left of Christ. He leans in to place his

a

left arm on Christ's chest and encircles his shoulder with the right. With this disingenuous embrace, the perfidious apostle identifies Christ to the soldiers. The drama of the moment is emphasized by the compression of its actors. The overlapping figures are evocative of the "great multitude"—the armed soldiers and servants of the high priest—said to be attendant at the scene. Christ, holding a book in his left hand, recoils from Judas, while at the same time extending his right arm to heal the ear of the high priest's servant Malchus, which Peter, in his anger, has just cut off.[4] Behind Peter another apostle appears alongside a figure wearing an elaborate headband, possibly identifying him as a member of the high priest's entourage.

Following his arrest, Jesus is tied to a column and flogged.[5] This relief representing the Flagellation is preserved as three separate pieces. Christ appears behind the column to which his hands and feet have been bound. The flanking figures of the flagellants, with their gaze firmly fixed on Christ, have been stopped in mid-motion as they pivot their torsos in the opposite direction, preparing to strike once more. The dramatic tripartite composition, with its strong vertical lines, is strictly symmetrical; an equilibrium

b

c

is achieved by the diagonal lines of Christ's crossed arms and the horizontal lines of the ground and the figures' belts.

In the fourth relief, the Entombment, the body of Christ is laid in a sarcophagus. The descriptions of the four Gospel accounts are conflated:[6] Nicodemus, holding an unguent jar, stands over the body of Christ, accompanied by Joseph of Arimathea, who had obtained the body of Christ from Pontius Pilate and bought the linen shroud to cover it. A woman usually identified as the Virgin Mary holds Christ's head, while Mary Magdalen leans over the foot of the bier holding another unguent jar.[7] The sarcophagus itself stands on two thin columns. The surface of the copper is richly engraved with a floral pattern across the base, scrolling decoration at the necks of the garments, diaper patterning on the Magdalen's gown, and fleur-de-lis ornamentation on the Virgin's.

The four panels are allied not only in theme but also in style and decoration. They share the following characteristics: thin figures, elongated limbs, fixed gazes, moon-shaped faces, rounded gestures, and drapery that falls in slender, straight folds. The capital of the column of the Flagellation is identical to those that support the sarcophagus of Christ; the folds of the shroud of the Entombment resemble the pleating of the tablecloth of the Last Supper; the stacking of the figures in the Last Supper—with those behind only partly defined and those in the foreground rendered in

smaller scale—recurs on the relief of the Betrayal.

These fragments of an original ensemble have been identified by art historians for the last forty years—and labeled by the *Émaux méridionaux* (Gauthier 1987)—as "groupe des appliques du maître de la Passion" (group of appliqués by the Master of the Passion). Georg Swarzenski was the first to link the reliefs by style,[8] though he saw the Walters Betrayal as the work of a distinct hand. His discussion of the four was predicated upon their stylistic links to the relief of the Baptism of Christ preserved in Boston (cat. 122). He dated all the reliefs to the second quarter of the thirteenth century by comparing them with the chasse of Saint-Viance (cat. 118). In fact, the Limousin chasse of Saint-Viance (Corrèze) includes an enameled quatrefoil depicting a similarly composed Flagellation scene. In addition, the principal face of the chasse of Saint-Viance presents repoussé figures that can be compared with the Passion reliefs.

To test the implications of Swarzenski's association of the reliefs, the five plaques were exhibited together in Cleveland in 1967; the juxtaposition suggested to the curator that the Boston appliqué, because of stylistic and technical anomalies, did not come from the same ensemble as the others.[9]

The scale of the reliefs seems to indicate their provenance from an altar frontal with relief figures set against an enameled ground of the type described at Saint-Martial, Limoges; Bourganeuf; and Grandmont, as Rückert, fol-

d

lishing the stylistic relationship between the appliqués and the tomb effigy of John of France (cat. 146), dated after his death in 1248. It is clearly this monument that stands midway between the appliqué figures of the tabernacle of Cherves (cat. 98) and the appliqué reliefs grouped together here. The Ascension relief (cat. 116) is, however, distinct, and arguably earlier than the Passion reliefs assembled here, whereas the Baptism, the unique example incorporating enamel, seems slightly later.

GF and BDB

NOTES
1. Our thanks to Françoise Arquié-Bruley for research in this regard See Matthew 26: 20–29; Mark 14: 17–25; Luke 22: 14–21; John 13: 1–4.
2. Réau 1957, vol. 2, pp. 414–15.
3. Matthew 26: 47–52; Mark 14: 43–47; Luke 22: 14–21; John 18: 3–11.
4. This follows the account in John 18:10.
5. Matthew 27:30; Mark 15:19; Luke 23:16; John 19:1.
6. Matthew 27: 57–60; Mark 15: 43–46; Luke 23: 50–56; John 19: 38–42. Only John mentions the presence of Nicodemus; however, he alone makes no mention of women being present.
7. The women who were witnesses to the Entombment are variously identified in the Gospel accounts; none of them is said to be the Virgin Mary.
8. Following the written suggestion of Marie-Madeleine Gauthier.
9. William D. Wixom, oral communication. In 1976, Alain Erlande-Brandenburg, following M.-M. Gauthier's request, collected data at the Research Laboratory of the National Museums of France that confirmed the coherence of the reliefs preserved in France, despite their idiosyncrasies. These conclusions may be found in the Rapport du Laboratoire de recherche des musées de France, *Examen d'appliqués limousines en cuivre doré (milieu du XIIIe s.) à la demande de M.-M. Gauthier par S. Delbourgo, F. Drilhon, Ch. Lahanier, F. Françaix*, 1976, typescript.
10. For Saint-Martial, probably dated after 1183, see Gauthier 1987, no. 238, p. 200, with earlier literature; for Bourganeuf, see Texier 1843, pp. 82, 270; Rupin 1890, p. 100; for Grandmont, see Souchal 1963, pp. 50, 54, 56; Gauthier 1987, no. 246, p. 204, with earlier literature.
11. Otavsky 1972.
12. Archives Départementales de la Haute-Vienne, ms. I, Sem. 82. Transcribed by Rückert 1959, p. 7, and Souchal 1963, p. 51.

EXHIBITIONS: *Last Supper*: Vatican City 1963, no. 35; Cleveland 1967, pp. 130–33, no. IV-5; Paris 1979, no. 229.
*Betrayal*: Cleveland 1967, pp. 130–33, no. IV-6.
*Flagellation*: Paris 1865; Paris 1934; Cleveland 1967, pp. 130–33, no. IV-7; Paris 1979, no. 230.
*Entombment*: Paris 1865; Paris 1867, no. 1967; Cleveland 1967, pp. 130–33, no. IV-8.

BIBLIOGRAPHY: *Last Supper*: E. Du Sommerard 1883, no. 4994; Rupin 1890, p. 363, fig. 428; Swarzenski 1951, p. 24; Rückert 1959, p. 7, fig. 5; Gauthier 1967a, p. 119; Otavsky 1973, pp. 59–65, fig. 15; Gaborit-Chopin and Taburet 1981, no. 34; Swarzenski and Netzer 1986, p. 98.
*Betrayal*: Swarzenski 1951, p. 24; Gauthier 1967a, p. 119; Otavsky 1973, pp. 59–65, fig. 16; Swarzenski and Netzer 1986, p. 98.
*Flagellation*: A. Du Sommerard 1838–46, vol. 5, *Album*, chaps. V and IX, pl. XXXVIII-3; E. Du Sommerard 1883, no. 4993; Rupin 1890, pp. 364–65, fig. 430; Swarzenski 1951, p. 24; Rückert 1959, p. 7; Otavsky 1973, pp. 59–65, fig. 17; Swarzenski and Netzer 1986, p. 98.
*Entombment*: Darcel 1865b, p. 440; Giraud 1881, pl. III; Molinier 1903, no. 149, pl. VII; Swarzenski 1951, p. 24, n. 1; Hunter 1958; Rückert 1959, p. 7; Gauthier 1967a, p. 119; Otavsky 1973, pp. 59–65, fig. 18; Swarzenski and Netzer 1986, p. 98.

lowed by Gauthier and subsequent authors, has suggested.[10] Unfortunately, the descriptions of these altars are imprecise and late in date; none of them, for example, specifies the number of components or gives more than general subjects. These descriptions are, moreover, only representative of the types of ensembles that surely existed in other locations as well. Publications since Rückert have focused on the inclusion or exclusion of surviving appliqué reliefs from one or more reconstructed examples. In this, Otavsky is the most encyclopedic,[11] correctly grouping the reliefs associated here with a group of Marys at the Crucifixion and a Deposition in the Abegg Stiftung, Bern (fig. 121*a*), but also with the Boston Baptism (cat. 122), the centurion (cat. 120), and the Chaalis relief (cat. 61)—all as a part of a monumental cycle of scenes from the Old and New Testaments corresponding to the description of the altar of Grandmont made in the sixteenth century by Pardoux de La Garde (before 1591). He described the altar as illustrating "histoires du vieux et du nouveau testament, les treze apostres et autres sainctz, le tout eslevé en bosse et enrichi de petite pierrerie" (stories from the Old and New Testaments, the thirteen apostles and other saints, the whole raised in relief and enriched with small stones).[12] The theory is attractive in its simplicity, but it is difficult to reconcile with the stylistic differences apparent among the various appliqués.

Rückert's publication, however, was important in estab-

# 120. Centurion

Limoges, ca. 1240–50

Copper: repoussé, engraved, chased, and gilt; medium blue enameled beads

25.5 × 8.8 cm (10 × 3½ in.)

Ex colls.: Albert Germeau (in 1865, but not included in the sale of May 4–7, 1868). Gift of Nicolas Landau, 1976

Condition: Gilding might be modern.

Musée du Louvre, Paris; Département des Objets d'art (OA 10625)

The centurion is one of the appliqué figures formerly affixed to a large composite chasse belonging to Albert Germeau, exhibited in Paris in 1865, known by a photograph (fig. 120a) and especially by the engraving published in the journal *L'Art pour tous* in 1870. The center was occupied by a group representing the Descent from the Cross, to which a Holy Woman in a very different style was added; to the left were two Holy Women and to the right two figures: Saint John and this centurion.[1] The two Holy Women and the Descent from the Cross now belong to the Abegg Stiftung in Bern.[2] The similar dimensions[3] and the close stylistic relationship between the appliqués in Bern and in Paris, together with the Christ on the Cross formerly in the Spitzer Collection and today in the Musée de Picardie in Amiens (cat. 121), indicate that these elements originally belonged to a single ensemble. The similarities between the face of the centurion, on the one hand, and those of the Christs from Bern and Amiens, on the other, are particularly striking.

The overall characteristics of this style are those observed on the appliqués of the Passion (cat. 119): a strong sculptural treatment; a stylization of the drapery with deeply cut furrows terminating in "spoon shapes"; and wide, round faces on which the protruding beards and mustaches are, like the hair, finely engraved. The centurion can be more specifically compared to the Kiss of Judas (cat. 119b), which is undeniably one of the most beautiful reliefs of the group. They share the same animated quality in the repoussé work, which is stripped of all harshness, and the same subtlety in the engraved and chased areas. It is therefore highly probable that we are dealing here, as has been suggested many times,[4] with elements of one large ensemble, probably an altar decoration, dedicated to the Passion of Christ.

Otavsky has compared these appliqués to the tombs of the children of Saint Louis, John (d. 1248) and Blanche (d. 1243; see cat. 146), on one hand, and the stained-glass windows of Chartres—the great north rose window (between 1223 and 1236)—and those of the Sainte-Chapelle in Paris (between 1243 and 1248), on the other, which places the execution of these reliefs at about 1240 to 1250.

The centurion is also distinguished by the precise representation of several details typical of thirteenth-century military attire: the short conical helmet, the sword carried beneath the cloak, and the large buckler in the form of a triangular shield, bearing an escarbuncle of eight rays terminating in fleurs-de-lis, imaginary arms (see cat. 15). If helmets of this type were common in the first half of the century—they are found, for example, in the stained glass of the Charlemagne window at Chartres (ca. 1215–20)[5]—the triangular shield seems not to have become widespread until about mid-century: similar shields decorated with the same carbuncle can be found, for example, in the scenes of combat in the Book of David in the Maciejowski Bible, attributed to Paris or northern France about 1245 to 1250.[6]

ET-D

Notes

1. See Darcel 1865b; old photo in *Albums Maciet*, Paris, Bibliothèque de l'Union Centrale des Arts Décoratifs, and engraving in *L'Art pour tous* (April 15, 1870), no. 248, fig. 2204: the figure was placed before a nimbus bearing the inscription s[anctus] . ipolitus . mart[irum], which explains his erroneous identification with Saint Hippolytus. This Mosan nimbus was preserved in the Brummer Collection (see sale, Zurich, October 19, 1979, no. 215).
2. The group with the Holy Women had been acquired by Werner Abegg about 1930; that of the Descent from the Cross, having passed into the Salavin Collection, was sold (no. 47) in Paris on November 22, 1972, to the Abegg Stiftung, Bern. See Otavsky 1973, p. 37.
3. Height of the group of Holy Women, 24 cm (9½ in.); of the Descent from the Cross, 31 cm (12¼ in.); of the cross from Amiens, 43 cm (16⅞ in.),

Fig. 120a. Composite chasse. Albert Germeau Collection in 1865

Christ alone, 27.5 cm (10⅞ in.); see Otavsky 1973, and below, cat. 121.

4. See Otavsky 1973, p. 58; Swarzenski and Netzer 1986, p. 98.

5. Grodecki and Brisac 1984, no. 24, p. 244; Manhes-Deremble and Deremble 1993, pp. 308–9.

6. New York, Pierpont Morgan Library, ms. M 638, for example, fol. 39r and 41r; see Cockerell and Plummer 1975, pls. pp. 179 and 187; Avril 1995, p. 26.

EXHIBITION: Paris 1865, no. 604 (the ensemble of the chasse).

BIBLIOGRAPHY: Darcel 1865b, pp. 439–40; *L'Art pour tous* 9, no. 248 (April 15, 1870), fig. 2204; Otavsky 1973, p. 55, fig. 9, p. 56; "Nouvelles acquisitions," *La Revue du Louvre* 26, no. 4 (1976), p. 299, fig. 7; Swarzenski and Netzer 1986, pp. 98, 100.

# 121. Christ on the Cross

Limoges, ca. 1240–50

*Cross*: copper: engraved and gilt; champlevé enamel: green

*Christ*: copper: repoussé, engraved, chased, and gilt

*Cross*: 43 × 30 cm (16¹⁵⁄₁₆ × 11¹³⁄₁₆ in.)

*Christ*: 27.5 × 25.8 cm (10¹³⁄₁₆ × 10⅛ in.)

Ex colls.: Frédéric Spitzer, Paris (before 1890; sale, Paris, April 16–June 17, 1893, no. 282, pl. IX); Charles Davis, London (in 1897); Thomas G. Carmichael (sale, London, May 12, 1902, no. 65, ill.); A. Maignan (in 1906). Bequest of A. Maignan, 1927.

Condition: Wear to the gilding; some losses to enamel on cross.

Musée de Picardie, Amiens (MP 3063.524)

This Christ, from the former Spitzer Collection, does not appear on the reproductions of the composite chasse from the Germeau Collection (fig. 120*a*), but it is so similar to the Christ on the Descent from the Cross placed at the center of that chasse that their attribution to the same hand and their common origin cannot be doubted. The two large faces, with half-closed eyes, short beards, and fine, wavy mustaches, are practically identical. The same is true for the torsos, in the smooth relief that erases anatomical details; and for the loincloths, tied with a large knot on the right hip and draped to delineate the successive curves, which are deeply cut but hardly broken.

It is possible that this cross once decorated the center of the back of the Germeau chasse, which was never reproduced nor described. But because the chasse was a nineteenth-century pastiche on which Limousin appliqués were reused, this question concerns only the later history of these pieces.

Fig. 121*a*. Descent from the Cross. Abegg Stiftung, Bern

On the other hand, the face of the chasse (known from the engraving and the photograph) displayed other appliqué figures that flanked the Descent from the Cross now in the Abegg Stiftung in Bern (fig. 121*a*): a group of two Holy Women, also belonging to this museum, a Saint John, whose present whereabouts is unknown, and the centurion that entered the Musée du Louvre in 1976 (cat. 120). Otavsky has judiciously noted that the Holy Women and Saint John rest not on a thin, undulating band suggesting the ground like the ensemble of figures in the Deposition but on more solid bases of circular shape.[1] In all likelihood these figures were not originally placed in a straight line at the sides of the Descent from the Cross. The Saint John, in particular, would not fit as a companion to the figure of the Virgin. Therefore, the Holy Women and Saint John might have accompanied instead the Crucifixion now in the Musée de Picardie in Amiens.

The way in which the feet of Christ, resting on the suppedaneum, are placed immediately above the head of Adam emerging from his tomb may seem surprising. One can see, however, that on Limousin enameled crosses the skull or the small half-length figure representing Adam is often placed just below the feet of Christ. Such is the case, for example, on the cross with the name Garnerius (cat. 103). Furthermore, the figure in half relief appears perfectly adapted to the cross; the head, in particular, is placed precisely in front of the halo situated at the crossing. It thus seems likely that the cross and the Christ were made one for the other.

The decoration of this cross, which suggests a flowering branch, alludes to the theme of the Tree of Life. It is close to that of the cross placed behind the central group on the tabernacle of Cherves (cat. 98) and fits perfectly into the evolution of Limousin production. Otavsky compared this vegetal decoration to the one on the cross in the Crucifixion of the Psaltery of Robert of Lindesey, illuminated about 1220 to 1222[2]—a good comparison, although the branch of the manuscript is more floral and abundant. Some Continental examples could also be cited, especially on goldsmiths' works, such as the niello reverse of the cross of Clairmarais (ca. 1210–20), on which small trefoil leaves sprout regularly from the central branch.[3]

Like the centurion (cat. 120) and the other figures from the Germeau chasse attributable to the same hand, the Amiens Crucifixion can be dated to about 1240 to 1250.

ET-D

NOTES
1. Otavsky 1973, p. 54.
2. Ibid., pp. 57–58, fig. 13, p. 57. London, Society of Antiquarians, ms. 59, fol. 35.
3. Saint-Omer, Musée de l'Hôtel Sandelin (repository of the brotherhood of Notre-Dame des Miracles). See Saint-Omer 1992, no. 2, p. 39, ill.

EXHIBITIONS: London 1897, no. 58; Amiens 1967.

BIBLIOGRAPHY: Palustre and Molinier 1890, no. 75; Molinier 1901, no. 189, ill.; Migeon 1906, p. 14, fig. 33; Thoby 1953, no. 110; Otavsky 1973, pp. 54–56, fig. 11; Gauthier 1978c, p. 35, fig. 21; Lernout et al. 1992, no. 6, pp. 22–23.

# 122. The Baptism of Christ

Limoges, ca. 1250–60

Copper: repoussé, engraved, chased, scraped, and gilt; champlevé enamel: gray-blue; blue glass

36.8 x 21.2 x 2.8 cm (14½ x 8⁵⁄₁₆ x 1⅛ in.)

Ex colls.: Albert Germeau, prefect of the Haute-Vienne (sold May 5, 1868, no. 51); Countess Isabelle Dzialynska, Paris (in 1903); Prince Witold Czartoryski, Goluchow Castle, Poland; Adam-Louis Czartoryski; J. Pollack, Paris; [Wildenstein and Company, New York]

Condition: Gilding largely intact; some loss to enameling of water, replaced with plaster paste; repairs at the bottom and edges of the figure of Christ, at the center and near the top; break in lower left corner repaired with copper strip on back.

Courtesy, Museum of Fine Arts, Boston; Francis Bartlett Fund (50.858)

Exhibited in New York only.

John the Baptist steps forward into the river Jordan, suggested by a series of arched lines of gray-blue enamel and engraved fish that cover the lower torso of Jesus. The Baptist is identifiable by his animal-skin garment;[1] he touches Jesus on the upper arm with his left hand, and in his right he holds a gilt vessel from which a torrent of water pours down over Jesus' head. His body naked, Jesus stands immobile, his right hand raised in blessing and his left covering the lower part of his torso.

This scene of the Baptism of Christ was surely part of a larger cycle, either of the Life of Christ or of the Baptist himself. Its scale suggests that it came from an altar frontal and, in such a context, along with other surviving appliqués (cat. 119d), it has been associated with the destroyed altars of Bourganeuf, Saint-Martial, and Grandmont. In this regard, however, it is important to note the stylistic distinctions between the Boston relief and others: only the Boston relief is enameled; it is the only surviving appliqué of this scale that does not depict a scene from the Passion of Christ; the more abstracted planes of the faces suggest a later date; and the reverse of the plaque is less thoroughly and deeply worked.

The Boston relief is the only example in Limoges enameling of the representation of this scene. However, portrayals in other media, especially illuminated manuscripts, suggest that a figure is missing from the other side of Christ—the angel who ministered to him at the close of the ceremony.[2] An illumination from the Sacramentary of Saint-Étienne, Limoges, shows the Baptist pouring water from a vessel suspended over Christ's head.[3] A cycle of miniatures in the Pierpont Morgan Library, New York, traditionally—albeit not demonstrably—associated with Saint-Martial, Limoges, shows the angel holding Christ's tunic.[4]

It is tempting to follow the suggestion that the Boston relief comes from the altar at Bourganeuf. According to Texier and Rupin,[5] that altar represented scenes from the life of Saint John the Baptist, patron of the Hospitalers and therefore of the commandery of Bourganeuf, but the procès-verbal, done at Bourganeuf in October 1672, following improvements to the building, which Texier cited in full, mentions an "image of Saint John the Baptist, our Patron, in relief in wood" but does not describe the subject represented on the "retable of enameled copper, with figures in relief."[6]

BDB

Notes

1. See Matthew 3:1–17; Mark 1:4–11; Luke 3:21–22; John 1:26–34.
2. Squilbeck 1966–67, pp. 69–116.
3. Paris, BNF, ms. lat. 9438, fol. 24; Gaborit-Chopin 1969, fig. 179.
4. M44; see Swarzenski 1951, p. 21, fig. 4.
5. Texier 1843, p. 82; Rupin 1890, p. 100.
6. Texier 1843, p. 270: "l'image de St. Jean-Baptiste nostre patron, en relief de bois . . . un retable de . . . cuivre émaillé, garni . . . de figures en relief."

Exhibitions: Paris 1865, no. 617; Paris 1880, no. 2; Cleveland 1967, pp. 130–33, no. IV-4.

Bibliography: Darcel 1865b, pp. 439–40; Giraud 1881, pl. 3; Garnier 1886, pp. 426–27; Rupin 1890, pp. 358–59, fig. 423; Molinier 1903, no. 148, p. 39, pl. 6; Swarzenski 1951, pp. 17–25, fig. 1; *Bulletin of the Museum of Fine Arts, Boston* 1957, pp. 80–81, no. 32; Hunter 1958, p. 29; Rückert 1959, p. 7; Le Goff 1964, p. 432, fig. 162; Swarzenski 1969, p. 488, no. 9; Otavsky 1973, p. 59, fig. 14; Gauthier 1978c, p. 35; Boston 1981, no. 37; Steinberg 1983, pp. 136–37, fig. 154; Swarzenski and Netzer 1986, no. 32, pp. 98–101.

# 2

## *The realm of chivalry*

# 123. The Coffret of Saint Louis

Limoges, 1234–37 (1236?)

Openwork copper medallions: repoussé, engraved, and gilt; copper medallions with enamel: engraved and gilt; champlevé enamel: black; lapis, medium, and lavender blue; red, and white; blue glass beads; tin sheet over wood core; green paint

14 × 36.5 × 19 cm (5½ × 14⅜ × 7½ in.)

COATS OF ARMS: The forty-six extant escutcheons bear twenty-three different arms. The identifications proposed below are those of Hervé Pinoteau:[1]

1. Azure semy-de-lis Or (France Ancient)
2. Gules a Castle Or (Castile)
3. Checky Or and Azure a Bordure Gules over all a Canton Ermine (Dreux-Brittany)
4. Azure two Barbels addorsed Or (Bar)
5. Gules a Chain in cross saltire and orle Or dimidiated by Azure a Bend Argent cotised Or (Navarre-Champagne)
6. Argent gerattie of Crosses a Cross Potent Or (Jerusalem)
7. Gules three Lions passant guardant in pale Or (England)
8. Bendy Azure and Or a Bordure Gules (Burgundy Ancient)
9. Checky Azure and Or a Bordure Gules (Dreux)
10. Or three Torteaux Gules a Label of five points Azure (Courtenay with a brisure of cadency)
11. Azure a Bend Argent cotised Or (Champagne)
12. Or a Lion rampant Sable queue fourchy (Flanders)
13. Gules a Cross cletchy and pometty Or voided of the field (Toulouse)
14. Or a Cross Gules between four Eagles Azure (Montmorency)
15. Gules a Lion rampant Argent queue fourchy (Montfort-l'Aumary)
16. Bendy Vair and Gules (Coucy)
17. Gyronny Argent and Gules (Beaumont-en-Gâtinais)
18. Gyronny Argent and Gules a Label of five points Azure (the above with a brisure of cadency)
19. Checky Gules and Or a Chief Argent fretty Azure (Roye)
20. Barry Argent and Azure (La Marche-Lusignan)
21. Gules two Bars Or (Harcourt)
22. Gules three Buckles Or (Malet)
23. Or a Fess Gules between three Torteaux

PROVENANCE: The abbey church of Notre-Dame du Lys, where it held four bones of Saint Louis and his hairshirt, given by King Philip the Fair (see the inventory recorded in 1678, published in Ganneron 1855, pp. 1–2). Transferred to the parish church of Dammarie-les-Lys after the suppression of the monastery (1793). "Discovered in this church" in 1853 in "one of the two large chasses situated at the right and the left of the high altar," where it had been placed in 1838 (Ganneron 1855). Acquired in 1858.

CONDITION: The paint has been restored; original paint preserved only under the medallions; crystal cabochons missing from two of the four corners.

Musée du Louvre, Paris; Département des Objets d'art (MS 253)

The coffret is composed of a beech-wood core, sheathed with tin, and ornamented with crystal cabochons and medallions, gilded and enameled in the Limousin technique. The medallions are decorated either with human figures and animals or with coats of arms. The tin is covered with green paint, and the core, which in its present state is not entirely medieval, is abundantly ornamented with large, round-headed copper studs.

Twenty-five openwork medallions cover the front, sides, and lid of the coffret. The most prominent of these contains the locking mechanism. The elongated body of a basilisk stretches across the top of the lid; a thin metal strip with a latch at the end issues from a hinge in the monster's mouth and locks securely on the left side of the central medallion of the upper register. The opening for the keyhole appears in the center of this medallion, between the intertwined bodies of two other basilisks. The scenes depicted on these openwork medallions are varied: no two are identical. Although the most common theme is that of animals in combat, an assortment of fantastic beasts—basilisks, griffins, chimeras—and realistic creatures—lions, deer, and hounds—appear intertwined in a variety of poses. Men appear in five of the openwork medallions: Samson wrestles with a lion, two men are slaying monsters, another returns from the hunt with a stag slung over his shoulder, and, in the lower right corner of the principal face, a musician plays a troubadour fiddle. The variety of the openwork scenes is furthered by a two-headed eagle, a lion, a rampant fantastic beast, and two flowers.

Eleven medallions, with scenes in reserve set against blue enameled grounds, adorn the back of the coffret. Two of these medallions portray confronted animals. The other nine present scenes of pageantry, chivalry, or the hunt: one of these depicts a falconer on horseback; another shows an amorous couple embracing; and on another a man plays a troubadour fiddle while a woman dances with clappers (a medieval percussion instrument) in her hands. The central medallion shows a king—wearing a crown, sitting on a throne, and holding a scepter—greeting one of his subjects.

In a definitive study of the heraldic evidence presented on the coffret, Hervé Pinoteau identified each of the forty-six surviving coats of arms (there seem to have been fifty-three originally), examining the forms, the contours, the distribution, the places of honor, the sequences, the repetitions, and the different sizes. The dates of death, of marriage, of conferrings of knighthood, of heraldic changes, and of changes in title or of function of each of the persons concerned establishes the date of 1234–37 for the coffret, and possibly even the summer of 1236, far earlier than was previously believed. The study further revealed that each device was that of either a relative of Louis IX or a member of his court, while the most important positions were reserved for the arms of Louis himself. His arms not only are the largest in size but also appear eleven times on the

Principal face

Reverse

coffret; only the arms of his mother, Blanche of Castile, appear more frequently (in twelve separate places).

The precise date of the Saint Louis coffret, determined by its heraldic program, offers an important chronological point of reference for the history of Limousin enamelwork in the thirteenth century. The decorative program of the openwork medallions (on the front, the sides, and the lid) and the enameled medallions (on the back) is evocative of the programs displayed on coffrets created about 1220–30, such as that of Cardinal Bicchieri (cat. 88), which was made before 1227. But the less secure handling of the engraving, the somewhat hasty execution, and the summary rendering of facial features are hallmarks of a style that is still dependent on early Gothic art. Twenty years later the same qualities are manifest on the Aachen coffret (fig. 9) of 1258.[2] Such continuity of style and technique in the mid-thirteenth century testifies to the force of a conservative tradition in Limoges work of this period.

Although the coats of arms have now been fully examined, questions about the coffret's original destination remain to be determined. Did it once contain a solemn document, relics, or a precious object? Ganneron's initial publication (1855) of the coffret retraced its history to the abbey church of Notre-Dame du Lys. Having been deposited in the parish church of Dammarie-les-Lys after the destruction of the monastery in 1793, the coffret was discovered in 1853 within a chasse at the altar.[3] Ganneron identified it with a reliquary recorded in an inventory of 1678 as containing four bones and the hair shirt of Louis IX, presented to the abbey of Notre-Dame du Lys by Philip the Fair after the canonization of his grandfather in 1297.[4]

Philip the Fair did indeed distribute relics of his grandfather—including an arm bone and several finger bones to the abbey of Lys—after the opening of the saint's tomb in 1306/08.[5] But how can the history of the relics be reconciled with the coats of arms presented on the coffret? Such an object, bearing the royal arms, would have served as a fitting container for these holy relics. If Philip the Fair presented the coffret to the abbey, it was not he who had it made, as literature before Pinoteau suggests.[6] Either the coffret was already housed at the abbey of Notre-Dame du Lys, or it came into Philip's possession by inheritance when he ascended the throne in 1285. These questions have yet to be resolved.

It is important to note that William of Saint-Pathus, the confessor of Saint Louis's widow, Marguerite of Provence, states that relics of both Blanche and Louis were housed within the abbey's walls before Louis's canonization in 1297. In his early-fourteenth-century Miracles of Saint Louis, William describes a Sister Clemence of Sens who, in 1278, made a pilgrimage to an *escrinet* (coffret) in which hair shirts and other penitential articles of "discipline" belonging to Saint Louis were kept, hoping to gain the saint's intercession to relieve an eye malady.[7] Although William's language describing the *escrinet* is vague, it is possible to assume that the Louvre coffret and the *escrinet* are one and the same, which would place its arrival at the abbey sometime before 1278.

Another possible explanation for the coffret's appearance at the abbey of Notre-Dame du Lys is that it was originally a donation of King Louis himself. From its foundation in 1244—sponsored by the pious Louis and his mother, Blanche of Castile—this affluent abbey profited greatly from royal beneficence.[8] The arms of both founders appear many times on the coffret, with Louis's receiving the most prominence. Certainly, an object as richly decorated as the Saint Louis coffret would have been a fitting gift from the king to the abbey at Lys, either at the time of its foundation

Left and right sides

362

in 1244 or upon an important occasion such as the interment of Blanche of Castile's heart there in 1253. Although heraldic evidence has provided a date for the coffret, the exact circumstances of its arrival at the abbey have yet to be determined.

BDB and MP

NOTES

1. Pinoteau 1983, pp. 97–130.
2. See Vaivre 1974, pp. 97–124.
3. The parish priest, very excited by his discovery, showed the coffret to Aufauvre and Fichot, who were in the region in 1853, doing research for their *Monuments de Seine-et-Marne* (Paris 1858); they were the first to authenticate its antiquity.
4. Ganneron 1855, pp. 1–2. This theory was accepted by subsequent scholars; see Grésy 1854. Because of this provenance, Napoleon III acquired the coffret in 1858 for the enormous sum of 25,000 francs and presented it to the Musée des Souverains.
5. Pinoteau 1983, p. 98.
6. See especially Darcel 1891, pp. 566–68.
7. Guillaume de Saint-Pathus 1931, pp. 70–75. Fragments of one of Saint Louis's hair shirts are currently housed in the church of Saint-Alpais, Melun; they were on display in the Paris exhibition (1970, no. 43).
8. See Ganneron 1855, pp. 59–66.

EXHIBITIONS: London 1932, no. 1060; Paris 1960, no. 167; Paris 1970, no. 83.

BIBLIOGRAPHY: Ganneron 1855; Grésy 1854, pp. 637–47; Aufauvre and Fichot 1858, pp. 30–36; Gautier 1858, p. 855; Barbet de Jouy 1858, no. 35; Darcel 1891, pp. 566–68, no. 940; Rupin 1890, pp. 437–40, figs. 486–88; Marquet de Vasselot 1914, no. 69; Guillaume de Saint-Pathus 1931; Vaivre 1974, pp. 97–124; Pinoteau 1978–79, pp. 77–78; Pinoteau 1983, pp. 97–130.

Lid

# 124. Twelve Medallion

Limoges, ca. 1240–60

Copper: engraved and gilt; champlevé enamel: medium and light blue and white

Diams.: 9 cm (3½ in.)

Ex colls.: Alexis Berg; J. Pierpont Morgan, London and New York

Condition: Losses to gilding.

The Metropolitan Museum of Art, New York; Gift of J. Pierpont Morgan, 1917 (17.190.2144–2155)

Each medallion presents figures reserved against a blue enamel ground enlivened with foliate decoration. Each has four holes for attachment,[1] surely to a large coffret similar to, or larger than, the Bicchieri coffret (cat. 88). The medallions may have been used in combination with others ornamented with fantastic animals as on the Bicchieri coffret, or with heraldic motifs as on the coffret of Saint Louis (cat. 123) or the series in an Italian private collection.[2]

Fig. 124*a*. Coffret. Museo Leone, Vercelli

Fig. 124*b*. Composite triptych. The Metropolitan Museum of Art, New York

The style of the medallions, with their sketchy drawing and freely designed foliate grounds, appears as early as the 1220s, on the coffret of Cardinal Bicchieri. A later date is suggested for the Metropolitan roundels by comparison with the Saint Louis coffret, which is also summary in its treatment of figures and less precise than the Bicchieri coffret in its definition of foliation. Designs related by subject and composition can also be found on works produced for religious use, such as the chasse of Saint Viance, with its narrative account of the miraculous domestication of a wild bear (cat. 118).

The twelve medallions include chivalric and hunting scenes and images of fantastic animals. Although generically related in theme, they do not appear to constitute a coherent program. This is not surprising, however; the small roundels on the coffret in the Museo Leone, Vercelli (fig. 124*a*) are likewise diverse in subject.

Several of the medallions, including the one with the fighting men, have trial engraving marks on the reverse. These are not, however, for identifiable designs; nor do they constitute a numbering system. In 1949, the twelve medallions were removed from a triptych composed of medieval and nineteenth-century elements (fig. 124*b*). This pastiche was assembled by Alexis Berg, who signed the supporting wood panel of the central plaque: "Fait par Alexis Berg, [restau]rateur d'objets d'art et d'antiquités l'an 1854" (Made by Alexis Berg, restorer of works of art and antiquities in the year 1854).[3]

BDB

Notes

1. The attachment holes of one (.2149) are oriented differently, but their common engraving, style, and palette reinforce their common provenance.
2. Castronovo 1992, ill. p. 217.
3. At present it is not known where Berg was working. Charles T. Little (oral communication) has observed that the die stamps found on the Metropolitan triptych recur on a number of composite pieces inspired by medieval art, suggesting that Berg was a prolific restorer. As some works by Berg were in the Spitzer Collection, a link with either Paris or Aachen may be posited.

Bibliography: Unpublished.

# 125. Gemellion: Offering to a Queen

Limoges, second third of the 13th century

Copper: stamped, engraved, and gilt; champlevé enamel: medium and lavender blue, turquoise, and red

Diam. 21 cm (8¼ in.)

Ex colls.: Abbé Fauvel (in 1729); Pierre Révoil. Acquired in 1828

Condition: Wear, tarnishing, and numerous losses to the enamel; gilding almost completely lost, certainly because of prolonged use.

Musée du Louvre, Paris; Département des Objets d'art (MRR 171)

Exhibited in Paris only.

A gemellion is a type of basin used when washing the hands. As the word itself indicates (*gemellus* means twin), gemellions were made in pairs, and the water was poured from one to the other—as seen, for example, in certain illuminations representing the washing of Pilate's hands during the Passion.[1] Some gemellions are furnished with a spout on the outside to allow for the flow of water (cats. 131, 132). The term *gemellion* first appeared in Carolingian times: the *Liber pontificalis* records eight (*gemiliones* VIII) among the gifts of Pope Gregory IV (r. 827–44) to the church of Saints Callistus and Cornelius.[2] However, in the Middle Ages, these objects were usually called basins (*pelves* or *bacini*). They must have been sufficiently common from the twelfth century onward since the monk Theophilus told of their fabrication: *pelves quibus aqua in manis funditur . . .* (basins from which water is poured out on the hands).[3] These basins, it seems, usually were made of silver, as were those given to Notre-Dame in Paris by several bishops in the twelfth and thirteenth centuries, such as Peter Lombard (d. 1160) and William III (d. 1240).[4] Nonetheless, the mention of basins of Limousin work is not uncommon in thirteenth- and fourteenth-century texts; "*bacinos* from Limoges" figure among the gifts of the prior Helyas to the cathedral of Rochester (before 1222)[5] and "*duo bacini qui sunt de opere Lemovitico*" (two *bacini* that are of Limoges work) are listed in the inventory of the possessions of the bishop of Toulouse, Foulques, who died in 1231.[6] The form and the enameled decoration of these Limousin gemellions have sometimes been compared with those of Eastern objects—in particular the basin from the Tiroler Landesmuseum Ferdinandeum in Innsbruck;[7] the medallions distributed over its surface are enhanced with cloisonné-enamel decoration on copper, which would suggest the influence of Islamic models in the creation of this type of object. However, it should also be noted that copper basins of the same shape, with engraved decoration, were made in great numbers in Europe in the twelfth century.[8]

Gemellions, as well as candlesticks (see cat. 33), probably were used in both liturgical[9] and secular contexts. Since the decoration on most Limousin gemellions consists of secular scenes or coats of arms—the two motifs are sometimes combined as well[10]—perhaps they were intended for secular use. The Limousin gemellions generally have come down to us in rather poor condition, due undoubtedly to their prolonged use as water containers and—on occasion, for those that were or still are preserved in churches—as collection plates.[11]

This gemellion belonged to Abbé Fauvel and subsequently was given to the Musée du Louvre as part of the Révoil Collection; the earliest to have been published, it was included in the *Monuments de la monarchie françoise* by Bernard de Montfaucon (1729). Its iconography has baffled all scholars. The central scene depicts a female figure offering a cup to a seated queen. In the six medallions on the periphery are young seated and crowned figures; the ones with scepters are probably kings. Without a doubt, this gemellion exemplifies the taste for courtly scenes evoking the refined life led by kings or important individuals, rather than the precise representation of a historic or literary episode—a taste especially exploited in the decoration of Limousin gemellions, since forty-two examples with this theme were catalogued by Marquet de Vasselot (1952).

The style of the figures, with their full faces and their draperies falling in deep, curving folds, links the object to the group of Limousin works that adopted the forms of early Gothic art; this would suggest a date at the beginning of the thirteenth century for the gemellion. However, the somewhat summary and repetitive quality of the execution argues for a comparison with mid-thirteenth-century works—in particular, the medallions on the coffret in Aachen (fig. 9) or the coffret of Saint Louis (cat. 123).

ET-D

NOTES

1. Notably in the *Bibles moralisées* from the early thirteenth century, for example, folio 410 of the manuscript at BNF, Paris, ms. lat. 11560; see A. Laborde 1912, pl. 265.
2. Duchesne 1955, vol. 2, p. 80.
3. Theophilus 1961 and 1986, p. 137.
4. Guérard 1850, p. 60: *bacini duo argentei* (two silver basins); and p. 38: *bacinos argenteos* (silver basins).
5. Lehmann-Brockhaus 1955–60, vol. 2, no. 3752, p. 397.
6. Catel 1633, pp. 901–2.
7. Berlin 1989, no. 1/255, fig. 607. The dating and attribution of the Innsbruck basin have been questioned in recent literature. See Redford 1990, pp. 119–35.
8. Particularly in the region of Rhein-Westfalen; see Weitzmann-Fiedler 1981; Paris 1995, no. 7.
9. During the Mass, for the ritual of the Washing of the Hands; see Durand 1854, vol. 2, pp. 174–75; Cabrol 1907, cols. 104–6. See also the very detailed description of the ritual of the Washing of the Hands before the pontifical Mass given in the *Ordo Romanus* XIV, drawn up in the fourteenth century; see *PL* 78, cols. 1147–48, cited by Marquet de Vasselot (1952, p. 11).
10. Among the 129 gemellions indexed by Marquet de Vasselot (1952), only fifteen have a religious decoration.
11. For example, the one in the church at Cambronne (Oise), today in the Musée Vivenel, Compiègne (Oise); see Marquet de Vasselot 1952, no. 55.

BIBLIOGRAPHY: Bernard de Montfaucon vol. 1 (1729), p. 348, pl. 32; Millin 1792, p. 11, pl. 42; L. Laborde 1852, no. 55; Courajod 1886, pp. 10–11; Rupin 1890, pp. 547–48, fig. 612; Darcel 1891, no. 134; Marquet de Vasselot 1914, no. 80; Marquet de Vasselot 1952, no. 22, pp. 37–38.

# 126. Gemellion: Knight

Limoges, third quarter of the 13th century

Copper: formed, engraved, stippled, and gilt; champlevé enamel: medium blue, red, and white

Diam.: 27 cm (10⅝ in.)

Ex colls.: Arthur Sachs, Cambridge, Mass. (in 1932); [Joseph Brummer, New York (purchased, 1946)]

Condition: Scattered losses to gilding and enamel overall, especially in "pie-crust" edge.

The Metropolitan Museum of Art, New York; The Cloisters Collection, 1947 (47.101.40)

Exhibited in New York only.

At the center of this enameled basin, a knight with shield and banner rides a sleek horse across a field of white enamel enlivened by reserved foliate ornament. Surrounding him are standing figures of warriors in mail, also reserved against an enameled ground of blue. The back of the plate is engraved with a pattern of six lobes and fleurs-de-lis.

Shallow metal bowls such as this served as hand basins. This example, which has no spout, was used to catch the water poured from the other basin of the set, whose wall would have been pierced to form a small drain that fed into a spout shaped like an animal's mouth (see cats. 131, 132).

Gemellions were used both by priests at the altar[1] and by noblemen at table,[2] as the emphatically secular decoration of this example suggests. Still, the presence of a secular subject does not preclude the use of gemellions in an ecclesiastical context.[3] Knights on horseback appear on a number of surviving gemellions.[4]

The delicate engraving of the figures around the rim calls to mind that of the figures on the box excavated at Saint-Martial (cat. 127). A number of the subjects illustrated on gemellions are common to medallions on the coffret of Saint Louis (cat. 123), suggesting a general date for gemellions beginning in the second quarter of the thirteenth century, which coincides with inventory citations.

The Metropolitan gemellion is remarkable among surviving examples for the quality of its engraving and for its good condition, especially since many gemellions were apparently used later as offering plates, and their metal and enamel were worn away by the coins dropped into them.[5]

BDB

Notes
1. Braun 1932, p. 550, pl. 111, nos. 429, 431.
2. Rupin 1890, p. 545.
3. See the discussion by Marquet de Vasselot 1952, pp. 15–16.
4. See the catalogue in Marquet de Vasselot 1952, pp. 23ff.
5. Ibid., p. 7, n. 1, and cat. 125 above.

Exhibitions: Cambridge, Mass., 1931; Los Angeles and Chicago 1970, no. 63; Yokohama 1989, p. 6, no. 7.

Bibliography: Ross 1932a, pp. 9–13.

# 127. Box with Courtly Scenes

Limoges, second half of the 13th century

Copper: engraved and gilt; champlevé enamel: ultramarine blue

H.: 6.1 cm (2⅜ in.); diam. of base: 5.1 cm (2 in.)

PROVENANCE: Found in Limoges in 1837 during an excavation on the site of the abbey of Saint-Martial.

EX COLL.: Victor Gay, Paris (before 1887). Gift of the friends of Victor Gay to the Louvre, 1909

CONDITION: Cover is lost; missing areas in the upper edge; 2 mm gap at the join; some gilding still remains.

Musée National du Moyen Âge, Thermes de Cluny, Paris; lent by the Département des Objets d'Art of the Louvre since 1949 (OA 6279)

The provenance of this small, cylindrical box, with its secular decoration—which unfortunately has become hard to read—is of twofold interest: first, because, along with another fragment,[1] it remains one of the rare extant champlevé enamels ever found buried in the soil of Limoges, the city where these were manufactured by the thousands in the Middle Ages, and second, because it was found on the very site of the famous Limousin abbey of Saint-Martial,[2] one of the cultural centers of Aquitaine.

The two parts that make up the body of the box were cut from a single sheet of copper (2.5 to 3 mm thick): a rectangle was formed into a cylinder for the vertical surface and into a disk for the bottom.[3] The missing cover, which was probably flat, overlapped the box itself by a few millimeters and rested on the molding still visible at the neck of the box. The outside of the cylinder is ornamented with eight gilt figures in reserve on a medium-blue background covered with a rinceau pattern: only through its linear progression as set forth in 1887 in the *Glossaire archéologique* by Victor Gay (see fig. 127a) can a continuous reading of the four scenes be accomplished. The same amorous couple is represented in four stages of an encounter. In each episode the young people are depicted standing, face to face, and dressed the same way, with certain details very close; the only variations are the poses, the attributes of the young man, and a few details of the young woman's costume or hairstyle.

In keeping with the overall significance accorded these scenes by Maurice Ardant and Ernest Rupin, who regarded them as preliminary events preceding the couple's engagement, and, following Ardant and Rupin's reading of their sequence, Markus Müller attempted to interpret the four scenes with greater precision in his thesis on the representations of courtly love in France in works of art of the thirteenth and fourteenth centuries. Müller sees them as a graphic unfolding of the stages leading up to an amorous conquest, following courtly ritual, as it was codified in medieval literary treatises. According to these discourses of courtly love, Müller identifies the four phases as follows: 1) the *Encounter*; 2) the *Homage to submission*, the "*immixtio manum*" of the medieval texts; 3) the *Gift*; 4) the *Kiss*. The scenes thus would be read counterclockwise—that is to say, contrary to the linear progression schematic proposed by Gay, beginning with the scene on the far right, followed by that on the left, and concluding with the two in the middle.

If, as its decoration suggests, the primary purpose for which this box was made was a secular one, its function nonetheless remains mysterious. Because of the place where it was discovered, some scholars believed that it might have contained relics or hosts (Ardant), while others more convincingly considered it an "engagement box"

369

Fig. 127*a*. Sequence of images (after Gay 1887, p. 168)

(Rupin). Although none of these uses is a certainty, they are all possibilities, and each might have been the case at one time or another. The box perhaps contained makeup, ointments, jewels, written tokens of love, or other "relics," and probably all or some of these things. On the other hand, the numerous but all-too-brief references to boxes (*boîtes*) in medieval texts—*boete, boeste, bussola, boueste, boîtelette, boistes*—scarcely help us to identify their original purpose, contents, or form.[4]

One other small, round box, contemporaneous with the present one, may be compared with it, and likewise attests to the production in Limousin workshops of this type of secular object. Now in the Museum of Fine Arts, Boston, it has been identified by Lillich as a seal box because of its shape and iconography. On one side of its flat cover it bears an escutcheon, and on the other side a pair of lovers similar to the couple depicted here.[5]

GF

NOTES

1. An enameled handle of a crosier found in Saint-Augustin in Limoges; see Gauthier 1987, no. 59.
2. The box was found at the site where, one hundred and twenty-five years later, Gauthier, Perrier, Blanchon, and Maury excavated the tombs of Saint Martial and Saint Valerie; see Gauthier 1962b, pp. 205–48.
3. The absence of any visible system of assembly allows one to think that the two parts were soldered together. The gap, two millimeters in length between the edges of the vertical surface—which is hard to explain—may also indicate the disappearance of a soldered join that would have made the meeting of the two edges invisible.
4. See Gay 1887–1928, vol. 1, pp. 167–71.
5. Boston, Museum of Fine Arts (49. 470; William F. Warden Fund); see Lillich 1986, pp. 239–51.

BIBLIOGRAPHY: Ardant 1838b, pp. 109–10; Ardant 1838a; Gay 1887–1928, vol. 1, fig. p. 168; Rupin 1890, p. 574, figs. 646, 647; Migeon 1909, p. 426, fig. 424(3); Marquet de Vasselot 1914, no. 52; Lillich 1986, p. 242; Müller 1994, pp. 89–101.

# 128. Pyx

Limoges, ca. 1260–80

Copper: engraved, chased, and gilt; champlevé enamel: lapis blue, turquoise, brick red, and white

H.: 8.2 cm (3¼ in.); diam.: 6.6 cm (2⅝ in.)

COAT OF ARMS: Or à fleur-de-lis Argen

EX COLL.: Edme Durand. Acquired in 1825.

CONDITION: Significant wear to gilding.

Musée du Louvre, Paris; Département des Objets d'art (MR 2656)

Exhibited in Paris only.

As Longpérier (1842) has already noted, the decoration of this pyx—escutcheons bearing a fleur-de-lis and stylized castles probably derived from those on the arms of Castile—places this object in the group using an ornamental style that combined the emblems of the kingdoms of France and Castile. This trend was popular after the marriage in 1200 of the French king Louis VIII to Blanche, the daughter of the king of Castile. The fleurs-de-lis here, however, are silver on a gold shield rather than gold on blue, the colors of the arms of the kings of France. The borders of many of the stained-glass windows in the Sainte-Chapelle in Paris (1241–48)[1] are among the oldest examples of this decoration, which ornaments the robes of John and Blanche of France (cat. 146) made at Limoges about the same time. In fact, these motifs recur throughout the entire second half of the thirteenth century.[2]

Longpérier also noted that the fleurs-de-lis are of the same design as those on the counterseal of Louis VIII (1223–26)[3] in which the upper part of the escutcheon bearing the fleurs-de-lis is similarly rounded. This type of fleur-de-lis appears also on the first counterseal of Louis IX, used from 1126 to 1150 but not thereafter.[4] The escutcheon rounded at the top is likewise found in the twelfth and the early thirteenth century.[5]

The design of long, straight leaves is, on the other hand, found on other pyxes decorated with armorial motifs or busts of angels, such as the one in the Musée du Louvre[6] in which the style of the angels suggests a date in the third quarter of the thirteenth century. The restrained range of colors supports such dating. The shapes of the escutcheon and the fleur-de-lis here are thus archaic.

ET-D

NOTES

1. See Grodecki 1959, for example, window O, pl. 18, or C, pl. 76.
2. For example, the chevet of La Trinité at Vendôme, about 1280; see Grodecki and Brisac 1984, fig. 154, p. 162.
3. Pastoureau 1979, fig. 18, p. 40.
4. Paris 1970, no. 21; Dalas 1991, nos. 74bis, 75bis, 76bis, pp. 154–56.
5. See the seals of Pierre de Courtenay (1184) and the future Louis VIII (1211), ill. in Pastoureau 1979, figs. 17, 18, p. 40.
6. OA 937; see Rupin 1890, fig. 270, p. 208.

EXHIBITIONS: Vatican City 1963, no. 135; Paris 1979, no. 210.

BIBLIOGRAPHY: Longpérier 1842, pp. 155–56, ill.; L. Laborde 1852, no. 50; Courajod 1888, no. 119, p. 53; Rupin 1890, pp. 205–6, fig. 262; Darcel 1891, no. 130; Marquet de Vasselot 1914, no. 101.

# 129. Pyx

Limoges, second half of the 13th century

Copper: engraved and gilt; champlevé enamel: medium blue, red, and white

H.: 7.9 cm (3⅛ in.); diam.: 6.8 cm (2⅝ in.)

EX COLL.: Michael Friedsam, New York (in 1931)

CONDITION: The cross finial is modern.

The Metropolitan Museum of Art, New York; The Friedsam Collection, Bequest of Michael Friedsam, 1931 (32.100.284)

Exhibited in New York only.

Although the coats of arms that decorate pyxes may indicate that the objects were presented by a particular family or were destined for its use, heraldic decoration was also used as a strictly ornamental device (see Pastoureau, above p. 339, and cat. 128). Given the simplicity of the motifs here and the decorative quality of the colors, that seems to have been the case with this example and related ones, such as those preserved in Budapest,[1] Detroit,[2] and Hartford (fig. 18). The decoration evokes the medieval world of chivalry, which explains the popularity of this type of object among nineteenth-century collectors.

BDB

NOTES
1. Kovács 1968, fig. 35.
2. Barnet 1988, p. 17, fig. 2, from the Lord Londesborough Collection (see Fairholt 1857, pl. XXI).

EXHIBITION: Cambridge, Mass., 1975, pp. 65–69.

# 130. Censer

Limoges, second half of the 13th century (ca. 1270?)

Copper: pierced, engraved, and gilt; champlevé enamel: medium blue, green, yellow, red, and white

H.: 19 cm (7½ in.); diam.: 13 cm (5⅛ in.)

COATS OF ARMS: 1. Gules a Cross Saltire and Orle of Chains linked with an Annulet Or stoned Vert; impaled by Azure a Bend Argent cotised potent counter-potent Or (Navarre-Champagne). 2. Gules a Lion Or (Champlitte?).

EX COLLS.: Baron Max von Goldschmidt-Rothschild, Frankfurt am Main; [Rosenberg and Stiebel, New York]

CONDITION: Overall wear to enamel and gilding; losses to enamel; suspension loop on top is a replacement.

The Metropolitan Museum of Art, New York; The Cloisters Collection, 1950 (50.7.3ab)

The censer is composed of two principal elements: a hemispherical cup set on a hexagonal base and a hemispherical cover surmounted by a "roof" with keyhole-shaped openings. Lugs paired at the intersection of the top and bottom halves served as guides for the chains used to suspend and swing the censer. Four openwork medallions, each with a fantastic creature biting its own tail, have been applied to the lid. The ring at the top probably replaced a finial similar to the one preserved on the Limoges enameled censer in Nuremberg.[1] A chain attached to this finial would have allowed the censer to be opened and replenished from an incense boat (navette) used in conjunction with it.

The burning of incense was an integral part of Church ceremony in the Middle Ages, used in the Mass and the Daily Office as well as in other ceremonies, including the blessing of candles and palms and of the faithful themselves.[2] Consequently, censers were part of standard liturgical equipment, their fabrication described by goldsmiths such as Theophilus[3] and their symbolism elucidated by writers such as William Durandus, bishop of Mende in the thirteenth century, for whom the incense itself was an emblem of prayer and the censer a metaphor for the Body of Christ.[4] Given their frequent—and somewhat rough—use, it is not surprising that few enameled examples of censers and navettes survive.

The Metropolitan's piece is particularly unusual in having heraldic decoration. The hexagonal base has two sets of arms, each repeated three times. One is Navarre-Champagne, first used by Theobald I, count of Champagne in 1201, king of Navarre in 1234 (d. 1253).[5] These arms appear on the coffret of Saint Louis (cat. 123) and recur frequently in Limoges work. The second may represent the

seigneurs de Champlitte, viscomtes de Dijon, although it may simply be a stock coat of arms with a lion charge.[6]

The notation "zu G-R 112" appears in white ink under the foot.

BDB

NOTES
1. Rupin 1890, p. 535, fig. 599.
2. Barraud 1860, p. 666.
3. Theophilus refers to both a cast and a repoussé censer. See Hawthorne and Smith 1976, pp. 130–38.
4. *Rationale divinorum officiorum* 1484, "De incenso benedicendo e thuribulum mittedo."
5. Pinoteau 1983, p. 100.
6. See *Europäisches Stammtafeln* 1978, vol. 3, chart 348. I am grateful to Elliot Nesterman for his suggestions.

EXHIBITIONS: Los Angeles and Chicago 1970, no. 62.

BIBLIOGRAPHY: Frazer 1985–86, p. 21, ill. 15.

# 131. Armorial Gemellion

Limoges, third quarter of the 13th century

Copper: formed, engraved, and gilt; champlevé enamel: medium blue, turquoise, and white

Diam.: 23 cm (9⅟₁₆ in.)

HERALDRY: 1. Argent semy of crosslets a cross potent or.

EX COLLS.: [Altounian (sold 1929)]; [Joseph Brummer, New York, 1929 (sold 1949, pt. 1, no. 720, ill. p. 192)]

CONDITION: Gilding almost entirely worn away; loss of definition of engraving through wear; losses to enamel.

The Metropolitan Museum of Art, New York; Rogers Fund, 1949 (49.56.8)

The arms of the Latin Kingdom of Jerusalem appear at the center of this gemellion. Surrounding the arms are four lunettes with men in combat, armed with shields and clubs. Filling the space between each lunette is an image of a three-towered castle in reserve. The back of the plate is engraved with fleurons at the center of a star pattern and fleurs-de-lis at the terminals.

Originally part of a pair, this was the bowl used to pour water over the hands from the small "gargoyle," or animal-head spout, set into the side.

Although worn from use, the engraving of the lunettes recalls the medallions with secular ornament made for the coffret of Saint Louis (cat. 123) and others now isolated from their original contexts (cat. 124). Coats of arms appear at the centers of numerous other gemellions[1] and are the principal decorative motif on scores of others.[2] This is the unique example that shows the arms of the Latin Kingdom of Jerusalem at the center of the basin; the arms do figure, among others, on a gemellion in the Musée National du Moyen Âge, Paris.[3]

Like the Limoges candlesticks excavated in the Holy Land or the pyxes with Crusader imagery (fig. 18),[4] this gemellion testifies to the dialogue between the Latin Kingdom of Jerusalem and the Limousin, first evidenced in the late twelfth century (see "Opus lemovicense," above, p. 40).

BDB

NOTES
1. Marquet de Vasselot 1952, nos. 17–19, 27–31, 33–37, 50, 52, 54, 70, 72–75, 77–79, 81–83.
2. Ibid., nos. 100–129.
3. Ibid., no. 115.
4. Hartford, Conn., Wadsworth Atheneum (1949.165); Cambridge, Mass., 1975, pp. 78–79, no. 12.

BIBLIOGRAPHY: Unpublished.

# 132. Armorial Gemellion

Limoges, second half of the 13th century

Copper: stamped, engraved, and gilt; champlevé enamel: medium blue, turquoise, brick red, and white

Diam. 22.6 cm (8⅞ in.)

COATS OF ARMS:

In the center: Azure semy-de-lis Or (for France), impaling Gules three Castles Or (for Castile): Alphonse de Poitiers

On the periphery: Bendy Or and Gules: Turenne; Paly Or and Gules: Provence; Barruly Azure and Argent: Lusignan

EX COLL: Bequest of Arthur de Marsy, 1901.

CONDITION: Gilding almost entirely lost; overall wear and many losses to enamel.

Musée du Louvre, Paris; Département des Objets d'art (OA 5529)

Exhibited in Paris only.

This gemellion is equipped with a spout and pierced with six holes to allow for the flow of water. The decoration is made up almost entirely of coats of arms, inscribed within circles with enameled grounds of turquoise in the center and white on the periphery; there, standing female figures with outstretched arms alternate with the escutcheons, which they seem to be presenting.

Heraldic decoration, alone or combined with other themes, is found on more than three quarters of the extant gemellions. On most of them, as Marquet de Vasselot noted,[1] the coats of arms seem to have been employed as decorative motifs, without actual heraldic significance. Such is the case with this gemellion, decorated in the center with an escutcheon of France impaling Castile, the arms of Alphonse of Poitiers, count of Toulouse (d. 1271),[2] and, on the periphery, with several of the escutcheons most often seen on Limousin works: Lusignan, Turenne, and Provence[3] (see cats. 134, 136). The style of the female figures between the escutcheons, who are very summarily executed, retains characteristics of early Gothic art, while the rinceaux in reserve, a weak and rather repetitive design, appear to have been widespread, especially in the second half of the thirteenth century (see cat. 137). The rinceau decoration that surrounds the escutcheons, forming a long sideways S, is, furthermore, close to that on the armorial medallions of the Longpont coffret (cat. 133): the arrangement of the decoration on this gemellion, it would seem, was influenced by that found on secular coffrets of this type.

This gemellion may be compared with others that are decorated in the center with the arms of Alphonse of Poitiers and, on the periphery, with various escutcheons being presented by female figures—notably, the example in

the Kunstgewerbemuseum in Schloss Köpenick, Berlin.[4] The characteristics of the decoration, such as the inclusion of the arms, lead us to propose a date of execution in the second half of the thirteenth century, perhaps after 1271, if one accepts that the arms of Alphonse of Poitiers would not have been used as ornamentation during his lifetime.[5]

ET-D

NOTES
1. Marquet de Vasselot 1952, pp. 97–98.
2. Pinoteau and Le Gallo 1966, p. 12.
3. For the armorial bearings of the Lusignan, see Eygun 1938, pp. 217–20, pls. XIII, XIV: the seals of Hugh X (1243) to Hugh XIII (1281); for those of Turenne, see Bosredon and Rupin 1886, pp. 27–37: the seals of Raymond IV (1211 or 1214, 1225) to Raymond VII (1290 and 1297?); for those of Provence, see Pinoteau and Le Gallo 1966, p. 8.
4. Inv. 74-414; see Marquet de Vasselot 1952, no. 117.
5. See Dionnet 1994, p. 112.

EXHIBITIONS: Paris 1960, no. 41; Paris 1964, no. 136.

BIBLIOGRAPHY: "Le Comte de Marsy. . .," 1901, p. liv; Marquet de Vasselot 1952, no. 116; Beaure d'Augères 1959 and 1960, p. 360 and pl. II, p. 362.

# 133. The Coffret of Blessed John of Montmirail

Limoges, ca. 1270

Copper: engraved, stippled, and gilt; champlevé enamel: sky blue, gray, green, dark red, and white; red and brown leather over wood core

15 × 78.7 × 17.5 cm (5⅞ × 31 × 6⅞ in.)

COATS OF ARMS:

*Lid*

1, 9. Bendy Or and Azure a Bordure Gules (Burgundy Ancient)[1]
2, 4, 13. Gules three Castles Or (Castile)
3, 8, 14. Azure semy-de-lis Or (France Ancient)
5, 12. Vair
6. Azure a Lion rampant Or *impaling* Bendy Or and Gules[2]
7, 11, 16. Azure semy-de-lis Or *impaling* Gules semy of Castles Or (Alphonse of Poitiers)[3]
10. Barry Vair and Gules (Coucy)[4]
15. Chevrony Or and Gules (Chateaugontier?)

*Front*

1. similar to no. 1 of the lid (Burgundy Ancient)
2, 5. similar to no. 7 of the lid (Alphonse of Poitiers)
3, 4. similar to no. 3 of the lid (France Ancient)
6. similar to no. 6 of the lid (Guy VI of Limoges)
7. Paly Or and Gules (Provence)
8. Checky Or and Azure a Bordure Gules (Dreux, simplified)
9. similar to no. 2 of the lid: Castile
10, 14. Barruly Argent and Azure a Bordure Gules (Partenay-l'Archevêqué)
11. Azure a Lion rampant Or
12. Barry Or and Gules
13. Barruly Argent and Azure a Label Gules (with an error in the barruly: Lusignan, Count of Eu)
15. Paly Or and Gules on a Chief Azure three Bezants
16. Gules three Crescents Argent

*Back*

1. Bendy Or and Gules (Turenne)
2. Or a Cross recercely Gules (Aubusson)
3, 4. similar to no. 3 of the front (France)
8. Barruly Argent and Azure (Lusignan,[5] Count of La Marche)
9. Vair a Fess Gules
10. Barry wavy Argent and Gules (Rochechouart)

*Left Side*

1. similar to no. 6 of the lid and the face (Guy VI of Limoges)
2. similar to no. 12 of the front (Turenne)
3. similar to no. 2 of the cover and the face (Alphonse of Poitiers)
4. Per Fess in chief Argent a Lion rampant Gules and in base Azure a Fess Argent
5. similar to no. 15 of the lid

*Right Side*

1. similar to no. 8 of the back (Lusignan)
2. similar to no. 3 of the cover and of the face, and 4 of the back (France Ancient)
3. Paly Or and Azure

PROVENANCE: Mentioned for the first time at the abbey of Longpont (Aisne) in 1639. Preserved during the French Revolution by the lay sacristan of Longpont Abbey, Lebeau, who was also mayor; preserved since 1804 in the Church of Saint-Sébastien, Longpont. Classified, Monuments Historiques, 1901

CONDITION: Restored between 1855 and 1859 (see Larigaldie 1909, p. 157), at which time there were fifty-three medallions (Corneaux

1879, p. 102), although Poquet mentions only forty-nine in 1869; a "fragment of a plaque in yellow-enameled copper" was found in 1888; fifty medallions present now; lid separated from the coffret during World War I reattached; some border strips replaced; wear to gilding; lock plate replaced.[6]

Treasury of the Abbey, Longpont (Aisne)

The oblong coffret is decorated on each face by coats of arms inscribed on small gilt-copper medallions. On the front there are sixteen medallions and a lock plate; sixteen medallions also decorate the lid, along with a looped handle and the hasp of a lock in the form of an enameled serpent. Five medallions appear on the left side; two are missing from the right side, so only three medallions appear there. Ten medallions adorn the back, with one missing. The edges are covered with strips of stamped gilt copper.

The coffret of Longpont, with its decoration of secular medallions encircled by gilt-copper bosses applied to a parchment skin, belongs to a tradition established as early as the coffret from Conques (cat. 7) and represented throughout the twelfth and thirteenth centuries by a few key surviving examples. Among these, the coffret of Saint Louis (cat. 123) provides an important point of comparison. On both, the palette includes a dark opaque red and a chalky white. Both coffrets display a number of the same arms, including those of France, Burgundy, Toulouse, Dreux, Lusignan, and of Alphonse of Poitiers, brother of Saint Louis. The arms of Champagne—for Duke Hugh IV—and those of Viscount Guy VI of Limoges (d. 1263) figure as well on the coffret from the treasury of Aachen (fig. 9), dated to 1258 by de Vaivre (1974).

Corneaux (1888) recognized the arms of the royal family of Saint Louis and of eminent figures of this period (1250) among those represented on the coffret.

According to Alain-Charles Dionnet, who has graciously

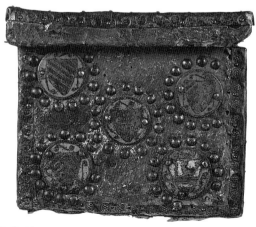

Left side

Lid

Principal face

Reverse

written to us with his interpretation, this coffret brings together—in addition to the arms of Saint Louis, of his brother (Alphonse of Poitiers), and of his wife (Marguerite of Provence)—those of principal feudal lords of the North (Burgundy, Dreux, Coucy) and those of the barons of Poitou and Aquitaine with subject families of the Limousin on the reverse: Turenne, Lusignan-La Marche, Lusignan-comtes d'Eu, Aubusson, Rochechouart, and, possibly, Berry (because of the wavy vair pattern). Dionnet suggests that the coffret could have been commissioned by the viscount of Limoges, destined for Alphonse of Poitiers, to "contain the homage of the barons of Poitou and Aquitaine who rose in insurrection at the instigation of the count of La Marche and of Angoulême" in 1241 and who were defeated at Taillebourg in 1242. The coffret then would have been made in 1242. The object may have come into the possession of King Philip III the Bold upon the death of Alphonse of Poitiers (1271) and subsequently been given by the king to the abbey of Longpont, which enjoyed his patronage.

This attractive theory does not, however, take into account all the coats of arms displayed on the coffret and, above all, appears contradictory to the style of its decoration. A comparison of these enamels with those of the Louvre coffret and the one at Aachen (fig. 9) suggests that the Longpont coffret was produced at a later date. Its decoration is limited to coats of arms that are often simplified and even summarily reproduced; the ornamentation in the form of a limited, and very sober, rinceau pattern, along with the somewhat dull colors of the enamels, with their dominant tones of blue-gray and medium blue, indicates a date of no earlier than the middle of the thirteenth century.

Coats of arms from the Limousin, including those of France, Alphonse of Poitiers, Turenne, Coucy, and Lusignan, are accorded important positions within the program of the Longpont coffret's heraldic decoration. The considerable repetition of several of these coats of arms—the arms of France appear nine times, those of Castile six times, those of Burgundy and of Guy VI of Limoges three times each—raises doubts as to the serious heraldic content of these arms, which were more likely used for a decorative rather than a dynastic or personal purpose.[7]

As for its use, the coffret of John of Montmirail can be considered in relation to comparable coffrets. Decorated with heraldic and secular ornamentation, the coffret of Saint Louis (cat. 123) was used as a reliquary for the bones and hair shirt of Louis IX at Notre-Dame du Lys. The relics of Cardinal Guala Bicchieri were likewise preserved in a coffret (cat. 88) at Sant'Andrea in Vercelli, the cathedral which also housed the remains of Blessed Amadeus IX, duke of Savoy. Similarly, the coffret of Longpont contains the relics of John of Montmirail, a knight in the service of Philip Augustus who decided late in life to dedicate himself to a completely spiritual existence. He first established a hospital for the sick near his castle. In 1210, after the completion of his children's education, he left his home to enter the Cistercian foundation at Longpont, near Soissons. During his years there the abbey church was under construction. John of Montmirail died in 1217, ten years before its completion. Following his death, numerous cures were effected through his intercession.[8]

In 1236, the monks of Longpont sought to have John of Montmirail canonized; perhaps to further this initiative, his *Vita* was compiled about 1241. After 1250 the abbot asked for authorization to transfer the relics of John to the choir of the abbey church. Authorization for this process was accorded by the chapter of Cîteaux in 1253, at the same time that permission to erect a monument was given. Not coincidentally, Marie of Montmirail, a daughter of John, began to make frequent visits to the church and to maintain candles at the tomb. In 1252, and then again in 1261, his son Matthew is recorded as making donations of wax to keep candles burning at his father's tomb. In 1271, Marie decided to be interred there as well, and also made a gift of candles in 1272, the year of her death. The tradition was maintained by Enguerrand IV of Coucy, the son of Marie, who in turn made donations of wax to the abbey in 1267 and 1270.[9] As both Marie and and her son were champions of the cult of their ancestor (in the same spirit of Philip the Fair in regard to Saint Louis), it is reasonable to suggest that it was following the initiative of the members of the Montmirail family that the relics of the Blessed John were placed in the coffret now preserved at Longpont.

BDB

NOTES
1. See Pinoteau and Le Gallo 1966, p. 38 and pl. XIV.
2. See Vaivre 1974, pp. 109–12.
3. Pinoteau and Le Gallo 1966, p. 12 and pl. IV.
4. See Pinoteau 1983, pp. 102–3.
5. Ibid., pp. 103–4. We wish to thank Alain-Charles Dionnet for his help in identifying the arms.
6. The style of the lock plate is close to that on the medallions produced by the Maison André in imitation of the series from Biella (see cat. 91).
7. This was also the conclusion put forth by Dionnet (1994, p. 111), concerning a gemellion decorated with arms that was sold in Lille (March 21, 1993, no. 11). See also in this volume Pastoureau, above, p. 339.
8. See *Vie des Saints* 1950, vol. 9, pp. 613–15.
9. On the cult of John of Montmirail, see Muldrac 1652; Larigaldie 1909; Dimier 1960–61.

EXHIBITIONS: Paris 1965, no. 95; Paris 1970, no. 45; Senlis 1987, no. 8.

BIBLIOGRAPHY: Muldrac 1652, pp. 186, 466; Boitel 1859, pp. 479, 486, 628, 636, 679–84; Poquet 1869, pp. 71–75; Corneaux 1888; Rupin 1890, p. 442; Larigaldie 1909, pp. 87–88, 156–57; Lefèvre-Pontalis 1912, p. 419; Dimier 1960–61, pp. 182–91; Dimier and Montesquiou, n.d., pp. 87–88; de Vaivre 1974, pp. 102–4, figs. 13–16; Mathieu 1975, pp. 78–79, ill. p. 97; Dionnet 1994, pp. 103, fig. 9, 105, 109.

# 134. Pair of Traveling Candlesticks

Limoges, end of the 13th or beginning of the 14th century

Copper: formed, engraved, and gilt; champlevé enamel: medium blue, red, and white

32.100.285: H.: 27.5 cm (10⅞ in.); 32.100.286: H.: 28.5 cm (11¼ in.)

Ex colls.: Gaspar, Jean, and Stéphan Bourgeois Frères, Cologne (sale, October 19–27, 1904, no. 374); Michael Friedsam, New York

The Metropolitan Museum of Art, New York; The Friedsam Collection, Bequest of Michael Friedsam, 1931 (32.100.285, .32.100.286)

Coats of arms:[1]
1. Azure semy-de-lis Or (France Ancient)
2. Azure a Fleur-de-lis Or or Azure three Fleurs-de-lis Or (France badly interpreted)
3. Bendy Or and Azure a Bordure Gules (Burgundy Ancient)
4. Gules an Escarbuncle Or dimidiated by Azure a Bend Argent cotised Or (Navarre/Champagne)
5. Azure two Barbels addorsed Or (Bar without the semy of Cross-crosslets fitchy Or)
6. Barry Argent and Azure (Lusignan)
7. Gules three Castles Or (Argent?) impaling Azure three Fleurs-de-lis Or (Alphonse de Poitiers badly interpreted)

Notes
1. I wish to thank Robert Theo Margelony for identification of the arms.
2. The basins are probably gemellions (see cats. 126, 131). See Gay 1887–1928, vol. 2, p. 80.
3. Rupin 1890, p. 524.
4. Hildburgh (1920, pp. 132–35) also mentions a pair in the Wallace Collection, London. Ducatel sale, Paris, April 21–26, 1890, no. 44; Flannery Sale, no. 42.
5. Rupin 1890, p. 524.
6. Baker 1981, pp. 336–38.

Bibliography: Unpublished.

Each candlestick consists of a number of elements threaded on a central stem: three legs, four cylinders separated by three knops, a bobeche, and a pricket with clips to support the taper. The legs, threaded one after the other, can be rotated on the stem, allowing the candlesticks to lie flat for storage or travel. Each foot is decorated with two or three coats of arms taken from the standard repertory of Limoges enamels (see cats. 123, 133). The combination does not correspond to a particular genealogy, and there are numerous errors in the drawing and in the colors.

Apparently such candlesticks were used in a secular context and in association with other Limoges objects; the will of Guillaume d'Ercuis (d. 1329) refers to "duobus bacinis cupreis de opere Lemovicensi cum candalabris ejusdem operis" (two basins of copper of Limoges work with candlesticks of the same work).[2]

Candlesticks of this type are preserved in public collections—at Dijon and in the Victoria and Albert Museum, London, for example[3]—and occasionally appear at auction.[4] One pair is preserved at Rocamadour, an important medieval pilgrimage site south of Limoges.[5] Others have been excavated, including a pair at Grave Priory, Bedfordshire, one of six English dependencies of Fontevrault.[6]

BDB

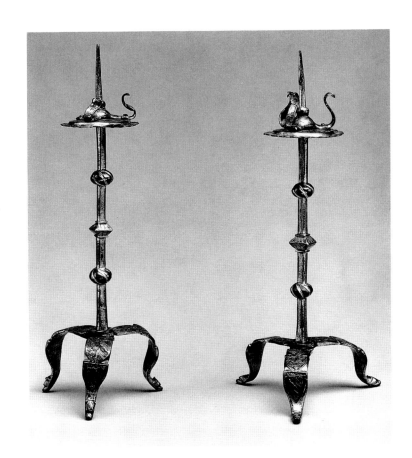

379

# 135. Six Candlesticks

Limoges, last third of the 13th century

Copper: engraved and gilt; champlevé enamel: medium blue, brick red, and white

H.: 23; 21.5; 19; 16.5; 14; 10.5 cm (9; 8½; 7½; 6½; 5½; 4⅛ in.); diam.: 10; 9.5; 9.3; 8.7; 8.3; 7.8 cm (3¹⁵⁄₁₆; 3¾; 3⅜; 3⅜; 3¼; 3⅛ in.)

COATS OF ARMS: Azure semy-de-lis Or (France)

Azure a Bend Argent Cotised Or (simplified Champagne)

Bendy Or and Gules (Turenne)

EX COLL.: Edme Durand. Acquired in 1825.

CONDITION: Considerable wear to the gilding.

Musée du Louvre, Paris; Département des Objets d'art (MR 2661–2665)

The six candlesticks, of a type whose regularly decreasing size and hollow spikes allow them to stack perfectly to facilitate their transport, were known as "traveling candlesticks." This ensemble, which is apparently complete, seems the most important of its kind to have survived; it can be compared to the one, made up of five pieces, preserved in the Österreichisches Museum für angewandte Kunst in Vienna.[1] Composed only of a base and a tall, pointed spike—and also distinguished by the armorial decoration on the bases—these candlesticks are of a type popular in the second half of the thirteenth and the early fourteenth century. Although the bases are usually pyramidal or, more rarely, in the form of a truncated cone,[2] the circular, convex shape here is a rare variant, the choice of which is probably explained by the need to stack the candlesticks easily while preserving another effective aspect: the base also serves as a bobeche to catch the wax.

As on most candlesticks of this type, the armorial escutcheons are purely decorative. The escutcheons of France, Champagne, and Turenne are, moreover, among the most frequently used by the Limousin workshops in the second half of the thirteenth century (cats. 132, 133). The use of only three colors—medium blue, brick red, and white—was also widespread in Limousin production during this period. The form of the escutcheons suggests a date of about 1270–80 (oral communication from Michel Pastoureau). The rather conservative character of the decorative elements used by the Limousin workshops leads us to place the execution of these six candlesticks in the last third of the thirteenth century.

ET-D

NOTES
1. Inv. Em 438 a–e.
2. François (1991, p. 93) counted almost thirty, all with a hexagonal base. Among the candlesticks with slightly concave truncated-cone bases, we might note the one in the Louvre, which was found in Villelouet (Loir-et-Cher). (See Paris 1945, no. 258.)

EXHIBITION: Paris 1964, no. 137.

BIBLIOGRAPHY: L. Laborde 1852, nos. 58–63; Courajod 1888, no. 121 (2660–2665), pp. 53–54; Rupin 1890, pp. 521–22, figs. 584–90; d'Allemagne 1891, pp. 120–22; Darcel 1891, nos. 155–60; Marquet de Vasselot 1914, nos. 55–60; François 1991, pp. 92–93.

# 3

*The strength of tradition*

# 136. Cruet

Limoges, mid-13th century

Copper: formed, engraved, and gilt; champlevé enamel: medium and light blue, turquoise, green, red, and white; glass cabochon

H.: 16.3 cm (6⁷⁄₁₆ in.)

CONDITION: Wear to gilding; losses to enamel; foot, handle, and spout are replacements.

EX COLLS.: John, 2nd Marquess of Breadalbane (in 1857); Pitt-Rivers; Dr. Philip Nelson, Liverpool; [F. A. Drey (in 1947)]; [Joseph Brummer, New York (purchased 1947)]

The Metropolitan Museum of Art, New York; The Cloisters Collection, 1947 (47.101.39)

Exhibited in New York only.

The pear-shaped vessel was one of a pair made to contain the wine and water that the priest combined during the celebration of the Eucharist. The body of the cruet is decorated with scrolling vines set on either side of narrow gilt ovals in rather tightly organized circular patterns. The tapering neck has a thin band of alternating lozenge and oval ornament with a band of cloudlike decoration above and below. At the knop a band of reserved scrolling foliate decoration is repeated in enamel near the lip. Standing on a circular foot, the cruet has a thin S-shaped spout secured by an arm that protrudes from the body and, opposite the spout, a thin, ear-shaped handle. The flat lid, decorated with a glass cabochon, is hinged just above the handle.

The profile of the cruet seems to depend on Eastern Christian and Islamic vessels adapted for use in major church treasuries in the West, including, for example, the treasury of San Marco, Venice, and the abbey of Grandmont.[1] The cruet made at Abbot Suger's order for the abbey of Saint-Denis, which incorporates a sardonyx Byzantine vase, is very similar in form.[2] (The long-necked spout set low on the bowl, common to both cruets, served to control the flow of liquid.)[3]

Few Limoges cruets survive. Nelson published only seven: the example in his own collection (now the Metropolitan cruet), the Bibliothèque Nationale (cat. 137), two in Hungary, one in Florence, one excavated in Ireland and preserved in Belfast, and another in the subsequently dispersed collection at Goluchow Castle (Poznań), Poland. Three burettes, two of which may form a pair, are preserved in the Museo de Olat (Catalonia).[4] Another, not identified as one of these, appeared at auction in 1992 (Sotheby's, London, December 6–9, lot 7). Among surviving cruets of Limoges enamel, the Metropolitan's is the only example that does not have religious imagery.

The decoration of cruets other than the Metropolitan's example is comparable both in design and execution to Limoges pyxes. The enameling of the Metropolitan's example is more elegant and arguably slightly earlier in date.

In the earliest illustrations of the cruet (in Waring 1857 and Rupin 1890), there is no lid. Curiously, in the engraving of 1859 published by Didron, the cruet has an embossed lid and a finial. Though the spout and the handle are replacements of before 1857, the forms approximate those found on other cruets. The bases found on the cruets in Florence and, formerly, at Goluchow Castle are probably replacements, as is the Metropolitan's.

BDB

NOTES

1. New York 1984, nos. 30, 31, with additional comparative illustrations; Paris 1984, nos. 30, 31. The cruet from Grandmont is wider at the top than at the bottom; see Paris 1965, no. 368, pl. 73.
2. Montesquiou-Fezensac and Gaborit-Chopin 1973, vol. 3, pl. 22; New York 1981, no. 26, Paris 1991, no. 29a.
3. Ewers with spouts set low on the body appear only later in Islamic examples, first in glass and subsequently in metal. Similar objects appear in Persian manuscripts of the fifteenth century but do not survive. (Oral communication, Linda Komaroff.)
4. These are apparently excavated pieces. A photograph is preserved in the Stohlman Archives, Princeton University.

EXHIBITIONS: London 1857b; Los Angeles and Chicago 1970, no. 59; Binghamton, N.Y., 1975.

BIBLIOGRAPHY: London 1857a, p. 26, "Vitreous Art," pl. VIII, no. 2; Didron 1859, p. 154; Rupin 1890, pp. 528–29, fig. 593; Nelson 1938, pp. 49–54, pl. XXI; Rorimer 1948, p. 247, ill.; Rorimer 1963, p. 142, fig. 70; Frazer 1985–86, pp. 10–11, ill.; Theuerkauff-Liederwald 1988, p. 175.

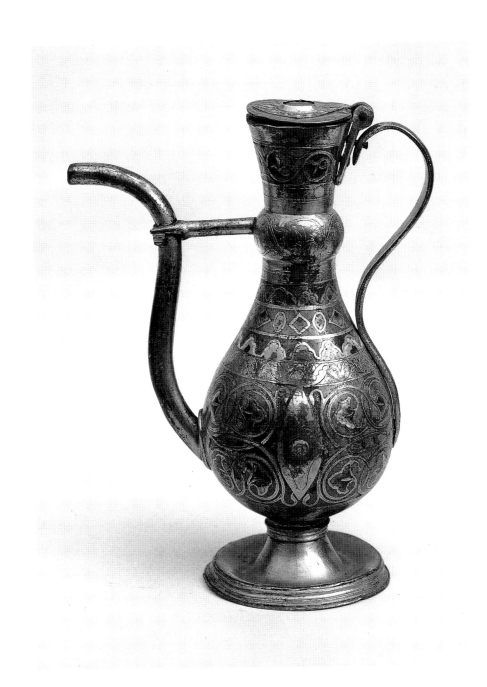

# 137. Cruet

Limoges, mid-13th century

Copper: engraved and gilt; champlevé enamel: medium blue

14.8 × 12.5 cm (5⅞ × 4⅞ in.)

PROVENANCE: From the treasury of Saint-Denis (not included in the inventory because of its low commercial value). Deposited in the Cabinet des Antiques of the Bibliothèque Nationale de France in 1794.

CONDITION: Base is missing; knop on the cover broken; body dented, probably the result of one or several falls. Substantial losses to the enamel, especially on the body. Handle and spout resoldered.

Bibliothèque Nationale de France, Paris; Cabinet des Médailles (no. 412)

Exhibited in Paris only.

During Mass, cruets were used in pairs to hold the water and the wine (see cat. 136), which were then mixed in a chalice. In the earliest extant lists of liturgical objects—those describing the gifts of fourth- and fifth-century popes to various churches in Rome—certain vessels called *amae*, which certainly were used for this purpose, often are mentioned after chalices but before patens. For example, among the items bequeathed by Pope Silvester (r. 315–35) to Saint Peter's basilica were: ". . . calices aureos III [. . .] calices argenteos XX [. . .] amas aureas II [. . .] amas argenteas V. . . ."[1]

In the oldest medieval texts one sometimes still finds the terms *amae* or *amulae* for cruets, but they are also referred to as *phialae* (vials), *vasa* (vessels), *vascula* (small vessels), *poti* (pots), *urceoli* (ewers), *buretae* (cruets), and, most often, *ampullae*[2]—a less precise term that nevertheless evokes their form and their relatively small size. We can cite from the earliest and most explicit descriptions those objects that were sent in 1137 by Bishop Werner to the cathedral of Münster: "calicem et duas ampullas ad servandum vinum et aquam,"[3] or the ones included in the bequest of Bishop Peter Lombard (d. 1160) to the cathedral of Notre-Dame in Paris: "duobus vasculis argenteis ad ministrandum aquam et vinum."[4]

In the Middle Ages, cruets were made of various materials. Sometimes they incorporated vessels from antiquity of hard stone or Fatimid rock crystal, such as the examples in the treasury of San Marco in Venice or the cruet from the treasury of Grandmont, formerly in Saint-Georges-des-Landes and then in Milhaguet; the last example was fitted with a gilt-copper mount executed by Limousin goldsmiths in the thirteenth century.[5] Cruets most often seem to have been silver, like those bequeathed by Peter Lombard to Notre-Dame, mentioned above. Furthermore, silver is the metal cited by the monk Theophilus in chapter LVIII of *On Divers Arts* when he describes the making of a cruet.[6]

However, the most luxurious were gold; thus, the sum of money left to Notre-Dame by Bishop Ranulfus in 1288 was used for the fabrication of "duas buretas aureas."[7]

Cruets are rarely mentioned among works from Limoges, yet the 1238 inventory of the treasury of Trier Cathedral lists one: "Ampullas duas operis de Lemugis."[8]

Like the majority of other extant examples, the cruets executed in Limoges, which are very similar, are in the form of small vases with rounded bodies, long necks, handles, and curved spouts—a type probably Eastern in origin (see cat. 136). The decoration most often, as here, is composed of busts of angels—a motif favored by Limousin enamelers to adorn liturgical objects (cats. 69, 71–74).

Of the surviving Limousin cruets, the one from Saint-Denis is not only the most famous but also among the most carefully executed; it is closest to the cruet in the Museo Nazionale del Bargello in Florence.[9] The engraved angels resemble those on many works made during the second and third quarters of the thirteenth century. The combination, in the rinceau decoration, of some enameled florets and others in reserve, with pointed leaves that form "hooks," recalls, for example, the ornamentation of the small chasse of Saint Fausta in the Musée de Cluny,[10] which could not have been executed before 1240 or 1260 at the latest—which is why we suggest dating the Saint-Denis cruet to about the middle of the thirteenth century.

ET-D

NOTES
1. Duchesne 1955, vol. 1, p. 176. "Three gold chalices . . . , twenty silver chalices . . . , two gold vases . . . , five silver vases."
2. See the examples given in Nelson 1938, p. 49.
3. See "one chalice and two ampullae for serving wine and water," Bischoff 1967, no. 139, pp. 141–42, l. 8.
4. "Two small silver vessels for serving wine and water." See Guérard 1850, p. 60.
5. Treasury of San Marco in Venice: cruets of rock crystal and sardonyx, see Paris 1984, nos. 31, 34; Milhaguet cruet: see Paris 1965, no. 368, pl. 73; Gauthier 1983, no. 14, p. 36 (stolen in 1981 from the Musée Municipal de l'Évêché, Limoges).
6. See Theophilus 1961 and 1986, p. 109.
7. ". . . Pecunie summam . . . quam . . . convertimus in duas buretas aureas, ad ministrandum vinum et aquam in missa" (A sum of money . . . which . . . we exchanged for two gold cruets for serving wine and water at Mass), in Guérard 1850, p. 185.
8. "Deux ampoules de l'oeuvre de Limoges" (Two ampullae of Limoges work); see Bischoff 1967, no. 91, p. 96, l. 22.
9. Nelson 1938, pp. 52–53. The example sold at Sotheby's, London, on July 8, 1865 (no. 5), has similarities to the BNF example that are at the

very least surprising (with the exception of the engraved bust of an angel, the style of which is suspicious). It may have been made on the basis of nineteenth-century engravings.

10. Inv. Cl 2827, see Rupin 1890, pp. 385–86, fig. 441; Gauthier 1972b, p. 188.

EXHIBITIONS: Limoges 1948, no. 103; Vatican City 1963, no. 102; Paris 1991, no. 43.

BIBLIOGRAPHY: Way 1845, p. 168; Didron 1859, p. 153 and pl.; *L'Art pour tous*, no. 420 (Dec. 15, 1877), p. 1677, figs. 3664–68; Rupin 1890, pp. 528–29; Rohault de Fleury 1883–89, vol. 4, pl. CCCXXVII; Duret 1932, p. 230, fig. 207; Nelson 1938, pp. 49–54, and fig. 1; Gauthier 1950, p. 54 and pl. 59; Montesquiou-Fezensac and Gaborit-Chopin 1973, p. 123, pl. 108A.

# 138. Pyx

Limoges, mid-13th century

Copper: engraved and gilt; champlevé enamel: medium blue, turquoise, and white

H.: 10.8 cm (4¼ in.); diam.: 6.9 cm (2¹¹⁄₁₆ in.)

INSCRIBED: IH[ESU]S

EX COLL.: Michael Friedsam, New York (in 1931)

CONDITION: The cross finial is modern.

The Metropolitan Museum of Art, New York; The Friedsam Collection, Bequest of Michael Friedsam, 1931 (32.100.282)

Exhibited in New York only.

The Christogram is set in reserve against a disk of white enamel. This decoration is frequently found on Limousin pyxes[1] and is indicative of its function, recalling the form found on molds for the host in the fourteenth and fifteenth centuries.[2]

The limited palette, dominated by the medium blue ground, is common to that of most preserved examples and cannot be earlier than the second quarter of the thirteenth century—the period at which documentary sources make reference to pyxes of Limoges work (see cat. 74).

BDB

# 139. Pyx

Limoges, second quarter of the 13th century

Copper: engraved and gilt; champlevé enamel: blue-black; dark, medium, and light blue; turquoise, green, yellow, red, and white

H.: 9.5 cm (3¾ in.); diam.: 6.49 cm (2⁹⁄₁₆ in.)

EX COLL.: Michael Friedsam, New York (in 1931)

CONDITION: The cross finial is modern.

The Metropolitan Museum of Art, New York; The Friedsam Collection, Bequest of Michael Friedsam, 1931 (32.100.281)

Exhibited in New York only.

Angels set in medallions, like those of this example, are the most common form of decoration of Limousin pyxes (see cat. 74), probably an allusion to the Eucharist as the "bread of angels."[1]

Hosts reserved for individual communicants were small in size: Honorius of Autun, in the twelfth century, compared them to *deniers*, coins that were about 2.5 centimeters (1 in.) in diameter.[2] This pyx, like a number of other examples, has a cupule set in the interior similar to the ones found in Eucharistic doves, another common form of Eucharistic reserve in Limoges work, perhaps the type to which William of Salisbury refers (see cat. 74). Apparently meant to facilitate removal of the wafers, such cupules show that a given pyx could hold only a small number of hosts.

The rich palette of this pyx is consistent with a dating in the first decades of the thirteenth century, though the rather awkward engraving may indicate a slightly later date.

BDB

NOTES
1. See Rupin 1890, pp. 208–10, figs. 272, 273, 276; pyxes preserved at Saint-Hilaire-Foissac (Corrèze), in the Musée d'Auxerre, and in the treasury of the cathedral of Moutiers.
2. See Schulte 1910, especially p. 491.

EXHIBITIONS: Los Angeles and Chicago 1970, no. 58, p. 128; Cambridge, Mass., 1975, pp. 65–69.

NOTES
1. See cat. 69 and Rohault de Fleury 1883–89, vol. 5, p. 92.
2. Rupin 1890, p. 226.

# 140. Dedicatory Plaque from an Altar

Limoges, 1267

Copper: engraved and gilt; champlevé enamel: medium blue, green, red, and white

8.5 × 35.4 cm (3⅜ × 13¹⁵⁄₁₆ in.)

INSCRIBED: HOC . ALTARE CO[N]SACRATV[M] : EST : AB : AYMERICO : LEM [OVICENSI]/: EP[ISCOP]O : IN : HONORE : S[AN]C[T]E : ×[CRUCIS] : ET : BEATE : MARIE : VIRGINIS:/ ET B[EAT]I ANDREE : AP[OSTO]LI : ET : B[EAT]I : LAVRIANI : M[ARTI]R[IS] ATQ[UE] PO[N]TIFIC /IS : ET : B[EAT]I : NICHOLAI : EP[ISCOP]I : ET C[ON]F[ESSORI]S : ET . BEATE : KATHERINE/ VIRGINIS : ET : M[ARTI]R[S]I : ET OM[N]IVM : S[AN]C[T]O[RUM] : Q[U]INTO : NONAS : MAII:/ ANNO : AB : INCARNACIONE : D[OMIN]NI : MCCLXVII : T[EM]P[O]R[E] : WI[LLEMI] : P[RI]OR[IS]. (This altar is consecrated by Aymeric, bishop of Limoges, in honor of the Holy Cross and of the Blessed Virgin Mary and of Blessed apostle Andrew and of Blessed Nicholas, bishop and confessor, and of Blessed Catherine, virgin and martyr, and all the saints, the fifth day of the nones of May [May 3], in the year of the incarnation of the Lord, 1267, in the time of the prior William.)

Laborde in 1852 and by Darcel in 1857; later, it was said to have come from the church at Genouillac—as, for example, in the catalogue of the Germeau sale of 1868. In 1883, however, Molinier stated that "no one knows today which Genouillac [sic] is meant here." Indication of actual provenance, the priory of the Artige, was published by Gauthier (1960), as a result of the discovery by Decanter, in the Departmental Archives of Haute-Vienne, of a description recorded in 1586 by Martial Gay, lord of Nexon and of Champagne and presiding judge of the court of Limoges. Fleeing from a plague epidemic, he had, in fact, taken refuge at the Artige, where he had uncovered several inscriptions, among them that on the present plaque.

The text notes precisely that the dedication took place "in the time of prior William," who was undoubtedly either

PROVENANCE: From the priory of the Artige (Haute-Vienne), where the plaque is mentioned, and its inscription copied, in 1586 by Martial Gay (Cahier de notes in Chartrier de Nexon, Archives Départementales de la Haute-Vienne, 26 J, file 13, fol. 4r)

EX COLLS.: Albert Germeau (in 1852); (sale, Paris, May 4–7, 1868, no. 61); Countess Isabelle Dzyalinska, Paris (in 1883), then Goluchów Castle, Poland (in 1903). Acquired in 1942.

CONDITION: Traces of corrosion; missing enamel replaced by inserted plugs, in the inscription and on the left side, around the figures and their halos. Dent on the head of the Virgin, which makes her features unreadable.

National Museum, Warsaw (deposited at the Poznań Museum) (52 M 987 MN)

Exhibited in Paris only.

The principal decoration on this plaque is a six-line inscription commemorating the dedication of an altar by Aymeric de la Serre, bishop of Limoges, on May 3, 1267.

The fact that this is one of only four dated inscriptions known to us in the history of Limousin enamelwork—and that it is the oldest among them (see cats. 152, 156, 157)—confers on this modest work an importance much greater than that suggested by its artistic quality alone. Known since the middle of the nineteenth century, the plaque was first published, without mention of its provenance, by

William Gaillard, recorded in 1264, or William of Croizille, who died in 1285.[1]

The plaque is the only extant enameled-copper work of this type made in Limoges. Its modest appearance probably was in keeping with that of the priory for which it was made. Nothing indicates that it was part of the decoration of some more important object.

Three of the saints mentioned in the inscription are represented at the ends of the plaque: the Virgin and Saint Andrew are at the left, and Saint Catherine is submitting to torture on the wheel, at the right. The style of the plaque is still generally linked to early Gothic art, which prevailed during the first half of the thirteenth century; it is particularly apparent here in the treatment of Saint Catherine's robe, but the stylized drapery and the reduction of the facial features reveal the plaque's later execution, confirmed by the inscription.

ET-D

NOTE

1. Senneville 1900, p. 414. On the order of the Artige, see Becquet 1970.

EXHIBITION: Paris 1865, no. 649.

BIBLIOGRAPHY: L. Laborde 1852, pp. 64–65; Darcel 1857, p. 114; Darcel 1865b, p. 518; Molinier 1883, pp. 213–14; Rupin 1890, pp. 136–37, fig. 209, p. 137; Molinier 1903, no. 147, p. 39; Gauthier 1960b; Gauthier 1976a, p. 185; Favreau et al. vol. 2, 1978, p. 197.

# 141. Chasse of Saint-Aurelian of Limoges

Limoges, end of the 13th century

Copper: engraved and gilt; champlevé enamel: black, ultramarine blue, turquoise, green, yellow, red, gray, and white

28 × 21.7 × 11 cm (11 × 8½ × 4⅜ in.)

INSCRIBED: s[anctus] martin // s[anctus] sesatoris // s[anc]ta katerina

PROVENANCE: Documented in the church of Saint-Aurelian in Limoges since 1723. Sold to the prefect of Haute-Vienne, Albert Germeau,[1] before 1855. Acquired by the Musée de Cluny at the sale of the Germeau Collection (Paris, May 1–7, 1868, no. 52).

CONDITION: The ends of the crest are broken off. End with door has lost a large amount of enamel; considerable wear to the gilding.

Musée National du Moyen Âge, Paris; Thermes de Cluny (Cl 8668)

Principal face

Right side

Its established provenance makes this chasse of special interest: the inventory of 1723 of the chapel of Saint-Aurelian in Limoges, published by Tintou in 1979, provides the following description: "a small chasse of enameled copper on which is written on one side Saint Catherine, an image of Saint Peter in the opening of the door, and on the other side the figures of Saint Aurelian and Saint Cessateur."[2] Built between 1471 and 1475 by the butchers' guild of Limoges as an annex to the church of Saint-Cessateur, which was situated outside the town walls in a secluded area, the chapel of Saint-Aurelian was to house the relics and reliquaries of this ancient church in order to ensure their safety. Since the eighth century, the church of Saint-Cessateur had contained the tomb of its patron saint, Cessateur, or Cessadre, the twenty-ninth bishop of Limoges (who died in 742), as well as the body of Saint Aurelian, which was brought there at a later date, as Levet (1994) has shown.

The reliquary, in the form of a building that is rectangular in plan, consists of a coffret resting on four feet and surmounted by a sloping roof with a crest. The chasse, made entirely of metal without an underlying wood core, is composed solely of thick copper plates assembled by means of a system of tenon and mortise joints; hence it is relatively heavy (almost three kg [6.6 lbs.]) for its size. This method of metal assembly was widely employed in Limousin workshops from the mid-thirteenth century—in particular for certain chasses whose roofs open and for the bases of pyxes. Each of the two faces is formed from a single copper plate, bent to mark the transition to the roof. To our knowledge, this aspect of the construction remains unique. On each end is a plaque, one of which has a door to provide access to the relics. The treatment of the four sides is identical: silhouetted, engraved, and gilt figures in reserve stand out against an overall background of ultramarine blue enamel, covered with decorative motifs, sometimes in reserve and sometimes enameled. The execution is rapid and somewhat careless, and the quality of the enameling is mediocre. From the personalized decoration and the history of this chasse we can deduce that its initial contents, which have since disappeared, were without any doubt the relics of Saint Cessateur—and perhaps those of other saints as well. The similar treatment of the two faces makes it impossible to distinguish the principal face from the lesser one, as can usually be done with Limousin chasses. However, the inscriptions and the imagery of the face on which the two bishops are depicted would suggest that it is of greater importance.

The scenes on both faces are developed in continuous fashion, from the lower part to the roof, set against a rinceau-patterned background with large palmette flowers in graduated shades of red, green, and yellow, or of black, blue, and white. On one face, two haloed bishops, standing and in profile, identified by a horizontal inscription as Saint Martin and Saint Cessateur, face each other; both wear miters and liturgical vestments and hold crosiers, emblematic of their episcopal rank. On the other face is the Martyrdom of Saint Sebastian: at the left, the emperor Diocletian, seated on his throne, orders the martyrdom of the saint with his raised right hand; at the right, a nimbed Saint Sebastian, his body pierced with the arrows of two archers, is tied to a column. On one of the gabled ends, Saint Catherine, identified by an inscription, is depicted standing, wearing a halo and a crown; on the other end, Saint Peter, with a large key, stands at the entrance to Paradise on the door, occupying his symbolic and traditional place, in Limoges, as guardian of the relics.[3] The only other known representation of Saint Martin, bishop of Tours, on a Limousin reliquary is that on the chasse from Laguenne (Corrèze) and now in the Musée Thomas Dobrée in Nantes:[4] on the principal face, Saint Martin, also identified by an inscription, and Saint Calminius surround Christ in Majesty. The structure, dimensions, and different methods of execution of the Saint Calminius chasse certainly appear to have little in common with the chasse of Saint Aurelian, but its rinceau ornamentation with large poly-

chrome palmette flowers in reserve might suggest that it inspired the reliquary described here. This type of luxuriant rinceau pattern, which serves as the background of enameled plaques, first appeared about 1230 in the decoration of tabernacles in Cherves (see cat. 98) and in Chartres. The use of this polychrome floral rinceau ornamentation, which also may be seen on the small chasse of Saint Fausta now in the Musée de Cluny[5]—datable, like the chasse from Nantes, to the middle of the thirteenth century—would prevail in simplified forms, as the present chasse attests, until the close of the thirteenth century.

GF

Reverse

Left side

NOTES
1. Leniaud 1979, pp. 363–72.
2. ". . . Une petite châsse de cuivre en émail òu est escrit, à coste santa Caterina, à ouverture de la porte d'image de S. Pierre de l'autre costé les figures de saint Aurélien [sic] et Saint Cessateur." *Inventaire de la chapelle Saint-Aurélien de Limoges, 1723* (Limoges, Archives Départementales de la Haute-Vienne, 4 E. 5-30).
3. On Saint Peter as "porter of paradise" on Limousin chasses, see Texier 1857, col. 311.
4. Inv. 896-1-22; see Costa 1961, vol. I, pp. 32–35, no. 35, fig.
5. Inv. Cl 2827; E. du Sommerard 1883, no. 4499.

EXHIBITIONS: Limoges 1948, no. 121; Vatican City, 1963, no. 100.

BIBLIOGRAPHY: *Bulletin archéologique*, 1842–43, p. 146; Texier 1842, p. 168; Ardant 1848, p. 134; Texier 1851, pp. 199–200, no. 137; Ardant 1855, pp. 52–54; Ardant 1865, p. 117; E. du Sommerard 1883, no. 4508; Rupin 1890, pp. 394–95, fig. 449; Tintou 1974, p. 118; Levet 1994, pp. 148–51.

# 142. Almond-Shaped Reliquary

Limoges, after 1255, probably about 1260–80

Copper: engraved, chased, and gilt

16.7 × 10.7 cm (6⅝ × 4¼ in.)

INSCRIBED:

Face:

DE TVNICA : BEATI . FRANCISCI : C[ON]F[ESSORIS] : ET . DE . CAPILLIS : EI[US] / [DE] CAPILL[IS] . B[EAT]E . CLARE . V[IR]G[IN]IS : ET . DE . TVNICA : ET DE VELLO EI[A] / DE TVNICA : BEATI : AN / TONII CONFESOR[IS] (Of the tunic of blessed Francis, confessor, and his hair. Hair from the blessed Clare, virgin, of her tunic and her veil. Of the tunic of blessed Anthony, confessor.)

Reverse:

S[AN]C[T]I MARCI[A]LIS : AP[OSTO]LI . S[AN]C[T]I . LAVRENCII : M[A]R[TIR]IS : S[AN]C[T]I : BL / ASII : EP[ISCOP]I : ET M[A]R[TIR]IS S[AN]C[T]I GEORGII : [MA]R[TIR]IS // S[AN]C[T]I NICHOLAI E[PISCOP]I . S[AN]C[T]I XPOFORI . (CHRISTOFORI) . M[A]R[TIR]IS // S[AN]C[T]E : PRISCE . VIRG[INIS] : S[AN]C[T]I / SILVANI : M[A]R[TIR]IS // S[AN]C[T]E MARINE : VIRG[INIS] // S[AN]C[T]E CATERINE : VIRG[INIS] ET M[A]R[TIR]IS // S[AN]C[T]E : PETRONILL[E] / S[AN]C[T]E VALERIE . VIRG[INIS] . ET . MARTIRIS S[AN]C[T]I : PARDUL : FI : [CON]F[ESSORIS] / DE TVNICA B[EAT]I : IAC / OBI AP[OSTO]LI (Of Saint Martial, apostle. Of Saint Lawrence, martyr. Of Saint Blaise, bishop and martyr. Of Saint George, martyr. Of Saint Nicholas, bishop. Of Saint Christopher, martyr. Of Saint Prisca, virgin. Of Saint Sylvan, martyr. Of Saint Marina, virgin. Of Saint Catherine, virgin and martyr. Of Saint Petronilla. Of Saint Valerie, virgin and martyr. Of Saint Pardoux, confessor. Of the tunic of blessed James, apostle.)

PROVENANCE: Mentioned for the first time at the collegiate church of Saint-Martin, Brive, in 1882. Classified, Monuments Historiques, 1891

Reverse

CONDITION: The present support is modern. Around the single hole on the bottom of the plaque, on the reverse, there is the trace of a trefoil shape, which probably corresponds to a former attachment. Extensive wear to the gilding on the reverse, principally on the periphery.

Collegiate Church of Saint-Martin, Brive (Corrèze)

This modest object is an example of those reliquaries that contained a great many different relics, reassembled in two small compartments measuring 4.4 and 4.9 centimeters (1¾ and 1⅞ in.) in diameter. Mentioned in the inscriptions on the principal side, along with various saints whose cults were quite widespread, are various local saints (Pardoux, Valerie, and Martial) as well as several "recent" saints (the founders of the minor orders, Francis, Clare, and Anthony of Padua). The iconography honors both these Franciscan and "local" saints, since Saint Valerie, presenting her head to Saint Martial, occupies the upper register, while Saint Clare and Saint Francis are represented below.

The unusual shape of this reliquary is surprising. Indeed, we know of no other example, in Limoges or elsewhere, of a reliquary composed, like this one, of a single metal plaque with one or several compartments containing relics presented directly to the gaze of the faithful. The trace of a trefoil-shaped element on the lower tip of the plaque and comparison with other known reliquaries support the hypothesis that this object perhaps was originally a shutter articulated by a hinge, which opened onto a more important object whose structure, including a wood core and metal revetments, is suggested by the reliquary of Saint Francis of Assisi in the Louvre (cat. 101). Furthermore, it is highly probable that the compartments, which are of different shapes and poorly adapted to their support, initially were not affixed to the almond-shaped plaque; the large rivets, whose round, ungilded heads disrupt the engraved decoration on the other side, seem, in fact, to prove this.

The style of the decoration is somewhat archaic as well as naive: the composition is very summary and the four figures rapidly sketched in. Their thin, long-limbed silhouettes; their small facial features; and the V-shaped fold formed by Saint Martial's robe are indications of the undoubtedly rather late date of execution, during the thirteenth century, which is further confirmed by the representation of Saint Clare, who died in 1253 and was canonized in 1255, and the presence of her relics. These figures, moreover, are not unlike those on the dedicatory plaque from an altar in the priory of Artige, dated 1267, now belonging to the Muzeum Narodowe, Warsaw (cat. 140), which supports our dating of the Brive plaque to between 1260 and 1280. The reliquary

from the collegiate church of Saint-Martin in Brive, a rare
example of the evolution of Limousin production after the
middle of the thirteenth century, is also striking testimony
to the rapid spread in the Limousin of the cults of Saint
Francis, Saint Clare, and Saint Anthony, whose orders were,
in fact, firmly established in the region early on: a
Franciscan house was founded in Brive as early as 1226–27
by Saint Anthony, who stayed at the hermitage of Grottes in
1227, situated near the city gates.[1] This establishment
housed a relic of Saint Anthony, which was transferred to
the church of Saint-Martin at Brive in 1791. The convent of
Poor Clares, founded in 1242, was joined to the Benedictine
abbey of Bonnesaigne in 1759, and closed in 1791. It is,
therefore, highly possible that, as has been recently proposed, the reliquary is from one of these establishments.[2]

ET-D

NOTES

1. See Roy de Pierrefitte 1864, pp. 149–50; Nadaud, published in Lecler
   1903, pp. 709–10; Paris 1993, no. 3.
2. Dedieu, in Paris 1994, no. 3.

EXHIBITIONS: Tulle 1887, no. 59; Paris 1965, no. 394; Paris 1993, no. 3.

BIBLIOGRAPHY: Rupin 1882, pp. 420–24; Molinier 1887a, no. 59, pp.
524–26, ill. p. 525; Rupin 1890, pp. 134–36; Poulbrière 1891, p. 19; Forot
1913, pp. 178ff.; Pérol 1948, p. 84 and ill. p. 91; Favreau et al. 1978, vol. 2,
no. 29, pp. 35–38; Boehm 1989, p. 141, fig. 40.

# 143. Quatrefoil Reliquary

Limoges, last third of the 13th century

Copper: punched, engraved, and gilt; five large rock-crystal cabochons; small colored-glass cabochons (originally fifty-eight); wood core

21.5 × 17.7 × 4.3 cm (8½ × 7 × 1¹¹⁄₁₆ in.)

PROVENANCE: Sale, Paris, Hôtel Drouot, November 16, 1988 (no catalogue). Gift of the Société des Amis du Louvre, 1989

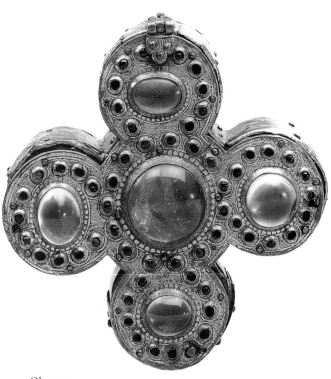

Obverse

CONDITION: Original wood core and original setting preserved. Copper foil covering the core on the obverse, under the hinged and detachable plaque, missing, except for a few small fragments still attached to the wood. Twenty-nine small cabochons missing. Gilding partly worn. Restoration on the copper plaque covering the lower right-hand edge.

Musée du Louvre, Paris; Département des Objets d'art (OA 11233)

Exhibited in Paris only.

This quatrefoil reliquary belongs to a small group of objects with the same form, faithfully modeled after the reliquaries created in Limousin workshops to house the relics of Saint Francis of Assisi (cat. 101). Indeed, in form, in construction, and in the decoration on the obverse, these reliquaries are identical. The wood core, in which spaces were cut to hold the relics, was covered with two plaques: the first, carved

and then attached with rivets, and a second, mounted on hinges and set with five large cabochons of rock crystal.

The abandonment of all enamel decoration, the iconography, and the style of the reverse all distinguish it from this model. This, moreover, is the only element in which the ornamentation varies from one work to another. Depicted on the Louvre reliquary are Christ in Majesty, as well as the four evangelists seated at their desks—an iconographic innovation in Limoges, for the evangelists usually are represented by their symbols, as on the example (with a base) in the Musée de Cluny.[1]

A reliquary of this type, decorated, on the reverse, with the engraved figure of an angel, was still preserved in the hospital in Limoges at the end of the nineteenth century;[2] it was recently identified at the Brooklyn Museum in New York, but lacking the original revetment of the reverse and mounted on a modern base.[3] Another example, in the Walters Art Gallery, Baltimore, is decorated on both sides with crystal cabochons ringed by smaller cabochons of colored glass.[4]

These objects, accordingly, take as their point of departure the reliquaries of Saint Francis of Assisi but in a simplified form, using only goldsmithing techniques on gilt copper. Works modest in quality, they reveal how the Limousin workshops were able to simplify the production of a type of object originally created for more exceptional relics. The present reliquary, acquired by the Louvre in 1989, also attests to the evolution of a Limousin style and iconography. The representation of the evangelists sitting at their desks, surrounding Christ in Majesty seated on a throne on which is depicted a candlestick—an allusion to the vision of the Apocalypse (4: 2–8)—while frequently encountered in the thirteenth century, was a novelty in Limoges.

The style—characterized by a clear-cut attenuation of the silhouettes and faces of the figures; by the draperies, with their unusual, V-shaped folds; and especially by the long lines that suggest both the poses and outlines of the figures and the pleats of their garments—derives from the artistic currents that developed in the Île-de-France during the second half of the thirteenth century. In the treatment of backgrounds—with their small quatrefoils in reserve—and the other stylistic peculiarities, the reliquary may be compared with the one in the treasury of the church of Saint-Martin in Brive (cat. 142). If one takes these considerations into account, a date is suggested clearly after the middle—and probably during the final third—of the thirteenth century for the present reliquary.

ET-D

392

Reverse

The reliquary open

NOTES

1. Taburet-Delahaye 1989, no. 30.
2. Rupin 1890, figs. 548, 549.
3. Taburet-Delahaye 1989, no. 30 n. 1, figs. 2, 3, p. 102.
4. Inv. 53139; see ibid., fig. 1, p. 102. The backs of other examples, known only through photographs, such as the one in a private collection in Barcelona in 1924, are unknown (photo documentation of the Amattler Institute, Barcelona).

EXHIBITION: Paris 1990, no. 4.

BIBLIOGRAPHY: "Chronique des amis" 1990.

# 144. Crosier of the Dauphin Humbert

Limoges, end of the 13th century

Copper: stamped, repoussé, engraved, chased, and gilt; colored glass cabochons

35.4 × 14.5 cm (13¹⁵⁄₁₆ × 5¾ in.)

PROVENANCE: Said to have been found in 1793 in the tomb of the dauphin of Viennois, Humbert II (d. 1355), in the Dominican church of Paris.

Ex COLL.: Pierre Révoil, Paris (before 1811). Acquired in 1828.

CONDITION: Extensive wear to the gilding and corrosion of the copper, certainly caused by burial. Eighteen of the cabochons decorating the shaft and volute are missing (only the mounting or a part of it is preserved); of the remaining eight, two are broken.

Musée du Louvre, Paris; Département des Objets d'art (MRR 812)

Exhibited in Paris only.

A Gothic theme par excellence, the Crowning of the Virgin appeared in a group of Limousin crosiers executed during the second quarter of the thirteenth century.[1]

Certain motifs of this crosier, such as the finely engraved foliage suggesting ferns, were already widespread during the first half of the thirteenth century (cat. 98). But the flattened knop with lozenge-shaped bosses—in this case decorated with engraved fleurs-de-lis, an innovation in Limousin production—would be difficult to imagine occurring before the middle of the thirteenth century.[2] Further, the small, round faces and the rather summary V-shaped folds of the drapery recall, in the full relief, the treatment of engraved works such as the Brive and Louvre reliquaries (cats. 142, 143).

This crosier may be compared particularly with the one in the Germanisches Nationalmuseum of Nuremberg, which was sketched by Eugène Delacroix before 1863, when it was in the Soltykoff Collection.[3] The example from the Révoil Collection was painted by Claude Bonnefond, a student of Révoil's, probably about 1810–13.[4]

According to the inventory of the Révoil Collection (1828), the crosier was found in 1793 in the tomb of the dauphin Humbert of Viennois, in the Dominican church at Paris. The dauphin Humbert II of Viennois was responsible for the accession of this province to France; in 1349, he abandoned all his possessions in favor of the future king, Charles V; from that time, the oldest son of the king of France bore the title dauphin until his accession to the throne. Named patriarch of Alexandria by the pope, Humbert retired to a Dominican monastery. In his will of May 25, 1355, he asked to be buried in the church of the Parisian monastery of that order;[5] his tomb there was described and reproduced by Millin (1792): it was covered with a great slab of copper engraved with a representation of the deceased, with his episcopal attributes accompanied by his armorial bearings.[6]

ET-D

NOTES

1. See the examples in the Museo Civico Medievale, Bologna, and the Louvre; Marquet de Vasselot 1941, nos. 103, 104.
2. This type of knop is found on a small number of Limousin crosiers that all seem to date from later than the middle of the thirteenth century, with the exception of the example now in the Walters Art Gallery, Baltimore, whose authenticity is not without question.
3. See Marquet de Vasselot 1936.
4. Louvre, Département des Peintures, inv. 2695: the painting comes from the Révoil Collection. It was painted by Bonnefond during his apprenticeship with Révoil, which allows for a dating of about 1810–13. See Paris 1979–80, no. 276. I heartily thank V. Pomarède, conservator in the Département des Peintures, for written communication in this connection.
5. See Allard (seventeenth century), published by Gariel 1970, pp. 432–86.
6. Millin 1792, pp. 35–41 and pl. IV.

EXHIBITIONS: Paris 1963, no. 69; Paris 1979, no. 213.

BIBLIOGRAPHY: Barraud and Martin 1856, p. 87, fig. 107, p. 88; Darcel 1891, no. 718; Marquet de Vasselot 1914, no. 77; Enlart 1916, p. 360; Marquet de Vasselot 1941, no. 115; Garmier 1980, p. 55.

# 145. Appliqué Figure: Wise Man

Limoges, ca. 1310–20

Copper: repoussé, engraved, and gilt; lapis blue enameled beads

H.: 14 cm (5½ in.)

EX COLL.: Charles-Jules Maciet. Gift of Charles-Jules Maciet, 1903

CONDITION: Wear to the gilding and on the surface of the copper, notably on the face, probably caused by burial.

Musée du Louvre, Paris; Département des Objets d'art (OA 5933)

This elegant appliqué figure is a rare example of a Limousin relief that presents the best qualities of the refined style developed by artists at the time of Philip the Fair. The elongation of the figure; the supple and natural fall of the mantle, whose border sketches a long, sinuous line; the elegance of the gesture; and the attenuation of the features of the face, with its thin mouth, short, fine, nose, and small eyes set in shallow sockets are all characteristic of this art. The parallel with the figure of Guy de Meyos kneeling before Saint Louis (cat. 152) makes it possible to attribute the piece to the same Limousin milieu of the early fourteenth century. The more pronounced slimness of the silhouette may suggest, however, a slightly later date in this case.

ET-D

BIBLIOGRAPHY: Marquet de Vasselot 1914, no. 108.

OBIIT NOBILI

VS · GVIDO · DE

R POST OCCA

MILLESIMO

TO · O · VVO · FE

APELLIO · G

ARGONESO

# VI

Gilded Images:
Sacred and
Funerary
Sculpture


13th–14th century

The manufacture of tombs appears comparatively late in the development of Limousin enamelwork. There is an illustrious precedent, the effigy of Geoffrey Plantagenet (cat. 15), dating to the years just after his death in 1151. Yet this exceptional work commemorating the military prowess of the count of Anjou and Maine, who is represented with drawn sword, is part of a funerary monument whose original disposition is still imperfectly understood. In all likelihood, the effigy was originally placed vertically, as it was still to be seen in the seventeenth century, for the likeness has neither the dimensions nor the posture befitting a recumbent tomb figure. Furthermore, its attribution to Limousin workmanship remains a matter of debate. The same can be said for the enameled ornamentation of the tomb of Ulger, bishop of Angers (d. 1148), now lost but partially known through the albums of watercolors of tomb ensembles in French churches recorded by Roger de Gaignières in the eighteenth century.[1] Again we have an enameled funerary monument, yet its shape suggests a very sizable reliquary, whose principal side and sloping roof, the only elements visible inside the cathedral, were rhythmically divided by an arcade enclosing images of Christ, surrounded by apostles and prophets, and of Ulger himself, accompanied by Angevin canons.

It is not until the middle of the thirteenth century that we find Limousin enamelers executing tomb decorations with effigies laid out on the sarcophagi, in keeping with the format that first appeared in Europe at the end of the eleventh century. The earliest such Limousin effigies were created in relief; there were no preceding flat enameled representations. Indeed, the enameled sheets would have been too fragile to be placed on the ground and too difficult to see had they been affixed to the body of a sarcophagus. During the first quarter of the thirteenth century Limousin enamelers first of all mastered and then widely adopted the fabrication of small appliqué figures (of thirty centimeters [about twelve inches] at most), executed in repoussé and destined to decorate reliquaries or altars.[2] Only at the end of the second quarter and toward the middle of the thirteenth century did the artists attempt to hammer sheets of copper on a larger scale to create effigies or head reliquaries whose dimensions in both cases approximated life size.

This delay may also have been due to the rarity of large-scale sculpted human figures in the Limousin during the eleventh, twelfth, and thirteenth centuries; influenced by the wave of Romanesque style originating around Toulouse in the Midi, this region produced predominantly ornamental sculpture, such as the capitals carved with plant forms of the former abbey church of Saint-Martial of Limoges[3] and the archstones of the abbey of Sainte-Marie de la Règle.[4] The three masculine heads now in the Musée Municipal de l'Évêché in Limoges, probably from statues originally placed within an arcature against the facade of the city's cathedral, constitute a notable exception.[5] Similarly, funerary sculpture is very weakly represented. The museum at Limoges, however, does have an effigy of an abbot or a bishop from Saint-Augustin-lès-Limoges: the deceased, sculpted in low relief, is laid out on a bed of state,[6] the frame of which is girded with a broad rinceau. The head, with open eyes, stands out against a large pillow.[7] The monument, traditionally dated to the end of the twelfth century, has little in common with the general conception of tombs produced in the Limousin several decades later. Only a small number of stone or granite tombs of the thirteenth century attest to the limited activity of Limousin tomb-makers working in these materials. One such memorial, the effigy of a woman in the church of Saint-Laurent-les-Églises (Haute-Vienne), depicts the deceased lying with her hands folded across her breast, wearing a gown, a long mantle, and a coif;[8] the costume as well as the style evoke the second half of the thirteenth century and are thus contemporaneous with the known effigies produced in the Limousin. However, the four floral bosses, placed two by two at the woman's head and feet, may be interpreted as a reference to a bed of state, a theme not found among tombs of Limoges work.

Our knowledge of these tombs is based on three types of sources. First, there is the visual evidence of the five monuments that escaped, at least partially, the vicissitudes that caused the destruction of the rest. The oldest of these examples are the tomb of Bishop Mauricio of Burgos (d. 1238), placed in the choir of the cathedral of Burgos (list, no. 1), and the effigies of two children of Saint Louis, Blanche of France (d. 1243) and John of France (d. 1248), previously buried at Royaumont (cat. 146) but whose tomb effigies were installed in the ambulatory of the church of Saint-Denis in the nineteenth century. Two other monuments are somewhat more recent, that of William II of Valence (d. 1288) in Westminster Abbey (list, no. 16; for list, see appendix, p. 434) and that of Blanche of Champagne, countess of Brittany (d. 1285), which was taken from the abbey of la Joie (Morbihan) during the nineteenth century and is now in the Département des Sculptures of the Musée du Louvre (cat. 151). To these one may add the funerary masks of Herbert Lanier and his wife, Alès (cat. 150), the only surviving fragments of two full-length effigies produced in the Limousin near the end of the thirteenth century, from the priory of la Papillaye near Angers.

A second source of information regarding these monuments is found in the watercolor paintings of the albums of Roger Gaignières, which depict images of about ten lost tombs.[9] The third source comprises written materials, which are of two types. A few archival documents attest to commissions made in Limoges, such as the famous payment to John of Limoges for the tomb of Gauthier Merton, bishop of Rochester (list, no. 11); the receipt for payment given by the knight William Le Borgne to the heirs of John II, duke of Brittany, concerning the effigy of Blanche of Champagne, commissioned in Limoges by the duke (cat. 151); and a letter from the prior of Grandmont on the subject of the payment and transportation, from Limoges, of the tomb of Thibaut IV of Champagne (list, no. 4). Finally, a Norman text, the Chronique des comtes d'Eu (Chronicles of the counts of Eu), mentions the commission made in Limoges by John of Brienne for the execution of his tomb as well as that of his mother, destined for the Cistercian abbey of Foucarmont (list, no. 17). To this indisputable evidence we can add early references to and descriptions of tombs—collected predominantly in the nineteenth century by specialists in Limousin enamel—that occasionally permit us to identify works of Limousin manufacture.

Indeed, early in the nineteenth century, metal effigies, whether made in the Limousin or not, became of interest to scholars. In 1855 Arcisse de Caumont devoted several pages to them in the *Bulletin monumental*, in an article based essentially on his knowledge of the Gaignières watercolors; however, he also cited copper tomb plates and fourteenth century effigies, probably Parisian.[10] That same year, Maurice Ardant, in his book on Limoges enamelers, assembled most of the archival sources mentioning tombs commissioned in Limoges during the thirteenth and fourteenth centuries.[11] In 1857 Abbé Texier, writing on the topic "enameled tombs," drew up a catalogue featuring the effigy of Geoffrey

Plantagenet (cat. 15), the tomb of Aymeric Guerry (list, no. 9), and the monuments of the counts of Champagne, as well as a list of metal tombs, compiled from the Gaignières albums, that specified not only the works embellished with enamel but also monuments whose copper material was mentioned by Gaignières.[12] These juxtapositions of Limousin and non-Limousin monuments, in conjunction with the bibliographic authority of Abbé Texier on matters of enamelwork, were no doubt partially responsible for some of the overly generous attributions to Limoges in more recent literature. Viollet-le-Duc, in the "tomb" entry of his *Dictionnaire raisonné de l'architecture française* published in 1868, also elaborated on metallic tombs, but his preoccupations were formal and archaeological, and he did not mention the Limousin origins of certain monuments that he discussed.[13] In 1890 Rupin published the most complete manuscript then written on tombs of Limoges work, relying heavily on monographs devoted to individual monuments.[14] Marquet de Vasselot, in 1898, also cited most of the Limousin tombs in the course of his work on the funerary masks in the Musée d'Angers and the Musée du Louvre.[15]

More recently, during her study of the enamels of Grandmont, published in 1963, Souchal became interested in sepulchres of Limoges work, evidence, she considered, of the precocious aptitude of Limousin enamelers for works in relief, which led her to early datings for funerary sculpture.[16] And Gauthier, working toward the publication of the *Corpus des émaux méridionaux* (forthcoming), assembled the records of Limousin copper tombs and organized the material into lists, for which the present author is greatly beholden. The current lists, however, cannot be regarded as exhaustive, and expanding them requires the analysis of a great number of old sources, which can be done only with the utmost care, under the constraints of working in geographically isolated areas and with tremendous vigilance in the reading of the documents. The following information must therefore be considered provisional, open to further development as well as critical rereadings.

It is not always easy to distinguish tombs of Limousin production from other tombs in copper and in relief, even when there is graphic evidence. If one trusts the Gaignières albums, monuments of Limousin manufacture were usually placed on a stone block; only the effigy of William Roland (cat. 147) has a tomb plaque that sits on four cylindrical feet. Most often, the sides of the wood chest placed on the stone base were not heavily decorated; the tombs of laypersons, however, were frequently embellished with coats of arms, easily translated into enamel, the number of which varied. The funerary monument of Alix of Thouars and Yolanda of Brittany (cat. 149), which bore a considerable quantity of such decoration, was also demarcated with lions, set on the corners of the chest. Following this example, one might suggest a similar placement for the copper lions that, according to Pardoux de la Garde, surrounded the sepulchre of Aymeric Guerry (list, no. 9), and for the lions curiously positioned, during Gaignières's time, at the back of the tomb of Bishop Michael of Villoiseau (cat. 148). At least two tombs received distinctive ornamentation on their sides: the one of John of Brienne and his mother (list, no. 17), with its images of Christ, the Virgin, Saint Denis, and Saint Paul near the figures' heads; and the tomb of Cardinal Peter of la Chapelle-Taillefer (list, no. 19), the coffret of which was surrounded with copper appliqué figures, placed within an arcade to create a funerary procession.

Seen from above, Limousin tombs all presented a large rectangular enameled border of ornamental foliage, punctuated either with armorial bearings, on lay sepulchres, or with cabochons or small inscribed figures in quatrefoils, on ecclesiastical monuments. Candles were sometimes placed along this border; from the description by Beaumesnil of the tomb of the cardinal Peter of la Chapelle-Taillefer, we can specify that these candles could be detached. The epitaph was usually inscribed in enameled letters on the inside bevel of the frame. The texts that we know were drawn up in Latin verse, abundantly praising the virtues of the deceased, the sole exception being the sober inscription in French that once accompanied the effigy of Blanche of Champagne.

Constructed of several pieces of copper with abutted or slightly overlapping edges, the effigies were affixed to a wood core by nails. The metal figures, worked in relief and repoussé, were gilded; enamel was used only locally, generally by means of small plaques attached to the clothing, as can be seen on the tomb of William II of Valence (list, no. 16), where just the belt, the baldric, and two small escutcheons fastened to the shoulders were enameled; filigree decorated the sleeves and the fillet of the knight.

The effigies were laid on flat beds recessed below the borders. When this ground was enameled—and it assuredly was not on the tomb of Juhel III of Mayenne (list, no. 8)—it received a decoration of ornamental foliage or heraldic motifs. On the most highly finished tombs, such as those of the children of Saint Louis (cat. 146) or that of Philip of Dreux, bishop of Beauvois (list, no. 10), appliqué figures—angels or monks reciting prayers—surrounded the reclining likenesses. The effigy itself customarily rested its feet on a support, an animal or a simple console; several tombs, however, such as those of William Roland (cat. 147) and Claude Renaud (list, no. 12), lacked this element. The gestures of the figures varied: hands joined in prayer, hands crossed on the breast, hands holding attributes. The heads, excluding those of Philip of Dreux (list, no. 10) and the children of Saint Louis (cat. 146), rested on a pillow enameled with floral or heraldic motifs. Most of the illustrations of the Gaignières albums depict the effigies with closed eyes. But the accidents of conservation are such that among the seven faces of extant Limousin figures, four have their eyes open—the children of

Saint Louis, Blanche of Champagne, and William of Valence—while Mauricio of Burgos and the Lanier couple have their lids closed. Although its theological significance remains unclear, the iconography of funerary sculptures with closed eyes, witnessed as early as the beginning of the thirteenth century, was no doubt intended to represent the sleep of death during the wait for the Last Judgment. The occurrence of both customs, open and closed eyes, on Limousin monuments suggests that the commissioner of the tomb could indicate his preference to the Limousin tombmakers.

These characteristics distinguish the effigies of Limousin work from those of bronze or cast-copper alloys created in other artistic centers. In northern France during the thirteenth century, effigies in copper were generally presented under a trefoil arch, often treated as an architectural motif, following an iconography that derived ultimately from antiquity.[17] Such an arch never appears on tombs of Limousin manufacture, and, accordingly, a certain number of tombs have been excluded from the following corpus of Limousin tombs.

The oldest known effigy produced in the Limousin seems to date to the second quarter or mid-thirteenth century, as the tombs of Bishop Mauricio of Burgos (d. 1238, list, no. 1) and those of the children of Saint Louis (cat. 146) still bear witness. The funerary monuments of Blanche and John of France, who died in 1243 and 1248, respectively, are comparable stylistically to the groups of appliqué figures in the Musée de Cluny, Paris; Baltimore; and Minneapolis (cat. 119), for which Karel Otavsky has proposed a dating contemporary with the Sainte-Chapelle (ca. 1242–48).[18] Among the lost monuments, the old textual descriptions and the illustrations in the Gaignières albums suggest that although the subjects' dates of death preceded the mid-thirteenth century—as was the case for Gerald IV, bishop of Cahors (d. 1199, list, no. 2), for Philip of Dreux (d. 1217, list, no. 10), and for the knight Juhel III of Mayenne (d. 1220, list, no. 8)—it was not unusual for delays to push back the execution of the tombs to the middle or third quarter of the thirteenth century.

The second half of the thirteenth century saw the greatest success for tombs of Limoges work, with about fifteen documented monuments dispersed geographically from England to Navarre. The use of enamels allowed the legible and indelible reproduction of armorial bearings at a time when heraldry was spreading. This no doubt accounts for some of the popularity of Limousin enamelers even as the quality of heraldic motifs tended to diminish with the execution of rather repetitive works such as gemellions or candlesticks, which were also suitable for armorial decoration. During the fourteenth century the execution of tombs of Limoges work dissipated and then disappeared altogether; it is in the Limousin itself that one encounters the last enameled funerary monument, that of Hugh Roger (d. 1363) in Saint-

Germain-les-Belles (list, no. 24).

Obviously costly works, reserved for the elite, the tombs made of Limoges work were often created to shelter the remains of bishops, including Gerald IV, bishop of Cahors, and Ayimeric Guerry, archbishop of Lyons, at Grandmont (list, nos. 2 and 9); Philip of Dreux, bishop of Beauvais (list, no. 10); and Michael of Villoiseau, bishop of Angers, in the Dominican monastery of that city (cat. 148). They were particularly favored by the ducal family of Brittany, which erected them at the abbey of Villeneuve, at the Franciscan monastery in Nantes, at the abbey of la Joie near Vannes, and at the Fontevrault priory of Belhomer. This frequently noted phenomenon reflected the converging alliances of the Bretons.[19] Peter Mauclerc was a member of the Dreux family, which manifested a predilection for metal tombs as far back as the early thirteenth century.[20] Peter's son John I, duke of Brittany, married Blanche of Champagne, whose father, Thibaut IV, was buried in an enameled mausoleum commissioned in Limoges (list, no. 4), while her great-grandfather and grandfather, the counts of Champagne, Henry I the Generous and Thibaut III, lay in Saint-Étienne de Troyes under monuments of gilt silver embellished with enamels. The sepulchres of the children of Saint Louis may have also inspired the choice of a Limousin tomb for Robert of Brittany, a son of John I and Blanche of Champagne, who died young in 1259 (list, no. 5). The family tradition was again reinforced by the marriage, in 1274, of Arthur of Brittany, son of John II, to Marie of Limoges, heir of the viscount of Limoges. Furthermore, additional tombs continued to be commissioned from Limoges after this date by the ducal family and their relatives, including monuments in the abbeys of Villeneuve, Belhomer, and la Joie.

Among the motives that might have led certain powerful figures to commission Limousin tombs, or that might explain why they benefited from them after their deaths, is the existence of ties to the order of Grandmont, as some scholars have suggested. Indeed, the list of founders of Grandmont priories includes Bishop Philip of Dreux (list, no. 10) and Juhel III of Mayenne (list, no. 8).[21] Many of those who established such priorates, however, were not interred under monuments produced in the Limousin, notably the count of Champagne, Henry I the Generous. Other connections with the Limousin, long forgotten, must also have played a role in their selection.

Generally speaking, these expensive tombs—the sum of 450 livres paid to William le Borgne for the tomb of Blanche of Champagne seems to have been the total cost of the work—were intended to honor important figures and were situated in preferential positions in the churches, most often in the choir. A substantial number of the tombs of Limoges work honored the founders or benefactors of abbeys (Grandmont, Fontaine-Daniel, Evron, Villeneuve, Foucarmont, la Joie), collegiate churches (la Chapelle-Taillefert, Saint-Germain-les-Belles), monasteries (the Dominican order at Angers and Rieux), or priories (la Papillaye in Anjou). The mapping of known Limousin tombs reveals an area of diffusion concentrated essentially in the western half of France, which no doubt made it easy to dispatch monuments from Limoges and perhaps allowed enamelers to travel as well. Whatever the case, transportation of these works did not present the same difficulties as did the movement of solid or stone works, since the different plates of ornamentation and the effigy itself could be sent disassembled, thus reducing the burden and allowing for its division into a number of packages or crates.

BÉATRICE DE CHANCEL-BARDELOT

NOTES

1. See Gauthier 1987, pp. 106–10, no. 103.
2. See Gauthier 1992, p. 97.
3. Marcheix and Perrier 1969, pp. 65–71, nos. 142–57.
4. Ibid., pp. 78–83, nos. 162–73.
5. Ibid., pp. 72–77, nos. 158–60 (fragments of splay statues of a Roman portal); I owe the suggestion of the decoration of the arcature to Jean-René Gaborit.
6. On this subject, see Gaborit 1988, who proposes a dating to the third quarter of the twelfth century.
7. Bauch 1976, p. 37 and n. 108, with a dating to the third quarter of the twelfth century.
8. Laborderie 1946, pp. 86, 88.
9. Adhémar 1974; for reproductions in color of the Gaignières plates regarding most of the Limousin tombs, see Vaivre 1988a.
10. De Caumont 1855.
11. Ardant 1855.
12. Texier 1857.
13. Viollet-le-Duc 1868.
14. Rupin 1890, pp. 158–64.
15. Marquet de Vasselot 1898.
16. Souchal 1963, pp. 322–29.
17. Panofsky 1964, pp. 52, 55.
18. Otavsky 1973.
19. Copy 1986, p. 39; Copy 1993.
20. See, for comparison, the tombs of Robert II, count of Dreux, and Peter Mauclerc at Saint-Yved de Braine.
21. Gaborit, manuscript notes.

# 146. Tomb Effigies of John and Blanche of France

Limousin, ca. 1250

Copper: engraved and gilt; appliqués: repoussé, engraved, and gilt; champlevé enamel: black, medium blue, turquoise, green, yellow, red, and white; turquoise enameled beads; wood core

H.: 105.5 cm (41½ in.; John: 104 cm [41 in.], Blanche: 105.5 cm [41½ in.]); total width: 126.5 cm (49¾ in.; John: 53.5 cm [21⅛ in.], Blanche: 55 cm [21¹¹⁄₁₆ in.]); total depth: 13 cm (5⅛ in.; John: 6.5 cm [2⁹⁄₁₆ in.], Blanche: 5.5 cm [2³⁄₁₆ in.])

INSCRIBED: HIC JACET : JOHANNES : EXCELLENTISSIMI . LVD[OVICI REGIS FRANCORVM FILIVS QVI IN ETATE INFANCIE MIGRA]VIT AD XPM ANNO GRACIE : MILLESIMO : DVCENTESIMO : QVADRAGESIMO : SEPTIMO : SEXTO : IDVS : MARCII (Here lies John, son of the most excellent Louis, king of France, who, as a child, rejoined Christ in the year of the Lord 1247, the sixth of the ides of March [March 10, 1247]")

[HIC JACET BLANCA EXCELLENTISSIMI LVDOVICI REGIS FRAN]CORVM PRIMOGENITA QVE IN ETATE INFAN[CIE MIGRAVIT AD XPM ANNO GRACIE MILLESIMO DVCENTESIMO QUADRAGESIMO TERCIO, III KAL. MAII] (Here lies Blanche, oldest daughter of the most excellent Louis, king of France, who, as a child, rejoined Christ in the year of the Lord 1243, the third of the calends of May [April 29, 1243])

COATS OF ARMS: John: France, Castile, Provence; Blanche: Provence, France

PROVENANCE: Abbey of Royaumont (until 1791); the abbey church of Saint-Denis (1791–94); Musée des Monuments Français, then the abbey of Saint-Denis (ca. 1820)

CONDITION: John of France: The lion's tail has been remade; the plaque to the left of the prince's head has been partially reenameled(?); part of the inscription redone (between brackets in the above transcription); border remade.

Blanche of France: Head, greyhound, and background plaques framing the head and feet remade; major part of the inscription (between brackets in the above transcription) and border remade.

Abbey church, Saint-Denis

These two Limousin tombs of John of France (d. 1248) and of his sister Blanche of France (d. 1243), now joined, were similarly executed. The borders are made up of enameled plaques with blue grounds, decorated with vegetal rinceau patterns. Once, small circular plaques alternated with these and contained either busts of angels or armorial escutcheons; today, only a few of the escutcheons remain. Then came the epitaphs, now partially restored, executed in red letters between two blue enameled fillets. This border is situated on the same plane as the background plaques (six for the effigy of John, because of the small plaque between his ankles, and five for that of Blanche) and is ornamented with rinceaux terminating in floral buds. Set off against this ground are the effigies of a prince and a princess: one copper plaque in very high relief for the body and another plaque for the head. The prince—standing on a walking lion that gracefully raises its head and its enameled eyes toward

him—holds in his left hand a scepter topped with a fleur-de-lis and salutes with his right hand. His chubby face is framed by longish locks, molded like a wig, upon which rests a diadem decorated with turquoise enameled beads. His slightly staring eyes are also enameled. Like her brother, the princess is dressed in a supple tunic, under a robe with a lozenge pattern containing fleurs-de-lis and Castilian castles. She holds a fold of her robe in her left hand and a bowl in her right. On the background plaques surrounding the effigies are the silhouettes of four monks who are reading and two angels bearing censers.

In its original state, known through the Gaignières albums,[1] each of these effigies was placed atop a stone sarcophagus decorated with ornamental arcades, and enclosed within a tomb recess containing heraldic paintings on the lateral surfaces and the vault, while on the back there was a full-length painted portrait of the royal children. At the dawn of the eighteenth century, only a few areas of the metal ornamentation were damaged: the plaque situated to the left of John of France's head and the corner corresponding to the border had suffered losses; the effigy of Blanche of France was missing its head, but its feet still rested on a dragon. Some of the armorial escutcheons from the area around the princess's tomb had disappeared. In 1760, the tomb recess situated below the grand arcades surrounding the choir of the abbey church of Royaumont was destroyed to allow for the installation of a choir screen, resulting in further damage to the enameled decorations.[2] In August 1791, the tombs in Royaumont Abbey were opened in the presence of Dom Germain Poirier, and the decoration that was deemed worthy was transported to the abbey church of Saint-Denis;[3] the tomb plaques of John and Blanche of France left their second resting place in March 1794 [germinal year III] to become part of the Musée des Monuments Français—founded by Alexandre Lenoir in the Petits-Augustins—whence they were transferred to Saint-Denis during the Restoration.[4] They were repaired in the third quarter of the nineteenth century by Viollet-le-Duc and the enameler Le Gost, and reinstalled in two imitation thirteenth-century-style tomb recesses, the ceremony commemorated in photographs.[5] Re-created at that time were the missing medallions with angels, armorial escutcheons, plaques for the border epitaphs, the head of Blanche of France, and appliqué figures of monks and angels surrounding the royal children, and a greyhound was placed at Blanche's feet (instead of the original dragon). The original support—which was on two levels, one for the border and the other, recessed, for the background, and both joined by the bevel of the inscription—was abandoned in favor of a

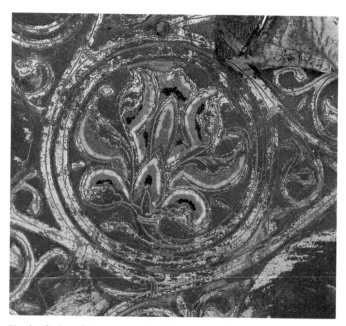

Tomb of John of France, enameled fleuron (detail)

flat support. The entire operation was facilitated by a total replacement of the border plaques and some of those containing the epitaphs. However, since John's funerary inscription was more complete than Blanche's, it was necessary to add copper inserts in two of the corners to compensate for the gaps created by the flattening of the support. A molding engraved with heraldic decorations was added to the outer edge of each of the funerary plaques. In 1955, the Monuments Historiques decided to remove the nineteenth-century restorations.[6] The Toulouse restoration workshop was assigned the task of joining the two plaques, following an arrangement perhaps inspired by the drawings in the Gaignières albums of the tomb of Alix of Thouars and Yolanda of Brittany, formerly in the abbey of Villeneuve (cat. 149), or by the tomb of Louis and Philippe, sons of the count of Alençon, formerly in Royaumont and now in the Musée de Cluny.[7] The appliqué figures, the medallions, and the escutcheons restored under the supervision of Viollet-le-Duc were detached from the tomb. The removal of the tomb recesses since then has allowed for a better view of the enameled plaques and of the effigies.

In spite of the vicissitudes they suffered, these two works are precious evidence in more than one respect: they can be dated approximately, and the fact that they are identical in concept suggests that they were commissioned simultaneously, after the death of the young prince (1248), and probably before the Treaty of Paris, the signing of which, in 1258, resulted in a new alliance between the Limousin and England. The plaques thus serve as an important chronological milestone in the production of Limoges enamels, and in particular of objects in relief. The workmanship of the head of the young prince, with his very round face and his wiglike hair, has been compared to the group of appliqué Passion figures (cat. 119). The wide eyes, with their enameled corneas and pupils, are rarely seen on Limousin statuettes; a few later works, such as the Virgin in Breuil-au-Fa or the one in La Sauvetat (cat. 156), provide other related examples. The large enameled fleurons on the background plaques, clearly more complex than those surrounding the apostles of Grandmont (cat. 58), are comparable to the fleurons on the Cherves tabernacle (cat. 98). They are accompanied by scrolling vines terminating in copper trefoil leaves—a stylistic innovation that would be developed, to the detriment of the enameled fleurons, on later works.

The two tomb plaques are among the earliest Limousin tombs, which is confirmed by the frontality and the slightly awkward mass of the silhouettes of the two royal children. They belong, as well, to the earliest known group of effigies of children—those of the sons and daughters of Saint Louis, who died at a young age and were buried at Royaumont—and they postdate by only a few years the tomb effigy of Philip Dagobert,[8] the brother of Saint Louis, who died in 1233 or 1234. Most often, the material used for these tombs was stone; the recourse to Limousin production seems to have been unique to the royal family and its entourage. It is true that the passing of the Limousin from French to English rule, in 1258, might have deterred the princely clientele from using the goldsmiths and enamelers of the region, thus benefiting the founders of bronze tombs and the copperplate engravers active in northern France. However, the commissioning of metal tombs was not unusual for the royal family: the tomb of Philip Augustus, dating to before 1264—which has been lost since the fifteenth century—was of silver,[9] as were those of Louis VIII and of Saint Louis, executed later in the second half of the thirteenth century, while several princely sepulchres were decorated with engraved-copper slabs. Also worthy of mention here is the tomb of the two children of Louis VIII, Alphonse and John of France, who were buried in the collegiate church of Saint-Louis in Poissy.[10] The representations of the two Prince Johns, both of whom hold scepters, are so close that it led de Montfaucon to identify the subjects as the same person.[11] The presence of this attribute in the hands of children who were not reigning monarchs may be interpreted as a sign that they were members of the royal family, following the example on the tombs of queens, who were also frequently endowed with this attribute.[12] Its inclusion reinforces the pomp of these funerary representations, enhanced here by the brilliance of the metal and the glitter of the enamels.

BC–B

NOTES

1. Paris, BNF, Département des Estampes, Pe 1a, fols. 25, 26, 29, 30; see Adhémar 1974, nos. 189, 190.
2. Pessiot 1994, p. 31.
3. Lauer 1908, p. 252; Vitry and Brière 1925, pp. 106–7.
4. Courajod 1878, pp. 77–78, no. 532, p. 79, no. 551, p. 181, no. 21.
5. Vitry and Brière 1925, p. 138. Photograph nos. 86892 and 86893, Direction du Patrimoine, Bureau des Objets Mobiliers, carton Saint-Denis, tombeaux.
6. Direction du Patrimoine, Bureau des Objets Mobiliers, dossier of the abbey church of Saint-Denis.
7. Pessiot 1994, pp. 29, 31, 32, 35, figs. 2, 4, 5, 8.
8. Sauerländer 1972, pp. 139–40; Erlande-Brandenburg 1975, pp. 164–65.
9. Erlande-Brandenburg 1975, p. 162.
10. Adhémar 1974, no. 258; Erlande-Brandenburg 1975, pp. 92–93, and n. 65.
11. Bernard de Montfaucon, vol.2 (1730), pp. 160–62.
12. Erlande-Brandenburg 1975, pp. 122–23.

EXHIBITIONS: Paris 1968b, no. 407; Royaumont 1970, nos. 136, 136 bis; Paris 1970, no. 32.

BIBLIOGRAPHY: Bernard de Montfaucon, vol. 2 (1730), pp. 160–62 and pls. XXVII, no. 4, and XXVIII; Millin 1791, pl. XI and pp. 12–13 of the entry on Royaumont; Willemin 1839, vol. 1, pp. 58, 59, pls. 91–93; Guilhermy 1848, pp. 164–67; Viollet-le-Duc 1868, p. 62; Viollet-le-Duc 1871–72, p. 222; Guilhermy 1875, pp. 156–58; Rupin 1890, pp. 159–60 and fig. 239; Lauer 1908, pp. 257–60; Vitry and Brière 1925, pp. 137–38, nos. 21, 22; Rückert 1959; Erlande-Brandenburg 1975, pp. 117, 119, 123; Bauch 1976, pp. 74–75; Gauthier 1977; Gauthier 1988, p. 107.

Lion at the feet of John of France (detail)

# 147. Tomb of William Roland (watercolor)

Limoges, third quarter of the 13th century

Copper: engraved and gilt; champlevé enamel: red and blue; cabochons(?); beads; wood core

INSCRIBED: MORES MAGNORVM SVPERANS ET DOCTOR EORVM, / TVTOR CVNCTORVM CLERI, PLEBIS, MONARCHORVM, / LVX PREDICTORVM DEBELLATORQVE MALORVM,[1] / JVSTVS, DEVOTVS, GRATVS,[2] SAPIENS, MODERATVS, / PRESVL GVILLELMVS ROLAND JACET HIC TVMVLATVS. / HORVM SVNT TESTES[3] PARCVS[4] CIBVS, ASPERA VESTIS, / COPIA MVLTARVM QVAS FVNDEBAT LACRYMARVM / DVM FLENS MISSARVM TRACTARET SANCTA SACRARVM. / CELLA PHISIS, LEGIS ARCHARVM CANONIS ARCHA, / LVX ET FORMA GREGIS, PATER ET PATRIE PATRIARCHA, / TERROR ERAT PRAVIS, JVSTIS APPLAVSIO SVAVIS.[5] / ISTI CELORVM DET GAVDIA REX SVPERORVM / CHRISTE SERVORVM MERITIS PRECIBVSQVE SVORVM (Surpassing the morals of the great and teaching them, / Protector of all, clerks, the common man, monks, / Their light and the enemy of the wicked, / Just, pious, grateful, wise, temperate, / The bishop William Roland lies buried here. / Attesting to this are his scant nourishment, his poor clothing, / The abundance of tears he shed / While he celebrated the holy mysteries sacred to the Mass. / Nature's sanctuary, coffer of the coffer of the right canon, / Light and model for the flock, father and patriarch of his country, / He was the terror of the wicked, the sweet approbation of the just. May the Lord give him the joys of the heavens, / O Christ, thanks to the merits of his followers and of his own prayers.)

PROVENANCE: Disappeared in the eighteenth century; known from a watercolor executed for Roger de Gaignières in the seventeenth century. Once in the middle of the sanctuary of the Cistercian abbey church of Notre-Dame of Champagne (Sarthe).

Bibliothèque Nationale de France, Paris (ms. lat. 17036, fol. 173)

The enameled copper tomb of the bishop of Le Mans William Roland (d. 1260) is raised above the ground and rests on six cylindrical feet. A large frame made up of copper plates enameled in blue and with gilded rinceau decoration surrounds the effigy. In the corners are square copper plaques containing four large beads. On the border, copper quatrefoils decorated with small, silhouetted mitered figures alternate with large cabochons—or, rather, red and blue enameled ornaments, an oval standing out from the elongated polylobed form below. Within this frame, a red enameled rim outlines the beveled border, which bears the epitaph. The ground of the compartment inside the frame in which the effigy rests is shaded in blue, but instead of suggesting an enameled background, the intention seems to have been to give the impression of a uniform copper plaque that is softly lit. The prelate is shown stretched out, his head resting on a cushion enameled with a red checkered pattern incorporating blue flowers; his eyes, which seem to be open—according to the watercolor in the Department of Manuscripts in the Bibliothèque Nationale—are depicted as closed in the copy of the drawing now in Oxford. In his left hand he clutches a book, while his right hand, folded across his chest, would probably have held the staff of a then-missing crosier. His feet do not rest on anything, but the relationship between the size of the effigy and the length of the compartment in which he rests precludes the inclusion, originally, of an animal or a decorative support.

Like many Limousin tombs, William Roland's was situated in the choir of a church, specifically the abbey church in Champagne, where it was placed before the main altar—a fact provided by Le Corvaisier de Courteilles;[6] Chappée rediscovered the site during the excavations undertaken in 1895–96, but he did not find any furnishings inside the sepulchre.[7]

This bishop of Le Mans inspired no biographies. Le Corvaisier de Courteilles identified him as the son of Hector Rolandi, financial adviser to the count of Le Mans; he was canon, then cantor, of the cathedral, and bishop of Le Mans from 1256. According to Gaignières, he traveled to Rome and died shortly after his return to Le Mans.[8] However, Le Corvaisier de Courteilles wrote that he fell ill in Genoa, and Chappée believed that he died there. The date of Roland's death is not certain and according to scholars occurred between 1258 and 1261. While Chappée dated the consecration of the abbey church in Champagne to 1269, citing this same year as the date of the transfer of the body of William Roland, Coutard proposed 1260 as the date, thus overlooking the supposed delay between the death of the prelate and his burial. Whatever the case, the appearance of the tomb accords with a date in the third quarter of the thirteenth century: the alb, the checked tunic, the chasuble decorated with floral motifs within circles, the miter, and the cushion are very similar to those on the tomb of Michael of Villoiseau, bishop of Angers, who died in 1260 (see cat. 148); the long mustaches and the short beards are seen on tomb effigies from the middle or the third quarter of the thirteenth century, as well as on the reliquary head in the Kunstgewerbemuseum, Berlin, dated to the mid-thirteenth century.[9]

Although a Limousin tomb effigy of a bishop is not surprising in the province of Maine, where at least three other tombs with enameled decoration have been confirmed to exist—including those of Geoffrey Plantagenet and the knights Juhel de Mayenne and Claude Renaud—the little that we know about the personality of William Roland gives no indication of a relationship with Limoges or an explanation of the choice of a Limousin production technique for his sepulchre.

BC-B

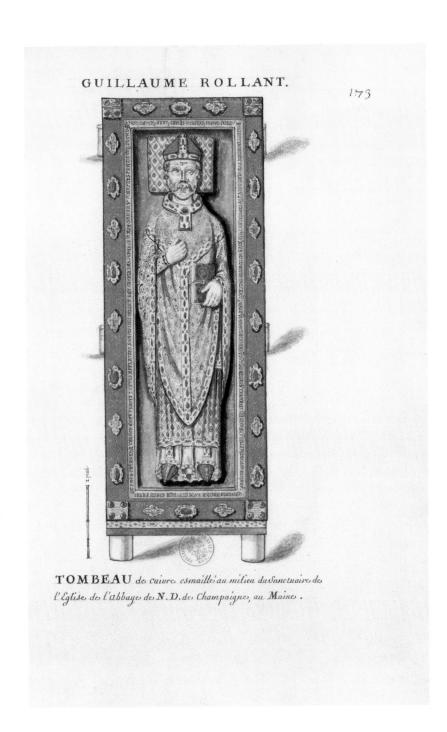

TOMBEAU *de cuivre, esmaillé au milieu du sanctuaire de l'Eglise de l'abbaye de N.D. de Champaigne, au Maine.*

NOTES

1. The first three lines of the epitaph are given by Le Corvaisier de Courteilles, by the *Gallia christiana*, and by Chappée, but do not appear on the summary of the inscription in the Gaignières albums.

2. Word given in the Gaignières album and by Chappée; the *Gallia christiana* gives *castus*.

3. The *Gallia christiana* transcribes: *hujus erat testis*; Gaignières gives *Horum sunt testis*, corrected by Chappée, whom we follow with *testes*.

4. Chappée gives *parcies*.

5. These three lines appear only in the transcription of the Gaignières album.

6. Le Corvaisier de Courteilles 1648, p. 519.

7. Chappée 1897, p. 369.

8. Paris, BNF, ms. lat. 17036, fol. 171.

9. See Rückert 1959; Boehm 1990, pp. 191–96.

BIBLIOGRAPHY: *Gallia christiana*, vol. 14, col. 401; Le Corvaisier de Courteilles 1648, pp. 516–20; Caumont 1855; Hucher 1856, pp. 206–7; Texier 1857, col. 1404; Chappée 1897, pp. 352–53; Coutard 1897, pp. 187–91; Cordonnier-Dètrie 1971, pp. 9–40; Adhémar 1974, no. 267 bis; Vaivre 1988b, pp. 60, 64.

GRAPHIC SOURCES: Gaignières albums, Oxford, Bodleian Library, ms. Gough-Gagnières 14 (18359); copy in Paris, BNF, Département des Estampes, Pe 1g, fol. 219; Adhémar 1974, no. 267; Paris, BNF, Département des Manuscrits, ms. lat. 17036, fol. 173; Adhémar 1974, no. 267 bis.

# 148. Tomb of Michael of Villoiseau (watercolor)

Limoges, third quarter of the 13th century

Copper: gilt; champlevé enamel: blue, green, and red; cabochons; beads; wood core

INSCRIBED: VILLA CREAVIT AVIS MICHAELEM SIC TVMVLATVM / ET DEDIT ANDEGAVENSIS DOMINVS SIBI PONTIFICATVM. / DOCTRINA CLARVS, CVNCTIS DVLCEDINE CHARVS, / ECCLESIAE TVTOR, PRIMVM PROBITATE SEQVTOR. / SVMME DEVS COELI, QVI LAESVS CVSPIDE TELI / CLAMASTI TER HELI, VENIAM DES HVIC MICHAELI. / FVNVS AVIS VILLAE MICHAELIS SIC CAPE : MILLE ANNIS / MISCE BIS CENTVM TRIGINTAQVE BIS. (The Lord created Michael Villoiseau, buried here, / And gave him the bishopric of Angers, / Famous for his doctrine, dear to all for his mild manner, / Protector of the church, and above all its guardian through his honesty. / God on high, who, wounded by the point of a spear, three times called "Eli," show mercy to this Michael: / remember the day of the death of Michael of Villoiseau in the year one thousand / two hundred and thirty two.)

PROVENANCE: This work, sold and dismantled in 1725, is known from a watercolor executed for Roger de Gaignières in the seventeenth century. Formerly in the choir of the church of the Dominican convent of Angers (Maine-et-Loire), protected by a wood case.

Bibliothèque Nationale de France, Paris (ms. lat. 17030, fol. 103)

This watercolor of the tomb of Michael of Villoiseau, bishop of Angers (r. 1240–60), presents an abstracted view of the border framing the bevel and the compartment in which the effigy was placed—a border known only from a drawing by the seventeenth-century Angevin scholar Jacques Bruneau de Tartifume: twelve candlesticks alternated with other elements rendered simply as ovals, but the exact location of the epitaph is not indicated either in the documentation in the Gaignières album or in Bruneau de Tartifume's sketch. The Gaignières album details the ornamentation of the bevel, "enriched by many damascenes with the figures of several bishops and religious"[1] depicted within arcades. In the background, decorated with a checkered pattern incorporating floral motifs, are three lions and an angel, as well as the effigy of Michael of Villoiseau; his head rests on a cushion, his eyes are closed, and he holds a book and his crosier. The tip of the staff of the crosier pierces the neck of the dragon at the prelate's feet.

Many of the features of this monument set it apart from other Limousin tombs: the inclusion of the engraved figures on the bevel is unusual, and the surface that supports the effigy generally is covered either with plain copper plaques (see the tomb of Juhel de Mayenne, Appendix, list, no. 8), or with plaques decorated with enameled rinceau patterns (see the tombs of John of France, cat. 146, and of Robert of Brittany, Appendix, list, no. 5), or with plaques enameled with heraldic imagery (see the tomb of Yolanda of Brittany, cat. 149); here, the plaque with simple engraved motifs is shown in a highly unusual rose color. Only the drawing in the Gaignières album attests to the presence of enameling on the cushion that supports the head of the bishop and on his vestments, which is very similar to the decoration on the cushion of the tomb effigy of the bishop of Le Mans William Roland (d. 1260; see cat. 147) and on his dalmatic and his chasuble.

In spite of its peculiarities, it is very likely that the tomb of Michael of Villoiseau might be included in the list of monuments of Limousin production. The ties between this prelate and a Grandmontine of La Haye-aux-Bonshommes (Maine-et-Loire, commune of Avrillé), emphasized by Port,[2] reinforces the probability of a commission executed in Limoges. The inclusion of copper lions, pointed out by Bruneau de Tartifume and depicted by the artist of the Gaignières album, also represents a point in common with the tomb of an archbishop of Lyons, Aymeric Guerry (d. 1257), who was buried at Grandmont itself in a monument of Limoges work (Appendix, list, no. 9). However, Michael of Villoiseau, a prelate who jealously guarded his episcopal privileges relative to the abbeys and collegiate churches in his diocese, preferred his tomb to be in the Dominican convent church in Angers—whose founding he had supported, and which was situated in close proximity to the cathedral—rather than in La Haye-aux-Bonshommes.

Like many tombs of this sort, that of Michael of Villoiseau underwent successive modifications. Bruneau de Tartifume observed it in the choir of the Dominican church in Angers, where it was usually covered with a wood case; he noted that several of the precious stones "as large as pigeons' eggs," and the beads decorating the vestments had been removed, as were the copper lions in each of the four corners. The tomb, originally slightly raised off the ground, was, in 1723, lower than the level of the new flagstone pavement;[3] then, suddenly, in 1725 the copper from the tomb was sold to a founder, and the monument was replaced by a marble slab at floor level, engraved with a new epitaph. According to a handwritten note in the margin of a collection by Claude Pocquet de Livonnière (1684–1762), the Angevin scholar kept the head and the crosier of the effigy in his study.[4] While these two relics have never been found, the excavations of 1892, on the other hand, recovered the ring, chalice, and funerary paten of the prelate, as well as his copper crosier. Now preserved along with the rest of the findings in the Musée des Beaux-Arts in Angers, these objects attest to Michael of Villoiseau's preference for Limousin workmanship.

BC-B

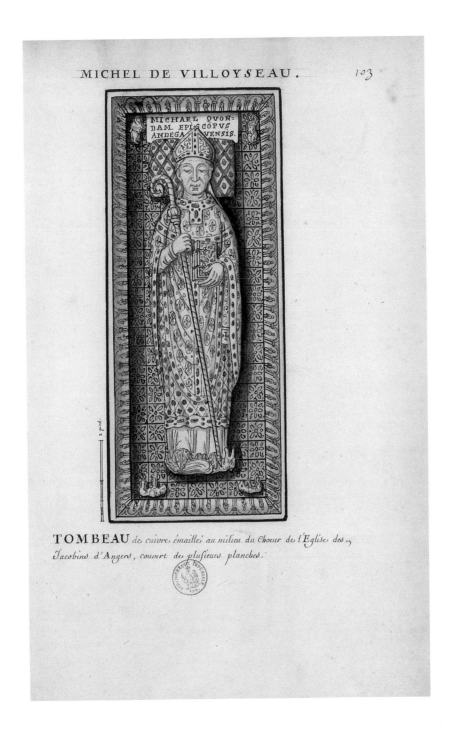

TOMBEAU *de cuivre émaillé au milieu du Choeur de l'Eglise des Jacobins d'Angers, couvert de plusieurs planches.*

NOTES

1. Bruneau de Tartifume 1623 and 1977, p. 93.

2. Port 1874–78, vol. 3 (Villoiseau).

3. Farcy 1877, p. 26.

4. Angers, Bibliothèque Municipale, ms. 691 (624), fol. 704.

BIBLIOGRAPHY: Bruneau de Tartifume 1623 and 1977, vol. I, pp. 91–94; Péan de la Tuilerie 1776, pp. 90–92; Caumont 1855; Texier 1857, col. 1404; Farcy 1877, pp. 24–26; Rupin 1890, p. 161; Farcy 1892; Thorode 1897, p. 332; Farcy 1905; Marquet de Vasselot 1941, p. 73; Lévesque 1961, pp. 24–26; Vaivre 1988a, pp. 61, 64.

GRAPHIC SOURCES: Drawing of Bruneau de Tartifume (1977 ed., pp. 92–94) and albums of Gaignières, Paris, BNF, ms. lat. 17030, fol. 103 (Bouchot 1891, no. 6661; Adhémar 1974, no. 266 bis); Oxford, Bodleian Library, ms. Gough-Gaignières 14 (18359) and copy in Paris, BNF, Est. Pe Ig, fol. 192 (see Bouchot 1891, no. 2721; Guibert m.d., pl. I-99, and Adhémar 1974, no. 266).

# 149. Tomb of Alix of Thouars and of Yolanda of Brittany (watercolors)

Limoges, ca. 1250 and 1272–80

INSCRIBED: *Epitaph of Alix of Thouars:*

PRAESENTI TVMBAE PAR SIMPLICITATE COLVMBAE / CORPORE SVBMISSA, HALIZ, BRITONVM COMITISSA. / INTER OPES HVMILIS ITA VIXIT, QVOD SIBI VILIS / MVNDVS ERAT PRIDEM, LICET ARRIDIRET EIDEM. / TANDEM FINITA FELICI FVNERE VITA / FRATRES HVJVS ALIT CONVENTVS NOBILIS HALIZ. (Haliz, countess of Brittany, like a dove in her simplicity, her body placed below this tomb, lived humbly among riches, in such a way that for a long time she found the world despicable, in spite of its charms for her. Having completed her life with a happy end, the noble Haliz nurtured the brothers of this monastery.)

*Epitaph of Yolanda of Brittany, countess of La Marche:*
PETRI DE BRANA BRITONVM DVCIS, HIC IOLANA / NOBILIS EST PROLES, TEGIT HAEC QVAM FVLGIDA MOLES, / MARCHENSIS FLORE COMITATVS FLORVIT ORAE. / IN TANTO DECORE DOMINI FLAGRAVIT AMORE. / INTER OPES MODICA, SAPIENS, PIETATIS AMICA, / CONSTANS, MVNDIFICA, CVM CORPORE MENTE PVDICA, / MITIS, SECVNDA VIRTVTIBVS, ORE JOCONDA, / PER TOTVM MVNDA, PATIENS, SERMONE FACVNDA, / SANCTA MARIA DEI MATER PIA, CLARA DIEI / STELLA, SVAE DET ET PARARE LOCVM REQVIEI. / DIE DOMINICA POST FESTVM BEATI DYONISII OBIIT / DOMINA IOLENDIS COMITISSA DE MARCHIA ET ENGOLISMENSIS, ANNO DOMINI MCCLXXII. (Here lies Yolanda, daughter of Peter of Braine, duke of Brittany, covered by this splendid monument; the county of La Marche shone with the flower of her life. In great glory she burned with the love of God. Sober among riches, and wise, she combined modesty of body and soul; gentle, inclined toward virtue, agreeable, pure, patient, and eloquent in everything; may Holy Mary, pious Mother of God, clear morning star, find her a place to rest. Yolanda, countess of La Marche and Angoulême, died on the Sunday following Saint Denis in 1272.)

COATS OF ARMS:
On the Alix of Thouars side:
*1.* Or (?) a Bend Argent; *2.* Gules three Pallets Or (Argent?); *3.* Azure semy-de-lis Or *(France Ancient)*; *4.* Gules a Bend Or (Argent?); *5.* Gules a Cross clechy pometty and voided Or *(Botherel)*; *6.* Argent five Bars Azure a Label Gules *(on the engraving of the Lonineau work: Barruly Argent and Azure a Label Gules; 7.* Or (Argent?) a Dolphin Azure; *8.* Gules a Lion rampant Or *(Vitré)*; *9.* Or three Pallets Gules on a Chief of the first a Label of the second; *10.* Or three Pallets Gules; *11.* Vair *(Lohéac)*; *12.* Azure a Lion rampant Or; *13.* same as nine; *14.* Bendy Or and Gules *(Limoges or Turenne)*; *15.* Or (Argent?) a Lion rampant Gules *impaling* Or (Argent?) an Estoil of eight points dimidiated Azure; *16.* Or a Lion rampant Azure *impaling* Bendy Or and Azure; *17.* Azure a Lion rampant Or *impaling* Bendy Or and Gules; *18.* Or (Argent?) a Cross fourchy Gules; *19.* Quarterly 1 and 4 Azure a Fleur-de-lis Or 2 and 3 Argent a Fleur-de-lis Azure; *20.* Barry Argent and Azure a Fess Gules; *21.* Azure a double-headed Eagle displayed Or; *22.* Checky Or and Azure a Bordure Gules over all a Canton Ermine *(Pierre Mauclerc)*; *23.* Azure a Lion rampant Or debruised by a Bend Argent; *24.* Or (Argent?) a Chevron Gules between three Acorns (?) Sable *(on the engraving in the Lobineau work: Azure a Chevron Gules between three Pineapples [i.e., pinecones] Or (?)); 25.* Or a Lion rampant within an Orle Gules; *26.* Argent three Chevronels Gules *(Machecoul)*; *27.* Or three Pallets Gules on a Chief of the first a Label of the second; *28.* Argent three Bars dancetty Azure; *29.* same as 26 *(Machecoul)*; *30.* Or (Argent?) a Lion rampant Gules a Bordure compony Or and Sable; *31.* same as 11; *32.* Azure a Bend Argent cotised Or

*(Champagne) impaling* Or a Bend Argent; *33.* same as 22 *(Pierre Mauclerc)*; *34.* Gules two Lions passant Or *(Anjou Ancient or Pouancé)*; *35.* Azure three Lions rampant Or *(?) impaling* Bendy Or and Gules *(Limoges or Turenne)*; *36.* Or (Argent?) a Lion Gules within an Orle embattled on the inner edge Sable; *37.* same as 6 *(?)*; *38.* Or Martlets in orle a Canton Gules *(Chemillé)*

On the Yolanda of Brittany side:
*1.* Gules a Pallet between ten Billets Or; *2.* Argent three Bars Azure debruised by a Bend Argent; *3.* Gules two Lions passant Or *(Pouancé or Anjou Ancient)*; *4.* Or semy-de-lis Azure a Canton Gules *(Thouars)*; *5.* Lozengy Or and Gules; *6.* Azure semy-de-lis Or *(France Ancient)*; *7.* Gules three Lions passant guardant in pale Or *(England)*; *8.* Checky Or and Azure a Bordure Gules a Canton Ermine *(Pierre Mauclerc)*; *9.* Gyronny of six Gules and Vair; *10.* Or (Argent?) five Bars Azure; *11.* same as eight *(Pierre Mauclerc)*; *12.* Argent four Bars Azure over all three Lions rampant Gules; *13.* same as 7 *(England)*; *14.* Azure three Bars Argent debruised by a Bend Gules; *15.* Checky Or and Gules *impaling* Or (Argent?) three Bars Gules; *16.* Azure three (?) Or; *17.* Azure an Eagle displayed Or; *18.* Azure a Lion rampant Or; *19.* same as 12; *20.* Lozengy Or and Gules a Pale Or; *21.* Or (Argent?) four Bars Azure; *22.* Argent three Bars dancetty Azure debruised by a Bend Gules; *23.* Bendy Or (Argent?) and Gules; *24.* same as 18; *25.* Or a Cross Gules; *26.* Gules a Lion rampant Or *(Vitré?) impaling* Or three Bars Gules; *27.* Gules three Fleurs-de-lis Or; *28.* Argent four Bars Azure; *29.* Bendy Or and Gules *(Limoges or Turenne)*; *30.* Or Martlets in orle a Canton Gules *(Chemillé)*; *31.* same as 8 and 11; *32.* Barry Vair and Gules; *33.* Or a Lion rampant within an Orle embattled on the inner edge Gules; *34.* Barry Argent and Azure a Label Gules; *35.* same as 29; *36.* same as 8, 11, and 31; *37.* Gules a Cross clechy pometty and voided Or *(Botherel)*; *38.* Azure semy-de-lis Or *(France Ancient)*

PROVENANCE: Lost work, known from three seventeenth-century watercolors executed for Roger de Gaignières and an engraving by N. Pitau after a drawing by Jean Chaperon (at the beginning of the eighteenth century). Tomb formerly in the abbey church of Villeneuve (Loire-Atlantique), dismantled during the Revolution.

Bibliothèque Nationale de France, Paris; Département des Estampes (Pe 11 c, fols. 78, 79)

Folio 78 exhibited in Paris; folio 79 exhibited in New York.

Of all the Limousin tombs known through reproductions, that of Alix of Thouars, duchess of Brittany (d. 1221), and of her daughter Yolanda (d. 1272) is the most lavish. Furthermore, the Gaignières albums contain three views of the monument, thereby underscoring its significance.

The work is complex. The tomb rests on a large, rectangular stone base. The wood sarcophagus is covered with copper plaques decorated with enameled rinceau patterns. Horizontal bands of decoration bearing medallions with

# Villeneuue

*Le mesme Tombeau veu du costé d'Alix Comtesse de Bretagne.*

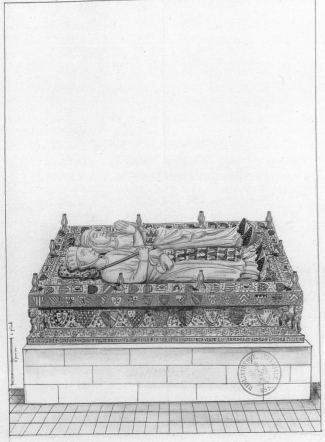

Folio 78

# Villeneuue

*Le precedent Tombeau veu du costé d'Ioland de Bretagne.*

Folio 79

411

busts of angels and a row of armorial shields extend around all four sides of the tomb, interrupted by four lions at the corners. Above is a wide, quadrangular border likewise containing coats of arms on its surface; the escutcheons on this border, which forms a frame around and between the effigies, are grouped in twos and alternate with plaques decorated with cabochons and baluster-shaped candlesticks. At the left, the duchess of Brittany, Alix, lies on an enameled ground of rinceau decoration; her feet are supported by a gadroon-ornamented base. She wears a robe and a tunic, as well as a mantle lined with vair; her head, resting on an enameled cushion and framed by two appliqué figures of incense-bearing angels, is coiffed with a type of barbette; her eyes are open; and she holds a scepter, a symbol of power. Her epitaph covers the length of the lowest border of the coffin, just above the stone base. To the right, Alix's daughter Yolanda of Brittany is represented with her eyes closed and her hands joined; she lies on a checkered background bearing the blazons of France, England, and, no doubt, Dreux, and her feet rest on a polygonal base, likewise ornamented with escutcheons. Dressed in a robe with decorated cuffs, she is draped, like her mother, in a mantle lined with vair, a flap of which, tucked under her right elbow, falls in conical folds. A wimple and widow's veil surround her face. On either side of her pillow, two kneeling angels perfume the deceased with incense; according to folio 79, these would have been seated angels, as one finds on certain stone monuments of the second quarter of the thirteenth century,[1] and not appliqué figures, more common on Limousin monuments.

This brilliant copper funerary monument is not surprising for a member of the ducal family of Brittany; Peter Mauclerc, husband of Alix of Thouars—whose burial place in Braine is marked with a bronze flagstone—belonged to a family whose predilection for metal tombs is apparent in this abbey in Champagne, and his descendants were faithful patrons of the Limousin enamelers during the second half of the thirteenth century. On the other hand, one might question, as Viollet-le-Duc did from 1868 on, the date and the homogeneity of a tomb made for two women, mother and daughter, who died more than fifty years apart, one in 1221 and the other in 1272. The sepulchre was in the choir of the Cistercian abbey church of Villeneuve, not far from Nantes, a monastery founded in 1200 by Alix's mother, Constance, duchess of Brittany, who made it her last home. The church was consecrated in 1223, and it was on that date that the sepulchres of Constance, her spouse, Guy of Thouars, and their daughter Alix were erected. The effigy of Alix of Thouars is represented dressed in a costume with a ribbed barbette also seen on monumental sculpture of about 1220–30: it is worn by an Old Testament female figure on the north porch of the cathedral of Chartres[2] and by one of the damned on the tympanum over the north arm of the

transept of the cathedral of Reims.[3] The head covering is also the same as that of Agnes of Baudement (d. 1207), countess of Dreux, who is buried in Saint-Yved in Braine.[4] This type of headdress, although somewhat modified, is still encountered in works from the second half of the thirteenth century. A fine example is seen on the engraved tomb slab of Catherine of Brittany (d. 1278) in the Dominican monastery in Paris:[5] it affords a glimpse of the cloth that covers Catherine's skullcap, which is filled out by the hair gathered above her ears, and, with the loop of cloth stretched around her chin, it creates a triangle around her face. However, the head coverings of Alix of Thouars, Agnes of Baudement, and the Old Testament woman depicted on the cathedral of Chartres, on the contrary, accentuate the ovals of their faces and are earlier in date. The watercolorist depicted the drapery of Alix of Thouars's mantle falling in V-shaped folds along her left side, evoking mid-thirteenth-century sculpture. This would therefore suggest that the effigy of Alix of Thouars was not executed until thirty years or so after she died—and not at the time of the consecration of the church in Villeneuve and of the burial of the duchess, which took place shortly after her death; the effigy, shown holding a scepter to symbolize the power over the duchy of Brittany that she is passing on to her husband, Peter Mauclerc, might have been commissioned at the time of the latter's death, in 1250; this is all the more likely since the duke chose to be buried in Braine in Champagne, near his parents, and not in the duchy. The effigy is set off against a background with enameled rinceau decoration, which is compatible with this hypothesis, since one finds similar ornamentation on the tombs of the children of Saint Louis in Royaumont, which are very close in date.

The effigy of Yolanda of Brittany is distinguished from that of her mother in several respects: her eyes are closed; the conical folds under her arm bear witness to a later style; and, lastly, the background studded with armorial escutcheons is a stylistic feature of the second half of the thirteenth century, when one also finds it on works produced outside of Limoges, as on the tomb of Marie of Bourbon (d. 1274) in Braine. The execution of this effigy therefore may have been carried out shortly after the death of Yolanda, in 1272, and the common base must have been made at the same time. A difference in the decoration of the border at the feet of the two women perhaps indicates a partial reuse of elements that belonged to Alix of Thouars's first monument: the small enameled plaques, alternately quatrefoil and lozengelike in shape, were placed on the border on her side, while small armorial shields were added to the profusion of escutcheons. In addition, the curious placement of Alix's epitaph probably also attests to this reworking of the tomb.

Compared with the monument of Robert of Brittany (d. 1259, see Appendix, list, no. 5), Alix's grandson, the

Villeneuve tomb is characterized by a proliferation of heraldic decoration: added to the coats of arms borne by the deceased, their parents, and their spouses are those of the noble families of Brittany, as well as such others as the arms of France and England. The identifications are all the more difficult since a comparison of the escutcheons depicted in the Gaignières plates with those in published engravings of Dom Lobineau's work reveals very few that are identical, and thus bodes ill for the veracity of these documents. The neighboring tomb of Andrea of Vitré (d. 1288), likewise in the abbey church of Villeneuve, also contains extensive heraldic decoration, indicating that its date is quite close to that of the ducal family's two monuments. Executed for the most part, as we have seen, at the time of Yolanda's death, this double tomb must have been commissioned by John I or one of his kinsmen; among the historical events that might have coincided with this commission are the treaty, concluded in 1274, announcing the marriage of Arthur of Brittany, grandson of John I of Brittany, to Marie of Limoges—which perhaps created closer ties between Brittany and the Limousin.

BC-B

NOTES

1. Tomb of Philip Dagobert, formerly in Royaumont, today in Saint-Denis, dated about 1235 (Sauerländer 1972, fig. 159 and pp. 139–40); see also the tombs of the Ote d'Artois, and of Louis and Philip d'Alençon, also erected in Royaumont (Pessiot 1994).
2. Sauerländer 1972, fig. 98.
3. Ibid., fig. 239.
4. Teuscher 1990, pp. 149–71; Adhémar 1974, no. 235.
5. Adhémar 1974, no. 360.

BIBLIOGRAPHY: Dubuisson-Aubenay 1636 and 1898, vol. 2, pp. 153–55; Lobineau 1707, p. 217; Bernard de Montfaucon, vol. 2 (1730), p. 167 and pl. XXXII, no. 3; Morice 1750–54, vol. 1, pp. 149–50; Texier 1857, col. 1403 (erroneous localization in Braine); Viollet-le-Duc 1868, pp. 63–64; Viollet-le-Duc 1871–72; Rupin 1890, p. 162; Copy 1986, pp. 40, and 268, no. 115; Vaivre 1988a, pp. 62–64; Gauthier 1988, p. 108.

GRAPHIC SOURCES: Gaignières album: Oxford, Bodleian Library, ms. Gough-Gaignières 1 (18346) and traced copy Pe I, fol. 99 (Bouchot 1891, no. 1961 and Adhémar 1974, no. 323); Paris, BNF, Est. Pe II c, fols. 78, 79 (Bouchot 1891, nos. 4951, 4952); Oa 9 fol. 83 (Bouchot 1891, no. 97) and Oa 10 fol. 12 (Bouchot 1891, no. 142); engraving by N. Pitau after a drawing by J. Chaperon, published by Lobineau (1707).

# 150. Masks of Herbert and Alès Lanier

Limoges, end of the 13th century

a. Mask of Herbert Lanier (d. 1290)

Copper: hammered, engraved, and gilt

36 × 23 × 14.5 cm (14⅛ × 9 × 5¾ in.)

PROVENANCE: Fragment of an effigy, from the church of the priory of Notre-Dame de La Papillaye (Maine-et-Loire, commune of Angers) until the Revolution

EX COLLS.: P.-A. Mordret (ca. 1858); sale, Paris, April 20–23, 1881, no. 235; Bligny, Gavet, Tollin, Isaac de Camondo (in 1897); bequest of Isaac de Camondo, 1911

Musée du Louvre, Paris; Département des Objets d'art (OA 6485)


b. Mask of Alès, wife of Herbert Lanier (d. end of the 13th century)

Copper: hammered, engraved, and gilt

29.5 × 16.5 × 11 cm (11⅝ × 6½ × 4⅜ in.)

PROVENANCE: Fragment of an effigy, from the church of the priory of Notre-Dame de La Papillaye (Maine-et-Loire, commune of Angers) until the Revolution

EX COLLS.: P.-A. Mordret (ca. 1858); sale, Paris, April 20–23, 1881, no. 235; acquired for the Musée Saint-Jean of Angers

Musées d'Angers (MA III R 368)

The two works, made from thick sheets of hammered, engraved, and gilt copper, depict the elongated faces and features of a man and a woman, respectively. The head and the neck of the man are entirely hidden by a cowl that rounds off the forms; its lower edge, as well as the border ringing the opening surrounding the face, contains a band decorated with lozenge-shaped motifs. The man's high, bulging forehead is topped with an even fringe of hair, and the eyes are closed, beneath a linear brow ridge. A play of wrinkles around the globelike eyes humanizes the metallic face of the effigy. The nose, long and fine, is framed by the curves of the cheeks and surmounts a mouth with thin lips surrounded by a beard and a closely cropped mustache. The features of the woman are very similar to those of her companion: closed eyes, a long nose with a slightly swollen bridge, thin lips, prominently rounded cheeks and forehead. The differences between the two works are in the details, such as the engraving of the eyebrows and the less precise and sensitive execution of the woman's mask, where the wrinkles were omitted around the eyes.

The head covering of the woman consists of a band placed high on the forehead, from which a drape of fabric falls, cradling the chin, while the neck is veiled by a wimple. Strands of wavy hair are visible at the temples. When the effigy was intact, the head covering no doubt was similar to those on the tombs of Yolanda of Brittany (see cat. 149) and Eustacia of Vitré in the abbey of Villeneuve and, consequently, must have been suitable for a high-ranking laywoman. The cowl, likewise, was worn by a certain number of laymen in the middle and second half of the thirteenth century, as attested by the watercolors in the Gaignières albums depicting the tomb of Robert Alneti (who died at the end of the twelfth century—but who is buried beneath a flagstone with a trefoil arcade, under a Gothic arch, which one does not encounter before the second quarter of the thirteenth century), formerly in Saint-Nicolas in Angers,[1] as well as in the scenes on the tympanum of the north arm of the transept of the church of Notre-Dame in Semur-en-Auxois (Côte-d'Or), of about 1240–50, where Abbanes, the steward of the king of India,[2] who received Saint Thomas, is depicted twice, in a long robe, his head enveloped in a cowl, whose lateral folds recall those of the Limousin mask. Also similar to the mask is Abbanes's short fringe of hair, which forms a row of bangs beneath the upper edge of the head covering.

Mask of Herbert Lanier

Mask of Alès Lanier

The effigies to which these two masks most likely belong can be identified as those of a bourgeois from Angers, Herbert Lanier—known through various financial documents of 1287–90, and the founder of Notre-Dame de La Papillaye, a priory of canons regular of Saint Augustine, who died in January or March 1290—and his wife, Alès. Early references to their tombs, situated until the Revolution in the choir of the church of the priory of La Papillaye a few kilometers from Angers, mention a copper sepulchre, with the effigies depicted in the round and surrounded with the arms of the Lanier family. The only difficulty with this identification is that the historian Barthélemy Roger claimed to have seen Herbert Lanier, in the seventeenth century, "fully armed, his helmet at his feet."[3] Indeed, this author appears to have wished to reinforce the noble pretensions of the Laniers and Lasniers of the seventeenth century by endowing their ancestor with military attire hardly compatible with the cloth cowl worn by the mask in the Musée du Louvre and absent from all other descriptions of the tomb. On the other hand, various eyewitnesses agreed with Barthélemy Roger about the presence of "escutcheons of his arms around his tomb, which is of gilded copper and very sumptuous for the time."[4] These escutcheons recall those on the tomb of Juhel de Mayenne at the abbey of Fontaine-Daniel, where the coats of arms of the Mayenne were repeated on the border, while the coffin bore the arms of related families; they also recall the heraldic decoration on the tombs of the ducal family of Brittany, which were contemporary with the Angevin tomb in the priory of La Papillaye (cat. 149).

The two Limousin masks find parallels in a small series of reliquary heads, several of which are included in this exhibition (see cats. 153, 154). The mask of Herbert Lanier evokes, in certain respects, the reliquary head with eyes likewise closed, formerly in the Figdor Collection and now in the Kunstgewerbemuseum, Berlin.[5] However, the latter is earlier, for its style is similar to that of the appliqué figures in the Passion group, dated to about the middle of the thirteenth century (cat. 119). The mask of Alès is almost contemporaneous with the two reliquary heads of the female companions of Saint Ursula, whose faces, weaker in execu-

tion, present the same elongated type, with a large nose, lips surmounted by a deep labio-nasal groove, and, most probably, the same technique of execution (cat. 153).

The method of production is easily discerned if one examines the reverse sides of the masks: the repoussé technique was employed, the sheet of copper, originally very thick (more than 1 mm), hammered, probably using a die. The hammer marks are easily visible in the X-radiographs of the two works; the flattening out of the metal caused accidents on the nose of the mask of Alès and on the lip of the mask of Herbert Lanier, after which the reverses were reinforced by adding small pieces of copper. The front sides were then polished by abrasion to obtain smooth volumes and engraved, before being gilded. Traces of gold on the mask of Alès bear witness to the very fine quality of the gilding, but the condition of the two works—the brown of the copper shows through in the broad planes of the forehead, cheeks, and chin—undoubtedly alters somewhat the perception of the forms as they originally were conceived by the Limousin enamelers and goldsmiths.

What is striking about the monuments of Herbert Lanier and his wife, Alès, is the revelation that, among the clientele of the Limousin tomb makers, there were wealthy bourgeois, elite townspeople who were sufficiently familiar with the bishops and the noble families to the point of following their example in commissioning tombs of Limousin manufacture.

BC-B

NOTES
1. Adhémar, 1974, no. 162.
2. Sauerländer 1972, fig. 291.
3. Roger 1852, pp. 274–75.
4. Ibid.
5. Rückert 1959, fig. I-2; Boehm 1991, pp. 191–96.

EXHIBITIONS (a): Paris 1968b, no. 405; Moscow–Leningrad 1980, no. 24; (b): Tours 1891; Châteaugontier 1892; Paris 1900, no. 1628; Paris 1962, no. 163; Paris 1968b, no. 404; Paris 1970, no. 41; Poitiers 1976, no. 2; Angers 1992, no. 33.

BIBLIOGRAPHY: Godard-Faultrier 1884, no. 3183 (b); Marquet de Vasselot 1898; Vitry 1914, pp. 463–64, with fig. (a); Gauthier 1950, p. 55; Rückert 1959, pp. 5–6, fig. 4 (a); Chancel 1991.

# 151. Tomb Effigy of Blanche of Champagne

Limoges, ca. 1306

Copper: repoussé, engraved, and champlevé (mask, hand, belt, cushion), and gilt (mask, cushion); copper, alloy of copper, zinc, and tin: hammered, stamped, and nailed (clothing); wood core (walnut), roughly shaped and made up of two superimposed thicknesses, at the level of the bust

205 × 24 × 50 cm (80¾ × 9½ × 19¾ in.)

INSCRIBED (on the belt; incomplete): [F]LERE : SI : QUOD : FUERIT : MISERER[E] (Cry if there is something to have pity about [?])

PROVENANCE: From the former abbey of the Cistercian nuns of La Joie Notre-Dame, Hennebont (Morbihan), which became national property and was sold on December 12, 1792. It remained in the partially destroyed abbey, property of the successive owners of that site.

EX COLL.: Alfred Ramée, Rennes (unknown date, after 1864). Acquired in 1873. Deposited in the Musée des Thermes et de l'Hôtel de Cluny between 1948 and 1988.

CONDITION: The left hand is missing; the dog has lost its muzzle; missing areas in the revetment: on the dog; on part of the scapular and the mantle, to the right; and on the lower part of the cushion. The gilding is almost all gone. Restoration done by Antoine Amarger in 1995, repairing the dent on the forehead.

Musée du Louvre, Paris; Département des Sculptures (OA 2082 and RF 2605)

Blanche of Champagne, also known as Blanche of Navarre, daughter of Thibaut IV, count of Champagne and of Brie and king of Navarre, wife of John I, duke of Brittany, died on August 11, 1283, at the Château de Hédé, near Rennes. The following year, she was buried in the choir of the abbey church of La Joie Notre-Dame, which she founded.

Her tomb was commissioned in Limoges by her son John II of Brittany, sometime between October 1286, the date of his accession to the dukedom, and November 18, 1305, the date of his death. This is confirmed by a receipt discovered and subsequently published by Benjamin Fillon in 1852: "May everyone know that we, Guillaume le Borgne, Knight, have had and received from the executors of our dear lord of fond memory, John, former duke of Brittany, count of Richemond, deceased, four hundred and fifty livres for the construction of the tomb and the sepulchre of our dear lady, may God keep her soul, my lady Blanche his mother, former duchess of Brittany, which our dear lord had ordered made in Limoges during the time that he lived. And we had the ten exequaturs delivered to those who made the said tomb . . . due the sum of the aforementioned four hundred and fifty livres. To Venes, our seal in testimony. The day . . . of the year of the Lord thirteen hundred and five."[1]

The monument was next mentioned by Dubuisson-Aubenay, in 1636, then by the abbess of La Joie in an account of the possessions of her convent, drawn up in 1725.

Later, the "tomb and the statue" appeared on the list of movable objects that, of course, were excluded from the sale, during the Revolution, of the lands and buildings of the abbey, which took place December 12, 1792.

Without knowledge of the specifics, we may assume, nevertheless, that the tomb effigy must have remained in the buildings of the abbey, which became private property and were partially demolished. The effigy was noted as being there in 1847 and in 1864. However, to scholars, who had lost sight of it, the monument was considered to have disappeared until 1872, the year in which it reappeared, in the possession of the collector Alfred Ramée, for some time magistrate in Rennes, who sold it to the Louvre in 1873.

Several like examples in the Gaignières albums (see cats. 147–49) and Dubuisson-Aubenay's text provide some idea of the monument's base, which was destroyed in the first half of the nineteenth century: a wood pedestal "covered with copper plaques bearing the arms of Dreux-Brittany and Navarre impaling Champagne," the arms of the deceased depicted on "escutcheons composed of copper plaques mounted on the sepulchre." As color was the essential element of a blazon, these escutcheons undoubtedly were enameled—to which the text of the 1725 account, indicating a work "in color," would have made a reference. The coffin still bore the epitaph, which was transcribed by Dom Lobineau: "Here lies the high and mighty Lady Blanche of Navarre, wife of John the First, duke of Brittany, who founded this abbey in the year MCCLX and was buried in the habit of the order in the year MCCLXXXIV."

In the effigy, Blanche likewise is dressed in the habit of a Cistercian nun, with a scapular and a full mantle, a wimple, a veil, and a chin strap—the head covering of a widow or a nun, whose coquettish arrangement, with the hair gathered at the temples and little wisps peeking through the veil, here is meant to designate a laywoman. The position of the letters on the belt may be surprising, but this certainly reflects the original intention.

A comparison with other rare, extant Limousin funerary objects that are apparently contemporaneous—such as the masks of Herbert and Alès Lanier (ca. 1290; see cat. 150) or the effigy of William II of Valence (d. 1288), in Westminster Abbey (fig. 10)—reveals the faithfulness of the present work to the stylistic traditions of the Limousin workshops. We find the same characteristic elements in these works as in Blanche of Champagne: the narrow face; thin and slightly aquiline nose, with pinched nostrils; sharply outlined ridges of the brows; rather broad upper lip, with a well-defined labio-nasal groove. The eyes are prominent, like those of the English prince and, also like his, open—a detail peculiar to

French thirteenth-century tomb sculptures that corresponds to a certain concept of the afterlife in which the deceased is already depicted as one of the elect. However, other elements make this effigy of Blanche of Champagne an exceptional and singular work: a touch of elegance and a search for naturalism—certainly not based on portraiture, since the duchess was nearly sixty years old at the time of her death—that is clearly perceptible in the asymmetry of the open mouth or in the hint of mischief that marks the features. Above all, there is an attempt, here, at sculptural effect, a tactile sense, and a subtlety of modeling, which evokes the medium of sculpture and links this Limousin effigy to certain carved marble effigies from the Île-de-France: that of Marguerite d'Artois (d. 1311), in Saint-Denis,[2] or the sculpture, presumed to be of Marguerite de Dampierre, in the Louvre.[3]

One can better evaluate the extent of the stylistic evolution revealed here if one refuses to follow currently accepted opinion, deducing from the lack of any mention of the tomb in the will of John II of Brittany, drawn up in September 1302, that it was completed before that date. One need only reexamine the terms of the receipt of "those who make it" and the amount of the sum delivered, which was sufficiently high to correspond to the total cost of the tomb, to conclude that the effigy was executed during the year 1306 and, perhaps, was begun immediately upon the death of the duke.[4]

The inordinate length of the body is surprising. Its rigidity is nonetheless broken by the position of the slightly flexed left leg, which, once again, possibly derives from Parisian statuary of the early fourteenth century—to which the statue of Pierre d'Alençon, for example, originally from Poissy (and now in the Musée de Cluny, Paris),[5] bears witness. This movement felicitously upsets the symmetry, shifts the scapular, and creates several folds whose relative awkwardness contrasts with the technical accomplishment of the face.

This particular quality is due to an uneven mastery of the technical processes employed, which were studied recently by Antoine Amarger; his conclusions are as follows: mask, hand, belt, and cushion were not cast, as has often been asserted, but formed by the repoussé technique from thick sheets of copper (2 mm for the mask, more than 1 cm for the hands), then modeled by the careful removal of the material. For the hands, this required a complex procedure to give volume to the fingers and the palm, created by folding the metal sheet—a task almost comparable to that of sculpture. In order to cover the tears, four pieces of metal were attached to the reverse of the mask (three are still preserved)—an indication of the difficulty of the technical process, which is in perfect accord with Limousin practice and was carried out with undeniable expertise.

The procedure employed to create the body was also common in Limousin workshops: the use of thin sheets of

418

metal formed directly over a wood core. A study of the attempts at volume is interesting, as they were perhaps inspired by sculpture: for example, the folds on the left leg or, on the same side, the fall of the mantle, which curves and folds in on itself, revealing the lining. A certain originality is also found in the choice of materials: copper for the scapular, and an alloy of copper, zinc, and tin for the robe and the mantle.[6] This variety probably was intended to introduce nuances of color.[7] However, the execution shows relatively little care: there are nearly seventy pieces of metal, some of which are quite small. The division of labor within a single workshop seems obvious here. One cannot dismiss, as well, the possibility of later restorations.

A last remark is in order: although prepared for enameling, the cushion, with its checkered pattern enclosing floral motifs inscribed within circles, and the belt with its prettily drawn letters, were never decorated with color, for, contrary to the information supplied by Barbet de Jouy, no trace of enamel is visible, even under a microscope. Furthermore, this agrees with the findings of Dubuisson-Aubenay, who describes the "engraved copper" decoration on the pillow, where, curiously enough, he saw coats of arms. Was this for reasons of economy? Or incompletion of hasty work? The death of the man who commissioned the effigy might, under the circumstances, have played a not insignificant role.

The work retains its part of the mystery. It appears, in any case, to be essential to the understanding of Limousin production—the only extant female effigy among a supremely limited group. It also attests to the extent of the spread of this art form and to the preponderant place that the political circumstances and the play of familial alliances ensured it in Upper Brittany, from the middle of the thirteenth to the early fourteenth century, before the production of effigies in marble and stone. This development followed the marriage in 1275 of Arthur II, grandson of Blanche of Champagne, to the heiress of the viscount of Limoges, and was also the result of a tradition of metal tombs, eventually Limousin, well established in the house of Dreux, to which Duke John I belonged, and in Blanche's family in Champagne. Such a climate, recently reexamined by J.-Y. Copy, allowed for the importation of many Limousin works. The tomb effigy of Blanche of Champagne remains the only tangible testimony to this phenomenon.

FB

NOTES

1. Archives Départementales of the Loire-Atlantique, E 21. This text was verified and corrected against the original by M.-Ch. Remy, curator of the Archives Départementales of the Loire-Atlantique and by C. Leroy, curator of the Musée Thomas Dobrée, Nantes, whom I thank.

2. Erlande-Brandenburg 1987, fig. 123, p. 153.

3. Musée du Louvre, Département des Sculptures, *Catalogue sommaire illustré des Sculptures*, vol. 1, *Moyen Âge* (forthcoming).

4. The poor condition of the text renders certain passages illegible. The term "due the sum of the aforementioned four hundred and fifty livres" leaves one free to imagine installment payments (see Marquet de Vasselot [1941, p. 111], who alone imagines a completion in 1306).

5. Sandron 1991, pp. 82–83.

6. Physico-chemical composition of the four samples (in percentages), determined by the analysis undertaken in 1995 by the ICPI laboratory in Lyons. Sample 1: covering of the body; 2: covering of the right wrist; 3: metal from the fingers; 4: metal from the mask

|   | Copper | Lead | Zinc | Tin | Weight (mg) | Total |
|---|--------|------|------|-----|-------------|-------|
| 1 | 82 | 0.58 | 16 | 3.6 | 119.2 | 102 |
| 2 | 93 | 0.91 | 5 | 3.8 | 15.3 | 103 |
| 3 | 96.7 | 0.61 | 0.25 | 1.78 | 19.4 | 99.3 |
| 4 | 98.2 | 0.14 | 0.05 | 1.02 | 126.8 | 99.4 |

7. Corrosion and accumulated layers of wax obliterated this possible initial plan.

BIBLIOGRAPHY: Dubuisson-Aubenay 1636 and 1898, vol. 2, pp. 73–74; Lobineau 1707, pp. 400, 450; Morice 1742–46, vol. 1, col. 1205; Cayot-Delandre 1847, pp. 472–73, n. 1; Fillon 1852–53, vol. 2, pp. 129–30; Rosenzweig 1863, pp. 28–29; Clément de Ris 1873, pp. 122–23 (engraving showing the stump of the left hand, now missing); Barbet de Jouy 1874, no. 70 bis; Rupin 1890, pp. 155–67; Le Méné 1903, pp. 123–84; Delage and Courtot 1923, pp. 5–30; Aubert and Beaulieu 1950, no. 264; Copy 1986, pp. 8, 33, 39, 41, 48, 50.

# 152. Funerary Plaque of Guy de Meyos

Limoges, 1307

Copper: engraved, silver-plated, and gilt; champlevé enamel: lapis blue, blue-green, red, and white

32 × 20 cm (12⅝ × 7⅞ in.)

INSCRIBED: + OBIIT : NOBILIS : CLERICUS : DOM / INUS : GUIDO : DE : MEYIOS : DIE : SABB / ATI : POST : OCCULI : MEI : ANNO : DOM / INI : MILLESIMO : TRICENTESIMO : S / EXTO : QUI : FECIT : CONSTRUI : ISTA / M : CAPELLAM : ET : SEPULTUS : IACET :/: HIC : REQUIESCAT : IN : PACE : AMEN : (Noble cleric, master Guy de Meyos, died Saturday after the Oculi Sunday, in the year of our Lord thirteen hundred and six. He built this chapel and is buried here. May he rest in peace. Amen)

COATS OF ARMS: Or three Lions rampant Gules

PROVENANCE: This piece was supposedly found on the property of the Templar Commandery of Civray (Vienne), about 1880.

EX COLLS.: Métayer; Gaillard de La Dionnerie, Poitiers (in 1883; sale, Paris, November 30–December 2, 1903, no. 23); Frédéric Engel-Gros, Ripaille castle (Haute-Savoie; sale, Paris, May 30–June 1, 1921, no. 235). Gift of the children of Frédéric Engel-Gros, 1921.

CONDITION: Very worn, certainly because of prolonged burial. Surface of the copper corroded, notably on the face of Guy de Meyos. Losses to the enamel, principally at the top left. An appliqué depicting the Hand of God was once set in the medallion over the figure of Guy de Meyos. The head of Saint Louis is misshapen and pushed far to the right, probably as the result of a violent blow from a tool.

Musée du Louvre, Paris; Département des Objets d'art (OA 7495)

The plaque is composed of two parts. The actual epitaph comprises a long inscription in Latin, the letters enameled in red divided over seven lines and separated by continuous strokes of blue enamel. Above, the deceased, Guy de Meyos, dressed in a voluminous hooded robe and flanked by an escutcheon bearing his arms, kneels before a haloed king, whose robe is strewn with fleurs-de-lis on an azure field; the latter is surely Saint Louis (r. 1226–70; canonized in 1297), as Molinier had proposed. In the upper part, the Hand of God, in a cruciform medallion, designates the kneeling cleric.

This representation, made barely ten years after the canonization of the king of France, is of exceptional interest. It is one of the oldest depictions of the haloed king, and, furthermore, bears the exact date of the Saturday following Oculi Sunday, 1306—that is to say, after the third Sunday of Lent, 1307, since the year began then at Easter and not on January 1. In addition, this commemorative image seems to suggest a personal link between the king—or, at the least, the Capetian monarchy that he embodies—and the kneeling cleric, to whose coat of arms he points with his right hand.

This work, unique in Limousin production in its conception, its iconography, and its dated inscription, was published for the first time in 1883, very shortly after its discovery, but with no indication of its provenance. This was only revealed in 1935, by Bobe. This writer—who made it known that he was an eyewitness to the discovery—indicated that the plaque had come to light on the grounds of the Order of Knights Templar in Civray, then the property of M. Métayer, during the digging of a trench, "about 1885."[1] The correct date of the discovery was "about 1882," as Galopaud (1944) said, since the first publication of the plaque was in 1883: the object would have been accompanied at the time by "two other objects aedicular in form—reliquaries, no doubt—which have not been preserved." Furthermore, the actual poor state of preservation of the plaque would have been due not only to its lengthy burial but also to the treatment it received, since the owner had had "the ill-conceived idea of rubbing it with sand."[2]

The site of the discovery therefore must have been the chapel founded by Guy de Meyos to house his tomb, as indicated in the enameled inscription on the plaque. The plaque is thus one of the rare Limousin works to be precisely dated, and, what is more, by an inscription (cats. 140, 156, 157). However, the importance of this epitaph in the context of Limousin production does not reside only in the historical circumstances of its execution and its discovery. It is also of supreme interest in demonstrating that, at the beginning of the fourteenth century, certain Limousin artists knew how to adapt themselves to the evolving style and to assimilate the principal characteristics of the art developed in the Île-de-France under Philip the Fair—seen here in the elegant and fluid drapery of the cleric and the fine-featured faces and short, softly waved hair of the two figures. The work, which seems to capture in metal a personal bond between Guy de Meyos and the recently canonized king, is, furthermore—as J. Durand noted[3]—precious testimony to the success of the expansion of the Capetian monarchy only a few years before the Order of Knights Templar was suppressed by Philip the Fair (1314).

ET-D

NOTES
1. A confirmation of this can be found in a mention by Faye (1849, p. 461): "I have heard that at the back of a cave was the tomb of one of the commanders with an inscription, but I looked for it in vain." On the commandery of Civray, known through rare documents, of which the oldest dates from 1184, see Faye 1849, pp. 74–77, and Galopaud 1944, pp. 304–5.
2. Galopaud 1944, p. 305.
3. In Senlis 1987.

EXHIBITIONS: London 1932, no. 576 l; Paris 1960, no. 178; Senlis 1987, no. 11; Tokyo 1991, no. 40.

BIBLIOGRAPHY: Gaillard de La Dionnerie and Lasteyrie 1883; Molinier 1883, pp. 94–95; Rupin 1890, p. 167; Molinier 1891, p. 189; Molinier 1901, pp. 192–93, fig. p. 193; Migeon 1921; Marquet de Vasselot and Prinet 1921; Ganz 1925, no. 19, pp. 434–35, pl. 137; Bobe 1935, pp. 264–65, with pl.; Marquet de Vasselot 1941, p. 111; Galopaud 1944, pp. 301–7, fig.; Gauthier 1976a, p. 191.

# 153. Reliquary Heads

a. Limoges, fourth quarter of the 13th century

Copper: formed, chased, scraped, engraved, and gilt

33 x 18 cm (13 x 7 1/16 in.)

INSCRIBED (on base at center front): HIC EST CAPVT VNI[US]. D[E]. VNDECIM MIL[L]IBVS VIRGINVM ET MARTIRV[M] (Here is the head of one of the eleven thousand virgins and martyrs).

EX COLLS.: Bouvier, Amiens (formed ca. 1840/50; sold December 8–16, 1873, no. 262); Aimé Desmottes, Lille and Paris (sale, Paris, March 19–23, 1900, no. 39); Albert Maignan (in 1906)

CONDITION: Losses to gilding; two feet of base missing (noted 1897); proper left foot repaired; hinges replaced except proper left and center back bottom; dentilated edge partly lost; traces of polychromy on lips; relics preserved.

Musée de Picardie, Amiens (Inv. no. 357.189.51)

b. Limoges, fourth quarter of the 13th century

Copper: formed, chased, scraped, engraved, and gilt

34 x 21 cm (13 3/8 x 8 1/4 in.)

INSCRIBED (on base at center front): HIC EST CAPVT VNI[US] DE VNDECIM MIL[L]IB[US] VIRGINVM ET MARTIRV[M] (Here is the head of one of the eleven thousand virgins and martyrs).

CONDITION: Loss of gilding at hairline; break at inner corner, proper right eye; lower hinge at proper right replaced, proper left hinge replaced.

Classified, Monuments Historiques, June 25, 1891

Collegiate Church of Saint-Martin, Brive (Corrèze)

Inscriptions on the Amiens and Brive reliquaries identify each as containing the head of a companion of Saint Ursula. With the head in Paris (cat. 154), which bears no inscription but clearly represents a sister image, they are part of a group of seven Limousin gilt-copper reliquary heads.[1]

In both the Amiens and the Brive reliquaries, the virgin martyr has softly flowing hair parted in the center, almond-shaped eyes, a long nose that is broad at the tip, and a thickened neck. A thin, plain band suggesting a diadem encircles the head, whose crown is pierced in the form of a cross. Both heads have a splayed base; a flattened tab foot, surmounted by an incised roundel representing a half-length angel, is at each corner. Across the front, the identifying inscription is engraved in three lines of Gothic capitals. Each of the heads is composed of three pieces of copper: the front and back halves are soldered along a vertical seam, which is reinforced with metal and solder on the inside. A third sheet of copper extends from the crown of the head to the diadem set just over the forehead and is secured by hinges at the two sides and the center back. The bases of the Amiens and Brive examples are formed from a single sheet of copper. A dentilated band helps to secure the head to the base and to obscure the joint.

The legend of Saint Ursula is documented by Geoffrey of Monmouth in *Historia . . . Regum Britanniae* and elaborated in *The Golden Legend*.[2] The king of England, a pagan, wished to see his son marry Ursula, a British princess renowned for her beauty and wisdom. Not wishing to marry an idol worshiper, Ursula placed discouraging conditions upon the betrothal. She demanded a fleet of ships and a three-year engagement so her new fiancé would have time to study the doctrines of the Church. She also required ten virgins as companions and, for their moral support, another one thousand virgins each for herself and the original ten. Thus, eleven thousand and eleven virgins set off on a journey to Rome. The sight of this flock was so wondrous that it attracted many followers, including the converted queens of Sicily and Constantinople as well as Pope Cyriacus.[3]

On their return journey, at Cologne, Etherius, Ursula's betrothed, found the virgins by following the command of an angel. But the entire group was surrounded and martyred outside the walls of the city by an army of Huns. After all had been slaughtered save Ursula, the king of the Huns offered to spare her if she would become his wife; when she denied him, he shot her with an arrow from his own hand.

In 1106, when the citizens of Cologne decided to expand the city walls, they discovered a mass of human remains dating from the Roman era. Assuming the dead to be the martyred virgins, they named the place *ager Ursulanus* (the field of Ursula). The devotion to Saint Ursula and her companions spread throughout northern Europe, aided by gifts of their relics such as an entire body presented to King Louis IX of France about 1250. Relics had reached the Limousin in 1181, following the trip to Cologne of a delegation of Grandmontain monks sponsored by their abbot, William of Treignac.

In Limousin work, the devotion to Saint Ursula is manifest in these three reliquary heads, a fourth from the Finoelst Collection,[4] and a chasse (cat. 114).

Abbé Texier identified the head at Brive as the one of Saint Essence given to the Ursulines at Brive when the relics of Grandmont were dispersed in 1790 after the suppression of the monasteries.[5] Although the abbey possessed relics of the companions of Saint Ursula, including seven skulls, the documents concerning their dispersal specify that the heads of these martyrs were distributed without reliquaries.[6] Thus, although the Ursuline community at Brive was among the recipients, it could not have received the Brive reliquary at that date, as Texier supposed.[7]

a

b

Sixteenth-century inventories of Grandmont (1566–67, 1575) record heads in silver coupes (paired hemispheres); these may have replaced medieval reliquaries of gilt copper. The common collection histories of the Louvre and Amiens examples within fifty years of the French Revolution and the pattern of devotion to the Ursuline martyrs as a group—not only at Cologne but at Grandmont and elsewhere—suggest that these reliquaries were made for use in a single church.

J.-J. Marquet de Vasselot was the first to relate head reliquaries in gilt copper to copper tombs produced in Limoges. Comparisons of the heads at Amiens, Brive, and Paris with the tomb-monuments of Blanche of France (d. 1243) (cat. 146) and William of Valence in Westminster Abbey (see fig. 10 and de Chancel-Bardelot, Appendix, list, no. 16), made between 1288 and 1296, suggest a date for the three reliquaries in the last quarter of the thirteenth century, since the facial features of the reliquaries resemble the sleek, uninterrupted line of brow and nose and the almond-shaped eyes of the Westminster figure. The engraved figures of angels in roundels set at the corners of the bases of the Amiens and Brive reliquaries are in the style that typically decorates small enameled or gilt-copper reliquary caskets and pyxes of Limoges manufacture of the mid-thirteenth century. A chasse in the Metropolitan Museum (32.100.288) is representative. The engraved lines defining the figures and drapery of the half-length angels are characteristically nervous, the facial features summarily rendered, and the ground enlivened with an overall engraved pattern.

BDB

NOTES
1. Boehm 1991, pp. 136–37.
2. Jacques de Voragine 1967, pp. 294–98.
3. The nineteenth Pope after Saint Peter. He was stricken from all records after he offended the Holy See by relinquishing his title in order to follow the virgins.
4. Present location unknown; Finoelst Sale 1927, no. 342.
5. Texier 1857, col. 894. These relics were given to the Ursulines at Eymoutiers, Beaulieu, Brive, and Ussel.
6. Ardant 1848, p. 74.
7. The museum in Brive preserves the relic received by the Ursulines in 1790; it is kept in a silver reliquary made in 1877. See Moser-Gautrand and Vidalo-Borderie (1983), who noted the two reliquaries preserves in the museum and in the collegiate church of Saint-Martin.

EXHIBITIONS (a): Amiens 1860, p. 182, no. 1133; Brussels 1880, p. 25, no. 158; Tokyo and Kyoto 1972, no. 56; (b): Paris 1900, no. 1699; Paris 1965, no. 395, pp. 214–15.

BIBLIOGRAPHY (a): Marquet de Vasselot 1898, pp. 266–67, fig. 4; Migeon 1906, pp. 14–16, fig. 246; Rückert 1959, figs. 11, 13; Bessard 1981, pp. 96–97; Boehm 1991, pp. 184–90; Chancel 1991, p. 38, fig. 19; Chancel 1992, pp. 30–32, no. 10; Falk 1994, p. 155, no. 13; (b): Texier 1857, col. 894; Molinier 1887a, p. 518, no. 48 (published with catalogue of Tulle exhibition, but not in exhibition); Rupin 1890, p. 450, fig. 500; Marquet de Vasselot 1898, p. 267; Guibert 1901, p. 14; Braun 1940, p. 414; Rückert 1959, p. 13, fig. 10; Souchal 1966, p. 210, fig. 4, p. 208; Favreau et al. 1978, vol. 2, pp. 38–39, no. 30, pl. VI, fig. 11; Camille 1989, pp. 275–76, fig. 145 (called head from Billac church); Boehm 1991, pp. 197–99; Chancel 1991, p. 38; Chancel 1992, pp. 31–32; Falk 1994, p. 155, no. 14.

# 154. Reliquary Head

Limoges, last quarter of the 13th century

Copper: formed, chased, scraped, engraved, and gilt; pearl

H.: 33 cm (13 in.); diam. of base: 15.2 cm (6 in.)

Ex colls.: Bouvier, Amiens (formed ca. 1840/50; sale, Amiens, December 8–16, 1873, no. 264); Aimé Desmottes, Lille and Paris (sale, Paris, March 19–23, 1900, no. 40); Octave Homberg (in 1904)

Condition: Substantial loss to gilding; vertical tear in metal at proper left back; hinges replaced at left and right.

Musée du Louvre, Paris; Département des Objets d'art (OA 6120)

This head of a female saint extends only to the middle of the neck. Her softly flowing hair is parted at the center, and a diadem with a single central pearl encircles her head at the line slightly above the forehead. She has almond-shaped eyes, a rather broad nose, full jaw, and an enigmatic smile. At the crown of the head, the copper is pierced with a cross surrounded by circles. The head stands on a splayed base with a central knop and a dentilated edge with incised lines. The reliquary can be opened by means of hinges placed at either side and at the back of the diadem.

The construction of the head is identical to that of the Amiens and Brive examples (cat. 153), but this piece differs in the construction of the base, which is formed of two pieces of metal, the circular foot and the upper cylinder. The Louvre reliquary differs also in the engraving, the waving of the hair, and the shape of the mouth. This suggests that the reliquaries were made in the same workshop with reference, but not strict adherence, to a common model.

Although the provenance of this reliquary and the identity of the saint are not known with certainty, it is probable that the head represents one of the virgin-martyr companions of Saint Ursula. Like the head from Amiens, the Louvre head can be traced to the Bouvier Collection in the late nineteenth century. Its closest parallel is the reliquary formerly in the Finoelst Collection, Paris (sold 1927, present location unknown), which is similarly placed on a splayed base (reminiscent of the chalicelike base of some Limousin statuettes of the Virgin)[1] and which also bears no inscription.

It is possible that all four surviving heads were made for use as an ensemble in a single church. This is suggested first by the remarkable survival of such closely related heads and by the common collection histories of the Louvre and Amiens examples, within fifty years of the French Revolution.

BDB

Note
1. Rückert 1959, p. 10.

Exhibitions: Possibly Amiens 1860, no. 1133 ("quatre reliquaires en forme de buste"); Brussels 1880, p. 25, no. 159.

Bibliography: Marquet de Vasselot 1898, p. 268, fig. 5; Migeon 1904b, pp. 42–43, ill. 36; Rückert 1959, p. 10, fig. 7; Bessard 1981, pp. 96–97; Boehm 1991, pp. 270–73; Chancel 1992, p. 31; Falk 1994, pp. 153–54, no. 12.

# 155. Virgin and Child

Limoges, last third of the 13th century

Copper: repoussé, engraved, and gilt, cut out and nailed; champlevé enamel: lapis and light blue, red; gems; wood core (walnut)

43.5 × 20 cm (17⅛ × 7⅞ in.)

INSCRIBED (on the inside part of the plinth; red enameled letters between blue lines): AVE MARIA GRACIA PLENA (Hail Mary, full of grace)

EX COLLS.: The count of La Béraudière (sale, Paris, May 18–30, 1885, no. 395); Baron Oppenheim; [Jacques Seligmann, Paris (ca. 1906)]; J. Pierpont Morgan, London and New York

CONDITION: Scepter that the Virgin carried and right arm of Child have disappeared; revetment of the plinth, on the inside, has been partially removed; gems missing on the brooch.

The Metropolitan Museum of Art, New York; Gift of J. Pierpont Morgan, 1917 (17.190.124)

About forty extant Limousin statuettes of the enthroned Virgin Mother, with or without wood cores, have been recorded up until now.[1] Many of them, in which the throne was hollowed out at the back and the opening provided with a door, were long considered reliquaries until Marie-Madeleine Gauthier convincingly proposed that they were actually tabernacles, designed to be placed on the altar.

The frontal poses of the Virgin Mary and the Christ Child, the gestures, and the attribute of the Child, who holds a closed book in his left hand and very likely was lifting his right hand in blessing, are all features shared by traditional Romanesque carved-wood Virgins in Majesty. However, the off-center position of the Child, who is seated on his mother's left knee, while characteristic of the series of Limousin statues, reflects the stylistic evolution taking place in the second half of the twelfth century.

This faithfulness to an outdated iconographic type was accompanied by a completely new sculptural experimentation in the Limousin workshops of the thirteenth century, which were faced with the prospect of creating images in the round as well as with the desire to rival sculpture. Certainly, great care was taken in depicting decoration on clothing: the fur linings of cloaks, the borders of garments finely engraved with a variety of motifs, the pendants of enameled belts, luxurious jewels, and majestic lobed crowns with engraved rinceau ornamentation. Yet, what is essential here remains the emphasis on smooth planes and strongly modeled volumes. The monumental sculpture that flowered at the time in Bordeaux, Charroux, and Limoges was, perhaps, the source of these preoccupations; small ivory sculptures, especially, could have served as models,[2] resulting in noticeable awkwardness that is apparent, among other places, in the verticality and the stiffness of the folds.

The pentagonal throne, the front of which bears simple engraved decoration, is adorned with arcades and pierced with keyhole-shaped openings—a common motif in Limoges—and topped with finials now missing their terminal elements. The iconography of the images on the enameled revetment at the sides and back deviates somewhat from that usually followed by Limousin artists: represented on the sides of the bench are the Angel and the Virgin of the Annunciation, the visual symbols of the mystery of the Incarnation, and the apostles Peter and Paul, an evocation of the Church with which Mary—the Mother of God, its source, and its emblem—is closely associated. For here, the two themes have been united. The first is suggested simply by the inscription: AVE MARIA GRACIA PLENA. The second is represented by the figure of Saint Peter, wearing the conical, pontifical miter and holding his keys. Peter is situated at the door to the tabernacle, surrounded by the other apostles who, all curiously unbearded, are shown in threes, under two rows of superimposed arches situated on the sides of the throne. The silhouettes of the apostles are in reserve on a ground of enamel (lapis blue, light blue, and red), with a rinceau decoration that is likewise in reserve. Saint Peter himself is set off against a lapis-blue ground decorated with polychrome fleurons.

The design of the engraved figures, as well as the slightly summary quality of the enameled decoration, the fusion of iconographic themes, and the style of the figures argue for a date in the last third of the thirteenth century. The present work should be compared with several statuettes preserved in the church of Breuil-au-Fa (Haute-Vienne), in the Victoria and Albert Museum, London (formerly in the collection of Dr. Hildburgh), and in the convent of Santa Clara in Huesca (Spain), the Aragonese province that, in 1955, Hildburgh proposed as the place of origin of most of these images. However, we choose to abandon his theory here in favor of a Limousin provenance.

FB

NOTES
1. The list made by Marie-Madeleine Gauthier regarding the framework of the corpus of southern enamels was published by her in 1993; see Gauthier 1993, pp. 95–100, and 133–37, on the function of these works.
2. See, for example, the Virgin dated between 1240 and 1250 from the former Homberg Collection, now in the Art Institute of Chicago. See Randall 1993, no. 1.

EXHIBITION: Paris 1900, no. 2520.

BIBLIOGRAPHY: Molinier 1904, no. 133, p. 57; "The Oppenheim Collection" 1906, p. 229; Hildburgh 1955, pp. 120–21; Gauthier 1968b, pp. 68–72, 84–86.

Reverse

427

# 156. Virgin and Child

Limoges, 1319

Copper: repoussé, engraved, and gilt, cut out and nailed; champlevé enamel: lapis blue, turquoise, red, and white; gems; wood core

53 × 26 × 22.5 cm (20⅞ × 10¼ × 8⅞ in.)

INSCRIBED (framing the door, on the reverse of the throne; blue enameled letters between red lines): + DOMINVS / HODO : DE MO / NTE ACVTO / HOSPITALAR / IVS PRIOR A / LVERNHIE : F / ECIT : FIER / I HANC YMAG / INEM : AD H / ONOREM B / EATE : VI / RGINIS : GL / ORIOSE : A / NNO DOMINI / MILLESIM / O TRIENTE / SIMO DEC / IMO NONO : / DOMI-NVS / DEVS : JESVS / XPS : PER SV / AM SA[N]C[TA]M MI / SERICORDIAM CVSTODIAT : / EVM : IN VITAM ETERNAM : AMEN (The lord Odon of Montaigu, hospitaler, prior of Auvergne, had this image made in honor of the blessed and glorious Virgin in the year of the Lord 1319. May the Lord God Jesus Christ by his blessed mercy keep it for eternal life. Amen)

PROVENANCE: Given in 1319 to the church of La Sauvetat by Odon de Montaigu, grand prior of the province of Auvergne of the Hospitalers of Saint John of Jerusalem.

CONDITION: Scepter of the Virgin missing; a few fleurons of her crown missing, as well as one of the shafts terminating in knobs that decorate the corners of the throne; small area in the revetment of the plinth missing on the inside; the crown of the Child and the cap of hair of the Virgin have been remade.

Church of Saint John the Baptist, La Sauvetat (Puy-de-Dôme)

Among Limousin statuettes of the enthroned Virgin Mother (see cats. 52, 155), the one from La Sauvetat is the only dated work in the series and is exceptional on account of the inscription on the back of the throne. It provides us with a record of the date of the work and the name of the donor, who is better known now since de Vaivre (1992) identified his effigy in the church of Olloix—once the seat of the commandery to which La Sauvetat belonged—and discovered a seal with his arms, a charged fess on a checky field. This coat of arms, carved in the stone at La Sauvetat, allows us to attribute to the head prior the erection or the modification of the dungeon and certain buildings in the fortified village, acquired in 1324 from Guigue VIII, dauphin of Viennois. References to Odon of Montaigu, invested with his office between 1309 and 1312, appear in texts until 1344.

The inscription, furthermore, has the merit of perhaps substantiating Gauthier's hypothesis that Marian statuettes whose thrones have an opening at the back, closed off by a door, were not reliquaries, as had long been believed, but tabernacles,[1] for the prayer with which the inscription ends, addressed to Christ, seems to echo the one recited in the Middle Ages during the distribution of Communion.[2]

The iconographic type still follows the tradition of Romanesque Virgins in Majesty—the figures are hieratic and frontal—and the Christ Child holds a book and makes a ges-

ture of blessing. Nonetheless, the Child is slightly detached from his mother, on whose left knee he sits, his pose consti-tuting a concession to the stylistic evolution that took place from the second half of the twelfth century on, and which appears to be characteristic of the Limousin group.[3]

The style of the statuette—the Virgin composed of about fifteen separate pieces of copper and the Child of about a half dozen—derives from the artistic developments in the thirteenth century and demonstrates the progress made in the direction of elegance and naturalism. The folds of the drapery swell and become supple, creating large peaks on the right side and between the legs of the Virgin; the borders of the robes fall in scroll-like configurations, and a flap of the Virgin's mantle flows in a long and fluid diagonal toward her slightly turned right foot. This mantle, with its gracefully cadenced drapery, envelops the Virgin's slender silhouette, except for her arms, which are clearly visible. On the back, the short veil falls, somewhat clumsily, into many folds. All these elements seem to derive, not without some delay, from contemporary designs for wood and ivory stat-ues of the second half—or even the middle—of the thir-teenth century.

This relatively successful attempt to rival sculpture is curiously combined with a perhaps conscious archaism apparent in the full and inexpressive faces, with their large blue enameled eyes and black dot to designate the pupils, endowing the figures with a fascinating gaze like that of the Virgin of Breuil-au-Fa (Haute-Vienne),[4] which is close to the present example in La Sauvetat.

The decorative details were carefully executed: an engraved band delineates the edges of the robes, and the motif on the reverse of the Virgin's mantle simulates vair. The pendant on Mary's belt is enameled in blue, and her shoes are attractively decorated with blue, red, and turquoise enamel. The enameling of the throne is of a more common type.

This throne, preceded by a semicircular platform, is sup-ported by four feet in the form of lions' paws, simply fash-ioned of bent and engraved strips of metal. It is decorated with the traditional keyhole-shaped cutouts and, at the cor-ners, with twisted shafts topped with knobs. At the sides, the enamelers depicted Saint Peter holding his keys and Saint Paul clasping the sword of his martyrdom; the two saints are often included in Limousin representations of the Virgin Mary to suggest the Church, with which she is so closely associated in medieval theology and symbolism. These small images were summarily indicated, without great concern for standard iconographic convention, since the two apostles appear beardless and shod, while, in addition, Saint Paul has a very thick head of hair.[5] On the door of the tabernacle an angel points to the dedicatory inscription. All of these silhouetted figures, in reserve, are set off against a ground of blue enamel strewn with fleurs-de-lis—a decorative motif without any precise heraldic significance.

Hildburgh compared this work with the group surrounding the Metropolitan Museum's Virgin and Child (see cat. 155), which he presumed to be originally from the Aragonese province of Huesca in Spain. Even if certain similarities outweigh the obvious differences, that the La Sauvetat Virgin is a product of the Limousin workshops is nonetheless incontestable. Situated chronologically almost at the end of the series of statuettes of the Virgin Mary, it bears witness, in its fine qualities as well as in its weaknesses, to both the stylistic evolution and the attachment to artistic stereotypes of the Limousin workshops, on the eve of their extinction.

FB

NOTES
1. Gauthier 1968b, pp. 68–71.
2. Jungmann 1954, pp. 288–90.
3. For example, the wood statues of Rosay-sur-Lieure (Eure) or of Taverny (Val-d'Oise).
4. Paris 1965, no. 359.
5. The use of inadequate models could have, perhaps, justified the iconographic peculiarities that Marie-Madeleine Gauthier (1993a) suggests interpreting as a desire to express, by their youth, the renewal tied to the jubilee of the year 1300.

BIBLIOGRAPHY: Guelon 1882; Palustre 1883, pp. 497–504; Rupin 1890, pp. 471–73; Hildburgh 1955, pp. 122–23; Vaivre 1992, pp. 577–614; Gauthier 1993a, pp. 87–137, especially pp. 124–27; Gauthier 1993b, pp. 121–36.

Reverse

# 157. Head of Saint Ferreolus

Limoges (Aymeric Chrétien), 1346

Copper: formed, chased, scraped, engraved, and gilt; champlevé enamel: turquoise, green and red; semiprecious stones

60 x 32 cm (23⅝ x 14⅞6 in.)

INSCRIBED (on amice at center back): D[OMI]N[US] : GVIDO : DE : BRVGERIA : P[A]RO/CHIA : S[AN]C[T]I : MARTINI : VET[ER]IS : CA/P[E]LL[ANU]S : ISTI[US] : ECC[LES]IE : DE ANEX/O[N]IO : FECIT : FIERI : LEM[OVICIS] : HOC/CAPVT : IN HONORE : B[EAT]I FERR/EOLI : PONTIFICIS + EGO : AY/MERICVS XPI[STI]ANI : AVRIFA/BER : DE CASTRO : LEM[OVICENSI] : FECI/HOC OPVS : LEM[OVICIS] : ANNO : D[OMI]NI M[I]LL[ESIM]O : CCCXL SEXTO DE P[RE]CEPTO : D[I]C[T]I: D[OMI]NI GVIDO[N]IS/ DE /BRVGERIA (Lord Guido de Brugières of the parish of Saint-Martin-le-Vieux, chaplain of the church at Nexon, had this head made in Limoges in honor of the blessed pontiff Ferreolus; I, Aymeric Chrétien, goldsmith of the château of Limoges, made this work at Limoges in the year of Our Lord one thousand three hundred forty-six at the order of Lord Guido de Brugières.)

PROVENANCE: Known from its inscription to have been ordered by Guido de Brugières of the parish of Saint-Martin-le-Vieux, chaplain of the church at Nexon. Classified, Monuments Historiques, June 20, 1891.

CONDITION: Losses to gilding; damage to enameling of inscription; vertical break in miter at proper right; lappets of miter broken off; some stones missing from miter and parure of amice; relics inside in post-medieval silver coupe.

Parish Church of the Beheading of Saint John the Baptist, Nexon (Haute-Vienne)

The shoulder-length image of Saint Ferreolus rests on a low, flat, polylobed base set on four clawed feet. The bishop-martyr is bearded and mustached, with softly curling hair, short bangs, almond-shaped eyes, thin mouth, fine narrow nose, and strongly delineated brows. He wears an amice and a miter decorated with engraved foliate bands and cabochons. On the circlet of the miter is an antique intaglio. The miter additionally has engraved, beaded decoration on the upper edges and, on the front, two enameled quatrefoils with three-quarter-length angels reserved against a green ground. On the reverse, two quatrefoils with addorsed and regardant birds are also reserved against a green ground. Engraved copper strips imitative of lappets were once suspended from the hinged tabs at the left and right of the center back of the miter. A large inscription plate, the letters enameled in blue and green, is set at the center back of the amice; the copper at the center front is worked in imitation of the bunched fabric of the amice. A small corpus of Christ is set at the center front of the base, which has a repeating flame decoration outlining its edge. The reliquary may be opened at the miter, which is hinged at either side.

The history of Saint Ferreolus was first chronicled by Gregory of Tours.[1] A sixth-century bishop of Limoges, Ferreolus was particularly noted for restoring the church of Saint-Martin at Brive. According to Bonaventure de Saint-Amable,[2] the saint's relics were brought to Nexon following the Norman invasions, and Geoffrey de Vigeois (d. 1185) also recorded that they were in the church at Nexon;[3] in the early fourteenth century Bernardus Guidonis noted that they were in a *capsa* (chasse),[4] which, according to Collin,[5] was given by the local noble family, Las Tours.

The reliquary of Saint Ferreolus, although one of a group of extant Limousin examples in gilt copper (see cats. 153, 154), is the only one dated and signed by its artist. It is the latest preserved example of copper sculpture produced in Limoges in the Gothic period.

Alain Erlande-Brandenburg (1983) suggested that the reliquary may be a composite work assembled in the mid-fourteenth century using an earlier mask. Yet he gave no comparative earlier examples to support the suggestion, and the rendering of the features, hair, and beard accords logically with a mid-fourteenth-century date. The scale and sophistication of the head recall the Limoges relief sculpture for enameled tombs, especially the head from the effigy of Blanche of Champagne of about 1306 (cat. 151), in such details as the curling locks of hair at the temples.

In a manner typical of Limoges workshop technique, engraved lines on the base of the head of Saint Ferreolus indicate the placement points for the attached elements.

Similar lines at the shoulder and neck indicate the placement of the amice. The miter is similarly engraved for the setting of the enameled quatrefoils. Such guidelines may indicate that some elements were separately worked by an apprentice for later assembly, especially since the bust slightly overlaps its outline on the base.

The quatrefoils of the miter, typical of the vocabulary of Limoges work in the decoration with figures of angels, are remarkable for the use of green enamel, which also appears in areas of the inscription. The range of green tones in Limoges work begins to vary from about the mid-thirteenth century, suggesting some experimentation.

BDB

NOTES
1. *History of the Franks* 1974, p. 395; *Vita S. Aridi*, sects. 10 and 29, in Gregory of Tours 1936–65, pp. 310–13; PL. 71, cols. 1127, 1138.
2. Bonaventure de Saint-Amable 1676–85, vol. 3, col. 215.
3. *Chronique de Geoffrey* 1864, chap. 15, p. 23; Ballot and Dautrement 1945–50, p. 18.
4. Labbé 1657, vol. 2, p. 632.
5. Collin 1672, p. 406.

EXHIBITIONS: Paris 1867, no. 93; Limoges 1886, pp. 60–61, no. 61, pl. XXXXIII; Paris 1900, no. 1678, 1693bis; Limoges 1948, p. 72, no. 151; Paris 1965, no. 370, pl. 138; Paris 1981, pp. 240–41, no. 194, ill.; Paris 1993, pp. 132–33.

BIBLIOGRAPHY: Ardant 1848, p. 140; Verneilh 1863, pp. 39–42; Linas 1867 and 1868, pp. 285–86; Guibert 1886, p. 223; Rupin 1890, pp. 175–77, figs. 251–53, pl. XXI; Marquet de Vasselot 1897, p. 266; Fayolle 1898, p. 299; Guibert 1901, p. 14; Demartial 1923, pp. 439–40, ill.; Kovács 1964b, p. 63, no. 21; Erlande-Brandenburg 1983, p. 109; Boehm 1991, pp. 254–60.

# Appendixes

# Tombs of Limoges Work

BÉATRICE DE CHANCEL-BARDELOT

## Extant and lost tombs of confirmed Limousin production

1. Ca. 1238. Burgos, cathedral. Tomb of Dom Mauricio, bishop of Burgos (d. 1238).

Bibl.: Rupin 1890, p. 164; Roulin 1899, pp. 487–95; Leguina 1912, pp. 11–37; Duran and Ainaud de Lasarte 1956, p. 66; Gauthier 1977, pp. 258–59; Gauthier 1988, p. 106. Exh.: Paris 1968, p. 255, no. 395.

The tomb of the bishop Mauricio of Burgos is one of the outstanding remaining testaments to the contribution Limousin enamelers made to the funerary arts of the thirteenth century. We might, however, question the homogeneity of the copper revetment, particularly that of the face, which is depicted in the sleep of death. The contraction of the eyebrows and the forehead and the somewhat tired wrinkles that mark the lower part of the prelate's face correspond to sculptural elements of the second quarter of the thirteenth century, such as the figure of Saint Denis(?) on the left portal of the west facade of Reims cathedral (Sauerländer 1971, fig. 210); the frown on the bishop's face also recalls a king on the south facade of the transept of the same cathedral (Sauerländer 1972, fig. 265). Nonetheless the very beautiful, carefully rendered mask is far removed from the geometricizing tendencies evident in the faces of reliquary heads or of the few surviving Limousin effigies.

2. Mid-13th century(?). Grandmont (Haute-Vienne, near Limoges arrondissement, Ambazac canton, commune of Saint-Sylvestre), monastery church. Tomb of Gerald IV, known as Hector, bishop of Cahors (d. 1199). Damaged during the Wars of Religion. Disappeared at the end of the eighteenth century.

Documentary sources: Pardoux de la Garde, Archives Départementales de la Haute-Vienne, ms. 1, Sém. 81, fols. 128v, 129: "Dans le choeur des religieux au mylieu de ladite esglise est un fort beau sepulchre de cuyvre doré esmaillé, par dessus lequel est l'effigie d'un evesque élevé en bosse (fol. 129) tenant une crosse de la main droite et de la senestre un livre sur lequel sont escriptz et gravez les vers que s'ensuyvent" (In the choir in the middle of the aforementioned church is a very handsome sepulchre of enameled gilt copper, above which is the effigy of a bishop in relief [fol. 129] holding a cross in his right hand and in his left a book upon which the following verses are written and engraved). Levesque 1662, p. 200: "Tandem anno quem Pardulphus non annotavit, post laudabilem vitae conversationem, obiit in Grandimonte et in medio choro juxta precedentem archiepiscopum sub simili mausolaeo sepultus est eidem piissimi versus leonini affiguntur, sed notandi praesertim decem qui compendio se vitam referunt" (In a year which Pardulphus did not record, after a praiseworthy life, he died and was buried in the middle of the choir next to the former archbishop in a similar sepulchre engraved with the same Leonine verses but with ten further verses that record his life).
Bibl.: Texier 1842 and 1843, pp. 152–53; Texier 1850 and 1851, p. 169; Rupin 1890, pp. 160–61; Lecler 1907–9, pp. 84–86; Souchal 1963, pp. 323–24, 329; Favreau et al. 1978, pp. 211–12, no. 104.

Gerard IV, bishop of Cahors, may have been the beneficiary, if we ascribe the execution of his funerary monument to the date of his death, of one of the earliest and finest effigy tombs of Limoges work. Souchal suggests that this is the case, plausibly emphasizing that the community of Grandmont probably did not tarry in honoring the memory of its benefactor; she thus places the work about 1209 (the year Gerard died, according to abbé Texier). The year of his death, however, should be viewed as a guide to an approximate dating. We prefer to hypothesize that the monument described by Pardoux de la Garde and compared by Levesque to the memorial for Aymeric Guerry was erected about the same time as Guerry's tomb, in the mid-thirteenth century, perhaps replacing an earlier cenotaph.

1

3. Mid-13th century. Royaumont (Val-d'Oise, Montmorency arrondissement, Viarmes canton, commune of Asnières-sur-Oise), Cistercian abbey; later moved to the abbey of Saint-Denis. Tomb of John (d. 1243) and Blanche of France (d. 1248) (cat. 146).

4. Ca. 1253–67. Pamplona, cathedral. Tomb of Thibaut le Chansonnier—Thibaut IV of Champagne, Thibaut I, king of Navarre (d. 1253). Disappeared in 1276.

Documentary source: Martène 1717, vol. 1, cols. 1124–25: letter of 1267 from the prior of Grandmont to Thibaut V of Champagne, delivered by John "Chatelas" (from Chaptelat), burgher of Limoges, asking where the tomb should be transported (the Pamplona cathedral) and requesting payment for the tomb.
Bibl.: Favyn 1612, pp. 298, 325; Ardant 1855, p. 83; Texier 1857, cols. 1400–1402; Daunou and Naudet 1860 [1840], pp. 507, 508; Arbellot 1888, p. 243; Rupin 1890, p. 161; Mâle 1908, pp. 447–48; Taittinger 1987, pp. 313–14.

Having died in Pamplona, Thibaut I of Navarre was buried there in the cathedral of Santa Maria, as was his son Henry, who was king of Navarre between 1271 and 1274. Relating the troubles that arose in the country in 1276, chroniclers mention the pillaging of the cathedral and, according to William of Nangis, the destruction of the gilt-copper tomb of King Henry I. One can presume that Thibaut I's sepulchre met the same fate. There is no precise record of how this monument looked.

5. Ca. 1259. Nantes (Loire-Atlantique), Franciscan monastery church choir. Tomb of Robert of Brittany (d. 1259), son of John I, duke of Brittany, and Blanche of Champagne. Disappeared during the eighteenth century.

5

435

8

Graphic source: Gaignières album, Paris, Bibliothèque Nationale, Département des Estampes, Pe 11c, fol. 74 (Bouchot 1891, no. 4947; Adhémar 1974, no. 265).
Documentary source: Dubuisson-Aubenay 1636, pp. 73–74: "Pierre dure élevée de deux pieds sur terre, longue de cinq et large environ de deux, portant au dessus une autre plus courte et plus estroite de bois, eslevée encore prez d'un pié, couverte d'une lame de cuivre émaillé à fleurs et creuse comme une bière, portant une statue d'enfant gisant, vêtu d'une chemise dorée et ayant une calotte ou béguin sur le tête; icelle sepulture ou bière portant au dessus ladite statue, le tout émaillé de diverses couleurs et tout autour est escrit en lettre noire d'émail . . . . Puis, tout autour de l'oreiller, aussy de cuivre émaillé à fleurs, sur lequel sa teste repose, il y a escrit aussy en esmail noir: *Qui obiit IIII idus febroarii, anno Domini MCCLIX*" (Hard stone raised two feet off the ground, five long and about two wide, above it another shorter and narrower of wood, also more than a foot high, covered with a thin plate of copper enameled with flowers and hollow like a coffin, containing an effigy of a child, clothed in a gild`ed chemise and wearing a cap or hood on the head; this sepulchre or bier carrying the aforementioned statue, the whole enameled with various colors and inscribed all around in black enamel letters . . . . Next, all around the pillow, likewise enameled with flowers, upon which the head rests, inscribed in black enamel is: *Qui obiit IIII idus febroarii, anno Domini MCCLIX* [who died on the fourth of the ides of February (i.e., February 9) A.D. 1259]).
Bibl.: Lobineau 1707, vol. 2, col. 432; Morice 1750–54, col. 980; Grégoire 1882, p. 32; Brault 1925, p. 178; Copy 1986, pp. 40, 50, 269, no. 135; Vaivre 1988b, pp. 61, 64; Copy 1993, pp. 385–401.

6. Ca. 1260. Angers (Maine-et-Loire), Jacobin monastery. Tomb of Michael of Villoiseau, bishop of Angers (d. 1260) (cat. 148). Disappeared in 1728.

7. Ca. 1260. Champagne (Sarthe, Le Mans arrondissement, Sillé-le-Guillaume canton, commune of Rouez), Cistercian abbey of Notre-Dame. Tomb of William Roland, bishop of Le Mans (d. 1260) (cat. 147). Disappeared.

8. Third quarter of the 13th century. Fontaine-Daniel (Mayenne, Mayenne arrondissement and canton, commune of Saint-Georges-Buttavent), Cistercian abbey. Tomb of Juhel III of Mayenne (d. 1220). Disappeared between 1772 and 1776.

Graphic source: Gaignières album, Oxford, Bodleian Library, Gough-Gaignières ms. 14 (18359); Paris copy, Bibliothèque Nationale, Département des Estampes, Pe 1g, fol. 200 (Bouchot 1891, no. 2729; Adhémar 1974, no. 51).
Documentary sources: Anonymous, seventeenth century (Guyard de la Fosse 1850, p. 52): "On y voit son tombeau, avec la figure couchée, de lumière dorée, de la forme de ceux des anciens seigneurs . . . la tête nue et les mains jointes . . . . Quant à ses pieds, ils sont appuyés sur le dos d'un chien. . . . Sa cotte d'armes . . . dénote aussi qu'il était guerrier, ainsi qu'il est certifié par son épitaphe, écrite en lettres gothiques" (One saw his tomb, with his sleeping figure, gilded by the light, fashioned like those of the old

lords . . . the head bare, the hands joined. . . . As for his feet, they rest on the back of a dog. . . . His coat of arms . . . indicates that he was also a warrior, which is affirmed by his epitaph, written in Gothic letters). Pierre-François Davelu, *Répertoire topographique et historique du Maine*, ca. 1775, Le Mans, Bibliothèque Municipale, ms. B 471 (according to abbé Angot; passage not found in the manuscript).
Bibl.: Guyard de la Fosse 1850, p. 52; Texier 1857, col. 1404; Rupin 1890, p. 161; Leblanc 1892, p. 71; Grosse-Dupéron and Gouvrion 1896, pp. 3, 290–91; Angot 1900–1902; Vaivre 1988a, pp. 60, 64; Vaivre 1988b, p. 84; Mussat 1988, pp. 140–41, 148–49.

The tomb of Juhel de Mayenne certainly dates after 1243, the year of the consecration of the abbey church of Fontaine-Daniel. The large armorial shield hanging from the knight's belt, the Mayenne coat of arms repeated on the frame, and the small escutcheons placed on the walls of the sarcophagus evidence the monument's role in the development of heraldry on Limousin tombs. According to Mussat, the presence of the Avaugour family coat of arms is an additional indication that the sepulchre should be dated in the third quarter of the thirteenth century.

9. Third quarter of the 13th century. Grandmont (Haute-Vienne, Limoges arrondissement, Ambazac canton, commune of Saint-Sylvestre), monastery church. Tomb of Aymeric Guerry, archbishop of Lyons (d. 1257). Damaged during the Wars of Religion. Disappeared at the end of the eighteenth century.

Documentary sources: Pardoux de la Garde, Archives Départementales de la Haute-Vienne, ms. 1 Sém. 81, fol. 132: "Dans ladite esglise au choeur, tirant vers le grand autel, est un aultre sepulchre, semblable a l'aultre, de cuyvre doré esmaillé; est tout entouré de lions de cuyvre doré. Sur ycellui est l'effigie et pourtraict en bosse bien eslevé d'un archevesque comte de Lyon et primat des Gaules. L'épitaphe duquel est engravée dans le cuyvre esmaillé autour dudit sepulchre en vers latine comme s' ensuit . . ." (In the aforementioned church in the choir, near the main altar, is another sepulchre, similar to the other, of enameled and gilt copper; it is surrounded by lions of gilt copper. On it is the effigy and portrait in full relief of an archbishop, count of Lyons and primate of the Gauls. The epitaph engraved on the enameled copper around the sepulchre in Latin is as follows . . .); Levesque 1662, p. 199: "Cui in medio choro appositum est mausolaeum cum effigie aurichalco expressa gemmis et leonisis (quae ultimo seculo sublata sunt a Sangermano, ut ibidem dictum). Hujus sicut et sequentium virorum epitaphia refert Pardulphus hoc modo" (to whom is situated, in the middle of the choir, a sepulchre with an effigy of copper decorated with gems and accompanied by lions, depicting a mitered bishop, with thin plates[?] and other wonderful things [which are held up during the last century by Saint Germain(?) as they say there]. Pardulphus reports his epitaph and those of following people as follows . . .). The small copper pieces (laminulis et aliis curiosis) were removed by the soldiers of Saint-Germain-Beaupré during the Wars of Religion.
Bibl.: Texier 1842 and 1843, pp. 153–54; Texier 1850 and 1851, p. 165; Texier 1857, col. 1399; Rupin 1890, pp. 160–61; Arbellot 1892, pp. 59–70; Lecler 1907–9, pp. 456–58; Hugon 1931, p. 478; Souchal 1963, p. 324; Favreau et al. 1978, pp. 215–16, no. 107; Galland 1994, pp. 140–46.

10. Ca. 1272–74. Beauvais (Oise), cathedral, "à gauche du grand autel dans le choeur" (to the left of the main altar, in the choir). Tomb of Philip of Dreux, bishop of Beauvais (d. 1217). Disappeared during the eighteenth century.

Graphic source: Gaignières albums, Oxford, Bodleian Library, Gough-Gaignières ms. 1 (18346); Paris copy, Bibliothèque Nationale, Département des Estampes, Pe 1, fol. 76 (Bouchot 1891, no. 1938; Guibert n.d., pl. 203; Adhémar 1974, no. 62); Paris, Bibliothèque Nationale, Département des Manuscrits, ms. Clairambault 645, fol. 45 (Bouchot 1891, no. 7232; Adhémar 1974, no. 61).
Bibl.: Louvet 1631–35, vol. 2, pp. 360–61; Texier 1857, col. 1402; Desjardins 1865, pp. 12–14, 238, 253; Viollet-le-Duc 1871–72, vol. 2, p. 225; Rupin 1890, p. 158; Souchal 1963, pp. 325–26; Murray 1989, p. 53.

Philip of Dreux was first buried in the church of la Basse-Oeuvre in Beauvais. In 1274 his tomb was moved to the new choir of the cathedral completed in 1272; at this time it received an epitaph. According to the inscription as transcribed by Louvet, the remains of the bishop Roger de Champagne, who died in 1022, were added to those of Philip of Dreux, which explains Desjardins's uncertainty and lack of coherence concerning the identification of the prelate represented by the enameled effigy. The treatment of the figure, which is very similar to that of William Roland (d. 1260), confirms the date furnished by the transfer of the prelate's remains.

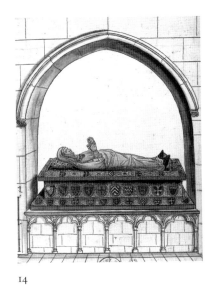

10                                    12                                                          14

11. Ca. 1276. Rochester, cathedral. Tomb of Gauthier of Merton, bishop of Rochester (d. 1276). Disappeared.

Documentary source: Highfield 1954, p. 137, no. 111: Several payments by the executors of the will. 40 pounds sterling and 6 deniers for construction and transportation; 40 pounds sterling and 8 deniers to draw up the plan and oversee the execution; 10 pounds sterling and 8 deniers tournois to go get the tomb in Limoges and bring it back to Rochester.
Bibl.: Ardant 1855, p. 83; Texier 1857, cols. 999, 1059; Arbellot 1888, p. 243; Rupin 1890, pp. 162–63; Marquet de Vasselot 1941, pp. 15–16; Gauthier 1988, p. 106.

12. Ca. 1277. Evron (Mayenne, Laval arrondissement), Benedictine abbey of Notre-Dame. Tomb of Claude Renaud, lord of Evron (d. 1277). Disappeared at the end of the eighteenth century.

Graphic source: Gaignières album, Oxford, Bodleian Library, Gough-Gaignières ms. 14 (18359); Paris copy, Bibliothèque Nationale, Département des Estampes, Pe 1g, fol. 205 (Bouchot 1891, no. 2734; Adhémar 1974, no. 356).
Documentary source: Pierre-François Davelu (ca. 1775), Le Mans, Bibliothèque Municipale, ms. B 471, p. 444: "On voit dans cette église près du grand autel le tombeau de Renaud, vicomte de Champagne, mort en 1277" (In this church, near the main altar, one sees the tomb of Renaud, viscount of Champagne, who died in 1277).
Bibl.: Gérault 1838, pp. 25, 70–71; Texier 1857, col. 1404; Rupin 1890, p. 161; Angot 1900–1902; Lefèvre-Pontalis 1903, p. 28ff.

The effigy with closed eyes on this tomb is quite similar to that of Juhel de Mayenne. The main differences are the enameled plates of ornamental foliage upon which Renaud's figure lies, the knight's bare hands, and, on the chest of the sarcophagus, the presence of medallions with busts of angels instead of armorial bearings. The very small portion of the framing of the tomb that still existed during Gaignières's day does not reveal if this area received any heraldic ornamentation, in accordance with the arms decorating the shield and tunic.

13. Mid-13th century, ca. 1272–80. Villeneuve (Loire-Atlantique, Nantes arrondissement, Aigrefeuille-sur-Moine canton, commune of Le Bignon), Cistercian abbey. Tombs of Alix of Thouars (1199–d. 1221) and her daughter Yolanda of Brittany (d. 1272) (cat. 149). Disappeared.

14. Ca. 1288. Villeneuve (Loire-Atlantique, Nantes arrondissement, Aigrefeuille-sur-Moine canton, commune of Le Bignon), Cistercian abbey. Tomb of Andrea or Eustacia de Vitré, daughter of Andrew III of Vitré and Thomassa of Pouancé, married Oliver

of Machecoul in 1269 (d. 1288). Disappeared at the end of the eighteenth century.

Graphic source: Gaignières album, Paris, Bibliothèque Nationale, Département des Estampes, Pe 2, fols. 37, 38 (Bouchot 1891, nos. 3832, 3833; Adhémar 1974, no. 446).
Documentary source: Dubuisson-Aubenay 1636, p. 156.
Bibl.: Dubuisson-Aubenay 1636, pp. 156–58; La Nicollière 1859–61, p. 268; Copy 1986, pp. 33, 41, 268, no. 116; Vaivre 1988a, p. 64.

This tomb belonged to a relative of the ducal family of Brittany (the deceased's father's first wife was Catherine of Thouars, sister of Alix of Thouars, duchess of Brittany; the deceased's husband, Oliver of Machecoul, was the son of Pierre Mauclerc, duke of Brittany). The figure of the woman is depicted with closed eyes, and the monument is embellished with many armorial escutcheons, quite similar in conception to the double tomb, also in the abbey of Villeneuve, of Alix of Thouars and Yolanda of Brittany (cat. 149).

15. Ca. 1290. Angers, priory of Notre-Dame de La Papillaye. Tombs of Herbert Lanier (d. 1290) and his wife Alès (cat. 150). Disappeared at the end of the eighteenth century.

16. Ca. 1292. London, Westminster Abbey. Tomb of William of Valence, earl of Pembroke (d. 1296).

Bibl.: Texier 1857, cols. 999, 1059, 1405; Rupin 1890, pp. 168–69; Gauthier 1972b, pp. 192–95, 376–77, as between 1288 and 1296; Gauthier 1988b, pp. 109–10 (with mention of an unpublished study of this tomb by Gauthier, given as a talk in London in 1986).

On this monument, the only preserved example of a knight's tomb made of Limoges work, the figure is represented clothed in a tunic over armor, a heraldic shield attached to the belt; a hood of mail covers the head, which is crowned with a diadem. The gloved hands are joined, the eyes are open. The copper revetments inside the *cuvette* and the tomb frame have disappeared. The ornamentation is refined, with filigree bands embellishing the clothing. The face presents the geometricizing tendencies and a certain hardening of the features characteristic of the evolution of Limousin modeling at the end of the thirteenth century.

17. Ca. 1294. Foucarmont (Seine-Maritime, Dieppe arondissement, Blangy-sur-Bresle canton), Cistercian abbey of Saint John the Evangelist. Tombs of John of Brienne (d. 1294) and his mother (d. 1260). Lost during the Wars of Religion.

437

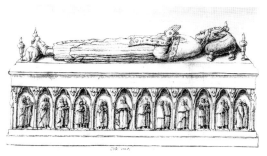

19

Documentary source: "Chronique des comtes d'Eu," in *Recueil 1738–1904*, vol. 23 (1876), pp. 444–46.
Bibl.: Guibert 1879, p. 196.

These tombs were commissioned from Limoges by John of Brienne during his lifetime, completed and transported by John des Forges. They were "de fin cuyvre et y a deux ymaiges enlevez tres grands . . . et est tout l'ouvraige qui est suroré d'or fin" (of fine copper with two figures in high relief . . . and the whole work is gilded with fine gold). At the heads of the effigies were their names and enameled armorial bearings and "ymaiges eselez des princes" (distinct images of princes) as well as images of Christ, the Virgin, Saint Denis, Saint Peter, and Saint Paul. In addition, "lesdites tombes sont closes en manière de huche et ont esté empirées moult durant les guerres" (these tombs are enclosed in a wooden framework and have been much damaged during the wars).

18. Beginning of the 14th century. Abbey of la Joie (Morbihan, Lorient arrondissement, canton and commune of Hennebont), later, Musée du Louvre. Tomb of Blanche of Champagne, duchess of Brittany (d. 1285) (cat. 151).

19. First quarter of the 14th century. La Chapelle-Taillefert (Creuse, Guéret arrondissement and canton), collegiate church. Tomb of Cardinal Peter of la Chapelle-Taillefert (d. 1312). Disassembled and transported to Guéret between 1763 and 1769; disappeared during the Revolution.

Graphic source: P. Beaumesnil, drawings, about 1769–70 (Paris, Bibliothèque Nationale, Département des Estampes, Ve 883–4, pp. 6–9, and Ro 86033–35).
Documentary source: Ibid.
Bibl.: Texier 1842 and 1843, p. 156; Ardant 1855, p. 86; Texier 1857, cols. 1060–65, 1402; Rupin 1890, p. 167; Hugon 1931, pp. 471–78; Lacroix 1988; Lacroix 1993.

20. Ca. 1310. Poitiers (Vienne), monastery church of the Jacobins. Tomb of Guy of Lusignan, count of La Marche and Angoulême (d. 1307 or 1310). Probably disappeared during the Wars of Religion.

Documentary source: Bouchet 1524, fol. 7 of part four: "L'an 1310. Au dit an alla de vie a trespas Guy de Luzignen, comte de la Marche et d'Engoulesme en la ville de Poictiers; et fut enterré en l'église des frères prescheurs où l'on veoit encore de present sa riche sépulture enlevée en bosse de cuivre doré et richement esmaillée" (The year 1310. Guy of Lusignan, count of La Marche and Angoulême, in the town of Poitiers, departed this life; he was buried in the church of the preaching brothers where one still can see his opulent sepulchre raised in relief in gilt and splendidly enameled copper).
Bibl.: Crozet 1942, p. 69, no. 284; Copy 1986, p. 42.

The son of Hugh XII le Brun, count of La Marche and Angoulême, Guy of Lusignan was the grandson of Yolanda of Brittany, who was buried beneath a tomb of Limoges work in Villeneuve (cat. 149), and Hugh XI le Brun, count of La Marche and Angoulême. This monument was most likely the victim of the pillaging of the Jacobin monastery of Poitiers in March 1559

or of the occupation of the town by Protestants in 1562 (on these troubles, see Barbier 1891, pp. 3–4, n. 2).

21. Ca. 1321–22. Maure-de-Bretagne (Ille-et-Vilaine, Redon arrondissement), parish church. Tomb of Thomas of Anast, bishop of Quimper (1321–22; d. 1322). Disappeared after 1623.

Documentary source: Proceedings of 1623 (ed. Guillotin de Corson 1884 and 1887): "Un ancien tombeau hault et enlevé [de terre], sur lequel il y une effigie en cuivre doré, d'un evesque qu'on dist avoir esté de Cornouaille, environné de plusieurs escussons d'esmail et entre iceux les armes pleines et escartelées de Maure" (An old tomb, high and raised [from the ground], on which there is an effigy in gilt copper, of a bishop whom they say was from Cornwall, surrounded by several enamel escutcheons and, between them, the full and quartered arms of Maure).
Bibl.: Guillotin de Corson 1884, vol. 5, p. 158; Guillotin de Corson 1887, p. 178; Banéat 1973, vol. 2, p. 349; Copy 1981, pp. 175–76, no. 94; Copy 1986, pp. 58, 268, no. 123.

22. After 1327. Tombs of William of Harcourt, Blanche of Avaugour, John of Harcourt, and Jeanne of Thorigny. Disappeared.

Documentary source: Paris, Archives Nationales, JJ 65, fols. 152v, 153r. Testament of William of Harcourt, dated July 18, 1327 (Delage 1918, p. 278): Two tombs "hautes et levées de l'Oeuvre de Limoges" (high and raised of Limoges work) for William of Harcourt and Blanche of Avaugour, his wife, and two others "auxi de l'oeuvre de Limoges qui seront par terre et sanz enleveure" (also of Limoges work, which would be on the ground, not raised) for John of Harcourt, his son who died young, and Jeanne of Thorigny, his first wife.
Bibl.: Delage 1918, pp. 275–81.

These tombs were destined for the collegiate church of la Saussaye but were perhaps placed in the church of Notre-Dame du Parc at Harcourt; they are mentioned neither by the historians of the Harcourt family nor by the chroniclers of the collegiate church of la Saussaye.

23. Ca. 1329. Rieux (Haute-Garonne, Muret arrondissement), Dominican monastery. Tomb of Arnaud Frédet, bishop of Couserans (d. 1329). Disappeared during the Revolution.

Documentary sources: Blaise Binet, *Mémoire*, Paris, Bibliothèque Nationale, Département des Manuscrits, collection Languedoc, vol. 20, fol. 159ff.: "Le couvent des Jacobins possède le tombeau d'Arnaud Frédet, jacobin, évêque de Couserans, mort en 1329; il est situé dans le choeur; il est d'airain, peint en émail de différentes couleurs et posé sur un socle de pierre de taille, qu'il s'était fait faire de son vivant et sur lequel il est représenté habillé pontificalement, la mitre sur la tête qui est appuyée sur un carreau, les mains croisées sur la poitrine, la crosse passée dans le bras gauche, ayant à ses pieds un gros animal qui a des ailes et des serres d'aigle, et toutes les pièces d'accompagnement d'airain émaillé, de même que les écussons de ses armes qui ornent le pourtour du tombeau. L'inscription sépulcrale, qui est en latin . . ." (The Jacobin monastery possesses the tomb of Arnaud Frédet, Jacobin, bishop of Couserans, who died in 1329; it is situated in the choir; it is of brass, painted in enamel of different colors and positioned on a freestone base, which he had made while he was alive and upon which he is depicted dressed pontifically, miter on his head, which is supported by a square cushion, hands crossed on his breast, crosier carried in his left arm, at his feet a large animal with the wings and talons of an eagle, and all the trimmings in enameled brass, including the escutcheons and coats of arms that decorate the periphery of the tomb. The inscription, which is in Latin . . .). Letter from the Rieux district to the Revolution (published by Contrasty): "Un sarcophage en cuivre émaillé qui est placé derrière le grand autel de l'église des Dominicains de Rieux; il repose sur un dé en maçonnerie avec une inscription en lettres gothiques et des écussons de même métal; il représente l'évêque du Couserans . . . l'épaisseur du cuivre est d'environ 1 ligne et paraît de bonne qualité" (A sarcophagus in enameled copper situated behind the main altar of the Dominican church of Rieux; it rests on a stonework block with an inscription in Gothic letters and escutcheons of the same metal; it depicts the bishop of Couserans . . . the thickness of the copper is about one *ligne* and appears to be of good quality).
Bibl.: Martin 1917, pp. 190–93; Contrasty 1936, p. 55.

24. Ca. 1363. Saint-Germain-les-Belles (Haute-Vienne, Limoges arrondissement), collegiate church. Tomb of Hugh Roger, bishop of Tulle, cardinal (d. 1363). Disappeared during the Revolution.

Documentary source: Paris, Bibliothèque Nationale, *Armoires de Baluze*, vol. 21, p. 48 (Fage 1885): "fait de bronze doré esmaillé, de longueur de neuf pieds, trois de largeur et deux de hauteur, soustenu par un berceau de boys d'un pied de hauteur" (made of enameled, gilt brass, nine feet long, three feet wide, and two feet high, held up by a wooden support one foot high), with armorial bearings, effigy, covered with ornaments of enameled scenes; border with coats of arms and figures; base with figures in relief. Baluze 1717, pp. 195–96: Hugh Roger died in 1363, at the monastery of Montolieu (Carcassonne diocese), where he was buried. Later his body was brought to the Limousin where it was buried "in ecclesia Sancti Germani de Manso Sereno, vulgo *Maseré* aud procul Tutela, in monumento aereo elegantissimo quod etiamnum superest. Testamentum enim condiderat, in quo edixerat uti ex facultatibus et bonis suis constitueretur collegium canonicorum saecularium in aedibus suis apud Villam Novam Avenionensis diocesis. Sed executores illius testamenti mutavere locum et mentem fundatoris transtulerunt ad diocesim Lemovicensem, ad ecclesiam videlicet parrochialem Sancti Germani de Manso Sereno, ubi collegium illud institutum fuit, uti pluribus a nobis dictum est in Notis ad vitas paparum Avenionensium" (in the church of Saint-Germain de Manso Sereno, commonly called Maseré, not far from Tulle, is an exquisite brass monument which one can still see. By his testament he [Hugh Roger] established from his wealth and his properties a college of secular canons in his church at Villeneuve in the diocese of Avignon. But the executors of his testament, changing the location and the founder's intention, moved this undertaking to the diocese of Limoges to the parish church of Saint-Germain de Manso Sereno, where this college of canons was established as the Writings on the Lives of the Avignon Popes repeatedly tell us).
Bibl.: Fage 1885, pp. 51–57; Linas 1886d, pp. 83–84; Linas 1886, pp. 150–53; Poulbrière 1886, p. 155; records of the meetings, April 27 1937, *BSHAL* 77 (1938), pp. 111–12.

## Tombs without effigies made in the Limousin

Among the shapes a funerary monument might take in the Middle Ages, a tomb with an effigy was but one; churches and religious buildings also housed simple engraved slabs, or plaques set into the walls, bearing epitaphs and, sometimes, figural representations or decorative or heraldic motifs. Only rare examples of works of this type by Limousin enamelers survive. To these we add a mention of the Belhomer monument, by the eighteenth century already too deteriorated for its original appearance to be determined.

25. Belhomer (Eure-et-Loire, Nogent-le Rotrou arrondissement, La Loupe canton, commune of Belhomer-Guéhouville), priory of Fontevrault. Tomb of a child(?) of Arthur II of Brittany and Yolanda of Dreux, who died at an early age (perhaps Blanche, born in 1300). Partially dismantled before 1711. An inscription allows us to date the tomb between 1294 (the year Arthur of Brittany married Yolanda of Dreux) and 1305 (the year Arthur became duke of Brittany).

Documentary source: "CXVII. Monument trouvé au prieuré de Belhomer," *Mémoires pour l'histoire des sciences et des beaux-arts recueillis par l'ordre de son altesse sérénissime monseigneur prince souverain des Dombes, août 1711* ("CXVII. Monument found in the priority of Belhomer," Report for the history of sciences and fine arts gathered on the order of His Most Serene Highness the Lord Sovereign Prince of Dombes, August 1711), "Monument. . ." 1711, pp. 1365–70: "On y a trouvé un tombeau qui était autrefois magnifique. Il était de bronze doré et émaillé. Un misérable qui a voulu s'enrichir en dépouillant ce tombeau en a enlevé tout ce qu'il a pu, et il n'est resté que deux écussons et une lame de cuivre qui environnait la base du tombeau. Les deux écussons sont de figure antique, c'est à dire presque en triangle. De ces deux écussons, l'un représente les armes de Montfort. . . . L'autre représente les armes de Dreux-Bretagne. . . . Sur la lame de cuivre on lit ces paroles: *domini jadis aliter dicti Arturi de Britannia, vice comitis Lemovecensis et dominae Iolandis* . . . ." (A tomb has been found that was once magnificent. It was of gilt and enameled brass. A wretch who wanted to enrich himself by despoiling this tomb has removed all that he could, and there remain only two escutcheons and a sheet of copper surrounding the base of the tomb. These two escutcheons are of an old type, almost triangular in form. One of the two escutcheons represents the Montfort arms . . . . The other represents the Dreux-Brittany arms . . . . On a sheet of copper are these words: "of the aforesaid lord, also called Arthur of Brittany, vice earl of Limoges, and of the lady Yolanda" [The article identifies these personages as Arthur II, duke of Brittany, and his wife Yolanda of Dreux and dates the inscription between the marriage of Arthur and Yolanda (1294) and Arthur's accession to the duchy of Brittany (1305); it concludes that the tomb is probably that of one of the couple's children.] "It could only be the tomb of Blanche, their fourth daughter, who, according to history, died young. But since Jeanne, their eldest, was born in

1298, it is unlikely that Blanche was born in 1300, and if she was not born until 1310, the monument could not have had anything to do with her. But we are conjecturing too much and should not do so; we should content ourselves with awaiting the memoirs of the lords of Belhomer").
Bibl.: Poulbrière 1886, pp. 154–58; Rupin 1901, pp. 601–6.

Most likely this was a tomb with an effigy, as it had heraldic decoration in keeping with what we know of such monuments commissioned by the ducal family of Brittany.

26. Civray (Vienne). Funerary plaque of Guy de Meyos (d. 1307). Musée du Louvre (cat. 152).

27. Limoges (Haute-Vienne), Jacobin monastery, cloister. Tomb of Isabelle of Ventadour (d. 1278), wife of Guy de Meyos. Disappeared during the eighteenth century.

Documentary sources: Testament (Nadaud 1880): "Isabelle de Ventadour, par souci de discrétion, souhaitait que sa tombe située dans le choeur de l'église fût recouverte de cuivre" (Isabelle of Ventadour desired that her tomb be covered in copper, so as not to disturb the comings and goings in the church choir). *Memoralia pro conventu lemovicensi* (Douais 1892): "In muro portae claustri cernitur in lamina aenea episcopus indutus pontificalibus et officium sepulturae faciens pro femina, fratribus nostris stipatus. Hae fuit domina Elisabeth de Ventodorio, quae obiit anno Domini M.CC.LXXVIII nonis octobris; ad cujus sepulchrum hi versus leguntur . . ." (In the wall of the cloister entrance one can see on a copper plate the figure of a bishop dressed in the garments of his office which is used as a tomb for a woman, surrounded by our brothers. She was Lady Elizabeth of Ventadour who died in the year of Our Lord 1278, on the nones [the seventh] of October, over which tomb one can read these verses . . . ). Archives Départementales de la Haute-Vienne, no. 4406: foundation of anniversaries by Isabelle de Ventadour, 1277.
Bibl.: Texier 1850 and 1851, p. 197; Nadaud 1880, pp. 252–53; Poulbrière 1886, pp. 154–55; Douais 1892, pp. 45, 305; Favreau et al. 1978, pp. 126–27.

28. Saint-Léonard-de-Noblat (Haute-Vienne, Limoges arrondissement), priory of Augustinian canons regular of l'Artige-Grande. Plaque bearing the epitaph of Agnes, mother of Brother Arnaud of Rozier de l'Artige (d. 1253) and plaque bearing the epitaph of prior Peter des Prés (d. 1313). Disappeared.

Documentary source: Martial de Gay, lord of Nexon, 1586 manuscript (Archives Départementales de la Haute-Vienne, no. 26 J 13).
Bibl.: Becquet 1963b, p. 99, appendix 3.

## Lost tombs possibly made in the Limousin

Angers (Maine-et-Loire), cathedral. Tomb of William of Beaumont, bishop of Angers (d. 1240). Disappeared in 1747.

Graphic sources: Bruneau de Tartifume 1623 and 1977, p. 58; Gaignières album, Oxford, Bodleian Library, Gough-Gaignières ms. 14 (18359); Paris copy, Bibliothèque Nationale, Département des Estampes, Pe 1f, fol. 59 (Bouchot 1890, no. 2588; Guibert n.d., pl. I–23; Adhémar 1974, no. 183); Angers copy, Musée d'Angers, inv. MA VI R 985–1; Paris, Bibliothèque Nationale, Département des Manuscrits, ms. lat. 17030, fol. 93 (Bouchot 1890, no. 6660). Documentary sources: Angers, Bibliothèque Municipale, ms. 618, vol. 1, p. 255; ms. 656 (sale to a founder); ms. 666 (cleaning of the tomb); Angers, Archives Départementales du Maine-et-Loire, Lehoreau, book V.
Bibl.: Bruneau de Tartifume 1623 and 1977, pp. 57–59; Péan de la Tuilerie 1776 and 1879, p. 82 (n.); Texier 1857, col. 1404; Caumont 1855; Port 1874–78, pp. 252–53; Farcy 1877, pp. 22–24; Vaivre 1988b, p. 64.

The tomb, with no enamelwork, was given a rectangular frame used for monuments of Limousin work. The flat border, which usually was enameled, here instead has an epitaph. The inside of the cuvette has an engraved decoration of fleurs-de-lis in a lozenge pattern reminiscent of some cast-metal tomb slabs (for example, the tomb of Charles the Bald, formerly in Saint-Denis). The prelate's unbearded face is very close to that on the tomb of Michael of Villoiseau, also a bishop of Angers, as seen in the Gaignières

Angers, Tomb of William of Beaumont

of Canteloup, archbishop of Bordeaux, must be celebrated on the seventh of the Kalends of April [March 25]. For this we have eleven pounds of income owed by the recorded feudatories. He was buried near the main altar, at the right corner between two pillars, in a sepulchre whose top bears a brass figure). Paris, Bibliothèque Nationale, Département des Manuscrits, ms. lat. 17024 (Gaignières album), fol. 112bis: "Est sepultus juxta magnum altare inter duo pilaria, ubi est imago de letone" (He was buried near the main altar, between two pillars, where there is a brass figure). Lopès 1668 and 1882, vol. 1, p. 217: "Sur lequel tombeau est son image de pierre, qui estoit couverte de lames de cuivre doré" (On this tomb is his image in stone, which was covered in sheets of gilt copper).
Bibl.: Lopès 1668 and 1882, vol. 2, p. 266; Courteault 1925, p. 125; Gardelles 1963, p. 19, n. 37, and p. 194.

Lopez's reference to a stone figure covered with sheets of gilt copper is unusual, since metal effigies generally had wooden cores. The attribution of the tomb to the Limousin is possible but not proven.

Casteljaloux (Lot-et-Garonne, Nérac arrondissement), Franciscan convent. Tomb of Amanieu VII of Albret. Commissioned about 1341 by Bernard-Ezi II of Albret. Disappeared during the Wars of Religion.

Documentary source: Archives Départementales des Pyrénées-Atlantiques, E31: First testament of Bernard-Ezi, lord of Albret and Nérac, March 1341 (précis in Raymond 1867, p. 8), "ordonnant que de belles sépultures soient faites à sa mère, à son père, à sa soeur, dame de Bergerac, à son frère Guitard, vicomte de Tartas, mais que la tombe de son père soit la plus belle et de laiton bien travaillé, comme on le fait à Limoges" (ordering that fine sepulchres be made for his mother [Rose of Bourg], his father [Amanieu VII], his sister, lady of Bergerac [Martha, wife of Richard of Pons, lord of Bergerac], and his brother Guitard, viscount of Tartas, but that the tomb of his father be the finest and of well-wrought brass, such as they make in Limoges).
Bibl.: Samazeuilh 1860, pp. 89–90, 102–3, 234; Raymond 1867, pp. 6, 8, 10; Luchaire 1873, p. 34.

The Franciscan convent of Casteljaloux had been, since the time of Amanieu VI, grandfather of Bernard-Ezi II, the site of the Albret family sepulchre, and in his testament Amanieu VII of Albret selected it as his resting place (Archives Départementales des Pyrénées-Atlantiques, E21, between 1307 and 1338). It is thus conceivable that Bernard-Ezi II respected this choice and took charge himself of the wills of his family concerning their tombs, as his second testament, dated 1358 (Archives Départementales des Pyrénées-Atlantiques, E37) does not repeat the arrangements of the first. The sad fate of these Albret family monuments, following the sacking of the Franciscan convent in 1568, does not permit us to learn more about this tomb, which may well have been made in Limousin.

albums (see no. 6 in this appendix). Despite this similarity it is not possible to assert with certainty that this tomb was of Limousin manufacture.

Ca. 1292. Beauvais (Oise), abbey church of Saint-Lucien. Tomb of John Cholet of Nointel (d. 1292). Pilfered about 1358 and 1628; disappeared during the eighteenth century.

Graphic sources: Gaignières album, Oxford, Bodleian Library, Gough-Gaignières ms. 9 (18354); Paris copy, Bibliothèque Nationale, Département des Estampes, Pe 11, fol. 14 (Bouchot 1891, no. 3525; Guibert n.d., pl. 219; Adhémar 1974, no. 436), with caption: "Il est à costé gauche du grand autel de l'église de l'abbaye de S. Lucien de Beauvais. Il estoit fort orné d'esmail et de cuivre doré. On assure que sa figure estoit d'argent; elle n'est plus que de bois peint. Sur le bord du tombeau est escrit son épitaphe en lettres esmaillees d'or sur fond rouge. . . . Le bord au dessous de cette escriture était tout autour de compartimens de ses armes esmaillées sur du cuivre que l'on a osté et mis autour du retable de l'autel. Il en reste seulement du costé de la teste contre le mur et aux pieds" (It is to the left of the main altar of the abbey church of Saint Lucian of Beauvais. It was heavily decorated with enamel and gilt copper. People say that the face was of silver; now it is only painted wood. Around the edge of the tomb his epitaph is inscribed in gold enameled letters on a red ground. . . . The border beneath this writing was girded with compartments bearing his arms enameled on copper, which have been removed and placed around the altar's retable. The only remaining enamel is beside the head against the wall and at the feet).
Bibl.: Louvet 1631–35, vol. 1, pp. 391–92; Texier 1857, col. 1403; Rupin 1890, p. 161.

According to Louvet, the massive silver effigy of the prelate was sold about 1358, after the ravages of the Hundred Years War. The tomb's embellishments and ornaments were removed in 1628 and placed on the main altar of the abbey church. The Limousin provenance of the tomb is not certain.

Bétête (Creuse, Guéret arrondissement, Châtelus-Malvaleix canton).

Documentary source: Archives Départementales de la Creuse, H 529, vol. 14, record of April 10, 1791: "Un vieux mausolée de bois et cuivre" (An old wood and copper mausoleum).
Bibl.: Louradour 1977, p. 9.

This monument may be the remains of a Limoges-work tomb.

Bordeaux (Gironde), cathedral. Tomb of Arnaud III of Canteloup (1306–d. 1332), archbishop of Bordeaux. Noted in 1785; disappeared, probably during the Revolution or between 1803 and 1812, during work conducted by Combes, architect of the department.

Documentary source: "Obituaires . . ." 1878, pp. 65–66: Necrology, no. XLI, March 26: "Anniversarium domini Arnaldi de Cantelupo, archiepiscopi Burdegale, debet fieri VIIo kalendis aprilis. Pro quo habemus XI libras census, debitas per subscriptos feudatorios. Et est sepultus juxta magnum altare, in dextro cornu, inter duo pilaria, in sepultura operate cum ymagine desuper[de letone]" (The anniversary of Lord Arnaud

Limoges (Haute-Vienne), Jacobin monastery. Tomb of Bernard Gui, bishop of Lodève (d. 1331). Disappeared during the Wars of Religion.

Documentary source: *Brevis chronica* 1657, pp. 512–17; Quétif and Echard 1719, pp. 576ff.: "Frater Bernardus Guidonis, ordinis Fratrum Praedicatorum, episcopus Lodovensis, qui in presbyterio ecclesiae Fratrum Lemovicensium cum ad sinistram altaris in decenti ac eminenti tumulo fabrefacto ex letone requiescit" (Brother Bernard Gui, of the Order of Preachers [Dominicans], bishop of Lodève, who rests in the presbytery of the Friars' church in Limoges, to the left of the altar, in an appropriate and remarkable monument well wrought of brass).
Bibl.: Poulbrière 1886, p. 155; Arbellot 1896, p. 16.

Bernard Gui, who died on December 30, 1331, at the château Lauroux, today in the canton of Lodève (Hérault), had a solemn funeral in his bishopric of Lodève. Then, according to his wishes, his body was transported to Limoges to be buried in the church of the Dominicans (today the church of Sainte-Marie), to the left of the main altar, beneath a tomb raised off the ground and made of brass or, more probably, copper. According to Quétif and Echard, this monument was stolen when the Dominican church was pillaged during the religious disturbances of the sixteenth century. There is a strong likelihood that, given its location, the tomb of Bernard Gui was of Limousin manufacture.

Pamplona, cathedral. Tomb of Henry I, king of Navarre (d. 1274). Disappeared in 1276.

Documentary source: Nangis 1860, pp. 507, 508: "Ils mirent le comble à leur scélératesse en portant des mains sacrilèges sur la tombe en cuivre doré du roi Henri, qui reposait dans l'église de Santa Maria; croyant qu'elle était d'or, ils la mirent en pièces" (They reached the height of their wickedness by laying their sacrilegious hands on the gilt-copper tomb of King Henry, which rested in the church of Santa Maria; believing it to be of gold, they tore it to pieces).

Son of Thibaut I, king of Navarre, who himself had a Limousin tomb (see no. 4 in this appendix), Henry I was probably buried beneath a similar monument. However, its early disappearance, two years after the king's death, makes it impossible to know anything more about it.

Sauve-Majeure (Gironde, Bordeaux arrondissement, Créon canton, commune of La Sauve), Benedictine abbey. Tomb of Amanieu of Gresinhac, archbishop of Auch (1226–d. 1242). Disappeared after 1672.

Documentary sources: Dulaura, Bordeaux, Bibliothèque Municipale, ms. 1871, pp. 311–13, and description p. 350: "Son effigie est au dessus de ce sépulcre; elle est de bois, revêtue de plaques de cuivre bien façonné et doré et le représentent de petites statues" (His effigy lies on top of this sepulchre [in the choir of the church]; it is made of wood, covered with sheets of copper, well made and gilded, and there are small statues). Archives Départementales de la Gironde, H17, record of bishop's visitations, 1605 and 1608; Bertrand Pagès, Bordeaux, Bibliothèque Municipale, ms. 770. Bibl.: Cirot de la Ville 1845, pp. 316, 345; Thomas 1910, pp. 44–54; Gardelles 1979, pp. 183–90; Favreau et al. 1979, no. 39, pp. 133–34.

The attribution of this tomb to Limousin is quite possible, even probable.

## *Tombs known through graphic or written sources whose association with Limousin production is improbable or impossible to demonstrate*

Braine (Marne), collegiate church of Saint-Yved. Tomb of Robert II, count of Dreux (d. 1218). Disappeared during the Revolution.

Graphic sources: Gaignières albums, Oxford, Bodleian Library, Gough-Gaignières ms. 1 (18346); Paris copy, Bibliothèque Nationale, Département des Estampes, Pe 1, fol. 75 (Bouchot 1891, no. 1937: "tombe de cuivre en cuvette" [basin-shaped copper tomb]; Guibert n.d., pl. I–275; Adhémar 1974, no. 65); Oa 9, fol. 40 (Bouchot 1891, no. 61). Bibl.: Texier 1857, col. 1402; Souchal 1963, pp. 325–26; Teuscher 1990, pp. 171–74.

Although cited by the historians of Limousin enamels, who call it simply "tombe du cuivre" (copper tomb), the monument of Robert II of Dreux can be excluded from the list of tombs of Limousin production because of its affinities with other cast tombs in the Paris basin (such as the almost contemporary one of Evrard of Fouilloy [d. 1222] in the Amiens cathedral).

Tomb of Peter of Dreux, known as Mauclerc, duke of Brittany (d. 1250). Disappeared during the Revolution.

Graphic sources: Gaignières albums, Oxford, Bodleian Library, Gough-Gaignières ms. 1 (18346); Paris copy, Bibliothèque Nationale, Département des Estampes, Pe 1, fol. 98 (Bouchot 1891, no. 1960; Guibert n.d., pl. I–277; Adhémar 1974, no. 230); Oa 9, fol. 81 (Bouchot 1891, no. 95). Bibl.: Texier 1857, cols. 1402–3; Souchal 1963, pp. 327–28; Teuscher 1991, pp. 179–85.

Although cited in works devoted to Limousin enamels, the tomb of Peter of Dreux, known as Mauclerc, seems to have been excluded from the list of tombs of Limousin production and likened instead to works in cast metal. Teuscher suggests it was commissioned in 1238 and executed in the following years.

Tomb of Mary of Bourbon, countess of Dreux (d. 1274). Disappeared during the Revolution.

Graphic sources: Gaignières albums, Oxford, Bodleian Library, Gough-Gaignières ms. 1 (18346); Paris copy, Bibliothèque Nationale, Département des Estampes, Pe 1, fols.

Braine, Tomb of Peter of Dreux

Braine, Tomb of Mary of Bourbon

78–80 (Bouchot 1891, no. 1940–42; Guibert n.d., pls. I–280, I–281; Adhémar 1974, no. 340—fol. 78: the effigy and figures at tomb's base; fol. 79: figures at tomb's side; fol. 80: ink transcription of inscriptions; Oa 10, fol. 13 (Bouchot 1891, no. 143). Bibl.: Texier 1857, col. 1402; Rupin 1890, p. 162; Souchal 1963, pp. 327–28; Gauthier 1988, pp. 107–8; Teuscher 1991, pp. 189–97.

Cited in works devoted to Limousin enamels, the tomb of Mary of Bourbon bore inscriptions on a red or blue enameled ground and armorial bearings that were similarly enameled. The place of production can be located in a region other than the Limousin, perhaps in Paris, where enamelers are confirmed to have worked during the reign of Saint Louis. The molded frame of the plaque on which the effigy rests and the trefoil arch that shelters it are not typical of usual tombs of Limousin production.

Cherlieu (Haute-Saône, Vesoul arrondissement, Vitrey canton, commune of Montigny-lès-Cherlieu), Cistercian abbey. Tomb of Alix of Méranie (d. 1279). Disappeared between 1694 and 1751.

Bibl.: Clerc 1840–46, p. 457, no. 4: "Un compte du receveur de la comtesse Mahaut, belle-fille d'Alis, à la date de 1306, parle d'une tombe en cuivre doré, faite récemment à Paris pour Alis et conduite à Cherlieu" (A receipt from the countess Mahaut, daughter-in-law of Alis, dated 1306, mentions a gilt-copper tomb, recently made in Paris for Alis, and brought to Cherlieu). Besson 1847, pp. 50–51; Chatelet 1855, pp. 298–326.

According to ancient sources cited by nineteenth-century historians, the copper tomb of Alix of Méranie was made in Paris in 1306 and was commissioned by Mahaut of Artois. It bore a statue in gilt copper reclining on a wooden bed. This figure disappeared in the first half of the eighteenth century, and the remaining metal was recast as copper lamps for the abbey choir.

Corbeil (Essone). Tomb of Ingeburge of Denmark, queen of France (d. 1236). Moved in 1736 and lost during the Revolution.

Graphic sources: Gaignières albums, Paris, Bibliothèque Nationale, Département des Estampes, Oa 9, fol. 36 (Bouchot 1891, no. 58); Oxford, Bodleian Library, Gough-Gaignières ms. 2 (18347); Paris copy, Bibliothèque Nationale, Département des Estampes, Pe 1a, fol. 22 (Bouchot 1891, no. 1994; Adhémar 1974, no. 166: "tombe de pierre au milieu du choeur de Saint-Jean-en-l'Île, commanderie proche Corbeil" [metal tomb in the middle of the choir of Saint-Jean-en-l'Île, a commandery near Corbeil]). Bibl.: Pinard 1852, pp. 565–75, esp. p. 573; Erlande-Brandenburg 1975, pp. 93, 114, 181.

The Gaignières caption describes this monument as a "stone tomb," but in fact it seems to have been a work in metal, probably cast; indeed the presence of an inscription giving the maker's name—magister Hugo de Plagliaco—can be compared to several other examples of signatures on cast-metal slabs from the first half of the thirteenth century; one of them, covering the remains of Barthélemy of Roye (d. after 1221), treasurer of

Evreux, Tomb of Jean of La Cour
d'Aubergenville

France, buried at Joyenval (Bouchot 1891, no. 3988; Adhémar 1974, no. 155), is signed Hugo de Plalli (Plailly, Oise, Senlis arrondissement and canton), which seems to be the French form of the name of the same caster. Among the funerary monuments reproduced in the Gaignières albums, the yellow copper tomb of Eudes of Sully, the bishop of Paris, who died in 1208 (Bouchot 1891, nos. 4532, 6792; Adhémar 1974, no. 55), and that of Robert III of Dreux in Braine (Bouchot 1891, no. 1939; Adhémar 1974, no. 151, incorrectly captioned "tombe de pierre plate" [tomb of flat stone]), also bear signatures of casters.

## Evreux (Eure), cathedral. Tomb of Jean of La Cour d'Aubergenville (d. 1256), bishop of Evreux.

Graphic sources: Gaignières albums, Paris, Bibliothèque Nationale, Département des Manuscrits, ms. lat. 17034, fol. 73 (Bouchot 1891, no. 6722; Adhémar 1974, no. 253); Oxford, Bodleian Library, Gough-Gaignières ms. 8 (18353); Bibliothèque Nationale copy, Département des Estampes, Pe 1d, fol. 72 (Bouchot 1891, no. 2340).
Bibl.: Texier 1857, col. 1404.

The monument of Jean of La Cour d'Aubergenville is cited by Texier as a "magnifique tombeau en cuivre, en relief [du] XIIIe siècle" (magnificent copper tomb in relief from the thirteenth century). It bears a trefoil arch and a plaque which has an engraved ground and a lozenge pattern containing fleurs-de-lis. It is captioned in the Gaignières albums as a "tombeau de bronze" (bronze tomb), a description preferred by the present author.

## Fontevraud (Maine-et-Loire, Saumur arrondissement, Montsoreau canton), abbey. Tomb of Isabelle of Angoulême, queen of England (d. 1246).

Effigy of polychromed wood, which was probably never revetted with metal.

## Jard (Seine-et-Marne, Melun arrondissement and canton, commune of Voisenon), Augustinian abbey. Tomb of Jean of Melun, bishop of Poitiers (d. 1257). Disappeared.

Graphic source: Gaignières albums, Paris, Bibliothèque Nationale, Département des Manuscrits, ms. lat. 17042, fol. 197 (Bouchot 1891, no. 6825; Adhémar 1974, no. 254).

Like the tomb of Jean of La Cour d'Aubergenville, that of Jean of Melun seems to have been a work of cast metal, similar to the monument of Charles the Bald in the abbey of Saint-Denis.

## Jouy (probably Jouy-le-Châtel or Jouy-en-Brie, Seine-et-Marne, Provins arrondissement and canton, commune of Chenoise), Cistercian abbey. Tomb of Simon of Beaulieu, archbishop of Bourges (d. 1294). Disappeared.

Graphic sources: Gaignières albums, Paris, Bibliothèque Nationale, Département des Manuscrits, ms. lat. 17024, fol. 143 (Bouchot 1891, no. 6595; Adhémar 1974, no. 443); Oxford, Bodleian Library, Gough-Gaignières ms. 10 (18355); Paris copy, Bibliothèque Nationale, Département des Estampes, Pe 10, fol. 36 (Bouchot 1891, no. 3661).
Bibl.: Rupin 1890, p. 161; Vaivre 1988a, pp. 64–65.

This tomb, bearing witness to chromatic development and whose effigy is sheltered in pointed equilateral arching, seems to mark the stage at the end of the thirteenth century of the evolution of bronze funerary monuments of royal usage. In the early 1200s that of Eudes of Sully marked the beginning of this series, which includes the slab of Evrard of Fouilloy from about 1225 in Amiens cathedral and at mid-century the tombs of Charles the Bald, Jean of La Cour d'Aubergenville, and Jean of Melun. Although the Gaignières caption reads "tombeau de cuivre émaillé" (tomb of enameled copper), this work can be set apart from tombs of Limousin fabrication.

## Maubuisson (Val-d'Oise, Pontoise arrondissement, Saint-Ouen-l'Aumône canton and commune), Cistercian abbey church. Tomb of Blanche of Castille, queen of France (d. 1258). Disappeared in November 1790.

Documentary source: Lebrun-Desmarettes 1718, p. 263: copper tomb in the middle of the choir of Maubuisson.
Bibl.: Guilhermy 1848, p. 237; Erlande-Brandenburg 1975, pp. 114, 165.

The tomb of Blanche of Castille, in copper or more likely in bronze, bore an effigy in relief of the queen dressed in the habit of a Cistercian nun. Revolutionary inventories recorded the tomb's weight as 3,300 pounds (1.5 metric tons), which leads us to believe it was a cast-metal work.

## Mirebeau (Vienne), collegiate church of Notre-Dame. Tomb of Maurice of Blason, bishop of Poitiers (d. between 1201 and 1214 or about 1216). Disappeared during the Wars of Religion.

Documentary source: Poitiers, Bibliothèque Municipale, ms. of D. Fonteneau, vol. 67, p. 415 (copy of a 1618 account): "C'est dans la commune créance que Maurice, évêque de Poitiers, a son tombeau au milieu du choeur de ladite église collégiale; et quelques anciens rapportent avoir vu, auparavant les premiers troubles de ceux de la religion prétendue, au milieu dudit choeur, sur un tombeau de pierre, la figure de l'évêque –étendu sur icelui avec la mitre et la crosse, le tout étant de bois" (It is commonly believed that the tomb of Maurice, bishop of Poitiers, is in the middle of the choir of the collegiate church; and some ancients [anciens] report having seen, before the first of the so-called religious troubles, in the middle of the said choir, a stone tomb on which the figure of the bishop was stretched with his miter and crosier, all made of wood."
Graphic source: Gaignières albums, Paris, Bibliothèque Nationale, Département des Manuscrits, ms. lat. 17042, fol. 143: Maurice of Blason buried in Notre-Dame in Mirebeau.
Bibl.: Thibeaudeau 1839, vol. 1, p. 397; Crozet 1942, p. 60, no. 238.

The wood effigy of Maurice of Blason disappeared in the sixteenth century; like that of Isabelle of Angoulême in the Fontevrault abbey, it may have had a simple ornamentation of painted polychrome.

## Paris, Notre-Dame cathedral. Tomb of Eudes of Sully, bishop of Paris (d. 1208).

Graphic sources: Gaignières albums, Paris, Bibliothèque Nationale, Département des Estampes, Pe 11a, fol. 26 (Bouchot 1891, nos. 4532, 6792; Adhémar 1974, no. 55) and Bibliothèque Nationale, Département des Manuscrits, ms. lat. 17040, fol. 93 (Bouchot 1891, no. 6793).
Documentary source: Lebrun-Desmarettes 1718, p. 254: "tombe de cuivre" (copper tomb).
Bibl.: Rupin 1890, pp. 158–59; Souchal 1963, pp. 325–27; Erlande-Brandenburg 1975, p. 115; Teuscher 1991, p. 178.

The yellow copper effigy in relief of Eudes of Sully is depicted with closed eyes and is a particularly early example of this convention (if the funerary monument is in fact contemporary with the prelate's death). The work bears an inscription that includes the name of the maker—*Stephanus de Boisse me fecit*—as do a number of bronze slabs from the thirteenth century. Rupin proposed a connection between this name and that of a Limousin family. The place-name Boisse is, however, found in Saintonge, Périgord, Vendée, and the Aveyron (that is, mostly in western France), and the use of the name as a patronym does not necessarily imply a tie to the Limousin.

In fact, the tomb of Eudes of Sully belongs to a small group of bronze funerary monuments produced in Île-de-France, which are occasionally signed.

## Prières (Morbihan, Vannes arrondissement, Muzillac canton, commune of Billiers), Cistercian abbey. Tomb of Jean I, duke of Brittany (d. 1286). Disappeared.

Documentary source: P. Le Baud, *Chronicques et ystoires des Bretons . . .* (1922), vol. 4, p. 92: "Sur la fosse duquel fut collocqué ung tombeau précieux d'or, d'argent et de pierres très riches et à l'envyron petites ymages d'argent ouvrées subtillement avecques l'épitaphe qui s'ensuit . . ." (Over the grave there was a precious tomb of gold, silver, and very valuable stones around which were small silver images finely worked with the following epitaph . . .).
Bibl.: Copy 1981, vol. 2, pp. 436–37, no. 230; Copy 1986, pp. 33, 41, 293, no. 3.

Copy interprets the materials in Pierre Le Baud's text as gilt copper and enamels. This is possible, but the tomb of the duke of Brittany could be different from those of the members of his entourage, made of metals more precious than the copper used for a son who died young, Robert of Brittany (no. 5 in this appendix), for his parents (nos. 13 and 14 in this appendix), or even for the duchess (no. 18 in this appendix).

## Provins (Seine-et-Marne), general hospital. Tomb of the heart of Thibaut V, count of Champagne and king of Navarre (d. 1270).

Bibl.: Rupin 1890, pp. 161–62; Rohan-Chabot 1939, vol. 2, pp. 207–8.

The cover's copper plaques are decorated with hybrid beings that are not in the repertory of Limousin enamelers of the second half of the thirteenth century. These creatures can, however, be related to the fantasies in margins of manuscripts from eastern and northern France. The empty spaces, left for armorial bearings, were probably once filled with enameled plaques, whose place of production is very difficult to determine despite the known links between Thibaut V and Limoges.

## Saint-Denis (Seine-Saint-Denis), abbey church. Tomb of Charles the Bald (d. 877). Disappeared in 1792.

Graphic sources: Du Tillet, Paris, Bibliothèque Nationale, Département des Manuscrits, ms. fr. 2848, fol. 42; Bonfons 1586–88, vol. 2, pl. p. 24; Bonfons 1605, pl. p. 62. Gaignières albums, Oxford, Bodleian Library, Gough-Gaignières ms. 2 (18347); Bibliothèque Nationale copy, Département des Estampes, Pe 1a, fol. 12 (Bouchot 1891, no. 1984; Adhémar 1974, no. 256).
Bibl.: Texier 1857, col. 1403; Viollet-le-Duc 1868, p. 60 of the "tomb" article; Vitry and Brière 1925, pp. 107–8, 134; Erlande-Brandenburg 1975, pp. 114, 153.

The tomb of Charles the Bald fits into the series of cast bronze tombs from Île-de-France. The rich orphreys decorating the king's cloak, which were interpreted by Viollet-le-Duc as plaques of encrusted enamel, were rightly compared by Erlande-Brandenburg to those of the effigy of Evrard of Fouilloy (d. 1222) in the Amiens cathedral. According to Erlande-Brandenburg, the work, which was 95 inches (2.4 meters) long, was already in Saint-Denis when the funeral of Philip Augustus took place in 1223.

## Tomb of Marguerite of Provence (d. 1295). Disappeared in 1792.

Graphic sources: Peiresc, Carpentras, Bibliothèque Inguimbertine, Peiresc ms. X, fol. 515 (Erlande-Brandenburg gives ms. 17791, fols. 515, 516); Gaignières albums, Paris, Bibliothèque Nationale, Département des Estampes, Oa 9, fol. 67 (Bouchot 1891, no.

Troyes, Tomb of Henry I, count of Champagne

81); Oxford, Bodleian Library, Gough-Gaignières ms. 2 (18347); Paris copy, Bibliothèque Nationale, Département des Estampes, Pe 1a, fol. 24 (Bouchot 1891, no. 1996; Adhémar 1974, no. 451); *Monuments de la monarchie*, vol. 3, pl. XXVI-4.
Bibl.: Texier 1857, col. 1403; Erlande-Brandenburg 1975, pp. 114, 167, no. 97.

The tomb of Marguerite of Provence was a metal sheet, and its inclusion on Texier's list can be justified only by the use of a coppery material.

## Troyes (Aube), Saint-Étienne collegiate church. Tombs of Henry I, count of Champagne (d. 1181), and Thibaut III, count of Champagne (d. 1201). Disappeared during the Revolution.

Documentary sources: Duhalle, Troyes, Bibliothèque Municipale, ms. copy by Finot; Camuzat 1610, fols. 329, 330.
Bibl.: Arnaud 1837, pp. 29–34; Texier 1857, cols. 1399–402; Gaussen 1861; Roserot 1945–48, vol. 3, entry on Troyes, p. 1602 (with earlier bibliography); Jottrand 1965; Bauch 1976, pp. 34–35.
Graphic reconstructions: Preparatory documents, Troyes, Bibliothèque Municipale (ms. 2398) and the Musée des Beaux-Arts. Lithograph by M.-C. Fichot, from a drawing by A.-F. Arnaud, published in Arnaud 1837: tomb of Henry I. Gaussen 1861: tomb of Thibaut III.

There is a solid historiographic tradition linking the counts of Champagne and the order of Grandmont. However, Duhalle's very detailed description of the counts' tombs and the few enameled plaques from these monuments, now in the Musée de Troyes and the treasury of the cathedral, do not allow us to regard them as the work of Limousin enamelworkers. Furthermore, the structure of the tomb of Henry I was very unusual, the effigy being sheltered inside a latticework arcade. A similar arrangement is found in the tomb of Saint Stephen of Obazine (Corrèze), perhaps inspired by the Champagne example, but from a later date, toward the mid-thirteenth century. The tomb of Thibaut III may be echoed in that of Marie of Bourbon in Braine—both depict figures related to the deceased around the coffer. The effigies were covered in silver leaf, which may have been worked by the local goldsmiths. The extant enamel plaques, in chromatic range and treatment of decorations, are more fruitfully compared to enamels from the Meuse.

443

# Techniques and Materials in Limoges Enamels

ISABELLE BIRON, PETE DANDRIDGE, AND MARK T. WYPYSKI,
WITH THE COLLABORATION OF MICHEL VANDEVYVER

## Metal Compositions

Although systematic identification of alloy composition was not within the scope of this study, representative samples were taken from eight objects and analyzed by energy dispersive X-ray spectrometry (EDS) to detect major and minor elements. All were nearly pure copper; the minor elements detected were iron, lead, nickel, and arsenic, none of which appear to be deliberate additions. The apparent absence of lead in some of the analyses may be due either to segregation within the alloy or to the fact that the minimum detection limit for lead is approximately 0.5 percent with EDS.

### Table 1. Metal Analyses

| Cat. No. | Acc. No. | Location | Cu | As | Ni | Ag | Fe | Pb |
|---|---|---|---|---|---|---|---|---|
| 6 | 17.190.426 | Plaque | 96.9 | | | 0.4 | 0.8 | 1.9 |
| 8f | 17.190.690 | Medallion | 99.1 | | | | 0.2 | 0.7 |
| 43 | 17.190.783 | Plaque | 98.4 | | | | 0.1 | 1.4 |
| | | Rivet (Lion) | 99.9 | | | | 0.1 | |
| 58b | 17.190.123 | Plaque | 98.7 | 0.4 | | | 0.1 | |
| | | Appliqué | 99.5 | 0.7 | | | 0.6 | |
| 79 | 17.190.790 | Appliqué | 99.1 | | | | 0.4 | 0.5 |
| 96 | 17.190.789 | Appliqué | 98.8 | | | | 0.4 | 0.8 |
| 105 | 17.190.344 | Breast | 99.4 | 0.4 | | | 0.2 | |
| | | Beak | 99.8 | | | | 0.2 | |
| | | Leg | 97.8 | 1.7 | 0.4 | | 0.1 | |
| | | Base Plate* | 99.5 | | 0.3 | | 0.2 | |
| NE | 17.190.744 | Rivet (Eagle)* | 99.9 | | | | 0.1 | |

* : later manufacture
NE : not exhibited

## Solder Compositions

Samples of solder from seven different objects were analyzed by EDS and were shown to be predominantly copper/tin alloys with the tin ranging from ten to twenty percent, with some lead unintentionally introduced along with the copper (see note 26 for a discussion of cat. 6). The liquidus temperature for the binary alloys would fall between 1020°C and 900°C, with the lead depressing the temperature somewhat further. The presence of phosphorous in the sample from the dove (cat. 105) is indicative of a modern solder and seems to confirm the fact that mechanical deformation of the chest necessitated not only its reshaping and partial regilding along the seam but also the resoldering of the tabs that key the two halves of the dove together.

### Table 2. Solder Analyses

| Cat. No. | Acc. No. | Location | Cu | Sn | Pb | P | Ag |
|---|---|---|---|---|---|---|---|
| 6 | 17.190.426 | Join of overlaid plaques | 85.0 | | | | 15.0 |
| 48 | 17.190.757 | Christ: seam at crown | 78.0 | 20.0 | 1.3 | | |
| 58b | 17.190.123 | Plaque: corner seam | 90.0 | 10.0 | | | |
| 105 | 17.190.344 | Breast: interior tab | 94.0 | 5.0 | | 1.0 | |
| 109 | 32.100.289 | Volute: seam at spine | 77.0 | 22.0 | 1.0 | | |
| 111 | 17.190.835 | Shaft: seam | 82.0 | 16.0 | 0.3 | | |
| 129 | 32.100.284 | Dish: vertical seam | 92.0 | 8.0 | 0.1 | | |

## Enamel Compositions

Tables 3 and 4 give the analyses of the enamels from fifty-three objects, in relative weight percentages of the common oxide forms for each element. Table 3 gives the EDS results from the Metropolitan Museum (MMA); ion beam analyses from the Laboratoire de Recherche des Musées de France (LRMF) are presented in table 4. The results in both tables are organized by object in approximate chronological order, identified by the catalogue numbers and museum accession numbers. The enamel colors are noted with the following abbreviations: WHI = white opaque; DBO = dark blue opaque; DBT = dark blue translucent; BLU = medium blue opaque; LBL = light blue opaque; TUR = turquoise opaque; GRN = green opaque; DGR = dark green opaque; LGR = light green opaque; GRT = green translucent; YEL = yellow opaque; RED = red opaque; FLS = flesh-colored opaque; AME = amethyst translucent; BLK = black translucent; CLR = clear uncolored; *CLR* = a clear glass cabochon, not an enamel. The less-than sign (<) for several of the antimony oxide and tin oxide results in tables 3 and 5 indicates that the overall amount present is below the EDS minimum detection limits, but its presence was confirmed by high-magnification analysis of the crystalline opacifiers.

445

# Table 3. EDS Analyses of Enamels (MMA)

| Cat. No. | Acc. No. Object | Date | Color | $Na_2O$ | MgO | $Al_2O_3$ | $SiO_2$ | $P_2O_3$ | $SO_3$ | Cl | $K_2O$ | CaO | $TiO_2$ | $Cr_2O_3$ | MnO | $Fe_2O_3$ | CoO | NiO | CuO | ZnO | $As_2O_3$ | $SnO_2$ | $Sb_2O_3$ | PbO |
|---|---|---|---|---|---|---|---|---|---|---|---|---|---|---|---|---|---|---|---|---|---|---|---|---|
| NE | 17.190.688 | 10th c. | WHI | 15.1 | 1.1 | 2.4 | 65.7 | nd | 0.2 | 0.4 | 0.6 | 7.4 | nd | nd | 0.2 | 0.4 | nd | nd | nd | nd | nd | nd | 5.4 | nd |
| | Luna | | DBT | 17.0 | 0.6 | 2.7 | 68.0 | nd | 0.4 | 0.9 | 0.7 | 7.0 | 0.1 | nd | 0.6 | 1.0 | 0.1 | nd | 0.2 | nd | nd | nd | 0.9 | nd |
| | Plaque | | TUR | 17.9 | 0.5 | 2.5 | 68.8 | nd | 0.4 | 1.3 | 0.6 | 5.2 | 0.1 | nd | nd | 0.5 | nd | nd | 1.9 | nd | nd | nd | 1.0 | nd |
| | Carolingian | | GRN | 17.4 | 0.7 | 2.4 | 66.8 | nd | nd | 1.0 | 0.9 | 5.8 | 0.1 | nd | 0.3 | 0.9 | nd | nd | 1.6 | nd | nd | nd | <0.6 | 2.6 |
| | | | YEL | 17.3 | 0.5 | 2.2 | 67.4 | nd | nd | 1.2 | 0.6 | 5.8 | 0.1 | nd | 0.4 | 0.6 | nd | nd | nd | nd | nd | nd | <0.6 | 3.9 |
| | | | RED | 15.5 | 0.9 | 2.9 | 58.0 | nd | nd | 1.1 | 1.0 | 7.1 | 0.1 | nd | 0.3 | 1.6 | nd | nd | 2.6 | 0.6 | nd | 0.5 | nd | 7.0 |
| 4 | 1983.38 | early 12th c. | WHI | 15.0 | 2.4 | 1.9 | 66.1 | nd | 0.3 | 0.6 | 0.6 | 6.6 | 0.1 | nd | nd | 0.4 | nd | nd | nd | nd | nd | nd | 5.0 | nd |
| | Roundel | | DBO | 13.8 | 0.5 | 1.8 | 64.6 | nd | 0.4 | 0.6 | 0.7 | 6.9 | 0.1 | nd | 0.6 | 1.6 | 0.4 | nd | 1.3 | nd | nd | nd | 6.5 | nd |
| | | | TUR | 16.4 | 0.5 | 2.0 | 65.0 | nd | 0.4 | 1.1 | 0.6 | 4.6 | 0.1 | nd | nd | 0.6 | nd | nd | 4.5 | nd | nd | nd | 2.5 | nd |
| | | | GRN | 15.1 | 0.6 | 2.2 | 60.8 | nd | nd | 0.6 | 0.8 | 7.0 | 0.1 | nd | 1.8 | 0.8 | nd | nd | 3.9 | nd | nd | nd | 2.5 | 4.0 |
| 6 | 17.190.428 | early 12th c. | TUR | 12.5 | 0.7 | 1.7 | 63.5 | nd | 0.3 | 0.7 | 0.7 | 5.7 | nd | nd | 0.2 | 0.5 | nd | nd | 5.5 | nd | nd | nd | 6.0 | 0.50 |
| | Plaque | | AME | 16.4 | 1.5 | 2.7 | 67.0 | nd | 0.3 | 0.9 | 0.9 | 7.3 | 0.2 | nd | 2.5 | 1.1 | nd | nd | nd | nd | nd | nd | nd | 0.7 |
| 8h | 17.190.692 | 1110–1130 | DBO | 14.9 | 0.5 | 2.3 | 64.4 | nd | nd | 0.7 | 0.7 | 6.9 | nd | nd | 0.8 | 1.2 | 0.3 | nd | 0.5 | nd | nd | nd | 4.1 | 2.6 |
| | Medallion | | TUR | 18.3 | 0.5 | 2.0 | 62.6 | nd | nd | 1.3 | 0.5 | 6.0 | 0.1 | nd | nd | 0.8 | nd | nd | 2.8 | nd | nd | nd | 2.9 | 0.7 |
| 8i | 17.190.693 | 1110–1130 | GRN | 17.9 | 0.6 | 2.0 | 65.8 | nd | nd | 1.2 | 0.7 | 6.3 | 0.1 | nd | 0.1 | 0.8 | nd | nd | 2.3 | nd | nd | nd | <0.6 | 3.4 |
| | Medallion | | | | | | | | | | | | | | | | | | | | | | | |
| 10 | 17.190.695 | ca. 1150 | WHI | 16.4 | 3.5 | 1.8 | 65.1 | nd | 0.4 | 0.7 | 0.5 | 6.7 | 0.1 | nd | nd | 0.4 | nd | nd | nd | nd | nd | nd | 3.9 | nd |
| | Chasse | | LBL | 15.0 | 0.5 | 2.1 | 65.6 | nd | nd | 0.7 | 0.8 | 7.3 | 0.1 | nd | 0.5 | 0.9 | 0.10 | nd | 0.4 | nd | nd | nd | 3.5 | 1.8 |
| | | | RED | 0.9 | 3.5 | 1.5 | 47.3 | 5.40 | 0.3 | 0.3 | 18.3 | 16.4 | 0.1 | nd | 0.2 | 2.1 | nd | nd | 1.7 | 0.4 | nd | nd | nd | 0.5 |
| 10 | 17.190.710 | ca .1150 | TUR | 17.7 | 0.6 | 1.8 | 62.9 | nd | 0.5 | 1.1 | 0.5 | 4.3 | 0.1 | nd | nd | 0.7 | nd | nd | 3.8 | nd | nd | nd | 3.8 | nd |
| | Chasse | | GRN | 15.0 | 0.7 | 2.5 | 61.2 | nd | nd | 1.1 | 0.7 | 5.0 | 0.1 | nd | 0.4 | 0.7 | nd | nd | 2.5 | nd | nd | nd | <0.6 | 10.2 |
| 62 | 17.190.409a | 1185–1195 | WHI | 15.1 | 1.3 | 2.3 | 57.6 | nd | nd | 0.6 | 0.8 | 7.7 | 0.10 | nd | 0.5 | 0.6 | nd | nd | 0.1 | nd | nd | 2.8 | 6.1 | 4.3 |
| | Cross | | LBL | 14.1 | 0.4 | 2.5 | 63.0 | nd | nd | 0.5 | 0.6 | 7.2 | 0.10 | nd | 0.3 | 0.8 | 0.1 | nd | 0.7 | nd | nd | 0.6 | 7.5 | 1.8 |
| | | | DBT | 17.2 | 0.4 | 2.5 | 68.3 | nd | nd | 0.9 | 0.7 | 5.9 | nd | nd | 0.6 | 1.2 | 0.2 | nd | 0.4 | 0.2 | nd | nd | 1.7 | 1.0 |
| NE | 17.190.409c | 1185–1195 | WHI | 15.9 | 2.2 | 1.9 | 66.2 | nd | 0.4 | 0.8 | 0.7 | 7.0 | nd | nd | 0.1 | 0.5 | nd | nd | 0.1 | nd | nd | nd | 4.1 | nd |
| | St. Matthew | | DBO | 16.6 | 0.6 | 2.1 | 65.1 | nd | 0.3 | 0.8 | 0.7 | 6.6 | nd | nd | 1.0 | 1.1 | 0.3 | nd | 0.4 | nd | nd | nd | 4.0 | 2.0 |
| | Plaque | | TUR | 17.9 | 0.4 | 1.8 | 67.6 | nd | 0.2 | 1.2 | 0.5 | 5.0 | nd | nd | 0.1 | 0.6 | nd | nd | 3.4 | nd | nd | nd | 2.3 | 0.6 |
| | | | YEL | 16.3 | 0.4 | 2.1 | 64.7 | nd | nd | 1.0 | 0.70 | 6.0 | nd | nd | 0.3 | 0.5 | nd | nd | 0.4 | nd | nd | nd | <0.6 | 6.8 |
| | | | RED | 16.4 | 0.8 | 2.4 | 61.8 | nd | nd | 0.7 | 1.00 | 6.9 | 0.10 | nd | 0.9 | 5.4 | nd | nd | 1.0 | nd | nd | 0.9 | nd | 2.6 |
| 63a | L51.10.2 | 1185–1195 | DBO | 15.8 | 0.5 | 2.0 | 65.6 | nd | 0.3 | 0.8 | 0.7 | 7.0 | nd | nd | 0.5 | 1.2 | 0.2 | nd | 0.3 | nd | nd | nd | 3.2 | 1.4 |
| | Plaque | | | | | | | | | | | | | | | | | | | | | | | |
| 47 | 17.190.812 | 1185–1200 | WHI | 13.6 | 1.2 | 1.5 | 56.9 | nd | nd | 0.7 | 1.6 | 6.2 | nd | nd | 0.4 | 0.5 | nd | nd | 0.8 | nd | nd | 3.6 | 3.1 | 9.3 |
| | Plaque | | DBO | 13.7 | 2.2 | 1.5 | 56.5 | nd | nd | 0.7 | 1.3 | 6.1 | nd | nd | 0.4 | 1.6 | 0.3 | nd | 0.2 | nd | nd | 3.2 | 2.0 | 9.8 |
| | | | LBL | 11.6 | 2.7 | 0.7 | 49.3 | nd | nd | 0.5 | 1.8 | 4.2 | nd | nd | 0.1 | 1.5 | 0.2 | nd | 0.1 | nd | nd | 7.2 | nd | 19.8 |
| | | | RED | 13.6 | 1.9 | 1.8 | 58.1 | 0.60 | nd | 1.1 | 2.1 | 8.9 | 0.1 | nd | 0.3 | 1.8 | nd | nd | 2.4 | nd | nd | 0.5 | nd | 6.4 |
| 67 | 17.190.761 | 1190–1200 | WHI | 15.0 | 1.7 | 2.2 | 63.5 | nd | nd | 0.7 | 0.7 | 7.1 | nd | nd | 0.2 | 0.5 | nd | nd | 0.6 | nd | nd | nd | 6.1 | 1.2 |
| | Plaque | | BLU | 14.9 | 0.4 | 2.0 | 65.7 | nd | nd | 0.7 | 0.5 | 5.7 | nd | nd | 0.6 | 1.6 | 0.3 | nd | 0.4 | nd | nd | nd | 4.8 | 2.2 |
| | | | GRN | 17.7 | 0.4 | 2.1 | 67.6 | nd | nd | 1.2 | 0.5 | 5.7 | nd | nd | 0.3 | 0.7 | nd | nd | 2.2 | nd | nd | nd | 0.6 | 1.8 |
| | | | YEL | 16.5 | 0.5 | 2.0 | 64.8 | nd | nd | 1.0 | 0.7 | 5.3 | 0.1 | nd | 0.1 | 1.1 | nd | nd | 0.1 | nd | nd | nd | 0.9 | 8.2 |
| | | | RED | 15.2 | 1.5 | 2.6 | 60.0 | nd | nd | 1.0 | 1.3 | 7.6 | 0.1 | nd | 0.5 | 2.9 | nd | nd | 1.6 | nd | nd | 0.6 | nd | 6.0 |
| 31 | 17.190.513 | 1190–1200 | WHI | 15.6 | 1.2 | 1.7 | 63.7 | nd | 0.3 | 0.6 | 0.6 | 7.4 | nd | nd | 0.2 | 0.5 | nd | nd | 0.8 | nd | nd | nd | 6.8 | nd |
| | Chasse | | BLU | 15.6 | 0.7 | 2.0 | 65.1 | nd | nd | 1.0 | 1.0 | 6.8 | 0.1 | nd | 0.5 | 1.1 | 0.2 | nd | 1.1 | nd | nd | nd | 2.5 | 2.5 |
| | | | TUR | 19.1 | 0.7 | 2.0 | 64.0 | nd | 0.2 | 1.6 | 0.6 | 4.1 | 0.1 | nd | nd | 0.9 | nd | nd | 4.4 | nd | nd | nd | 2.7 | 0.5 |
| | | | GRN | 13.2 | 0.6 | 2.1 | 65.8 | nd | nd | 0.9 | 0.5 | 7.3 | nd | nd | 0.4 | 0.7 | nd | nd | 1.5 | nd | nd | nd | <0.6 | 7.0 |
| | | | RED | 13.1 | 2.1 | 2.1 | 57.2 | nd | nd | 1.1 | 1.9 | 7.8 | 0.2 | nd | 0.4 | 1.8 | nd | nd | 2.5 | nd | nd | nd | nd | 8.8 |
| | | | BLK | 17.0 | 1.5 | 2.0 | 64.7 | nd | 0.3 | 1.1 | 1.1 | 7.6 | 0.1 | nd | 0.5 | 5.0 | nd | nd | 0.2 | nd | nd | nd | nd | nd |
| 43 | 17.190.783 | ca. 1200 | DBO | 15.1 | 0.6 | 2.0 | 66.9 | nd | 0.2 | 0.8 | 0.6 | 6.8 | 0.1 | nd | 0.8 | 1.3 | 0.3 | nd | 0.8 | nd | nd | nd | 2.9 | 1.2 |
| | Book cover | | BLU | 14.1 | 0.5 | 2.0 | 64.3 | nd | 0.2 | 0.7 | 0.6 | 8.0 | nd | nd | 0.5 | 1.2 | 0.2 | nd | 0.6 | nd | nd | nd | 4.5 | 3.0 |
| | | | GRN | 13.8 | 0.5 | 2.0 | 65.2 | nd | nd | 0.9 | 0.4 | 7.1 | 0.1 | nd | 0.4 | 0.7 | nd | nd | 1.7 | nd | nd | nd | <0.6 | 7.1 |
| 86 | 17.190.798 | 1210–1220 | WHI | 15.0 | 1.5 | 2.0 | 63.5 | nd | nd | 0.7 | 0.7 | 7.2 | nd | nd | 0.2 | 0.5 | nd | nd | 0.2 | nd | nd | nd | 5.9 | 3.3 |
| | Book cover | | BLU | 15.0 | 0.6 | 2.0 | 65.8 | nd | 0.2 | 0.8 | 0.7 | 7.0 | 0.1 | nd | 0.6 | 1.2 | 0.2 | nd | 0.4 | nd | nd | nd | 3.4 | 2.0 |
| | | | GRN | 15.4 | 0.6 | 2.0 | 60.9 | nd | nd | 0.9 | 0.7 | 5.7 | 0.1 | nd | 0.4 | 0.6 | nd | nd | 3.6 | nd | nd | 0.5 | 0.8 | 8.2 |
| | | | YEL | 16.9 | 0.7 | 2.0 | 63.7 | nd | nd | 1.2 | 0.6 | 5.3 | 0.1 | nd | 0.2 | 0.8 | nd | nd | 0.1 | nd | nd | nd | 0.7 | 8.4 |
| | | | RED | 14.3 | 1.6 | 2.3 | 57.8 | nd | nd | 0.8 | 1.4 | 7.4 | 0.2 | nd | 0.5 | 3.7 | nd | nd | 1.9 | nd | nd | 0.6 | nd | 8.0 |
| 70b | 41.100.184 | 1200–1210 | DBO | 16.9 | 0.6 | 2.3 | 65.2 | nd | nd | 0.9 | 0.7 | 6.7 | nd | nd | 0.5 | 1.4 | 0.2 | nd | 1.0 | 0.3 | nd | nd | 2.6 | 1.2 |
| | Tabernacle | | TUR | 17.4 | 0.8 | 1.9 | 62.2 | nd | nd | 1.3 | 0.6 | 5.9 | 0.1 | nd | nd | 0.8 | nd | nd | 4.1 | nd | nd | 0.8 | 2.2 | 1.6 |
| | | | GRN | 17.5 | 0.4 | 1.9 | 67.6 | nd | nd | 1.3 | 0.6 | 5.5 | nd | nd | 0.3 | 0.5 | nd | nd | 3.5 | nd | nd | 0.5 | 0.6 | 0.9 |
| | | | YEL | 16.7 | 0.5 | 2.0 | 62.6 | nd | nd | 1.1 | 0.6 | 5.5 | nd | nd | 0.3 | 0.9 | nd | nd | 0.6 | nd | nd | nd | 1.0 | 8.0 |
| | | | RED | 14.4 | 1.6 | 2.3 | 58.6 | nd | nd | 0.9 | 1.3 | 7.5 | 0.1 | nd | 0.5 | 3.3 | nd | nd | 1.8 | 0.1 | nd | 0.7 | nd | 6.0 |
| 81 | 17.190.833 | 1200–1220 | WHI | 14.1 | 1.4 | 2.0 | 60.5 | nd | 0.2 | 0.7 | 1.0 | 7.5 | nd | nd | 0.3 | 0.5 | nd | nd | 0.2 | nd | nd | 1.7 | 5.6 | 3.7 |
| | Crosier | | DBO | 15.0 | 0.6 | 2.8 | 64.8 | nd | nd | 0.7 | 0.7 | 7.2 | 0.1 | nd | 0.8 | 1.4 | 0.3 | nd | 0.5 | nd | nd | nd | 3.5 | 2.5 |
| | | | GRN | 13.6 | 0.6 | 2.5 | 63.3 | nd | nd | 1.0 | 0.5 | 6.8 | nd | nd | 0.5 | 0.8 | nd | nd | 2.3 | 0.2 | nd | nd | 0.6 | 6.6 |
| | | | YEL | 16.3 | 0.5 | 2.5 | 61.8 | nd | nd | 1.1 | 0.6 | 5.6 | 0.1 | nd | 0.3 | 0.8 | nd | nd | 0.9 | nd | nd | nd | 0.6 | 9.0 |
| | | | RED | 14.5 | 1.6 | 2.9 | 58.2 | nd | nd | 0.8 | 1.2 | 7.3 | 0.2 | nd | 0.5 | 3.3 | nd | nd | 2.1 | nd | nd | 0.7 | | 6.0 |
| 83 | 14.40.703 | 1200–1220 | BLU | 14.9 | 0.5 | 2.0 | 64.8 | nd | nd | 0.7 | 0.6 | 6.7 | 0.1 | nd | 0.7 | 1.3 | 0.3 | nd | 0.4 | nd | nd | nd | 4.4 | 2.2 |
| | Chasse | | LBL | 14.8 | 0.6 | 2.2 | 66.2 | nd | 0.4 | 0.7 | 0.7 | 7.6 | 0.1 | nd | 0.3 | 1.0 | 0.1 | nd | 0.3 | nd | nd | nd | 4.7 | nd |
| | | | DBT | 13.1 | 3.1 | 1.0 | 67.0 | nd | 0.2 | 0.8 | 2.7 | 9.6 | 0.1 | nd | 0.9 | 0.9 | 0.2 | nd | 0.1 | nd | nd | nd | nd | nd |
| | | | GRN | 13.7 | 0.6 | 2.2 | 63.5 | nd | nd | 0.9 | 0.6 | 7.3 | nd | nd | 0.5 | 0.8 | nd | nd | 2.0 | 0.2 | nd | nd | 0.6 | 7.2 |
| | | | YEL | 15.2 | 0.6 | 2.3 | 62.8 | nd | nd | 0.8 | 0.7 | 6.1 | nd | nd | 0.3 | 0.7 | nd | nd | 0.4 | nd | nd | nd | 0.8 | 9.2 |
| 89 | 17.190.794 | ca. 1210 | BLU | 12.3 | 2.8 | 1.0 | 49.4 | nd | nd | 0.6 | 1.9 | 4.6 | nd | nd | 0.1 | 1.4 | 0.4 | nd | nd | nd | nd | 7.0 | nd | 19.0 |
| | Medallion | | RED | 14.9 | 1.7 | 3.0 | 60.9 | nd | nd | 1.0 | 1.4 | 7.3 | 0.2 | nd | 0.3 | 5.1 | nd | nd | 1.5 | nd | nd | 0.5 | nd | 2.4 |
| 102 | 17.190.694 | 1218–1225 | WHI | 15.4 | 2.5 | 1.7 | 63.5 | nd | 0.3 | 0.8 | 0.7 | 7.8 | nd | nd | nd | 0.6 | nd | nd | 1.4 | nd | nd | nd | 4.0 | 0.7 |
| | Plaque | | BLU | 13.2 | 1.7 | 1.6 | 65.9 | nd | nd | 0.8 | 1.8 | 7.7 | nd | nd | 0.8 | 1.0 | 0.3 | nd | 0.8 | nd | nd | 1.0 | 1.6 | 1.6 |
| | | | TUR | 17.3 | 1.4 | 1.6 | 63.9 | nd | nd | 1.0 | 2.0 | 5.8 | 0.1 | nd | 1.0 | 0.9 | nd | nd | 2.4 | nd | nd | 1.7 | 1.1 | 1.4 |
| | | | GRN | 16.3 | 0.6 | 2.3 | 66.7 | nd | nd | 1.2 | 0.7 | 6.5 | 0.1 | nd | 1.2 | 0.7 | nd | nd | 3.0 | nd | nd | nd | 0.6 | 2.2 |
| | | | YEL | 16.7 | 0.6 | 2.5 | 65.4 | nd | nd | 1.1 | 0.7 | 5.4 | 0.1 | nd | 0.1 | 0.8 | nd | nd | 0.7 | nd | nd | nd | 0.6 | 6.6 |
| | | | RED | 12.2 | 2.0 | 2.0 | 63.7 | nd | nd | 0.7 | 1.9 | 8.1 | 0.1 | nd | 1.3 | 3.3 | nd | nd | 1.7 | nd | nd | 0.5 | nd | 2.3 |
| 93 | 17.190.854 | ca. 1220 | WHI | 14.7 | 0.8 | 2.0 | 65.0 | nd | 0.4 | 0.7 | 0.7 | 7.1 | nd | nd | 0.2 | 0.6 | nd | nd | 0.3 | nd | nd | nd | 6.9 | nd |
| | Plaque | | BLU | 15.5 | 0.6 | 2.2 | 65.4 | nd | nd | 0.7 | 0.6 | 7.0 | nd | nd | 0.8 | 1.3 | 0.3 | nd | 0.3 | nd | nd | nd | 3.2 | 2.2 |
| | | | GRT | 15.2 | 0.6 | 2.3 | 62.3 | nd | nd | 1.0 | 0.6 | 6.4 | 0.1 | nd | 0.6 | 0.8 | nd | nd | 2.6 | nd | nd | nd | nd | 6.8 |
| | | | GRN | 13.5 | 0.7 | 2.0 | 60.0 | nd | nd | 1.0 | 0.5 | 6.8 | nd | nd | 0.4 | 1.0 | nd | nd | 2.2 | 0.1 | nd | nd | <0.6 | 11.5 |
| | | | YEL | 16.2 | 0.4 | 2.1 | 62.5 | nd | nd | 0.9 | 0.6 | 5.2 | 0.1 | nd | 0.3 | 0.9 | nd | nd | 0.2 | nd | nd | nd | 0.8 | 10.6 |
| | | | RED | 15.3 | 3.0 | 4.4 | 59.2 | 0.40 | nd | 1.2 | 1.8 | 6.7 | 0.3 | nd | 0.2 | 5.4 | nd | nd | 1.6 | nd | nd | nd | nd | 0.6 |
| NE | 17.190.509 | 1200–1250 | WHI | 14.0 | 2.8 | 1.0 | 52.3 | nd | nd | 0.8 | 1.7 | 5.0 | nd | nd | 0.2 | 0.5 | nd | nd | 0.1 | nd | nd | 6.9 | nd | 15.2 |
| | Situla | | DBO | 12.2 | 2.8 | 1.0 | 48.8 | nd | nd | 0.6 | 1.8 | 4.5 | nd | nd | 0.2 | 1.5 | 0.3 | nd | 0.2 | nd | nd | 7.3 | nd | 19.2 |
| | | | BLU | 14.1 | 0.5 | 2.2 | 65.0 | nd | nd | 0.8 | 1.1 | 6.6 | 0.1 | nd | 0.6 | 1.3 | 0.2 | nd | 0.5 | 0.2 | nd | 0.6 | 2.9 | 1.8 |
| | | | TUR | 14.1 | 2.0 | 0.9 | 51.3 | nd | nd | 0.8 | 1.6 | 4.7 | nd | nd | 0.1 | 0.4 | nd | nd | 2.2 | nd | nd | 6.4 | nd | 16.4 |
| | | | GRN | 13.7 | 0.6 | 2.2 | 63.2 | nd | nd | 0.9 | 0.5 | 7.2 | 0.1 | nd | 0.4 | 0.7 | nd | nd | 1.9 | 0.2 | nd | nd | <0.6 | 8.2 |
| | | | YEL | 16.9 | 0.6 | 2.1 | 62.7 | nd | nd | 1.0 | 0.6 | 6.2 | 0.1 | nd | 0.4 | 0.8 | nd | nd | 0.3 | nd | nd | <0.5 | <0.6 | 8.4 |
| | | | RED | 15.6 | 2.0 | 3.2 | 59.1 | 0.50 | nd | 0.9 | 1.6 | 6.8 | 0.1 | nd | 0.5 | 3.4 | nd | nd | 1.7 | nd | nd | 0.6 | nd | 4.2 |
| NE | 17.190.781 | 1200–1250 | WHI | 13.9 | 1.6 | 1.8 | 56.9 | nd | nd | 0.7 | 1.1 | 6.7 | nd | nd | 0.3 | 0.5 | nd | nd | 0.1 | nd | nd | 3.3 | 4.4 | 7.8 |
| | Cross | | DBT | 16.3 | 0.7 | 2.2 | 65.3 | nd | 0.2 | 0.9 | 0.7 | 6.8 | 0.1 | nd | 0.9 | 1.8 | 0.3 | nd | 0.6 | nd | nd | nd | 2.5 | 1.0 |
| | | | TUR | 17.1 | 1.6 | 1.8 | 59.8 | nd | nd | 1.2 | 0.8 | 4.8 | 0.1 | nd | 0.1 | 1.2 | nd | nd | 2.7 | nd | nd | 2.5 | 1.4 | 5.6 |
| | | | GRN | 13.8 | 0.6 | 2.5 | 64.1 | nd | nd | 0.9 | 0.5 | 6.9 | nd | nd | 0.4 | 1.1 | nd | nd | 1.3 | 0.2 | nd | nd | 0.8 | 6.6 |
| 92b | 17.190.795 | before 1227 | BLU | 15.8 | 4.9 | 1.4 | 48.6 | nd | nd | 0.7 | 1.7 | 7.3 | 0.1 | nd | 0.1 | 1.8 | 0.4 | nd | 0.1 | nd | nd | 5.3 | nd | 10.8 |
| | Roundel | | | | | | | | | | | | | | | | | | | | | | | |
| 58b | 17.190.123 | ca. 1231 | WHI | 13.5 | 2.6 | 1.0 | 48.5 | nd | nd | 0.6 | 2.0 | 4.4 | nd | nd | 0.2 | 0.4 | nd | nd | nd | nd | nd | 7.7 | nd | 18.5 |
| | Plaque | | BLU | 12.5 | 2.7 | 1.5 | 53.4 | nd | nd | 0.8 | 2.5 | 5.3 | nd | nd | 0.5 | 1.1 | 0.2 | nd | 0.5 | 0.3 | nd | 5.5 | nd | 12.4 |
| | | | LBL | 14.4 | 0.5 | 2.1 | 62.6 | nd | 0.3 | 0.6 | 0.7 | 8.3 | 0.1 | nd | 0.4 | 1.2 | 0.2 | nd | 0.4 | nd | nd | 0.6 | 7.6 | 0.6 |
| | | | TUR | 15.1 | 2.5 | 1.5 | 46.6 | nd | nd | 0.7 | 1.6 | 4.9 | nd | nd | 0.1 | 0.7 | nd | nd | 2.1 | nd | nd | 7.4 | nd | 17.2 |
| | | | GRN | 18.0 | 0.6 | 2.2 | 65.4 | nd | nd | 1.2 | 0.8 | 5.6 | 0.1 | nd | 0.4 | 0.8 | nd | nd | 2.8 | nd | nd | nd | 0.6 | 2.2 |
| | | | YEL | 17.3 | 0.6 | 2.3 | 65.3 | nd | nd | 1.2 | 0.6 | 5.6 | 0.1 | nd | 0.3 | 0.7 | nd | nd | 0.3 | nd | nd | nd | 0.6 | 5.6 |
| 105 | 17.190.344 | 1215–1235 | TUR | 17.3 | 2.5 | 1.3 | 48.6 | nd | nd | 1.0 | 2.4 | 6.0 | nd | nd | 0.2 | 0.7 | nd | nd | 1.2 | nd | nd | 5.7 | nd | 13.2 |
| | Dove | | YEL | 17.9 | 0.6 | 2.1 | 66.0 | nd | nd | 1.3 | 0.7 | 5.5 | 0.1 | nd | 0.2 | 0.5 | nd | nd | 0.1 | nd | nd | nd | 0.7 | 5.0 |

NE : not exhibited      nd : not detected

*continued on next page*

## Table 3. EDS Analyses of Enamels (MMA) *continued*

| Cat. No. | Acc. No. Object | Date | Color | $Na_2O$ | MgO | $Al_2O_3$ | $SiO_2$ | $P_2O_3$ | $SO_3$ | Cl | $K_2O$ | CaO | $TiO_2$ | $Cr_2O_3$ | MnO | $Fe_2O_3$ | CoO | NiO | CuO | ZnO | $As_2O_3$ | $SnO_2$ | $Sb_2O_3$ | PbO |
|---|---|---|---|---|---|---|---|---|---|---|---|---|---|---|---|---|---|---|---|---|---|---|---|---|
| 115 | 1974.228.1 | 1235–1245 | WHI | 12.5 | 2.5 | 1.2 | 49.1 | nd | nd | 0.6 | 1.8 | 5.8 | nd | nd | 0.5 | 0.6 | nd | nd | 0.2 | nd | nd | 6.7 | nd | 17.9 |
| | Chasse | | LBL | 10.1 | 2.4 | 1.3 | 51.1 | nd | nd | 0.6 | 2.4 | 4.7 | nd | nd | 0.2 | 0.7 | nd | nd | 2.4 | nd | nd | 6.7 | nd | 16.3 |
| | | | DBO | 13.7 | 2.7 | 1.4 | 50.5 | nd | nd | 0.8 | 1.7 | 6.5 | nd | nd | 0.3 | 1.0 | 0.2 | nd | 0.5 | 0.4 | nd | 5.9 | nd | 14.4 |
| | | | GRN | 10.5 | 2.3 | 1.2 | 46.1 | nd | nd | 0.6 | 1.7 | 5.8 | 0.1 | nd | 0.5 | 3.4 | nd | nd | 2.9 | nd | nd | 7.2 | nd | 17.8 |
| | | | RED | 12.9 | 2.5 | 2.0 | 66.1 | nd | 0.2 | 1.1 | 1.7 | 6.9 | 0.2 | nd | 0.1 | 5.7 | nd | nd | 1.4 | nd | nd | nd | nd | nd |
| | | | CLR | 1.5 | 6.2 | 1.7 | 47.4 | 5.6 | nd | 0.4 | 17.5 | 16.5 | 0.1 | nd | 1.1 | 0.6 | nd | nd | 0.6 | nd | nd | nd | nd | nd |
| 155 | 17.190.124 | 1250–1275 | DBO | 13.5 | 3.1 | 1.3 | 51.7 | nd | nd | 0.8 | 1.5 | 6.1 | nd | nd | 0.6 | 1.3 | 0.2 | nd | 0.5 | 0.3 | nd | 5.0 | nd | 13.8 |
| | Virgin and Child | | | | | | | | | | | | | | | | | | | | | | | |
| NE | 35.75 | 1250–1300 | WHI | 12.9 | 2.7 | 1.5 | 50.8 | nd | nd | 0.7 | 1.4 | 5.7 | nd | nd | 0.6 | 0.7 | nd | nd | 0.2 | nd | nd | 6.5 | nd | 17.6 |
| | Gemellion | | DBO | 15.7 | 3.0 | 1.5 | 49.0 | nd | nd | 0.8 | 1.7 | 5.6 | 0.1 | nd | 0.2 | 1.8 | 0.3 | nd | 0.9 | 0.6 | nd | 5.7 | nd | 13.4 |
| | | | TUR | 13.2 | 2.8 | 1.4 | 48.3 | nd | nd | 0.7 | 1.4 | 5.4 | nd | nd | 0.2 | 0.8 | nd | nd | 1.5 | nd | nd | 7.6 | nd | 17.2 |
| 129 | 32.100.284 | 1250–1300 | WHI | 14.0 | 2.8 | 1.7 | 51.9 | nd | nd | 0.9 | 1.6 | 6.6 | nd | nd | 0.3 | 0.7 | nd | nd | 0.1 | nd | nd | 5.3 | nd | 15.0 |
| | Pyx | | LBL | 13.7 | 2.8 | 1.4 | 51.1 | nd | nd | 0.7 | 2.3 | 7.5 | nd | nd | 0.3 | 1.3 | 0.2 | nd | 0.7 | 0.3 | nd | 3.9 | nd | 13.0 |
| | | | RED | 16.3 | 4.4 | 2.0 | 56.8 | nd | nd | 0.9 | 2.2 | 8.3 | 0.1 | nd | 0.6 | 4.6 | nd | nd | 1.4 | nd | nd | nd | nd | 2.6 |
| 124 | 17.190.2146 | 1240–1260 | BLU | 13.9 | 3.4 | 1.4 | 51.7 | nd | nd | 0.8 | 1.5 | 5.9 | 0.1 | nd | 0.6 | 1.2 | 0.2 | nd | 0.5 | 0.2 | nd | 4.8 | nd | 14.4 |
| | Medallion | | | | | | | | | | | | | | | | | | | | | | | |
| 131 | 49.56.8 | 1250–1275 | WHI | 13.3 | 2.7 | 1.3 | 49.4 | nd | nd | 0.7 | 1.9 | 5.8 | nd | nd | 0.5 | 0.6 | nd | nd | 0.3 | nd | nd | 6.1 | nd | 17.8 |
| | Gemellion | | DBO | 16.2 | 2.7 | 1.3 | 50.3 | nd | nd | 0.8 | 1.8 | 6.4 | nd | nd | 0.3 | 1.1 | 0.2 | nd | 1.0 | 0.4 | nd | 4.8 | nd | 12.7 |
| | | | TUR | 11.8 | 3.0 | 1.3 | 47.6 | nd | nd | 0.7 | 1.7 | 6.2 | 0.1 | nd | nd | 0.6 | nd | nd | 3.6 | nd | nd | 5.3 | nd | 17.6 |
| 134 | 32.100.285 | ca. 1300 | WHI | 11.7 | 2.6 | 1.6 | 51.1 | nd | nd | 0.6 | 1.7 | 5.4 | nd | nd | 0.8 | 0.7 | nd | nd | 0.6 | nd | nd | 6.0 | nd | 17.6 |
| | Candlestick | | LBL | 12.9 | 2.7 | 1.4 | 53.3 | nd | nd | 0.9 | 1.7 | 6.3 | nd | nd | 0.7 | 1.1 | 0.2 | nd | 0.3 | 0.2 | nd | 4.8 | nd | 13.0 |
| | | | RED | 12.9 | 2.4 | 2.7 | 63.5 | 0.6 | nd | 0.9 | 2.0 | 7.2 | 0.1 | nd | 0.7 | 5.5 | nd | nd | 2.0 | nd | nd | nd | nd | 0.5 |

NE : not exhibited nd : not detected

## Table 4. Ion Beam Analyses of Enamels (LRMF)

| Cat. No. | Identification | Date | Color | $Na_2O$ | MgO | $Al_2O_3$ | $SiO_2$ | $P_2O_3$ | $SO_3$ | Cl | $K_2O$ | CaO | $TiO_2$ | $Cr_2O_3$ | MnO | $Fe_2O_3$ | CoO | NiO | CuO | ZnO | $As_2O_3$ | $SnO_2$ | $Sb_2O_3$ | PbO |
|---|---|---|---|---|---|---|---|---|---|---|---|---|---|---|---|---|---|---|---|---|---|---|---|---|
| 3 | OA 6273 | ca. 1100 | DBO | 18.78 | 0.22 | 1.55 | 55.23 | 0.09 | 0.52 | 0.79 | 0.77 | 7.73 | 0.001 | 0.000 | 0.55 | 2.10 | 0.93 | 0.05 | 3.31 | 0.04 | 0.21 | 0.06 | 7.73 | 0.22 |
| | Plaque | | FLS | 41.43 | 0.00 | 1.66 | 41.43 | 0.00 | 0.18 | 0.45 | 0.83 | 6.22 | 0.000 | 0.001 | 0.04 | 0.83 | 0.02 | 0.01 | 0.41 | 0.02 | 0.03 | 0.00 | 3.94 | 0.41 |
| | | | TUR | 17.59 | 0.29 | 1.66 | 53.74 | 0.00 | 0.56 | 1.32 | 1.47 | 6.84 | 0.001 | 0.002 | 0.00 | 0.68 | 0.03 | 0.03 | 11.72 | 0.09 | 0.09 | 0.20 | 2.74 | 0.49 |
| | | | GRN | 23.69 | 0.13 | 1.71 | 56.60 | 0.00 | 1.16 | 2.15 | 1.18 | 7.90 | 0.001 | 0.001 | 0.13 | 1.97 | 0.06 | 0.04 | 0.00 | 0.12 | 0.07 | 0.26 | 1.05 | 1.97 |
| 7 | Medallion 21 | 1110–1130 | LBL | 8.71 | 0.31 | 2.24 | 66.45 | 0.32 | 0.37 | 1.41 | 1.08 | 9.97 | 0.002 | 0.000 | 0.66 | 1.25 | 0.55 | 0.02 | 0.46 | 0.01 | 0.02 | 0.04 | 4.15 | 1.91 |
| | | | GRN | 9.21 | 0.36 | 2.33 | 64.61 | 0.43 | 0.00 | 1.81 | 1.22 | 8.61 | 0.000 | 0.000 | 0.30 | 0.95 | 0.03 | 0.03 | 1.90 | 0.01 | 0.02 | 0.17 | 1.46 | 6.46 |
| | | | YEL | 8.91 | 1.45 | 3.40 | 68.03 | 0.34 | 0.00 | 1.70 | 1.45 | 7.23 | 0.001 | 0.001 | 0.17 | 0.47 | 0.00 | 0.01 | 0.14 | 0.01 | 0.01 | 0.05 | 0.87 | 5.70 |
| | | | DBO | 9.47 | 0.52 | 2.93 | 64.62 | 0.26 | 0.65 | 1.21 | 1.12 | 9.05 | 0.002 | 0.000 | 0.43 | 1.72 | 0.30 | 0.01 | 0.78 | 0.01 | 0.06 | 0.07 | 6.63 | 0.11 |
| | | | WHI | 13.90 | 4.72 | 0.00 | 53.11 | 0.05 | 0.00 | 1.18 | 17.70 | 0.30 | 0.001 | 0.008 | 0.07 | 0.59 | 0.00 | 0.00 | 0.26 | 0.01 | 0.05 | 0.07 | 7.44 | 0.47 |
| 7 | Medallion 23 | 1110–1130 | GRN | 5.71 | 1.30 | 4.00 | 70.00 | 0.30 | 0.20 | 1.60 | 1.30 | 6.00 | 0.001 | 0.002 | 0.28 | 1.10 | 0.01 | 0.02 | 2.00 | 0.00 | 0.00 | 0.25 | 0.90 | 5.00 |
| | | | YEL | 2.48 | 1.01 | 3.68 | 77.21 | 0.14 | 0.01 | 1.56 | 1.38 | 4.60 | 0.001 | 0.001 | 0.28 | 0.60 | 0.00 | 0.01 | 0.11 | 0.01 | 0.02 | 0.06 | 0.74 | 6.16 |
| 7 | Medallion 24 | 1110–1130 | BLU | 8.87 | 0.15 | 1.79 | 67.14 | 0.36 | 0.00 | 0.90 | 1.43 | 9.49 | 0.001 | 0.000 | 0.49 | 1.25 | 0.50 | 0.02 | 0.54 | 0.01 | 0.03 | 0.04 | 5.10 | 1.79 |
| | | | GRN | 10.96 | 0.10 | 1.52 | 59.72 | 0.43 | 0.00 | 0.87 | 1.25 | 7.60 | 0.001 | 0.000 | 0.27 | 1.30 | 0.02 | 0.04 | 2.71 | 0.02 | 0.04 | 0.33 | 1.85 | 10.86 |
| | | | WHI | 15.12 | 0.23 | 1.35 | 60.00 | 0.24 | 0.20 | 0.90 | 1.40 | 11.50 | 0.001 | 0.000 | 0.13 | 0.65 | 0.01 | 0.00 | 0.10 | 0.01 | 0.04 | 0.06 | 7.70 | 0.30 |
| 7 | Medallion 7 | 1110–1130 | LBL | 11.07 | 0.07 | 1.49 | 59.66 | 0.39 | 0.57 | 1.55 | 1.49 | 11.70 | 0.002 | 0.000 | 0.80 | 1.66 | 0.50 | 0.02 | 0.80 | 0.01 | 0.03 | 0.07 | 5.74 | 2.29 |
| | | | WHI | 11.27 | 0.23 | 1.35 | 64.17 | 0.25 | 0.79 | 1.58 | 1.35 | 11.37 | 0.002 | 0.001 | 0.15 | 0.59 | 0.00 | 0.00 | 0.18 | 0.01 | 0.07 | 0.06 | 6.19 | 0.34 |
| | | | YEL | 10.87 | 0.10 | 1.42 | 66.40 | 0.24 | 0.16 | 1.85 | 1.96 | 7.65 | 0.001 | 0.002 | 0.22 | 0.69 | 0.00 | 0.01 | 0.22 | 0.01 | 0.01 | 0.09 | 0.98 | 7.08 |
| | | | TUR | 11.03 | 0.11 | 1.34 | 65.00 | 0.28 | 0.56 | 2.47 | 2.58 | 8.18 | 0.001 | 0.000 | 0.04 | 0.83 | 0.04 | 0.04 | 3.36 | 0.06 | 0.03 | 0.22 | 3.36 | 0.34 |
| | | | DBO | 9.13 | 0.15 | 1.37 | 63.79 | 0.34 | 0.79 | 1.28 | 1.28 | 9.81 | 0.002 | 0.000 | 0.51 | 1.86 | 0.90 | 0.04 | 0.98 | 0.03 | 0.17 | 0.07 | 7.16 | 0.20 |
| | | | GRN | 11.00 | 0.10 | 1.58 | 63.00 | 0.36 | 0.00 | 1.70 | 1.30 | 8.10 | 0.003 | 0.002 | 0.30 | 1.10 | 0.03 | 0.04 | 2.30 | 0.02 | 0.01 | 0.26 | 1.30 | 7.40 |
| 7 | Medallion 4 | 1110–1130 | DBO | 9.97 | 0.11 | 1.41 | 62.81 | 0.32 | 0.70 | 1.19 | 1.09 | 8.77 | 0.002 | 0.000 | 0.55 | 1.99 | 0.90 | 0.04 | 1.08 | 0.03 | 0.18 | 0.06 | 8.34 | 0.32 |
| | | | LBL | 11.39 | 0.08 | 1.20 | 59.75 | 0.27 | 0.72 | 1.31 | 1.31 | 10.76 | 0.002 | 0.000 | 0.68 | 1.67 | 0.59 | 0.02 | 0.84 | 0.01 | 0.04 | 0.07 | 7.17 | 2.03 |
| | | | TUR | 10.99 | 0.11 | 1.45 | 67.05 | 0.19 | 0.89 | 1.90 | 1.45 | 6.71 | 0.002 | 0.002 | 0.01 | 0.89 | 0.02 | 0.04 | 3.58 | 0.05 | 0.04 | 0.34 | 3.69 | 0.45 |
| | | | WHI | 8.95 | 0.35 | 1.67 | 70.20 | 0.22 | 0.66 | 1.23 | 0.90 | 9.65 | 0.002 | 0.000 | 0.03 | 0.44 | 0.01 | 0.00 | 0.31 | 0.01 | 0.04 | 0.05 | 5.18 | 0.04 |
| | | | GRN | 9.32 | 0.19 | 1.78 | 70.29 | 0.28 | 0.61 | 1.59 | 1.41 | 8.43 | 0.001 | 0.001 | 0.28 | 0.75 | 0.02 | 0.02 | 2.06 | 0.03 | 0.03 | 0.09 | 0.75 | 1.97 |
| 8j | OA 6280 | 1110–1130 | DBO | 17.36 | 0.61 | 1.74 | 61.27 | 0.08 | 0.00 | 1.42 | 1.02 | 7.15 | 0.000 | 0.001 | 0.71 | 1.33 | 0.22 | 0.02 | 0.51 | 0.01 | 0.01 | 0.03 | 6.13 | 0.82 |
| | Medallion | | LBL | 8.45 | 0.34 | 2.25 | 67.57 | 0.00 | 0.00 | 0.04 | 1.01 | 11.26 | 0.001 | 0.002 | 0.02 | 1.24 | 0.47 | 0.03 | 0.45 | 0.01 | 0.02 | 0.00 | 4.50 | 1.69 |
| | | | TUR | 23.61 | 0.11 | 1.12 | 50.58 | 0.28 | 1.18 | 1.82 | 1.46 | 4.50 | 0.001 | 0.002 | 0.11 | 0.56 | 0.03 | 0.03 | 11.24 | 0.06 | 0.08 | 0.22 | 2.81 | 0.79 |
| | | | LGR | 28.83 | 0.16 | 1.94 | 51.82 | 0.04 | 0.79 | 0.55 | 0.97 | 4.53 | 0.000 | 0.001 | 0.10 | 0.97 | 0.02 | 0.04 | 6.48 | 0.04 | 0.03 | 0.00 | 1.62 | 0.00 |
| | | | DGR | 11.35 | 0.46 | 2.45 | 69.03 | 0.02 | 0.82 | 1.16 | 1.53 | 4.60 | 0.000 | 0.002 | 0.00 | 0.61 | 0.03 | 0.02 | 7.67 | 0.02 | 0.02 | 0.15 | 0.00 | 0.00 |
| | | | WHI | 16.22 | 3.34 | 1.91 | 66.77 | 0.08 | 0.50 | 0.95 | 1.00 | 8.59 | 0.000 | 0.000 | 0.10 | 0.48 | 0.02 | 0.01 | 0.19 | 0.01 | 0.04 | 0.00 | 0.00 | 0.00 |
| NE | OA 10026 | 1170–1180 | DBO | 16.75 | 1.90 | 1.01 | 44.67 | 0.24 | 0.00 | 0.71 | 1.79 | 3.91 | 0.001 | 0.000 | 0.00 | 2.01 | 0.44 | 0.05 | 0.45 | 0.01 | 0.74 | 7.26 | 0.03 | 18.99 |
| | Plaque | | TUR | 17.23 | 0.53 | 0.93 | 26.50 | 0.90 | 0.61 | 1.57 | 1.39 | 8.61 | 0.001 | 0.002 | 0.00 | 0.93 | 0.02 | 0.06 | 13.25 | 0.03 | 0.08 | 8.22 | 0.03 | 18.55 |
| | | | GRN | 18.79 | 0.43 | 1.66 | 57.82 | 0.13 | 0.35 | 1.04 | 1.30 | 7.23 | 0.001 | 0.001 | 0.00 | 1.30 | 0.03 | 0.03 | 4.34 | 0.05 | 0.04 | 0.20 | 0.80 | 4.34 |
| 14 | OA9485 | 1170–1190 | GRN | 13.28 | 1.17 | 2.47 | 56.33 | 0.41 | 3.17 | 1.80 | 1.12 | 6.90 | 0.030 | 0.000 | 0.26 | 1.08 | 0.01 | 0.00 | 2.15 | 0.48 | 0.00 | 0.34 | 0.62 | 6.01 |
| | Candlestick base | | BLU | 16.18 | 1.34 | 2.35 | 55.97 | 0.36 | 0.45 | 1.20 | 1.12 | 8.63 | 0.027 | 0.000 | 0.70 | 1.78 | 0.27 | 0.11 | 0.46 | 0.20 | 0.00 | 0.25 | 3.11 | 3.33 |
| | | | DBO | 13.66 | 1.11 | 2.52 | 61.87 | 0.31 | 1.46 | 1.51 | 0.89 | 8.14 | 0.024 | 0.000 | 0.62 | 1.29 | 0.45 | 0.18 | 0.81 | 0.15 | 0.00 | 0.00 | 3.60 | 1.96 |
| | | | RED | 17.19 | 2.66 | 1.96 | 43.45 | 0.58 | 2.61 | 1.43 | 1.82 | 7.86 | 0.031 | 0.000 | 0.34 | 1.39 | 0.01 | 0.00 | 1.81 | 0.48 | 0.00 | 0.21 | 0.24 | 10.63 |
| 25 | OA 8205 | 1180–1190 | DBT | 17.97 | 1.65 | 4.10 | 61.42 | 0.34 | 0.73 | 1.66 | 0.80 | 7.04 | 0.019 | 0.000 | 0.56 | 1.26 | 0.65 | 0.32 | 0.56 | 0.09 | 0.16 | 0.02 | 1.67 | 0.32 |
| | Book cover— | | GRN | 20.18 | 2.18 | 4.34 | 52.90 | 0.20 | 0.00 | 1.35 | 0.52 | 5.09 | 0.028 | 0.000 | 0.30 | 0.71 | 0.01 | 0.00 | 1.46 | 2.07 | 0.00 | 0.32 | 0.51 | 6.93 |
| | Crucifixion | | DGR | 21.44 | 2.05 | 3.50 | 48.08 | 0.35 | 0.00 | 1.27 | 0.82 | 4.81 | 0.035 | 0.000 | 0.38 | 1.14 | 0.02 | 0.09 | 3.18 | 0.20 | 0.00 | 0.58 | 0.81 | 7.05 |
| | | | BLK | 17.62 | 1.42 | 4.29 | 60.12 | 0.30 | 0.70 | 1.99 | 1.22 | 7.91 | 0.056 | 0.000 | 0.34 | 4.87 | 0.03 | 0.00 | 0.53 | 0.09 | 0.00 | 0.02 | 0.32 | 0.52 |
| | | | RED | 17.25 | 1.85 | 3.91 | 63.91 | 0.37 | 1.24 | 1.81 | 1.02 | 7.11 | 0.023 | 0.000 | 1.97 | 0.99 | 0.00 | 0.00 | 0.25 | 0.08 | 0.00 | 0.00 | 0.09 | 0.09 |
| | | | BLU | 15.50 | 1.08 | 3.31 | 61.76 | 0.25 | 1.63 | 1.21 | 0.77 | 7.40 | 0.022 | 0.000 | 0.39 | 1.28 | 0.19 | 0.00 | 0.32 | 0.08 | 0.00 | 0.08 | 4.76 | 1.14 |
| 25 | OA 8205 | 1180–1190 | DBT | 19.82 | 2.14 | 4.46 | 57.10 | 0.18 | 0.70 | 1.66 | 1.00 | 7.77 | 0.021 | 0.000 | 0.71 | 1.44 | 0.38 | 0.27 | 0.49 | 0.09 | 0.00 | 0.04 | 1.63 | 0.23 |
| | Book cover—Border | | RED | 22.67 | 1.65 | 3.73 | 54.86 | 0.17 | 0.77 | 1.98 | 1.10 | 6.70 | 0.027 | 0.000 | 0.36 | 5.03 | 0.02 | 0.00 | 0.18 | 0.10 | 0.00 | 0.00 | 0.38 | 0.31 |
| 25 | OA 8205 | 1180–1190 | DBO | 15.75 | 1.22 | 3.08 | 62.62 | 0.27 | 1.22 | 1.72 | 0.92 | 7.80 | 0.023 | 0.000 | 0.75 | 1.68 | 0.62 | 0.00 | 0.53 | 0.18 | 0.00 | 0.00 | 2.57 | 0.37 |
| | Book cover—Plaque | | RED | 17.27 | 1.46 | 3.40 | 62.53 | 0.20 | 0.87 | 2.30 | 0.95 | 6.14 | 0.035 | 0.000 | 0.41 | 5.35 | 0.03 | 0.00 | 0.13 | 0.07 | 0.00 | 0.00 | 0.48 | 0.25 |
| | | | GRN | 15.05 | 1.02 | 2.74 | 58.87 | 0.15 | 0.00 | 1.73 | 0.56 | 7.07 | 0.028 | 0.000 | 0.46 | 0.75 | 0.01 | 0.00 | 1.99 | 2.82 | 0.00 | 0.00 | 0.49 | 6.74 |
| NE | MRR 249 | ca. 1200 | TUR | 17.55 | 0.00 | 2.51 | 62.70 | 0.00 | 0.00 | 1.08 | 1.50 | 6.27 | 0.000 | 0.000 | 0.00 | 1.00 | 0.03 | 0.02 | 2.88 | 0.09 | 0.05 | 0.25 | 3.13 | 0.44 |
| | Book cover | | DBO | 16.63 | 1.11 | 2.83 | 61.58 | 0.11 | 0.20 | 0.79 | 0.99 | 7.39 | 0.000 | 0.000 | 0.62 | 1.35 | 0.81 | 0.04 | 0.49 | 0.02 | 0.20 | 0.06 | 4.06 | 1.35 |
| | | | GRN | 20.87 | 0.14 | 1.39 | 55.66 | 0.15 | 1.62 | 2.34 | 1.39 | 7.65 | 0.001 | 0.001 | 0.42 | 1.04 | 0.03 | 0.02 | 2.78 | 0.05 | 0.03 | 0.21 | 1.04 | 5.57 |
| 65 | OA 7284 | 1185–1195 | WHI | 16.58 | 0.24 | 1.54 | 55.65 | 0.55 | 0.71 | 0.82 | 0.71 | 7.81 | 0.002 | 0.015 | 0.59 | 1.18 | 0.01 | 0.00 | 2.37 | 0.02 | 0.03 | 0.12 | 9.12 | 0.18 |
| | Cross | | DBO | 18.94 | 0.30 | 1.67 | 56.07 | 0.21 | 0.46 | 1.09 | 0.61 | 7.58 | 0.001 | 0.006 | 0.61 | 2.27 | 0.37 | 0.02 | 2.27 | 0.01 | 0.02 | 0.08 | 5.00 | 2.12 |
| | | | TUR | 22.48 | 0.32 | 1.45 | 56.21 | 0.16 | 0.44 | 1.33 | 0.64 | 6.42 | 0.001 | 0.006 | 0.03 | 0.80 | 0.04 | 0.03 | 3.69 | 0.07 | 0.08 | 0.32 | 4.18 | 0.48 |
| 92a | OA 9476 | before 1227 | DBO | 18.11 | 4.71 | 1.93 | 43.34 | 0.37 | 0.00 | 1.36 | 1.44 | 5.16 | 0.034 | 0.000 | 0.05 | 1.34 | 0.29 | 0.00 | 0.61 | 4.98 | 1.67 | 8.35 | 0.00 | 11.92 |
| | Medallion | | | | | | | | | | | | | | | | | | | | | | | |
| 59 | Les Billanges | 1220–1230 | DBO | 11.33 | 2.72 | 2.24 | 55.43 | 0.40 | 0.00 | 1.09 | 1.39 | 4.24 | 0.008 | 0.000 | 0.33 | 1.15 | 0.21 | 0.01 | 0.51 | 1.43 | 0.07 | 7.00 | 0.03 | 9.96 |
| | Reliquary | | | | | | | | | | | | | | | | | | | | | | | |
| 101 | OA 4083 | after 1228 | DBO | 14.06 | 1.05 | 2.76 | 61.69 | 0.22 | 0.96 | 1.18 | 1.11 | 7.39 | 0.029 | 0.000 | 0.60 | 1.45 | 0.47 | 0.16 | 0.47 | 0.05 | 0.00 | 0.00 | 4.19 | 1.38 |
| | Reliquary | | GRN | 14.00 | 1.35 | 3.09 | 62.17 | 0.22 | 2.51 | 2.62 | 1.46 | 6.09 | 0.028 | 0.000 | 0.39 | 0.82 | 0.01 | 0.00 | 2.21 | 0.20 | 0.00 | 0.30 | 0.51 | 3.02 |
| | | | YEL | 12.75 | 1.44 | 3.22 | 66.87 | 0.24 | 0.00 | 2.05 | 1.12 | 5.09 | 0.026 | 0.000 | 0.21 | 0.95 | 0.01 | 0.00 | 0.27 | 0.08 | 0.00 | 0.23 | 0.77 | 7.22 |
| 135 | MR 2663 | Last 3rd | DBO | 13.39 | 1.81 | 2.13 | 31.95 | 0.41 | 5.71 | 1.08 | 2.72 | 6.80 | 0.035 | 0.000 | 0.72 | 2.00 | 0.29 | 0.00 | 2.78 | 3.59 | 0.00 | 13.99 | 0.03 | 2.31 |
| | Candlestick | 13th c. | RED | 2.32 | 1.44 | 1.73 | 37.40 | 1.93 | 2.71 | 0.88 | 2.62 | 11.02 | 0.040 | 0.000 | 0.49 | 2.22 | 0.05 | 0.00 | 3.53 | 1.00 | 0.00 | 7.82 | 0.00 | 8.23 |
| 152 | OA 7495 | 1307 | TUR | 15.75 | 0.14 | 0.55 | 27.39 | 0.20 | 0.67 | 1.03 | 2.19 | 4.11 | 0.001 | 0.003 | 0.03 | 0.68 | 0.04 | 0.10 | 9.59 | 0.06 | 0.04 | 9.59 | 0.00 | 27.39 |
| | Plaque | | DBO | 12.82 | 0.64 | 0.90 | 25.63 | 0.52 | 0.84 | 0.67 | 3.20 | 2.56 | 0.000 | 0.002 | 0.38 | 2.31 | 0.94 | 0.11 | 14.10 | 0.70 | 0.03 | 8.33 | 0.13 | 24.35 |
| | | | WHI | 13.47 | 0.13 | 0.27 | 20.21 | 0.22 | 0.65 | 1.01 | 2.16 | 2.02 | 0.000 | 0.003 | 0.03 | 1.35 | 0.02 | 0.09 | 5.39 | 0.03 | 0.00 | 18.86 | 0.00 | 33.68 |
| 157 | Nexon | 1346 | BLU | 9.80 | 1.17 | 3.46 | 45.50 | 0.13 | 0.37 | 0.41 | 4.92 | 8.00 | 0.049 | 0.000 | 0.48 | 1.46 | 0.04 | 0.31 | 8.32 | 0.77 | 0.58 | 8.52 | 0.02 | 4.84 |
| | Reliquary | | GRN | 7.73 | 1.30 | 0.32 | 56.31 | 0.03 | 0.36 | 0.08 | 11.45 | 1.55 | 0.040 | 0.000 | 0.20 | 1.53 | 0.96 | 0.58 | 5.67 | 6.50 | 0.17 | 4.79 | 0.03 | 0.39 |

NE : not exhibited.

## Later Enamels

During the course of our technical examinations of the objects in the exhibit, we tried to be aware of possible alteration or production at a later time. The revival of the medieval aesthetic in the nineteenth century resulted in a proliferation of enameled objects in "the style of Limoges." Given the popularity of medieval objects among collectors during the nineteenth and twentieth centuries, it would not be unusual for copies to be included with original material. Of greater concern were those objects—intended to deceive—made by artists or restorers well versed in the stylistic vocabulary of the medieval world. The André studio in Paris undertook extensive campaigns in the nineteenth century to restore medieval objects, including pieces made in Limoges, that often included replication of the originals (see cat. 8). Written records in the Monuments Historiques describe the fabrication of a replacement medallion for the coffret of Abbot Boniface from Aveyron during the course of a restoration program in the nineteenth century (cat. 7; see entry for a full description). By drawing on the large body of original material in the exhibition and the identifiable pieces of later manufacture, it was possible to develop a set of technical and analytical criteria against which questioned pieces could be evaluated and identified.

Visual observation can be used to delineate the degree of color homogeneity and the particle size of the enamel. The industrial production of enamels, beginning in the nineteenth century, produced finely powdered enamels of uniform color, as opposed to the earlier coarsely powdered, heterogeneous mixtures (see fig. 21). The possible regilding of an original surface can often be discerned with the naked eye by the detection of gilding within scratches or across worn surfaces.

X-ray radiography can indicate whether an object has been cast or wrought. Since the Limousin metalworker used hammered, wrought copper almost exclusively, the detection of a cast structure in an object normally formed from sheet would be cause for concern. The ability of radiography to reveal the tooling of the copper substrate below the glass can provide additional information regarding the carving or working that can be useful in characterizing different hands and possibly distinguishing a later addition from the originals in a series, as was the case for the replacement medallions on the coffret of Abbot Boniface.[1] Finally, X-rays serve as an indicator of the relative amount of lead used in the fabrication of the enamel. The amount of lead present in medieval enamels is low in comparison with most later glass compositions; therefore, the radiograph of a medieval Limoges enamel shows a higher degree of contrast between the enamel and the reserved metal relative to a nineteenth-century copy.

The analyses of the enamel compositions have been extremely useful in differentiating between original and later enamels (see tables 5 and 6). Two roundels from the Art Museum, Princeton University, a book cover and a base from a Eucharistic dove at the Metropolitan Museum, two replacement medallions from Abbot Boniface's coffret, and the courtly scenes from the Louvre were all found to be lead-alkali glasses with lead content of between twenty and forty percent, calcium levels of one to five percent, with potassium greater than calcium. Many of these had very little or no detectable aluminum, and only small amounts or no detectable magnesium. The opaque enamels, except for one yellow, all appear to contain the opacifying agent lead arsenate, with some tin found in several examples. The yellow enamel was found to contain a mixture of lead antimonate and lead-tin yellow. The blue enamels analyzed were tinted with cobalt, the turquoises and reds with copper oxides, and the greens with chromium, possibly with some lead-tin yellow. The use of lead arsenate as an opacifier and chromium green as a colorant are both eighteenth-century innovations; moreover, the compositions are very similar to those found on several nineteenth-century enameled gold objects analyzed at The Metropolitan Museum of Art.[2]

---

1. The markings of a drill bit are apparent in the radiographs in each of the circular cells around the perimeter of the replacement medallions.
2. M. T. Wypyski, unpublished analyses.

## Table 5. EDS Analyses of Later Enamels (MMA)

| Cat. No. | Acc. No. Object | Date | Color | Na$_2$O | MgO | Al$_2$O$_3$ | SiO$_2$ | P$_2$O$_5$ | SO$_3$ | Cl | K$_2$O | CaO | TiO$_2$ | Cr$_2$O$_3$ | MnO | Fe$_2$O$_3$ | CoO | NiO | CuO | ZnO | As$_2$O$_3$ | SnO$_2$ | Sb$_2$O$_3$ | PbO |
|---|---|---|---|---|---|---|---|---|---|---|---|---|---|---|---|---|---|---|---|---|---|---|---|---|
| 105 | 17.190.344 Dove (base) | 1235–1235* | BLU | 4.9 | nd | 0.5 | 47.5 | nd | nd | nd | 8.3 | 3.0 | nd | nd | 0.3 | 0.3 | 0.1 | nd | 0.8 | nd | 0.7 | 2.3 | ND | 29.5 |
| NE | 17.190.774 Book cover | 1150–1200* | WHI | 5.0 | nd | 0.4 | 40.6 | nd | nd | nd | 4.6 | 5.0 | nd | nd | 0.2 | 0.5 | nd | nd | nd | nd | 1.2 | ND | ND | 42.0 |
| | | | BLU | 3.2 | nd | 0.8 | 42.3 | nd | nd | nd | 5.3 | 3.0 | nd | nd | 0.8 | 0.2 | 0.6 | nd | 1.1 | nd | 1.0 | 0.7 | ND | 42.5 |
| | | | GRN | 5.4 | nd | nd | 49.6 | nd | nd | nd | 7.0 | 2.0 | nd | 0.2 | nd | 0.2 | nd | nd | 1.6 | nd | 0.7 | 0.5 | ND | 33.8 |
| | | | YEL | 4.2 | nd | nd | 38.3 | nd | nd | nd | 3.7 | 3.7 | nd | nd | nd | 0.2 | nd | nd | 0.4 | nd | ND | 1.0 | <0.6 | 49.0 |
| | | | CLR | 7.2 | nd | nd | 56.8 | nd | nd | nd | 8.1 | 1.0 | nd | nd | 0.1 | 0.1 | nd | nd | | nd | ND | ND | ND | 27.8 |
| NE | Art Museum, Princeton, Y1950-14 Roundel | 1200–1250* | BLU | 4.1 | nd | nd | 36.4 | nd | nd | nd | 6.0 | 1.5 | nd | nd | 0.2 | 0.5 | 3.4 | 0.8 | nd | nd | 3.2 | ND | ND | 43.0 |
| | | | TUR | 6.7 | nd | 0.3 | 51.6 | nd | nd | nd | 11.2 | 3.0 | nd | nd | 0.1 | 0.3 | nd | nd | 2.8 | nd | 3.8 | ND | ND | 19.0 |
| | | | RED | 7.4 | nd | 0.6 | 48.4 | nd | nd | nd | 3.8 | 4.6 | nd | nd | 0.4 | 2.4 | nd | nd | 2.9 | nd | 0.7 | ND | ND | 29.2 |
| NE | Art Museum, Princeton, Y1930-170 Roundel | 1200–1250* | BLU | 4.4 | nd | nd | 37.6 | nd | nd | nd | 4.9 | 1.5 | nd | nd | 0.2 | 0.2 | 2.5 | 0.2 | 0.3 | nd | 3.1 | ND | ND | 44.5 |

NE : not exhibited          nd : not detected          * Composition does not agree with given date.

## Table 6. Ion Beam Analyses of Later Enamels (LRMF)

| Cat. No. | Acc. No. Object | Date | Color | Na$_2$O | MgO | Al$_2$O$_3$ | SiO$_2$ | P$_2$O$_5$ | SO$_3$ | Cl | K$_2$O | CaO | TiO$_2$ | Cr$_2$O$_3$ | MnO | Fe$_2$O$_3$ | CoO | NiO | CuO | ZnO | As$_2$O$_3$ | SnO$_2$ | Sb$_2$O$_3$ | PbO |
|---|---|---|---|---|---|---|---|---|---|---|---|---|---|---|---|---|---|---|---|---|---|---|---|---|
| 7 | Medallion 10 | 1110–1130* (1878?) | DBT | 4.77 | 0.70 | 3.00 | 60.00 | 0.53 | 0.00 | 0.60 | 6.50 | 4.20 | 0.001 | 0.003 | 0.15 | 0.10 | 1.42 | 0.09 | 0.15 | 0.02 | 0.59 | 0.06 | 0.00 | 17.00 |
| | | | LBL | 8.71 | 1.10 | 3.00 | 53.00 | 0.90 | 0.00 | 0.40 | 6.00 | 2.50 | 0.000 | 0.001 | 0.01 | 0.26 | 0.11 | 0.15 | 0.03 | | 1.47 | 0.10 | 0.00 | 22.00 |
| | | | LBL | 15.58 | 1.01 | 1.90 | 50.00 | 0.75 | 0.00 | 0.25 | 3.00 | 2.50 | 0.003 | 0.004 | 0.00 | 0.08 | 0.03 | 0.13 | 1.65 | 0.04 | 1.81 | 0.08 | 0.00 | 21.00 |
| | | | GRN | 10.75 | 1.70 | 2.50 | 49.00 | 1.10 | 0.00 | 0.30 | 3.30 | 2.30 | 0.004 | 0.019 | 0.00 | 0.03 | 0.14 | | 1.64 | 0.05 | 4.52 | 0.14 | 0.00 | 22.00 |
| | | | WHI | 0.94 | 0.92 | 3.70 | 60.50 | 0.63 | 0.00 | 0.17 | 6.70 | 4.00 | 0.002 | 0.000 | 0.01 | 0.11 | 0.02 | 0.09 | 0.23 | 0.02 | 0.51 | 0.13 | 0.00 | 21.00 |
| 7 | Medallion 29 | 1110–1130* (1878?) | BLU | 9.15 | 1.20 | 3.00 | 50.00 | 0.90 | 0.00 | 0.45 | 6.50 | 2.80 | 0.001 | 0.003 | 0.02 | 0.09 | 0.32 | 0.10 | 0.50 | 0.02 | 1.82 | 0.00 | 0.00 | 23.00 |
| | | | GRN | 11.26 | 1.50 | 2.50 | 38.00 | 1.20 | 0.00 | 0.40 | 2.30 | 1.09 | 0.002 | 0.024 | 0.00 | 0.09 | 0.01 | 0.12 | 0.80 | 0.04 | 5.87 | 0.09 | 0.00 | 54.50 |
| | | | WHI | 8.40 | 0.70 | 3.20 | 53.00 | 0.50 | 0.00 | 0.46 | 6.50 | 3.90 | 0.000 | 0.003 | 0.00 | 0.13 | 0.00 | 0.06 | 0.25 | 0.02 | 0.71 | 0.06 | 0.00 | 22.00 |
| 127 | OA 6279 Box | 1250–1300* (before 1900) | BLU | 8.12 | 1.21 | 1.24 | 30.01 | 0.23 | 0.00 | 0.35 | 3.90 | 1.14 | 0.035 | 0.064 | 0.31 | 0.21 | 0.16 | 0.13 | 1.34 | 0.35 | 21.33 | 3.52 | 0.13 | 24.33 |
| | | | GRN | 8.24 | 1.32 | 1.19 | 28.18 | 0.21 | 0.00 | 0.46 | 3.27 | 0.96 | 0.036 | 0.145 | 0.27 | 0.24 | 0.03 | 0.00 | 1.19 | 0.59 | 24.99 | 1.74 | 0.02 | 24.69 |

* Composition does not agree with given date.

# Table 7. Munsell Notations for Enamel Colors (MMA Objects)

The hue is designated by the first set of numbers and the letters, while the value and chroma are defined, respectively, by the second set of numbers.

| Cat. No. | Acc. No. | Object | Date | Light Blue | Medium Blue | Dark Blue | Light Green | Medium Green | Dark Green | Red (opaque) | Red (translucent) | Turquoise | Yellow | Other |
|---|---|---|---|---|---|---|---|---|---|---|---|---|---|---|
| 4 | 1983.38 | Roundel | ca. 1100 | 10BG 4/4 | 7.5PB 3/6 | 7.5PB 2/4 | | 10GY 4/4 | | | | 2.5B 5/6 | 2.5GY 5/4 | 2.5YR 6/2 Flesh |
| 6 | 17.190.426 | Plaques | ca. 1100 | 2.5PB 6/4 | 7.5PB 3/8 | 7.5PB 2/4 | | 2.5G 4/6 | | | | 5B 5/4 | | |
| | 17.190.427 | | | 2.5PB 6/4 | 7.5PB 3/8 | 7.5PB 2/4 | | 5G 4/6 | | 10R 4/8 | | 2.5B 5/6 | | |
| | 17.190.428 | | | 2.5PB 6/4 | 7.5PB 3/8 | 7.5PB 2/4 | | 2.5G 4/6 | | | | 2.5B 5/6 | | |
| | 17.190.429 | | | 2.5PB 6/4 | 7.5PB 3/8 | 7.5PB 2/4 | | 2.5G 4/6 | | 2.5YR 3/6 | | 2.5B 5/6 | | |
| 8 | 17.190.689 | Roundels | 1100–1130 | 5PB 4/6 | | | 7.5PB 2/8 | 7.5GY 5/6 | 5G 3/4 | | | | | |
| | 17.190.690 | | | | | | 7.5PB 2/6 | 10GY 5/6 | 5G 3/4 | | | 5B 4/6 | | |
| | 17.190.691 | | | 5PB 4/6 | 7.5PB 4/6 | | | 7.5GY 5/6 | 5G 3/4 | | | 5B 4/6 | | |
| | 17.190.692 | | | 5PB 4/6 | | | 7.5PB 2/6 | 10GY 5/6 | 5G 3/4 | | | 5B 4/6 | | |
| | 17.190.693 | | | 5PB 4/6 | 7.5PB 4/6 | | | 10GY 4/6 | 5G 3/4 | | | 5B 4/6 | | |
| 10 | 17.190.685–687,695,710–711 | Chasse | ca. 1150 | | 5PB 3/4 | 5PB 2/4 | 10GY 5/6 | 10GY 5/6 | | 10R 3/6 | | | | |
| 24 | 17.190.514 | Chasse | 1180–1190 | 5PB 4/6 | 7.5PB 3/6 | 7.5PB 2/6 | 7.5GY 5/4 | 2.5G 4/4 | | 10R 3/6 | | 2.5B 5/6 | 5Y 7/6 | 5PB 2/4 v dark blue |
| 26 | 17.190.773 | Cross | 1180–1190 | 5PB 5/10 | | | 7.5PB 2/10 | 2.5G 6/8 | 2.5G 3/6 | 7.5R 3/10 | | 2.5B 5/10 | 7.5Y 8/12 | |
| 27 | 17.190.523 | Chasse | 1180–1190 | 5PB 4/6 | 7.5PB 3/6 | 7.5PB 2/6 | | 10GY 4/4 | 5G 3/4 | 10R 3/6 | 10RP 2/2 | 2.5B 5/4 | 5Y 7/6 | |
| 28 | 41.100.155 | Chasse | ca. 1190 | 5PB 4/4 | 7.5PB 3/6 | 7.5PB 2/6 | | 10GY 4/4 | | 10R 3/6 | | 2.5B 5/4 | 5Y 7/6 | |
| 31 | 17.190.513 | Chasse | 1190–1200 | 5PB 5/8 | | 5PB 2/6 | 10GY 5/8 | | | 7.5R 3/8 | 5R 2/1 | 7.5B 4/8 | 7.5Y 4/10 | |
| 43 | 17.190.783 | Book cover | ca. 1200 | 7.5PB 4/12 | 7.5PB 4/8 | 7.5PB 2/6 | 10GY 5/8 | | 10G 3/4 | 7.5R 3/8 | 5R 2/2 | 2.5B 4/6 | 5Y 8/10 | |
| 46 | 17.190.779 | Plaque | 1185–1200 | 5PB 4/6 | 7.5PB 3/4 | | | 10GY 4/4 | | 10R 3/8 | | 5B 6/6 | 5Y 7/6 | |
| 47 | 17.190.812 | Plaque | 1185–1200 | 2.5PB 6/2 | 5PB 3/4 | | 10GY 5/4 | | | | | 10BG 6/4 | 5Y 8/6 | |
| 48 | 17.190.757 | Book cover | 1185–1210 | 10B 6/2 | 5PB 3/4 | 5PB 2/4 | 10GY 5/4 | | | 7.5R 3/6 | | | 5Y 7/4 | |
| 58 | 17.190.123 | Plaque | ca. 1231 | 5PB 4/4 | 5PB 3/6 | 5PB 2/4 | | | 5G 3/4 | 10R 3/6 | | 2.5B 5/4 | 7.5Y 7/6 | |
| NE | 17.190.409b | Plaques | 1185–1195 | 7.5PB 4/6 | 7.5PB 3/6 | 7.5PB 2/6 | 7.5GY 5/6 | | | 10R 3/6 | | 5B 4/6 | 5Y 7/6 | |
| | 17.190.409c | | | 7.5PB 4/6 | | 7.5PB 2/6 | 7.5GY 5/6 | | | 10R 3/6 | | 5B 4/6 | 5Y 7/6 | |
| 63a | L.51.10.2 | Plaque | 1185–1195 | 5PB 5/6 | 7.5PB 3/8 | 7.5PB 2/4 | | 10GY 4/4 | 10G 2/4 | 10R 3/6 | 5R 2/2 | 7.5B 5/6 | 5Y 7/8 | |
| 63b | 17.190.771 | Plaque | 1185–1195 | 5PB 5/6 | | 7.5PB 2/4 | | 10GY 4/4 | 10G 2/4 | 10R 3/6 | | 7.5B 5/6 | 5Y 7/8 | |
| 64 | 17.190.772 | Plaque | 1185–1195 | 5PB 5/6 | | 5PB 2/6 | | 10GY 4/4 | 10G 3/4 | 10R 3/6 | | 2.5B 5/6 | 2.5Y 7/8 | |
| 66 | 17.190.759 | Plaque | 1190–1200 | 5PB 4/4 | 5PB 3/4 | 7.5PB 2/4 | | 2.5G 4/4 | | 10R 3/6 | | 10BG 5/4 | 5Y 8/8 | 2.5R 4/2 Mauve |
| 67 | 17.190.761 | Plaque | 1190–1200 | 5PB 4/4 | | 7.5PB 2/4 | 10GY 5/4 | | 2.5BG 2/4 | 10R 3/6 | | | 5Y 7/6 | |
| NE | 17.190.770 | Plaque | 1190–1200 | 5PB 5/6 | 7.5PB 3/6 | 7.5PB 2/6 | | 10GY 4/4 | | 2.5YR 3/4 | | 5B 5/6 | 7.5Y 7/6 | |
| 70b | 41.100.184a | Tabernacle | 1200–1210 | 2.5PB 6/4 | 7.5PB 3/6 | 7.5PB 2/4 | | 10GY 4/4 | 2.5BG 2/4 | 10R 3/6 | | 2.5B 5/6 | 5Y 7/6 | 5PB 2/2 v dark blue |
| 72 | 17.190.853 | Coffret | 1200–1220 | | 5PB 3/4 | 7.5PB 2/6 | | 7.5GY 4/4 | 5G 3/4 | 2.5R 3/6 | | 2.5B 4/4 | 7.5Y 6/6 | |
| 83 | 14.40.703 | Chasse | 1200–1220 | 5PB 4/6 | 7.5PB 3/8 | 7.5PB 2/4 | 10GY5/4 | | | 2.5YR 4/6 | | 2.5B 4/4 | 5Y 8/8 | |
| 86 | 17.190.798 | Book cover | 1210–1220 | 7.5PB 4/6 | | 7.5B 2/4 | | 2.5G 4/6 | | 7.5R 2/6 | | 2.5B 4/6 | 10Y 6/8 | |
| 89 | 17.190.794 | Medallion | ca. 1210 | | 5PB 3/6 | | | | | 10R 3/6 | | | | |
| 92b | 17.190.795 | Medallions | before 1227 | | 5PB 3/4 | | | | | | | | | |
| | 17.190.796 | | | | 5PB 3/4 | | | | | | | | | |
| 93 | 17.190.854 | Plaque | ca. 1220 | 5PB 5/4 | 5PB 2/6 | 7.5PB 2/4 | 7.5GY 5/4 | | 2.5BG 3/4 | 2.5YR 3/6 | | | 5Y 8/6 | |
| 98 | 17.190.735 | Tabernacle | 1220–1230 | | 5PB 3/4 | | 7.5GY 5/4 | | | 7.5R 3/6 | | 7.5BG 6/4 | 10Y 6/6 | |
| 102 | 17.190.694 | Plaque | 1218–1225 | | | 5PB 2/4 | | 7.5G 4/4 | | 7.5R 3/6 | | 2.5PB 2/4 | 5Y 7/6 | |
| 105 | 17.190.344 | Dove | 1215–1235 | | 5PB 3/3 | 10B 2/1 | 10GY 4/4 | | | 7.5R 3/4 | | 10BG 5/2 | 10Y 6/6 | |
| 109 | 32.100.289 | Crosier | 1220–1230 | 7.5BG 5/2 | 2.5PB 4/4 | 5PB 3/4 | 10GY 4/4 | | | 2.5R 3/6 | | 5B 4/4 | 5Y 7/6 | |
| 113 | 32.100.290 | Appliqués | 1200–1225 | | 5PB 3/4 | | | | | | | 5B 4/4 | 5Y 7/6 | |
| | 32.100.291 | | | 2.5PB 5/2 | 5PB 3/4 | | 7.5G 4/4 | | | | | 5B 4/4 | 5Y 7/6 | |
| 115 | 1974.228.1 | Chasse | 1235–1245 | | 2.5PB 3/4 | 5PB 2/8 | | 5G 5/6 | | 7.5R 3/10 | | 5PB 5/8 | | |
| 130 | 50.7.3a&b | Censer | ca. 1270 | | 5PB 3/3 | | | 10GY 3/4 | | 7.5R 3/6 | | 5B 5/6 | | |
| 134 | 32.100.285 | Candlesticks | ca. 1300 | | 5PB 3/4 | | | | | 7.5R 2/6 | | | | |
| | 32.100.286 | | | | 5PB 3/4 | | | | | 7.5R 2/6 | | | | |
| 138 | 32.100.281 | Pyx | 1225–1250 | 2.5PB 6/6 | 5PB 3/10 | 5PB 2/6 | 10GY 5/8 | | | 7.5R 4/12 | | 5B 6/6 | | 2.5PB 2/2 v dark blue |
| NE | 17.190.781 | Plaque | 1200–1250 | 5PB 5/4 | 5PB 3/4 | 5PB 2/4 | 10GY 5/4 | | 10G 3/4 | 10R 3/6 | | 2.5B 5/4 | | |

NE : not exhibited

# Bibliography

(The *Bulletin de la Société archéologique et historique du Limousin* is abbreviated as *BSAHL;* the *Patrologia latina* as *PL*)

**Adhémar 1974.** J. Adhémar. "Les Tombeaux de la collection Gaignières, dessins d'archéologie du XVIIe siècle." *Gazette des Beaux-Arts* 84, no. 2 (1974), pp. 1–192.

**Agricola 1950.** Georgius Agricola. *De re metallica.* Trans. Herbert Clark Hoover and Lou Henry Hoover. New York, 1950.

**Agricola 1987.** Georgius Agricola. *De re metallica.* Ed. and trans. A. France-Lanord. Thionville, 1987.

**Alexander 1970.** J.J.G. Alexander. *Norman Illumination at Mont St. Michel 966–1100.* Oxford, 1970.

**Allard 1864.** In Gariel 1864 and Gariel 1970

**D'Allemagne 1891.** Henri-René d'Allemagne. *Histoire du luminaire depuis l'époque romaine jusqu'au XIXe siècle.* Paris, 1891.

**D'Allemagne 1928.** H.-R. d'Allemagne. *Les Accessoires du costume et du mobilier depuis le treizième jusqu'au milieu du dix-neuvième siècle.* 3 vols. Paris, 1928.

**Allier 1832.** Achille Allier. *Esquisses bourbonnaises.* Moulins, [1832]. New edition, Marseille, 1979.

**Allier 1833.** A. Allier. *L'Ancien Bourbonnais (Histoire, monumens, moeurs, statistique).* Continuation by A. Michel. Engravings under the supervision of A. Chenavard. Vol. 1. Moulins, 1833.

**Amiens 1860.** *Exposition provinciale: Notice des tableaux et objets d'art, d'antiquité et de curiosité.* Exh., Hôtel de Ville. Amiens, 1860.

**Amiens 1967.** *Les Objets d'art religieux anciens.* Exh. cat., Maison de la Culture. Amiens, 1967. No catalogue.

**Amiranashvili 1962.** Shalva I. Amiranashvili. *Les Émaux de Géorgie.* Trans. François Hirsch. Paris, 1962.

**Amouric and Foy 1985.** H. Amouric and D. Foy. "Histoire des techniques et sources documentaires, méthodes d'approches et expérimentation en région méditerranéenne." *Cahiers du GIS* 7 (1985), pp. 157–71.

**Amouric and Foy 1991.** H. Amouric and D. Foy. " 'De la salicorne aux soudes factices'— Mutations techniques et variation de la demande." In *L'Évolution des techniques, est-elle autonome? Colloque d'Aix-en-Provence, Espace Mejanes, 17 novembre 1989,* ed. M. C. Amouretti and G. Comet, pp. 40–57. Cahiers d'Histoire des Techniques, 1.

Marseille, 1991.

**A. Andersson 1956.** A. Andersson. "Chalices and Patens: Early Fourteenth-Century Examples in Sweden." *The Connoisseur,* September 1956, pp. 112–16.

**B. M. Andersson 1976.** Britt-Marie Andersson. "Les Trésors d'émaux limousins en Suède médiévale." *BSAHL* 103 (1976), pp. 107–36.

**B. M. Andersson 1980.** B.-M. Andersson. *Émaux limousins en Suède: Les Châsses, les croix.* Antikvariskt Arkiv, 69. Stockholm, 1980.

**Angers 1985.** *Saint-Aubin d'Angers du VIe au XXe siècle.* Exh., Association culturelle du département du Maine-et-Loire.

**Angers 1992.** *Souvenir de musée: Le 150e anniversaire du musée Saint-Jean (1841–1991).* Exh., Musée des Beaux-Arts. Angers, 1992.

**Angot 1900–1902.** Abbé A.-V. Angot. *Dictionnaire historique, topographique et biographique de la Mayenne.* Laval, 1900–1902.

**Annual Report 1989–90.** The Metropolitan Museum of Art. *Annual Report of the Trustees.* New York, 1990.

**Anselme 1726–33.** Père Anselme. *Histoire généalogique et chronologique de la maison royale de France, des pairs, grands officiers de couronne et de la maison du Roy et des anciens barons du royaume.* 9 vols. Paris, 1726–33.

**Arbellot 1873.** Chanoine François Arbellot. "Chronique (trouvaille à N.-Dame de la Règle)." *BSAHL* 22 (1873), pp. 161–62.

**Arbellot 1887.** F. Arbellot. "Châsse émaillée de l'église de Bellac." *BSAHL* 34 (1887), pp. 21–27.

**Arbellot 1888.** F. Arbellot. "L'Oeuvre de Limoges." *BSAHL* 35 (1888), pp. 237–49.

**Arbellot 1892.** F. Arbellot. "Aimeric Guerrut, archevêque de Lyon." *BSAHL* 40 (1892), pp. 59–70.

**Arbellot 1896.** F. Arbellot. "Étude biographique et bibliographique sur Bernard Guidonis, évêque de Lodève." *BSAHL* 45 (1896), pp. 5–44.

**Arborio Mella 1856.** C. E. Arborio Mella. *Cenni Storici sull'Abbazia di Sant'Andrea Apostolo in Vercelli dal 1200 al 1857.* Turin, 1856.

**Arborio Mella 1883.** C. E. Arborio Mella. "La Cassa già di deposito delle ossa del cardinale Guala Bicchieri." *Atti della Società piemontese di Archeologica e Belle Arti* 4 (1883), pp. 256–62.

**Ardant 1838a.** Maurice Ardant. "Médailles et monnaies trouvées à Saint-Martial de Limoges." *Mémoires de la Société royale des Antiquaires de France,* n. s., 14 (1838), p. 172.

**Ardant 1838b.** M. Ardant. "Médailles et monnaies trouvées dans l'emplacement qu'occupait l'église Saint-Martial de Limoges." *Bulletin de la Société royale d'Agriculture, des Sciences et Arts du Limousin* 16, no. 3 (July 1838), pp. 109–10.

**Ardant 1848.** M. Ardant. *Des Ostensions: Origine de ces solennités religieuses, dates des principales, détails sur leurs cérémonies, les reliques et les reliquaires.* Limoges, 1848.

**Ardant 1855.** M. Ardant. *Émailleurs et émaillerie de Limoges.* Isle, 1855.

**Ardant 1865.** M. Ardant. *Limoges et le Limousin. Guide de l'étranger: 2eme partie.* Paris, 1865.

**Arminjon et al. 1995.** Catherine Arminjon et al. *Émaux limousins du Moyen Âge: Corrèze, Creuse, Haute-Vienne.* Limoges, 1995. English edition trans. Jennie Rabb.

**Armoires de Baluze.** *See* Fage 1885

**"Armorial" 1872.** "Armorial des évêques de Limoges et de Tulle." *BSAHL* 21 (1872), pp. 133–49.

**Armorial général 1926–54.** *Armorial général de J. B. Rietstap.* With supplement by Victor and Henri Rolland. 7 vols. The Hague, 1926–54.

**Arnaud 1837.** A.-F. Arnaud. *Voyage archéologique et pittoresque dans le département de l'Aube et dans l'ancien diocèse de Troyes.* Troyes, 1837.

**Arnaud 1844.** A.-F. Arnaud. *Notice sur les objets trouvés dans plusieurs cercueils de pierre à la cathédrale de Troyes.* Troyes, 1844. Published extract from *Mémoires de la Société d'agriculture, sciences et arts de l'Aube* 12 (1844–45).

**Arnaud d'Agnel 1904.** G. Arnaud d'Agnel. "Le Trésor de l'église d'Apt (Vaucluse)." *Bulletin archéologique du Comité des travaux historiques et scientifiques,* 1904, pp. 329–35.

**Arquié-Bruley 1981.** Françoise Arquié-Bruley. "Un Précurseur: Le Comte de Saint-Morys (1772–1817), collectionneur d'antiquités nationales (2e partie)." *Gazette des Beaux-Arts* 96 (February 1981), pp. 61–77.

**Arquié-Bruley 1983.** F. Arquié-Bruley. "Les Monuments français inédits (1806–1839) de N. X. Willemin et la découverte des Aniquités nationales." *Revue d'art canadienne/Canadian Art Review* 10, no. 2 (1983), pp. 139–56.

**Arquié-Bruley 1984.** F. Arquié-Bruley. "Un Fonds iconographique rapporté par Millin d'Italie, au Cabinet des Estampes. Quelques dessins d'objets médiévaux." *Bulletin de la Société nationale des Antiquaires de France,* 1984, pp. 193–96.

**Arti del Medio Evo 1989.** *See* Florence 1989

**Art of Medieval Spain 1993.** *The Art of Medieval Spain, A. D. 500–1200.* Exh. cat., The Metropolitan Museum of Art. New York, 1993.

**L'Art rétrospectif 1886.** *See* Guibert and Tixier 1886

**Art Treasures 1987.** *See* New York 1987

**Art Treasures Exhibition, Manchester.** *See* London 1857b

**Ashbee 1967.** Benvenuto Cellini. *The Treatises of Benvenuto Cellini on Goldsmithing and Sculpture.* Trans. C. R. Ashbee. New York, 1967.

**Atlas 1857.** *See* Weerth 1857

**Aubert 1933.** Marcel Aubert. *L'Église Saint-Sernin de Toulouse.* Paris, 1933.

**Aubert 1939 and 1954.** M. Aubert. *L'Église de Conques.* Paris, 1939. 2d ed. Paris, 1954.

**Aubert and Beaulieu 1950.** M. Aubert and M. Beaulieu. *Musée national du Louvre: Description raisonnée des sculptures.* Vol. 1, *Moyen Âge.* Paris, 1950.

**Aubert et al. 1958.** M. Aubert et al. *Le Vitrail français.* Paris, 1958.

**Aubrun 1981.** Michel Aubrun. *L'Ancien Diocèse de Limoges des origines au milieu du XIe siècle.* Clermont-Ferrand, 1981.

**Aufauvre and Fichot 1858.** Amédée Aufauvre and Charles Fichot. *Les Monuments de Seine-et-Marne.* Paris, 1858.

**Auriol and Rey 1930.** Le chanoine Achille Auriol and Reymond Rey. *La Basilique Saint-Sernin de Toulouse.* Toulouse, 1930.

**d'Avout 1847.** L. d'Avout. "Rapport de la commission chargée de l'examen de la Bible de Souvigny et de sa prétendue restauration." *Bulletin de la Société d'émulation de l'Allier* 1 (1847 [1846–50]), pp. 196–200.

**Avranches 1991.** *L'Enluminure romane au Mont Saint-Michel, Xe–XIIe siècles.* Exh., Bibliothèque municipale. Avranches, 1991.

**Avril 1983.** François Avril. "Les Arts de la couleur." In Avril, Barral i Altet, Gaborit-Chopin 1983.

**Avril 1995.** F. Avril. *L'Enluminure à l'époque gothique 1200–1420.* Paris, 1995.

**Avril, Barral i Altet, Gaborit-Chopin 1983.** F. Avril, Xavier Barral i Altet, and Danielle Gaborit-Chopin. *Le Monde roman 1060–1220: Les Royaumes d'Occident.* Paris, 1983.

**Aymard and Malégue 1857.** A. Aymard and M. Hippolyte Malégue. *Album photographique d'archéologie religieuse.* Paris, 1857.

**Babelon 1946.** Jean Babelon. *L'Orfèvrerie française.* Paris, 1946.

**Baker 1981.** Evelyn Baker. "The Medieval Travelling Candlesticks from Grove Priory, Bedfordshire: SP923227." *Antiquaries Journal* 61, no. 2 (1981), pp. 336–38.

**Ballot and Dautrement 1945–50.** M. Ballot and Léon Dautrement. *Histoire de la Corrèze et du Bas-Limousin.* Paris, 1945–50.

**Baluze 1717.** E. Baluze. *Historiae Tutelensis libri tres.* Paris, 1717.

**Banéat 1973.** P. Banéat. *Le Département d'Ille-et-Vilaine: Histoire, archéologie, monuments.* 4 vols. 3d ed. Paris, 1973.

**Barbet de Jouy 1858.** H. Barbet de Jouy. *Notice du Musée des souverains.* 2d ed. Paris, 1858.

**Barbet de Jouy 1874.** H. Barbet de Jouy. *Musée national du Louvre: Description des sculptures du Moyen Âge, de la Renaissance, et des temps modernes.* Paris, 1874.

**Barbier 1891.** A. Barbier. "Chroniques de Poitiers aux XVe et XVIe siècles. Deuxième partie: Première guerre civile à Poitiers (1562)." *Mémoires de Société des Antiquaires de l'Ouest* 14 (1891), pp. 1–222.

**Barbier de Montault 1876.** Xavier Barbier de Montault. "Les Gants liturgiques." *Bulletin monumental* 42 (1876), pp. 401–67, 649–72, 777–809.

**Barbier de Montault 1881.** X. Barbier de Montault. "Inventaire de la basilique royale de Monza." *Bulletin monumental* 47 (1881), pp. 145–86.

**Barbier de Montault 1886a.** X. Barbier de Montault. "Authentiques des XIIIe, XIVe et XVe siècles récemment découverts à la cathédrale d'Albi." *Mémoires de la Société archéologique du Midi de la France* 14 (1886), pp. 177–228.

**Barbier de Montault 1886b.** *See* Palustre and Barbier de Montault 1887

**Barbier de Montault 1887.** *See* Palustre and Barbier de Montault 1886

**Barbier de Montault 1894.** X. Barbier de Montault. "Les Émaux champlevés du musée de Poitiers." *Bulletin de la Société scientifique, historique et archéologique de la Corrèze* 16 (1894), pp. 521–34.

**Barbier de Montault 1897a.** X. Barbier de Montault. "Le Trésor liturgique de Cherves en Angoumois." *Bulletin et mémoires de la Société archéologique et historique de la Charente,* ser. 6, 7 (1897), pp. 81–257.

**Barbier de Montault 1897b.** X. Barbier de Montault. *Le Trésor liturgique de Cherves en Angoumois.* Angoulême, 1897. Reprint in book form of Barbier de Montault 1897a.

**Barbier de Montault 1898a.** X. Barbier de Montault. "Les Crucifix émaillés d'Angoulême." *Bulletin de la Société scientifique, historique et archéologique de la Corrèze* 20, no. 2 (1898), pp. 165–78.

**Barbier de Montault 1898b.** X. Barbier de Montault. "Un Crucifix habillé du XIIIe siècle." *Bulletin de la Société scientifique, historique et archéologique de la Corrèze* 20, no. 1 (1898), pp. 573–83.

**Barbier de Montault 1898c.** X. Barbier de Montault. "La Généalogie du crucifix du trésor de Cherves, d'après ses congénères d'Angoulême et de Paris au XIIIe siècle." *Revue de l'Art chrétien* 9 (1898), pp. 217–24.

**Barbier de Montault 1898d.** X. Barbier de Montault. *Traité d'Iconographie chrétienne . . . dessins par M. Henri Nodet.* 2 vols. Paris, 1890; also Paris, 1898.

**Barcelona 1961.** *Arte romanico et Saint-Jacques-de-Compostela.* Exhibition sponsored by the Counsel of Europe. Barcelona, 1961.

**Barnet 1988.** Peter Barnet. "Medieval Limoges Enamels in the Detroit Institute of Arts." *Bulletin of the Detroit Institute of Arts* 63, nos. 3 and 4 (1988), pp. 17–26.

**Baron 1972.** *See* Tokyo and Kyoto 1972

**Barral i Altet 1983.** X. Barral i Altet. "Architecture, sculpture et mosaïque." In Avril, Barral i Altet, and Gaborit-Chopin 1983.

**Barrau 1875.** E. de Barrau. "Découverte d'un reliquaire à Conques." *Revue religieuse de Rodez et de Mende,* May 14, 1875, pp. 237–38.

**Barraud 1858.** Abbé Pierre-Constant Barraud. "Notice archéologique et liturgique sur les ciboires." *Bulletin monumental* 24 (1858), pp. 396–442, 561–637.

**Barraud 1860.** Abbé P.-C. Barraud. "Notice archéologique et liturgique sur l'encens et les encensoirs." *Bulletin monumental* 26 (1860), pp. 389–421, 501–36, 621–68.

**Barraud and Martin 1856.** P.-C. Barraud and A. Martin. "Le Bâton pastoral." *Mélanges d'archéologie, d'histoire et de littérature* 4 (1856), pp. 161–256. Also published under the same title, in book form, in Paris, 1856.

**Barrera and Velde 1989.** J. Barrera and B. Velde. "A Study of French Medieval Glass Composition." *Journal of Glass Studies* 31 (1989), pp. 48–54.

**Barrière 1977.** Bernadette Barrière. *L'Abbaye cistercienne d'Obazine en Bas-Limousin: Les origines, le patrimoine.* Tulle, 1977.

**Barrière 1984.** B. Barrière. *Limoges.* Coll. Atlas historique des villes de France. Bordeaux, 1984.

**Barrière 1986.** B. Barrière. "L'Abbaye Saint-Martial de Limoges au Moyen Âge." In *Les Limousins en quête de leur passé,* pp. 25–38. Limoges, 1986.

**Barrière 1989.** B. Barrière. "Limoges du Ve au XIIIe siècle." In *Histoire de Limoges,* ed. L. Pérouas. Toulouse, 1989.

**Barrière 1994.** B. Barrière. "Le Limousin médiéval." In *Atlas du Limousin,* pp. 42–47. Limoges, 1994.

**Barruol 1967.** G. Barruol. *Provence romane.* Vol. 2, *Haute-Provence.* La Pierre-qui-Vire, 1967.

**Bastard 1857.** A. de Bastard. "Rapport fait à la section d'archéologie le 28 juillet 1856, sur une crosse du XIIe siècle trouvée dans l'église de Trion, arrondissement de Nogent-le-Retrou." *Bulletin du Comité de la langue et de l'histoire de la France* 4 (1857), pp. 401–912.

**Bauch 1976.** K. Bauch. *Das mittelalterliche Grabbild Figürliche Grabmäler des 11 bis 15 Jahrhunderts in Europa.* Berlin–New York, 1976.

**Baudot and Chaussin 1935–59.** R. R. P. P. Baudot and Chaussin. *Vie des Saints et des bienheureux selon l'order du calendrier avec l'histoire des fêtes.* 13 vols. Paris, 1935–59.

**Bautier 1992.** R.-H. Bautier. *Le Musée Jacquemart-André à Chaalis: Historique de l'abbaye royale de Chaalis et guide-itinéraire des collections du musée.* N. p., 1992.

**Beani 1912.** Gaetano Beani. *La Chiesa pistoiese dalla sua origine ai tempi nostri.* Pistoia, 1912.

**Beard 1929.** Charles R. Beard. "A 'St. Ursula' Bowl, in the Possession of Capt. E. G. Reidy." *The Connoisseur,* February 1929, pp. 83–90.

**Beaulieu 1968.** M. Beaulieu. "Les Gants liturgiques en France au Moyen Âge." *Bulletin archéologique du Comité des travaux historiques et scientifiques,* 1986, pp. 137–53.

**Beaure d'Augères 1959 and 1960.** M. and M. Beaure d'Augères. "Les Armoiries de quelques gémellions limousins des XIIIe et XIVe siècles." *BSAHL* 87, no. 1 (1959), pp. 246–47, and 87, no. 2 (1960), pp. 349–82.

**Beauséjour 1891.** E. de Beauséjour. *Le monastère de Luxeuil, L'Église abbatiale: Étude historique et archéologique.* Besançon, 1891.

**Becker, Schorsch, Williams, Wypyski 1994.** L. Becker, D. Schorsch, J. Williams, and M. Wypyski. "Technical and Material Studies: Enamel Enkolpion." In *The Art of Late Rome and Byzantium in the Virginia Museum of Fine Arts,* ed. Anna Gonosova and Christine Kondoleon, pp. 410–15 (appendix 2). Richmond, Va., 1994.

**Becquet 1956.** Dom Jean Becquet. "Le Bullaire de l'ordre de Grandmont." *Revue Mabillon,* no. 184 (1956), pp. 82–93; no. 185 (1956), pp. 156–68; no. 186 (1956), pp. 189–201.

**Becquet 1958.** Dom J. Becquet. "La Règle de Grandmont." *BSAHL* 87, no. 1 (1958), pp. 9–36.

**Becquet 1960.** Dom J. Becquet. "La Première Crise de l'ordre de Grandmont." *BSAHL* 89 (1960), pp. 283–324.

**Becquet 1963a.** Dom J. Becquet. "Bibliothèque des écrivains de l'ordre de Grandmont." *Revue Mabillon* 53 (1963), pp. 59–79.

**Becquet 1963b.** Dom J. Becquet. "Aux origines du prieuré de l'Artige, chef-d'ordre canonial en Limousin." *BSAHL* 90 (1963), pp. 85–100.

**Becquet 1968.** *Scriptores ordinis Grandimontensis.* Ed. Dom J. Becquet. Corpus Christianorum. Continatio mediaevalis, 8. Turnhout, 1968.

**Becquet 1970.** Dom J. Becquet. "L'Ordre de l'Artige." *BSAHL* 97 (1970), pp. 83–142.

**Becquet 1975.** Dom J. Becquet. "Les Sanctuaires dédiés à saint Thomas de Cantorbéry en Limousin." In *Thomas Becket: Actes du colloque international de Sédières, 19–24 août 1973,* ed. Raymonde Foreville, pp. 159–61. Paris, 1975.

**Becquet 1977–81.** Dom J. Becquet. "Les Évêques de Limoges aux Xe, XIe et XIIe siècles." *BSAHL* 104 (1977) to 108 (1981).

**Becquet 1979–82.** Dom J. Becquet. *Paroisse et paroissiens au XIIe siècle: Structures et mentalités (1979–1982).* Typescript. Archives départementales de la Haute-Vienne et Bibliothèque municipale de Limoges.

**Becquet 1985a.** Dom J. Becquet. "Grandmont." In *Dictionnaire d'histoire et de géographie ecclésiastique.* vol. 16, cols. 414–26. 1985. Paris, 1912–.

**Becquet 1985b.** Dom J. Becquet. "Vie canoniale en France aux Xe–XIIe siècles." *Variorum Reprints.* London, 1985.

**Becquet 1987.** Dom J. Becquet. "L'Institution épiscopale à l'époque romane (Xe–XIIe siècles)." *Institution et les pouvoirs dans les églises de l'antiquité à nos jours: Miscellanea historiae ecclesiasticae, VIII, Colloque de Strasbourg, septembre 1983,* ed. Bernard Vogler, pp. 85–87. Bibliothèque de la Revue d'histoire ecclésiastique, 72. Brussels–Louvain, 1987.

**Becquet 1990.** Dom J. Becquet. "Pierre Coral et la Chronique de Saint-Martial au XIIIe siècle." *BSAHL* 118 (1990), pp. 31–51.

**Becquet 1991.** Dom J. Becquet. "Les Saints dans le culte en Limousin au Moyen Âge." *BSAHL* 119 (1991), pp. 26–59.

**Belaubre 1986.** J. Belaubre, with Evelyne Robert. *Histoire numismatique et monétaire de la France médiévale: De la période carolingienne à Charles VII.* Paris, 1986.

**Benoît 1927.** F. Benoît. "Inscription carolingienne et objets provenant de l'abbaye de Montmajour." *Bulletin de la Société nationale des Antiquaires de France,* 1927, pp. 110–14.

**Benoît 1928.** F. Benoît. *L'Abbaye de Montmajour.* Paris, 1928.

**Benoît 1951.** F. Benoît. "Introduction." In *Catalogue général de manuscrits des bibliothèques publiques de France,* 59. Supplement: Aix, Arles, Avignon. Paris, 1951.

**Berlin 1989.** *Europa und der Orient 800–1900.* Exh., Berliner Festspiele. Berlin, 1989.

**Bernard 1965.** M. Bernard. *Répertoire de manuscrits médiévaux contenant des notations musicales.* Vol. 1, Bibliothèque Sainte-Geneviève, Paris. Paris, 1965.

**Bernard de Montfaucon.** *See Monuments de la monarchie française*

**Bertaux 1910.** Émile Bertaux et al. *L'Exposition rétrospective de Saragossa.* Exh. cat. Paris, 1910.

**Bertaux 1912.** E. Bertaux. *Quelques pièces de la collection Claudius Côte.* Lyon, 1912.

**Berthelé and Drochon 1892.** J. Berthelé and Abbé Drochon. *Paysages et monuments du Poitou photographiés par J. Robuchon.* Vol. 10. Paris, 1892.

**Bertrand 1893.** A. Bertrand. "Notes sur la

Bible de Souvigny." *Bulletin-Revue de la Société d'émulation et des Beaux-Arts du Bourbonnais* 1 (1893), pp. 262–69.

**Bertrand 1903.** A. Bertrand. "Un Ornement de bronze provenant de la Bible de Souvigny." *Bulletin de la Société d'émulation du Bourbonnais* 11 (1903), pp. 339–41.

**Bessard 1981.** Bella Bessard. *Il Tesoro: Pellegrinaggio ai corpi santi e preziosi della cristianità.* Milan, 1981.

**Besson 1847.** L. Besson. *Mémoire historique sur l'abbaye de Cherlieu.* Besançon, 1847.

*Bibliotheca Sanctorum 1961–70. Bibliotheca Sanctorum.* 12 vols. Rome, 1961–70.

*Bildkataloge des Kestner-Museums Hannover 1966.* Ferdinand Stuttmann. *Mittelalter: Bronze, Email, Elfenbein.* Vol. 8 of *Bildkataloge des Kestner-Museums Hannover.* Hannover, 1966.

**Bilimoff 1977.** Michèle Bilimoff. "Boucles de ceintures d'émail champlevé limousines." *BSAHL* 104 (1977), p. 178.

**Bilimoff 1979.** M. Bilimoff. "Saint Michel dans les crosses limousines, styles et chronologie." *Actes du 102e Congrès national des Sociétés Savantes, Limoges, 1977,* pp. 37–51. Paris, 1979.

**Bimson 1981.** M. Bimson. "A Preliminary Survey of Two Groups of Twelfth-Century Mosan Enamels." *Annales du 8e Congrès internationale d'Étude historique du Verre, Londres–Liverpool, 18–25 septembre 1979,* pp. 161–64. Liège, 1981.

**Binghamton, New York, 1975.** *Islam and the Medieval West.* Exh. cat., Center of Medieval and Early Renaissance Studies. New York, 1975.

**Birch n. p.** *Vita sanctissimi Marcialis apostoli.* Ed. W. de Gray Birch. N. p., 1877.

**Bischoff 1967.** B. Bischoff. *Mittelalterliche Schatzverzeichnisse. Erster Teil: von der Zeit Karls des Grossen bis zur Mitte des 13. Jahrhunderts.* Munich, 1967.

**Biscoglio 1995.** Frances M. Biscoglio. " 'Unspun' Heroes: Iconography of the Spinning Woman in the Middle Ages." *Journal of Medieval and Renaissance Studies* 25, no. 2 (Spring 1995), pp. 163–76.

**Blanchet and Dieudonne 1912–36.** Adrien Blanchet and A. Dieudonne. *Manuel de Numismatique française.* 4 vols. Paris, 1912–1936.

**Bobe 1935.** A. Bobe. *Histoire de Civray.* Paris, 1935.

**Blumenthal 1943–45.** *See New York 1943*

**Boehm 1989.** *See Yokohama 1989*

**Boehm 1991.** Barbara Drake Boehm. *Medieval Head Reliquaries of the Massif Central.* Ann Arbor, Mich., 1991. Microfilm.

**Boehm 1992.** Barbara D. Boehm. "A Pair of Limoges Candlesticks in The Cloisters Collection." In *The Cloisters: Studies in Honor of the Fiftieth Anniversary,* ed. Elizabeth C. Parker and Mary B. Shepard, pp. 147–65. New York, 1992.

**Boehm 1993.** Entry, in New York 1993, no. 134.

**Boinet 1929.** A. Boinet. *La Cathédrale de Bourges.* 2d ed. Paris, 1929.

**Boinet 1948.** A. Boinet. "Exhibition of Limousin Enamels at the Museum of Limoges." *The Connoisseur,* December 1948, pp. 75–82.

**Boitel 1859.** Abbé Boitel. *Histoire du bien-heureux Jean surnommé l'Humble, seigneur de Montmirail-en-Brie.* Paris, 1859.

**Bonaventure de Saint-Amable 1676–85.** Bonaventure de Saint-Amable. *Histoire de saint Martial, apôtre des Gaules et principalement de l'Aquitaine et du Limousin . . .* 3 vols. Clermont-Ferrand and Limoges, 1676–85.

**Bonfons 1586–88.** Pierre Bonfons. *Les Antiquitez, croniques et singularitez de Paris . . . par Gilles Corrozet . . . augmentées par N. B. [Nicolas Bonfons] parisien.* 2 vols. in 1. Paris, 1568–88.

**Bonfons 1605.** P. Bonfons. *Les Fastes, antiquitez et choses plus remarquables de Paris.* Paris, 1605.

**Bonnélye [1864].** *See Chronique de Geoffroy*

*Book of Pontiffs 1989. The Book of Pontiffs (Liber pontificalis): The Ancient Biographies of the First Ninety Bishops to A. D. 715.* Trans. Raymond Davis. Liverpool, 1989.

**Bordeaux 1971.** *Bordeaux—2000 ans d'histoire.* Exh., Musée d'Aquitaine. Bordeaux, 1971.

**Borenius 1932.** Tancred Borenius. *St. Thomas Becket in Art.* London, 1932.

**Borenius 1933.** T. Borenius. "Some Further Aspects of the Iconography of St. Thomas of Canterbury." *Archaeologia* 83 (1933), pp. 171–86.

**Bosredon and Rupin 1886.** P. de Bosredon and E. Rupin. *Sigillographie du Bas-Limousin.* Brive, 1886.

**Boston 1940.** *Arts of the Middle Ages.* Exh. cat., Museum of Fine Arts, Boston. 1940.

**Boston 1981.** *Masterpieces from the Boston Museum.* Exh. cat., Museum of Fine Arts, Boston. 1981.

**Botineau.** *See Geoffrey de Vigeois, Chronique*

**Bouchet 1524.** J. Bouchet. *Les Annales d'Acquitaine.* Paris, 1524.

**Bouchot 1891.** H. Bouchot. *Inventaire des dessins exécutés pour Roger de Gaignières et conservés aux départements des Estampes et des Manuscrits.* 2 vols. Paris, 1891.

**Bouillet 1892 and 1905.** A. Bouillet. *Le Trésor de Conques.* Mâcon, 1892. New edition 1905.

**Bouillet and Servières 1900.** A. Bouillet and L. Servières. *Sainte Foy: Vierge et martyre.* Rodez, 1900.

**Bouillet 1846.** J.-B. Bouillet. *Statistique monumentale du département du Puy-de-Dôme.* 2d ed. Clermont-Ferrand, 1846.

**Bourgeois Frères Sale 1904.** [Stephan Bourgeois.] *Catalogue des objets d'art et de haute curiosité: composant la collection Bourgeois frères et dont la vente aura lieu à Cologne . . . du 19 au . . . 27 octobre, 1904. La vente sera faite par le ministère de Mr. Krings . . . elle sera dirigée par Mr. H. Lempertz.* Cologne, 1904.

**Bourret 1880.** [Joseph Christian Ernest] Bourret. *Procès-verbaux authentiques et autres pièces concernant la reconnaissance des reliques de sainte Foy vierge et martyre.* Rodez, 1880.

**Boussard 1956.** J. Boussard. *Le Gouvernement d'Henri II Plantagenêt.* Paris, 1956.

**Bouvier Sale 1873.** *Catalogue des objets d'art et de curiosité . . . tableaux anciens, composant la Collection de M. Bouvier, d'Amiens . . . vente Hôtel Drouot, 8–13, et 15–16 décembre.* Paris, 1873.

**Boze 1820.** Elzear Boze. *Histoire de l'église d'Apt.* Apt, 1820.

**Brandt 1976.** Karl Heinz Brandt. "Erzbischofsgräber im Bremer St.-Petri-Dom." *Zeitschrift für Archäologie des Mittelalters* 4 (1976), pp. 7–28.

**Branner 1977.** Robert Branner. *Manuscript Painting in Paris During the Reign of St. Louis: A Study in Styles.* Berkeley, Calif., 1977.

**Branner 1989.** R. Branner. *The Cathedral of Bourges and Its Place in Gothic Architecture.* New York and Cambridge, Mass., 1989.

**Brault 1925.** F. Brault. "Le Couvent des Cordeliers de Nantes: Étude historique." *Bulletin de la Société archéologique et historique de Nantes et de la Loire-Inférieure* 65 (1925), pp. 165–93.

**Braun 1924.** Joseph Braun. *Der christliche Altar in seiner geschichtlichen Entwicklung.* 2 vols. Munich, 1924.

**Braun 1932.** J. Braun. *Das christliche Altargerät in seinem Sein und in seiner Entwicklung.* Munich, 1932.

**Braun 1940.** J. Braun. *Die Reliquiare des christlichen Kultes und ihre Entwicklung.* Freiburg-im-Breisgau, 1940.

**Breck and Rogers 1925.** Joseph Breck and Meyric R. Rogers. *The Pierpont Morgan Wing: A Handbook.* New York, 1925.

**Breck and Rogers 1929.** J. Breck and M. R. Rogers. *The Pierpont Morgan Wing: A Handbook.* 2d ed. New York, 1929.

*Brevis chronica 1657. Brevis chronica de vita episcopi Lodovensis.* Ed. Philippe Labbé. 1657.

**Briend 1993.** C. Briend. *Les Objets d'art: Guide des collections,* Musée des Beaux-Arts de Lyon. Paris, 1993.

**Brill 1969.** R. H. Brill. "The Scientific Investigation of Ancient Glasses." In *Proceedings of the 8th International Congress on Glass,* July 1–8, 1968, vol. 1, pp. 47–68. 2 vols. London, 1969.

**Brill 1988.** R. H. Brill. "Scientific Investigations of the Jalame Glass and Related Finds." In *Excavations at Jalame: Site of a Glass Factory in Late Roman Palestine,* ed. Gladys D. Weinberg, pp. 257–94. Columbia, Mo., 1988.

**Brill, Barag, and Oppenheim 1970.** R. H. Brill, Dan Barag, and A. Leo Oppenheim. "The Chemical Interpretation of the Texts." In *Glass and Glassmaking in Ancient Mesopotamia: An Edition of the Cuneiform Texts Which Contain Instructions for Glassmakers with a Catalogue of Surviving Objects,* ed. A. Leo Oppenheim, Robert H. Brill, Dan Barag, and Axel von Saldern, pp. 105–28. Corning, N.Y., and London, 1970.

**Brisac 1974.** C. Brisac. "Le Sacramentaire manuscrit 63 de la Bibliothèque municipale de Clermont Ferrand, nouvelles données sur l'art figuré à Clermont-Ferrand autour de 1200." *Bulletin historique et scientifique de l'Auvergne* 86 (1974), pp. 303–15.

**Brisset 1949.** Abbé G. Brisset. "Notes sur l'exposition des émaux champlevés du musée de l'Eveché de Limoges (1948)." *BSAHL* 83 (1949), pp. 97–102.

**Brouillet 1885.** Pierre-Amédée Brouillet. *Notice des tableaux, dessins, gravures, statues, objets d'art anciens et modernes, curiosités . . . composant les collections de la ville de Poitiers.* 2 vols. Poitiers, 1884–85.

**Brown 1983.** Katharine R. Brown. "Roundel with an Elder of the Apocalypse." *The Metropolitan Museum of Art: Notable Acquisitions, 1982–83.* New York, 1983.

**Brown and Blin-Stoyle 1959.** M. A. Brown and A. E. Blin-Stoyle. "A Sample and Analysis of British Middle and Late Bronze Age Material Using Optical Spectrometry." *Proceedings of the Prehistoric Society* 25 (1959), pp. 188–208.

**Brummer Sale 1949.** Joseph Brummer Collection. Sale cat. New York: Parke-Bernet Galleries, 23 April. New York, 1949.

**Brummer Collection.** Marie-Madeleine S. Gauthier. *The Ernest Brummer Collection.* Part I, *Medieval, Renaissance and Baroque Art.* Zurich, 1979.

**Brun and Pernot 1992.** N. Brun and M. Pernot. "The Opaque Red Glass of Celtic Enamels from Continental Europe." *Archaeometry* 34, no. 2 (1992), pp. 235–52.

**Bruneau de Tartifume 1623 and 1977.** J. Bruneau de Tartifume. *Histoire d'Angers contenant ce qui est remarquable en tout ce qui estoit anciennement la ville d'Angers.* Angers, 1623. Ed. T. Cuvrays. Brussels, 1977.

**Brunswick 1995.** *Heinrich der Löwe und seine Zeit: Herrschaft und Repräsentation des Welfen 1125–1235.* Exh., Herzog Anton Ulrich–Museum, 1995.

**Brussels 1880.** *Exposition rétrospective de Bruxelles.* Brussels, 1880.

**Brussels 1882.** *L'Art ancien à l'Exposition nationale belge.* Brussels, 1882.

**Brussels 1958.** *Imago Christi.* Exposition universelle, Pavillon du Saint-Siège. Brussels, 1958.

**BSAHL 1885.** "Séance du 24 avril 1883." *BSAHL* 30 (1885), p. 328.

**Buckton 1985.** David Buckton. " 'Necessity the Mother of Invention' in Early Medieval Art." *Transactions of the Third Canadian Conference of Medieval Art Historians,* pp. 1–6. London, Canada, 1985.

**Bull 1993.** Marcus Bull. *Knightly Piety and the Lay Response to the First Crusade: The Limousin and Gascony, c. 970–c. 1130.* Oxford, 1993.

*Bulletin archéologique 1842–43.* "Coffret émaillé dans l'église des Bouchers de Limoges." *Bulletin archéologique du Comité historique des arts et monuments* 2 (1842–43), p. 146.

**Bulletin of the Museum of Fine Arts, Boston 1957.** "Georg Swarzenski, 1876–1957 (A Commemorative)." *Bulletin of the Museum of Fine Arts,* Boston 55, nos. 301–2 (1957), pp. 53–119.

**Burger 1930.** W. Burger. *Abendländische Schmelzarbeiten.* Berlin, 1930.

**Burin 1985.** E. Burin. "Réflexions sur quelques aspects de l'enluminure dans l'Ouest de la France au XIIe siècle: Le manuscrit latin 5323 de la Bibliothèque nationale." *Bulletin monumental* 143, no. 3 (1985), pp. 209–25.

**Burlington Fine Arts Club 1897.** *See* London 1897

**Bynum 1986.** Caroline Walker Bynum. *Holy Feast and Holy Fast.* Berkeley, Ca., 1986.

**Cabrol 1907.** Dom Fernand Cabrol. "Ablutions." In Cabrol and Leclercq 1907–53, vol. 1, pt. 1, cols. 103–11 (1907).

**Cabrol 1925.** F. Cabrol. "Huiles saintes." In Cabrol and Leclercq 1907–53, vol. 6, pt. 2, cols. 2779–90 (1925).

**Cabrol and Leclercq 1907–53.** Dom F. Cabrol and Dom Henri Leclercq, with Henri Marrou. *Dictionnaire d'archéologie chrétienne et de liturgie.* 15 vols. Paris 1907–53.

**Cahier 1867.** C. Cahier. *Caractéristiques des saints.* In *L'Art populaire.* 2 vols. Paris, 1867.

**Cahiers grandmontains** 1990–95. *Cahiers grandmontains,* nos. 1–11. 1990–95.

**Cahn 1967.** Walter Cahn. *The Souvigny Bible: A Study in Romanesque Manuscript Illumination.* Ph.D. diss., New York University, 1967. Ann Arbor, Michigan, 1970. Microfilm.

**Cahn 1980.** W. Cahn. "Autour de la Bible de Lyon: Problèmes du roman tardif dans le centre de la France." *Revue de l'Art* 47 (1980), pp. 11–20.

**Cahn 1981.** W. Cahn. "Note on a Roman Enamel in Moulins." *Gesta* 20, no. 1 (1981), pp. 155–59.

**Cahn 1982.** W. Cahn. *Romanesque Bible Illumination.* Ithaca, N.Y., 1982. French translation: *La Bible romane.* Fribourg–Paris, 1982.

**Caley 1926.** E. R. Caley. "Leydon Papyrus X: An English Translation with Brief Notes." *Journal of Chemical Education* 3 (1926), pp. 1149–66.

**Calmon 1931.** J. Calmon. "Présentation du trésor de la cathédrale de Cahors." *Revue religieuse de Cahors et de Rocamadour,* no. 36 (1931), pp. 486–89.

**Cambridge, Mass., 1931.** "Art works loaned by Arthur Sachs." Fogg Art Museum. Cambridge, Mass., 1931. No catalogue.

**Cambridge, Mass., 1975.** *Eucharistic Vessels of the Middle Ages.* Exh. cat., Busch-Reisinger Museum, Harvard University. Cambridge, Mass., 1975.

**Camille 1989.** Michael Camille. *The Gothic Idol: Ideology and Image-making in Medieval Art.* Cambridge, 1989.

**Campbell 1979.** M. L. Campbell. "Scribe faber lima: A Crozier in Florence Reconsidered." *The Burlington Magazine* 121, no. 915 (1979), pp. 364–69.

**Camuzat 1610.** N. Camuzat. *Promptuarium sacrarum antiquitatum Tricassinae diocesis.* Troyes, 1610.

**Carr 1978.** Carolyn K. Carr. "Aspects of the Iconography of Saint Peter in Medieval Art of Western Europe to the Early Thirteenth Century." Ph.D. diss., Case Western Reserve University, 1978.

**Castelnau d'Essenault 1864 [1863].** J.-M. de Castelnau d'Essenault. "Mémoires sur l'exposition d'objets anciens faite à l'occasion du Congrès." *Congrès archéologique de France, XXXe session, Rodez-Albi-Le Mans, 1863,* pp. 151–62. Paris, 1864.

**Castronovo 1992.** Simonetta Castronovo. "Il tesoro di Guala Bicchieri cardinale di Vercelli." In *Gotico in Piemonte,* ed. Govanni Romano, pp. 166–76. Torino, 1992.

**Catedrales 1981.** *See* Madrid 1981

**Catel 1633.** G. de Catel. *Mémoires de l'histoire du Languedoc.* Toulouse, 1633.

**Caudron 1975.** Simone Caudron. "Les Châsses de Thomas Becket en émail de Limoges." *Thomas Becket: Actes du colloque international de Sédières, 19–24 août 1973,* ed. Raymonde Foreville, pp. 233–41. Paris, 1975.

**Caudron 1976.** S. Caudron. "Émaux champlevés de Limoges et amateurs britanniques du XVIIIe siècle." *BSAHL* 103 (1976), pp. 137–68.

**Caudron 1977.** S. Caudron. "Connoisseurs of Champlevé Limoges Enamels in Eighteenth-Century England." In *Collectors and Collections,* pp. 9–23. The British Museum Year Book, 2. London, 1977. The English version of Caudron 1976.

**Caudron 1979.** S. Caudron. "Les Émaux champlevés méridionaux dans les collections du Victoria et Albert Museum." *Le Limousin, Études archéologiques: Actes du 102e Congrès national des sociétés savantes: Archéologie et Histoire de l'Art, Limoges, 1977,* pp. 53–68. Paris, 1979.

**Caudron 1993.** S. Caudron. "Les Châsses reliquaires de Thomas Becket émaillées à Limoges: Leur géographie historique." *BSAHL* 121 (1993), pp. 55–82.

**Caumont 1850.** Arcisse de Caumont. *Abécédaire ou rudiment d'archéologie.* Paris, 1850.

**Caumont 1855.** A. de Caumont. "Statues tombales en cuivre doré ou émaillé des XIIe, XIIIe et XIVe siècles." *Bulletin monumental* 1 (1855), pp. 459–71.

**Cayot-Delandre 1847.** F.-M. Cayot-Delandre. *Le Mobihan, son histoire et ses monuments.* Vannes–Paris, 1847.

**Cellini 1967.** *See* Ashbee 1967

**Chabaneau 1895.** C. Chabaneau. "Mémoire du Consulat." Supplement to the *Revue des langues romanes,* ser. 4, 8 [38] (1895), pp. 1–260.

**Chabouillet 1861.** A. Chabouillet. *Description des antiquités et objets d'art composant le cabinet de M. Louis Fould.* Paris, 1861.

**Chailley 1960.** J. Chailley. *L'École musicale de Saint-Martial de Limoges jusqu'à la fin du XIe siècle.* Paris, 1960.

**Chancel 1991.** Béatrice de Chancel. "Les Gisants démasqués: Les Masques de gisants du XIIIe siècle du Musée d'Angers et du Louvre." *Revue du Louvre et des musées de France* 41, no. 3 (1991), pp. 30–43.

**Chancel 1992.** *See* Lernout et al. 1992

**Chappée 1896.** J. Chappée. *Le Tombeau de Jean de Chanlay évêque du Mans à l'abbaye de Preuilly.* An extract from *Revue archéologique et historique du Maine.* Mamers, 1896.

**Chappée 1897.** J. Chappée. "Les Sépultures de l'abbaye de Champagne et les fouilles de 1895–1896." *Revue historique et archéologique du Maine* 41 (1897), pp. 348–77.

**Châteaugontier 1892.** *Exposition d'art rétrospectif.* Exh., Châteaugontier, 1892.

**Chatelet 1855.** P. F. Chatelet. "Les Monuments de l'abbaye de Cherlieu (Haute-Sâune)." *Académie des Sciences de Besançon,* 1855, pp. 298–326.

**Chicago 1970.** Donald F. Rowe, S. J. *Enamels: The XII to the XVI Century: A Special Exhibition at the Martin D'Arcy Gallery of Art, February 2 through March 2, 1970.* Exh. cat, The Martin D'Arcy Gallery of Art, Loyola University of Chicago. Chicago, 1970.

**Christensen 1952.** Erwin O. Christensen. *Objects of Medieval Art from the Widener Collection.* Washington, D.C., 1952.

**Chronicle of Conques.** *See* Desjardins 1879

**"Chronique" 1884.** "Chronique." *Revue bourbonnaise* 1 (1884), p. 291.

**Chronique de Geffroy [1864].** *Chronique de Geoffroy, prieur de Vigeois.* Translated by François Bonnélye. Tulle, [1864].

**"Chronique des amis" 1990.** "Chronique des amis du Louvre. Un reliquaire du XIIIe siècle." *La Revue du Louvre et des musées de France,* no. 2 (1990), p. i.

**Chronique des Arts 1970.** "La Chronique des arts." *Gazette des Beaux-Arts* 75 (February 1970), pp. 1–127.

**Chronique des comtes d'Eu 1876.** *Chronique des comtes d'Eu.* In *Recueil des historiens des Gaules et de la France,* vol. 23, ed. J. N. de Wailly, L. V. Delisle, and C.-M. G. B. Jourdain, pp. 439–48. Paris, 1876.

**Chroniques des comtes d'Anjou 1913.** *See* Halphen and Poupardin 1913

**Chroniques de Saint-Martial de Limoges.** *See* Duplès-Agier 1874

**Cirot de la Ville 1845.** Cirot de la Ville. *Histoire de l'abbaye et congrégation de la Grande-Sauve.* Vol. 2. Bordeaux, 1845.

**Civrays 1977.** *See* Bruneau de Tartifume 1623 and 1977

**Clément and Guitard 1900.** Chanoine Silvain Clément and A. Guitard. *Vitraux de Bourges: Vitraux du XIIIe siècle de la cathédrale de Bourges.* Bourges, 1900.

**Clément de Ris 1859.** L. Clément de Ris. *Les Musées de province.* Vol. 1. Paris, 1859.

**Clément de Ris 1873.** L. Clément de Ris. "Statue de Blanche de Champagne, duchesse de Bretagne." *Revue des Sociétés savantes des départements,* ser. 5, 6 (1873), pp. 122–23.

**Clerc 1840–46.** E. Clerc. *Essai historique sur la Franche-Comté.* Besançon, 1840–46.

**Cleveland 1967.** *Treasures from Medieval France,* ed. William D. Wixom. Exh. cat., The Cleveland Museum of Art. Cleveland, 1967.

**Clouzot 1934.** A. Clouzot. *Les Arts du métal.* Paris, 1934.

**Cochet 1868 and 1875.** J.-B. Cochet. *Catalogue du musée d'antiquités de Rouen.* Rouen, 1868. 2d ed. Rouen, 1875.

**Cockerell and Plummer 1975.** S. C. Cockerell and J. Plummer. *Old Testament Miniatures.* New York, 1975.

**Collection Spitzer 1890.** *La Collection Spitzer: Antiquité—Moyen Âge—Renaissance.* 6 vols. Paris, 1890–93.

**Collin 1672.** Abbé Jean Collin. *Histoire sacrée de la vie des saints principaux et autre personnes plus vertueuses qui ont pris naissance, qui ont vécu, ou qui sont en vénération particulière en divers lieux du diocèse de Limoges.* Limoges, 1672.

**Collins 1972.** Fletcher Collins, Jr. *The Production of Medieval Church Music-Drama.* Charlottesville, Va., 1972.

**Cologne 1982.** *Die heiligen drei Könige—Darstellung und Verehrung: Katalog zur Ausstellung des Wallraf-Richartz-Museums in der Josef-Haubrich-Kunsthalle.* Exh. cat. Cologne, 1982.

**Cologne 1985.** *Ornamenta Ecclesiae.* Exh., Joseph-Haubrich-Kunsthalle. Cologne, 1985.

**Cologne and Brussels 1972.** *Rhein und Maas: Kunst und Kultur 800–1400.* Exh. cat., Schnütgen Museum, Kunsthalle, Cologne, and Königlichen Museen für Kunst und Geschichte, Brussels, 1972. French edition: *Rhin-Meuse: Art et culture 800–1400.*

**Compositiones variae 1920.** *Compositiones Variae: A Classical Technology.* Edited by J. M. Burnam. Boston, 1920.

**"Le Comte de Marsy . . ." 1902.** "Le comte de Marsy, sa mort, ses funérailles." *Bulletin de la Société historique de Compiègne* 10 (1901), pp. xxxiii–lvi.

**Congrès archéologique de France 1890.** "Excursion à Brionne, au Bec-Hellouin et à Beaumont-le-Roger." *Congrès archéologique de France, LVIe session. Séances générales tenues à Évreux . . . en 1889 par la Société française d'archéologie,* pp. 99–121. Paris, 1890.

**Contrasty 1936.** J. Contrasty. *Histoire de la cité de Rieux-Vovestre et de ses évêques.* Meaux, 1936.

**Conway 1915.** William Martin Conway. "The Abbey of Saint-Denis and Its Ancient Treasures." *Archaeologia, or Miscellaneous Tracts Relating to Antiquity,* 2d ser., 66 (1915), pp. 103–58.

**Coo 1965.** J. de Coo. "L'Ancienne Collection Micheli au musée Mayer van den Bergh: L'Inventaire dressé par lui-même." *Gazette des Beaux-Arts* 61 (1965), pp. 345–70.

**Cook 1923.** W. S. Cook. "The Earliest Painted Panels in Catalonia (II)." *The Art Bulletin* 6, no. 2 (1923), pp. 31–63.

**Copy 1981.** J.-Y. Copy. *Les Tombeaux en Haute-Bretagne aux XIVe et XVe siècles: Synthèse et notices.* Thesis, Université de Rennes. Rennes, 1981. Typescript.

**Copy 1986.** J.-Y. Copy. *Art, société et politique au temps des ducs de Bretagne: Les Gisants haut-breton.* Paris, 1986.

**Corblet 1859.** Abbé Jules Corblet. "Les Chandeliers d'église au Moyen Âge." *Revue de l'Art chrétien* 3 (1859), pp. 17–57.

**Corblet 1885a.** Abbé J. Corblet. *Histoire dogmatique, liturgique et archéologique du sacrement de l'Eucharistie.* 2 vols. Paris, 1885.

**Corblet 1885b.** Abbé J. Corblet. "Des Vases et ustensiles eucharistiques." *Revue de l'Art chrétien*, ser. 3, 2 (1885), pp. 154–63, 427–38; n. s., 3 (1885), pp. 53–64, 311–28; n. s., 4 (1886), pp. 46–61.

**Cordonnier 1961.** P. Cordonnier. "La grande enluminure du manuscrit de Pline l'Ancien (XIIe siècle)." *Revue historique et archéologique du Maine*, ser. 2, 41 (1961), pp. 101–10.

**Corneaux 1879.** Abbé Corneaux. *Longpont et ses ruines.* Soissons, 1879.

**Corneaux 1888.** Abbé Corneaux. *Communication faite par M. l'abbé Corneaux à la Société des Antiquaires de France, dans sa séance du 13 juillet 1887, au sujet d'une plaque en cuivre jaune émaillé provenant de la châsse de Jean de Montmirail, à Longpont.* Nogent-le-Rotrou, 1888.

**Corpus des inscriptions 1978.** *See* Favreau et al. 1978

**Corpus des inscriptions 1979.** *See* Favreau et al. 1979

**Corpus des inscriptions 1984.** *See* Favreau et al. 1984

**D. Costa 1961.** D. Costa. *Catalogue des Musées départementaux de la Loire-Atlantique.* Vol. 1 (12th to 16th centuries). Musée Thomas Dobrée. Nantes, 1961.

**G. Costa 1981.** G. Costa. "Le Trésor de Saint-Sernin de Toulouse." *Monuments historiques de France* 115 (July-August 1981), pp. 73–88.

**G. Costa 1993.** G. Costa. "Le Trésor de la cathédrale de Cahors." *Congrès archéologique de France (1989): Quercy* 147 (1989), pp. 79–85.

**Cottineau 1935–39.** L. H. Cottineau. *Répertoire topo-bibliographique des abbayes et prieurés.* 3 vols. Mâcon, 1935–39. Vols. 1 and 2 (1935–39); vol. 3 by Grégoire Poras (1970).

**Cougny 1875.** G. de Cougny. "Chronique." *Bulletin monumental* 41 (1875), pp. 479–81.

**Councils and Synods 1964.** *Councils and Synods with Other Documents Relating to the English Church.* Vol. 2, pt. 1 (1205–1265); pt. 2 (1265–1313). Ed. F. M. Powicke and C. R. Cheney. Oxford, 1964.

**Courajod 1878.** L. Courajod. *Alexandre Lenoir, son journal et le Musée des Monuments français.* Vol. 1. Paris, 1878.

**Courajod 1886.** L. Courajod. *La Collection Révoil du musée du Louvre.* Caen, 1886. An extract from the *Bulletin monumental* (1886).

**Courajod 1888.** L. Courajod. "La Collection Durand et ses séries du Moyen Âge et de la Renaissance au musée du Louvre." *Bulletin monumental* 54 (1888), pp. 326–80. Also reprinted in book form in Caen, 1888.

**Courajod and Molinier 1885.** L. Courajod and E. Molinier. *Donation du baron Ch. Davillier: Catalogue des objets exposés au musée du Louvre.* Paris, 1885.

**Courteault 1925.** P. Courteault. "Les Sépultures anciennes de la cathédrale Saint-André." *Revue historique de Bordeaux et du département de la Gironde* 18 (1925), pp. 113–31; 176–89; 241–50.

**Coutard 1897.** A. Coutard. "L'Évêque Guillaume Roland (1260), à propos des fouilles de Champagne." *Province du Maine* 5 (1897), pp. 187–91.

**Craddock 1985.** Paul T. Craddock. "Medieval Copper Alloy Production and West African Bronze Analyses—Part I." *Archaeometry* 27 (1985), pp. 17–41.

**Craplet 1962.** B. Craplet. *Auvergne romane.* La Pierre-qui-Vire, 1962.

**Crozet 1942.** René Crozet. *Textes et documents relatifs à l'histoire des arts en Poitou: Moyen Âge, début de la Renaissance.* Poitiers, 1942.

**Crozet 1961.** R. Crozet. "Sur un détail vestimentaire féminin du XIIe siècle." *Cahiers de civilisation médiévale* 4 (1961), pp. 55–56.

**Cullity 1978.** Bernard D. Cullity. *Elements of X-ray Diffraction.* 2d ed. Reading, Mass., 1978.

**"Le Curieux" 1923.** "Carnet d'un curieux." *Renaissance de l'art français et des industrie de luxe,* p. 530. 1923.

**Czerwenka 1867.** Bernhard F. Czerwenka. *Die Khevenhüller: Geschichte des Geschlechtes mit besonderer Berücksichtigung des XVII. Jahrhunderts. Nach archivalischen Quellen.* Vienna, 1867.

**Dabrowska 1995.** E. Dabrowska. Forthcoming in *BSAHL* 123 (1995).

**Dalas 1991.** M. Dalas. *Archives nationales: Corpus des sceaux français du Moyen Âge.* Vol. 2, *Les Sceaux des rois et de régence.* Paris, 1991.

**Dandridge and Wypyski 1992.** Peter Dandridge and Mark T. Wypyski. "Preliminary Technical Study of Medieval Limoges Enamels." *Materials Issues in Art and Archeology III: Proceedings of the Materials Research Society Symposium,* ed. Pamela B. Vandiver et al., vol. 267, pp. 817–26. Pittsburgh, 1992.

**Darcel 1854.** Alfred Darcel. "Le Ciboire d'Alpais." *Annales archéologiques* 14 (1854), pp. 5–16.

**Darcel 1856.** A. Darcel. "Trésor de Conques: Les Autels portatifs." *Annales archéologiques* 16 (1856), pp. 77–89.

**Darcel 1857.** A. Darcel. "Plaque en cuivre doré et émaillé (1267) appartenant à M. Germeau." *Bulletin du Comité de la langue, de l'histoire et des arts de la France* 4 (1857), pp. 113–14.

**Darcel 1865a.** A. Darcel. "Une Encyclopédie des arts industriels du Moyen Âge et de la Renaissance." *Gazette des Beaux-Arts* 19, no. 2 (August 1865), pp. 120–34, and no. 3 (September), pp. 248–71.

**Darcel 1865b.** A. Darcel. "Union central des Beaux-Arts appliqués à l'industrie, Musée rétrospectif: Le Moyen Âge et la Renaissance." *Gazette des Beaux-Arts* 2 (1865), pp. 427–45; 507–33.

**Darcel 1867.** A. Darcel. *Notices des Émaux et de l'orfèvrerie.* Musée du Moyen Âge et de la Renaissance; série D. Paris, 1867.

**Darcel 1868.** A. Darcel. "Notes sur quelques Émaux anciens envoyés à l'Exposition universelle." *Gazette des Beaux-Arts* 24, no. 1 (1868), pp. 368–84.

**Darcel 1874.** *See* Darcel and Basilewsky

**Darcel 1878a.** A. Darcel. "Le Moyen Âge et la Renaissance au Trocadéro." *Gazette des Beaux-Arts,* ser. 2, 18, no. 4 (October 1878), pp. 520–75; for "Les Émaux champlevés," see pp. 565–75.

**Darcel 1878b.** A. Darcel. "Les Émaux champlevés." In *L'Art ancien,* ed. Beaumont et al., pp. 248–58. Vol. 2 of *Exposition universelle 1878.*

**Darcel 1883.** A. Darcel. *Notice des Émaux et de l'orfèvrerie.* Musée du Moyen Âge et de la Renaissance; série D. Paris, 1883.

**Darcel 1884.** A. Darcel. *L'Exposition rétrospective de Rouen.* Rouen, 1884.

**Darcel 1891.** A. Darcel. *Notice des Émaux et de l'orfèvrerie.* Musée national du Moyen Âge et de la Renaissance (Louvre); série D. 4th ed. with supplement by E. Molinier. Paris, 1891.

**Darcel and Basilewsky 1874.** A. Darcel and A. Basilewsky. *Collection Basilewsky: Catalogue raisonné précédé d'un essai sur les arts industriels du Ier au XVIe siècle.* 2 vols. Paris, 1874.

**Daunou and Naudet 1860 [1840].** In *Recueil* 1738–1904, vol. 20, ed. J. Naudet and P.-C.-F. Daunou. Paris, [1840].

**Decanter 1960.** Jacques Decanter. "L'Inventaire de l'église du Prieuré d'Aureil en 1542." *BSAHL* 80 (1960), pp. 383–92.

**De Castris 1981.** *See* Naples 1981

**Decoux-Lagoutte 1891.** E. Decoux-Lagoutte. "Hommes illustres de Treignac, I: Guillaume de Treignac VIe prieur de Grandmont." *Bulletin de la Société des lettres, sciences et arts de la Corrèze* 13 (1891), pp. 271–79.

**Delage 1918.** Franck Delage. "Note sur quatre tombes de l'Oeuvre de Limoges (1327)." *BSAHL* 83, no. 2 (1950)

**Delage 1941.** F. Delage. "Tombes découvertes dans les ruines des Bénédictins de Limoges." *BSAHL* 79 (1941), pp. 54–60.

**Delage and Courtot 1923.** F. Delage and P.-L. Courtot. "Statue tombale de Blanche de Champagne, duchesse de Bretagne." *BSAHL* 70 (1923), pp. 5–30.

**Delisle and Berger 1916.** *Recueil des actes de Henri II, roi d'Angleterre et duc de Normandie.* Ed. L. Delisle and E. Berger. Vol. 1. Paris, 1916.

**Demartial 1923.** André Demartial. "L'Orfèvrerie émaillée de Limoges." *Congrès archéologique de France: LXXXIVe session tenue à Limoges en 1921,* pp. 431–43. Paris, 1923.

**Demartial 1932.** A. Demartial. "Vente Homberg." *BSAHL* 74, no. 1 (1932), p. xxvi (minutes of meeting).

**Demay 1880a.** G. Demay. *Le Costume au Moyen Âge d'après les sceaux.* Paris, 1880.

**Demay 1880b.** G. Demay. *Inventaire des sceaux de la Normandie.* Paris, 1880.

**Demus 1970.** O. Demus. *La Peinture murale romane.* Paris, 1970.

**C. and J.-P. Deremble 1993.** *See* Manhes-Deremble and Deremble 1993

**Dereux and François 1992.** D. Dereux and G. François. "A propos d'une crosse en émaux de Limoges du XIIIe siècle découverte à Andres (Pas-de-Calais)." *Revue du Nord-Archéologie* 74, no. 296 (1992), pp. 189–206.

**Deschamps and Thibout 1951.** Paul Deschamps and Marc Thibout. *La Peinture murale en France: Le haut Moyen Âge et l'époque romane.* Paris, 1951.

**Deschamps and Thibout 1963.** P. Deschamps and M. Thibout. *La Peinture murale en France au début de l'époque gothique (1180–1380).* Paris, 1963.

**Desjardins 1865.** G. Desjardins. *Histoire de la cathédrale de Beauvais.* Beauvais, 1865.

**Desjardins 1879.** G. Desjardins. *Cartulaire de l'abbaye de Conques en Rouergue.* Paris, 1879.

**Desmottes Sale 1900.** *Objets d'art et de haute curiosité du Moyen-Âge et de la Renaissance . . . composant la collection de feu M. Desmottes.* Hôtel Drouot, Paris, 1900.

**Deyres 1985.** M. Deyres. *Maine roman.* La Pierre-qui-Vire, 1985.

**Dez 1934.** G. Dez. "Vieux émaux." In *Musée de Poitiers.* Poitiers, 1934.

**Dictionnaire d'archéologie chrétienne 1914.** *See* Cabrol and Leclercq 1907–53.

**Dictionnaire d'histoire 1912–.** *Dictionnaire d'histoire et de géographie ecclésiastique.* General eds. Alfred Baudrillart, Albert Vogt, and Urbain Rouzies. 24 vols. Paris, 1912–.

**Didron, 1859.** Adolphe Napoléon Didron (ainé). *Manuel des oeuvres de bronze et d'orfèvrerie du Moyen Âge . . . dessins de L. Gaucherel; gravures de E. Mouard.* Paris, 1859.

**Dimier 1960–61.** Anselme Dimier. "Le Bienheureux Jean de Montmirail, moine de Longpont." *Mémoires de la Fédération des Sociétés Savantes de l'Aisne* 7 (1960–61), pp. 182–91.

**Dimier 1964.** A. Dimier. *Les Moines bâtisseurs.* Paris, 1964.

**Dimier and Montesquiou n. d.** A. Dimier and F. de Montesquiou. *Longpont, abbaye cistercienne.* Paris, n. d.

**Dionnerie.** *See* Gaillard de La Dionnerie

**Dionnet 1994.** Alain-Charles Dionnet. "Deux gémellions limousins du XIIIe siècle: Interprétations héraldiques." *BSAHL* 122 (1994), pp. 94–112.

**Distelberger et al. 1993.** Rudolf Distelberger, Alison Luchs, Philippe Verdier, and Timothy H. Wilson. *Western Decorative Arts.* Part 1, *Medieval, Renaissance, and*

*Historicizing Styles Including Metalwork, Enamels, and Ceramics.* The Collections of the National Gallery of Art, Systematic Catalogue. Washington, D.C., 1993.

**Douais 1888.** Celestin Douais. "Deux reliquaires de l'église Saint-Sernin, à Toulouse," *Revue de l'Art chrétien* 38 (1888), pp. 154–69.

**Douais 1892.** C. Douais. "Les Frères prêcheurs de Limoges." *BSAHL* 40 (1892), pp. 261–363.

**Douais 1904.** C. Douais. *Trésor et reliques de Saint-Sernin de Toulouse: 1, Les Inventaires (1246–1657).* Vol. 2 of *Documents sur l'ancienne province de Languedoc.* 3 vols. Paris, 1901–6.

**Drouault 1905.** R. Drouault. "Monographie du canton de Saint-Sulpice-les-Feuilles (suite)." *BSAHL* 55 (1905), pp. 241–79.

**Du Boys 1855.** A. Du Boys. "Inventaire du trésor de l'ordre de Grandmont, 1666." *BSAHL* 6 (1855), pp. 5–72.

**Dubuisson-Aubenay 1636 and 1898.** Dubuisson-Aubenay. [François Nicolas Baudot]. *Itinéraire de Bretagne en 1636, d'après le manuscrit original.* Ed. Léon Maître and Paul de Berthou. Nantes, 1898–1902.

**Du Cange 1883–87.** Charles, seigneur Dufresne Du Cange. *Glossarium mediae et infimae latinitatis.* 10 vols. Niort, 1883–87.

**Du Cange 1840–50.** Charles, seigneur Dufresne Du Cange. *Glossarium mediae et infimae latinitatis.* 7 vols. Paris, 1840–50.

**Du Chesne 1636–49.** André Du Chesne. *Historiae Francorum scriptores coaetanei . . .* 5 vols. Paris, 1636–49.

**Duchesne 1892.** L. Duchesne (Mgr). "Saint Martial de Limoges." *Annales du Midi* 4 (1892), pp. 289–330.

**Duchesne 1910.** L. Duchesne. *Fastes épiscopaux de l'ancienne Gaule.* Paris, 1910.

**Duchesne 1955 and 1981.** L'Abbé L. Duchesne. *Le Liber pontificalis: Texte, introduction et commentaire.* 3 vols. Reprint of 1955 edition. Paris, 1981.

**Dugasseau 1864.** C. Dugasseau. *Notice des tableaux composant le musée du Mans precédé d'une notice historique.* Le Mans, 1864. Reissued under the title *Catalogue du musée de peinture et d'histoire naturelle du Mans precédé d'une notice historique.* Le Mans, 1892, and Le Mans, 1905.

**Du Mège 1836–37.** Alexandre Du Mège. "Mémoires sur quelques châsses ou reliquaires . . . conservés dans les églises du Midi de la France." *Mémoires de la Société Archéologique du Midi de la France* 3 (1836–37), pp. 307–36.

**Duplès-Agier 1874.** Henri Duplès-Agier, ed. *Chroniques de Saint-Martial de Limoges.* Pub. d'après les manuscrits originaux pour la Société de l'Histoire de France. Paris, 1874.

**Duran and Ainaud de Lasarte 1956.** Austin Duran y Sanpere and Juan Ainaud de Lasarte. *Escultura gotica.* Madrid, 1956.

**G. Durand 1484.** Guillaume Durand. *Rationale divinorum officiorum.* Strasbourg, 1484.

**G. Durand 1854.** G. Durand. *Rational ou Manuel des divins offices.* Ed. C. de

Barthélémy. Paris, 1854.

**Durand and Nougaret 1992.** Geneviève Durand and Jean Nougaret, eds. *L'Ordre de Grandmont, art et histoire: Actes des Journées d'Études de Monpellier, 7 et 8 octobre 1989.* Monpellier, 1982.

**Durandus 1906.** Guillaume Durand. *The Symbolism of Churches and Church Ornaments: A Translation of the First Book of the Rationale divinorum officiorum, Written by William Durandus.* Ed. and trans. John Mason Neale and Benjamin Webb. London, 1906.

**J. Durand 1985.** Jannic Durand. Entry in *Moyen Âge, orfèvrerie, Musée du Louvre, nouvelles acquisitions du département des objets d'art, 1980–1984.* Exh. cat., Musée du Louvre. Paris, 1985.

**J. Durand 1987.** In Senlis 1987

**Duret 1932.** D. Duret. *Mobilier, vases, objets et vêtements liturgiques: Étude historique.* Paris, 1932.

**Durliat 1986.** Marcel Durliat. *Saint-Sernin de Toulouse à la fin de l'époque romane.* Toulouse, 1986.

**Durliat 1989.** M. Durliat. "La Signification des majestés catalanes." *Cahier archéologique* 37 (1989), pp. 69–95.

**A. Du Sommerard 1838–46.** A. Du Sommerard. *Les Arts au Moyen Âge.* 5 vols. Paris, 1838–46.

**E. Du Sommerard 1883.** E. Du Sommerard. *Musée des Thermes et de l'Hôtel de Cluny: Catalogue et description des objets d'art de l'Antiquité, du Moyen Âge et de la Renaissance.* Paris, 1883.

**Dussieux 1841.** L. Dussieux. *Recherches sur l'histoire de la peinture en émail, dans les temps anciens et modernes et spécialement en France.* Paris, 1841.

**Duthuit 1929.** G. Duthuit. "La Donation Martin le Roy: Orfèvrerie, émaillerie." *Bulletin des musées de France* 10 (1929), pp. 4–16.

**Duveen 1935.** James Henry Duveen. *Collections and Recollections: A Century and a Half of Art Deals.* London, 1935. American title: *Art Treasures and Intrigue.*

**Edge and Paddock 1988.** David O. Edge and John Miles Paddock. *Arms and Armor of the Medieval Knight: An Illustrated History of Weaponry in the Middle Ages.* New York, 1988.

*L'École de Limoges.* Marie-Madeleine S. Gauthier. *L'École de Limoges 1190–1216.* Vol. 2 of Gauthier, *Émaux méridionaux.* Forthcoming.

*Émaux 1948.* See Limoges 1948

**England 1986.** Pamela England. "A Technical Investigation of Medieval Enamels." In *Catalogue of Medieval Objects: Enamels and Glass,* ed. Hanns Swarzenski and Nancy Netzer, pp. xviii–xxvi. The Museum of Fine Arts, Boston, 1986.

*English Romanesque Art.* See London 1984.

**Enlart 1916.** C. Enlart. *Manuel d'archéologie française.* Vol. 3, *Le Costume.* Paris, 1916.

**Enlart 1927–28.** C. Enlart. "L'Émaillerie cloi-

sonnée à Paris sous Philippe le Bel." *Fondation Eugène Piot, Monuments et mémoires* 29 (1927–28), pp. 1–197.

**Eraclius.** *See* Merrifield 1967

**Erdmann 1953.** Kurt Erdmann. "Arabische Schriftzeichen als Ornamente in der abendlandischen Kunst des Mittelalters." *Akademie der Wissenschaft und der Litteratur* 47 (1953), pp. 465–513.

**Erlande-Brandenburg 1975.** Alain Erlande-Brandenburg. *Le Roi est mort: Étude sur les funérailles, les sépultures et les tombeaux des rois de France jusqu'à la fin du XIIIe siècle.* Paris, 1975.

**Erlande-Brandenburg 1983.** A. Erlande-Brandenburg. *L'Art gothique.* Paris, 1983.

**Erlande-Brandenburg 1987.** A. Erlande-Brandenburg. *Le Monde gothique: La Conquête de l'Europe, 1260–1380.* Paris, 1987.

**Ernst and Heusinger 1963.** Konradt Ernst and Christian von Heusinger. *Die Wiegendrucke des Kestner-Museums.* Hannover, 1963.

*España sagrada 1763.* Enrique Flórez. *España sagrada: Theatro geographico-historico de la iglesia de España.* 51 vols. Madrid, 1747–1879. Vol. 17, *De la Santa Iglesia de Orense en su estado antiguo y presente.* Madrid, 1763.

**Etaix and Vregille 1970.** R. Etaix and B. Vregille. "Les Manuscrits de Besançon, P.-F. Chifflet et la Bibliothèque Bouhier." *Scriptorium* 24 (1970), pp. 27–39.

*Europäisches Stammtafeln 1978.* *Europäisches Stammtafeln: Stammtafeln zur Geschichte der europäischen Staaten.* 15 vols. Marburg, 1978.

**Evans 1948.** J. Evans. *Art in Medieval France.* London, 1948.

*Exposition rétrospective 1900.* See Molinier and Marcou.

*Exposition—Union centrale des Arts décoratifs appliqués à l'industrie 1880.* See Paris 1880

*Exposition—Union centrale des Beaux-arts appliqués à l'industrie 1867.* See Paris 1865

*Exposition universelle 1867.* See Paris 1867

*Exposition universelle 1878.* See Paris 1878

*Exposition universelle 1900.* See Paris 1900

**Eygun 1938.** F. Eygun. *Sigillographie du Poitou.* Poitiers, 1938.

**Fage 1885.** R. Fage. "Le Tombeau du cardinal de Tulle à Saint-Germain-les-Belles." *BSAHL* 33 (1885), pp. 51–57.

**Fairholt 1857.** Frederick W. Fairholt. *Miscellanea graphica: Representations of Ancient, Medieval, and Renaissance Remains in the Possession of Lord Londesborough.* Intro. Thomas Wright. London, 1857.

**Falk 1994.** Birgitta Falk. "Bildnisreliquiare: zur Entstehung und Entwicklung der metallenen Kopf-, Büsten-, und Halbfigurenreliquiare im Mittelalter." *Aachener Kunstblätter* 59 (1991–93 [1994]), pp. 99–238.

**Falke 1931.** Otto von Falke. "Romanische Emailarbeiten von Limoges." *Pantheon* 8 (July–December 1931), pp. 282–85.

**Fanjoux 1847.** G. Fanjoux. "Essai paléographique et archéologique sur la Bible de Souvigny." *Bulletin de la Société d'émulation de l'Allier* 1 (1846–50 [1847]), pp. 353–70.

**Farcy 1877.** L. de Farcy. *Notices archéologiques sur les tombeaux des évêques d'Angers.* Angers, 1877.

**Farcy 1892.** L. de Farcy. "Redécouverte du tombeau de Michel de Villoiseau évêque d'Angers." *Revue de l'Anjou* 25 (1892), pp. 181–84.

**Farcy 1897.** L. de Farcy. "Le Tombeau d'Ulger à Angers (extrait de la semaine religieuse d'Angers)." *Revue de l'Art chrétien* (1897), pp. 87–88.

**Farcy 1905.** L. de Farcy. "Épaves." *Revue de l'Art chrétien* 58 (1905), pp. 188–90.

**Farnier 1913.** R. Farnier. *La Condition juridique des personnes et des biens dans l'ordre de Grandmont.* Limoges, 1913.

*Fastes du Gothique 1981.* See Paris 1981

**Favreau et al. 1978.** Robert Favreau and Jean Michaud. *Limousin: Corrèze, Creuse, Haute-Vienne.* Poitiers, 1978. Vol. 2 of *Corpus des inscriptions de la France médiévale.* General ed. Edmond-René Labande. 17 vols. Paris, 1974–.

**Favreau et al. 1979.** R. Favreau, B. Leplant, and J. Michaud. *Dordogne, Gironde.* Poitiers, 1979. Vol. 5 of *Corpus des inscriptions de la France.* Paris, 1974–.

**Favreau et al. 1984.** R. Favreau, J. Michaud, and B. Leplant. *Aveyron, Lot, Tarn.* Paris, 1984. Vol. 9 of *Corpus des inscriptions de la France.* Paris, 1974–.

**Favreau 1988.** R. Favreau, ed., with G. Pon. *Le Diocèse de Poitiers.* Paris, 1988.

**Favyn 1612.** A. Favyn. *Histoire de Navarre.* Paris, 1612.

**Faye 1849.** L. Faye. *Notes historiques sur la ville de Sivrai.* Poitiers, 1849. An extract from the *Bulletin de la Société des Antiquaires de l'Ouest.*

**Fayolle 1897.** G. de Fayolle. "Le Trésor de l'Église de Saint-Nectaire." *Congrès archéologique de France, LXIIe session, séances générales tenues à Clermont-Ferrand en 1895,* pp. 292–306. Paris and Caen, 1897.

**Fayolle 1921.** G. de Fayolle. "Oeuvres d'orfèvrerie et d'émaillerie limousines exposées à Tulle." *Congrès archéologique de France, LXXXIVe session, Paris, 1921,* pp. 333–46. Paris, 1923.

**Fayolle 1924.** G. de Fayolle. "Oeuvres d'orfèvrerie en Auvergne." *Congrès archéologique de France, LXXXVIIe session, tenue à Clermont-Ferrand, 1924,* pp. 433–49. Paris, 1925.

**Férotin 1897.** D. Marius Férotin. *Recueil des chartes de l'abbaye de Silos.* Paris, 1897.

**Ferrari 1964.** Sabatino Ferrari. "La Badia di San Baronto (Sulle tracce d'un'opera d'arte smarrita." In *Chiese romaniche e moderne in Pistoia e Diocesi,* pp. 57–70. Pistoia, 1964.

**Fillitz 1969.** Hermann Fillitz, with Anton von Euw and others. *Das Mittelalter.* Vol. 1. Berlin, 1969.

**Fillitz and Pippal 1987.** Hermann Fillitz and

Martina Pippal. *Schatzkunst: Die Goldschmiede- und Elfenbeinarbeiten aus Österreichischen Schatzkammern des Hochmittelalters.* Salzburg, 1987.

**Fillon 1852–53.** B. Fillon. "Tombeau de Blanche, duchesse de Bretagne. Lettre de M. B. Fillon au directeur des Archives de l'art français." In *Archives de l'art français,* vol. 3. Paris, 1852–53. Documents, *Recueil de documents inédits relatifs à l'histoire des arts en France,* ed. Philippe de Chennevières, vol. 2, pp. 129–30. Paris, 1852–53.

**Fillon 1868.** B. Fillon. "Sépultures des anciens abbés de Nieul-sur-l'Autise (Vendée)." *Bulletin monumental* (1868), pp. 923–25.

**Fingerlin 1971.** Ilse Fingerlin. *Gürtel des hohen und späten Mittelalters.* Munich and Berlin, 1971.

**Finoelst Sale 1927.** *Catalogue de la collection A. Finoelst de Paris, primitifs des écoles italienne, allemande, française et de Barcelone . . . émaux limousins . . . ivoires, sculptures, tissus . . . meubles français . . . vente galerie Georges Giroux . . . 26 et 27 sept.,* rédigé par Joseph Destrée. Brussels, 1927.

**Flannery Sale 1983.** *The Thomas F. Flannery Jr. Collection.* Sotheby's, November 30, December 1, December 12, 1983. London, 1983. (Another element of the sale on March 6, 1984.)

**Flavigny 1992.** L. Flavigny. "Médaillon avec Osée." In *Musée des Antiquités, Rouen: De l'Egypte ancienne à la Renaissance rouennaise,* ed. A Dantet and P. Périn. Rouen, 1992.

**Florence 1989.** *Arti del Medio Evo e del Rinascimento: Omaggio ai Carrand, 1889–1989: Museo nazionale del Bargello, 20 marzo–25 guigno 1989.* Exh. cat. Florence, 1989.

**Fontaine 1931.** Georges Fontaine. "Les Émaux limousin." *L'Illustration,* December 5, 1931. Unpaginated.

**Fontaine 1937.** G. Fontaine. "La Collection Personnaz." *Bulletin des Musées de France* 9 (July 1937), pp. 101–3.

**Forbes 1972.** R. J. Forbes. *Studies in Ancient Technologies.* 9 vols. 2d ed. Leiden, 1964–.

**Forestié 1885.** J. Forestié. Communication to the session of April 8, 1885. *Bulletin archéologique du Comité des travaux historiques et scientifiques,* 1885, p. 184.

**Foreville 1975.** Raymonde Foreville. "Le Culte de saint Thomas Becket en Normandie: Enquête sur les sanctuaires anciennement placés sous le vocable du martyr de Canterbury." In *Thomas Becket: Actes du colloque international de Sédières, 19–24 août 1973,* ed. Raymonde Foreville, pp. 347–69. Paris, 1975.

**Foreville 1976.** R. Foreville. "La Diffusion du culte de Thomas Becket dans la France de l'ouest avant la fin du XIIe siècle." *Cahiers de civilisation médiévale* 19 (1976), pp. 347–69.

**Forot 1907.** Victor Forot. *La Verité sur la colombe eucharistique volée à Laguenne (Corrèze) par la bande des antiquaires Thomas et cie* [signed:] *V. Forot décembre 1907.* Brive [Tulle], n. d. [1907].

**Forot 1913.** V. Forot. "Catalogue raisonné des richesses monumentales et artistiques du département de la Corrèze: Les objets d'art

de nos églises." *Bulletin de la Société des Lettres, Sciences, et Arts de la Corrèze* 25 (1913), pp. 54–87. Also reprinted, under the same title, in book form in Paris, 1913.

**Forot 1924.** V. Forot. "Le Trésor de Gimel." *Bulletin de la Société scientifique, historique et archéologique de la Corrèze* 46, no. 3 (1924), pp. 246–59.

**I. H. Forsyth 1972.** Ilene H. Forsyth. *The Throne of Wisdom: Wood Sculptures of the Madonna in Romanesque France.* Princeton, 1972.

**W. H. Forsyth 1946.** William H. Forsyth. "Mediaeval Enamels in a New Installation." *The Bulletin of The Metropolitan Museum of Art* 4, no. 9 (May 1946), pp. 232–39.

**Foucart-Borville 1987.** Jacques Foucart-Borville. "Essai sur les suspenses eucharistiques comme mode d'adoration privilégié du Saint Sacrement." *Bulletin monumental* 145 (1987), pp. 267–89.

**Foucart-Borville 1990.** J. Foucart-Borville. "Les Tabernacles eucharistiques dans la France du Moyen Âge." *Bulletin monumental* 148 (1990), pp. 349–81.

**Foy 1985.** D. Foy. "Essai de Typologie des verres médiévaux d'après les fouilles provençales et lanquedociennes." *Journal of Glass Studies* 27 (1985), pp. 18–71.

**Foy 1990.** "Les Matières premières et le combustible." In Danièle Foy, *Le Verre médiéval et son artisanat en France méditerranéenne,* pp. 29–55, 407–19. Marseille, [1988].

**Foy and Sennequier.** In Rouen 1989

**Francesco d'Assisi 1982.** *Francesco d'Assisi, Storia e arte.* Milan, 1982.

**François 1855.** René François. "La façon de l'émaillerie recueillie des anciens émailleurs." In *Émailleurs et émaillerie de Limoges,* ed. Maurice Ardant, pp. 27–34. Isle, 1855.

**François 1991.** G. François. "Chandeliers civils armoriés du XIVe siècle acquis par le musée municipal de Limoges en 1990." *BSAHL* 119 (1991), pp. 83–96.

**Frazer 1985–86.** Margaret English Frazer. "Medieval Church Treasures." *Metropolitan Museum of Art Bulletin* 43, no. 3 (Winter 1985–86), pp. 1–56.

**Freestone 1990.** I. C. Freestone. "Laboratory Studies of the Portland Vase." *Journal of Glass Studies* 32 (1990), pp. 103–07.

**Freestone 1991.** I. C. Freestone. "Looking into Glass." In *Science and the Past,* ed. Sheridan Bowman, pp. 37–56. London, 1991.

**Freestone 1992.** I. C. Freestone. "Theophilus and the Composition of Medieval Glass." In *Materials Issues in Art and Archeology III: Proceedings of the Materials Research Society Symposium,* ed. Pamela B. Vandiver et al., vol. 267, pp. 739–45. Pittsburgh, 1992.

**Freestone 1993.** I. C. Freestone. "Compositions and Origins of Glasses from Romanesque Champlevé Enamel." In *Northern Romanesque Enamel,* pp. 37–45. Vol. 2 of *Catalogue of the Medieval Enamels in the British Museum.* Edited Neil Stratford. London, 1993.

**Freestone, Bimson, and Buckton 1990.** I. C. Freestone, M. Bimson, and D. Buckton. "Compositional Categories of

Byzantine Glass Tesserae." In *Annales du 11e Congrès de l'Association internationale pour l'Histoire du Verre, Bâle 29 août–3 septembre 1988,* pp. 271–79. Amsterdam, 1990.

**Freiburg 1972.** *Art Médiéval de France: Collections du musée Thomas Dobrée de Nantes.* Exh. Freiburg, 1972.

**Frolow 1941.** A. Frolow. "Deux Inscriptions sur les reliquaires byzantins." *Revue archéologique* 18 (1941), pp. 233–42.

**Frolow 1961.** A. Frolow. *La Relique de la vraie croix: Recherches sur le développement d'un culte.* Paris, 1961.

**Frolow 1965.** A. Frolow. *Les Reliquaires de la vraie croix.* Paris, 1965.

**Gaborit 1972.** See Québec–Montréal 1972

**Gaborit 1976a.** Jean-René Gaborit. "L'Autel majeur de Grandmont." *Cahiers de civilisation médiévale* 19, no. 3 (July–September, 1976), pp. 231–46.

**Gaborit 1976b.** J.-R. Gaborit. "L'Autel majeur de Grandmont." *Bulletin de la Société des Antiquaires de France,* 1976, pp. 31–33.

**Gaborit 1979.** J.-R. Gaborit. "Un Groupe de la descente de croix au musée du Louvre." *Fondation Eugène Piot, Monuments et Mémoires* 62 (1979), pp. 149–83.

**Gaborit 1988.** J.-R. Gaborit. "Sur un lit de parade: Essai d'interprétation d'un motif funéraire." In *La Figuration des morts dans la chrétienté médiévale jusqu'à la fin du premier quart du XIVe siècle: Colloque, 26–28 mai 1988, Abbaye royale de Fontevraud,* ed. Roger Grégoire, pp. 117–23. N. p., 1989.

**Gaborit-Chopin 1968.** Danielle Gaborit-Chopin. "Les Dessins d'Adhémar de Chabannes." *Bulletin archéologique du Comité des travaux historiques et scientifiques,* n. s., 3 (1967) [1968], pp. 163–225.

**Gaborit-Chopin 1969.** D. Gaborit-Chopin. *La Décoration des manuscrits à Saint-Martial de Limoges et en Limousin du IXe au XIIe siècle.* Paris and Geneva, 1969.

**Gaborit-Chopin 1972.** D. Gaborit-Chopin. "Deux émaux limousins." *La Revue du Louvre et des musées de France* (1972), pp. 205–11.

**Gaborit-Chopin 1978.** D. Gaborit-Chopin. *Ivoires du Moyen Âge.* Fribourg, 1978.

**Gaborit-Chopin 1983a.** D. Gaborit-Chopin. "Chronique des amis du Louvre, février–avril 1983." *La Revue du Louvre et des musées de France* 33, no. 1 (1983), p. i (following p. 84).

**Gaborit-Chopin 1983b.** D. Gaborit-Chopin. "Les Arts précieux." In Avril, Barral i Altet, Gaborit-Chopin 1983.

**Gaborit-Chopin 1988.** D. Gaborit-Chopin. "Nicodème travesti: La Descente de croix d'ivoire du Louvre." *Revue de l'Art* 81 (1988), pp. 31–44.

**Gaborit-Chopin 1991.** D. Gaborit-Chopin, in D. Alcouffe and Gaborit-Chopin. "Une Dation de remarquable objets médiévaux." *La Revue du Louvre et des musées de France* 4 (1991), pp. 7–10.

**Gaborit-Chopin 1992.** D. Gaborit-Chopin. "La Croix de la Roche-Foulques." *Bulletin de la Société nationale des antiquaires de*

France, 1992, pp. 408–25.

**Gaborit-Chopin and Lahanier 1982.** D. Gaborit-Chopin and Christian Lahanier. "Étude scientifique de la plaque émaillée de Geoffroy Plantagenêt." *Annales du Laboratoire de Recherche des Musées de France,* pp. 7–27. Paris, 1982.

**Gaborit-Chopin and Taburet 1981.** D. Gaborit-Chopin and E. Taburet. *École du Louvre: Objets d'art.* Paris, 1981.

**Gaillard de La Dionnerie and Lasteyrie 1883.** Gaillard de La Dionnerie and R. de Lasteyrie. [spécimens d'émaillerie limousine.] *Bulletin du Comité des travaux historiques et scientifiques,* 1883, pp. 11–12.

**Galbreath 1942.** D. L. Galbreath. *Manuel du blason.* Lausanne, 1942.

**Galland 1994.** B. Galland. *Deux archevêchés entre la France et l'Empire: Les Archevêques de Lyon et les archevêques de Vienne du milieu du XIIe siècle au milieu du XIVe siècle.* Rome, 1994.

**Gallego Lorenzo 1989.** See Lorenzo 1989

**Galletti 1753.** L. Galletti. *Notizie istoriche del monastero, di San Baronto della diocesi Pistoiese.* N. p., 1753.

*Gallia christiana.* Denis de Sainte-Marthe. *Gallia christiana in provincias ecclesiasticas distributa.* 13 vols. Paris, 1715–25; vols. 14–16, ed. B. Haureau. Paris, 1856–65.

**Galopaud 1944.** E. Galopaud. "Un des très beaux objets d'art du musée du Louvre." *Bulletin et mémoires de la Société archéologique et historique de la Charente,* 1944, pp. 301–7.

**Gams 1873.** P.-B. Gams. *Series episcoporum ecclesiae catholicae, quotquot innotuerunt a beato Petro apostolo.* Ratisbonne, 1873.

**Ganneron 1855.** Edmond Ganneron. *La Casette de saint louis, roi de France donnée par Philippe le Bel à l'Abbaye du Lis.* Paris, 1855.

**Ganz 1925.** P. Ganz. *La Collection Engel-Gros.* Paris, 1925.

**Gardelles 1963.** J. Gardelles. *La Cathédrale Saint-André de Bordeaux: Sa place dans l'évolution de l'architecture et de la sculpture.* Bordeaux, 1963.

**Gardelles 1979.** J. Gardelles. "Reliquaires et objets d'art médiévaux à la Sauve-Majeure." In *Saint-Emilion-Libourne, La Religion populaire en Acquitaine: Fédération historique du Sud-Ouest, 29e Congrès, 1977,* pp. 183–90. Bordeaux, 1979.

**A. Gardner 1954.** Albert ten Eyck Gardner. "Beckford's Gothic Wests." *Bulletin of The Metropolitan Museum of Art,* n. s., 13, no. 2 (October 1954), pp. 41–49.

**J. S. Gardner 1897.** See London 1897

**Gariel 1864.** Hyacinthe Gariel. "Oeuvres diverse." *Bibliothèque historique et littéraire du Dauphiné.* Vol. 1. Grenoble, 1864.

**Gariel 1970.** H. Gariel, ed. *Dictionnaire historique, chronologique, géographique, généalogique, héraldique, juridique, politique et botanique du Dauphiné, par Guy Allard; publié pour la première fois et d'après le manuscrit original . . .* Grenoble, 1864. Reprint. Geneva, 1970.

**Garmier 1980.** J.-F. Garmier. "Le Goût du Moyen Âge chez les collectionneurs lyonnais du XIXe siècle." *Revue de l'Art* 47 (1980), pp. 53–64.

**Garnier 1886.** Édouard Garnier. *Histoire de la verrerie et de l'émaillerie.* Tours, 1886.

**Gatty 1883.** C. T. Gatty. *Catalogue of Medieval and Later Antiquities Contained in the Mayer Museum.* Liverpool, 1883.

**Gauchery and Grossouvre 1966.** P. Gauchery and A. de Grossouvre. *Notre Vieux Bourges.* Bourges, 1966.

**Gaussen 1861.** A. Gaussen. *Portefeuille archéologique de la Champagne.* Bar-sur-Aube, 1861.

**Gauthier 1948.** *See* Limoges 1948

**Gauthier 1950.** Marie-Madeleine S. Gauthier. *Émaux limousins champlevés des XIIe, XIIIe et XIVe siècles.* Paris, 1950.

**Gauthier 1955.** M.-M. S. Gauthier. "La Légende de sainte Valérie et les émaux champlevés de Limoges." *BSAHL* 86 (1955), pp. 35–80.

**Gauthier 1956.** M.-M. S. Gauthier. "Une Crosse à grande fleur en émail champlevé de Limoges à la cathédrale de Poznán (Pologne)." *BSAHL* 86 (1956), pp. 283–86.

**Gauthier 1957.** M.-M. S. Gauthier. "Les Émaux champlevés limousins et l'Oeuvre de Limoges: Quelques problèmes posés par l'émaillerie champlevée sur cuivre en Europe méridionale du XIIe au XIVe siècle." *Cahiers de la Céramique et des arts du feu,* no. 8 (1957), pp. 146–67.

**Gauthier 1958a.** M.-M. S. Gauthier. "Les Décors vermiculés dans les émaux champlevés limousins et méridionaux: Aperçus sur l'origine et diffusion du motif au XIIe siècle." *Cahiers de civilisation médiévale* 1 (1958), pp. 349–69.

**Gauthier 1958b.** M.-M. S. Gauthier. "Émaux limousins champlevés." *L'Information d'histoire de l'art* 3 (1958), pp. 67–78.

**Gauthier 1960a.** M.-M. S. Gauthier. "Émaux et orfèvreries." In *Limousin roman,* ed. Jean Maury, Marie-Madeleine S. Gauthier, and Jean Porcher, pp. 280–91. La Pierre-qui-Vire, 1960. 2d ed. 1974.

**Gauthier 1960b.** M.-M. S. Gauthier. "La Plaque de dédicace émaillée datée 1267 aujourd'hui au musée national de Varsovie et les autels de l'Artige." *BSAHL* 87 (1960), pp. 333–48.

**Gauthier 1962a.** M.-M. S. Gauthier. "Los Esmaltes meridiónales en la Esposición internacional de arte románico en Barcelona y Santiago de Compostela." *Goya: Revista de Arte,* no. 48 (1962), pp. 400–408.

**Gauthier 1962b.** M.-M. S. Gauthier. "Première campagne de fouilles dans le 'Sepulchre' de Saint-Martial de Limoges." *Cahiers archéologiques* 12 (1962), pp. 205–48.

**Gauthier 1963a.** M.-M. S. Gauthier. "Le Frontal limousin de San Miguel de Excelsis et l'Oeuvre de Limoges." *Art de France,* no. 3 (1963), pp. 40–62.

**Gauthier 1963b.** M.-M. S. Gauthier. "Le Trésor de Conques." In *Rouergue roman,* Georges Gaillard et al., pp. 98–187. La Pierre-qui-Vire, 1963. 2d ed., 1975.

**Gauthier 1964a.** M.-M. S. Gauthier. "Notes et documents. Dossiers." *L'Information d'histoire de l'art* 9 (1964), pp. 78–83.

**Gauthier 1964b.** M.-M. S. Gauthier. "Observations préliminaires sur les restes d'un revêtement d'émail champlevé fait pour la confession de Saint Pierre à Rome." *BSAHL* 91 (1964), pp. 43–61.

**Gauthier 1964 (1967).** *See* Gauthier 1967

**Gauthier 1966.** M.-M. S. Gauthier. "Une Châsse limousine du dernier quart du XIIe siècle: Thèmes iconographiques, composition et essai de chronologie." *Mélanges offerts à René Crozet, Professeur à l'Université de Poitiers, Directeur du Centre d'Études Supérieures de Civilisation Médiévale, à l'occasion de son soixante-dixième anniversaire, par ses amis, ses collègues, ses élèves, et les membres du C.É.S.C.M.,* ed. Pierre Gallais and Yves-Jean Riou, vol. 2, pp. 937–51. 2 vols. Poitiers, 1966.

**Gauthier 1967a.** M.-M. S. Gauthier. "L'ange 'grand comme nature', jadis pivotant au sommet de la flèche de Saint-Pierre du Dorat." *BSAHL* 94 (1967), pp. 109–29.

**Gauthier 1967b.** M.-M. S. Gauthier. "Le Goût Plantagenêt et les arts mineurs dans la France du Sud-Ouest." In *Stil und Überlieferung in der Kunst des Abendlandes I, Akten des XXI internationalen Kongress für Kunstgeschichte, Bonn, 14–19 sept. 1964,* pp. 139–55. Berlin, 1967.

**Gauthier 1967c.** M.-M. S. Gauthier. "A Limoges Champlevé Book-Cover in the Gambier-Parry Collection." *The Burlington Magazine,* March (1967), pp. 151–57.

**Gauthier 1968a.** M.-M. S. Gauthier. "La Clôture émaillée de la confession de Saint Pierre au Vatican, lors du Concile de Latran, IV, 1215." In *Synthronon: Art et archéologie de la fin de l'Antiquité et du Moyen Âge. Recueil d'études par André Grabar et un groupe de ses disciples,* pp. 237–246. Paris, 1968.

**Gauthier 1968b.** M.-M. S. Gauthier. "Les 'Majestés de la Vierge,' limousines et méridionales au 13e siècle au Metropolitan Museum of Art de New York." *Bulletin de la Société nationale des Antiquaires de France* (1968), pp. 66–95.

**Gauthier 1968c.** M.-M. S. Gauthier. "Musée municipal de Limoges: Les collections d'émaux champlevés, acquisitions récentes." *La Revue du Louvre et des musées de France,* 1968, pp. 447–54.

**Gauthier 1968d.** M.-M. S. Gauthier. "Les Reliures en émail de Limoges conservées en France: recensement raisonné." In *Humanisme actif: Mélanges d'art et de littérature offerts à Julien Cain,* vol. 1, pp. 271–87. 2 vols. Paris, 1968.

**Gauthier 1969.** *See* Gauthier 1972c

**Gauthier 1970.** *See* Gauthier 1968b

**Gauthier 1972a.** M.-M. S. Gauthier. "L'Art de l'émail champlevé en Italie à l'époque primitive du gothique." In *Il Gotico a Pistoia: Nei suoi rapporti con l'arte gotica italiana. Atti del 20 Convegno internazionale di studi, Pistoia, 24–30 Aprile 1966,* pp. 271–93. Rome, 1972.

**Gauthier 1972b.** M.-M. S. Gauthier. *Émaux du Moyen Âge occidental.* Fribourg, 1972.

**Gauthier 1972c.** M.-M. S. Gauthier. "De la palette au style chez les émailleurs du Moyen Âge." In *Évolution générale et développements régionaux en histoire de l'art: Actes du XXIIe Congrès internationale de histoire de l'art, Budapest 1969,* ed. György Rózsa, pp. 621–35. 3 vols. Budapest, 1972.

**Gauthier 1973.** M.-M. S. Gauthier. "Colombe limousine prise aux rêts d'un 'antiquaire' benedictin a Saint-Germain-des-Pres, vers 1726." In *Intuition und Kunstwissenschaft: Festschrift für Hanns Swarzenski zum 70. Geburtstag,* ed. Peter Bloch et al., pp. 171–90. Berlin, 1973.

**Gauthier 1975a.** M.-M. S. Gauthier. "Une Crosse limousine datable de 1230 à 1240, découverte dans une tombe épiscopal de la cathédrale de Brême." *Bulletin de la Société nationale des Antiquaires de France,* pp. 165–80. Paris, 1975.

**Gauthier 1975b.** M.-M. S. Gauthier. "La Meurtre dans la cathédrale, thème iconographique médiévale." *Thomas Becket, Actes du Colloque international de Sédières, 19–24 août, 1973,* ed. Raymonde Foreville, pp. 247–53. Paris, 1975.

**Gauthier 1976a.** M.-M. S. Gauthier. "Les Inscriptions des émaux champlevés limousins du XIIe au XIVe siècle: Essai de classement de chronologie." *Bulletin de la Société National des Antiquaires de France,* 1976, pp. 176–90.

**Gauthier 1976b.** M.-M. S. Gauthier. "Reliques des Saint Innocent et châsse limousine au trésor de Saint-Denis." In *Essays in Honor of S. McKnight Crosby,* ed. Pamela Z. Blum, pp. 293–302. Published in *Gesta* 15, nos. 1 and 2 (1976).

**Gauthier 1977.** M.-M. S. Gauthier. "Antichi ripristini e restauri moderni su smalti e orificerie medioevali." In *Il Restauro delle opere d'arte, Pistoia, Centro italiano di Studi di Storia e d'Arte,* 1977, pp. 251–62.

**Gauthier 1978a.** M.-M. S. Gauthier. "La Croix émaillée de Bonneval au musée de Cluny." *Revue du Louvre et des musées de France* 4 (1978), pp. 267–85.

**Gauthier 1978b.** M.-M. S. Gauthier. "Un Croix médiévale entièrement émaillée, présumée provenir de l'abbaye de Bonneval." *Bulletin de la Société nationale des Antiquaires de France,* 1978, pp. 53–56.

**Gauthier 1978c.** M.-M. S. Gauthier. "Du tabernacle au rétable: Une innovation limousine vers 1230." *Revue de l'Art* 40–41 (1978), pp. 23–42.

**Gauthier 1979.** M.-M. S. Gauthier. "Art, savoir-faire médiéval et laboratoire moderne, à propos de l'effigie funéraire de Geoffroy Plantagenêt." *Académie des Inscriptions et Belles-Lettres, comptes-rendus des séances,* 1979, pp. 105–31.

**Gauthier 1981.** M.-M. S. Gauthier. "Un Style et ses lieux: Les ornements métalliques de la Bible de Souvigny." *Gesta* 20, no. 1 (1981), pp. 141–53.

**Gauthier 1982.** M.-M. S. Gauthier. "Reliquaires du XIIIe siècle entre le proche Orient et l'Occident Latin." In *Il Medio oriente e l'occidente nell'arte del XIII secolo,* ed. Hans Belting, pp. 55–69. Bologna, 1982.

**Gauthier 1983.** M.-M. S. Gauthier. *Les Routes de la Foi: Reliques et reliquaires de Jérusalem à Compostelle.* Fribourg, 1983.

**Gauthier 1987.** M.-M. S. Gauthier, with Geneviève François. *Émaux méridionaux: Catalogue international de l'oeuvre de Limoges.* Vol. 1, *L'Époque romane.* Paris, 1987.

**Gauthier 1988.** M.-M. S. Gauthier. "Naissance du défunt à la vie éternelle: Les tombeaux d'émaux de Limoges au XIIe et XIIIe siècle." In *La Figuration des morts dans la chrétienté médiévale jusqu'à la fin du premier quart du XIVe siècle: Colloque, 26–28 mai 1988, Abbaye royale de Fontevraud,* ed. Roger Grégoire, pp. 97–116. N. p., 1989.

**Gauthier 1990.** M.-M. S. Gauthier. "L'Atelier d'orfèvrerie de Silos à l'époque romane." In *El Romanico en Silos: IX centenario de la consagracion de la iglesia y claustro, 1088–1988,* pp. 377–95. Silos, 1990.

**Gauthier 1992.** M.-M. S. Gauthier. "Reflets de la spiritualité grandmontaine et sources de l'histoire médiévale au miroir des émaux méridionaux." In *L'Ordre de Grandmont, art et histoire: Actes des Journées d'Études de Monpellier 7 et 8 octobre 1989,* ed. Geneviève Durand and Jean Nougaret, pp. 91–105. Monpellier, 1992.

**Gauthier 1993a.** M.-M. S. Gauthier. "Le Décor des églises en France méridionale (XIIIe–mi XVe siècle)." *Cahiers de Fanjeaux* 28 (1993), pp. 87–137.

**Gauthier 1993b.** M.-M. S. Gauthier. "L'Image de la mère de Dieu sise à La Sauvetat: Statue de l'oeuvre de Limoges offerte en 1319 par Odon de Montaigu à la commanderie de Saint-Jean de Jérusalem au diocèse de Clermont." *BSAHL* 121 (1963), pp. 121–36.

**Gauthier and François 1981.** M.-M. S. Gauthier and Geneviève François. *Medieval Enamels: Masterpieces from the Keir Collection.* Ed. and trans. Neil Stratford. Exh. cat., The British Museum. London, 1981.

**Gauthier, Perrier, and Blanchon 1961.** M.-M. S. Gauthier, J. Perrier, and D. Blanchon. "Fouilles sur l'emplacement de l'abbaye de Saint-Martial." *BSAHL* 88 (1961), pp. 49–71.

**Gauthier forthcoming.** *See L'École de Limoges* and *L'Oeuvre de Limoges,* vols. 2 and 3 of *Émaux méridionaux*

**Gautier 1858.** T. Gautier. "La Cassette de saint Louis." *Le Moniteur,* July 7, 1858, p. 855.

**Gay 1887–1928.** Victor Gay. *Glossaire archéologique du Moyen Âge et de la Renaissance.* 2 vols. Paris, 1887–1928. Vol. 2 (1928) completed by Henri Stein.

**Gay 1909.** "La Collection Gay (Paris) aux Musées nationaux." *Gazette des Beaux-Arts* (1909), p. 20

**Gay Sale 1909.** *Catalogue des objets d'art et de haute curiosité du Moyen-Âge et de la Renaissance . . . provenant de l'ancienne collection de M. V. G[ay] . . . vente . . . hôtel Drouot 23–26 mars.* Paris, 1909.

**Geoffrey de Vigeois, *Chronique.*** Pierre Botineau, ed. *Geoffrey de Vigeois: Chronique.* Thesis, École des Chartes, 1964. Ed. and trans. by Botineau and B. Barrière forthcoming.

**Geoffrey de Vigeois.** *See* Chronique de Geoffroy 1864 (trans. Bonnélye)

**Geoffrey de Vigeois** *See* Labbé 1657 (editor)

**George and Yapp 1991.** Wilma B. George and

Brunsdon Yapp. *The Naming of the Beasts: Natural History in the Medieval Bestiary.* London, 1991.

**Gérault 1838.** Gérault, curé d'Evron. *Notice historique sur Evron, son abbaye et ses monuments.* Laval, 1838.

*Gesta consulum Andeguvorum* **1913.** *Gesta consulum Andeguvorum* (French title: *Gestes des comtes d'Anjou/Chroniques des comtes d'Angou*). *See* Halphen and Poupardin 1913

**Gibson and Wright 1988.** *Joseph Mayer of Liverpool, 1803–1886.* Ed. Margaret Gibson and Susan M. Wright. Occasional Papers (Society of Antiquaries of London): New Series, 11. London, 1988.

**Giradot 1859.** A. de Giradot. "Histoire et inventaire du trésor de la cathédrale de Bourges." *Mémoires de la Société Impériale des Antiquaires de France,* ser. 3, 24 (1859), pp. 192–272.

**Giradot and Durand 1849.** A. de Giradot and Hippolyte Durand. *La Cathédrale de Bourges.* Moulins, 1849.

**Giraud 1881.** Jean-Baptiste Giraud. *Les Arts du métal: Recueil descriptif et raisonné . . . des objets ayant figuré à l'Exposition de l'Union Centrale des Beaux-Arts.* Paris 1881.

**Giraud 1887 and 1897.** J.-B. Giraud. *Catalogue sommaire des musées de la ville de Lyon.* Lyon, 1887. 2d ed. Lyon, 1897.

*Glory of the Confessors* **1988.** Gregory of Tours. *Glory of the Confessors.* Trans. Raymond Van Dam. Liverpool, 1988. For Latin text, see *Gregorii Turonensis Opera.*

*Glory of the Martyrs* **1974.** Gregory of Tours. *Glory of the Martyrs.* Trans. Raymond Van Dam. Liverpool, 1988.

**Godard-Faultrier 1884.** V. Godard-Faultrier. *Inventaire du musée d'antiquités Saint-Jean et Toussaint.* Angers, 1884.

*Golden Legend* **1969.** *The Golden Legend of Jacobus de Voragine.* Ed. and trans. Granger Ryan and Helmut Ripperger. New York, 1941. Reprint. New York, 1969.

**Goldschmidt 1914.** A. Goldschmidt. *Die Elfenbeinskulpturen aus der Zeit der Karolingischen und Sächsischen Kaiser VIII–XI Jahrhundert.* Vol. 1. Berlin, 1914. New edition Berlin–Oxford, 1969.

**Goldschmidt and Weitzmann 1979.** A. Goldschmidt and K. Weitzmann. *Die Byzantinischen Elfenbein Skulpturen des X–XIII Jahrhunderts.* Vol. 1. 2d ed. Berlin 1979.

**Gómez-Moreno 1968.** *See* New York 1968

**Gomot 1872.** H. Gomot. *Histoire de l'abbaye royale de Mozat.* Paris, 1972.

**Gondelon 1886.** L. Gondelon. *L'Église de Mozac.* Riom, 1886.

**Goñi Gaztambide 1979.** José Goñi Gaztambide. *Historia de los obispos de Pamplona.* Vol. 1, Siglos IV–XIII. 10 vols. Pamplona, 1979–.

**Gonzague 1875.** F. L. de Gonzague. "Découverte de la date authentique de la châsse émaillée de Conques." *Revue religieuse de Rodez et de Mende,* June 18, 1875, p. 296.

**Grabar 1958.** André Grabar. *Ampoules de Terre Sainte (Monza, Bobbio).* Paris, 1958.

**Grabar 1968.** A. Grabar. *Christian Iconography: A Study of Its Origins.* Princeton, 1968.

**Graham and Clapham 1926.** Rose Graham and A. W. Clapham. "The Order of Grandmont and Its Houses in England." *Archaeologia* 25 (1926), pp. 159–210.

**Granboulan 1994.** A. Granboulan. "De la paroisse à la cathédrale: Une approche renouvelée du vitrail roman dans l'Ouest." *Revue de l'Art* 103 (1994), pp. 43–52.

**Gratuze and Barrandon 1990.** B. Gratuze and J.-N. Barrandon. "Islamic Glass Weights and Stamps: Analysis Using Nuclear Techniques." *Archaeometry* 32 (1990), pp. 155–62.

**Grégoire 1882.** P.-M. Grégoire. *État du diocèse de Nantes en 1790.* Nantes, 1882.

*Gregorii Turonensis Opera* **1885.** [Gregory of Tours] *Gregorii Turonensis Opera.* Ed. Wilhelm Arndt and Bruno Krusch. 2 vols. Monumenta Germaniae historica. Scriptorum rerum merovingicarum, 1. Hannover, 1884–85.

**Gregory 1980.** Cedric E. Gregory. *A Concise History of Mining.* New York, 1980.

**Gregory of Tours.** *See* individual titles (*Glory of the Confessors; Glory of the Martyrs; History of the Franks; Vita S. Aridi*)

**Gregory of Tours 1885.** *Miracula et opera minora [De gloria martyrum, De gloria confessorum].* Ed. Max Bonnet and B. Krusch. Vol. 1, part 2 of *Gregorii Turonensis Opera* 1885.

**Gregory of Tours 1963–65.** [Gregory of Tours.] *Histoire des Francs (Historia Francorum).* Trans. Robert Latouche. 2 vols. Paris, 1963–65.

**Grésy 1854.** Eugène Grésy. "Cassette de saint Louis dans l'église de Dammarie (Seine et Marne)." *Revue Archéologique* 10, part 2 (October 1853–March 1854), pp. 637–42.

**Grésy 1857.** E. Grésy. "Notice sur l'abbaye de Preuilly (Seine-et-Marne)." *Mémoires de la Société impériale des Antiquaires de France* 23 (1857), pp. 373–81.

**Grimouard de Saint-Laurent 1869.** Grimouard de Saint-Laurent. "Iconographie de la croix et du crucifix." *Annales archéologiques* 26 (1869), pp. 357–79.

**Grodecki 1959.** Louis Grodecki. *Les Vitraux de la Sainte-Chapelle et de Notre-Dame de Paris.* Paris, 1959.

**Grodecki 1977.** L. Grodecki, with Catherine Brisac and Claudine Lautier. *Le Vitrail roman.* Fribourg, 1977.

**Grodecki and Brisac 1984.** L. Grodecki and C. Brisac. *Le Vitrail gothique au XIIIe siècle.* Fribourg–Paris, 1984. Translated as *Gothic Stained Glass 1200–1300.* Ithaca, N.Y., 1985.

**Grosse-Dupéron and Gouvrion 1896.** A. Grosse-Dupéron and E. Gouvrion. *L'Abbaye de Fontaine-Daniel: Étude historique.* Mayenne, 1896.

**Guelon 1882.** Abbé P.-F. Guelon 1882. *Histoire de La Sauvetat-Rossille, chef-lieu d'une commanderie de Saint-Jean de Jérusalem en Auvergne.* Clermont-Ferrand, 1882.

**Guennol Collection 1975.** *See* Rubin 1975

**Guérard 1850.** B. Guérard. *Collection des cartulaires de France: Cartulaire de l'église Notre-Dame de Paris.* Vol. 6. Paris, 1850.

**Guibert n. d.** Louis Guibert. *Les Dessins d'archéologie de Roger de Gaignières.* 6 vols. Paris, n. d.

**Guibert 1876.** L. Guibert. "Destruction de l'ordre et de l'abbaye de Grandmont." *BSAHL* 24 (1876), pp. 344–61 and p. 378.

**Guibert 1877a.** L. Guibert. "Destruction de l'ordre de Grandmont: Appendice." *BSAHL* 25 (1876), pp. 33–384.

**Guibert 1877b.** L. Guibert. "Le Monastère de Balezis." *Almanach limousin* 19 (1877), pp. 1–10.

**Guibert 1877c.** L. Guibert. *Une Page de l'histoire du clergé français au XVIIIe siècle: Destruction de l'ordre de l'abbaye de Grandmont.* Paris, 1877.

**Guibert 1879.** L. Guibert. "Tombeau de fabrication limousine à Foucarmont." *BSAHL* 27 (1879 [1880]), p. 196.

**Guibert 1885.** L. Guibert. "L'Orfèvrerie et les orfèvres de Limoges." *BSAHL* 32 (1885), pp. 35–116.

**Guibert 1886.** L. Guibert. "L'Orfèvrerie limousine et les émaux d'orfèvrerie à l'Exposition rétrospective de Limoges." *BSAHL* 33 (1886), pp. 179–236.

**Guibert 1887 and 1988.** L. Guibert. "La Société archéologique de Limoges à l'Exposition artistique de Tulle." Bulletin de la Société des letters, sciences, arts de la Corrèze (1887), pp. 409–37; reprinted in *BSAHL* 35 (1888), pp. 595–622.

**Guibert 1888a.** L. Guibert. "L'École monastique d'orfèvrerie de Grandmont et l'autel majeur de l'église abbatiale. Notice accompagnée des deux inventaires les plus anciens du trésor (1496–1515)." *BSAHL* 36 (1888), pp. 51–98. Also published, under the same title, in book form in Paris, 1888.

**Guibert 1888b.** Guibert 1886 reprinted in book form, under the same title, in Limoges, 1888.

**Guibert 1889.** L. Guibert. "Monuments historiques. Rapport de la commission chargée d'examiner à nouveau la liste des Monuments historiques et de dresser la nomenclature des objets mobiliers auxquels il y a lieu d'appliquer les articles 8 à 13 de la loi du 30 mars 1887." *BSAHL* 36 (1889), pp. 458–83.

**Guibert 1901.** L. Guibert. "Les Vieux émaux de Limoges à l'Exposition de 1900." *Bulletin de la Société des Lettres, Sciences et Arts de la Corrèze* 23 (1901), pp. 5–40.

**Guibert and Tixier 1886.** L. Guibert and Jules Tixier. *L'Art rétrospectif à l'Exposition de Limoges.* Limoges, 1886. 2d ed., 1889.

**Guilhermy 1848.** F. de Guilhermy. *Monographie de l'église royale de Saint-Denis, tombeaux et figures historiques.* Paris, 1848.

**Guilhermy 1875.** F. de Guilhermy. *Inscriptions de la France du Ve au XVIIIe siècle.* Vol. 2, *Ancien diocèse de Paris.* Paris, 1875.

**Guillaume de Saint-Pathus 1931.** Guillaume de Saint-Pathus. *Les Miracles de Saint Louis.* Ed. Percival B. Fay. Paris, 1931.

**Guillotin de Corson 1884.** A. Guillotin de Corson. *Pouillé historique de l'archevêché de Rennes.* Vol. 5. Rennes, 1884.

**Guillotin de Corson 1887.** A. Guillotin de Corson. "Les Seigneuries de la comtesse de Maure dans la Haute-Bretagne en 1623." *Revue historique de l'Ouest* 3 (1887), pp. 165–87, 307–14, 444–57, and 557–85.

**Gunner 1855.** W. H. Gunner. "Antiquities and Works of Art Exhibited." *The Archaeological Journal* 12 (1855), pp. 182–89.

**Gustavs 1993.** S. Gustavs. "Work Debris and Tools of a Germanic Silver/Bronzesmith and Glassworker from Klein Köris, southeast of Berlin." In *Outils et ateliers d'orfèvres des temps anciens,* ed. Christiane Eluère, pp. 197–202. Paris, 1993.

**Guyard de la Fosse 1850.** J.-B. Guyard de la Fosse. *Histoire des seigneurs de Mayenne et de ce qui s'est passé de plus considerable en cette ville.* Le Mans, 1850.

**Hahnloser and Brugger-Koch 1985.** H. Hahnloser and S. Brugger-Koch. *Corpus der Hartsteinschliffe des 12.–15. Jahrhunderts.* Berlin, 1985.

**Halphen and Poupardin 1913.** Louis Halphen and René Poupardin. *Chroniques des comtes d'Anjou et des seigneurs d'Amboise.* Paris, 1913.

**Harmignies 1983.** Roger Harmignies. "A propos du blason de Geoffroi Plantagenêt." In *Les Origines des armoires: IIe Colloque international d'héraldique, Bressanone–Brixen 5–9.X.1981,* ed. Hervé Pinoteau, Michel Pastoureau, and Michel Popoff, pp. 55–63. Académie internationale d'Héraldique. Paris, 1983.

**Hartford, Conn., 1950.** *See Religious Art* 1950

**Havard 1896.** H. Havard. *Histoire de l'orfèvrerie française.* Paris, 1896.

**Hawthorne and Smith 1976.** *See* Theophilus 1976

*Die heiligen drei Könige* **1982.** *See* Cologne 1982.

**Helleputte 1884.** G. Helleputte. "Matériaux pour servir à l'histoire des vases aux saintes huiles." *Revue de l'Art chrétien* 34 (April 1884), pp. 146–153.

**Henderson 1985.** J. Henderson. "The Raw Materials of Early Glass Production." *Oxford Journal of Archaeology* 4 (1985), pp. 267–91.

**Henderson 1991.** J. Henderson. "Chemical Characterization of Roman Glass Vessels, Enamels and Tesserae." In *Materials Issues in Art and Archeology II: Proceedings of the Materials Research Society Symposium,* ed. Pamela B. Vandiver et al., vol. 185, pp. 601–07. Pittsburgh, 1991.

**Héron 1965.** A. Héron. "Découverte de peintures murales à la chapelle Ste-Radegonde de Chinon." *Bulletin de Amis du vieux Chinon* 6, no. 9 (1965), pp. 481–88.

**Herrgott 1726.** Marquard Herrgott. *Vetus disciplina monastica, seu collecto auctorum Ordinis S. Benedicti maximam partem ineditorum . . .* Paris, 1726.

459

**Heyne 1922.** Hildegard Heyne. *Das Gleichnis von den Klugen und törichten Jungfrauen: Ein literarisch- ikonographische Studie zur altchristlichen zeit.* Leipzig, 1922.

**Highfield 1954.** J.R.L. Highfield, ed. *The Early Rolls of Merton College Oxford.* Oxford, 1954.

**Hildburgh 1920.** Walter L. Hildburgh, discussion in *Proceedings of the Society of Antiquaries,* March 18, 1920, pp. 129–40.

**Hildburgh 1936.** W. L. Hildburgh. *Medieval Spanish Enamels and Their Relation to the Origin and the Development of Copper Champleve Enamels of the Twelfth and the Thirteenth Centuries.* Oxford, 1936.

**Hildburgh 1955.** W. L. Hildburgh. "Medieval Copper Champlevé Enamelled Images of the Virgin and Child." *Archeologia* 96 (1955), pp. 116–58.

**Hildesheim 1993.** *Bernward von Hildesheim und das Zeitalter der Ottonen.* Exh., Dom und Diözesan Museum. Hildesheim, 1993.

**Hippolytus of Rome 1968.** Hippolyte de Rome. *La Tradition apostolique.* Ed. B. Botte. Paris 1968.

*Histoire des Francs 1963–65. See* Gregory of Tours 1963–65

*Historia Gaufredi ducis Normannorum. See* Marmoutier 1913

*History of the Franks 1974.* Gregory of Tours. *History of the Franks.* Trans. Lewis Thorpe. London, 1974. For the French translation, see *Histoire des Francs.*

**Hoberg 1944.** Hermann Hoberg, ed. *Die Inventare des päpstlichen Schatzes in Avignon 1314–1376.* Vatican City, 1944.

**Hoos 1988.** H. Hoos. "Kerzenleuchter aus acht Jahrhunderten." *Weltkunst* 58 (1988), pp. 46–47.

**Hospital 1979.** Françoise Hospital. "Les Inscriptions sur les croix dans l'oeuvre de Limoges." *Actes du 102e Congrès national des Sociétés savantes, Limoges, 1977: Section d'archéologie et d'histoire de l'art,* pp. 21–35. Paris, 1979.

**Hreglich and Verità 1986.** S. Hreglich and M. Verità. "Applications of X-ray Microanalysis to the Study and Conservation of Ancient Glasses." *Scanning Electron Microscopy* 1986, part 2, pp. 485–90.

**Hucher 1856.** E. Hucher. *Études sur l'histoire et les monuments du département de la Sarthe.* Le Mans–Paris, 1856.

**Hucher 1860.** E. Hucher. "L'Émail de Geoffroy Plantagenêt au musée du Mans." *Bulletin monumental* 7 (1871), pp. 669–93.

**Hucher 1878a.** E. Hucher. *L'Émail de Geoffroy Plantagenêt au musée du Mans reproduit en photochromie par le procédé Vidal et accompagné d'une dissertation sur l'origine et le but de cet émail.* Paris–Tours–Le Mans, [1878].

**Hucher 1878b.** E. Hucher. "Visite du Muséum à la préfecture du Mans." *Congrès archéologique de France, session XLVe, Le Mans,* 1878, pp. 46–68. Paris–Tours, 1879.

**Hugon 1931.** H. Hugon. "L'Épitaphe du cardinal de la Chapelle-Taillefer et les I. P. lemovici fratres." *BSAHL* 73 (1931), pp. 471–78.

**Huici and Juaristi 1929.** S. Huici and W. Juaristi. *El Santuario de San Miguel de Excelsis, Navarral y su retable esmaltado . . .* Madrid, 1929.

**Hunter 1958.** Sam Hunter. "A Plaque from Limoges." *The Minneapolis Institute of Arts Bulletin* 47, no. 3 (1958), pp. 29–33.

**Hutchison 1989.** Carole Hutchison. *The Hermit Monks of Grandmont.* Kalamazoo, Mich., 1989.

**Huyghe 1929.** R. Huyghe. "La Donation Martin le Roy." *Beaux-Arts,* no. 7 (1929), pp. 3–5.

*Icones. See* Philostratus

*Imagines 1614. Les Images ou tableavx de platte peinture des deux Philostrates sophistes grecs et les statues de Callistrates . . .* Trans. Blaise de Vigenère. Paris, 1614. Reprinted in Meyer 1895.

**Jacques de Voragine 1967.** Jacques de Voragine. *La Légende dorée.* Ed. and trans. J.-B. Roze, with H. Savon. 2 vols. Paris, 1967.

**Jalabert 1954.** Denise Jalabert. "Fleurs peintes à la voûte de la chapelle du Petit-Quevilly (Seine-Inférieure)." *Gazettes des Beaux-Arts* 43 (1954), pp. 5–26.

**Javakhishvili and Abrahamishvili 1986.** A. Javakhishvili and G. Abrahamishvili. *Jewellery and Metalwork in the Museums of Georgia.* Leningrad, 1986.

**Jeoufre 1859.** M. Jeoufre, curé de Saint-Viance. *La Vie miraculeuse de saint Vincentian . . . appelé vulgairement S. Viance.* Saint-Flour, 1859.

**Jeulin 1925–26.** P. Jeulin. "L'Ancien couvent des Cordeliers de Nantes de 1791 à 1925 (étude archéologique)." *Bulletin de la Société archéologique et historique de Nantes et de la Loire-Inférieure* 65 (1925), pp. 195–215, and 66 (1926), pp. 133–45.

**Johnson 1938–39.** Rozelle P. Johnson. *Compositiones variae, from Codex 490, Biblioteca Capitolare, Lucca, Italy; An Introductory Study.* Illinois Studies in Language and Literature, 23. Urbana, 1938–39.

*Joseph Mayer of Liverpool 1988. See* Gibson and Wright 1988

**Jottrand 1965.** M. Jottrand. "Les Émaux du trésor de la cathédrale de Troyes décoraient-ils les tombeaux des comtes de Champagne?" *Gazette des Beaux-Arts* 65 (1965), pp. 257–64.

**Jougla de Morenas 1939.** Henri Jougla de Morenas. *Grand Armorial de France, catalogue général des armoiries des familles nobles de France . . .* 6 vols. Paris, 1934–39.

**Jouhaud 1949.** L. Jouhaud. "Les Châsses de pacotille." *BSAHL* 83, no. 1 (1949), pp. 48–75.

**Jülich 1986–87.** Theo Jülich. "Gemmenkreuze: Die Farbigkeit ihres Edelsteinbesatzes bis zum 12. Jahrhundert." *Aachener Kunstblätter* 54–55 (1986–87), pp. 99–258.

**Jungmann 1954.** Josef-Andreas Jungmann. *Missarum sollemnia: Explication génétique de la Messe romaine.* 2 vols. Paris, [1964].

**Kauffmann 1975.** Claus M. Kauffmann. *Romanesque Manuscripts 1066–1190.* London and Boston, 1975.

**King 1965.** Archdate A. King. *Eucharistic Reservation in the Western Church.* New York, 1965.

**Kitzinger 1977.** Ernst Kitzinger. *Byzantine Art in the Making: Main Lines of Stylistic Development in Mediterranean Art, 3rd–7th century.* Cambridge, Mass., 1977.

**Köhler 1930.** W. Köhler. *Die karolingischen Miniaturen.* Vol. 1, *Die Schule von Tours.* Berlin, 1930.

**Kolias 1988.** T.-G. Kolias. *Byzantinische Waffen: Ein Beitrag zur Byzantinischen Waffenkunde von den Anfängen bis zur Lateinische Eroberung.* Vienna, 1988.

**Kondakov 1891.** N. Kondakov. *Oukazatel otdelenya srednikh vehoui epokhi Vozrozhdeniya* (Guide to the Department of Medieval and Renaissance Art). The State Hermitage Museum, St. Petersburg, 1891.

**Kovács 1964a.** Eva Kovács. "Le chef de Saint Maurice à la cathédrale de Vienne (France)." *Cahiers de civilisation médiévale* 7 (1964), pp. 19–26.

**Kovács 1964b.** E. Kovács. *Kopfreliquiäre des Mittelalters.* Budapest, 1964.

**Kovács 1968.** Eva Kovács. *Limosiner Email in Ungarn (Aus dem Ungarischen von Férénc Gottschlig.)* Budapest, 1968.

**Krautheimer 1980.** Richard Krautheimer. *Rome: Profile of a City, 312–1308.* Princeton, N. J., 1980.

**Krems-an-der-Donau 1964.** *Ausstellung Romanische Kunst in Österreich, veranstaltet von der Stadtgemeinde Krems-an-der-Donau, 21. Mai bis 25. Oktober 1964, Minoritenkirche Krems-Stein, Niederösterreich.* Exh. cat., Krems-an-der-Donau, 1964.

**Krysanovskaïa 1986.** M. Krysanovskaïa et al. *Decorative Arts in the Hermitage.* Leningrad, 1986.

**Krysanovskaïa 1990.** M. Krysanovskaïa. "Alexander Petrovich Basilewsky: A Great Collector of Medieval and Renaissance Works of Art." *Journal of the History of the Collections* 1, no. 2 (1990), pp. 143–55.

*Kunstdenkmäler 1857–80. See* Weerth 1857

**Labarte 1856.** J. Labarte. *Recherches sur la peinture en émail dans l'Antiquité et au Moyen-Âge.* Paris, 1856.

**Labarte 1865.** J. Labarte. *Histoire des Arts industriels au Moyen Âge et à l'époque de la Renaissance.* 4 vols., and *Album,* 2 vols. Paris, 1864–65. Vol. 3 (1865).

**Labbé 1657.** Philippe Labbé. *Nouae bibliothecae manuscriptorum librorum. Tomus primus: Historias, chronica, sanctorum, sanctarumque vitas, translationes, miracula, stemmata genealogica, ac familia antiquitatis, praesertim Franciae, monumenta . . . tomus secundus: Rerum Aquitanicarum, praesertim Bituricensium, vberrima collectio . . .* 2 vols. Paris, 1657.

**A. Laborde 1911–27.** Alexandre de Laborde (comte). *La Bible moralisée conservée à Oxford, Paris, et Londres: Reproduction intégrale du manuscrit du XIIIe siècle, accompagnée d'une notice par le comte A. de Laborde.* 5 vols. Paris, 1911–27.

**L. Laborde 1852 and 1853.** Léon de Laborde (marquis). *Notice des émaux exposés dans les galeries du Musée du Louvre.* Part 1: *Histoire et descriptions;* part 2: *Notice des émaux, bijoux et objets divers exposés dans les galeries du Musée du Louvre.* 2 vols. Paris, 1852–53.

**Laborderie 1946.** A. de Laborderie. *Quarante-six églises limousines.* Limoges, 1946.

**Lacrocq 1933.** L. Lacrocq. "Note" for the "Procès-verbeaux des séances du 28 juin 1933." *BSAHL* 74 (1933), p. xxxix.

**Lacroix 1988.** P. Lacroix. "La Double Énigme du tombeau du cardinal Pierre de la Chapelle-Taillefer." *Mémoires de la Société des sciences naturelles et archéologiques de la Creuse* 43 (1988), pp. 363–71.

**Lacroix 1993.** P. Lacroix. "Le Tombeau de Pierre de la Chapelle." *Les Annales de la Chapelle-Taillefert* 2, fasc. 1 (1993), pp. 5–97.

**Laffitte 1991.** M.-P. Laffitte. *Bibliothèque nationale. Reliures précieuses.* Paris, 1991.

**Lafontaine-Dosogne 1975.** J. Lafontaine-Dosogne. "L'Influence byzantine sur les émaux du Moyen Âge occidental." *Revue belge d'archéologie et d'histoire* 44 (1975), pp. 166–69.

**Lambert 1937.** A. Lambert. "Boniface." In *Dictionnaire d'histoire et de géographie ecclésiastique* 9 (1937), cols. 948–49.

**Landais 1961a.** Hubert Landais. "La Donation Côte au musée du Louvre." *La Revue du Louvre et des musées de France* 3 (1961), pp. 128–34.

**Landais 1961b.** H. Landais. "La Donation Mège." *La Revue du Louvre et des musées de France* 3 (1961), pp. 12–16.

**Landais 1976.** H. Landais. "Contribution à l'étude des origines de l'émaillerie limousine." *Fondation Eugène Piot, Monuments et Mémoires* 60 (1976), pp. 113–31.

**Landes 1983.** Richard Landes. "A Libellus from St. Martial of Limoges Written in the Time of Adémar de Chabannes (989–1034). Un faux à retardement," *Scriptorium* 37, no. 2 (1983), pp. 178–204.

**Landes 1994.** R. Landes. "Autour d'Adémar de Chabannes (1034)." *BSAHL* 122 (1994), pp. 23–54.

**Landes and Paupert 1991.** R. Landes and Catherine Paupert. *Naissance d'apôtre: La Vie de saint Martial de Limoges.* Brepols, 1991.

**La Nicollière 1859–61.** S. de La Nicollière. "Une Pierre tombale de l'abbaye de Villeneuve: Olivier de Machecoul, XIIIe siècle." *Bulletin de la Société archéologique de Nantes et du département de la Loire-Inférieure* 1 (1859–61), pp. 259–75.

**Lanmon and Whitehouse 1993.** D. Lanmon and D. Whitehouse. *The Robert Lehman Collection.* Vol. 11, *Glass,* pp. 250–54. New York, 1993.

**Lapauze 1925.** Henry Lapauze. *Catalogue sommaire des collections Dutuit: Notice historique sur les frères Dutuit.* Paris, 1925.

**Lapkowskaya 1971.** E. A. Lapkowskaya. *Applied Art of the Middle Ages in the Collection of the State Hermitage, Artistic Metalwork.* Moscow, 1971. Summaries in English, French, and German.

**Laporte 1988a.** J.-P. Laporte. "Le Tissu aux faisans nimbés de Jouarre: Description et étude stylistique." *Bulletin du Centre international d'étude des textiles anciens,* no. 66 (1988), pp. 15–25.

**Laporte 1988b.** J.-P. Laporte. *Le Trésor des saints de Chelles.* Chelles, 1988.

**Larigaldie 1909.** M. G. Larigaldie. *Chevalier et moine ou Jean de Montmirail connétable de France 1165–1217.* Paris, 1909.

**Larigauderie-Beigeaud.** M. Larigauderie-Beigeaud. *Recherches sur les prieurés grandmontains de Charente. Architecture et Histoire, XIIe–XVIIe siècles.* Ed. R. Favreau. Poitiers, 1994.

**Lasko 1964.** Peter Lasko. "A Notable Private Collection." *Apollo,* June 1964, pp. 464–73.

**Lasko 1972.** P. Lasko. *Ars Sacra, 800–1200.* Hammondsworth, England, 1972.

**C. de Lasteyrie 1901.** Charles de Lasteyrie. *L'Abbaye de Saint-Martial de Limoges: Étude historique, économique et archéologique.* Paris, 1901.

**F. de Lasteyrie 1859.** Ferdinand de Lasteyrie. *Notice sur la châsse de Saint Viance.* Brive, 1859.

**F. de Lasteyrie 1865.** F. de Lasteyrie. "Observations critiques sur le trésor de Conques et sur la description qu'en a donnée M. Darcel." *Mémoires de la Société impériale des Antiquaires de France* 28 (1865), pp. 48–68.

**R. de Lasteyrie 1884.** R. de Lasteyrie. [crois en cuivre émaillé de travail limousin.] *Bulletin de la Société nationale des Antiquaires de France,* 1884, pp. 192–95.

**Lauer 1908.** Philippe Lauer. "L'Abbaye de Royaumont." *Bulletin monumental* 72 (1908), pp. 215–68.

**Lavedan and Hugueney 1969.** P. Lavedan and J. Hugueney. "Bastides de l'Agenais: Villeneuve-sur-Lot-Vianne." *Congrès archéologique de France, session CXXVII, Agenais, 1969,* pp. 9–41. Paris, 1972.

**Le Baud 1922.** Pierre Le Baud. *Cronicques et ystoires des Bretons . . . Publié d'après le première rédaction inédite . . .* Ed. Charles de La Lande de Calan. Vol. 4 (1922). Rennes, 1907–.

**Leblanc 1892.** E. Leblanc. *L'Abbaye de Fontaine-Daniel, sa fondation et ses derniers jours.* Mayenne, 1892.

**Lebrun-Desmarettes 1718.** J.-B. Lebrun-Desmarettes. *Voyages liturgiques.* Paris, 1718.

**Lecler 1891.** André Lecler. "Les grandes châsses de Grandmont." *BSAHL* 38 (1891), pp. 173–75.

**Lecler 1901.** A. Lecler. *Histoire de Grandmont.* Paris, 1901.

**Lecler 1903.** A. Lecler. "Pouillé historique du diocèse de Limoges, manuscrit de l'abbé Joseph Nadaud, 1775." *BSAHL* 53 (1903), pp. 1–841.

**Lecler 1907–9.** Le chanoine A. Lecler. "Histoire de l'abbaye de Grandmont." *BSAHL* 57 (1907–8), pp. 129–71, 413–78; 58 (1908–9), pp. 44–94, 431–97; 59 (1909), pp. 14–66, 366–404. Also published in book form in Limoges, 1909.

**Lecler 1908.** A. Lecler. "La Châsse d'Ambazac." *BSAHL* 57, no. 2 (1908), pp. 565–79.

**Lecler 1910–11.** A. Lecler. "Histoire de l'abbaye de Grandmont." *BSAHL* 60 (1910–11), pp. 86–162 and 371–452.

**Lecler 1920.** A. Lecler. *Dictionnaire historique et géographique de la Haute-Vienne.* Limoges, 1920–26. Reprinted Marseille, 1976.

**Le Corvaisier de Courteilles 1648.** A. Le Corvaisier de Courteilles. *Histoire des évêques du Mans et de ce qui s'est passé de plus mémorable dans le diocèse pendant leur pontificat.* Paris, 1648.

**Ledru 1903.** A. Ledru. "Plainctes et doléances du chapitre du Mans en 1562." *Archives historiques du Maine* 3, no. 2 (1903), pp. 169–257.

**Le Feuvre and Alexandre 1932.** A. Le Feuvre and A. Alexandre. *Musée du Mans (Hôtel de Tessé). Catalogue du musée des arts. Peinture. Sculpture. Dessins. Gravures. Objets d'art.* Le Man, 1932.

**Lefèvre-Pontalis 1903.** E. Lefèvre-Pontalis. "L'Église abbatiale d'Evron." *Revue historique et archéologique du Maine* 54 (1903), pp. 5–43.

**Lefèvre-Pontalis 1912.** E. Lefèvre-Pontalis. "Septième excursion: Abbaye de Longpont." *Congrès archéologique de France, LXXVIIIe session tenue à Reims en 1911,* vol. 1, pp. 410–27. Paris and Caen, 1912.

*Legende dorée 1993. See Paris 1993*

**Le Goff 1964.** Jacques Le Goff. *La Civilisation de l'occident médiéval.* Paris, 1964.

**Le Goff 1968.** J. Le Goff. "Apostolat mendiant et fait urbain dans la France médiévale." *Annales, économies, sociétés, civilisations,* 1968, pp. 335–52.

**Leguina 1909.** Enrique de Leguina [D. Enrique de Leguina y Vidal]. *Esmaltes españoles: Los frontales de Orense, San Miguel, "in Excelsis," Silos y Burgos. Apuntes reunidos por D. Enrique de Leguina . . .* Madrid, 1909.

**Leguina 1912.** E. de Leguina. "La Estatua del Obispo D. Mauricio en la Catedral de Burgos." *Arte español,* 1912, pp. 11–37. Madrid.

**Lehmann-Brockhaus 1955-60.** Otto Lehmann-Brockhaus. *Lateinische Schriftquellen zur Kunst in England, Wales und Schottland, vom Jahre 901 bis zum Jahre 1307.* 5 vols. Munich, 1955–60.

**Lemaître 1989.** J.-L. Lemaître. *Mourir à Saint-Martial: La commération des morts et les obituaires à Saint-Martial de Limoges du XIe au XIIIe siècle.* Paris, 1989.

**Le Mans 1880.** *Exposition des Beaux-Arts: Section de l'art rétrospectif. Notice sommaire des objets d'art de la section rètrospective.* Le Mans, 1880.

**Le Mans 1980.** *Les Trésors du patrimoine sarthois de l'antiquité à nos jours.* 2 vols. Exh. cat., Abbaye de l'Epau. Le Mans, 1980.

**Lemasson 1886.** Abbé Lemasson. *Notice historique sur Savigny.* Saint-Lô, 1886.

**Le Méné 1903.** Abbé J.-M. Le Méné. "Abbaye de la Joie." *Bulletin de la Société polymathique du Morbihan* 49 (1903), pp. 123–84.

**Leniaud 1979.** J.-M. Leniaud. "Albert Germeau, un préfet collectionneur sous la monarchie de Juillet." *Actes du 103eme Congrès national des Société savantes, Nancy–Metz, 1978,* vol. 2, pp. 363–72. Paris, 1979.

**Leniaud 1983.** J.-M. Leniaud. "Pour une histoire des vols d'objets d'art: L'affaire et le vol de la châsse d'Ambazac." *Actes du 105e Congrès national des Société savantes, Caen, 1980,* pp. 375–84. Paris, 1983.

**Leningrad–Moscow 1980.** *L'Art décoratif d'Europe occidentale IXe–XVIe siècles d'après les collections du Musée du Louvre et du Musée de Cluny.* Exh., The State Hermitage Museum, Leningrad; Pushkin Museum of Fine Arts, Moscow, 1980–81.

**Lenoir 1821.** A. Lenoir. *Musée des Monuments français ou Description historique et chronologique des statues, bas-reliefs et tombeaux des hommes et des femmes célèbres . . . 8 vols.* Paris, 1821.

**Leone de Castris 1981.** Pierluigi Leone de Castris. "Lo smalto champlevé a Limoges." In *Medioevo e produzione artistica di serie: Smalti e avori gotici in Campania,* ed. Paola Giusti and Pierluigi Leone de Castris, pp. 13–30. Exh. cat., Museo Duca di Martina, Naples. Florence, 1981.

**Le Paige 1777.** A.-R. Le Paige. *Dictionnaire topographique, historique, généalogique et bibliographique de la province et du diocèse du Maine.* Vol. 2. Le Mans–Paris, 1777.

**Lernout et al. 1992.** Fr. Lernout et al. *Le Moyen Âge au Musée de Picardie.* Amiens, 1992.

**Leroux 1911.** A. Leroux. "La Légende de saint Martial dans la littérature et l'art anciens (suite)." *BSAHL* 60, no. 1 (1910) and 60, no. 2, pp. 353–66.

**Le Roy-Pierrefitte.** See Roy-Pierrefitte 1851 and Roy-Pierrefitte 1864

**Levesque 1662.** J. Levesque. *Annales ordinis Grandimontis.* Troyes, 1662.

**Lévesque 1961.** J.-D. Lévesque. *L'Ancien couvent des frères prêcheurs d'Angers.* Angers, 1961.

**Levet 1994.** J. Levet. "L'Église Saint-Aurélien à Limoges." *BSAHL* 122 (1994), pp. 130–54.

**Leydon Papyrus 1926.** See Caley 1926

**Liebgott 1986.** Niels-Knud Liebgott. *Middelalderens emaljekunst.* Copenhagen, 1986.

**Lièvre and Sauzay 1863.** E. Lièvre and A. Sauzay. *Musée impérial du Louvre: Collection Sauvageot.* 2 vols. Paris, 1863.

**Lillich 1986.** M. P. Lillich. "Gothic Heraldry and Name Punning: Secular Iconography on a Box of Limoges Enamel." *Journal of Medieval History* 12, no. 3 (1986), pp. 231–51.

**Lilyquist, Brill, Wypyski, and Koestler 1993.** C. Lilyquist, R. H. Brill, M. T. Wypyski, and R. J. Koestler. "Glass." In *Studies in Early Egyptian Glass,* ed. C. Lilyquist and R. H. Brill, with M. T. Wypyski, pp. 23–58. New York, 1993.

**Limoges 1886.** *L'Art rétrospectif, Exposition de Limoges 8 mai–22 août 1886.* Part 3, *Orfèvrerie, armes, et métaux.* Exh. cat., L'Hôtel de Ville. Limoges, 1886.

**Limoges 1948.** *Émaux limousins, XIIe, XIIIe, XIVe siècles.* Exh. cat., Musée Municipal. Catalogue by M.-M. S. Gauthier. Preface by Pierre Verlet. Limoges, 1948.

**Limoges 1992.** *Trésors d'émail: Catalogue des acquisitions, 1977–1992.* Exh. cat., Musée municipal de l'Evêché. Limoges, 1992.

**Limoges 1995.** *Splendeurs de Saint-Martial de Limoges au temps d'Adémar de Chabannes.* Exh., Musée municipal de l'Evêché. Limoges, 1995.

**Limouzin-Lamothe 1951.** Roger Lamothe-Limouzin. *Le Diocèse de Limoges des origines à la fin du Moyen Âge.* Strasbourg and Paris, 1951.

**Linas 1867 and 1868.** Charles de Linas. *L'Histoire du travail à l'Exposition universelle de 1867.* Paris, 1867; Arras and Lille, 1868.

**Linas 1881.** C. de Linas. *Émaillerie, métallurgie, toreutique, céramique: Les expositions rétrospectives, Bruxelles, Dusseldorf, Paris en 1880.* Paris–Arras, 1881.

**Linas 1883.** C. de Linas. "La Châsse de Gimel (Corrèze) et les anciens monuments de l'émaillerie." *Bulletin de la Société scientifique, historique et archéologique de la Corrèze* 5 (1883), pp. 245–71. Also published in book form, Paris, 1883.

**Linas 1884a.** C. de Linas. "Les Émaux de l'abbaye de Grandmont." *Revue de l'Art chrétien* 34, no. 2 (1884), pp. 341–43.

**Linas 1884b.** C. de Linas. "Une Plaque d'émail limousin et la châsse de saint Étienne, à Grandmont." *Revue de l'Art chrétien* 34 (1884), pp. 164–67.

**Linas 1885a.** C. de Linas. "Une nouvelle forme du nom Alpais." *Bulletin de la Société scientifique, historique et archéologique de la Corrèze* 7 (1885), pp. 112–13.

**Linas 1885b.** C. de Linas. *Oeuvres de Limoges conservées à l'étranger et documents relatifs à l'émaillerie limousine. Lettre à Ernest Rupin.* Paris, 1885. Also published under similiar title in the *Bulletin de la Société scientifique, historique et archéologique de la Corrèze* 7 (1885), pp. 47–130.

**Linas 1885 and 1886.** C. de Linas. "Les Crucifix champlevés polychromes, en plate peinture, et les croix émaillées." *Revue de l'Art chrétien* 35 (1885), pp. 453–78, and 36 (1886), pp. 62–70.

**Linas 1886a.** C. de Linas. "Émaillerie limousine: La croix stationale du musée diocésain de Liège et le décor champlevé à Limoges." *Bulletin de la Société d'art et d'histoire du diocèse de Liège* 4 (1886), pp. 1–35.

**Linas 1886b.** C. de Linas. "Les Émaux limousins de la collection Basilewsky à Saint-Petersbourg. Le Triptyque de la Cathédrale de Chartres." *Bulletin de la Société scientifique, historique et archéologique de la Corrèze* 8, no. 2 (1886), pp. 245–71.

**Linas 1886c.** C. de Linas. *Les Émaux limousins de la collection Basilewsky à Saint-Petersbourg. Le Triptyque de la Cathédrale de Chartres.* Paris, 1886. A reprint in book form of Linas 1886b.

**Linas 1886d.** C. de Linas. "La Tombe en cuivre émaillé du cardinal de Tulle." *Revue de* *l'Art chrétien* 36 (1886), pp. 83–84.

**Linas 1887.** C. de Linas. "Le Reliquaire de Pépin d'Aquitaine au trésor de Conques en Rouegue." *Gazette archéologique* 12 (1887), pp. 37–54 and 291–97.

**Lion-Goldschmidt 1956.** D. Lion-Goldschmidt. *Collection Adolphe Stoclet (première partie): Choix d'oeuvres appartenant à Madame Féron-Stoclet.* Pref. G. Salles. Brussels, 1956.

**Little 1985.** Charles T. Little. "A New Ivory of the Court School of Charlemagne." In *Studien zur mittelalterlichen Kunst 800–1250: Festschrift für Florentine Mütherich zum 70. Geburtstag,* pp. 11–28. Munich 1985.

**Lobineau 1707.** Dom G. A. Lobineau. *Histoire de Bretagne.* 2 vol. Paris, 1707. Reprint. Paris, 1973.

**London 1857a.** John B. Waring. *A Handbook to the Museum of Ornamental Art in the Art Treasures Exhibition [at Manchester] . . . To which is added The Armory by J. R. Planche; being a reprint of critical notices originally published in the "Manchester Guardian."* London, 1857.

**London 1857b.** *Catalogue of the Art Treasures of the United Kingdom collected at Manchester in 1857.* London, [1857].

**London 1862.** *Catalogue of the Special Exhibition of Works of Art of the Mediaeval, Renaissance, and More Recent Periods, on Loan at the South Kensington Museum, June 1862.* Ed. John C. Robinson. Revised ed. London, 1863.

**London 1874.** *See* London 1875

**London 1875.** John H. Pollen, ed.[?] *Catalogue of the Special Loan Exhibition of Enamels on Metal Held at the South Kensington Museum in 1874.* London, 1875.

**London 1897.** John S. Gardner, ed. *Burlington Fine Arts Club, Catalogue of a Collection of European Enamels from the Earliest Date to the End of the XVII Century.* London, 1897.

**London 1932.** *Commemorative Catalogue of the Exhibition of French Art, 1200–1900.* Exh. cat., Royal Academy of Arts. London, 1932.

**London 1962.** *An American University Collection: Works of Art from the Allen Memorial Art Museum, Oberlin, Ohio, under the Patronage of the American Ambassador His Excellency David K. E. Bruce.* Exh. cat., Iveagh Bequest, Kenwood, and London County Council. London, 1962.

**London 1978.** *British Heraldry.* Exh., British Museum. London, 1978.

**London 1981.** *See* Gauthier 1981

**London 1984.** *English Romanesque Art 1066–1200.* Ed. George Zarnecki et al. Exh. cat., Hayward Gallery. London, 1984.

**Longpérier 1842.** A. de Longpérier. "Description de quelques monuments émaillés du Moyen Âge." In *Le Cabinet de l'amateur.* Paris, 1842.

**Longpérier 1865.** A. de Longpérier. "Communication à la séance du vendredi 24 novembre." *Comptes-rendus des séances de l'Académie des inscriptions et belles-lettres,* 1865, p. 96.

**Longpérier-Grimoard 1866.** A. de Longpérier-Grimoard. "Rapport sur les fouilles dans le choeur de l'ancienne abbaye de Chaalis." *Bulletin de la Société nationale des Antiquaires de France,* 1866, p. 96.

**Lopès 1668 and 1882.** H. Lopès. *L'Église métropolitaine et primatiale Saint-André de Bordeaux . . .* 2 vols. Bordeaux, 1882–84. Edited and annotated edition based on a text first edited in 1668.

**Lorenzo 1989.** Josefa Gallego Lorenzo. "San Martin de Tours, San Marcial de Limoges y Santiago en el llamado 'frontal' de la catedral de Orense." In *Los Caminos y el arte: Actas de VI Congreso Español de Historia del Arte (Santiago de Compostela).* Vol. 2, *Caminos y viajes en al arte: Iconografía,* pp. 61–69. Santiago de Compostela, 1989.

**Los Angeles and Chicago 1970.** Vera K. Ostoia, ed. *The Middle Ages: Treasures from The Cloisters and The Metropolitan Museum of Art.* Exh. cat., Los Angeles County Museum and Art Institute of Chicago. Los Angeles and Chicago, 1970.

**Louradour 1977.** A. Louradour. "Reliquaires en Creuse au cours des siècles." *Mémoires de la Société des sciences naturelles et archéologiques de la Creuse* 39 (1977).

**Louvet 1631–35.** P. Louvet. *Histoire et antiquitez du païs de Beauvais.* 2 vols. Beauvais, 1631–35.

**Louvre forthcoming.** Musée du Louvre, Département des Sculptures. *Catalogue sommaire illustré des Scuptures.* Vol. 1, [The Middle Ages].

**Lowery, Savage, and Wilkins 1971.** P. R. Lowery, R. D. A. Savage, and E. L. Wilkins. "Scriber, Graver, Scorper, Tracer: Notes on the Experimenting in Bronze-working Technique." *Proceedings of the Prehistoric Society* 37, pt. 2 (1971), pp. 167–82.

**Luchaire 1873.** A. Luchaire. *Notice sur les origines de la maison d'Albret.* Pau, 1873.

**Lundstrom 1976.** A. Lundstrom. "Bead Making in Scandinavia in the Early Middle Ages." *Early Medieval Studies* 9 (1976), pp. 4–7.

**Mabillon 1707.** J. Mabillon. *Annales ordinis sancti Benedicti: occidentalium monachorum patriarchae.* Vol. 4. Paris, 1707.

**Madec 1975.** Goulven Madec. " 'Panis Angelorum' (selon les Père de l'Église, surtout S. Augustin)." In *Forma futuri: Studi in onore del cardinale Michele Pellegrino,* pp. 818–29. Turin, 1975.

**Madrid 1893.** *Las Joyas de la Exposición historico-europea de Madrid, 1892.* Exh. cat. Madrid, 1893.

**Madrid 1981.** *Catedrales de España.* 3 vols. Madrid, 1981.

**Maillé 1930 and 1939.** *See* Rohan-Chabot 1930 and Rohan-Chabot 1939

**Malbork 1994.** *Rzemiosło artystyczne zachodniej europy od XI do XVI wieku, ze zbiorów państwowego ermitażu w Sankt Petersburgu* (Western European artistic handicrafts from the eleventh to the sixteenth centuries). Exh. cat., Muzeum Zamkowe. Malbork, 1994.

**Mâle 1908.** Émile Mâle. *L'Art religieux de la fin du Moyen Âge en France: Étude sur l'iconographie du Moyen Âge et sur ses sources d'inspiration.* Paris, 1908.

**Mâle 1922.** É. Mâle. *L'Art religieux du XIIe siècle en France: Étude sur les origines de l'iconographie du Moyen Âge.* Paris, 1922.

**Mâle 1978.** E. Mâle. *Religious Art in France: The Twelfth Century: A Study of the Origins of Medieval Iconography.* Ed. Harry Bober. Trans. Marthiel Mathews. Princeton, 1978.

**Malinowsky 1873.** J. Malinowsky. "Découvertes archéologiques à Cahors en 1872." *Bulletin de la Société des études littéraires, scientifiques et archéologiques du Lot* 1 (1873), pp. 49–50.

**Mallay 1838.** A. Mallay. *Essai sur les églises romanes et romano-byzantines du département du Poy-de-Dôme.* Moulins, 1838.

**Mallé 1950 and 1951.** Luigi Mallé. "Antichi smalti cloisonné a champlevés dei secoli XI–XIII in raccolte e musei dei Piemonte." *Bolletino della Societa piemontese di Archeologia e Belle Arti* (1950), pp. 39–79; (1951), pp. 54–136.

**Mallé 1969.** L. Mallé. *Smalti—Avari del Museo d'arti antica.* Turin, 1969.

**Manchester 1857.** *See* London 1857b

**Manchester 1959.** *Romanesque Art, c. 1050–1200, from Collections in Great Britain and Eire.* Exh., City of Manchester Art Gallery. Manchester, 1959.

**Manhes-Deremble and Deremble 1993.** Colette Manhes-Deremble and J.-P. Deremble. *Les Vitraux narratifs de la cathédrale de Chartres: Étude iconographique.* Paris, 1993.

**Marbouty 1886.** Camille Marbouty. "Visite aux émaux anciens et modernes de Limoges." *BSAHL* 33, no. 2 (1886), pp. 77–88.

**Marcheix and Perrier 1969.** M. Marcheix and J. Perrier. *Guide du musée municipal: Collection archéologique.* Limoges, 1969.

**Marin de Carranrais 1877.** F. Marin de Carranrais. "L'Abbaye de Montmajour-lès-Arles: Étude historique d'après les manuscrits de D. Chantelou et autres documents inédits tirés des archives des Bouches-du-Rhône." *Revue de Marseille et de Provence* 23 (1877). Also published, under the same title, as a full-length book, in Marseille, 1877.

**Marmoutier 1913.** Jean de Marmoutier. *Historia Gaufredi ducis Normannorum et comitis Andegavorum.* French title: *Histoire de Geoffroi le Bel.* In Halphen and Poupardin 1913.

**Marquet de Vasselot 1898.** Jean-Joseph Marquet de Vasselot. "Quelques pièces d'orfèvrerie limousine." *Fondation Eugène Piot, Monuments et Mémoires* 4 (1898), pp. 257–269.

**Marquet de Vasselot 1905 and 1906.** J.-J. Marquet de Vasselot. "Les Émaux limousins à fond vermiculé (XIIe et XIIIe siècles)." *Revue Archéologique* 6 (1905), pp. 15–18; 231–45; 418–31. Also published, under the same title, in book form in Paris, 1906.

**Marquet de Vasselot 1906a.** J.-J. Marquet de Vasselot. *Catalogue raisonné de la collection Martin-le-Roy.* Vol. 1, *Orfèvrerie—Émaillerie.* Paris, 1906.

**Marquet de Vasselot 1914.** J.-J. Marquet de Vasselot. *Catalogue sommaire de l'orfèvrerie,*

et de l'émaillerie et des gemmes du Moyen-Âge au XVIIe siècle. Paris, 1914.

**Marquet de Vasselot 1917.** J.-J. Marquet de Vasselot. *Catalogue de la collection Arconati-Visconti.* Paris, 1917.

**Marquet de Vasselot 1921.** J.-J. Marquet de Vasselot. *Les Émaux limousins de la fin du XVe siècle et la première partie du XVIe: Étude sur Nardon Penicaud et ses contemporains.* Paris, 1921.

**Marquet de Vasselot 1923.** J.-J. Marquet de Vasselot. "La Donation Corroyer au musée du Louvre." *Bulletin des musées,* June 1923, pp. 149–51.

**Marquet de Vasselot 1932.** J.-J. Marquet de Vasselot. "Une Plaque de reliure limousine au musée du Louvre." *Fondation Eugène Piot, Monuments et Mémoires* 37 (1932), pp. 107–18. Also published, under the same title, in book form in Paris, 1932.

**Marquet de Vasselot 1936.** J.-J. Marquet de Vasselot. "Trois crosses limousines du XIIIe siècle dessinées par E. Delacroix." *Bulletin de l'Histoire de l'art français* 1936, pp. 137–47.

**Marquet de Vasselot 1938.** J.-J. Marquet de Vasselot. "Une Châsse limousine du XIIIe siècle léguée au Musée du Louvre." *Fondation Eugène Piot, Monuments et Mémoires* 36 (1938), pp. 123–36.

**Marquet de Vasselot 1941.** J.-J. Marquet de Vasselot. *Les Crosses limousines du XIIIe siècle.* Paris, 1941.

**Marquet de Vasselot 1952.** J.-J. Marquet de Vasselot. *Les Gémellions limousins du XIIIe siècle.* Paris, 1952. An extract from *Mémoires de la Société nationale des Antiquaires de France,* vol. 82.

**Marquet de Vasselot and Prinet 1921.** J.-J. Marquet de Vasselot and M. Prinet. [communication relative à une epitaphe en cuivre émaillé et doré.] *Bulletin de la Société nationale des Antiquaires de France,* 1921, pp. 239–41.

**Martène 1717.** D. E. Martène. *Thesaurus novus anecdotorum prodit nunc primum studio & opera domni Edmundi Martene & domni Usini Durand.* 5 vols. Paris, 1717.

**Martin 1909.** Henry Martin. *Psautier de saint Louis et de Blanche de Castille.* Paris, 1909.

**Martin 1917.** H. Martin. "Tombeau de cuivre émaillé de l'église des Dominicains de Rieux." *Bulletin archéologique du Comité des travaux historiques et scientifiques,* 1917, pp. lxxxii–lxxxiii, and pp. 190–93.

**Martin and Walker 1990.** J. Martin and L. Walker. "At the Feet of St. Stephen Muret: Henry II and the Order of Grandmont 'redivivus'." *Journal of Medieval History* 16 (1990), pp. 1–12.

**Martín Ansón 1984.** Maria Luisa Martín Ansón. *Esmaltes en España.* Madrid, 1984.

**Maryon 1971.** Herbert Maryon. *Metalwork and Enamelling.* New York, 1971.

**Mathieu 1975.** Marguerite-Robert Mathieu. *Montmirail en Brie, sa seigneurie et son canton.* Paris, 1975.

**Matrod 1906.** H. Matrod. *Deux émaux franciscains au Louvre: Les stigmates de saint François, leurs plus anciennes représentation connues.* Paris, 1906.

**Maupeou 1971.** C. de Maupeou. "Découverte d'un crosse dans l'église de Saint-Benoît (Vienne)." *Monument historique de la France* 17, no. 4 (1971), pp. 103–9.

**Maury, Gauthier, and Porcher 1960.** *See* Gauthier 1960

**Mazzanti 1920.** A. Mazzanti. "Note storiche illustrative delle Chiese e Parocchie della Diocesi pistoiese, S. Baronto antica chiesa abbaziale." *Il Monitore Diocesano* 11 (1920), fasicles 3–10.

**"Medieval Art and The Cloisters".** "Medieval Art and The Cloisters." In *Annual Report* 1989–90, pp. 28–29.

*Medieval Images 1978. See* New York 1978

**Ménégaux 1986.** C. Ménégaux. *Les Crosses en bronze, bronze doré, cuivre, cuivre doré et plomb du Xe au XIIIe siècle dans les collections publiques françaises.* Mémoire de l'École du Louvre, ed. D. Gaborit-Chopin, 1986. Typescript.

**Menendez Pidal forthcoming.** Faustino Menendez Pidal. "Armoriaux et décors brodées au milieu du XIIIe siècle." Forthcoming in *Cahiers du Léopard d'or* 8 (1996).

**Mérimée 1838.** Prosper Mérimée. *Notes d'un voyage en Auvergne.* Paris, 1838.

**Merrifield 1967.** "Incipit Primus et metricus liber Eraclii . . . De coloribus et artibus romanorum." In *Original Treatises on the Arts of Painting.* Ed. and trans. Mary P. Merrifield, vol. 1, pp. 182–257. 2 vols. Reprint. New York, 1967.

**Mexico City 1993.** *Tesoros medievales del museo del Louvre.* Exh., Museo del Palacio de Bellas Artes. Mexico City, 1993.

**Meyer 1895.** Alfred Meyer. *L'Art de l'émail de Limoges ancien et moderne: Traité pratique et scientifique.* Paris, 1895.

**Meyer, Meyer, and Wyss 1990.** Olivier Meyer, Nicole Meyer, and Michael Wyss. "Un Atelier d'orfèvre-émailleur récemment découvert à Saint-Denis." *Cahiers Archéologiques,* 38 (1990), pp. 81–94.

**Michel 1911.** Charles Michel. *Evangiles apocryphes.* Vol. 1, *Protoévangile de Jacques, Pseudo-Mathieu; Evangile de Thomas, Histoire de Joseph le Charpentier.* Ed. C. Michel. 2 vols. Paris, 1911–14.

*Middle Ages 1970. See* Los Angeles and Chicago 1970

**Migeon 1902.** Gaston Migeon. "La Collection Martin le Roy." *Les Arts,* no. 10 (1902), pp. 2–34.

**Migeon 1904a.** G. Migeon. "Les Enrichissements du département des objets d'art au musée du Louvre." *Gazette des Beaux-Arts* 1 (1904), pp. 188–200.

**Migeon 1904b.** G. Migeon. "Collection de M. Octave Homberg." *Les Arts,* no. 36 (1904), pp. 32–48.

**Migeon 1906.** G. Migeon. "La Collection de M. Albert Maignan, II." *Les Arts,* no. 59 (November 1906), pp. 3–16.

**Migeon 1909.** G. Migeon. "La Collection Victor Gay aux musées nationaux." *Gazette des Beaux-Arts* 4 (1909), pp. 408–32.

**Migeon 1918.** G. Migeon. "Enrichissements du Louvre pendant la guerre." *La Renaissance de l'art français et des industries de luxe,* no. 3 (May 1918), pp. 11–15.

**Migeon 1921.** G. Migeon. "Un Émail champlevé de Limoges au musée du Louvre." *La Revue de l'art ancien et moderne* 40 (1921), pp. 215–18.

**Migeon 1922.** G. Migeon. "Les Acquisitions du département des objets d'art du musée du Louvre." *Gazette des Beaux Arts* 64, no. 2 (1922), p. 134.

**Migne.** *See Patrologia latina*

**Millet 1916.** G. Millet. *Recherches sur l'iconographie de l'évangile aux XIVe, XVe et XVIe siècles d'après les monuments de Mistra, de la Macédoine et du Mont-Athos.* Paris, 1916.

**Millin 1791 and 1792.** A. Millin. Vol. 2 of *Antiquités nationales, ou, recueil de monumens pour servir à l'histoire générale et particulière de l'empire françois . . .* 5 vols. Paris, 1790–An VII [1799].

**Minneapolis 1966.** *Treasures from the Allen Memorial Art Museum.* Exh. cat., Minneapolis Institute of Arts. Minneapolis, 1966.

**Mognetti 1979.** E. Mognetti. "L'Abbaye de Montmajour." *Congrès archéologique de France, session CXXXIV, Pays d'Arles, 1976,* pp. 182–239. Paris, 1979.

**Molinier 1881.** Émile Molinier. "Note sur les origines de l'émaillerie française." *Le Cabinet historique,* 1881, pp. 90–106.

**Molinier 1882–88.** E. Molinier. "Inventaire du trésor du Saint Siège sous Boniface VIII (1295)." *Bibliothèque de l'École des Chartes* 43 (1882), pp. 281–310, 624–48; 45 (1884), pp. 31–57; 46 (1885), pp. 16–44; 47 (1886), pp. 646–67; 49 (1888), pp. 226–37.

**Molinier 1883.** E. Molinier. "Supplément au recueil des inscriptions du Limousin." In *Documents historiques bas-latins provençaux et français concernant principalement la Marche et le Limousin,* ed. L. Leroux, E. Molinier, and A. Thomas, pp. 90–346. Limoges, 1883.

**Molinier 1885.** E. Molinier. *Dictionnaire des émailleurs depuis le Moyen Âge jusqu'à la fin du XVIIIe siècle. Ouvrage accompagné de 67 marques et monogrammes.* Paris, 1885.

**Molinier 1886.** E. Molinier. "L'Exposition d'art rétrospectif de Limoges." *Gazette des Beaux-Arts* 34, no. 2 (1886), pp. 165–76.

**Molinier 1887a.** E. Molinier. "L'Orfèvrerie limousine à l'Exposition de Tulle en 1887, catalogue de l'Exposition." *Bulletin de la Société scientifique, historique et archéologique de la Corrèze* 9 (1887), pp. 469–537.

**Molinier 1887b.** E. Molinier. "Exposition rétrospective d'orfèvrerie à Tulle." *Gazette des Beaux-Arts,* ser. 2, 36, no. 2 (August 1887), pp. 148–56.

**Molinier 1889.** E. Molinier. *L'Émaillerie [conférence faite à l'Union centrale des Arts décoratifs en 1887].* Paris, 1889.

**Molinier 1890.** *See* Palustre and Molinier 1890

**Molinier 1891.** E. Molinier. *L'Émaillerie.* Paris, 1891.

**Molinier 1900.** E. Molinier. "L'Exposition rétrospective de l'art français: L'Orfèvrerie." *Gazette des Beaux-Art* 24, no. 2 (1900), pp. 160–72 and 349–65.

**Molinier 1901.** E. Molinier. *L'Orfèvrerie religieuse et civile du Ve à la fin du XVe siècle.* Paris, [1901]. Vol. 4 of *Histoire général des arts appliqués à l'industrie du Ve à la fin du XVIIIe siècle.* 6 vols. (vol. 6 by J. Guiffrey). Paris, 1896–1919.

**Molinier 1903.** E. Molinier. *Collections du château de Goluchów: Objets d'art du Moyen Âge et de la Renaissance.* Paris, 1903.

**Molinier 1904.** E. Molinier. *Collection du baron Albert Oppenheim, tableaux et objets d'art. Catalogue précédé d'une introduction par . . .* Paris, 1904.

**Molinier and Marcou 1900.** E. Molinier and Frantz Marcou. *Exposition rétrospective de l'art français des origines à 1800.* 2 vols. Paris, 1900.

**Molinier and Rupin 1887.** E. Molinier and E. Rupin. "Note sur différents objets d'art exposés à l'Exposition rétrospective de Tulle." *Bulletin de la Société nationale des Antiquaires de France,* 1887, pp. 256–62.

*Monnaies et medailles 1972. See* Paris 1972

*Monsters, Gargoyles and Dragons. See* South Hadley, Mass., 1977

**Montesquiou n. d.** *See* Dimier and de Montesquiou n. d.

**Montesquiou-Fezensac n. d.** Blaise de Montesquiou-Fezensac. *Abbaye de Longpont.* Paris, n. d.

**Montesquiou-Fezensac and Gaborit-Chopin 1973.** Blaise de Montesquiou-Fezensac and Danielle Gaborit-Chopin. *Le Trésor de Saint-Denis: Inventaire de 1634.* 3 vols. Paris, 1973–77.

**"Monument . . ." 1711.** "Monument trouvé au prieuré de Belhomer." *Mémoires pour l'histoire des sciences et beaux-arts,* 1711, pp. 1364–70.

*Monuments de la monarchie 1729.* R. P. dom Bernard de Montfaucon. *Les Monuments de la monarchie française, qui comprenant l'histoire de France, avec les figures de chaque règne . . .* 5 vols. Paris, 1729–33.

*Monuments . . . Guillaume Libri 1864. Monuments inédits ou peu connus, faisant partie du cabinet de Guillaume Libri, et qui se rapportent à l'histoire de l'ornementation chez differents peuple.* London, 1864.

**Morel 1952.** P. Morel. "A propos du nom d'Alpais." *BSAHL* 84 (1952–54), pp. 518–19.

**Morenas 1939.** *See* Jougla de Morenas 1939

**Morgan 1982.** Nigel Morgan. *Early Gothic Manuscripts, I: 1190–1250. A Survey of Manuscripts Illuminated in the British Isles,* 4. Ed. J.J.G. Alexander.

**Morice 1742–46.** Dom P.-H. Morice. *Mémoires pour servir de preuve à l'histoire ecclésiastique et civile de la Bretagne.* 3 vols. Paris, 1742–46. New edition, Paris, 1974.

**Morice 1750-54.** P. H. Morice and C. Taillandier. *Histoire ecclésiastique et civile de Bretagne.* Vol. 1 (Morice); Paris, 1750. Vol. 2 (Taillandier); Paris, 1754.

**Moschetti 1920.** A. Moschetti. "Uno Smalto limosino nel duomo di Monselice." *Dedalo* 1 (1920), pp. 40–46.

**Moscow 1956.** *Exposition de l'art français.* Exh. Moscow, 1956.

**Moscow and Leningrad 1990.** *Dekorativno-Prikladnoe Iskusstvo ot Pozdnei Antichnosti do Pozdnei Gotiki* (Decorative and Applied Arts from the Late Antique through the Late Gothic [works from the Metropolitan Museum of Art and the Art Institute of Chicago]). Exh. cat., Pushkin Museum of Fine Arts and The State Hermitage Museum. Moscow and Leningrad, 1990.

**Moser-Gautrand and Vidalo-Borderie 1983.** Claire Moser-Gautrand and Evelyn Vidalo-Borderie. "Le Chef reliquaire de Sainte-Essence conservé au Musée Ernest Rupin." *Bulletin de la Société scientifique, historique et archéologique de la Corrèze* 105 (1983), pp. 123–41.

**Muldrac 1652.** F. Antonio [Le P. F. Antoine] Muldrac. *Compendiosum abbatiae Longipontis suessionensis Chronicon in III part. distinctum.* Paris, 1652.

**Müller 1994.** M. Müller. "Ikonologische Studien zu französichen Minnedarstellungen des 13. und 14. Jahrhunderts," pp. 89–101. Ph.D. diss., University of Münster, 1994. Forthcoming, 1995.

**Munby 1972.** A.N.L. Munby. *Connoisseurs and Medieval Miniatures.* Oxford, 1972.

**Munich 1992.** *Schatzkammerstücke aus der Herbstzeit des Mittelalters: Das Regensburger Emailkästchen und sein Umkreis.* Ed. Reinhold Baumstark. Exh. cat., Bayerisches Nationalmuseum. Munich, 1992.

**Murray 1989.** S. Murray. *Beauvais Cathedral: Architecture of Transcendence.* Princeton, 1989.

*Musiques au Louvre 1994. Musiques au Louvre.* Paris, 1994.

**Mussat 1988.** A. Mussat. "Le Chevalier et son double." In *La Figuration des morts dans la chrétienté médiévale jusqu'à la fin du premier quart du XIVe siècle: Colloque, 26–28 mai 1988, Abbaye royale de Fontevraud,* ed. Roger Grégoire, pp. 137–54. N. p., 1989.

**Mütherich and Gaehde 1976.** Florentine Mütherich and Joachim E. Gaehde. *Carolingian Painting.* New York, 1976. French translation: *Peinture carolingienne.* Paris, 1977.

**Nadaud 1880.** J. Nadaud. *Nobiliaire du diocèse de la généralité de Limoges.* Vol. 4. Limoges, 1880.

**Namur 1969.** *Orfèvreries du trésor de la cathédrale de Namus.* Exh., Crédit communal de Belgique. Namur, 1969.

**Nangis 1860.** G. de Nangis. *Gesta Philippi III Francorum regis.* In Daunou and Naudet 1860 [1840].

**Nantes 1933.** *Exposition d'art religieux ancien et moderne à la Psalette.* Exh., Nantes, 1933.

**Naples 1981.** *Medioevo e produzione artistica di series: Smalti di Limoges e avori gotici in Campania.* Exh., Museo Duca di Martina. Naples, 1981.

**Naudet 1837.** C. Naudet. *Recueil d'objets d'art et de curiosité dessinés d'après nature par T. de Jomiont et J. Cagniet.* Paris, 1837.

Neale 1906. *See* Durandus 1906

Nelson 1938. Philip Nelson. "Limoges Enamels: Altar Cruets of the Thirteenth Century." *The Antiquaries Journal* 18, no. 1 (January 1938), pp. 49–54.

Netzer 1986. *See* Swarzenski and Netzer 1986

Nevers 1990. *Trésors cachés des églises de la Nièvre*. [Nevers], 1990.

Newman 1991. Richard Newman. "Materials and Techniques of the Medieval Metalworker." In *Catalogue of the Medieval Objects in the Museum of Fine Arts: Metalwork,* ed. Nancy Netzer, with technical essay and analyses by Richard Newman, pp. 18–41. Boston, 1991.

New York 1927. *Loan Exhibition of French Primitives (Limoges, 13th Century)*. Exh., Kleinberger Galleries. New York, 1927.

New York 1943. *Masterpieces in the Collection of George Blumenthal: A Special Exhibition*. Exh. cat., The Metropolitan Museum of Art. New York, 1943.

New York 1954. *Spanish Medieval Art. Loan Exhibition in Honor of Dr. Walter W. S. Cook Arranged by the Alumni Association, Institute of Fine Arts in Cooperation with The Metropolitan Museum of Art*. Exh. handbook, The Cloisters. New York, 1954.

New York 1968. Carmen Gómez-Moreno. *Medieval Art from Private Collections, A Special Exhibition at The Cloisters*. Exh. cat., The Metropolitan Museum of Art. New York, 1968.

New York 1970. *The Year 1200; A Centennial Exhibition at The Metropolitan Museum of Art*. Exh. cat. New York, 1970.

New York 1974. *Saints and Their Legends*. Pamphlet, Metropolitan Museum of Art. New York, 1974.

New York 1975. *Islam and the Medieval West*. Exh., Center of Medieval and Early Renaissance Studies. Binghamton, N.Y., 1975.

New York 1978. *Medieval Images*. Exh. cat., The Katonah Gallery (Katonah, N.Y., 1978) and the Aspen Center for Visual Arts (Aspen, Colo., 1979–80). New York, 1978.

New York 1981. *The Royal Abbey of Saint-Denis (1122–1151)* Ed. Sumner McKnight Crosby et al. Exh. cat., The Metropolitan Museum of Art. New York, 1981.

New York 1982. *The Vatican Collections: The Papacy and Art*. Exh. cat., Metropolitan Museum of Art. New York, 1982.

New York 1984. *The Treasury of San Marco, Venice*. Exh. cat., The Metropolitan Museum of Art. New York and Milan, 1984.

New York 1987. *Art Treasures of the Middle Ages*. Exh. cat., Michael Ward Gallery. New York, 1987.

New York 1993. *See* The Art of Medieval Spain

Nicard 1864. P. Nicard. [crosse émaillée.] *Bulletin de la Société impériale des Antiquaires de France*, 1864, p. 130.

Niewdorp 1979. H. Niewdorp. *Le Musée Mayer van den Bergh: Guide du visiteur*. Anvers, 1979.

Nikitine 1981. S. and M. Nikitine. "L'Émail Plantagenêt." In *La Cathédrale du Mans*, ed. A. Mussat. Paris, 1981.

Nodier, Taylor, and Cailleux 1829. Charles Nodier, J. Taylor, and Alphone de Cailleux. *Voyages pittoresques et romantiques dans l'ancienne France*. 18 vols. Paris, 1820–78. Auvergne, 2 vols., 1829–33.

I Normanni 1994. *I Normanni, popolo d'Europa 1030–1200*. Exh. cat., Palazzo Venezia. Rome, 1994.

"Obituaires . . ." 1878. "Obituaires de l'église Saint-André de Bordeaux." *Archives historiques de la Gironde* 18 (1878), pp. 1–241.

Les Objets d'art 1993. *Les Objets d'art, Moyen Âge et Renaissance, guide du visiteur*. Musée du Louvre. Paris, 1993.

Oddy 1981. W. A. Oddy. "Gilding through the Ages." *Gold Bulletin* 14, no. 2 (1981), pp. 75–79.

Oddy 1982. W. A. Oddy. "Gold in Antiquity: Aspects of Gilding and Assaying." *Journal of the Royal Society of Arts* 130, no. 5315 (1982), pp. 730–43.

Oddy 1988. W. A. Oddy. "The Gilding of Roman Silver Plate." In *Argenterie romaine et byzantine: Actes de la Table Ronde*, 11–13 octobre, 1988, ed. François Baratte, pp. 9–21. Paris, 1988.

Oddy 1991. W. A. Oddy. "Gilding: An Outline of the Technological History of the Plating of Gold onto Silver or Copper in the Old World." *Endeavour*, n. s., 15, no. 1 (1991), pp. 29–33.

Oddy, Bimson, La Niece 1981. W. A. Oddy, M. Bimson, and Susan La Niece. "Gilding Himalayan Images: History, Tradition and Modern Techniques." In *Aspects of Tibetan Metallurgy*, ed. W. A. Oddy and W. Zwalf, pp. 87–101. British Museum Occasional Papers, 15. London, 1981.

Oddy, La Niece, and Stratford 1986. W. A. Oddy, Susan La Niece, and Neil Stratford. *Romanesque Metalwork: Copper Alloys and Their Decoration*. London, 1986.

L'Oeuvre de Limoges. M.-M. S. Gauthier. *L'Oeuvre de Limoges et le marché européen*. Vol. 3 of Gauthier, *Émaux méridionaux*. Forthcoming.

Öhler 1995. N. Öhler. *The Medieval Traveller*. Trans. Caroline Hiller. Woodbridge, Suffolk, Eng., 1989. Reprinted Rochester, New York, 1995.

Oklahoma City 1985. *Songs of Glory: Medieval Art from 900–1500*. Exh., Oklahoma Museum of Art. Oklahoma City, Okla., 1985.

"The Oppenheim Collection" 1906. "The Oppenheim Collection at South Kensington Museum." *The Burlington Magazine* 9 (1906), pp. 227–34.

L'Ordre de Grandmont 1992. *See* Durand and Nougaret 1992

Otavsky 1973. Karel Otavsky. "Zu einer Gruppe von Kupferreliefs aus dem 13. Jahrhundert." *Artes Minores: Dank an Werner Abegg*, ed. Michael Stettler and Mechthild Lemberg, pp. 37–74. Bern, 1973.

Ottawa 1972. *L'Art et la Cour/Art and the Court: France and England from 1259–1328.*

Exh., National Gallery of Canada. Ottawa, 1972.

Oursel 1975. R. Oursel. *Haut-Poitou roman*. La Pierre-qui-Vire, 1975.

Pala d'Oro 1994. *La Pala d'Oro*. Ed. H. R. Hahnloser and R. Polacco. Venice, 1994.

Palustre 1883. Léon Palustre. "La Vierge de La Sauvetat (Puy-de-Dôme)." *Bulletin monumental* 11 (1883), pp. 497–504.

Palustre 1891. L. Palustre. *Album de l'exposition rétrospective de Tours*. Tours, 1891.

Palustre and Barbier de Montault 1886. L. Palustre and X. Barbier de Montault. *Le Trésor de Treves*. Paris, n. d. [1886]. Volume 1 of Palustre's *Mélanges d'art et d'archéologie*. Paris, 1886–87.

Palustre and Barbier de Montault 1887. L. Palustre and X. Barbier de Montault. *Orfèvrerie et émaillerie limousines. 1re partie: Pièces exposées à Limoges en 1886*. Paris, n. d. [1887]. Volume 2 of Palustre's *Mélanges d'art et d'archéologie*. Paris, 1886–87.

Palustre and Molinier 1890. L. Palustre and Émile Molinier. *L'Orfèvrerie religieuse*. Vol. 1 of *La Collection Spitzer*. Paris and London, 1890.

Panofsky 1964. Erwin Panofsky. *Tomb Sculpture: Four Lectures on Its Changing Aspects from Ancient Egypt to Bernini*. London, 1964. Reissued, New York, 1992.

Pardiac 1860. J.-B. Pardiac. "Notice archéologique et iconographique sur sainte Ursule." *Revue de l'Art chrétien* 4 (1860), pp. 353–77.

Paris 1865. *Exposition de 1865. Union central des Beaux-arts appliqués à l'industrie. Musée rétrospectif*. Paris, 1867.

Paris 1867. *Exposition universelle de 1867 à Paris. Catalogue général. Histoire du travail et monuments historiques*. Paris, [1867].

Paris 1878. E. de Beaumont et al. *Exposition universelle de 1878: Les Beaux-Arts et les arts décoratifs*. General ed. Louis Gonse. 2 vols. Paris, 1879. (Vol. 2 titled *L'Art ancien à l'Exposition de 1878; see Darcel 1878b*.)

Paris 1880. *Exposition de l'Union centrale des Arts décoratifs appliqués à l'industrie*. [Germain Bapst.] *Le Musée rétrospectif du metal à l'Exposition de l'Union centrale des Beaux-Arts, 1880*. Paris, 1881.

Paris 1884. *Exposition de l'Union centrale des Arts décoratifs. Catalogue*. Paris, 1884.

Paris 1889. *Exposition universelle de 1889: Les Beaux-Arts et les arts décoratifs. L'Art français rétrospectif au Tocadéro*. Ed. E. Bonnafée et al.; general eds. Louis Gonse and Alfred de Lostalot, Paris, 1889 [or 1890].

Paris 1900. *Exposition universelle de 1900. Catalogue illustré officiel de l'exposition rétrospective de l'art français, des origines à 1800*. Petit-Palais. Paris, 1900.

Paris 1913. *Exposition d'objets d'art du Moyen Âge et de la Renaissance*. Exhibition organized by the marquise de Granay, Chez Jacques Seligman, ancien Hôtel de Sagan. Paris, 1913.

Paris 1934. *La Passion du Christ dans l'art français*. Exh., Musée de sculpture comparée du Trocadéro, Sainte-Chapelle. Paris, 1934.

Paris 1935. *L'Orfèvrerie et le bijou d'autrefois*. Exh., Galerie Mellerio. Paris, 1935.

Paris 1937. *Chefs-d'oeuvre de l'art français*. Exh. cat., Palais national des Beaux-Arts. 2 vols. Paris, 1937.

Paris 1945. *Nouvelles acquisitions*. Exh., Musée du Louvre. Paris, 1945.

Paris 1947. *Reliure originale*. Exh., Bibliothèque nationale. Paris, 1947.

Paris 1954. *Manuscrits à peinture du VIIe au XIIe siècle*. Exh., Bibliothèque nationale. Paris, 1954.

Paris 1957. *Chefs-d'oeuvre romans des musées de province*. Exh., Musée du Louvre. Paris, 1957.

Paris 1958. *Byzance et la France médiévale: Manuscrits du IIe au XVIe siècle*. Exh., Bibliothèque nationale. Paris, 1958.

Paris 1960. *Saint Louis à la Sainte Chapelle*. Exh., Archives de France. Paris, 1960.

Paris 1962. *Cathédrales, sculptures, vitraux, objets d'art, manuscrits des XIIe et XIIIe siècles*. Exh., Musée du Louvre. Paris, 1962.

Paris 1963. *Notre-Dame de Paris 1163–1963*. Exh., Sainte-Chapelle. Paris, 1963.

Paris 1964. *Emblèmes, totems, blasons*. Exh., Musée Guimet. Paris, 1964.

Paris 1965. *Trésors des églises de France*. Intro. Jean Taralon. Exh. cat., Musée des Arts Décoratifs. Paris, 1965.

Paris 1967. *Vingt ans d'acquisitions au Musée du Louvre, 1947–1967*. Exh., Orangerie des Tuileries. Paris, 1967–68.

Paris 1968a. *Rome à Paris*. Exh., Petit-Palais. Paris, 1968.

Paris 1968b. *L'Europe gothique, XIIe–XIVe siècles*. Exh. cat., Musée du Louvre, Pavillon de Flore. Paris, 1968.

Paris 1970. *La France de Saint Louis*. Exh. cat., La Conciergerie. Paris, 1970.

Paris 1972. *Monnaies et medailles racontent l'histoire de France*. Exh. cat., Hôtel de la Monnaie. Paris, 1972.

Paris 1979. *Le "Gothique" retrouvé avant Viollet-le-Duc*. Exh. cat., Hôtel de Sully. Paris, 1979.

Paris 1980a. *Cinq années d'enrichissement du patrimoine national, 1975–1980*. Exh., Grand-Palais. Paris, 1980.

Paris 1980b. *La Vie mystérieuse des chefs-d'oeuvre: La science au service de l'art*. Exh., Grand-Palais. Paris, 1980.

Paris 1981. *Les Fastes du Gothique: Le siècle de Charles V*. Exh. cat., Galeries nationales du Grand Palais. Paris, 1981.

Paris 1983. *See* Avril, Barral i Altet, and Gaborit-Chopin 1983

Paris 1984. *Le Trésor de Saint-Marc*. Exh. cat., Grand-Palais. Paris, 1984.

Paris 1985. *Nouvelles acquisitions du Département des objets d'art, 1980–1984*. Exh. cat., Musée du Louvre. Paris, 1985.

Paris 1990. *Nouvelles acquisitions du Département des objets d'art, 1985–1990*. Exh.

cat., Musée du Louvre. Paris, 1990.

**Paris 1991.** *Le Trésor de Saint-Denis.* Exh., Musée du Louvre. Paris, 1991.

**Paris 1992.** *Byzance: L'Art byzantin dans les collections publiques françaises.* Exh., Musée du Louvre. Paris, 1992.

**Paris 1993.** *Legende dorée du Limousin: Les saints de la Haute-Vienne.* Exh. cat., Musée du Luxembourg. Paris, 1993.

**Paris 1995.** *Nouvelles acquisitions du Département des objets d'art, 1990–1994.* Exh. cat., Musée du Louvre. Paris, 1995.

**Paris, Arras, and Lille 1867.** *See* Linas 1867 and 1868

**Parkhurst 1952.** Charles Parkhurst. "Preliminary Notes on Three Early Limoges Enamels at Oberlin." *Allen Memorial Art Museum Bulletin* 9, no. 3 (Spring 1952), pp. 92–96.

**Pastè and Arborio Mella 1907.** Romualdo Pastè and Federico Arborio Mella. *L'Abbazia di S. Andrea di Vercelli.* Vercelli, 1907.

**Pastoureau 1979.** Michel Pastoureau. *Traité d'héraldique.* Paris, 1979. New edition, Paris, 1993.

**Pastoureau 1982.** M. Pastoureau. "Le Bestiare héraldique médiévale." In *L'Hermine et le sinople. Études d'héraldique médiévale,* pp. 105–16. Paris, 1982.

*Patrologia Latina. Patrologia cursus completus: Series latina.* Ed. J.-P. Migne. 221 vols. Paris, 1844–64.

**Péan de La Tuilerie 1776 and 1869.** Péan de La Tuilerie. *Description de la ville d'Angers et de tout ce qu'elle contient de plus remarquable.* Angers, 1776. Ed., with annotations, by C. Port. Angers, 1869.

**Pérat 1911.** André Pératé. *Collections Georges Hoentschel: Introduction et notices.* Vol. 1, *Émaux du XIIe au XVe siècle.* Paris, 1911.

**Pérol 1948.** Pierre Pérol. "Nomenclature par communes des pièces d'orfèvrerie émaillées ou non constituant une partie du trésor artistique corrézien." *Bulletin de la Société scientifique, historique et archéologique de la Corrèze* 70 (1948), pp. 83–92.

**Pérol 1962.** P. Pérol. *Histoire religieuse de Brive-la-Gaillarde.* Brive, 1962.

**Perrier 1938.** A. Perrier. "Quelques notes sur l'orfèvrerie émaillée limousine à l'Exposition de 1937." *Bulletin de la Société des lettres, sciences et arts de la Corrèze* 78, no. 1 (1938), pp. 49–53.

**Pesche 1834.** J.-R. Pesche. *Dictionnaire topographique, historique et statistique de la Sarthe suivi d'une biographie et d'une bibliographie.* Vol. 3. Le Mans–Paris, 1834.

**Pessiot 1994.** M. Pessiot. "Un Tombeau oublié de la famille de saint Louis à Royaumont: Le tombeau d'Ote (1291), fils de Philippe d'Artois." *Revue du Louvre et des musées de France,* 4 (1994), pp. 29–36.

**Petrassi 1982.** Mario Petrassi. *Gli Smalti in Italia.* Rome, 1982.

**Philostratus.** *Philostratus, Imagine; Callistratus, Descriptions.* Trans. Arthur Fairbanks. London and New York, 1931.

**Pidal forthcoming.** *See* Menendez Pidal

**Pinard 1852.** T. Pinard. "Ancienne commanderie de Malte, de Corbeil." *Revue archéologique,* 1852, pp. 565–75.

**Pinoteau 1966.** *See* Pinoteau and Le Gallo 1966

**Pinoteau 1978–79.** Hervé Pinoteau. "La Date de la cassette de saint Louis." *Bulletin de la Société nationale des Antiquaires de France,* 1978–79, pp. 77–78.

**Pinoteau 1983.** H. Pinoteau. "La Date de la cassette de saint Louis: été 1236?" *Cahiers d'Héraldique* 4 (1983), pp. 97–130.

**Pinoteau and Le Gallo 1966.** H. Pinoteau and Cl. Le Gallo. "L'Héraldique de saint Louis et de ses compagnons." *Les Cahiers nobles* 27 (1976).

**Piolin 1856.** Dom Paul Piolin. *Histoire de l'église du Mans.* Vol. 6. Paris, 1856.

**Piot 1861–62.** E. Piot. "Émaillerie limousine du XIIIe siècle: La cassette de saint Louis, le ciboire de Warwick." *Le Cabinet de l'amateur,* 1861–62, pp. 103–6.

*Plainctes et doléances du chapitre du Mans. See* Ledru 1903

**Planché 1846.** J.-R. Planché. "Remarks on an Enamelled Tablet, Preserved in the Museum at Mans and Supposed to Represent the Effigy of Geoffroy Plantagenêt." *The Journal of The British Archaeological Association* 1 (1846), pp. 29–39.

**Pliny 1962.** *Pliny, the Elder. Natural History.* Trans. D. E. Eicholz. Vol. 10. London and New York, 1962.

**Poitiers 1976.** *Aliénor d'Aquitaine et son temps.* Exh. Poitiers, 1976.

**Pollen, ed., 1875.** *See* London 1875

**Poquet 1869.** Alexandre Eusèbe Poquet. *Monographie de l'abbaye de Longpont: Son histoire, ses monuments, ses abbés, ses personnages célèbres, ses sepultures, ses possessions territoriales.* Paris, 1869.

**Porcher 1959a.** Jean Porcher. In Pierre d'Herbécourt and J. Porcher, eds. *Anjou roman.* Saint-Leger-Vauban, 1959.

**Porcher 1959b.** J. Porcher. *L'Enluminure française.* Paris, 1959.

**Porcher 1959c.** J. Porcher. In Marguerite Vidal, Jean Maury, and J. Porcher, eds. *Quercy roman.* La Pierre-qui-Vire, 1959.

**Porcher 1968.** J. Porcher. "Les Manuscrits à peinture." In J. Hubert, J. Porcher, and W. F. Wolback. *L'Empire carolingien.* Paris, 1968.

**Port, ed. 1869.** *See* Péan de La Tuilerie 1776 and 1869

**Port 1874–78.** Celestin Port. *Dictionnaire historique, géographique et biographique de Maine-et-Loire.* 3 vols. Paris–Angers, 1874–78.

**Pottier 1866.** Abbé Pottier. "Objets d'orfèvrerie." *Congrès archéologique de France session XXXIIe, Montauban, Cahors et Guéret, 1865,* pp. 329–30. Paris, 1866.

**Poulbrière 1875.** J.-B. Poulbrière. "Promenades à Gimel." *Bulletin monumental* 41 (1872), pp. 530–36.

**Poulbrière 1886.** J.-B. Poulbrière. "Les Tombes en métal du Limousin." *Bulletin de la Société des lettres, sciences, et arts de la Corrèze* 8 (1886), pp. 154–58.

**Poulbrière 1891.** J.-B. Poulbrière. *L'Église Saint-Martin de Brive.* Paris, [1891].

**Poulbrière 1894.** J.-B. Poulbrière. "Les Inscriptions de la pierre tumulaire de Maschalx et de la châsse de Saint-Viance." *Bulletin de la Société scientifique, historique et archéologique de la Corrèze* 16 (1894), pp. 229–32.

**Poulbrière 1894–99.** J.-B. Poulbrière. *Dictionnaire historique, archéologique des pariosses du diocèse de Tulle.* 2 vols. Tulle, 1894–99.

**Prague–Bratislava 1978.** *L'Art du Moyen Âge en France.* Prague, Narodni Galerie. Prague, 1978–79; Bratislava. Exh. Slovenská národná galéria. Bratislava, 1979.

**Pressouyre 1990.** L. Pressouyre. *La Rêve cistercien.* Paris, 1990.

**Prosinec–Únor 1986–87.** *Stredoveké u melecké remeslo: Ze sbirek Uméleckoprumyslového musea v Praze.* Exh., Prosinec, 1986; Únor, 1987.

**Québec–Montréal 1972.** J.-R. Gaborit, ed. *L'Art français du Moyen Âge.* Exh. cat., Musée du Québec, and Musée des Beaux-Arts, Montréal. Paris, 1972.

**Quétif and Echard 1719.** R. P. F. Jacobus Quétif and R. P. F. Jacobus Echard. Vol. 1 of *Scriptores ordinis praedicatorum recensiti, . . .* 2 vols. Paris, 1719–21.

**Raible 1908.** Felix Raible. *Der Tabernakel einst und jetzt. Eine historische und liturgische Darstellung der Andacht zur aufbewahrten Eucharistie.* Freiburg im Breisgau, 1908.

**Randall 1993.** R. H. Randall. *The Golden Age of Ivory: Gothic Carvings in North American Collections.* New York, 1993.

*Rationale divinorum officiorum 1484.* Guillaume Durand. *Rationale divinorum officiorum.* Strasburg, 1484

**Raymond 1867.** P. Raymond. *Inventaire sommaire de Archives départementales antérieures à 1790: Basses-Pyrénées.* Vol. 4, *Archives civiles série E no. 1 à 1765.* Paris, 1867.

**Réau 1957.** L. Réau. *Iconographie de l'art chrétien.* 3 vols. Paris, 1955–59.

*Recueil 1738–1904. Recueil des historiens des Gaules et de la France . . .* Ed. Martin Bouquet et al. 24 vols. Paris, 1738–1904.

**Redford 1990.** S. Redford. "How Islamic is it? The Innsbruck Plate and its Setting." *Muquarnas* 7 (1990), pp. 119–35.

*Religious Art 1950. Religious Art of the Middle Ages and Renaissance in the Collection of the Wadsworth Atheneum.* Exh. cat., Wadsworth Atheneum. Hartford, Conn., 1950.

**Rennes–Avranches 1991.** *See* Avranches 1991

**Renouard 1811.** Pierre Renouard. *Essais historiques et littéraires sur la ci-devant province du Maine, divisés par époques.* 2 vols. Le Mans, 1811.

**Reusens 1886.** Chanoine E. H. J. Reusens. *Éléments d'Archéologie chrétienne.* 2 vols. 2d ed. Louvain, 1885–86.

**Reuterswärd 1970.** P. Reuterswärd. "Arons Stav." In *Kontakt Nationalmuset* (Stockholm, 1970), pp. 19–26.

**Reuterswärd 1975.** P. Reuterswärd. "Arons Stav: Ett Tokningsalternativ för reliefen i Grnbaeks kyrka." *Iconographische Post* 6, 4 (1975), pp. 9–12.

*Revue bourbonnaise 1884.* "Chronique." *Revue bourbonnaise* 1 (1884), p. 291.

*Revue du Louvre 1981.* "Don, par la Société des Amis du Louvre, de six médaillons à décor profane et cinq à décor floral . . . provenant du décor d'un coffret, Limoges. . . ." *Revue du Louvre et des Musées de France* 30, no. 4 (1981), p. 288 ("Les recentes acquisitions des musées nationaux").

*Rhein und Maas 1972. See* Cologne and Brussels 1972

**Rickard 1932.** T. A. Rickard. *Man and Metals: A History of Mining in Relation to the Development of Civilization.* 2 vols. New York, 1932.

**Rietstap 1861.** J. B. Rietstap. *Armorial général précédé d'un dictionnaire des termes du blason.* 2 vols. 1st ed. 1861. 2d ed. Gouda, 1884–87.

**Ripoud 1840.** A. Ripoud. *Annuaire de l'Allier pour 1840.* Moulins, 1840.

**Robinson 1862.** *See* London 1862

**Roccavilla 1905.** A. Roccavilla. *L'Arte nel Biellese.* Biella, 1905.

**A. Roger 1866.** A. Roger. *Émaux et montres de la collection Bouvier.* Amiens, 1866.

**B. Roger 1852.** Barthélemy Roger. "Histoire de l'Anjou." [Ed. A. Lemarchand.] *Revue de l'Anjou.* 1852.

**Rohan-Chabot 1930.** A. de Rohan-Chabot, marquise de Maillé. "L'Église cistercienne de Preuilly (Seine-et-Marne)." *Bulletin monumental* (1931), pp. 257–354.

**Rohan-Chabot 1939.** A. de Rohan-Chabot, marquise de Maillé. *Provins, les monuments religieux.* 2 vols. Paris, 1939.

**Rohault de Fleury 1883–89.** Charles Rohault de Fleury. *La Messe: Études archéologiques sur ses monuments . . . continuées par son fils.* 8 vols. Paris, 1883–89. Vol. 5 (1887), vol. 6 (1888), vol. 8 (1889).

**Rome 1963.** *See* Vatican City 1963

**Rome 1994.** *I Normanni, popolo d'Europa 1030–1200.* Exh., Palazzo Venezia. Rome, 1994.

**Rorimer 1948.** James J. Rorimer. "A Treasury at The Cloisters." *Metropolitan Museum of Art Bulletin,* n. s., 6, no. 9 (May 1948), pp. 237–60.

**Rorimer 1963.** J. J. Rorimer. *The Cloisters, the Building, and the Collection of Medieval Art in Fort Tryon Park.* 3d ed. New York, 1963.

**Rosedale 1904.** H. G. Rosedale. *St. Francis of Assisi According to Brother Thomas of Celano.* London, 1904.

**Rosenberg 1907.** Marc Rosenberg. *Geschichte der Goldschmiedekunst auf Technischer Grundlage.* Part 1a: *Niello bis zum Jahre 1000 nach Chr.* 3 vols. Frankfurt, 1910–24.

**Rosenzweig 1863.** L. Rosenzweig. *Répertoire*

archéologique du département du Morbihan. N. p. 1863.

**Roserot 1945–48.** A. Roserot. *Dictionnaire historique de la Champagne méridionale (Aube), des origines à 1790.* 3 vols. Langres, 1945–48.

**Ross 1932a.** Marvin Chauncey Ross. "An Enamelled Gemellion of Limoges." *Bulletin of the Fogg Art Museum* 2 (November 1932), pp. 9–13.

**Ross 1932b.** M. C. Ross. "Note sur un émail limousin de Saint Guillaume de Dongéon." *Notes d'Art et d'Archéologie* 3 (July 1932), pp. 36–39.

**Ross 1933a.** M. C. Ross. "La Croix Gaillard de La Dionnerie au Musée Métropolitain de l'Art à New-York." *Bulletin de la Société des Antiquaires de l'Ouest*, ser. 3, 9, no. 4 (1933), pp. 909–11.

**Ross 1933b.** M. C. Ross. "Le Devant d'autel émaillé d'Orense." *Gazette des Beaux-Arts* 140 (1933), pp. 272–78.

**Ross 1933c.** M. C. Ross. "Un Relicario esmaltado en Roda de Isabena." *Revista Zurita* (1933), pp. 3–9.

**Ross 1934.** M. C. Ross. "Un Motif hispano mauresque." *Revue de l'art ancien et moderne* (1934), pp. 87–89.

**Ross 1935.** M. C. Ross. "Two Fragments from a Limoges Bookcover." *Revue archéologique*, ser. 6, 6 (September 1935), pp. 73–77.

**Ross 1939.** M. C. Ross. "An Enameled Reliquary from Champagnat." In *Medieval Studies in Memory of A. Kingsley Porter*, ed. Wilhelm R. W. Koehler, vol. 2, pp. 467–77. 2 vols. Cambridge, Mass., 1939.

**Ross 1940.** M. C. Ross. "Gli Smalti del museo sacro vaticano." *The American Journal of Archaeology* (1940), pp. 280–81.

**Ross 1941.** M. C. Ross. "De Opere lemovicense." *Speculum* 16, no. 4 (October 1941), pp. 453–58.

**Ross 1942.** M. C. Ross. "Chrismatoires médiévaux de Limoges." *BSAHL* 79 (1942), pp. 341–44.

**Rouen 1975.** *Manuscrits normands XIe–XII siècles.* Exh., Bibliothèque municipale. Rouen, 1975.

**Rouen 1989.** *A Travers le verre du Moyen Âge à la Renaissance.* Exh., Musée départemental des Antiquités de Rouen. Rouen, 1989.

**Rouen–Caen 1979.** *Trésors des Abbayes normandes.* Exh. cat., Musée des Antiquités, Rouen, and Musée des Beaux-Arts, Caen. Rouen, 1979.

**Roulin 1899.** E. Roulin. "La Statue tombale de dom Mauricio évêque de Burgos (XIIIe siècle)." *Bulletin de la Société scientifique, historique et archéologique de la Corrèze* (1899), pp. 487–95.

**Roux 1889.** A. Roux. *Les Inscriptions de Saint-Viance.* Tulle 1889.

**Roux 1949.** A. Roux. *La Cathédrale d'Apt d'après des documents inédits.* Apt, 1949.

***Royal Abbey of Saint-Denis 1981.*** *See* New York 1981

***Royaumes 1983.*** *See* Avril, Barral i Altet, Gaborit-Chopin 1983

**Royaumont 1970.** *Saint Louis à Royaumont.* Exh., Abbaye de Royaumont. Royaumont, 1970.

**Roy-Pierrefitte 1851.** Abbé J.-B. L. Roy-Pierrefitte. *Histoire de la ville de Bellac (Haute-Vienne).* Limoges, 1851.

**Roy-Pierrefitte 1864.** Abbé J.-B. L. Roy-Pierrefitte. "Les Religieux de saint François d'Assise dans la Marche et le Limousin." *BSAHL* 14 (1864), pp. 137–62.

**Ruben et al. 1872.** E. Ruben, F. Achard, and P. Ducourtieux. *Annales manuscrites de Limoges . . . dites manuscrit de 1638.* Limoges, 1872.

**I. E. Rubin 1975.** Ida Ely Rubin, ed. *The Guennol Collection.* 3 vols. Exh. cat., The Metropolitan Museum of Art. New York, 1975.

**M. Rubin 1991.** Miri Rubin. *Corpus Christi: The Eucharist in Late Medieval Culture.* Cambridge, 1991.

**Rubinstein 1917.** Stella Rubinstein. "Some Limoges Reliquaries of the Late Twelfth Century." *Arts and Decoration* 7 (April 1917), pp. 304–6.

**Rubinstein-Bloch 1926.** *Catalogue of the Collection of George and Florence Blumenthal.* Ed. Stella Rubinstein-Bloch. 6 vols. Paris, 1926–30.

**Rückert 1959.** Rainer Rückert. "Beiträge zur limousiner Plastik des 13. Jahrhunderts." *Zeitschrift für Kunstgeschichte* 22, no. 1 (1959), pp. 1–16.

**Rupin 1881a.** Ernest Rupin. "Notices sur quelques objets d'émaillerie limousine accompagnées de planches et de dessins [séance du 2 avril 1881]." *Revue des Sociétés savantes des départements*, ser. 7, 4 (1881), pp. 236–253.

**Rupin 1881b.** E. Rupin. "Coffret en cuivre doré et émaillé, XIIIe siècle: Église de Saint-Viance (Corrèze)." *Bulletin de la Société scientifique, historique, archéologique de Corrèze* 3, no. 1 (1881), pp. 27–34.

**Rupin 1882.** E. Rupin. "Reliquaire en cuivre ciselé et doré, fin du XIIIe siècle, église de Saint-Martin de Brive (Corrèze)." *Revue des Sociétés savantes des départements*, ser. 7, 5 (1882), pp. 420–24.

**Rupin 1890.** E. Rupin. *L'Oeuvre de Limoges.* Paris, 1890.

**Rupin 1901.** E. Rupin. "Tombeau limousin en cuivre doré et émaillé du prieuré de Belhomer." *Bulletin de la Société scientifique, historique, archéologique de Corrèze* 23 (1901), pp. 601–06.

**Rutter 1823.** John Rutter. *Delineations of Fonthill and Its Abbey.* London, 1823.

**Ryan Sale 1933.** *Gothic and Renaissance Art: Collection of the late Thomas Fortune Ryan.* American Art Association, Anderson Galleries, November 23–25. New York, 1933.

**Sacken and Kenner 1866.** Eduard Freiherr von Sacken and Friedrich Kenner. *Die Sammlungen des K. K. Munz- und antiken-Cabinetes.* Vienna, 1866.

***Saint-Amable 176–85.*** *See* Bonaventure de Saint-Amable

**Saint-Omer 1992.** *Trésors des églises de l'ar-*rondissement de Saint-omer. Exh., Musée de l'Hôtel Sandelin. Saint-Omer, 1992.

**Salet 1946.** F. Salet. "Émaux champlevés des XIIe et XIIIe siècles." *Art et décoration* 3 (1946), pp. 177–84.

**Salet 1951.** F. Salet. "Nouvelle Doctrine sur la chronologie de Saint-Martial de Limoges." *Bulletin monumental* (1951), pp. 322–26.

**Salet 1958.** F. Salet. *Chefs-d'oeuvre des musées de France, Cluny: Les Émaux de Limoges.* Publications filmées d'art et d'histoire. Montrouge, 1958.

**Samazeuilh 1860.** J.-F. Samazeuilh. *Monographie de la ville de Casteljaloux.* Nérac, 1860.

**Sandron 1991.** D. Sandron. "Acquisitions. Paris, musée de Cluny: Statue de Pierre d'Alençon." *Revue du Louvre et des musées de France* 3 (1991), pp. 82–83.

**Santiago de Compostela 1990.** *Galicia no tempo: Monastery of San Martino Pinario, Santiago de Compostela.* Exh. cat.; exh. in the church and monastery of San Martino Pinario. Santiago de Compostela, 1990.

**Sauerländer 1970.** Willibald Sauerländer. *Gotische Skulptur in Frankreich 1140–1270.* Munich 1970.

**Sauerländer 1972.** W. Sauerländer. *Gothic Sculpture in France 1140–1270.* New York, 1972. Translation of Sauerländer 1970. French translation: *La Sculpture gothique en France 1140–1270.* Paris, 1972.

**Sauvel 1950.** T. Sauvel. "Les Manuscrits limousins: Essai sur les liens qui les unissent à la sculpture monumentale, aux émaux et aux vitraux." *Bulletin monumental* (1950), pp. 117–44.

**Sauvel 1951.** T. Sauvel. "Les Crucifix en robe longue dans le sud-ouest de la France." *Bulletin de la Société nationale des Antiquaires de France*, 1950–51, pp. 102–4.

**Sauzay 1861.** A. Sauzay. *Musée impérial du Louvre: Catalogue du musée Savageot.* Paris, 1861.

**Sayre and Smith 1961.** Edward V. Sayre and Ray W. Smith. "Compositional Categories of Ancient Glass." *Science* 143, no. 3467 (June 9, 1961), pp. 1824–26.

**Sayre and Smith 1967.** E. V. Sayre and R. W. Smith. "Some Materials of Glass Manufacturing in Antiquity." *Archaeological Chemistry: The 3d Symposium of Archaeological Chemistry*, ed. M. Levey, pp. 279–311. Philadelphia, 1967.

**Sayre and Smith 1974.** E. V. Sayre and R. W. Smith. "Analytical Studies of Ancient Egyptian Glass." *Recent Advances in the Science and Technology of Materials*, ed. A. Bishay, vol. 3, pp. 47–70. New York, 1974.

**Scarpellini 1982.** P. Scarpellini. In *Francesco d'Assisi*, pp. 91–126.

**Schäfke 1982.** Werner Schäfke. "Die Wallfahrt zu den Heiligen Drei Königen." In *Cologne 1982*, pp. 73–80.

**Schapiro 1954.** Meyer Schapiro. "Two Romanesque Drawings in Auxerre and Some Iconographic Problems." In *Studies in Art and Literature for Belle da Costa Greene*, ed. Dorothy Miner, pp. 331–49. Princeton, 1954.

**Schapiro 1977.** M. Schapiro. *Romanesque Art: Selected Papers.* New York, 1977.

**Schatzkunst Trier 1984.** *See* Trier 1984

**Schildhauer 1988.** Johannes Schildhauer. *The Hansa: History and Culture.* Trans. Katherine Vanovitch. Leipzig, 1988.

**Schiller 1966–68.** Gertrud Schiller. *Ikonographie der Christliche Kunst.* 2 vols. Gütersloh, 1966–68.

**Schiller 1971.** G. Schiller. *Iconography of Christian Art.* Trans. Janet Seligman. 2 vols. Greenwich, Conn., 1971.

**Schlumberger et al. 1943.** G. Schlumberger, F. Chalandon, and A. Blanchet. *Sigillographie de l'Orient latin.* Paris, 1943.

**Schnitzler, Bloch, and Ratton 1965.** Hermann Schnitzler, Peter Bloch, and Charles Ratton. *Email, Goldschmiede- und Metallarbeiten: Europaïsches Mittelalter.* Lucerne–Stuttgart, 1965.

**Schnütgen 1884.** A. Schnütgen. "Matériaux pour servir à l'histoire des vases aux saintes huiles." *Revue de l'Art chrétien* 34 (October 1884), pp. 454–62.

**Schnütgen 1887.** A. Schnütgen. "Eine neuentdeckte eucharistische Taube." *Jahrbücher des Vereins von Alterthumsfreunden im Rheinlande* 83 (1887), pp. 201–14.

**Schrader 1986.** J. L. Schrader. "A Medieval Bestiary." *Metropolitan Museum of Art Bulletin*, n. s., 44, no. 1 (Summer 1986), pp. 3–56.

**Schrader 1971.** J. L. Schrader. "Recent Acquisition: St. Michael and a Thirteenth-Century Bishop's Crozier." *Bulletin of the Museum of Fine Arts*, Houston (June 1971), pp. 42–45.

**Schramm 1956.** Percy E. Schramm. *Herrschaftszeichen und Staatssymbolik Beitrage zu ihrer Geschichte vom dritten bis zum sechzehnten Jahrhundert.* Vol. 3 (1956). Schriften der Monumenta Germaniae Historica, 13. Stuttgart, 1954–56.

**Schulte 1910.** A. J. Schulte. Entry for "Host." In *Catholic Encyclopedia*, ed. Charles G. Herbermann et al., vol. 8, pp. 489–97. New York, 1910.

**Schwarzwälder 1978.** Herbert Schwarzwälder. "Der Bischofsstab des Grabes no. 18 im Bremer Dom im Rahmen der Limousiner Krummstäbe mit der Verkündigungsdarstellung." *Bremisches Jahrbuch* 56 (1978), pp. 205–15.

**Sciolla 1980.** Gianni Carlo Sciolla. *Il Biellese d'al medioevo all'ottocento. Artisti, committenti, cantieri.* Photographs, Mario Serra. Turin, 1980.

***Scriptores ordinis Grandimontensis.*** *See* Becquet 1968

***Sculptures et objets 1993.*** *Sculptures et objets d'art précieux du VIe au XVIe siècle.* Ed. Étienne Bertrand. Exh. cat., Galerie Brimo de Laroussilhe. Paris, 1993.

**Seligmann 1961.** Germain Seligmann. *Merchants of Art: 1880–1960: Eighty Years of Professional Collecting.* New York, 1961.

**Senlis 1987.** *De Hugues Capet à Saint-Louis, les Capétiens et Senlis.* Exh. cat., Musée d'Art et d'Archéologie. Senlis, 1987.

**Senneville 1900.** G. de Senneville. "Cartulaires des prieurés d'Aureil et de l'Artige en Limousin." *BSAHL* 48 (1900), pp. 1–500.

**Servières 1878.** L. Servières. *Guide du pélerin à Sainte-Foy de Conques.* Rodez, 1878.

**Sheperd 1960.** D. Sheperd. "An Early Tiraz from Egypt." *Bulletin of the Cleveland Museum of Art* 47 (1960), pp. 7–14.

**Skubiszewski 1965.** P. Skubiszewski. "Romanskie Cyboria w Ksztalcie Czary z Nakrywa." *Rocznik Historii Sztuki*, 1965, pp. 7–46.

**Smith and Hawthorne 1974.** *Mappae clavicula: A Little Key to the World of Medieval Techniques.* Ed. and trans. Cyril Stanley Smith and John G. Hawthorne. Transactions of the American Philosophical Society, n. s., 64, part 4. Philadelphia, 1974.

**Sohn 1989.** A. Sohn. "Der Abbatiat Ademars von Saint-Martial de Limoges (1063–1114): Ein Beitrag zur Geschichte des Cluniacensischen Klösterverben." *Beiträge zur Geschichte des alter Mönchtums und Benediktänertums* 37 (1989).

**Soltykoff Sale 1861.** *Vente du prince Soltykoff. Catalogue.* Paris, 8 April–1 May, 1861.

**Souchal 1962.** Geneviève F. Souchal. "Les Émaux de Grandmont au XIIe siècle." *Bulletin monumental* 120 (1962), pp. 339–57.

**Souchal 1963.** G. F. Souchal. "Les Émaux de Grandmont au XIIe siècle." *Bulletin monumental* 121 (1963), pp. 41–64, 123–50, 219–35, 307–29.

**Souchal 1964a.** G. F. Souchal. "L'Émail de Guillaume de Treignac, sixième prieur de Grandmont (1170–1188)." *Gazette des Beaux-Arts*, ser. 6, 63 (February 1964), pp. 65–80.

**Souchal 1964b.** G. F. Souchal. "Les Émaux de Grandmont au XIIe siècle." *Bulletin monumental* 122 (1964), pp. 7–35, 129–59.

**Souchal 1966.** G. F. Souchal. "Les Bustes reliquaires et la sculpture." *Gazette des Beaux-Arts*, ser. 6, 67, (April 1966), pp. 205–16.

**Souchal 1967.** G. F. Souchal. "Autour des plaques de Grandmont, une famille d'émaux limousins champlevés de la fin du XIIe siècle." *Bulletin monumental* 125 (1967), pp. 21–71.

**Souchal 1976.** G. F. Souchal. "A propos d'un Lion champlevé limousin: La survivance d'un thème sassanide." *Gesta* 15, nos. 1 and 2, pp. 285–91.

**South Hadley, Mass., 1977.** *Monsters, Gargoyles and Dragons: Animals in the Middle Ages.* Exh. cat., Mount Holyoke College. South Hadley, Mass., 1977.

**South Kensington 1862.** *See* London 1862

**South Kensington 1874.** *See* London 1875

**Špaček 1971.** L. Špaček. "Les Émaux limousins médiévaux en Tchécoslovaquie." *BSAHL* 97 (1971), pp. 173–85.

**Spanish Medieval Art.** *See* New York 1954

**Speyer and Mainz 1992.** *Das Reich der Salier, 1024–1125. Katalog zur Ausstellung des Landes Rheinland-Pfalz.* Exh. cat., Historisches Museum der Pfalz, Speyer, Bischofliches Dom- und Diözesanmuseum, Mainz, and Romisch-Germanisches Zentralmuseum, Mainz. Sigmaringen, 1992.

**Spitzer Collection 1890.** *See* Palustre and Molinier 1890

**Spitzer Sale 1893.** *Resumé du catalogue des objets d'art et de haute curiosité antiques du Moyen Âge et de la Renaissance, composant l'importante et précieuse Collection Spitzer, dont la vente publique aura lieu à Paris. Catalogue de la Vente, Paris 17 avril–16 juin.* Paris, 1893.

**Spufford 1988.** Peter Spufford. *Money and its Use in Medieval Europe.* Cambridge, 1988.

**Squilbeck 1966–67.** Jean Squilbeck. "Le Jourdain dans l'iconographie médiévale du Baptême du Christ." *Bulletin des Musées Royaux d'Art et d'Histoire*, ser. 4, 38–39 (1966–67), pp. 69–116.

**The State Hermitage 1994.** *The State Hermitage, Masterpieces from the Museum's Collections.* 2 vols. London, 1994.

**Steenbock 1965.** F. Steenbock. *Der kirchliche Prachteinband im frühen Mittelalter, von den Anfangen bis zum Beginn der Gotik.* Berlin, 1965.

**Steinberg 1983.** Leo Steinberg. *The Sexuality of Christ in Renaissance Art and in Modern Oblivion.* New York, 1983.

**Stern 1954.** H. Stern. "La Mosaïque de l'église Saint-Genès de Thiers." *Cahiers archéologiques* 7 (1954), pp. 185–98.

**Stohlman 1934.** Frederick W. Stohlman. "Assembly Marks on Limoges Champlevé Enamels as a Basis for Classification." *The Art Bulletin* 16 (1934), pp. 14–18.

**Stohlman 1935.** F. W. Stohlman. "Quantity Production of Limoges Champlevé Enamels." *The Art Bulletin* 17 (1935), pp. 390–94.

**Stohlman 1950.** F. W. Stohlman. "The Star Group of Champlevé Enamels and Its Connections." *The Art Bulletin* 32 (1950), pp. 327–30.

**Stothard 1817.** C. A. Stothard. *The Monumental Effigies of Great Britain.* London, 1817.

**Stratford 1984.** Neil Stratford. "Three English Romanesque Enamelled Ciboria." *The Burlington Magazine* 126 (1984), pp. 204–16.

**Stratford 1993.** N. Stratford. *Catalogue of the Medieval Enamels in the British Museum.* Vol. 2, *Northern Romanesque Enamel.* London, 1993.

**Stuttmann 1966.** *See* Bildkatalogue des Kestner-Museums Hannover 1966

**Suau 1982.** J.-P. Suau. "La Face majeure de la grande croix émaillée de l'ancienne collection Loisel de la Rivière Thibouville (Eure) au Met de N. York?" *Connaissance de l'Eure*, nos. 44–45 (July 1982), pp. 58–73.

**Suau 1983.** J.-P. Suau. "Un nouveau témoignage sur le crucifix Loisel-Morgan du Metropolitan Museum of Art de New York." *Connaissance de l'Eure*, no. 59 (1983), pp. 26–27.

**Supino 1898.** I. B. Supino. *Catalogo del R. Museo nazionale di Firenze.* Palazzo del potesta. Rome, 1898.

**Swarzenski 1951.** Georg Swarzenski. "A Masterpiece of Limoges." *Bulletin of the Museum of Fine Arts* (Boston) 49 (1951), pp. 17–25.

**Swarzenski 1954.** Hanns Swarzenski. *Monuments of Romanesque Art: The Art of Church Treasures in North-Western Europe.* London, 1954.

**Swarzenski 1969.** H. Swarzenski. "A Medieval Treasury." *Apollo* 90 (1969), pp. 484–93.

**Swarzenski and Netzer 1986.** H. Swarzenski and Nancy Netzer, with a technical essay by Pamela England. *Catalogue of Medieval Objects: Enamels and Glass.* Boston, 1986.

**Taburet-Delahaye 1989.** Elisabeth Taburet-Delahaye. *L'Orfévrerie gothique (XIIIe–début XVe siècle) au musée de Cluny.* Paris, 1989.

**Taburet-Delahaye 1990.** E. Taburet-Delahaye. " 'Opus ad filum': L'ornement filigrané dans l'orfévrerie gothique du centre et du sud-ouest de la France." *Revue de l'Art* 90 (1990), pp. 46–57.

**Taburet-Delahaye 1992.** E. Taburet-Delahaye. "Acquisition, Moyen Âge: Paris, musée du Louvre, département des Objets d'art." *Revue du Louvre et des musées de France* 1 (1992), p. 91.

**Taburet-Delahaye 1994.** E. Taburet-Delahaye. "L'Oeuvre de Limoges." *Louvre: Trésors du Moyen Âge, Dossier de l'Art* 11 (Dec. 93–Jan. 94), pp. 16–25.

**Taittinger 1987.** C. Taittinger. *Thibaud le Chansonnier.* Paris, 1987.

**Taralon 1966.** Jean Taralon, with Roseline Maître-Devallon. *Les Trésors des églises de France.* Paris, 1966.

**Teuscher 1990.** A. Teuscher. *Das Prämonstratenserkloster Saint-Yved in Braine als Grablege der Grafen von Dreux zur Stifterverhalten und Grabmalgestaltung im Frankreich des 13. Jahrhunderts.* Bamberg, 1990.

**Texier 1842 and 1843.** Abbé Jacques-Rémi Texier. "Essai historique et descriptif sur les émailleurs et les argentiers de Limoges." *Mémoires de la Société des Antiquaires de l'Ouest* 9 (1842), pp. 77–347. Also published in book form, under the same title, in Poitiers, 1843.

**Texier 1842a.** J.-R. Texier. [châsse d'Ambazac.] *Bulletin archéologique publié par le Comité historique des arts et monuments* 1 (1842), p. 144.

**Texier 1843a.** J.-R. Texier. [intervention à propos de la châsse d'Ambazac.] *Bulletin archéologique publié par le Comité historique des arts et monuments* (1843), pp. 411–12.

**Texier 1846.** J.-R. Texier. "L'Autel chrétien. Autels émaillés." *Annales Archéologiques* 4 (1846), pp. 238–85.

**Texier 1850 and 1851.** J.-R. Texier. "Manuel d'épigraphie suivi du recueil des inscriptions du Limousin." *Mémoires de la Société des antiquaires de l'Ouest* 18 (1850), pp. 1–365. Also published in book form, under the title *Recueil des inscriptions du Limousin*, in Poitiers, 1851.

**Texier 1853.** J.-R. Texier. "L'Orfévrerie au XIIIe siècle." *Annales Archéologiques* 13 (1853), pp. 324–31.

**Texier 1855a.** J.-R. Texier. "Note sur le trésor de l'abbaye de Grandmont." *BSAHL* 1 (1855), pp. 73–75.

**Texier 1855b.** J.-R. Texier. "Les Ostensoirs en Limousin." *Annales archéologiques* 15 (1855), pp. 285–99.

**Texier 1857.** Abbé Texier. *Dictionnaire d'orfévrerie[sic], de gravure et de ciselure chrétiennes, ou de la mise en oeuvre artistique des métaux, des émaux et des pierreries.* Paris, 1857.

**Theophilus 1843 and 1977.** Théophile. *Essai sur divers arts.* Trans. C. de Lescalopier. Paris, 1843. Also a 1977 edition, edited by J. Laget and P. Daviaud.

**Theophilus 1961 and 1986.** *Theophilus, Presbyter. De diversis artibus/The Various Arts.* Ed. and trans. Charles Reginald Dodwell. London, 1961. 2d ed. Oxford and New York, 1986.

**Theophilus 1976.** Theophilus. *On Divers Arts.* Trans. John G. Hawthorne and Cyril Stanley Smith. Reprint. New York, 1976.

**Theuerkauff-Liederwald 1988.** Anna-Elisabeth Theuerkauff-Liederwald. *Mittelalterliche Bronze- und Messinggefäse: Eimer, Kannen, Lavabokessel.* Berlin, 1988.

**Thibeaudeau 1839.** Antoine Thibeaudeau. *Histoire du Poitou.* 3 vols. Niort, 1939–40.

**Thoby 1953.** Paul Thoby. *Les Croix limousines de la fin du XIIe siècle au début du XIVe siècle.* Paris, 1953.

**Thoby 1959.** P. Thoby. *Le Crucifix, des origines au Concile de Trente: Étude iconographique.* Nantes, 1959.

**Thomas 1910.** F. Thomas. "Notes sur les objets d'art mobilier ayant existé à l'abbaye de la Sauve-Majeure." *Société archéologique de Bordeaux*, 1910, pp. 44–54.

**Thomas d'Aquin 1646.** R. P. Fr. Thomas d'Aquin. *Histoire de la vie de saint Calmine duc d'Aquitaine . . .* Tulle, 1646.

**Thompson 1980.** F. W. Thompson. "A Graffito on the Society's Limoges Châsse." *The Antiquaries Journal* 60, no. 2 (1980), pp. 350–52.

**Thorode 1897.** E. L. Thorode. *Notice de la ville d'Angers.* Angers, 1897.

**Tintou 1979.** M. Tintou. "Inventaire des églises Saint-Aurélien et Saint-Cessateur de Limoges, en 1723." *BSAHL* 106 (1979), pp. 115–21.

**Titled Nobility of Europe 1914.** Melville Henry Massue, Marquis of Ruvigny and Raineval. *The Titled Nobility of Europe: An International Peerage, or "Who's Who," of the Sovereigns, Princes, and Nobles of Europe.* London, 1914. Reprint. London, 1980.

**Tokyo 1991.** *Portraits du Louvre.* Exh., National Museum of Western Art. Tokyo, 1991.

**Tokyo and Kyoto 1972.** Françoise Baron. *L'Art du Moyen Âge en France.* Exh. cat., National Museum of Western Art. Tokyo, 1972.

**Torrione 1947.** P. Torrione. "Gerolamo da Vespolate autore del coro di San Sebastiano." *Riviste Biellese* 1, no. 2 (1947), pp. 21–27.

**Toulouse 1989.** *De Toulouse à Tripoli: La puissance toulousaine au XIIe siècle (1080–1208).* Exh. cat., Musée des Augustins. Toulouse, 1989.

**Tours 1891.** *Exposition rétrospective de 1890.* Catalogue by L. Palustre: *Album de l'Exposition rétrospective de Tours, publié sous les auspices de la Société archéologique de Touraine.* Tours, 1891.

**Transactions 1878–79.** *Transactions of the Bristol and Gloucestershire Archaeological Society* 3 (1878–79), pp. 36–38.

**Treasury of San Marco.** *See* New York 1984 and Paris 1984

**Trésors des églises 1965.** *See* Paris 1965

**Trichaud 1854.** J.-M. Trichaud. *Les Ruines de l'abbaye de Montmajour d'Arles.* Arles, 1854.

**Trier 1984.** *Schatzkunst Trier.* Bischofliches Generalvikariat Trier. Exh. cat. Trier, 1984.

**Trocmé 1966.** S. Trocmé. "Remarques sur la facture des peintures murales de la chapelle Sainte-Radegonde à Chinon." *Bulletin de la Société des Amis du vieux Chinon* 6, no. 10 (1966), pp. 542–49.

**Trotignon 1991.** Françoise Trotignon. *Le Trésor de l'église de Saint-Marcel près d'Argenton (Indre).* La Chatre, 1991.

**Trouillart 1643.** P. Trouillart. *Mémoires des comtes du Maine.* Le Mans, 1643.

**Tulle 1887.** Émile Molinier. *Exposition rétrospective d'orfèvrerie limousine.* Exh. cat. Tulle, 1887. Published in *Bulletin de la Société scientifique, historique et archéologique de la Corrèze* 9 (1887), pp. 469–537.

**Turner 1956.** W.E.S. Turner. "Studies in Ancient Glasses and the Glass-making Process. Part V: Raw Materials and Melting Processes." *Journal of the Society of Glass Technology* 40 (1956), pp. 277T–300T.

**Turner and Rooksby 1961.** W. Turner and H. Rooksby. "Further Historical Studies Based on X-ray Diffraction Methods of the Reagents Employed in Making Opal and Opaque Glasses." *Jahrbuch des Romisch-Germanischen Zentralmuseums* (Mainz) 8 (1961), pp. 1–6.

**Utrecht 1939.** *Willibrod-Herdenking 739–1939.* Catalogus van de Tentoonstelling van Vroeg-Middeleeuwsche Kunst. Utrecht, 1939.

**Vaivre 1974.** Jean-Bernard de Vaivre. "Le décor héraldique de la cassette d'Aix-la-Chapelle." *Aachener Kunstblätter* 45 (1974), pp. 97–124.

**Vaivre 1988a.** J.-B. de Vaivre. "Les Dessins de tombes médiévales de la collection Gaignières." In *La Figuration des morts dans la chrétienté médiévale jusqu'à la fin du premier quart du XIVe siècle: Colloque, 26–28 mai 1988, Abbaye royale de Fontevraud,* ed. Roger Grégoire, pp. 60–96. N. p., 1989.

**Vaivre 1988b.** J.-B. de Vaivre. "Les Tombeaux levés pour Gaignières dans les provinces de l'Ouest à la fin du XVIIe siècle." *303, Recherches et Créations* 18 (1988), pp. 57–75.

**Vaivre 1992.** J.-B. de Vaivre. "Odon de Montaigu prieur d'Auvergne de l'Ordre de Saint-Jean de Jérusalem au XIVe siècle." *Académie des inscriptions et belles-lettres: Comptes-rendus des séances,* 1992, pp. 577–614.

**Vatican City 1963.** *Émaux de Limoges du Moyen Âge: Églises et musées de France.* Exh., Biblioteca Apostolica Vaticana. Vatican City. 1963.

**Verdier 1965.** Philippe Verdier. "Limoges Enamels from the Order of Grandmont." *The Walters Art Gallery Bulletin* 17, no. 8 (May 1965), pp. 1–3.

**Verdier 1972.** P. Verdier. "Arts at the Court of France and England (1259–1328)." *Apollo,* July 1972, pp. 20–31.

**Verdier 1980.** P. Verdier. *Le Couronnement de la Vierge: Les Origines et les premiers développements d'un thème iconographique.* Paris, 1980.

**Verdier 1982.** P. Verdier. "A Thirteenth-Century Reliquary of the True Cross." *The Bulletin of The Cleveland Museum of Art* 69 (March 1982), pp. 94–110.

**Vergnolle 1994.** E. Vergnolle. *L'Art roman en France: Architecture, sculpture, peinture.* Paris, 1994.

**Verità 1985.** M. Verità. "L'Invenzione del Cristallo muranese: Una Verifica analitica delle fonti storiche." *Rivista della Stazione sperimentale del Vetro* 15 (1985), pp. 17–22.

**Verità and Toninato 1990.** M. Verità and T. Toninato. "A Comparative Analytical Investigation of the Origins of Venetian Glassmaking." *Rivista della Stazione sperimentale del Vetro* 20 (1990), pp. 169–75.

**Verità, Basso, Wypyski, and Koestler 1994.** M. Verità, R. Basso, M. T. Wypyski, and R. J. Koestler. "X-ray Microanalysis of Ancient Glassy Materials: A Comparative Study of WDS and EDS Techniques." *Archaeometry* 36, no. 2 (1994), pp. 241–51.

**Verlet 1948.** *See* Limoges 1948

**Verlet 1949.** Pierre Verlet. [douze médaillons d'émail limousin, du XIIIe siècle.] *Bulletin de la Société nationale des Antiquaires de France,* 1949, pp. 223–24.

**Verlet 1950.** P. Verlet. "Donation Larcade. Disques émaillés de Limoges au Musée du Louvre." *Musées de France* (January–February 1950), pp. 6–7.

**Verneilh 1863.** F. de Verneilh. "Les Émaux d'Allemagne et les émaux limousins: Mémoire en réponse à M. le comte de Lasteyrie." *BSAHL* 13 (1863), pp. 5–48.

**Vernier 1923.** P. Vernier. *Musée des Antiquités de la Seine-Inférieure. Guide du visiteur.* Rouen, 1923.

**Verrier 1942.** Jean Verrier. *La Cathédrale de Bourges et ses vitraux.* Paris, 1942.

**Vie des Saints 1950.** *Vie des Saints et des bienheureux selon l'order du calendrier avec l'histoire des fêtes par les RR. PP. Bénédictins de Paris . . .* Vol. 9: "September." Paris, 1950. *See also* Baudot and Chaussin 1935–59

**La Vie miraculeuse 1859.** *See* Jeoufre 1859

**Villa-amil y Castro 1907.** José Villa-amil y Castro. *Coleccion de articulos en su mayoría sobre el Mobiliario litúrgico de las iglesias gallegas, en la Edad Media.* Madrid, 1907.

**Vinson 1971.** R.-J. Vinson. "Émaux rares choisis dans la collection Vaudecrane." *Connaissance des Arts,* no. 238 (1971), pp. 76–83.

**Viollet-le-Duc 1868.** E.-E. Viollet-le-Duc. *Dictionnaire raisonné de l'architecture française.* Vol. 9. Paris, 1868.

**Viollet-le-Duc 1871–72.** E.-E. Viollet-le-Duc. *Dictionnaire raisonné du mobilier français de l'époque carolingienne à la Renaissance.* Vol. 2, *Ustensiles.* Paris, 1871–72.

**Vita prolixior.** *See* Landes and Paupert 1991

**Vita S. Aridi.** Gregory of Tours. *Vita sancti Aridi abbatis.* PL, vol. 71, cols. 1119–50.

**Vita Stephani Muretensis 1968.** *Vita Stephani Muretensis.* Ed. Dom J. Becquet, in Becquet 1968.

**Vitry 1914.** P. Vitry. "La Collection Camondo: Les sculptures et les objets d'art du Moyen Âge et de la Renaissance." *Gazette des Beaux-Arts* 56, no. 1 (1914), pp. 461–68.

**Vitry and Brière 1925.** P. Vitry and G. Brière. *L'Église abbatiale de Saint-Denis et ses tombeaux: Notice historique et archéologique.* Paris, 1925.

**Vittori 1976.** Ottavio Vittori. *Four Golden Horses in the Sun.* Ed. and trans. James A. Gray. New York, 1976.

**Vittori 1978.** O. Vittori. "Interpreting Pliny's Gilding: Archaeological Implications." *La Revista di Archeologia* 2 (1978), pp. 71–81.

**Voragine.** *See* Jacques de Voragine 1967

**Wagner 1956.** A. R. Wagner. *Heralds and Heraldry in the Middle Ages.* 2d ed. London, 1956.

**Wainwright, Taylor, Harley 1986.** I. Wainwright, J. Taylor, and R. Harley. "Lead Antimonate Yellow." In *Artist's Pigments: A Handbook of Their History and Characters,* ed. R. L. Feller, pp. 220–25. Washington, D.C., and Cambridge, 1986.

**Waring 1857.** *See* London 1857a

**Washington 1993.** *See* Distelberger et al. 1993

**Watts 1924.** W. W. Watts. *Victoria and Albert Museum Department of Metalwork: Catalogue of Pastoral Staves.* London, 1924.

**Way 1845.** A. Way. "Decorative Processes Connected with the Art during the Middle Ages." *The Archaeological Journal* 2 (June 1845), pp. 150–72.

**Weerth 1857.** Ernst aus'm Weerth. *Kunstdenkmäler des christlichen Mittelalters in den Rheinlanden.* 3 vols. (of which the third is the *Atlas*). Leipzig, 1857–66. Another edition of the *Atlas* dated 1868.

**Weitzmann-Fiedler 1981.** Josepha Weitzmann-Fiedler. *Romanische gravierte Bronzeschalen.* Berlin, 1981.

**Weixlgartner 1932.** A. Weixlgartner, ed. *Führer durch di Albert-Figdor Stiftung.* Vienna, 1932.

**Westermann-Angerhausen 1973.** H. Westermann-Angerhausen. *Die Goldschmiedearbeiten der Trierer Egbertwerkstatt.* Trèves, 1973.

**Wharton 1984.** Glenn Wharton. "Technical Examination of Renaissance Metals: The Use of Laue Back Reflection X-ray Diffraction to Identify Electroformed Reproductions." *Journal of the American Institute of Conservation* 23 (1984), pp. 88–100.

**Widener 1935.** *Collection Joseph Widener: Inventory of the Objets d'Art at Lynnewood Hall, Elkins Park (Pennsylvania).* Philadelphia, 1935.

**Willemin 1839.** Nicolas-Xavier Willemin. *Monuments français inédits pour servir à l'histoire des arts depuis le VIe siècle jusqu'au commencement du XVIIe . . . , accompagné d'un texte historique et descriptif, par André Pottier.* 2 vols. Paris, 1839.

**Williams 1977.** J. Williams. *Manuscrits espagnols du Haut Moyen Âge.* Paris, 1977.

**Wilson 1985.** D. M. Wilson. *Der Teppich von Bayeux.* London–Frankfort–Berlin, 1985.

**Wixom 1967.** *See* Cleveland 1967

**Wixom 1969.** William D. Wixom. "A Manuscript Painting from Cluny." *Cleveland Museum of Art Bulletin* 56, no. 4 (April 1969), pp. 131–35.

**Wixom 1982.** W. D. Wixom. In New York, 1982, pp. 33, 108–9.

**The Year 1200.** *See* New York 1970

**Yokohama 1989.** *Treasures from The Metropolitan Museum of Art: French Art from the Middle Ages to the Twentieth Century.* Exh. cat., Yokohama Museum of Art, 1989.

**Young 1933.** Karl Young. *The Drama of the Medieval Church.* 2 vols. Oxford, 1933.

**Zaluska 1979.** Yolanta Zaluska. "La Bible limousine de la bibliothèque Mazarine." In *Le Limousin, Études archéologiques: Actes du 102e Congrès national des sociétés savantes: Archéologie et Histoire de l'Art, Limoges, 1977,* pp. 69–98. Paris, 1979.

**Zaluska 1991.** Y. Zaluska. *Manuscrits enluminés de Dijon.* Paris, 1991.

**Zuchold 1993.** Gerd-H. Zuchold. *Der "Klosterhof" des Prinzen Karl von Preussen im Park von Schloss Glienicke in Berlin.* Berlin, 1993.

**Zurich 1946.** *Meisterwerke aus Oesterreich.* Exh. cat., Kunstgewerbemuseum, Zürich 1946–47.

# Index

470

474

# Photograph Credits

Angers, Musées d'Angers: cat. 150b

Arles, Centre international de conservation du livre: *fig. 79a*

Baltimore, Walters Art Gallery: cat. 119b, *fig. 59a*

Berlin, Kunstgewerbemuseum: *fig. 25a, 37a, 78b*

Bern, Abegg Stiftung: *fig. 121a*

Boston, Museum of Fine Arts: cat. 122

Centre National de la Recherche scientifique, Institut de Recherche et d'histoire des textes: *fig. 34b, 55a*

Cleveland, Museum of Art: *fig. 8, 63a*

Copenhagen, Nationalmuseet: *fig. 115a*

Florence, Scala: cat. 8a-d, 56b, 58f

Hanover, Kestner Museum: *fig. 46b*

Hartford (Connecticut), Wadsworth Atheneum: *fig. 18*

Inventaire général-SPADEM: cat. 9, 16, 54, 55, 71, 118, 142, 153, 157, *fig. 4, 14, 118a* (photos: Magnoux, Rivière, Thibaudin), cat. 32 (photo: Plire), cat. 45, 156, *fig. 5, 30a* and *b* (photos: Choplain, Maston), cat. 133 (photo: Lefébure)

Kansas City, Keir Collection (on deposit, the Nelson-Atkins Museum of Art): *fig. 78a*

Le Mans, Musées du Mans: cat. 15

London, British Museum: *fig. 19a, 20a, 22a, 39b*; by Courtesy of the Dean and Chapter of Westminster Abbey: *fig. 10*

Minneapolis Institute of Art: cat. 119d

Munich, Bayerisches National Museum: *fig. 70a*

New York, The Metropolitan Museum of Art: Jacket/cover photo and cat. 4, 6, 8, 10, 13, 24, 26 27, 28, 31, 33, 36, 37, 38, 43, 46, 47, 48, 52, 58b, 62, 63, 64, 66, 67, 70b, 72, 79, 81, 83, 86, 89, 92b, 93, 96, 98, 102, 105, 109, 111, 113c and d, 115, 117, 124, 126, 129, 130, 131, 134, 136, 138, 139, 155; *fig. 11, 19, 23, 24, 25, 10a, 62a* and *b, 124b*

Oberlin College (Ohio), Allen Memorial Art Museum: cat. 53

Oxford, Ashmolean Museum: *fig. 39a*

Paris: Bibliothèque Mazarine: *fig. 54a*

Bibliothèque Nationale de France: cat. 147, 148, 149, *fig. 1a, 1b, 3a, 7b, 9a, 16a*, ill. p. 435(5), 436, 437, 438, 440, 441, 442

Bulloz: cat. 121, 153a

Laboratoire de Recherche des Musées de France (D. Bagault): cat. 7, *fig. 20, 21, 22, 26*

Musée du Louvre, documentation du Département des Objets d'art: *fig. 3, 6, 7a, 9b, 17a, 23a*, ill. p. 435(1)

Photothèque des Musées de la Ville de Paris: cat. 58d and e

Réunion des Musées Nationaux: cat. 3, 5, 8, 11, 14, 17, 21, 25, 29, 30, 34, 35, 39, 41, 42, 44, 49, 50, 58a, 61, 65, 68, 69, 70a, 73, 74, 76, 77, 78, 80, 82, 84, 85, 87, 90, 92, 94, 95, 97, 101, 106, 107, 108, 110, 112, 116, 120, 123, 125, 128, 132, 133, 135, 143, 144, 145, 146, 150a, 152, 154 (D. Arnaudet, M. Beck-Coppola); 151 (R. G. Ojeda); 12, 57, 99, 100, 104, 113a and b, 119a and c, 127, 141, *fig. 7, 33a, 38a, 101a* (H. Lewandowski, G. Blot)

Perpignan, Bibliothèque Municipale: *fig. 8b*

Poitiers, Musées de la Ville de Poitiers et de la Société des Antiquaires de l'Ouest: cat. 23 (Ch. Vignaud), *fig. 46a* (H. Plessis)

Poznan, Musée national Ryszard Ran: cat. 140

Prague, Uméleckoprumyslové Muzeum: cat. 60

Princeton (New Jersey), Stohlman Archives: *fig. 13, 16, 17, 51a, 52a, 124a*

Rouen, Musées Départementaux de la Seine-Maritime (Yohann Deslandes): cat. 2a

Saint Petersburg, State Hermitage Museum: cat. 20, 58c

Stockholm, Nationalmuseum: *fig. 49a, 50a*

Toledo (Ohio), The Toledo Museum of Art: *fig. 31a*

Troyes, Musée: ill. p. 443

Vienna, Museum für angewandte Kunst: *fig. 12a* and *b*

Washington, D.C., National Gallery of Art (Philip A. Charles): cat. 22

J.-L. Albert (Rodez): cat. 1, 7, *fig. 8a*

J.-L. Auriol (Toulouse): cat. 40

Joris Luyten (Antwerp): cat. 2b

F. Magnoux (Limoges): cat. 19

Georg Mayer (Vienna): cat. 56a, 92

Pascale Néraud (Paris): *fig. 120a, 127a*

Studio Basset (Caluire): cat. 18

Studio Gesell (Argenton-sur-Creuse): cat. 75

Studio Helga Photo (United States): cat. 51, 114